Springer Hydrogeology

Series Editor

Juan Carlos Santamarta Cerezal, San Cristóbal de la Laguna, Sta. Cruz Tenerife, Spain

The *Springer Hydrogeology* series seeks to publish a broad portfolio of scientific books, aiming at researchers, students, and everyone interested in hydrogeology. The series includes peer-reviewed monographs, edited volumes, textbooks, and conference proceedings. It covers the entire area of hydrogeology including, but not limited to, isotope hydrology, groundwater models, water resources and systems, and related subjects.

More information about this series at http://www.springer.com/series/10174

BIN TRAVERLER FORM

Cut By: Antoni owalker Qty 21 Date 10/20

Scanned By: ________ Qty ________ Date ________

Scanned Batch ID's

Notes / Exceptions

Robert G. Maliva

Anthropogenic Aquifer Recharge

WSP Methods in Water Resources Evaluation Series No. 5

Springer

Robert G. Maliva
WSP
Florida Gulf Coast University
Fort Myers, FL, USA

ISSN 2364-6454 ISSN 2364-6462 (electronic)
Springer Hydrogeology
ISBN 978-3-030-11083-3 ISBN 978-3-030-11084-0 (eBook)
https://doi.org/10.1007/978-3-030-11084-0

Library of Congress Control Number: 2019935494

This Springer imprint is published by the registered company Springer Nature Switzerland AG
The registered company address is: Gewerbestrasse 11, 6330 Cham, Switzerland

Preface

Many areas of the world now face water shortages of varying severity, and water scarcity is being exacerbated by a combination of population growth and economic development. Fresh groundwater resources are under particular stress because of their often high quality, lesser vulnerability to disruptions in supply from short-term climatic variations, and their wide geographic extent that allows for decentralized use. Climate change is expected to impact global precipitation patterns and recharge rates, although there is still considerable uncertainty as to local directions and magnitudes of change. The solution to water scarcity will necessarily involve both demand- and supply-side solutions. It will be increasingly important to find means to optimize the use of all available water resources.

Anthropogenic aquifer recharge (AAR) is broadly defined as human processes that increase the flux of water from the land or surface water bodies into underlying aquifers. AAR has varying degrees of planning and intent. Managed aquifer recharge (MAR), a major subset of AAR and main focus of this book, is characterized by the intentional use of aquifers to store and treat waters. AAR can be a secondary result of intentional activities, such as the use of infiltration basins for stormwater management and septic systems for wastewater disposal, and changes in land use and land cover. AAR also includes recharge from unintended processes, such as leakage from water and sewage mains.

MAR includes a broad range of water storage and treatment techniques of varying scales. In developed countries, MAR is usually implemented based primarily on economic considerations. For example, storage of water underground in aquifers can be substantially less expensive than the construction of above-ground reservoirs and tank systems. Similarly, riverbank filtration systems can be less expensive to construct and operate than conventional intakes and engineered filtration systems. Small-scale decentralized MAR is part of low impact development and green infrastructure, which aim to infiltrate and recharge runoff from rainfall close to its source. In developing countries and poor rural areas of newly industrialized countries, limited local technical and economic resources are important considerations prompting the implementation of MAR. Indirectly extracting water from a well located near a surface water body can be a simple, low-technology means to significantly reduce exposure to waterborne pathogens (with associated health benefits) compared to direct consumption of surface waters. A common denominator for MAR systems is that recharge rates (i.e., the amount of water that actually reaches an aquifer) and water quality changes in recharged water are highly dependent on local hydrogeology and hydrogeochemistry.

The performance of MAR systems depends on site-specific hydrogeological conditions, which may not be locally favorable. Successful design and operation of MAR systems thus requires consideration of the hydrogeological factors that impact the flow of water into aquifers and its transport and mixing with native groundwater after recharge. The chemical quality of recharged water depends upon various biogeochemical processes that occur as water infiltrates into and percolates through the soil and flows through an aquifer. Recharged water is typically in chemical disequilibrium with aquifer minerals and native groundwater, and a variety of either beneficial or detrimental fluid–rock interactions may occur. Adverse fluid–rock interactions include the leaching of metalloids (e.g., arsenic) and metals into stored waters. Beneficial processes include the filtration of suspended solids and complete or partial removal of pathogens and chemical contaminants. Where recharge is incidental or unplanned, there is an opportunity for greater water resource benefits by incorporation of planning and considering and addressing system risks.

Most MAR technologies are mature in that they have long histories of employment and their basic design and operational concepts are understood. However, considerable room exists for improved implementation by learning from historical experiences and innovation. A major deficiency in the technical literature on MAR and unmanaged AAR is a paucity of practicably available, up-to-date information on the historical performance of systems. Even less common are retrospective studies that examine the specific hydrogeological conditions responsible for either the good or poor performance of systems. There is often an understandable unwillingness to draw attention to the fact that an MAR system that one designed is performing poorly (i.e., not meeting expectations). However, little has been written on many systems that are performing quite well. Professional engineers and hydrogeologists involved in MAR projects often do not have the inclination, time, or incentive to write technical papers and give technical presentations. The performance of small-scale MAR systems, such as stormwater infiltration basins and permeable pavement systems, is seldom routinely evaluated after construction, unless there are obvious, serious performance issues. The under-reporting of MAR systems is hampering the successful implementation of the technologies because valuable lessons from historical experiences are being lost. Success stories increase confidence in MAR technologies, which can spur further implementation.

This book is intended to provide an overview of AAR practices and design and operational basics with a focus on hydrogeological and hydrogeochemical factors. It is based largely on a literature review of global practices and personal experiences. The book is not intended to be a "how to" manual but rather summarizes technical issues important for successful implementation of MAR and AAR and presents selected project experiences. Numerous references are provided to more detailed papers and books on key topics, which can provide a solid foundation for implementation of MAR technologies. In the USA, there is a plethora of federal, state, and local guidelines and construction standard documents on MAR and unmanaged AAR systems used for stormwater management. Some design guidelines and recommendations from these government sources are provided herein to illustrate normal design practices and considerations, but it is emphasized that the recommended values for specific design parameters vary between localities and that professionals involved in projects should be intimately familiar with local practices, experiences, and regulatory requirements.

MAR will undoubtedly play an increasing role toward achieving safer and more sustainable water supplies. Achieving the potential benefits of MAR will require knowledge of the various system types, their benefits and limitations, and the controls over system initial and long-term performance. Much can be learned from historical experiences and considerable opportunities still exist for innovation for improved implementation.

Fort Myers, USA

Robert G. Maliva

Contents

About the Author

Dr. Robert G. Maliva has been a consulting hydrogeologist since 1992 and is currently a Principal Hydrogeologist and Technical Fellow with WSP USA Inc. and a Courtesy Faculty Member at the U.A. Whitaker College of Engineering, Florida Gulf Coast University. He is currently based in Fort Myers, Florida. He specializes in groundwater resources development including alternative water supply, managed aquifer recharge, and desalination projects.

He completed his Ph.D. in geology at Harvard University in 1988. He also has a master's degree from Indiana University, Bloomington, and a BA in geological sciences and biological sciences from the State University of New York at Binghamton. Upon completion of his doctorate degree, he has held research positions in the Department of Earth Sciences at the University of Cambridge, England, and the Rosenstiel School of Marine and Atmospheric Science of the University of Miami, Florida. He grew up in New York City and attended Stuyvesant High School in Manhattan.

Dr. Maliva has also managed or performed numerous other types of water resources and hydrological investigations including contamination assessments, environmental site assessments, water supply investigations, wellfield designs, and alternative water supply investigations. He has maintained his research interests and completed studies on such diverse topics as Precambrian silica diagenesis, aquifer heterogeneity, precipitates in landfill leachate systems, carbonate diagenesis, and various aspects of the geology of Florida. He gives frequent technical presentations and has numerous peer-reviewed papers and conference proceedings publications on ASR, injection well and water supply issues, hydrogeology, and carbonate geology and diagenesis. He has authored or coauthored the books, *Aquifer Storage and Recovery and Managed Aquifer Recharge Using Wells: Planning, Hydrogeology, Design, and Operation*, *Arid Lands Water Evaluation and Management*, and *Aquifer Characterization Techniques*.

Chapter 1
Introduction to Anthropogenic Aquifer Recharge

1.1 Introduction

Groundwater resources are critical for global drinking water supply, and food and industrial production. It is estimated that groundwater provides almost 50% of the drinking water used worldwide and 43% of all water consumptively used for irrigation (Smith et al. 2016). Groundwater resources are being depleted in many areas of the world due to pumping at rates that greatly exceed the influx of water into aquifers, which is broadly referred to as recharge. The rate of global groundwater depletion has been estimated to be approximately 95 km^3/year for the period 1993–2008 based on hydrological modeling using information from wells and the GRACE satellites (Döll et al. 2014). Konikow (2015) reported that the estimated total depletion of groundwater in the United States during the 20th century is about 800 km^3 and that the depletion increased to almost 1,000 km^3 by 2008. The greatest groundwater depletions were reported to have occurred in the High Plains Aquifer, Gulf Coast Aquifer Systems, and Central Valley of California (Konikow 2015). The Central Valley has had the greatest depletion intensity, which is defined as the volumetric rate of depletion divided by the aquifer area.

Groundwater depletion ultimately results in the exhaustion of an aquifer as a useable water source. Aquifers will not be pumped completely dry, but rather extraction costs progressively increase so that the use of an aquifer becomes economically unviable. In addition, declining groundwater levels may induce adverse water quality changes, such as vertical and horizontal saline-water intrusion, land subsidence, decreased stream baseflow, and dehydration of wetlands. Environmental impacts of aquifer depletion are constraining groundwater use in some areas (e.g., parts of the High Plains of the west-central United States and the Central Valley of California) long before physical exhaustion of the resource may occur (Scanlon et al. 2012).

The global rate of groundwater use, and thus aquifer depletion, has increased over time due to population growth, economic development, climate change, and other factors. Groundwater will increasingly be needed to perform a stabilization role in mitigating fluctuations in the supply of surface waters (Tsur 1990). Surface

R. G. Maliva, *Anthropogenic Aquifer Recharge*,
Springer Hydrogeology, https://doi.org/10.1007/978-3-030-11084-0_1

water supplies may locally become more variable due to climate change. Modeled intensification of the water cycle is projected to result in more extreme floods interspersed with long-term droughts. It has been recognized that stabilization of water supplies is the greatest value of groundwater resources, as opposed to full-time supply. Increasing, or at least maintaining, groundwater resources is a critical issue for sustainable irrigation and other water uses in areas facing water scarcity. The need for underground storage of water will increase to mitigate the impacts of climate change (Scanlon et al. 2012).

Humans impact the recharge of aquifers in a myriad of ways. The impacts can be either positive or negative with respect to aquifer water budgets and groundwater depletion. Todd (1959) defined artificial recharge as "the practice of increasing by artificial means the amount of water that enters a ground-water reservoir." The term "artificial" is no longer in favor as it can be construed as indicating that the practice is in some manner unnatural. Morel-Seytoux (1985) defined "managed recharge" as "any process that facilitates transformation of surface water into ground water." Dillon (2005) subsequently introduced the term "management of aquifer recharge" as describing the "intentional banking and treatment of waters in aquifers." The term "management of aquifer recharge" has been superseded by "managed aquifer recharge" (MAR). MAR has been alternatively defined as the "the purposeful recharge of water to aquifers for subsequent recovery or environmental benefits" (Dillon 2009; Parsons et al. 2012). Human activities can also increase groundwater recharge in unplanned, unintentional, and unmanaged manners. The term "anthropogenic aquifer recharge" (AAR) is recommended herein to broadly describe increases groundwater recharge caused by human activities. MAR is a subset of AAR.

MAR includes methods intended to increase the volume of water in storage, such as recharge by infiltration and using wells. The logic of MAR is straightforward; excess water is captured and stored underground when available during wet seasons or low-demand periods, and later recovered for use during dry or high-demand periods. The solution to groundwater depletion is to bring aquifer water budgets back into balance by decreasing extractions, increasing recharge, or a combination of both. Although MAR is recognized to be an important tool for addressing water scarcity, it is not a substitute for demand management in resolving groundwater over extraction (Gale et al. 2006). Foster and Garduño (2013) noted that

> whilst 'managed aquifer recharge' should be encouraged, it is not usually the solution to groundwater resources imbalance and if pursued in isolation (rather than as part of a balanced suite of management measures) may merely result in increased demand.

MAR can be viewed as a means for optimizing the use of aquifers. Dillon et al. (2012) observed that the "Objectives of groundwater management relate to maximizing economic utility of aquifers while sustaining the environment and providing security for meeting human needs."

MAR technologies have a long history. For example, harvesting and storage of monsoon rains, including techniques that recharged local aquifers, have been continuously practiced in India since at least the third millennium BC (Agarwal and

Narain 1997). However, adoption of MAR as a decentralized water management solution has been slow, and actually fallen out of favor in some areas in preference to large-scale solutions, such as major dams and desalination systems. Lahr (1982) observed over 35 years ago that

> community after community throw up their hands in frustration over the dwindling availability of water to supply their ever-expanding population, yet they and their water supply consultants do not adequately investigate ways to integrate underground storage with highly variable surface-water supplies.

The decentralized nature and low technical requirements of some types of MAR are now viewed as important advantages. Small-scale MAR systems implemented on the local level in developing countries can have marked water supply and public health benefits. For example, extraction of water from rivers or streams using wells (i.e., riverbank filtration), as opposed to direct extraction, can result in a dramatic reduction in pathogen concentrations and associated incidence of disease. MAR technologies are economically attractive for rural areas of developing countries where they are a viable alternative to treatment works for medium-sized water supplies (Hofkes and Visscher 1986). While the capital costs of MAR schemes are comparable to those of more conventional water treatment systems, their operational and maintenance (O&M) costs are likely to be lower (Hofkes and Visscher 1986). The lower O&M financial and technical requirements for MAR systems are particularly important in some developing countries where many engineered systems have fallen into disrepair or have been abandoned because of a lack of resources.

MAR includes "environmentally sound technologies" (ESTs), which are defined to (UNEP 2008)

> encompass technologies that have the potential for significantly improved environmental performance relative to other technologies. Broadly speaking, these technologies protect the environment, are less polluting, use resources in a sustainable manner, recycle more of their wastes and products, and handle all residual wastes in a more environmentally acceptable way than the technologies for which they are substitutes.

Small-scale stormwater management techniques include elements of "green infrastructure," "low-impact development" (LID), and rainwater harvesting. The objective of decentralized stormwater management is to infiltrate water closer to the site of rainfall and reduce off-site runoff. Stormwater infiltration techniques fall under the umbrella of AAR.

Depending upon the qualities and chemistries of recharged and native groundwater, and the geochemical processes active during recharge and within aquifers, AAR systems can either improve or degrade water quality. For example, infiltration, percolation, and flow and residence within aquifers are intentionally used in soil aquifer treatment (SAT) systems to improve the quality of wastewater. SAT systems can be highly effective in reducing the concentrations of pathogenic microorganisms and many chemical contaminants. Riverbank filtration (RBF) systems also take advantage of natural contaminant attenuation processes to improve water quality. Alternatively, adverse fluid-rock interactions, such as the leaching of arsenic and trace metals, has impaired the quality of water recharged in some MAR systems.

1.2 Definitions

A variety of terms have been used, often inconsistently, to categorized MAR and AAR systems. The key aspects of MAR are that aquifer recharge is intentional and there is at least some control over the recharge process. The National Research Council (2008) Committee on Sustainable Underground Storage of Recoverable Water introduced the similar term "managed underground storage of recoverable water" (MUS), which denotes the "purposeful recharge of water into an aquifer system for intended recovery and use as an element of long-term water resources management." The Dillon (2005) definition of MAR is currently most widely used for purposeful aquifer recharge.

The Australian NRMMC, EPHC and NHMRC (2009) differentiated between unintentional recharge and unmanaged recharge (UMAR). The former includes unplanned recharge, such as from pipe leakage. UMAR includes intentional activities that have a primary disposal function in which recharge is incidental, such as discharges to septic system leach fields. UMAR and unintentional aquifer recharge include:

- infiltration of stormwater in retention basins
- leakage from potable water and sewer mains
- discharges to on-site sewage disposal and treatment systems (e.g., septic systems and cesspools)
- irrigation return flows
- leakage from canals
- discharge of wastewater to ephemeral streams.

The distinction between MAR and UMAR can be an exercise in semantics, especially where systems serve multiple intended purposes. For example, a stormwater infiltration basin could be categorized as either a managed or unmanaged system depending upon whether its primary purpose is either stormwater management or aquifer recharge. Stormwater infiltration basins also serve both aquifer recharge and water treatment functions.

A differentiating criterion between MAR and UMAR is whether the additional aquifer recharge serves an intended beneficial purpose. For example, increased recharge of stormwater to a shallow aquifer that is not used for water supply and does not support a groundwater-dependent environment (i.e., is purely a water disposal system) could have neutral or potentially adverse hydrological impacts, and would be categorized as UMAR. The same system that was constructed with the primary goal to augment local groundwater supplies or increase dry season baseflow in a nearby stream would be categorized as MAR. MAR includes technologies to:

- store water underground for later use
- improve the quality of water through natural contaminant attenuation process that occur during infiltration, groundwater flow, and underground storage
- protect the quality of existing fresh groundwater resources (e.g., control saline-water intrusion)

- prevent or mitigate adverse impacts associated with groundwater use, such as land subsidence, reduction in stream baseflow, and wetland dehydration.

AAR also includes land use/land cover (LULC) changes that increase net recharge. Bouwer (2002) included enhanced recharge, induced recharge, and incidental recharge in artificial recharge. Enhanced recharge, as defined by Bouwer (2002), consists mainly of vegetation management techniques, such as replacing deep-rooted vegetation with shallow-rooted vegetation or bare soil, and changing vegetation to types that intercept less rainfall with their foliage. Induced recharge is achieved by placing wells near streams, rivers or other surface water bodies to lower the water table and draw more water into an aquifer. Incidental recharge includes human activities that unintentionally increase recharge.

The categorization of AAR techniques and processes used herein is intended to facilitate their discussion. An important goal is to increase the management and control of both UMAR and unintentional recharge. Recharge may progress from unmanaged to managed by approximately accounting for human health and environmental risks (NRMCC, EPHC and NHMRC 2009) and accounting for site-specific hydrogeology and geochemistry in their design and operation.

1.3 MAR Techniques

MAR techniques vary both in their objectives and how recharge is performed or induced (Table 1.1). MAR is most commonly employed to either increase the volume of freshwater stored in an aquifer or reduce the rate of decline in water storage caused by excessive groundwater pumping. Increasing the volume of water in storage may have intended secondary benefits, such as controlling saline-water intrusion and environmental protection (e.g., maintaining or restoring stream baseflows and wetland hydroperiods). The second broad category of MAR techniques have a primary water treatment goal. Various natural contaminant attenuation processes are taken advantage of to improve the quality of recharged water. Some MAR systems have dual storage and treatment objectives.

Depending upon the system, recharge is performed by either applying water onto a land surface (surface spreading), subsurface discharge into the vadose zone using wells, galleries, and trenches, or by injection using wells into either confined or unconfined aquifers. Recharge may also be induced by pumping groundwater close to connected surface water bodies (induced recharge). Modifications of the land surface and stream channels, such as by the removal (or change of) vegetation and construction of dams and levees, are also used to intentionally increase aquifer recharge.

It has been advocated that the more precise term "injection" be replaced with the more general and innocuous-sounding term "recharge" (e.g., National Research Council 2008) because the former may have a negative public association with disposal wells. The term "injection" is retained in places herein because it clearly and specifically describes the process of emplacing water into an aquifer using wells.

Table 1.1 Main MAR objectives and recharge methods

Objectives
Local storage of freshwater for later recovery
Replenishment of aquifers to maintain or increase groundwater production
Protection of groundwater resources (salinity barriers)
Environmental benefits
Water treatment (improve the quality of recharged water)
Recharge Methods
Surface spreading
Vadose zone recharge
Injection or recharge wells
Induced infiltration
Modification of land surfaces and stream channels

Stormwater best management practices (BMPs) and low-impact development (LID) strategies include techniques that straddle the boundaries between managed and unmanaged aquifer recharge and between storage and treatment systems. The objective of these techniques is to mitigate the impacts of urban land development (and associated increase in imperviousness) by increasing local infiltration, decreasing runoff, and improving the quality of runoff and recharged water.

There is inconsistency in the literature in the names used for the various MAR system types, with both multiple names used for a given technique and given names applied to multiple techniques. Follows are summaries of the main MAR techniques and what are considered their mostly widely accepted definitions.

1.3.1 Water Storage-Type MAR Techniques

Water storage-type MAR techniques differ in how recharge is performed and where recovery is performed (Table 1.2). Depending upon system type, recharge is performed using:

- wells completed in the saturated zone
- dry wells, galleries, or trenches completed in the vadose zone
- infiltration basins or reservoirs
- modification of the land surface or channels to increase recharge
- spreading water onto existing land surfaces or channels that have conditions favorable for recharge.

Aquifer storage and recovery (ASR) is an increasingly used technique to locally store water underground. ASR was defined by Pyne (1995) as

> The storage of water in a suitable aquifer through a well during times when water is available, and the recovery of the water from the same well during times when it is needed.

Pyne's definition describes a large majority of ASR systems as a single dual-function injection and recovery well is typically a more economically efficient design than separate dedicated injection and recovery wells. However, in some circumstance, injection and recovery using different wells may be operationally preferred. A modified definition of ASR is (Maliva and Missimer 2010)

Table 1.2 Water storage-type MAR techniques

Technique	Description
Aquifer storage and recovery	Injection of freshwater into an aquifer and its later recovery using either the same well or, less commonly, a nearby well
Aquifer recharge using wells	Injection of water into an aquifer with the goal of increasing overall aquifer water levels; water may be recovered anywhere within the aquifer
Infiltration basins	Constructed basins into which water is diverted to recharge an underlying water-table aquifer
Dry wells, Infiltration galleries, pits, soakaways, and trenches	Shallowly excavated structures used for subsurface infiltration into the vadose zone
In-channel infiltration systems	Dams, check dams, and levees constructed in channels to back-up, spread, slow, and retain water to increase wetted area and the duration of inundation
Percolation tanks	Basins created in ephemeral streams to capture part of monsoon flows for direct use and to recharge the underlying aquifer (term is commonly used in India)
Recharge releases	Slow controlled release of water from surface reservoirs, used as sedimentation basins, into ephemeral streams to enhance recharge of a downstream shallow aquifer
Sand dams	Low dams constructed on ephemeral streams to capture sand and create artificial aquifers used to store storm flows
Surface flooding and ditch-and-furrow systems	Systems that recharge through sheet flow on land surfaces or off-channel shallow ditch and furrow systems
Land cover changes	Land cover modifications to increase net infiltration, such as phreatophyte vegetation removal
Infiltration-based LID techniques	Development practices and modifications designed to increase on-site infiltration to pre-development levels

> The storage of water in a suitable aquifer through a well during times when water is available, and the recovery of the same or similar quality water using a well during times when it is needed.

The above definition captures the essential, defining feature of ASR in that it involves the local storage of water within an aquifer with injection and recovery being performed using wells.

Injection wells are also used to increase the total volume of water stored within an aquifer, which is referred to as "groundwater banking." Typically freshwater is injected into an aquifer containing freshwater. Recharge may be performed either near the water source or at an alternative location that is preferred for hydrogeological or operational reasons. Recovery is performed using existing or newly constructed wells near the points of use or water distribution infrastructure. The aquifer is used to convey water from the recharge point to recovery areas.

Surface spreading tends to be the most efficient means for recharging shallow aquifers, especially if suitable land is economically available. Wells are more prone to clogging because the injected water volume passes through a much small area (the borehole wall) compared, for example, to the bottom and sides of infiltration basins. Infiltration may be performed using constructed facilities (infiltration basin complexes), channels modified by the construction of levees and dams, modified land surfaces, or controlled discharges to ephemeral stream (wadi or arroyo) channels. Surface-spreading systems are usually designed and operated to increase, or arrest the decline of, aquifer water levels. Water is recovered using production wells at distributed locations with the groundwater basin. Infiltration basins are also used for some treatment-type MAR systems (e.g., soil-aquifer treatment systems).

Where confining strata in the vadose zone impede percolation to the water table, recharge may be performed using dry wells, infiltration trenches and galleries, or other subsurface systems completed in the vadose zone. Infiltration trenches and galleries are also used where there is limited land available for infiltration basins.

Recharge can be enhanced through modifications of the land surface, such as by vegetation management and run-off retention and inducement (Bouwer 1989a). Trees and other deep-rooted vegetation may be replaced with grasses or other shallow-rooted vegetation. Phreatophyte control is practiced in the western United States to decrease evapotranspiration (ET) losses and increase recharge rates. Soils may be soil covered or treated to minimize local infiltration so as to concentrate runoff from small events (which would otherwise be lost to ET) for infiltration elsewhere. Urban land surfaces can be modified to increased infiltration rates and aquifer recharge. Low-impact development (LID) techniques includes development practices and modifications designed to maintain on-site infiltration rate at pre-development levels. LID includes infiltration-based rainwater harvesting technologies, such as contouring parcels to retain water.

A very full toolbox of techniques is available to increase aquifer. Perhaps the most extreme MAR technique ever proposed (and fascinating from an historical perspective) was the use of nuclear craters for groundwater recharge (Todd 1965). It was recognized that subsurface nuclear detonations create craters that have large storage volume and whose sides have enhanced permeabilities. It was concluded that blast and contamination hazards from the creation of craters by nuclear explosions are understood and that no technical problems exist that cannot be managed if proper precautions are observed (Todd 1965). An economic feasibility analysis indicated that MAR using nuclear explosives was economically competitive with other recharge options. It was reported that the costs of nuclear devices from the United States Atomic Energy Commission (AEC) were (in 1965 dollars) $350,000 for a 10 kiloton (KT) device and $600,000 for a 2,000 KT device, hence a strong economy of scale (Todd 1965). Times have changed!

1.3.2 Water Treatment-Type MAR Techniques

Treatment type-MAR techniques take advantage of natural contaminant attenuation processes to improve the quality of recharged waters. Natural aquifer treatment (NAT) processes include (Chap. 7):

- filtration at land surface, basin and trench floors and sides, and the wellbore surface
- straining and filtration processes as the water flows through unsaturated and saturated strata
- sorption onto mineral grains and crystals, and organic matter
- biogeochemical process in both aerobic and anaerobic environments
- pathogenic inactivation during storage caused by aquifer physicochemical conditions and predation.

Treatment-type MAR techniques (Table 1.3) vary depending upon whether recharge is performed either by direct application of water using wells or surface-spreading methods, or is induced by pumping groundwater near a surface water body. Aquifer storage transfer and recovery (ASTR) refers to the use of separate injection and recovery wells to enhance chemical and microbial contaminant attenuation (Rinck-Pfeiffer et al. 2006). Water quality improvement occurs by physical and biogeochemical processes that occur along the flow path from injection to recovery wells and by providing aquifer retention time for biodegradation processes to occur.

Soil-aquifer treatment (SAT) is a wastewater treatment technology that involves controlled application of treated wastewater to infiltration basins. Water quality improvement occurs during flow through both the vadose zone and phreatic (saturated) zone. The term SAT has been used loosely in the literature to refer to general processes by which infiltration through soil is used to treat wastewater. However, SAT was originally defined as systems in which the recharged sewage effluent is recovered and its geographic extent in an aquifer is controlled (Bouwer 1985, 1989b, 1991). The treated water is recovered using production wells located around or between the infiltration basins.

The term "aquifer recharge and recovery" (ARR) has a more limited usage and is less well-defined in the literature. It is broadly defined as systems in which impaired

Table 1.3 Treatment-type MAR techniques

Technique	Description
Aquifer storage transfer and recovery	Injection of water into an aquifer and its recovery using different, nearby wells to naturally treat water through filtration, sorption, and biodegradation processes
Soil-aquifer treatment	Infiltration of wastewater into shallow basins to improve its quality by vadose and saturated zone processes
Aquifer recharge and recovery	Recharge of surface water by land application and local recovery, usually using wells, to improve water quality
Dune filtration	Infiltration of water into sand dunes and its recovery with the goal of improving its quality
Bank filtration or riverbank filtration	Pumping of wells or galleries near a surface water body to induce infiltration and improve the quality of the surface water
Stormwater BMPs	Stormwater management techniques that use infiltration processes to improve water quality

(non-potable) water is recharged by land application and locally recovered. SAT and dune filtration are subsets of ARR.

Bank filtration (BF), which is also referred to as riverbank filtration (RBF) and induced infiltration, involves the pumping of water near a surface-water body to take advantage of the natural water quality improvements that occur as water flows across the sediment-water interface and through an aquifer. Pumping induces additional infiltration that would not otherwise occur. Groundwater is pumped using either vertical wells, galleries, or horizontal collector wells. Bank filtration is a long-established technology for treating surface water, with the first known system being a water collection tunnel constructed along the River Clyde (Glasgow) in 1810 (Huisman and Olsthoorn 1983).

Stormwater best management practices (BMPs) are defined by the U.S. Environmental Protection Agency (USEPA 1995) as

> A practice or combination of practices that are determined to be the most effective and practicable (including technological, economic, and institutional considerations) means of controlling point and nonpoint source pollutants at levels compatible with environmental quality goals.

Stormwater BMPs include techniques, such as bioretention facilities, infiltration basins and trenches, and permeable pavements, that are intended to provide both water treatment and aquifer recharge.

1.3.3 Salinity Barrier Systems

Saline-water intrusion is the vertical or horizontal migration of saline groundwater into part of an aquifer containing freshwater as the result of human activities, such as over-pumping of groundwater, reduction in groundwater recharge and levels by land development activities (drainage), or the destruction of natural barriers that separate fresh and saline waters (Chap. 21). Inland groundwater pumping can results in a landward hydraulic gradient at the interface between saline and fresh waters. A positive salinity barrier involves the creation of an artificial hydraulic mound between the saline-water interface and an inland wellfield by aquifer recharge using injection wells or surface-spreading techniques. Surface water and treated wastewater are most commonly used in positive salinity barriers.

1.4 MAR as an Adaptation to Water Scarcity and Climate Change

Global climate is changing through a combination of natural processes and anthropogenic influences. Global increases in temperature will affect water resources through an intensification of the hydrologic cycle and increased water demands. Greater temperatures will result in higher ET rates and, in turn, increased precipitation (P). The effects of the hydrological intensification will be uneven, with some regions receiving an increase in available water (P-ET), whereas others will experience drier conditions. Increases in temperature will result in more precipitation falling as rain rather than snow, with implications on the timing of runoff (Arnell 1999). Climate change is expected to result in changes in flood and drought frequency and intensity (Arnell 1999; Kundzewicz et al. 2008). Climate change will also impact the seasonal timing and flashiness of precipitation.

Areas where the impacts of future climate change on freshwater resources are a threat to sustainable development include (Kundzewicz et al. 2008; Green et al. 2011):

- southern Great Plains and southwestern United States
- Mexico
- Caribbean
- Mediterranean basin (southern Europe, North Africa, Levant)
- South Africa
- southwestern South America
- southern and western Australia.

The impacts of climate change on water resources must be placed in perspective. Large parts of the world's population are already experiencing water stress. Rising water demands from population growth and economic development will greatly outweigh greenhouse warming in causing increasing water stress on a global level

(Vörösmarty et al. 2000). Impacts of increasing groundwater extraction will be much greater than the impact of sea level rise and changes in groundwater recharge (Clifton et al. 2010; Ferguson and Gleeson 2012; Taylor et al. 2013).

Climate variability and change can affect groundwater through (Gurdak et al. 2009; Döll 2009; Clifton et al. 2010; Kumar 2012; Taylor et al. 2013):

- changes in precipitation patterns and an increase in the intensity and frequency of extreme events (flashiness)
- increased ET due to higher temperatures
- changes in available water (P-ET) and the amount of recharge and run-off
- changes in soil moisture
- changes in soil structure
- change in land use/land cover and thus infiltration rates
- change in diffuse recharge and discharge rates
- decreased duration of ephemeral streamflow and thus time for recharge
- increased salinity of coastal aquifers from sea level rise
- contraction of freshwater lenses on small islands
- increased recharge causing more pollutants to wash into aquifers
- increased water demands.

Coastal aquifers are vulnerable to saline-water intrusion and inundation (landward movement of the coastline; Ferguson and Gleeson 2012). The greatest vulnerability to saline-water intrusion occurs in areas with high population growth and low hydraulic gradients (<0.001; Ferguson and Gleeson 2012). Low-lying areas are also vulnerable to inundation.

Recharge is effected by the intensity, seasonality, frequency, and type of precipitation, in addition to the magnitude (average annual rate) of precipitation. Recharge is also affected by changes in land cover and soil properties (Clifton et al. 2010), and the location and type of recharge (diffuse versus focused). Various feedbacks may exist. For example, increased winter rainfall during ENSO (El Niño Southern Oscillation) events trigger rapid increases in vegetation productivity in deserts that preclude deep drainage below the root zone and groundwater recharge in inter-drainage (interfluve) areas (Scanlon et al. 2005). Additional infiltrated water during wet periods is rapidly removed by vegetation. Increased intensity of rainfall events and flashiness of streams could have either a positive or negative impact on focused recharged in channels, depending upon whether it results in an increased duration of inundation, and thus time for infiltration to occur. Holman (2006) emphasized the importance of feedbacks, and societal and economic responses in evaluating the impacts of climate changes. For example, temperature increases may impact the length of the local growing season, which, in turn, may affect the choice of crops. Crop selection can impact irrigation requirements and soil structure, and, in turn, groundwater recharge rates.

Soil moisture balance studies and empirical evidence suggest that there is a nonlinear relationship between rainfall and recharge, in which recharge is biased toward heavy rainfall events (>10 mm/day) that temporarily exceed high rates of ET (Taylor et al. 2012). A shift toward a pattern of more intense rainfall events favors groundwater recharge, suggesting a greater availability of groundwater that could be used as an adaptation to variability in surface water resources and reduced soil moisture resulting from climate change (Taylor et al. 2012).

Increased climate variability will increase the value of, and demands on, groundwater as a perennial source of freshwater for irrigation, industrial, and domestic uses (Kundzewicz and Döll 2009; Taylor et al. 2013). Groundwater can serve as a bridge between droughts. However, the use of groundwater as a buffer may not be sustainable in areas experiencing an increase in the frequency, intensity and duration of droughts. Aquifers that are already in an overdraft condition may not have a sufficient stored groundwater capacity to meet additional needs. The importance of groundwater as a buffer for managing increased variability of surface-water supplies and the value of MAR as a means to supplement groundwater supplies for use in droughts are being increasingly recognized (Kundzewicz and Döll 2009; Green et al. 2011; Van der Gun 2012; Scanlon et al. 2012; Taylor et al. 2013).

1.5 MAR Advantages and Disadvantages

The advantages of subsurface storage and treatment of water have been summarized in a number of papers and books including Huisman and Olsthoorn (1983), Bouwer (2002), Pyne (1995, 2005), Dillon (2005), and Maliva and Missimer (2010, 2012). The main advantages of MAR are:

- augmentation of groundwater supplies enhances or maintains the value of groundwater as a buffer to variations in surface water supplies
- aquifers have very large storage capacities
- subsurface storage avoids water losses due to evaporation
- a reduced threat of pollution and sabotage
- lesser land requirements than is needed for surface reservoirs
- lesser environmental impacts due to smaller system footprints
- impacts of over pumping of groundwater (e.g., land subsidence, impacts to groundwater-dependent ecosystems) can be ameliorated
- a high reliability of treatment (subsurface contaminant attenuation processes are not effected by mechanical breakdowns or power outages)
- lower costs and technical resources requirements.

Groundwater use, and associated aquifer depletion, has increased dramatically over the past 50 years due to increasing water demands for irrigation, domestic, and industrial uses, and because groundwater is a decentralized and year-round available source of often good quality water (or at least better quality than available surface water supplies). The most basic benefit of MAR is that it can contribute to continued availability of groundwater resources. Greater recharge can offset depletion caused by excessive pumping.

The latter two listed advantages are particularly important for developing countries and poor rural areas of recently industrialized countries where economic and technical resources may not be available to construct and adequately maintain and operate expensive engineered centralized water and wastewater treatment facilities. Decentralized bank filtration systems, for example, can provide improved water quality compared to direct use of surface water, with minimal operation and maintenance requirements. Natural contaminant attenuation processes that are taken advantage of in MAR systems operate continuously without human intervention, so water quality is not dependent upon continuous supervision and power supply.

The disadvantages and limitations of MAR systems lie in that they depend on natural aquifer systems whose properties may not be locally favorable for the practice. There will always be some uncertainty in predictions of performance in advance of actual system construction and testing. For example, the storage-zone used for some ASR systems turned out to be unfavorable for the recovery of stored water due to very high degrees of aquifer heterogeneity (Maliva and Missimer 2010). Adverse changes in water quality may occur due to fluid-rock interactions. Operational challenges, such as clogging, can also impact system performance and increase operation and maintenance requirements and associated costs.

The success of MAR systems designed for water storage requires a functional legal and regulatory framework to protect the rights of system owners and operators to the water they store. Contrarily, regulatory requirements can be so onerous as to make MAR uneconomical, particularly for small systems without an economy of scale.

MAR systems can have adverse hydraulic and water quality impacts. ASR systems result in aquifer drawdowns during recovery because of the absence of residual local pressure buildups from previous injection (Maliva and Missimer 2008). During recovery, the impact of an ASR system on local aquifer water levels is often the same as that of a solely extractive well, irrespective of previous injected water volumes. Adverse-fluid-rock interactions, such as arsenic and metals leaching, has adversely impacted the quality of water stored in some ASR systems, resulting in regulatory violations and restrictions on the use of recovered waters. Nutrient-rich wastewater recharged in MAR system could impact the water quality of surface water bodies into which it discharges upon recovery.

While MAR systems that harvest rainwater and surface water have been demonstrated to clearly have local benefits, such systems also need to be considered in the context of overall watershed water budgets (Glendenning et al. 2012). In particular, MAR systems involving surface water capture and storage can have upstream-downstream tradeoffs and surface-water/groundwater interactions. Upstream capture of water can adversely impact the availability water for downstream users (Glendenning et al. 2012). Rising water tables could also have the unintended feedback of prompting increases in irrigated area (Glendenning et al. 2012).

Kumar et al. (2006, 2008) proposed that the current and potential future benefits of MAR in India have been exaggerated. A key point made is that in "closed basins," defined as basins in which renewable water resources are being fully utilized (i.e., no "wasted" water is available that could be captured), water harvested in upper catchments for aquifer recharge decreases the availability of water for downstream users. The widespread implementation of MAR in parts of India was reported to have resulted in lower water levels in downstream reservoirs (Kumar et al. 2008). With respect to check dams, it was noted that there is a general belief that because these structures are small, they are benign, even though there may be many dams in a watershed. Kumar et al. (2006) observed that the economics of water harvesting cannot be worked out for structures on an individual basis, but instead systems need to be evaluated based on their incremental benefits on a basin scale. The concern is that water harvesting systems can function primarily to change the distribution of hydrological benefits rather than augment overall water availability.

MAR projects can impact groundwater dependent ecosystems and sensitive receptors, which include aquifer organisms, stygofauna (organisms that live in groundwater), wetland, riparian and terrestrial phreatophyte vegetation, and the flora and fauna of connected wetlands, streams, lakes, and marine environments (Dillon et al. 2009). The impacts include hydraulic and water quality hazards. Some MAR projects have the goal of restoring aquifer water levels toward historical, more natural conditions. However, over time, anthropogenically disturbed conditions can become the environmental status quo. For example, human activities can both drain and create new wetlands, and over the time the flora and fauna become adapted to the new conditions and could be harmed by a return to previous conditions. MAR systems can also impact springs and stream flows and water levels in cave systems.

Ecosystem health can be impacted by changes in the elevation of the water table (upwards or downwards), particularly if the rate of fall or rise exceeds that to which groundwater dependent flora and fauna can adapt (Dillon et al. 2009). The duration of changes in water levels is also an important factor. Changes in groundwater levels can also impact the geochemistry of wetlands, lakes, pools, and streams. Lowering of the water table can result in a change from anoxic to oxic conditions in soils. Rising water levels can result in water logging of the normally aerated root zone, and a transition to anoxic conditions. Waterlogging can also result in salinization of soils.

MAR is not a panacea to water management. Where conditions are hydrogeologically favorable, MAR techniques can be a valuable and cost-effective tool for better managing existing water resources. However, it is critical to appreciate that conditions may not be locally favorable for MAR and to consider MAR in the context of overall local aquifer or basin water budgets.

1.6 MAR System Performance and Impacts

MAR systems are constructed to provide certain intended benefits. Systems are ultimately evaluated based on the degree to which they provide targeted benefits and avoid unintended negative conditions. For example, ASR systems that store freshwater in brackish aquifers are evaluated in terms of their recovery efficiency, which is defined as the ratio of the volume of water that is recovered at a suitable quality to the volume of water recharged (Sect. 13.2). Recovery efficiency goals should be set at the start of a project and the system objectively evaluated based on whether the recovery efficiency goals are met. Similar performance goals can be set for other types of MAR systems. Infiltration basins are evaluated based on their infiltration rates, which can be either a numerical rate or, for stormwater systems, the infiltration of the runoff from a given frequency storm over a given time period.

Prathapar et al. (2015), with respect to MAR implementation in India, emphasized the difference between MAR performance and impacts, and that most investigations focus on one or the other and not both. Performance is defined as the accomplishment of a given task, measured against pre-set known standards, and include:

- storage capacity
- infiltration rate
- reduction in infiltration related to clogging
- recovery efficiency.

Impacts are categorized as primary, secondary, and tertiary and can be either beneficial of negative. Primary impacts include:

- ground water rise
- improvement in water quality
- reductions in downstream supply
- modification of an environmental flow regime.

Secondary impacts (resulting from primary impacts) include:

- additional agricultural production
- increased irrigated area
- change to higher value crops.

Tertiary impacts (resulting from secondary impacts) include:

- socio-cultural impacts of increased agricultural production
- increased incomes
- improvement in family livelihoods in recharge areas
- changes in the value of agricultural land.

Prathapar et al. (2015) emphasized that secondary impacts also depend on other variables (e.g., supply of high-quality seeds, labor, energy costs, and fertilizer) in addition to a rise in groundwater levels. Enabling conditions in addition to water prevailed in studied watersheds in India, which contributed to reported secondary and tertiary impacts. Prathapar et al. (2015) stressed that causal link between agricultural production and groundwater recharge movements in India remains unfounded.

The economics of MAR systems also need to be carefully considered (Sect. 8.4). The economic benefits of MAR projects should exceed their costs. Costs may be considered a performance criterion, in which system construction and O&M costs are evaluated relative to a project budget. A system that clogs frequently and incurs higher maintenance costs than anticipated would be considered a poorly performing system.

1.7 Basic Feasibility, Design, and Operational Issues

MAR system design involves consideration of the following issues (Gale et al. 2006; and others):

- availability of suitable water of sufficient quality and quantity for recharge
- aquifer storage space
- the most efficient means of introducing water into an aquifer
- mechanism of recovery of water
- whether the intervention is going to be effective in terms of meeting project-specific goals
- impacts on downstream and other aquifer users including the environment
- realization of community (stakeholders) expectations
- relative impacts (e.g., amount of additional water provided compared to water use or aquifer overdraft).

Additional or more specific issues include:

- regulatory feasibility and requirements (i.e., whether required governmental approvals can be obtained and stored water be protected)
- regulatory monitoring requirements and associated costs
- water quality changes between recharge and recovery
- O&M requirements, particularly for management of cogging
- human health risks; both the risks posed by an MAR system and the degree to which existing risks are ameliorated by a system
- economics of MAR versus other water storage and treatment options.

The performance of AAR systems, in terms of the realized augmentation of water supplies and the quality of the recharged and recovered water, highly depends on aquifer hydrogeology and the geochemical processes that occur during and after recharge.

The objective of the book is to provide an overview of the diversity of AAR techniques and processes and the hydrogeological and geochemical factors that affect their performance. This book is written from an applied perspective with a focus on taking advantage of global historical experiences, both positive and negative, as a guide to future implementation. Most AAR techniques are now mature technologies in that they have been employed for some time, their scientific background is well understood, and their initial operational challenges and associated solutions have been identified. However, opportunities exist for improved implementation and some recently employed and potential future innovations are presented.

References

Agarwal, A., & Narain, S. (1997). *Dying wisdom. Rise, fall and potential of India's traditional water harvesting systems*. New Delhi: India Centre for Science and Environment.

Arnell, N. W. (1999). Climate change and global water resources. *Global Environmental Change, 9,* S31–S49.
Bouwer, H. (1985). Renovation of wastewater with rapid-infiltration land treatment systems. In T. Asano (Ed.), *Artificial recharge of groundwater* (pp. 249–282). Stoneham, Massachusetts: Butterworth Publishers.
Bouwer, H. (1989a). Estimating and enhancing groundwater recharge. In M. L. Sharma (Ed.), *groundwater recharge* (pp. 1–10). Rotterdam: A.A. Balkema.
Bouwer, H. (1989b). Ground water recharge with sewage effluent. In A. I. Johnson & D. J. Finlayson (Eds.), *Artificial recharge of ground water; Proceedings of the International Symposium, Anahiem, California*, August 23–27 (pp. 170–185). New York: American Society of Civil Engineers.
Bouwer, H. (1991). Groundwater recharge with sewage effluent. *Water Science and Technology, 23,* 2099–2108.
Bouwer, H. (2002). Artificial recharge of groundwater: Hydrogeology and engineering. *Hydrogeology Journal, 10*(1), 121–142.
Clifton, C., Evans, R., Hayes, S., Hirji, R., Puz, G., & Pizarro, C. (2010). *Water and climate change: Impacts on groundwater resources and adaptation options*. Water Working Notes 25. Washington D.C.: Water Sector Board, Sustainable Development Network of the World Bank Group.
Dillon, P. (2005). Future management of aquifer recharge. *Hydrogeology Journal, 13,* 313–316.
Dillon, P. (2009). Water recycling via managed aquifer recharge in Australia. *Boletín Geológico y Minero, 120*(2), 121–130.
Dillon, P., Escalante, E. F., & Tuinhof, A. (2012). *Management of aquifer recharge and discharge processes and aquifer storage equilibrium* (Groundwater Governance Thematic Paper 4). Rome: GEF-FAO.
Dillon, P., Kumar, A., Kookana, R., Leijis, R., Reed, D., Parsons, S., et al. (2009). *Managed aquifer recharge—Risks to groundwater dependent ecosystems—A review*. Canberra: CSIRO Water for a Healthy Country National Research Flagship.
Döll, P. (2009). Vulnerability to the impact of climate change on renewable groundwater resources: A global-scale assessment. *Environmental Research Letters, 4,* 1–12.
Döll, P., Mueller Schmied, H., Schuh, C., Portmann, F. T., & Eicker, A. (2014). Global-scale assessment of groundwater depletion and related groundwater abstractions: Combining hydrological modeling with information from well observations and GRACE satellites. *Water Resources Research, 50*(7), 5698–5720.
Ferguson, G., & Gleeson, T. (2012). Vulnerability of coastal aquifers to groundwater use and climate change. *Nature Climate Change, 2,* 342–345.
Foster, S., & Garduño, H. (2013). Irrigated agriculture and groundwater resources—Towards an integrated vision and sustainable relationship. *Water Science and Technology, 67*(6), 1165–1172.
Gale, I. N., Macdonald, D. M. J., Calow, R. C., Neumann, I., Moench, M., Kulkarni, H., et al. (2006). *Managed aquifer recharge: An assessment of its role and effectiveness in watershed management*. Keyworth, Nottingham: British Geological Survey.
Glendenning, C. J., Van Ogtrop, F. F., Mishra, A. K., & Vervoort, R. W. (2012). Balancing watershed and local scale impacts of rain water harvesting in India—A review. *Agricultural Water Management, 107,* 1–13.
Green, T. R., Taniguchi, M., Kooi, H., Gurdak, J. J., Allen, D. M., Hiscock, K. M., et al. (2011). Beneath the surface of global change: Impacts of climate change on groundwater. *Journal of Hydrology, 405,* 532–560.
Gurdak, J. J., Hanson, R. T., & Green, T. R. (2009). Effects of climate variability and change on groundwater resources of the United States. U.S. Geological Survey Fact Sheet 2009-3074.
Hofkes, E. H., & Visscher, J. T. (1986). *Artificial groundwater recharge for water supply of medium-size communities in developing countries*. The Hague: International Reference Centre for Community Water Supply and Sanitation.
Holman, I. P. (2006). Climate change impacts on groundwater recharge—Uncertainty shortcomings, and the way forward? *Hydrogeology Journal, 14*(5), 637–647.

Huisman, L., & Olsthoorn, T. N. (1983). *Artificial groundwater recharge*. Boston: Pitman Advanced Publishing.

Konikow, L. F. (2015). Long-term groundwater depletion in the United States. *Groundwater, 53,* 2–9.

Kumar, C. P. (2012). Climate change and its impact on groundwater resources. *International Journal of Engineering and Science, 1*(5), 43–60.

Kumar, M. D., Ghosh, S., Patel, A., Singh, O. P., & Ravindranath, R. (2006). Rainwater harvesting in India: Some critical issues for basin planning and research. *Land Use and Water Resources Research, 6,* 1–17.

Kumar, M. D., Patel, A., Ravindranath, R., & Singh, O. P. (2008). Chasing a mirage: Water harvesting and artificial recharge in naturally water-scarce regions. *Economic & Policy Weekly,* 61–71.

Kundzewicz, Z. W., & Döll, P. (2009). Will groundwater ease freshwater stress under climate change? *Hydrological Sciences, 54*(4), 665–675.

Kundzewicz, Z. W., Mata, L. J., Arnell, N. W., Döll, P., Jimenez, B., Miller, K., et al. (2008). The implications of project climate change for freshwater resources and their management. *Hydrological Sciences, 53*(1), 3–10.

Lahr, J. H. (1982). Artificial ground-water recharge: A solution to many US water-supply problems. *Ground Water, 20*(3), 262–266.

Maliva, R. G., & Missimer, T. M. (2008). ASR, useful storage, and the myth of residual pressure. *Ground Water, 46,* 171.

Maliva, R. G., & Missimer, T. M. (2010). *Aquifer storage and recovery and managed aquifer recharge using wells: Planning, hydrogeology, design, and operation*. Houston: Schlumberger Corporation.

Maliva, R. G., & Missimer, T. M. (2012). *Arid lands water evaluation and management*. Berlin: Springer.

Morel-Seytoux, H. J. (1985). Conjunctive use of surface and ground waters. In T. Asano (Ed.), *Artificial recharge of groundwater* (pp. 35–67). Boston: Butterworth.

Natural Research Council. (2008). *Prospects for managed underground storage of recoverable water*. Washington, D.C.: National Academies Press.

NRMMC, EPHC, & NHMRC. (2009). *Australian guidelines for water recycling: Managing health and environmental risks (Phase 2), managed aquifer recharge*. Natural Resource Management Ministerial Council, Environment Protection and Heritage Council, and National Health and Medical Research Council.

Parsons, S., Dillon, P., Irvine, E., Holland, G., & Kaufman, C. (2012). *Progress in managed aquifer recharge in Australia* (Waterlines report 73). Canberra: National Water Commission.

Prathapar, S., Dhor, S., Tamma Rao, G., & Maheshwari, B. (2015). Performance and impacts of managed aquifer recharge interventions for agricultural water security: A framework for evaluation. *Agricultural Water Management, 159,* 165–175.

Pyne, R. D. G. (1995). *Groundwater recharge and wells*. Boca Raton: Lewis.

Pyne, R. D. G. (2005). *Aquifer storage recovery, a guide to groundwater recharge through wells*. Gainesville, Florida: ASR Systems.

Rinck-Pfeiffer, S., Pitman, C., & Dillon, P. (2006). Stormwater ASR in practice and ASTR under investigation in Salisbury, South Australia, in Recharge systems for protecting and enhancing groundwater resources. In *Proceedings of the 5th International Symposium on Management of Aquifer Recharge*, Berlin, Germany, June 11–16, 2005 (pp. 151–159). Paris: UNESCO.

Scanlon, B. R., Faunt, C. C., Longuevergne, L., Reedy, R. C., Alley, W. M., McGuire, V. L., et al. (2012). Groundwater depletion and sustainability of irrigation in the US High Plains and Central Valley. *Proceedings of the National Academy of Sciences, 109*(24), 9320–9325.

Scanlon, B. R., Levitt, D. G., Reedy, R. C., Keese, K. E., & Sully, M. J. (2005). Ecological controls on water-cycle response to climate variability in deserts. *Proceedings of the National Academy of Sciences, 102*(17), 6033–6038.

Smith, M., Cross, K., Paden, M., & Laban, P. (2016). *Spring—Managing groundwater sustainability*. Gland, Switzerland: International Union for Conservation of Nature (IUCN).

Taylor, R. G., Scanlon, B., Döll, P., Rodell, M., van Beek, R., & Wada, Y. (2013). Groundwater and climate change. *Nature Climate Change, 3,* 322–329.

Taylor, R. G., Todd, M. C., Kongola, L., Maurice, L., Nahozya, E., Sanga, H., et al. (2012). Evidence of the dependence of groundwater resources on extreme rainfall in East Africa. *Nature Climate Change, 3*(4), 374–378.

Todd, D. K. (1959). Annotated bibliography on artificial recharge of ground water through 1954. *U.S. Geological Survey Water-Supply Paper* 1477.

Todd, D. K. (1965). Nuclear craters for groundwater recharge. *Journal American Water Works Association, 57*(4), 429–436.

Tsur, Y. (1990). Stabilization role of groundwater when surface water supplies are uncertain: The implications for groundwater development. *Water Resources Research, 26*(5), 811–818.

UNEP. (2008). *Every drop counts. Environmentally sound technologies that have the potential for significantly improved environmental performance relative to other technologies.* Osaka: United Nations Environmental Programme.

USEPA. (1995). *Economic benefits of runoff control.* Washington, D.C.: U.S. Environmental Protection Agency Office of Water.

Van der Gun, J. (2012). *Groundwater and global change: Trends, opportunities and challenges.* Perugia, Italy: United Nations World Water Assessment Programme.

Vörösmarty, C. J., Green, P., Salisbury, J., & Lammers, R. B. (2000). Global water resources: Vulnerability from climate change and population growth. *Science, 289*(5477), 284–288.

Chapter 2
Hydrogeology Basics—Aquifer Types and Hydraulics

2.1 Introduction

Hydrogeology is the branch of science that broadly deals with the distribution, movement, and chemical properties of groundwater in the soil and rock of the Earth's crust. It addresses the flow of groundwater (i.e. hydraulics), solute transport, and water quality and chemical reactions in the groundwater environment (i.e., hydrogeochemistry). Managed aquifer recharge (MAR) and other anthropogenic aquifer recharge (AAR) systems involve some or all the main aspects of hydrogeology. A review of the historical performance of aquifer storage and recovery (ASR) systems revealed that system performance is dependent on local hydrogeological conditions, that the existence of unfavorable conditions is often evident early in a project, and that there may not be an engineered solution to compensate for adverse local hydrogeological conditions (Maliva and Missimer 2010). Hence, the design, operation, and evaluation of MAR systems should be based on a sound understanding of hydrogeological principles, local hydrogeological conditions, and how hydrogeology impacts MAR system performance. Uncertainties associated with MAR projects can be reduced through more thorough and sophisticated hydrogeological evaluations at the start of projects.

Aquifer types and hydraulics are addressed in detail in groundwater textbooks (e.g., Freeze and Cherry 1979; Lohman 1979; Domenico and Schwartz 1998; Fetter 2001; Schwartz and Zhang 2003; Todd and Mays 2005). Aquifer characterization methods were reviewed by Maliva (2016) and aquifer hydraulic testing methods and data interpretation were reviewed by Kruseman and deRidder (1970), Lohman (1979), Walton (1997), and Kasenow (1997, 2006). This chapter provides a summary of hydrogeology basics most applicable to AAR.

R. G. Maliva, *Anthropogenic Aquifer Recharge*,
Springer Hydrogeology, https://doi.org/10.1007/978-3-030-11084-0_2

2.2 Aquifer Types and Terminology

2.2.1 Aquifers, Semiconfining and Confining Units

An aquifer was defined by the U.S. Geological Survey (Lohman et al. 1972) as a

> formation, group of formations, or part of a formation that contains sufficient saturated permeable material to yield significant quantities of water to wells and springs.

Aquifers are bounded by either the water table or strata composed of markedly less permeable material. Bounding strata that have a sufficiently low permeability to effectively prevent flow are referred to as aquicludes or confining units. Bounding strata through which flow is retarded are referred to as aquitards or semiconfining units. Semiconfining units are considered to be leaky confining units.

There is no widely accepted quantitative threshold as to how much water a formation or part of a formation has to yield in order for it to be considered an aquifer. Clearly, this depends upon circumstances. A shallowly buried, low-permeability formation that can economically provide sufficient water to dispersed low-capacity wells used for domestic self supply would be considered an aquifer. The same formation might be considered a semiconfining unit if it were juxtaposed with a deep prolific aquifer used for large-capacity production wells. Similarly there is no set threshold of permeability between a confining and semiconfining unit. Furthermore, hydrostratigraphic units that are categorized as aquifers are commonly internally subdivided into relatively high-permeability flow zones and less permeable confining or semiconfining units.

From a technical perspective, the categorization or naming of strata is not especially relevant so long as the conceptual model of the local flow system is accurate and the properties of the strata of concern are adequately determined.

2.2.2 Unconfined, Semiconfined, and Confined Aquifers

Aquifers are categorized as either unconfined, confined, or semiconfined depending upon their relationship to the regional water table and the confining properties of overlying and underlying strata (Fig. 2.1). The water table is defined as the surface in a ground-water body that is at atmospheric pressure. The water table is the level at which water stands in wells that penetrate an unconfined aquifer far enough to hold standing water (Lohman et al. 1972). The water table is commonly referred to as the upper boundary of the saturated or phreatic zone. However, this definition, in some instances, is technically incorrect because a capillary fringe that is fully saturated, but under less than atmospheric pressure, may be present above the water table. The water table is located at the base of the capillary fringe rather than at its top.

Groundwater is produced from unconfined aquifers largely by the drainage (dewatering) of water from pore spaces and recharge is performed by filling pores. Confined

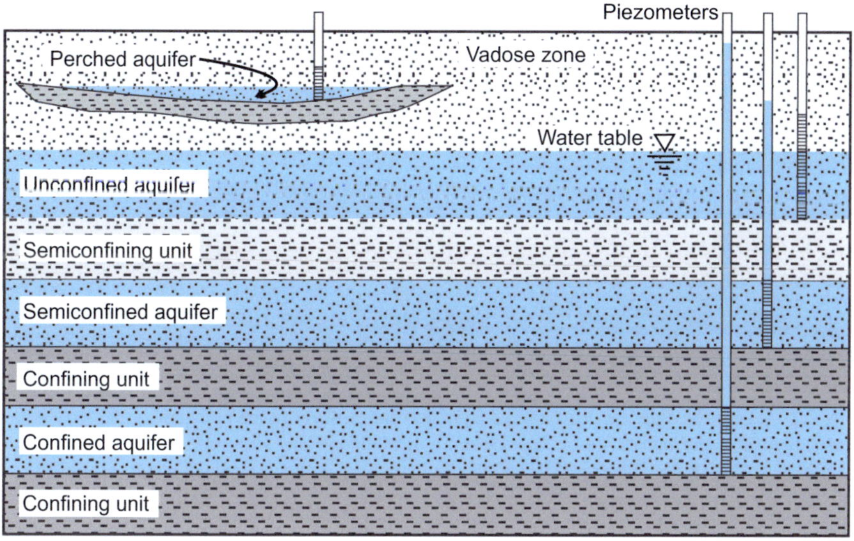

Fig. 2.1 Conceptual diagram of the main aquifer types and the relationship of their potentiometric surface elevations to the water table. The potentiometric surface of semiconfined and confined aquifers may be positioned below the water table, particularly where the aquifers are heavily exploited

and semiconfined aquifers are fully saturated and groundwater is stored and released by the expansion and compression of water and the aquifer in response to changes in pressure. A mapped surface of the elevation at which water rises in tightly cased wells is referred to as the potentiometric or piezometric surface. The potentiometric surface of an unconfined aquifer is approximately the water table. The potentiometric surface elevation may vary from the water table in wells that penetrate deeply into unconfined aquifers (not just the top) if an upward or downward component to groundwater flow exists (Lohman et al. 1972; Lohman 1979).

With a few exceptions, unconfined aquifers fall into the "semi-unconfined" aquifer type of Kruseman and de Ridder (1970), in that natural variations in sediment properties create a disparity in the flow of water between the vertical and horizontal directions. Vertical to horizontal anisotropies in permeability occur due to the interlayering of strata with different permeabilities. In practice, the term "semi-unconfined" is seldom used now as true isotropic conditions are seldom approached in unconfined aquifers.

Where the water table is located a large distance below land surface, permanent or temporary perched aquifers may occur. Perched aquifers contain groundwater under unconfined conditions that is separated from an underlying aquifer by an unsaturated zone. The top of a perched aquifer is a perched water table (Lohman et al. 1972), whose elevation is higher than that of the regional water table. Perched aquifers may occur where low hydraulic conductivity strata prevent or greatly retard the percolation of water to the regional water table. Groundwater in perched aquifers

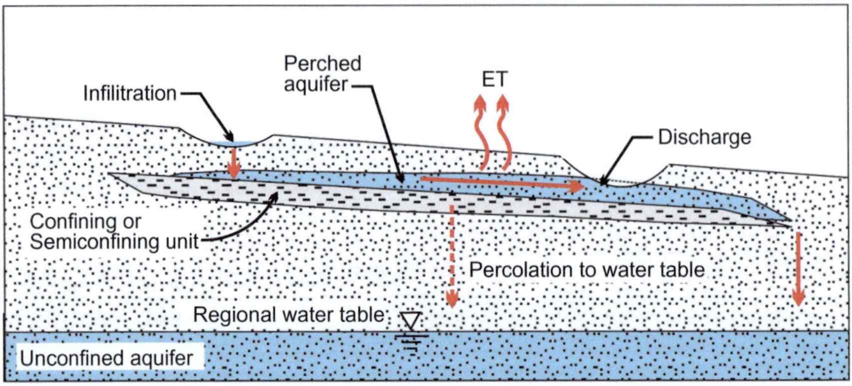

Fig. 2.2 Schematic diagram of a perched aquifer system. Water in the perched aquifer may either leak downwards through or around the confining or semiconfining strata and recharge the regional unconfined aquifer, flow downstream and be discharged, or be lost to evapotranspiration (ET)

may flow laterally in the direction of the slope of the confining strata. Recharge to the underlying regional aquifer may occur through slow leakage through the underlying semiconfining unit or through breaches in, or at the boundary of, the semiconfining unit (Fig. 2.2). For MAR systems the utilize land application for recharge, perched aquifer conditions can result in some, or most, of the recharged water not reaching the target aquifer or recharge occurring some distance away from the application site.

Confined aquifers, by definition, are bounded above and below by essentially impervious confining units. In semiconfined aquifers, also referred to as leaky aquifers, some leakage of water occurs into the aquifers through overlying or underlying strata (or both) when the aquifer is pumped. The potentiometric surface of confined and semiconfined aquifers occurs above the top of the aquifer.

Typically, "confining" strata are not completely impermeable. A confined aquifer is an idealized end-member whose characteristic hydraulic conditions are rarely met, although they may be approached in some aquifers. The degree of confinement is directly related to the vertical hydraulic conductivity and the thickness (i.e., leakance) of the bounding units.

The potentiometric surface of a confined aquifer during pumping continuously declines and does not reach an equilibrium condition. The pumping of semiconfined aquifers also causes the potentiometric surface to decline over time, but eventually the rate of leakage into a pumped aquifer reaches an equilibrium with the pumping rate and no further decline of the potentiometric surface occurs (Hantush and Jacob 1955; Hantush 1960; Walton 1960).

A working method for distinguishing between aquifers that are largely confined and aquifers that are semiconfined (leaky) is through the use of the Theis (1935) non-equilibrium solution (equation) for interpreting pumping test data. Logarithmic plots of drawdown versus time for confined aquifers plot on the Theis curve (assuming other method assumptions are met), whereas data from leaky aquifers eventually deviate from and plot below the Theis curve (Fig. 2.3).

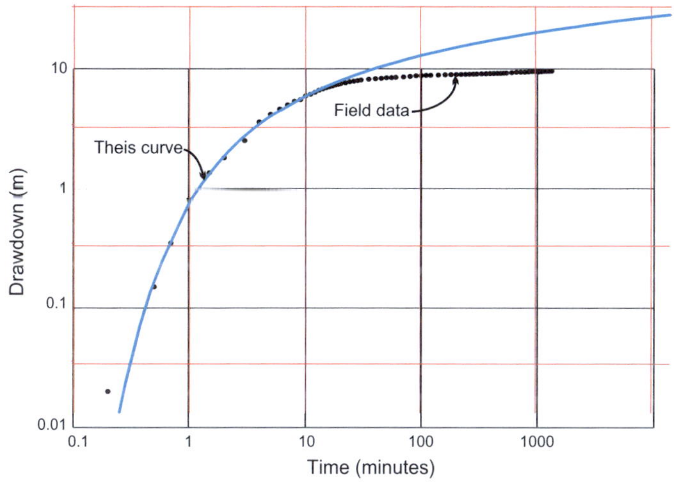

Fig. 2.3 Time-versus-drawdown data from a pumping test of a semiconfined aquifer plotted on a logarithmic scale. The Theis (1935) method is a curve-matching technique in which field data are superimposed on a type curve. Semiconfined conditions are evident by the data after about 15 min plotting below the Theis curve

2.2.3 *Porosity-Type Aquifer Characterization*

Groundwater flows through the pore spaces of aquifer sediment or rock. A variety of pore types may be present in a given aquifer. Most aquifers can be categorized based on whether their predominant porosity type is intergranular, fractures, or solution conduits. The porosity of unlithified sediments typically is entirely (or nearly entirely) intergranular porosity, which is the pore spaces between grains. Carbonate sediments may also contain significant intragranular porosity, such as pore spaces within hollow shell fragments. Incompletely cemented rock (e.g., porous sandstones and some limestone) may have predominately intergranular porosity.

Groundwater flow in hard-rock (very low porosity igneous and metamorphic rock) aquifers and well-indurated sedimentary rock is usually dominated by fractures because the matrix is essentially impermeable. Aquifer properties are largely controlled by the abundance, size (length and aperture width), orientation, and degree of interconnection of fractures. Groundwater flow in carbonate (limestone and dolomite) aquifers often occurs largely through secondary pores formed by the dissolution of the carbonate minerals. The size of the secondary pores in carbonate rocks is highly variable, ranging from solution enlarged fractures or bedding planes to large cave systems in some karst terrains.

A key difference between pore-types relevant to MAR is that intergranular porosity dominated aquifers tend to have high porosities, often in the 10–45% range. Fractured rock aquifers, on the contrary, usually have very low porosities (commonly $\leq$1%) and thus relatively low water storage capacity. Similarly, the main conduits system of karstic aquifers also typically constitute only a very small fraction of the total aquifer volume. Depending upon the karst system type, the main flow conduits may be hydraulically connected to varying degrees with the less permeable but more porous matrix.

Some additional types of aquifers are locally important. Extrusive volcanic rocks may have high porosity and permeability interflow zones. Carbonate aquifers may become recrystallized (e.g., replaced by dolomite) and have a high intercrystalline porosity.

2.2.4 Lithologic Aquifer Types

Aquifer rock type influences porosity, pore types, permeability, and geochemical reactivity. Hard-rock (crystalline) aquifers are usually composed of a dense mosaic of interlocking crystals of silicate minerals (e.g., quartz, feldspars, micas, amphiboles) with very low porosities. Sedimentary aquifers can be divided into two broad categories based on whether they are composed predominantly of either siliciclastic or carbonate sediment or rock. However, some aquifers are composed of intermixed or interbedded siliciclastic and carbonate rock. Clasts are fragments of pre-existing rock. Siliciclastic sedimentary rocks are composed predominantly of clasts of silicate minerals, of which quartz and feldspar are usually most common (Sect. 6.3). The intergranular space may be either open or filled to varying degrees with matrix or cement. The matrix of siliciclastic sediments and rock commonly consists of various clays minerals (e.g., illite, montmorillonite, kaolinite, chlorite) and silt-sized quartz. Common cements in siliciclastic rocks are calcite, clay minerals, iron minerals, and quartz. The porosity of siliciclastic rocks is highly variable ranging from primary values of commonly 35–45% in uncemented sands to close to 0% where their porosity has been largely occluded with cement.

Carbonate rocks, as the name implies, are composed predominantly of carbonate minerals, of which calcite ($CaCO_3$) and dolomite ($CaMg(CO_3)_2$) are most abundant by far. Carbonate rocks may be either clastic, in that they are composed of fragments of pre-existing carbonate rock (e.g., intraclasts and extraclasts) or shell (bioclasts), or they may form by local inorganic or biologically mediated precipitation (e.g., reefal rocks). Because of the variable sizes and shapes of grains, ranging from silt-sized particles to meter-scale coral heads, and biological stabilization (cementation) on the seafloor, carbonate sediments have an enormous range of depositional (primary) textures and fabrics, and associated hydraulic properties. Carbonate sediments are much more reactive under near surface physicochemical conditions than siliciclastic sediments and are thus more prone to a variety of alteration, dissolution, and precipitation processes that can change their porosity and permeability over time.

2.3 Aquifer Hydraulic Properties

Aquifers are hydraulically characterized by their ability to transmit water, which is quantified by their permeability, hydraulic conductivity, and transmissivity, and to store water, which is characterized by their porosity and storativity. Groundwater flow velocity and solute transport are also controlled by effective porosity and dispersivity values. Bulk aquifer properties (transmissivity, storativity, and confining strata leakance) are usually sufficient to evaluate the water level or pressure response of an aquifer to pumping. However, solute-transport is controlled by heterogeneities in porosity and permeability. Hence, MAR projects that are concerned with the recharge and recovery of a specific volume of water (e.g., storage of freshwater in a brackish aquifer), or for which travel path and time are of concern, require much greater attention to spatial variations in aquifer parameters (i.e., aquifer heterogeneity) than is needed, for example, in systems that involve recharging high-quality freshwater in freshwater aquifers.

2.3.1 Darcy's Law and Hydraulic Conductivity

The fundamental relationship in groundwater hydraulics is Darcy's Law, which describes the volumetric flow rate through a medium as a function of a property of the medium, the cross-section area of the flow path, and the pressure or hydraulic gradient along the flow path. Darcy's Law in one direction is generally expressed in differential and integrated form as

$$Q = -KA\left(\frac{dh}{dl}\right) \tag{2.1}$$

$$Q = -KA\left(\frac{\Delta h}{\Delta l}\right) \tag{2.2}$$

where

Q	volumetric flow or discharge rate (m^3/s)
K	is a constant of proportionality referred to as "hydraulic conductivity," which has the units of length over time (m/s)
$dh/\mathrm{d}l$, $\Delta h/\Delta l$	hydraulic gradient; change in head (Δh) with distance (Δl) along the flow path (dimensionless)
A	cross-sectional area of the flow path (m^2)

Specific discharge (q), which is also referred to as Darcy velocity, Darcy flux, and filtration velocity, is equal to the volumetric flow rate divided by the flow area:

$$q = \frac{Q}{A} \tag{2.3}$$

Specific discharge has the units of velocity (distance divided by time; m/s), but is not equal to the actual velocity of water flow. Average linear flow velocity (v) is inversely proportional to effective porosity (n_e):

$$v = \frac{Q}{An_e} = \frac{q}{n_e} = -\frac{K}{n_e}\left(\frac{dh}{dl}\right) \tag{2.4}$$

Effective porosity is the interconnected pore spaces in a rock or sediment through which water flows. Depending upon lithology, effective porosity may be significantly less than the total porosity. As porosity decreases, a correspondingly greater flow velocity is required for a given discharge rate (Q). Average linear velocity is important where solute transport and travel times and distances are of concern. For example, for a given hydraulic conductivity and hydraulic gradient, the travel time from a recharge site to a production well increases with increasing effective porosity. In the case of fractured rock aquifers, which typically have very low effective porosities, water flow and solute transport can be very rapid.

Hydraulic conductivity is an extrinsic property of a sediment or rock that depends on the properties of the fluid, particularly its temperature. Hydraulic conductivity is very commonly used in groundwater models and is expressed in groundwater technical reports and published papers as measured values for media without direct reference to the properties of the groundwater. This practice is usually acceptable because groundwater temperatures tend to be stable (i.e., vary over only a narrow range). However, temperature effects can be significant in MAR systems involving land surface application and where injected water has a different temperature than native groundwater.

Hydraulic conductivity also varies with direction (i.e., is anisotropic) in aquifers. Hydraulic conductivity in the vertical direction (K_z) is usually significantly less than that in the horizontal (x and y) directions (K_x and K_y). Horizontal directional anisotropy may also occur ($K_x \neq K_y$) in some aquifers. For example, fractured rock aquifers often have a preferred fracture orientation, and hydraulic conductivity can be much greater parallel to fractures than in the perpendicular direction.

Permeability is an intrinsic property of a rock or sediment in that it is not dependent on other variables or conditions. The relationship between hydraulic conductivity (K) and intrinsic permeability (k) is

$$K = \frac{k\rho g}{\mu} \tag{2.5}$$

where

ρ density (kg/m^3)
g gravitation acceleration (9.807 m/s^2)
μ dynamic viscosity (kg/(m s)

Dynamic viscosity is expressed in units of Pascal-seconds (Pa·s), which is equivalent to 1 kg/(m s), and in centipoise units (cP), which are equal to 1×10^{-3} kg/(m s). Permeability has units of length squared, with the SI unit being m^2. The unit of permeability commonly used in the oil and gas industry is the millidarcy (mD), which is equivalent to 9.869×10^{-16} ($\approx 1 \times 10^{-15}$ m^2) or 9.869×10^{-12} ($\approx 1 \times 10^{-11}$ cm^2). At 20 °C, 1 darcy is equal to about 9.61×10^{-6} m/s (0.831 m/d).

The dynamic viscosity of water is sensitive to changes in temperature, as expressed by the equation

$$\mu = 2.414x10^{-5} \cdot 10^{\left(\frac{247.8}{T-140)}\right)} \tag{2.6}$$

where T is temperature in degrees Kelvin. For example, a decrease in temperature from 30 to 20 °C results in a 25.6% increase in dynamic viscosity and a 20.4% decrease in hydraulic conductivity.

The effect of temperature on viscosity and hydraulic conductivity needs to be considered in systems in which there are significant temporal (e.g., seasonal) or spatial (e.g., depth-related) variations in temperature. The hydraulic conductivity of shallow aquifers, and thus infiltration rates, will be greater in the summer than in the winter, and greater during the day than during the night. Temperature effects on infiltration rates have been documented in natural stream channels (e.g. Ronan et al. 1998; Constantz et al. 1994) and stormwater infiltration basins (e.g., Braga et al. 2007). The temperature effect on viscosity, and thus hydraulic conductivity, may also be significant where cool water is injected into warmer aquifers. Aquifer hydraulic conductivity and transmissivity values obtained by pumping tests under one temperature regime should be corrected for temperature if they are to be used to model an aquifer under different temperature conditions.

2.3.2 *Transmissivity*

Transmissivity (T) is defined as the volumetric rate of water flow through a unit width (e.g., 1 m or 1 ft) of the cross-sectional area of an aquifer under a unit hydraulic gradient. Transmissivity has the units of length squared divided by time (m^2/d and ft^2/d). Transmissivity is also expressed in some older papers using the now obsolete units of gallons per day per foot (gpd/ft). In a horizontally layered aquifer, transmissivity is equivalent to the product of the average aquifer hydraulic conductivity $(\overline{K})$ and the aquifer thickness (b):

$$T = \overline{K}b \tag{2.7}$$

For an aquifer divided into "n" number of horizontal beds, transmissivity is calculated as

$$T = \sum_{i=1}^{n} K_i b_i \tag{2.8}$$

where K_i and b_i are the average hydraulic conductivity and thickness of each bed. As is the case for hydraulic conductivity, transmissivity is an extrinsic property of an aquifer and varies with temperature.

Transmissivity is a bulk property that quantifies the ability of aquifers to transmit water. Higher transmissivity values result in greater volumetric flow rates through an aquifer under a given hydraulic gradient. Aquifers with high transmissivities have lower drawdowns and injection pressures at given pumping and injection rates. Transmissivity values are typically measured by aquifer pumping tests. Calculation of transmissivity values from the hydraulic conductivity of individual beds usually results in a great underestimation of true aquifer values.

In unconfined aquifers, drawdowns in aquifer water levels decrease the saturated thickness and thus transmissivity of aquifers. As unconfined aquifers are depleted, their transmissivity decrease and drawdowns increase, even if pumping rates remain unchanged.

2.3.3 *Storativity*

The storage properties of aquifers is quantified using the storage parameters storativity, specific storage, and specific yield. Storativity (S), which is also referred to as storage coefficient, is defined as the volume of water that is released from a unit **area** of an aquifer (e.g., 1 m^2) under a unit decline (e.g., 1 m) of hydraulic head. Storativity is thus a dimensionless parameter. Specific storage (S_s) is defined as the volume of water that is released from a unit **volume** of an aquifer under a unit decline of hydraulic head. Specific storage has the units of the reciprocal of length (e.g., m^{-1}). The storativity of confined aquifers is the vertically integrated specific storage values, which for homogeneous aquifers is the product of their specific storage and thickness (b):

$$S = S_s b \tag{2.9}$$

The water released from storage in confined aquifers by pumping (or other pressure reductions) is produced by a combination of the compaction of the aquifer and expansion of the water. Specific storage is a function of the compressibilities of the water and the porous media:

$$S_s = \rho g(\alpha + n\beta) \tag{2.10}$$

where

ρ fluid density (kg/m^3)
g gravitational acceleration (9.8 m/s^2)
α compressibility of the porous media (m^2/N or Pa^{-1})
β compressibility of water (≈ 4.4 X 10^{-10} m^2/N)
n porosity (fractional)

More compressible materials, such as clays and unconsolidated sediments, have higher specific storage values than well-lithified rock.

Unconfined aquifers produce water primarily by gravity drainage, which is quantified by their specific yield (S_y). Specific yield is defined as the amount of water that will gravitationally drain from a unit area of an aquifer per unit change in head. The water that is retained in the aquifer (i.e., does not drain) is referred to as specific retention (S_r). The sum of specific yield and specific retention is the total porosity of a rock or sediment. The storativity of unconfined aquifers is the sum of their specific yield and the product of their average specific storage and thickness:

$$S = S_y + S_s b \tag{2.11}$$

In practice, the specific yield value is much greater than the specific storage (compressional) term and the later can usually be ignored.

The specific yields of unconfined aquifers are typically orders of magnitude greater than the storativity values of confined aquifers composed of the same material. The specific yield of porous granular sediments and rocks are usually in the 0.05 to 0.4 range, whereas storativity values are often on the order of 1×10^{-3} to 1×10^{-5}. A given drop in water level in an unconfined aquifer will result in the release of a much greater volume of water than would occur from the same magnitude pressure drop in a confined aquifer. A given rate of pumping will induce much greater drawdowns in confined aquifers than in unconfined aquifers. Similarly, recharge of a given volume of water will result in greater increases in pressure in confined aquifers. The rate of drawdown within a confined aquifer may decrease dramatically when the potentiometric surface falls below the top of the aquifer and the aquifer becomes unconfined.

Three porosity related terms are used to describe aquifers: total porosity, effective porosity and specific yield. Total porosity (often referred to as just porosity) is defined as the total pore volume of a rock divided by its total volume. Effective porosity refers to the interconnected pores through which water flows. Specific yield refers to the gravity-drainable porosity. In coarse-grained granular rock, total porosity, effective porosity and specific yield values are all close to each other.

The pressure response of aquifers to stresses (pumping or injection) is much more rapid than the rate of gravity drainage. Unconfined aquifers experience a delayed-yield phenomenon in which the initial water production is largely from depressuring (similar to the response of confined aquifers), followed by production from drainage. Fine-grained rocks may have a significant fraction of their total porosity filled with capillary bound water and, as a result, their specific yield may be considerably less than their total porosity. Some rocks contain pores that are isolated (not interconnected) and are thus not part of their effective porosity.

The gravity drainage process can take months or years to be completed in very fine-grained sediments (Prill et al. 1965; Johnson 1967). Hence with respect to MAR projects, an "operational specific yield" would be a more appropriate parameter, which is the gravitation drainage that occurs over the system operational time frame of interest. Maliva (2016) summarized techniques used to measure storativity, specific storage, and specific yields.

2.3.4 *Hydraulic Diffusivity*

Hydraulic diffusivity (α), which is also referred to as aquifer diffusivity, is defined as aquifer transmissivity divided by storativity:

$$\alpha = T/S \tag{2.12}$$

Hydraulic diffusivity is a not widely used parameter, but has relevance for MAR projects because it is directly related to the speed at which head (pressure) changes are propagated through porous media. Hydraulic diffusivity is positively related to transmissivity and inversely related to storativity. Pressure changes will propagate faster across an aquifer with a relatively high transmissivity and low storativity. For a given transmissivity, confined aquifers will have much greater hydraulic diffusivities than unconfined aquifers because of their much lower storativity values. Drawdowns from groundwater pumping will tend to be propagated much more slowly in unconfined aquifers than in confined aquifers.

Hydraulic diffusivity ties into an often misunderstood concept in MAR, which has been referred to as the "myth of residual pressure" (Maliva and Missimer 2008, 2010). Injection of water into confined aquifers, which inherently have high diffusivities, results in local increases in pressure (head) during injection. However, once injection is terminated, local pressure buildups quickly dissipate, in the same manner as drawdowns from pumping quickly dissipate once pumping is terminated. If the there is more than a short (usually several days) storage period, a local residual pressure increase will not persist until the time of recovery. Injection during wet periods will not ameliorate local impacts from dry season pumping.

Low hydraulic diffusivity values (such as may occur in unconfined aquifers) may result in a substantial time lag occurring between groundwater pumping or recharge in one part of a groundwater basin and hydraulic impacts elsewhere in the basin, such as changes in spring flows or river baseflow. The absence of adverse impacts to current pumping, therefore, is not conclusive evidence that continued pumping will not result in adverse impacts at some time in the future. The time lag might be taken advantage for beneficial purposes in MAR, for example, by strategically locating

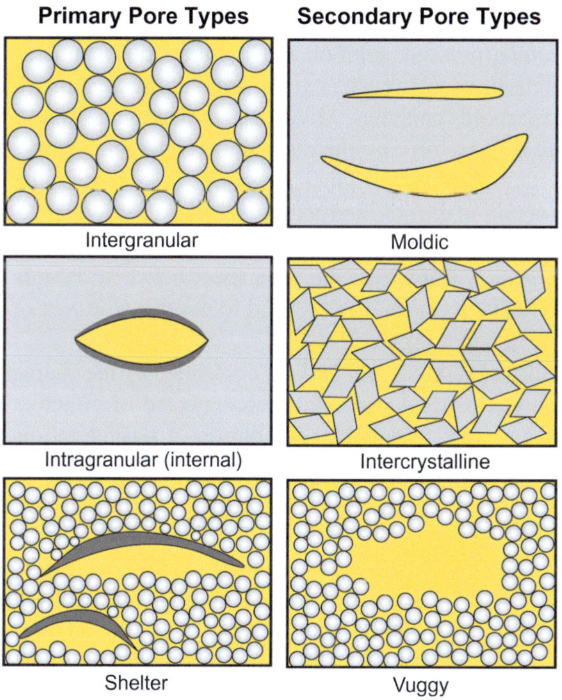

Fig. 2.4 Diagrams of common pore types (yellow) in sedimentary rock. Primary pores are present in the unaltered sediments. Secondary pores form after deposition as the result of dissolution and replacement processes. Secondary pores may form from the dissolution of single grains (moldic) or may be much larger than the size of grains (vuggy)

and timing recharge to benefit targeted groundwater-dependent ecosystems during dry seasons.

2.3.5 Porosity and Permeability

Porosity in sedimentary rocks can be subdivided into primary porosity and secondary porosity. Primary porosity is the original "fabric" or pattern of pores that is present in sediments immediately after deposition. Primary porosity includes intergranular (interparticle) porosity, which is the space between the grains, and intragranular porosity, which is pore spaces that occur within grains (e.g., space within a hollow shell; Fig. 2.4).

Secondary porosity is pore spaces that form after deposition as sediments and rock undergo diagenesis. Diagenesis is defined as the physical, chemical, and biological changes that take place in sediments and rock after they have been deposited, but before they enter the realm of metamorphism. Diagenetic processes that can reduce porosity include mechanical compaction (consolidation), chemical compaction (intergranular pressure solution and stylolitization), and cementation. Fracturing and dissolution processes can increase porosity. Secondary pores that form by dissolution include molds and vugs. Molds are formed by the dissolution of individual grains, whereas vugs form by the dissolution of larger volumes of rock. Large secondary pores include fractures, which are tabular or more irregular pores that form by the mechanical failure of rock, and conduits, which are elongate dissolutional features. Large conduits or caves may allow for the rapid (sometimes turbulent) flow of water. Intercrystalline porosity is the open space between crystals in a rock and is usually secondary. The term 'matrix' is used to describe the part of an aquifer that is neither fracture or conduit porosity.

The permeability of sedimentary rocks is controlled by the characteristics of their porosity, which includes total porosity, interconnected or effective porosity, pore-size distribution, and pore-throat-size distribution. A basic relationship in granular sediments and rocks (e.g., sands and sandstones) is that pore size, and thus permeability, generally increases as grain size increases (with consideration of sorting). Although permeability is correlated with grain and pore size, permeability is primarily controlled by the size of the pore throats that connect pores, rather than the size of the pores themselves. Pore throats are the constrictions that connect adjoining pores through which water flows. Under a given pressure gradient, the diameter, length, and shape of the pore throats are the principal controls over water flow, as opposed to the size of the larger pores. As a generalization, fine-grained rocks have small pore sizes and, in turn, smaller diameter pore throats and lower permeabilities.

Decreases in permeability, and thus infiltration and recharge rates, due to clogging are a major operational issue for MAR systems. Clogging is caused primarily by the obstruction of pore throats, which may be caused by the straining or filtration of sediment at the constrictions, gas bubbles becoming lodged in pore throats, occlusion of pore throats by cements, and biological growth (biofilm development).

Aquifers may contain two or more different types of porosity. Karstic aquifers have a matrix porosity and one or more generations of conduit or solution-enlarged fracture porosity (Fig. 2.5). In karst systems, groundwater flow is dominated by secondary porosity that has a very high hydraulic conductivity but often constitutes only a small fraction of the volume of the aquifer. On the contrary, the matrix often has a much lower hydraulic conductivity than secondary pores but contains most of the porosity and water storage in an aquifer. Dual-porosity conditions can have a large impact on the hydrogeochemistry of MAR systems because they can result in the juxtaposition of waters with different chemistries. Large, interconnected secondary pores may contain recharged water, whereas the adjacent matrix may still contain native groundwater. Diffusion between the matrix and secondary pores can adversely impact the quality of water stored in the secondary porosity. For example, saline native groundwater present with the matrix may "bleed" into freshwater stored in ASR systems, reducing the amount of water that can be recovered at an acceptable quality.

Fig. 2.5 (Top) Karstic limestone (Carboniferous age) in the Yorkshire Dales, near Malham, England. Extreme example of a dual-porosity system in which fracture apertures have been greatly widened by limestone dissolution. (Bottom) Karstic limestone exposure on Curacao. Cave is about 1 m high

2.3.6 Dispersivity

Hydrodynamic dispersion is the process by which solute "particles" are spread out parallel and transverse to the direction of average fluid flow (Freeze and Cherry 1979). It is essentially a more scientific term for mixing during fluid flow. Hydrodynamic dispersion includes molecular diffusion and mechanical dispersion. Molecular diffusion is the velocity independent flux of solute particles from areas of high to low concentrations. Mechanical dispersion is mixing caused by variations in fluid velocity. Variations in fluid flow velocity are the result of (1) velocity differences within a pore due to drag exerted by pore walls, (2) differences in pore sizes within a porous medium, and (3) differences in the length, branching, and interfingering of pore channels (i.e., tortuosity; Freeze and Cherry 1979).

Molecular diffusion and mechanical dispersion cannot normally be separated in groundwater systems and are instead combined into a single parameter called the hydrodynamic dispersion coefficient (*D*). Three hydrodynamic dispersion coefficients are defined based on their orientation with respect to the direction of groundwater flow. Longitudinal dispersion (D_L) occurs parallel to the direction of groundwater flow. Transverse or lateral dispersions (D_T) occurs perpendicular to the direction of flow on the horizontal plane, and vertical dispersion (D_V) occurs perpendicular to the direction of flow on the vertical plane.

Hydrodynamic dispersion coefficients are the sums of mechanical dispersion and coefficient of bulk diffusion (D^*), with the former being the product of the dispersivity value and average linear flow velocity in the principal direction of flow (υ_i):

$$D_L = \alpha_L \upsilon_i + D* \tag{2.13}$$

$$D_T = \alpha_T \upsilon_i + D* \tag{2.14}$$

$$D_V = \alpha_V \upsilon_i + D* \tag{2.15}$$

where α_L = the longitudinal dispersivity, α_T = the transverse dispersivity, and α_V = vertical dispersivity. Dispersivities have units of length (m or ft). Transverse and vertical dispersivity values are typically roughly an order of magnitude less than longitudinal dispersivity values within a given aquifer or aquifer zone.

From the above equations, it can be seen that hydrodynamic dispersion is dominated by diffusion as flow velocity approaches zero. At high flow velocities, mechanical dispersion is the dominant process and diffusion can be ignored. The ratio of advective to diffusive transport is commonly expressed using the dimensionless Perclet number. Diffusion is generally insignificant relative to mechanical dispersion at flow rates of 1 m/yr or greater (Apello and Postma 2005).

Dispersivity is an important variable in MAR systems in which solute-transport is a concern because it controls the mixing of recharged water and native groundwater. Dispersivity values are required for numerical solute-transport modeling, but the values have high degrees of uncertainty because they cannot be practically directly measured. In addition to the properties of geological media, dispersivity values are also dependent on scale; both the length of the flow path and aquifer thickness (Pickens and Grisak 1981; Molz et al. 1983; Gelhar 1986; Neumann 1990; Gelhar et al. 1992; Schulz-Makuch 2005).

Full-aquifer dispersivity values are controlled by the aquifer hydraulic conductivity distribution (i.e., degree of heterogeneity) and transverse migration between layers in response to hydraulic and concentration gradients (Pickens and Grisak 1981). Variations in hydraulic conductivity between aquifer layers can result in differences in solute concentrations between juxtaposed layers, which can cause increased dispersive and diffusive mixing.

Aquifer hydraulic data are seldom, if ever, sufficiently available to accurately calculate dispersivity values. In practice, dispersivity values used in numerical models are usually initially estimated based on rock or sediment types and adjusted, as needed, in the model calibration process. Dispersivity values needed to calibrate models depend on the degree to which aquifer heterogeneity is incorporated into the models. If heterogeneity is not adequately represented, then erroneously large dispersivity values may be required for model calibration (Konikow 2011). Hence, as aquifer heterogeneity is incorporated into models in greater detail and accuracy, smaller dispersivity values may be needed to calibrate models. Dispersivity values can thus be considered a parameter that captures unmodeled features of a system (Barnett et al. 2012).

2.4 Aquifer Heterogeneity

2.4.1 Types and Scales of Aquifer Heterogeneity

Aquifer heterogeneity refers to spatial variation in hydraulic, transport, and geochemical properties. Heterogeneity occurs at multiple scales both within and between beds, and heterogeneities of different scales are often superimposed upon one another. Anisotropy refers to the condition where properties vary with direction. All aquifers are heterogeneous and the degree of heterogeneity varies with scale. Aquifer heterogeneity can be caused by

- variations in the sediment composition and texture, such as grain size, shape, and sorting
- depositional environment or facies
- diagenesis
- structural geological process.

A fundamental challenge in aquifer characterization for MAR projects is developing a data collection and analysis approach that captures the scale of heterogeneity relevant to a specific project.

Aquifer heterogeneity can be categorized in terms of its type and scale. Layered heterogeneity refers to variations in properties in the vertical direction (between horizontal strata), whereas intralayer or lateral heterogeneity refers mainly to variations in the horizontal direction (within strata; Fig. 2.6). Layered heterogeneity in sedimentary aquifers is differences in aquifer properties between beds, bedsets, or formations. Layered heterogeneity occurs on multiple scales, and variations in properties may occur within a given layer. On a coarse-scale, a stratigraphic succession may be divided into aquifer and (semi)confining strata. Aquifers, in turn, may be divided into multiple zones, each assigned an independent transmissivity value.

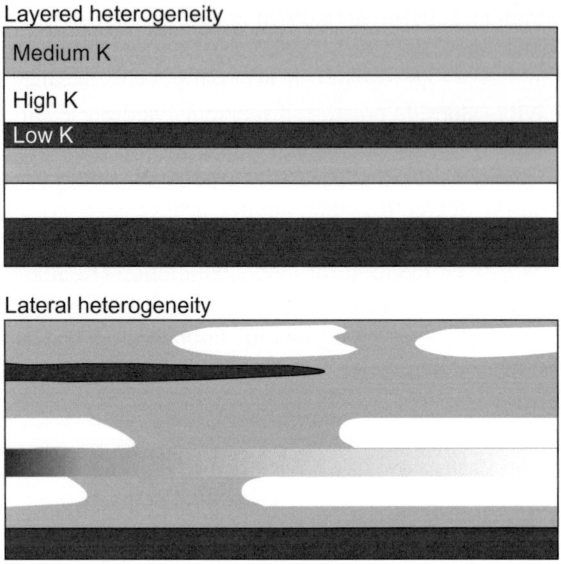

Fig. 2.6 Conceptual diagram of layered and lateral heterogeneity. Lateral heterogeneity may be due to discontinuities of sediment bodies or more gradual changes in properties with an aquifer layer. The data requirements to characterize lateral heterogeneity are much greater than that needed for layered heterogeneity

Intralayer (intrastratal) heterogeneity refers to compositional and hydraulic property variation within a given hydrostratigraphic unit. The heterogeneity is often the product of the three-dimensional interspersion within an aquifer or aquifer zone of bodies of sediment or rock with different hydraulic properties. Variations in depositional environment often result in spatial variations in grain size and thus hydraulic conductivity. For example, in fluvial aquifer system, channel sand bodies are usually coarser-grained and have lower silt and clay contents, and thus have higher hydraulic conductivities, than adjoining flood plain mud deposits. Aquifer heterogeneity is related to the three-dimensional distribution of channel sand and floodplain mud deposits.

Aquifer heterogeneity can also be the result of structural deformation features, such as folds and faults. Some rock types, such low-porosity quartzites and dolomites, are more brittle, and thus tend to have greater fracture densities than limestones (Stearns 1967; Nelson and Serra 1995; Domenico and Schwartz 1998). Preferentially fractured dolomite beds may be high-transmissivity flow zones (e.g., Maliva et al. 2002). Faults and fracture zones may also be the loci of interformational groundwater flow. Alternatively, faults can result in a compartmentalization of an aquifer if the fault (or more particularly fault gauge) acts as a permeability barrier or if displacement results in the juxtaposition of aquifer strata and confining strata. Dykes (dikes) are discordant, vertical or steeply dipping, tabular or sheet-like intruded bodies that cut across existing rocks. Depending on their properties, dykes can be barriers to groundwater flow.

Aquifer heterogeneity can impact the operation of MAR systems in numerous manners. Layered heterogeneity can impact the downward percolation of infiltrated water. Low hydraulic conductivity beds may retard vertical flow and cause perched aquifer conditions. Beds with high hydraulic conductivities can retain most of the infiltrated water and result in a high degree of lateral spreading. In MAR systems that use wells for recharge, high-transmissivity layers may receive most of the injected water, resulting in more rapid flow velocities, and greater lateral spreading and geographic extents of recharged water than would occur in a more homogeneous aquifer. ASR systems that store freshwater in brackish aquifers tend to have poor recovery of freshwater in highly heterogeneous aquifers in which flow is dominated by a thin high-transmissivity zone (Maliva and Missimer 2010). In MAR systems that rely upon aquifer residence time for the attenuation of pathogens and other contaminants, high degrees of aquifer heterogeneity can result in more rapid flow of recharged water to sensitive receptors and shorter residence times.

Spatial heterogeneities in hydraulic conductivity are relevant to the assessment and design of surface spreading systems in which system performance needs to be predicted from a limited number of infiltration tests. A key aquifer characterization issue is the degree to which testing programs capture aquifer heterogeneity that is relevant to system performance. Characterization of horizontal (spatial) aquifer heterogeneity is particularly challenging as the available data is usually limited to a small number of wells or borings. Techniques such as facies analysis and geostatistical modeling are available for better capturing of spatial heterogeneity (Maliva 2016).

2.4.2 Anisotropy

Aquifer anisotropy is directional differences in hydraulic conductivity or transmissivity. Stratified aquifers typically have large vertical (K_z) to horizontal (K_x, K_y) anisotropies because of differences in hydraulic conductivities between beds and finer-scale anisotropy within beds. The effective (average) horizontal and vertical hydraulic conductivity of a series of layers with a total thickness '*b*' are calculated using the equations

$$K_x = \sum_{i=1}^{n} \frac{K_{xi} b_i}{b} \tag{2.16}$$

$$K_z = \frac{b}{\sum_{i=1}^{n} \left(\frac{b_i}{K_{zi}} \right)} \tag{2.17}$$

where b_i is the thickness of layer "i" and K_{xi} and K_{zi} are the horizontal and vertical hydraulic conductivities of layer "*i*".

Effective vertical hydraulic conductivity is the weighted harmonic mean of the vertical hydraulic conductivity of each bed, whereas the effective horizontal hydraulic conductivity is the weighted arithmetic mean. Effective horizontal hydraulic conductivity, and thus transmissivity, is controlled largely by the most conductive beds, whereas, effective vertical hydraulic conductivity is controlled largely by the vertical hydraulic conductivity of the least conductive beds. Within aquitards, the greatest head decline may occur across a thin zone that provides most of the resistance to vertical flow (Bradbury et al. 2006). Vertical-to-horizontal anisotropy is thus largely a function of the difference in hydraulic conductivity between the most and least conductive beds.

Horizontal anisotropy is directional differences in hydraulic conductivity within a bed or aquifer (i.e., K_x and K_y are not equal). There are multiple causes of horizontal anisotropy. Horizontal anisotropy can be due to deposition fabrics, such as the orientation and connectivity of relatively high permeability sediment bodies (e.g., channel sands), or structural fabrics, such as a preferred orientation of open fractures.

Anisotropic aquifers are characterized with respect to transmissivity in terms of the principal directions (i.e., directions of maximum and minimum transmissivity) and magnitude of anisotropy (ratio of maximum to minimum values). Water will have a tendency to flow along the path of least resistance (i.e., direction of greatest hydraulic conductivity). Anisotropy in the horizontal direction can cause the predominant flow direction to deviate from the direction of the hydraulic gradient.

2.4.3 Connectivity

Interconnectedness of high-hydraulic conductivity units is of great importance in controlling groundwater flow and solute transport (Fogg 1986). One or more well-connected sands among a system of otherwise disconnected sands, for example, can completely alter a groundwater flow velocity field (Ritzi et al. 1994). Isolated transmissive units, on the contrary, may be largely isolated from regional flow systems. However, significant transport connectivity may not require complete connection of all zones of relatively high hydraulic conductivity (Bianchi et al. 2011). It has been documented that solutes can travel along preferential flow paths, leaking (jumping) from one hydraulic conductivity cluster to another, with transitions through low hydraulic conductivity zones (Bianchi et al. 2011).

Connectivity and associated aquifer heterogeneity, also depend upon the presence and continuity of low-permeability strata. Laterally continuous, low-permeability strata (e.g., shale units) may vertically compartmentalize an aquifer. Boundary conditions between sedimentary rock units (packets) are important features in determining effective reservoir and aquifer characteristics. The effective permeability of sand packets, for example, will be determined largely by the lower permeabilities of bounding units (Pryor 1973).

Evaluation of connectivity is critical to quantifying heterogeneity for hydrogeological investigations (Anderson 1997), and there is still the need to further develop and refine sedimentological techniques to identify and quantify connectivity among hydrofacies (Anderson et al. 1999). The challenge lies in extrapolating and interpolating one-dimensional facies or hydrofacies data from a limited number of wells into three-dimensional geological and numerical models (Webb and Anderson 1996).

References

Anderson, M. P. (1997). Characterization of geological heterogeneity. In G. Dagan & S. P. Neuman (Eds.), *Subsurface flow and transport: A stochastic approach* (pp. 23–43). Cambridge, UK: Cambridge University Press.

Anderson, M. P., Aiken, J. S., Webb, E. K., & Mickelson, D. M. (1999). Sedimentology and hydrogeology of two braided stream deposits. *Sedimentary Geology, 129*(3–4), 187–199.

Appelo, C. A. J., & Postma, D. (2005). *Geochemistry, groundwater and pollution* (2nd ed.). Leiden: AA Balkema.

Barnett, B., Townley, L. R., Post, V., Evans, R. E., Hunt, R. J., Peeters, L., Richardson S, Werner, A. D., Knapton, A., & Boronkay, A. (2012). *Australian groundwater modelling guidelines,* Waterlines Reports Series No. 82. Canberra: National Water Commission.

Bianchi, M., Zheng, C., Wilson, C., Tick, G. R., Liu, G., & Gorelick, S. M. (2011). Spatial connectivity in a highly heterogeneous aquifer: From cores to preferential flow paths. *Water Resources Research, 47*(5), W05524.

Bradbury, K. R., Gotkowitz, M. G., Hart, D. J., Eaton, T. T., Cherry, J. A., Parker, B. L., et al. (2006). *Contaminant transport through aquitards: Technical guidance for aquitard assessment.* Denver: American Water Works Research Foundation.

Braga, A., Horst, M., & Traver, R. G. (2007). Temperature effects on the infiltration rate through an infiltration basin BMP. *Journal of Irrigation and Drainage Engineering, 133*(6), 593–601.

Constantz, J., Thomas, C. L., & Zellweger, G. (1994). Influence of diurnal variations in stream temperature on streamflow loss and groundwater recharge. *Water Resources Research, 30*(12), 3253–3264.

Domenico, P. A., & Schwartz, F. W. (1998). *Physical and chemical hydrogeology* (2nd ed.). New York: Wiley.

Fetter, C. W. (2001). *Applied hydrogeology* (4th ed.): Englewood Cliffs, NJ: Prentice Hall.

Fogg, G. E. (1986). Groundwater flow and sand body interconnectedness in a thick multiple-aquifer system. *Water Resources Research, 22,* 679–694.

Freeze, R. A., & Cherry, J. A. (1979). *Groundwater*. Englewood Cliffs, NJ: Prentice-Hall.

Gelhar, L. W. (1986). Stochastic subsurface hydrology from theory to applications. *Water Resources Research, 22*(9S), 135S–145S.

Gelhar, L. W., Welty, C., & Rehfeldt, K. R. (1992). A critical review of data for field-scale dispersion in aquifers. *Water Resources Research, 38,* 1955–1974.

Hantush, M. S. (1960). Modification of the theory of leaky aquifers. *Journal of Geophysical Research, 65,* 3713–3725.

Hantush, M. S., & Jacob, C. E. (1955). Nonsteady radial flow in an infinite leaky aquifer. *Transactions of the American Geophysical Union, 36,* 95–100.

Johnson, A. I. (1967). *Specific yield—Compilation of specific yields for various materials.* U.S. Geological Survey Water Supply Paper 1662-D.

Kasenow, M. (1997). *Applied groundwater hydrology and well hydraulics.* Highlands Ranch, CO: Water Resources Publications.

Kasenow, M. (2006). *Aquifer test data: Evaluation and analysis*. Highlands Ranch, CO: Water Resources Publications.
Konikow, L. F. (2011). The secret of successful solute-transport modeling. *Ground Water, 49,* 144–159.
Kruseman, G. P., & de Ridder, N. A. (1970). *Analysis and evaluation of pumping test data*. Wageningen, Netherlands: International Institute for Land Reclamation and Improvement.
Lohman, S. W., et al. (1972). *Definition of selected ground-water terms—Revisions and conceptual refinements*. U.S. Geological Survey Water-Supply Paper 1988.
Lohman, S. W. (1979). *Ground-water hydraulics*. U.S. Geological Survey Professional Paper 708.
Maliva, R. G. (2016). *Aquifer characterization techniques*. Berlin: Springer.
Maliva, R. G., & Missimer, T. M. (2008). ASR, useful storage, and the myth of residual pressure. *Ground Water, 46,* 171.
Maliva, R. G., & Missimer, T. M. (2010). *Aquifer storage and recovery and managed aquifer recharge using wells: Planning, hydrogeology, design, and operation*. Houston: Schlumberger Water Services.
Maliva, R. G., Kennedy, G. P., Martin, W. K., Missimer, T. M., Owosina, E. S., & Dickson, J. A. D. (2002). Dolomitization-induced aquifer heterogeneity: Evidence from the Upper Floridan Aquifer, Southwest Florida. *Geological Society of America Bulletin, 114,* 419–427.
Molz, F. J., Güven, O., & Melville, J. G. (1983). An examination of scale-dependent dispersion coefficients. *Ground Water, 21,* 715–725.
Nelson, R. A., & Serra, S. (1995). Vertical and lateral variations in fracture spacing in folded carbonate sections and its relation to locating horizontal wells. *Journal of Canadian Petroleum Technology, 34,* 51–56.
Neuman, S. P. (1990). Universal scaling of hydraulic conductivities and dispersivities in geologic media. *Water Resources Research, 26,* 1749–1758.
Pickens, J. F., & Grisak, G. E. (1981). Scale-dependent dispersion in a stratified granular aquifer. *Water Resources Research, 17,* 1191–1211.
Prill, R. C., Johnson, A. I., & Morris, D. A. (1965). *Specific yield—Laboratory experiments showing the effect of time on column drainage*. U.S. Geological Survey Water-Supply Paper 1662-B.
Pryor, W. A. (1973). Permeability-porosity patterns and variations in some Holocene sand bodies. *American Association Petroleum Geologists Bulletin, 57*(1), 162–189.
Ritzi, R. W., Jr., Jayne, D. F., Zahradnik, A. J., Jr., Field, A. A., & Fogg, G. E. (1994). Geostatistical modeling of heterogeneity in glaciofluvial, buried valley aquifers. *Ground Water, 32,* 666–674.
Ronan, A. D., Prudic, D. E., Thodal, C. E., & Constantz, J. (1998). Field study and simulation of diurnal temperature effects on infiltration and variably saturated flow beneath an ephemeral stream. *Water Resources Research, 34*(9), 2137–2153.
Schwartz, F. W., & Zhang, Z. (2003). *Fundamentals of ground water*. New York: Wiley.
Schulze-Makuch, D. (2005). Longitudinal dispersivity data and implications for scaling behavior. *Ground Water, 43,* 433–456.
Stearns, D. W. (1967). Certain aspects of fracture in naturally deformed rocks. In R. E. Rieker (Ed.), *National science foundation advanced science seminar in rock mechanics* (pp. 97–118). Bedford, MA: Air Force Cambridge Research Laboratories.
Theis, C. V. (1935). The relation between the lowering of the piezometric surface and the rate and duration of discharge of a well using groundwater storage. *Transactions. American Geophysical Union, 16,* 519–524.
Todd, D. K., & Mays, L. W. (2005). *Groundwater hydrology* (3rd ed.). New York: Wiley.
Walton, W. C. (1960). *Leaky artesian aquifer conditions in Illinois*. Illinois State Water Survey Report of Investigation 39.
Walton, W. C. (1997). *Practical aspects of ground water modeling* (3rd ed.). Westerville, OH: National Ground Water Association.
Webb, E. K., & Anderson, M. P. (1996). Simulation of preferential flow in three-dimensional, heterogeneous conductivity fields with realistic internal architecture. *Water Resources Research, 32,* 533–545.

Chapter 3
Vadose Zone Hydrology Basics

3.1 Introduction

The vadose zone consists of the soil or rock located between land surface and the regional water table. The term "regional" is used to exclude local perched aquifers. The vadose zone includes the unsaturated zone and the capillary fringe, which is a usually thin saturated zone located above the water table in which water is held under less than atmospheric pressure (Fig. 3.1). Very fine-grained sediments and poorly sorted sediments with a very fine-grained component have relatively high capillary pressures and can thus have thick (>1 m) capillary fringes.

With the respect to aquifer recharge, the vadose zone is the interface between a surficial hydrological system and its underlying aquifer. The vadose zone is traditionally treated by hydrologists in regional-scale groundwater resources investigations in a lumped, "black box" fashion (Harter and Hopmans 2004). The key issue for groundwater resources evaluations is usually how much water passes through the vadose zone to recharge the underlying aquifer rather than the specific processes operative in the vadose zone. The vadose zone may be simply modeled through a rainfall-recharge relationship.

Processes that impact vertical and horizontal groundwater flow in the vadose zone have increasing importance as the depth to the water table, and thus vadose zone thickness, increases. Vadose zone processes can have great importance in managed aquifer recharge (MAR) systems that utilize surface spreading because the vadose zone is the location of many contaminant attenuation and other geochemical processes relevant to water quality. The vadose zone performs several main functions (Harter and Hopmans 2004):

- partitioning of precipitation (and applied water) into infiltration, runoff, evapotranspiration, interflow, and groundwater recharge
- storing and transferring water in the root zone

R. G. Maliva, *Anthropogenic Aquifer Recharge*,
Springer Hydrogeology, https://doi.org/10.1007/978-3-030-11084-0_3

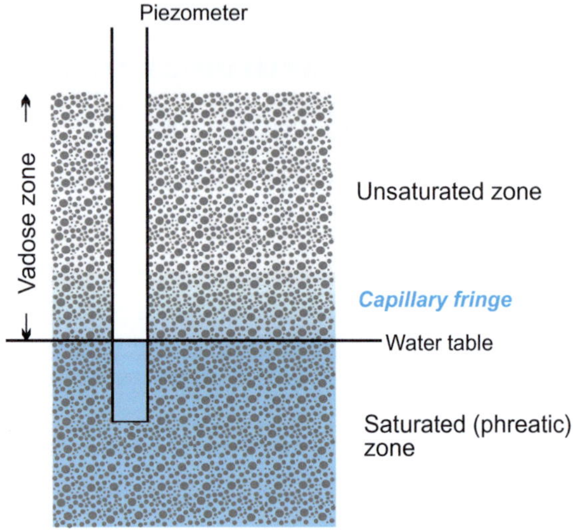

Fig. 3.1 Diagram of the relationship between the vadose zone, unsaturated zone, and water table. The lower part of the capillary fringe may be saturated but is above the water table

- storing and transferring water in the deep vadose zone between the root zone and water table
- storing, transferring, filtering, adsorbing, retarding, and attenuating solutes and contaminants before they reach the water table.

Salts, nutrients (e.g., nitrate), and chemicals present in surface water or applied to the land surface (e.g., fertilizers and pesticides) tend to accumulate in the vadose zone in semiarid and arid regions. Changes in recharge rates and water levels (both natural and anthropogenic) can mobilize salts and chemicals that accumulated in the vadose zone with associated impacts to groundwater and surface water quality.

There is an extensive literature on the hydrology of the vadose zone, including several dedicated textbooks (e.g., Stephens 1996; Selker et al. 1999) and chapters in general groundwater, soil sciences, agriculture, and contaminant hydrology texts (e.g., Bouwer 1978; Fetter 1999, 2001; Schwartz and Zhang 2003; Todd and Mays 2005). The mechanism and controls of the movement of contaminants from the unsaturated soil zone to underlying aquifers is of fundamental importance in the management and remediation of contamination.

Infiltration is the process by which surface water enters the soil (vadose zone). The term "percolation" is used to describe the downward movement of water through the vadose zone that occurs after infiltration. Precipitation that falls onto a land surface is partitioned between infiltration, runoff, and evapotranspiration (ET). Infiltrated water may either percolate to water table and enter an aquifer, be lost to ET, be stored in the vadose zone, or flow in the vadose zone to downgradient discharge areas. The flux of soil water across the water table into the phreatic zone is called "recharge." In arid and semiarid lands, vegetation is often highly effective in extracting water, and very little, if any, water that infiltrates between stream channels (i.e., in interfluve areas) reaches the water table.

Deep percolation refers to water that percolates past the root zone of plants (Bouwer 1978), after which it is much less susceptible to loss via ET. The similar term "net infiltration" is defined as the percolation flux that passes below the depth at which the rate of removal by ET becomes insignificant (Flint et al. 2002). Net infiltration is a function of precipitation, air temperature, root zone depth and root density, soil hydraulic properties and depths, and bedrock permeability (Flint et al. 2004). Under steady-state conditions, the net infiltration flux will be equal to the unconfined aquifer recharge rate unless some of the infiltrating water discharges before reaching the water table. For example, net infiltration water could enter a perched aquifer that discharges at a spring or a seep (Flint et al. 2002).

The deep percolation rate is approximately equal to the recharge rate with the caveat that in semiarid and arid lands with deep water tables, it may take an extremely long time for the small volume of water the passes through the root zone to reach the water table and, therefore, steady-state conditions may not exist.

The root zone in some areas may be very thick. Phreatophyte plants (which draw water from near the water table) can have very deep roots. Some members of the genera *Acacia* and *Boscia* (Jennings 1974; Canadell et al. 1996) were reported to have roots 10 s of meters deep. In the short-term (e.g., at the start of operation of an MAR system), some of the deep percolation may contribute to an increase in soil-water storage. However, over time, steady-state conditions may be approached in which the volume of water stored in the soil zone stabilizes.

3.2 Capillary Pressure

Capillary pressure has been defined "the pressure differential between two immiscible fluid phases occupying the same pores caused by interfacial tension between the two phases that must be overcome to initiate flow" (SPE International n.d.). A similar definition is that capillary pressure is the difference in pressure across the interface between two immiscible fluids, which are referred to as wetting and non-wetting phases:

$$P_c = P_{nw} - P_w \tag{3.1}$$

where (in Pa, N/m^2, or cm of water)

P_c capillary pressure
P_{nw} pressure of the non-wetting phase
P_w pressure of the wetting phase

With respect to the vadose zone, the wetting phase is water and the non-wetting phase is air. As the wetting phase, water tends to adhere to the solid surfaces of vadose zone sediments, whereas air, which is usually under atmospheric pressure, occupies the center of pores. Water in vadose zone, as the wetting phase, has a greater than atmospheric pressure, and thus the capillary pressure of water in the vadose zone is a negative value.

Capillary pressure in a tube can be calculated using the Young-Laplace equation:

$$P_c = \frac{2\gamma \cos\theta}{r} \tag{3.2}$$

where

P_c capillary pressure (Pa, N/m^2, $kg/(ms^2)$)
γ surface tension (N/m, J/m^2, kg/s^2)
θ wetting or contact angle (degrees)
r effective radius of the interface (tube) (m)

The wetting (contact) angle is the angle between the wetting and nonwetting phase where the two phases meet a surface. Both γ and θ are properties of the fluids and solid. Capillary pressure is inversely proportional to the radius of the interface, which in porous rock corresponds to pore throat size. Pore throats are constrictions that connect larger pores and behave essentially as small capillary tubes.

In the case of a vertical tube, surface tension will cause water to rise, while gravity will act to pull the water downward. For a vertical capillary tube with a radius "r" and in which water is the wetting phase, the height at which water would rise above static water level is calculated as

$$h = \frac{2\gamma \cos\theta}{\rho g r} \tag{3.3}$$

where

h height of water column (m)
g gravitational acceleration (9.807 m/s^2)
ρ density of water (≈1000 kg/m^3)

In the vadose zone, capillary pressure results in water being pulled into a formation. Fine-grained sediments, in general, have smaller pore sizes and pore throat diameters than coarser-grained sediments and thus greater capillary pressures. Unsaturated, very-fined grained materials (silts and clays) will tend to pull in and strongly hold water. During initial infiltration and percolation, water will tend to be preferentially drawn into finer-grained sediment and rock. Water that subsequently enters larger pores and pore throats will have a lower capillary pressure. Infiltration rates measured during the initial wetting of soils can be significantly greater than subsequent rates once soils become saturated. The effects of capillary pressure on infiltration rates need to be considered in the design and interpretation of the results of infiltration tests.

3.3 Soil-Water and Matric Potential

Soil-water potential (ψ (g, m, o)), also referred to as water potential and total water potential, is essentially the work water can do, per unit quantity, as it moves from its current state (elevation, temperature, pressure, chemical composition) to a reference state, which is soil saturated with pure water at an elevation defined to be zero. Soil-water potential is the sum of the gravitational potential (ψ (g)), matric potential (ψ (m)), and osmotic potential (ψ (o)):

$$\psi(g, m, o) = \psi(g) + \psi(m) + \psi(o) \tag{3.4}$$

Soil-water potential has the units of pressure (bars, atmospheres, kPa, or height of water).

The matric potential of soils is the component of the water potential due to the adhesion of water molecules to soil surfaces. Matric potential is the sum of capillary and adsorptive forces, and is always negative. It is greatest (most negative) in dry soils, approaching zero as a soil approaches saturation. Matric potential is also strongly related to soil texture and is greatest in very fine-grained material. Matric potential can be measured in the field using a tensiometer.

Osmosis is the process by which water flows across semi-permeable membranes from areas of low solute concentrations (high water potential) to areas of higher solute concentration (lower water potential). Within soils, clay layers may act as semi-permeable membranes. Osmotic potential is related to differences in salinity and is insignificant in most soil situations. Typically, osmotic potential is very small relative to matric potential.

Gravitational potential is related to differences in elevation (elevation head). Gravitational potential is usually also small relative to matric potential differences in dry vadose zone soils. Hence, soil-water potentials are approximately equal to matric potentials. Water within the vadose zone will move from areas of high matric potential to areas of low (more negative) matric potential.

The relationship between water potential and volumetric water content is illustrated by soil-water retention curves, which are also referred to as soil-moisture characteristic curves (Fig. 3.2). Water potential is zero where a soil is saturated and its volumetric water content is equal to its effective (interconnected) porosity. The shape of the soil-water retention curve depends on soil type and associated pore-size distribution.

Water potential increases as water is drained from the pores. In sands and gravels, drainage of water from large pores results in a relatively minor associated change in matric potential, which is reflected as a flat to gently sloping segment on the soil-water retention curve. Drainage of the remaining water from the very small pores and pore throats results in large increases in (more negative) matric potential.

The water content of a soil at a given water potential depends on the wetting history of the soil. Soil-water retention curves differ depending on whether the sample is being wetted or dried (Fig. 3.3). The difference in the relationship between soil

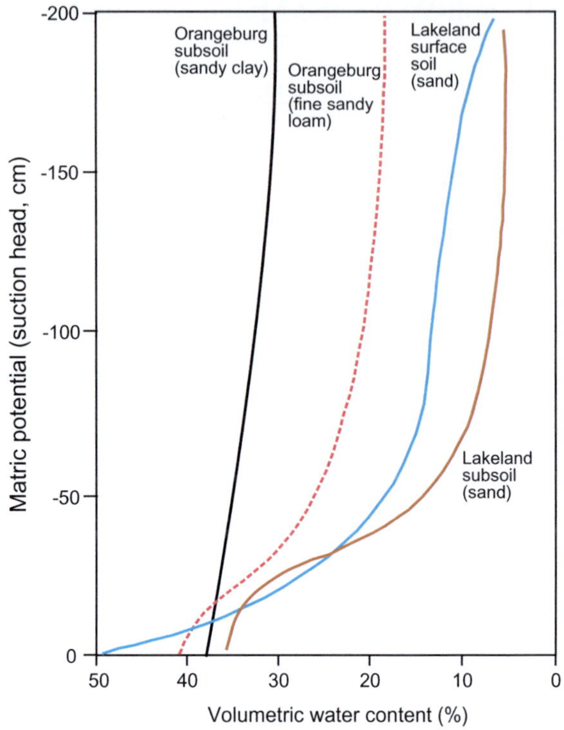

Fig. 3.2 Examples of soil-water retention curves for some Florida soils (source: Bouma et al. 1982)

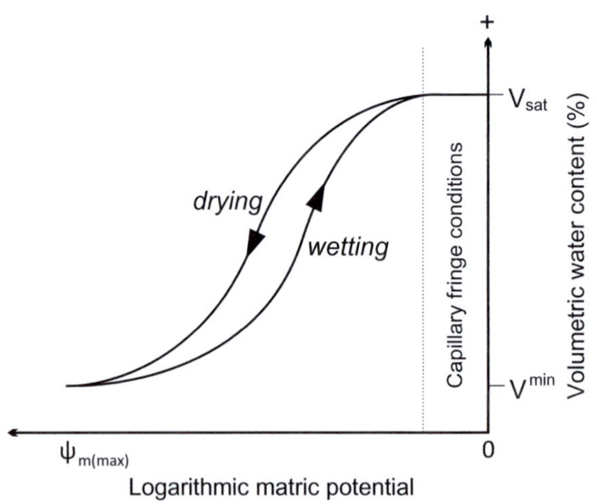

Fig. 3.3 Conceptual diagram illustrating the difference in soil-water retention curves between soils that are undergoing wetting and drying, which is referred to as hysteresis. Greater matric potential means a more negative pressure

water content and matric (water) potential under wetting and drying is referred to as "hysteresis." At a given water potential, the water content will be greater during drying than wetting. During drying, some water is retained in larger pores by smaller pore throats until the water potential at pore throats becomes sufficient to allow the water to drain.

3.4 Unsaturated Hydraulic Conductivity

The hydraulic conductivity of saturated sediments or rock is a function of their intrinsic permeability and the dynamic viscosity and density of the groundwater. The hydraulic conductivity of unsaturated sediments is also a function of the volumetric soil moisture content, rather than being a constant value. Unsaturated hydraulic conductivity increases with water content (Fig. 3.4). At low soil moisture contents, water is preferentially drawn into the smallest pores where it is tightly held by capillary pressure and thus flow is minimal. As soil moisture content increases, capillary pressure decreases (i.e., water is less tightly held) and greater flow rates occur under a given hydraulic gradient.

The relationship between unsaturated hydraulic conductivity and soil moisture content is dependent on soil properties, particularly grain types, pore size distribution, and texture (Bouwer 1964). Determination of site-specific unsaturated hydraulic conductivity values is further complicated by heterogeneities in the soil matrix and the presence, types, and degree of development of macroporosity. Unsaturated hydraulic conductivities can be estimated from the soil-water moisture retention function with at least one measurement of hydraulic conductivity at a known water content or from infiltration rates measured at several tensions (e.g., Brooks and Corey 1964;

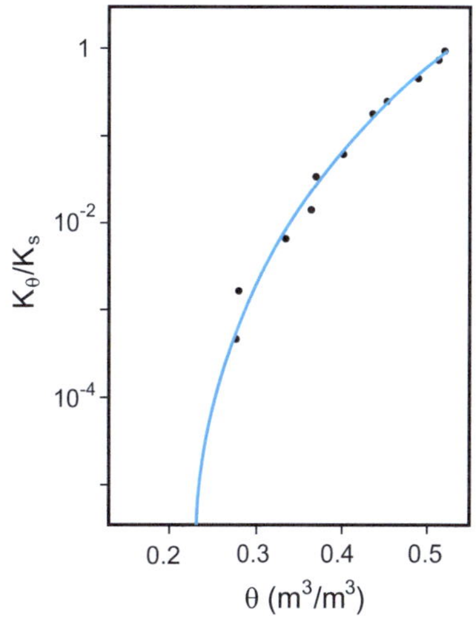

Fig. 3.4 Example of the relationship between soil moisture and relative hydraulic conductivity, which is defined as the ratio of measured hydraulic conductivity at various soil moistures to saturated hydraulic conductivity, for a loam soil (after Van Genuchten 1980)

Campbell 1974; Van Genuchten 1980; Ankney et al. 1991; Schaap and Leij 2000). Empirical data are required for each soil type to obtain soil-type specific hydraulic conductivity versus soil-moisture functions.

Under some circumstances, the relationship between hydraulic conductivity and grain size in unsaturated sediments may be the opposite of the relationship under saturated conditions. Finer-grain sediments may preferentially draw in water, and thus have higher volumetric water contents than nearby coarser-grained beds. The greater volumetric water content of the fine-grained sediments may give them a higher unsaturated hydraulic conductivity than drier, coarser-grained beds. Heilig et al. (2003), in a set of field experiments, demonstrated this relationship by showing that at a matric pressure of −280 cm, coarse sand had a hydraulic conductivity of 1.7×10^{-8} cm/s (4.8×10^{-4} ft/day) and fine sand had a corresponding hydraulic conductivity of 1.2×10^{-4} cm/s (3.4 ft/day).

3.5 Darcy's Equation for Unsaturated Sediments

One-dimensional vertical flow through the vadose zone can be expressed using a modification of Darcy's equation referred to as the Darcy-Buckingham equation:

$$q_Z = -K(\theta)\frac{\partial h}{\partial z} \tag{3.5}$$

where

q_z specific flow rate in the vertical (z) direction (m/d)
$K(\theta)$ hydraulic conductivity (m/d) at volumetric soil moisture θ
h total head (m)
z distance (m)
θ volumetric soil moisture content (unitless, m^3/m^3)

Volumetric soil moisture content is the ratio of the volume of water in a soil sample (V_w) to the total volume of the sample (V_t)

$$\theta = \frac{V_w}{V_t} \tag{3.6}$$

Within the vadose zone, total head is a function of both elevation (z) and water potential ($\psi(\theta)$):

$$h = z + \psi(\theta) \tag{3.7}$$

Equations 3.5 and 3.7 can be combined to relate flow rate to water potential:

$$q_Z = -K(\theta)\frac{\partial(z + \psi(\theta))}{\partial z} \tag{3.8}$$

$$q_Z = -K(\theta)(1 + \frac{\partial\psi(\theta)}{\partial z}) \quad (3.9)$$

The fundamental difficulty in applying the Darcy-Buckingham equation to actual projects is that the quantitative relationship between hydraulic conductivity and soil moisture content is poorly constrained and variable. Both hydraulic conductivity and soil-water potential are non-linear functions of water content (θ) and Eqs. 3.8 and 3.9 cannot be directly solved.

With respect to MAR projects, the Darcy Buckingham equation is important for providing a qualitative understanding of groundwater flow in the vadose zone. In practice, field determination of hydraulic conductivity and soil-water potential versus soil moisture content relationships is seldom performed for MAR projects. In numerical models of vadose zone processes, lithology-based generic relationships are often used. For example, in the USGS Unsaturated-Zone Flow package, unsaturated hydraulic conductivity values are obtained using the Brooks and Corey (1964) method (Niswonger et al. 2006). The values of the exponent (or lambda value) are usually estimated from sediment type.

3.6 Infiltration Theory

There are several terms applied, at times inconsistently, to infiltration processes. "Infiltration rate" refers to the velocity at which water enters the soil. Infiltration rates vary over time during an infiltration event, and are typically expressed in units of millimeters or inches per hour. "Infiltration capacity" is defined as the maximum rate at which water can be absorbed by a given soil per unit area under given conditions (Horton 1933). "Field capacity" or "field moisture capacity" is the amount of soil moisture or water held in a soil after excess water has drained away and the rate of downward movement has decreased. Horton's (1933) definition of the term "field moisture capacity" is roughly equivalent to specific retention. Field capacity usually refers to water held in soils after the rate of drainage slows (which usually takes place about 2–3 days after a rain or irrigation), whereas specific retention refers to water retained after gravity drainage is complete.

Infiltration will be most rapid in dry soils because of their greater capillary action (i.e., more negative soil-water potential), which acts to draw in water. However, the recharge rate in dry sediments will be less than the rate in pre-wetted sediments because water is retained under capillary pressure and the pore spaces must first be largely filled before significant percolation to the water table can occur.

Horton (1933) in his seminal paper considered the soil to act as a separating surface between infiltration and surface runoff. Rainfall will infiltrate into the soil at its infiltration capacity with excess water available for surface runoff. The initially infiltrated water will first make up any soil-moisture deficiency (i.e., soil moisture content below the field moisture capacity). Once the field moisture capacity is exceeded, some infiltrated water will be available for groundwater recharge. Horton (1933) also noted that only a fraction of the rainfall excess will actually become runoff into streams. A variable amount of the runoff will be captured in transit as surface detention and will either infiltrate into the soil or later evaporate.

After a rainfall or irrigation event, infiltration rates decrease over time in a regular, cyclical manner (Horton 1933; Jury and Horton 2004). The decrease in infiltration rates is caused by processes such as the packing of the soil surface, swelling of the soil, infiltration of fine materials into soil-surface openings, and the progressive filling of soil pores with water. Horton expressed this change in infiltration rate over time by the equation (Horton 1940)

$$f_t = f_c + (f_0 - f_c)e^{-\mathrm{kt}} \tag{3.10}$$

where

- f_t infiltration rate at time t (mm/h)
- f_0 initial or maximum infiltration rate (mm/h)
- f_c final or equilibrium infiltration rate after the soil has become saturated (mm/h)
- t time (h)
- k decay constant specific to the soil (h^{-1})

The total volume of infiltration (F_t; units = mm) after time t can be calculated as

$$F_t = f_c t + \frac{(f_0 - f_c)}{k}\left(1 - e^{-kt}\right) \tag{3.11}$$

Horton's equations provide a useful conceptual understanding of the infiltration process, but has the practical limitation that it is difficult to obtain accurate values for the parameters. Horton's model also does not consider macropore (conduit) flow.

The Green and Ampt (1911) method can be used to estimate infiltration rates in most soil types. The method is based on advancement during infiltration of a "wetting front" (Fig. 3.5), which is a sharp boundary between the upper saturated soil, with a soil moisture content (θ_s), and the underlying unsaturated soil, with a moisture content (θ_i). The Green-Ampt method is based on the assumption that the wetting front advances at the same rate with depth, soil in the wetted region has constant properties, and the soil-water potential below the wetting front does not change with time and depth (Jury and Horton 2004). The difference in soil moisture across the wetting front ($\theta_s - \theta_i$) is referred to as the soil moisture deficit ($\boldsymbol{\Delta\theta}$). The infiltration volume is equal to the product of $\Delta\theta$ and the distance the front moved since the start of infiltration (l).

The Green-Ampt equations for instantaneous infiltration rate and cumulative infiltration under conditions of negligible surface-water ponding are

$$f(t) = K_s\left[\frac{\psi_f \Delta\theta}{F(t)} + 1\right] \tag{3.12}$$

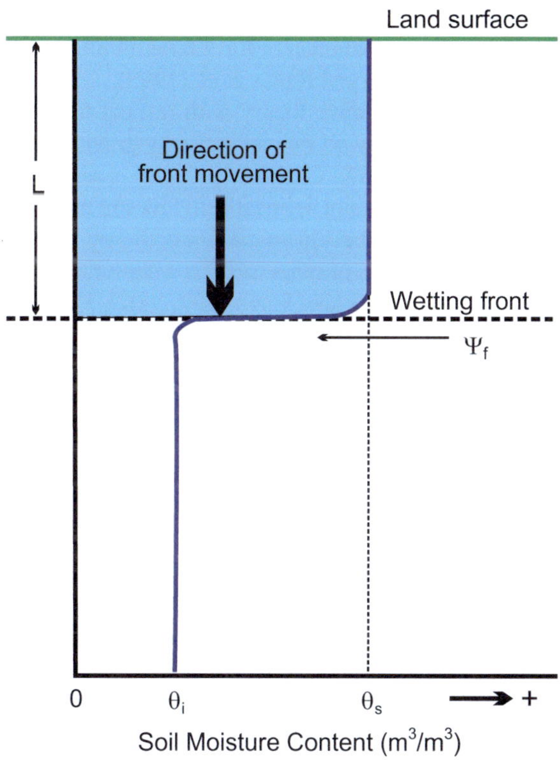

Fig. 3.5 Diagram of the Green-Amp equation parameters

$$F(t) = K_s t + \psi_f \Delta\theta \ln\left[1 + \frac{F(t)}{\psi_f \Delta\theta}\right] \quad (3.13)$$

where,

f(t) instantaneous infiltration rate at time "t" (mm/h)
F(t) cumulative infiltration at time "t" (mm)
K_s saturated vertical hydraulic conductivity (mm/h)
ψ_f soil-water potential (suction head) at the wetting front (mm)
$\Delta\theta$ change in soil moisture across the wetting front (dimensionless)

In order to solve the Green Ampt equation for the instantaneous infiltration rate or cumulative infiltration after a given time (t), data are needed for the variables K_s, $\Delta\theta$, and ψ_f. The variable being solved for, *F(t)*, is present in the equation itself, which precludes direct solution by algebraic manipulation. However, Eq. 3.13 can be solved by numerical techniques. Different values of *F(t)* can be systematically tried until the equation converges.

Saturated hydraulic conductivity is a property of soil composition and texture, which can be either directly measured or typical (generic) values for soil types may be used. Soil-water potential (suction head) is as function of soil type and volumetric water content and is graphically illustrated in water retention curves. The Green-Ampt infiltration parameters for a large number of soil types were determined by Rawls et al. (1983) and Rawls et al. (1993).

The kinematic wave theory with respect to unsaturated flow has been used to describe the downward flow of water by gravity drainage (Charbeneau 1984, 2000; Morel-Seytoux 1987). Capillary pressure is assumed to not impact infiltration. The method is therefore not applicable to low soil moisture conditions at which capillary flow dominates. The kinematic wave theory is mathematically expressed by the law of conservation of mass through a continuity equation and a flux-concentration relationship (Darcy's law; Singh 1997, 2002). The analytical solution of the kinematic wave theory can be used to model soil water distribution and recharge (Charbeneau 1984, 2000; Todd and Mays 2005).

3.7 Infiltration Controls

3.7.1 Introduction

Water naturally (e.g., rainfall) or anthropogenically applied (e.g., irrigation and MAR) to land surfaces is partitioned between infiltration, runoff, and evapotranspiration. Initially, water will more rapidly infiltrate into dry soils because of strong capillary action. As a soil fills with water, capillary forces diminish and ponding will occur if the soil cannot transmit water downward under gravity as rapidly as the application rate.

Conservation of mass considerations indicate that the rate of infiltration depends on the rates of water flow across and away from the soil-water interface. Once uppermost soils become saturated, water cannot infiltrate faster than the rate at which water flows away from the land surface. The bottleneck that controls the infiltration rate may be either the soil (or rock) at land surface, less permeable strata within the vadose zone, or the transmissivity of the water table aquifer. For example, low-permeability clay layers may retard percolation and eventually impede infiltration as overlying soils become saturated. Rapid infiltration can results in the formation of a hydraulic mound below the infiltration site (e.g., infiltration basin), which can reach land surface. Once the soil or rock below the infiltration site becomes fully saturated, the infiltration rate will be controlled by the rate of lateral flow away from the mound.

Vertical flow is dominated by gravity (capillary forces are minor) and the hydraulic gradient is one so long as the wetting front is above the water table. The hydraulic gradient may decrease dramatically once the wetting front reaches the water table. Further infiltration will be controlled by horizontal flow and the infiltration rate may slow dramatically (Bouwer 2002; Phillips and Kitch 2011). It is important to consider that vertical hydraulic conductivities (K_z) are often substantially less the horizontal hydraulic conductivities (K_h), and that the flow of water will seldom be straight down (Phillips and Kitch 2011).

Numerous factors influence infiltration rates including:

- type of soil (e.g., sand, loam, and clay) and its associated hydraulic properties
- soil vertical hydraulic conductivity
- moisture content of the uppermost soil
- soil moisture gradient with depth in the unsaturated or vadose zone
- slope of land surface
- vegetation or organic debris located at land surface
- presence of physical pathways in the surficial soil (e.g., deep mud cracks)
- presence of impervious crusts or coatings at land surface
- rate of rainfall versus the infiltration capacity of the soil
- degree that a given soil traps air between the wetted front and the water table
- pore-size distribution of the soil, which controls capillarity
- heterogeneity of the soil profile
- depth of ponding of water above land surface
- time.

3.7.2 Matrix and Macropore Recharge

Recharge is subdivided into matrix and macropore recharge. Matrix (interstitial) recharge occurs through the intergranular pore spaces of sediment or rock. Macropore recharge, on the contrary, occurs through zones of enhanced permeability. Macropores include features such as deep desiccation cracks, animal burrows, root tubes, pipes, fissures, and fractures (Stephens 1996; Wood et al. 1997; Wood 1999). The higher permeability of macropores results in much more rapid rates of infiltration and recharge compared to matrix recharge. Rain water is more likely to escape ET if it rapidly percolates through macropores, especially to below the root zone of the local vegetation.

Total recharge (recharge flux) is the sum of matrix and macropore recharge. Recharge often occurs simultaneously through both matrix porosity and macroporosity. Water that infiltrates into and percolates through macropores can be completely absorbed by the surrounding matrix before it reaches the water table (Stephens 1996). Macropores can also become clogged over time, which can result in the spatial distribution of deep soil water movement gradually changing over time (Stephens 1996).

Macropore recharge has several implications for MAR. Increased infiltration and recharge rates from macropores can be desirable for some MAR systems because it could allow for more water to be recharged during a limited period of water availability and thus allow for a decreased system footprint (recharge area). On the contrary, macropore recharge may be undesirable where vadose zone processes are being relied upon to improve water quality. Water could bypass or flow too rapidly through the vadose "treatment" zone. Macropores result in a heterogeneity that needs to be captured in the design of infiltration facilities. Infiltration testing procedures need to be carefully implemented so as to obtain data that are representative of a project site and not be biased either by the inclusion or exclusion of macropore recharge features. Additionally, the clogging potential of macropores also needs to be considered. Preferential clogging of macropores could result in system underperformance.

3.7.3 Surface Clogging Layers

A major operational issue for MAR systems that utilize surface spreading is the development of a low-permeability clogged layer that impedes infiltration (Chap. 11). Clogging may be the result of physical processes (e.g., deposition of very fine-grained sediment), biological activity, chemical precipitation, air or gas binding, or a combination of these processes. Physical clogging due to the deposition of very fine-grained sediments can be the result of suspended solids present in the applied water or the reworking (mobilization) and redeposition of fines already present in surficial soils. Clay swelling and dispersion can also contribute to the development of a surficial clogging layer.

Biological activity can contribute to clogging through the accumulation of organic matter (e.g., settling of algae), the growth of organisms, and the formation of biofilms (combination of cellular material and extracellular polysaccharides). Biological activity may also increase the pH of water and thereby promote the precipitation of calcium carbonate.

Infiltration rates are a function of both the vertical hydraulic conductivity of the clogging layer and the water (ponding) depth above the layer. Infiltration rates almost linearly increase with ponding depth in clogged basins. However, increases in ponding depth can compress clogging layers, making them less permeable (Bouwer 2002). As long as the water table is more than one meter below the bottom of a basin with a clogging layer, infiltration rates are unaffected by changes in groundwater level. Infiltration rates only decrease when the capillary fringe reaches the bottom of a basin (Bouwer 2002).

A key part of the design, operation, and maintenance of MAR systems is managing clogging. Measures may be taken to reduce the rate of clogging, such as pretreating the applied water. A variety of rehabilitation strategies are available for the different system types to restore infiltration capacity. For example, infiltration basins are often rehabilitated by periodic drying, harrowing (or other means) to disrupt the clogging layer, and physical removal of the clogging layer by scrapping. Clogging causes, impacts, and rehabilitation methods are further addressed in the Chap. 11 and sections on the different MAR system types.

3.7.4 Air Entrapment

As water infiltrates into and percolates through the vadose zone, air present in the pores is expelled (vented). Direct observations have demonstrated that air can inhibit infiltration when it is becomes entrapped ahead of the wetting front or is trapped as bubbles within the transmission zone (Constantz et al. 1988; Faybishenko 1999; Wang et al. 1998; Weir and Kissling 1992).

Air trapped below the wetting front will increase in pressure and eventually vent upward to allow water to flow deeper into the vadose zone. Purging of trapped air is controlled by the continuity of the wetting front, the grain-size distribution of the soil, and heterogeneities within the soil profile and corresponding variations in the distribution of hydraulic conductivity (Bouwer 1966). Air venting can cause a "finger flow" pattern within the vadose zone that may be independent of soil heterogeneity but usually occurs in concert with physical heterogeneities. Non-uniform early infiltration can isolate parts of a pore network by air encapsulation, thereby altering recharge rates (Carrick et al. 2010).

A more extreme effect of trapped air during infiltration is known as the "Lisse effect", in which the trapping and compression of air in the unsaturated zone after a rainfall events results in abnormally high water levels in shallow wells (Weeks 2002). Weeks (2002) concluded that the Lisse effect is likely more common than supposed and can produce an abnormal rise in water levels of about 0.10–0.55 m (0.3–1.8 ft). The abnormally high water levels in wells, if misconstrued as actual rises in the water table, could result in a substantial overestimation of groundwater recharge (Maliva and Missimer 2012).

3.7.5 Temperature Effects on Infiltration

Temperature impacts infiltration through its relationships with evaporation rates and hydraulic conductivity. Warmer temperatures during the summer can result in significant reductions in net infiltration rates because of more rapid evaporation of standing water and ET of shallowly infiltrated water in the upper vadose zone. Temperature affects the hydraulic conductivity of soils through its impact on the dynamic viscosity of water. Field and modeling studies have demonstrated that large diurnal variations in infiltration rates occur at some sites, which were attributed to the temperature-dependence of hydraulic conductivity (Jaynes 1990; Constantz et al. 1994; Ronan et al. 1998).

As temperature increases, the viscosity of water decreases and hydraulic conductivity increases, which for a given hydraulic gradient will result in more rapid infiltration. During a field study near Carson City, Nevada, both field data and modeling results indicate that infiltration rates (and thus stream losses) are greatest in the late afternoon when temperatures are the greatest and least in the morning when temperatures are lowest (Ronan et al. 1998). Infiltration rates in the Dan Region

(Sahfdan) Project (Israel) soil-aquifer treatment system were reported to be lower during the winter when the viscosity of water is greater (Icekson-Tel et al. 2003).

The magnitude of the temperature effect on recharge will depend largely on the magnitude of the diurnal and seasonal variations in temperature. The magnitude of diurnal stream temperature variations is inversely proportional to discharge, since the thermal mass of streams increases with increasing discharge (Constantz et al. 1994). The temperature effects on hydraulic conductivity, and thus stream losses, were thought to be greater than the potential effects of variations in ET rates.

3.8 Percolation and the Fate of Infiltrated Water

Aquifer recharge rates may be significantly less than infiltration rates. Water that enters the vadose zone can either reach the water table (and become recharge), be locally lost to ET, or flow laterally where it discharges to a stream or spring, becomes recharge, or is lost to ET at a downgradient location. Groundwater flow within the vadose zone is referred to as "interflow." Infiltrated water can also be stored in the vadose zone as soil moisture. Vadose zone processes, as would be expected, have a greater impact on recharge in more arid climates and where the vadose zone is thick (i.e., there is great depth to the water table).

Vegetation can be highly efficient at abstracting soil moisture. Plants in arid climates out of necessity tend to be very well adapted to capturing sparse water when available. Several studies of arid and rangeland systems have documented regions in which extraction of soil moisture within the root zone creates an upward water potential gradient below the root zone, which results in negligible current recharge in interdrainage (interfluve) areas (Andraski 1997; Izbicki et al. 2002; Scanlon et al. 2003, 2005; Stonestrom et al. 2007). Natural recharge in some arid regions appears to now occur only in areas where surface water flow is concentrated, such as in ephemeral river channels.

Infiltration tests performed in the Sultanate of Oman by Haimerl et al. (2002) illustrate some key relationships concerning infiltration and recharge by surface spreading. Tests were performed using an infiltration basin in which 11 tensiometers were installed at different depths. Groundwater levels were measured in an adjacent borehole. The arrival of the wetting front can be identified by the initiation of a decrease in water tension (suction). As would be expected, infiltration rates were greatest in dry soils, but actual recharge was greater in pre-wetted soil. The high matric potential for dry soil draws water into the soils, but the water tends to be retained. Pre-wetted soil requires a lesser infiltration volume until percolation to the water table (i.e., groundwater recharge) can occur.

A critical relationship is that surface-spreading systems with small water-loading (volume/area) rates may have large infiltration rates, but little or no recharge as the water is retained in the soil and subsequently lost to ET. Smaller wetted areas and longer application times can allow for greater recharge rates (Haimerl et al. 2002). From a design perspective, small recharge basins may result in greater recharge rates than would occur in a single large basin, for the same volume of applied water. Alternatively, large basins are often divided into multiple subbasins.

In areas with a deep water table, injection wells may be a more efficient method for groundwater recharge. However, groundwater recharge using basins may still be possible in areas with deep (>100 m; 328 ft) water tables. Izbicki et al. (2008) documented the testing of an infiltration pond in the western Mojave Desert (California) in which the water table was located 113–121 m (371–397 ft) below land surface. Migration of the wetting front was monitored using a series of tensiometers installed in a test borehole. It initially took three years for the infiltrated water to reach the water table. The time was reduced to about a year in subsequent tests due to the increased hydraulic conductivity from the residual increased water saturation from previous testing. Less additional water was also retained in the pre-wetted unsaturated zone. An important hydrogeological and operational issue is the presence of clay layers (paleosols) that limit the downward movement of water and enhance the spreading of water away from the infiltration area. The water at the wetting front also had a high salinity caused by the mobilization of salts, but the salinity quickly decreased to the concentrations in the infiltrating water.

Aquifer heterogeneity can greatly impact the rate and direction of groundwater flow in the vadose zone. A key issue in arid lands with thick unsaturated zones, is the effects of relatively impermeable strata, such as clays and caliche layers. These layers can cause the spreading of infiltrated water and result in more complex flow paths (Izbicki et al. 2002, 2007). Geological features in the vadose zone, such as slanted impermeable beds and higher transmissivity deposits (channel sands) can result in preferential flow directions.

Izbicki et al. (2007) documented an infiltration test performed on an alluvial fan in the Mojave Desert (near Victorville, California). Approximately 262,000 m^3 (69 million gallons) of water was infiltrated at the site, at which the water table is located about 130 m (427 ft) below land surface. Clay-rich layers impeded the downward flow of water and contributed to its lateral spreading. The rapid movement of small volume of water ahead of the wetting front was also observed, which may have been due to flow along preferential flow paths. For example, saturated conditions were observed to have developed atop a clay-layer 84.7 m (278 ft) below land surface while the main wetting from remained less than 60 m (197 ft) below land surface at that time. Izbicki et al. (2007) concluded that

> Given the time of travel and complex movement of water in thick, heterogeneous unsaturated zones, it may be unrealistic to assume that water infiltrated to depths below the root zone becomes groundwater recharge—especially where channel-abandonment processes on active alluvial fans may effectively strand water in the unsaturated zone before it reaches the water table.

Izbicki et al. (2002, 2007) noted that areas where natural recharge occur may be potential sites for artificial recharge because the unsaturated zone is pre-wetted and thick impermeable caliche layers are not present. In areas of natural recharge, chloride and other soluble salts (e.g., nitrate) may either have not accumulated or have been already leached from the unsaturated zone and thus have a lesser potential for degrading the quality of recharged water (Izbicki et al. 2002).

References

Andraski, B. J. (1997). Soil-water movement under natural-site and waste-site conditions: A multiple-year field study in the Mojave Desert, Nevada. *Water Resources Research, 33,* 1901–1916.

Ankeny, M. D., Ahmed, M., Kaspar, T. C., & Horton, R. (1991). Simple field method for determining unsaturated hydraulic conductivity. *Soil Science Society of America Journal, 55*(2), 467–470.

Bouma, J., Brown, R. B., & Rao, P. S. C. (1982). *Retention of water: Basics of soil-water relationships—Part II.* Gainesville, FL: Soil and Water Science Department, Florida Cooperative Extension Service, Institute of Food and Agricultural Sciences, University of Florida.

Bouwer, H. (1964). Unsaturated flow in ground-water hydraulics. *Journal of Hydraulics Division, American Society of Civil Engineers, 90*(HY5), 121–144.

Bouwer, H. (1966). Rapid field measurement of air entry value and hydraulic conductivity of soil as significant parameters in flow system analysis. *Water Resources Research, 2*(4), 729–738.

Bouwer, H. (1978). *Groundwater hydrology.* New York: McGraw-Hill.

Bouwer, H. (2002). Artificial recharge of groundwater: Hydrogeology and engineering. *Hydrogeology Journal, 10*(1), 121–142.

Brooks, R. H. & Corey, A. T. (1964). *Hydraulic properties of porous media*, Hydrology Papers No. 3. Fort Collins, CO: Colorado State University.

Campbell, G. S. (1974). A simple method for determining unsaturated conductivity from moisture retention data. *Soil Science, 117*(6), 311–314.

Canadell, J., Jackson, R. B., Ehleringer, J. R., Mooney, H. A., Sala, O. E., & Schluze, E.-D. (1996). Maximum rooting depth of vegetation types at the global scale. *Oecologia, 108*(4), 583–595.

Carrick, S., Buchan, G., Almond, P., & Smith, N. (2010). Atypical early-time infiltration into a structured soil near field capacity: The dynamic interplay between sorptivity, hydrophobicity, and air encapsulation. *Geoderma, 60,* 579–589.

Charbeneau, R. J. (1984). Kinematic models for soil moisture and solute transport. *Water Resources Research, 17*(3), 699–706.

Charbeneau, R. J. (2000). *Groundwater hydraulics and pollutant transport.* Upper Saddle River, NJ: Prentice Hall.

Constantz, J., Thomas, C. L., & Zellweger, G. (1994). Influence of diurnal variations in stream temperature on streamflow loss and groundwater recharge. *Water Resources Research, 30,* 3253–3264.

Constantz, J., Herhelrath, W. N., & Murphy, F. (1988). Air encapsulation during infiltration. *Soil Science Society of America Journal, 52,* 10–16.

Faybishenko, B. (1999). Comparison of laboratory and field methods for determining the quasi-saturated conductivity of soils. In M. T. Van Genuchten, F. J. Leij, & L. Wu (Eds.), *Characterization of the hydraulic properties of unsaturated porous media: Proceedings of the International Worship on Characterization and Measurement of Unsaturated Porous Media* (pp. 279–292). Riverside, California, October 22–24, 1997, University of California.

Fetter, C. W. (1999). *Contaminant hydrogeology* (2nd ed.). Upper Saddle River, NJ: Prentice Hall.

Fetter, C. W. (2001). *Applied hydrogeology* (4th ed.). Englewood Cliffs, NJ: Prentice Hall.

Flint, A. L., Flint, L. E., Kwicklis, E. M., Fabryka-Martin, J. T., & Bodvarson, G. S. (2002). Estimating recharge at Yucca Mountain, Nevada, USA, comparison of methods. *Hydrogeology Journal, 10,* 180–204.

Flint, A. L., Flint, L. E., Hevesi, J. A., & Blainey, J. B. (2004). Fundamental concepts of recharge in the desert Southwest: A regional modeling perspective. In J. F. Hogan, F. M. Phillips, & B. R. Scanlon (Eds.), *Groundwater recharge in a desert environment: The Southwestern United States, Water Science and Application Series* (Vol. 9, pp. 159–184). Washington, D.C.: American Geophysical Union.

Green, W. H., & Amp, G. A. (1911). Studies on soil physics. *Journal of Agricultural Science, 4*(1), 1–24.

Haimerl, G., Zunic, F., & Strobl, T. (2002). An infiltration test to evaluate the efficiency of groundwater recharge dams in arid countries. In V. P. Singh, M. Al-Rashed, & M. M. Sherif (Eds.), *Surface water hydrology* (pp. 313–322). Lisse: Swets & Zeitlinger.

Harter, T., & Hopmans, J. W. (2004). Role of vadose-zone processes in regional-scale hydrology: Review of opportunities and challenges. In R. A. Feddes & J. C. Van Dam (Eds.), *Unsaturated zone modeling: programs, applications and challenges* (pp. 179–208). Doordrecht: Kluwer.

Heilig, A., Steenhuis, T. S., Walter, M. Todd, & Herbert, S. J. (2003). Funneled flow mechanisms in layered soil: Field investigations. *Journal of Hydrology, 279,* 210–223.

Horton, R. E. (1933). The role of infiltration in the hydrologic cycle. *Transactions American Geophysical Union, 14,* 446–460.

Horton, R. E. (1940). An approach towards a physical interpretation of infiltration capacity. *Proceedings Soil Society of America, 5,* 399–417.

Icekson-Tal, N., Avraham, O., Sack, J., & Cikurel, H. (2003). Water reuse in Israel—the Den Region Project: evaluation of water quality and reliability of plant's operation. *Water Science and Technology: Water Supply, 3*(4), 231–237.

Izbicki, J. A., Flint, A. L., & Stamos, C. L. (2008). Artificial recharge through a thick, heterogeneous unsaturated zone. *Ground Water, 46,* 475–488.

Izbicki, J. A., Radyk, J., & Michel, R. L. (2002). Movement of water through the thick unsaturated zone underlying Oro Grande and Sheep Creek Washes in the western Mojave Desert, USA. *Hydrogeology Journal, 10*(3), 409–427.

Izbicki, J. A., Johnson, R. U., Kulongoski, J. & Predmore, S. (2007). Ground-water recharge from small intermittent streams in the western Mojave Desert, California. In D. A. Stonestrom, J. Constantz, T. P. A. Ferre, & S. A. Leake (Eds.), *Ground-water recharge in the arid and semiarid southwestern United States* (pp. 157–184). U.S. Geological Survey Professional Paper 1703G.

Jaynes, D. B. (1990). Temperature variations effects on field-measured infiltration. *Soil Science Society of America Journal, 54,* 305–311.

Jennings, C. H. M. (1974). *The hydrology of Botswana.* Ph.D. Thesis, University of Natal.

Jury, W. A., & Horton, R. (2004). *Soil physics.* Hoboken, NJ: Wiley.

Maliva, R. G., & Missimer, T. M. (2012). *Arid lands water evaluation and management.* Berlin: Springer.

Morel-Seytoux, H. J. (1987). Multiphase flow in porous media. In P. Novak (Ed.), *Developments in hydraulic engineering 4* (pp. 103–174). New York: Elsevier.

Niswonger, R. G., Prudic, D. E., & Regan, R. S. (2006). Documentation of the Unsaturated-Zone Flow (UZF1) *Package for modeling unsaturated flow between the land surface and the water table with MODFLOW-2005.* U.S. Geological Survey Techniques and Methods 6-A19.

Philips, C. E., & Kitch, A. A. (2011). A review of methods for characterization of site infiltration with design recommendations. In *Proceedings 43rd Symposium of Engineering Geology and Geotechnical Engineering.* Nevada: University of Las Vegas, March 2011.

Rawls, W. J., Brakensiek, D. L., & Miller, N. (1983). Green-Ampt infiltration parameters from soils data. *Journal of Hydraulic Engineering, 109,* 62–70.

Rawls, W. J, Ahuja, L. R., Brakensiek, D. L., & Shirmohammadi, A. (1993). Infiltration and soil movement. In D. R. Maidment (Ed.), *Handbook of hydrology* (pp. 5.1–5.51). New York: McGraw-Hill.

Ronan, A. D., Prudic, D. E., Thodal, C. E., & Constantz, J. (1998). Field study and simulation of diurnal temperature effects on infiltration and variably saturated flow beneath an ephemeral stream. *Water Resources Research, 34,* 2137–2153.

Scanlon, B. R., Keese, K., Reedy, R. C., Simunek, J., & Andraski, B. J. (2003). Variations in flow and transport in thick desert vadose zones in response to paleoclimatic forcing (0–90 kyr): Field measurements, modeling, and uncertainties. *Water Resources Research, 39*(7), 1179.

Scanlon, B. R., Reedy, R. C., Stonestrom, D. A., Prudic, D. E., & Dennehy, K. F. (2005). Impact of land use and land cover change on groundwater recharge and quality in the southwestern US. *Global Change Biology, 11,* 1577–1593.

Schaap, M. G., & Leij, F. J. (2000). Improved prediction of unsaturated hydraulic conductivity with the Mualem-van Genuchten model. *Soil Science Society of America Journal, 64*(3), 843–851.
Schwartz, F. W., & Zhang, H. (2003). *Fundamentals of ground water*. New York: Wiley.
Selker, J. S., McCord, J. T., & Keller, C. K. (1999). *Vadose zone processes*. Boca Raton, FL: CRC Press.
Singh, V. P. (1997). *Kinematic wave modeling in water resources, Environmental hydrology*. New York: Wiley.
Singh, V. P. (2002). Is hydrology kinematic? *Hydrologic Processes, 16,* 667–716.
SPE International (n.d.) Petrowiki. Capillary pressure. http://petrowiki.org/Capillary_pressure. Accessed April 2, 2017.
Stephens, D. B. (1996). *Vadose zone hydrology*. Boca Raton, FL: CRC Press.
Stonestrom, D. A., Prudic, D. E., Walvoord, M. A., Abraham, J. D., Stewart-Deaker, A. E., Glancy, P. A., Constantz, J., Lacniak, R. J., & Andrasji, B. J. (2007). Focused ground-water recharge in the Amargosa Desert Basin. In D. A. Stonestrom, J. Constantz, T. P. A. Ferré, & S. A. Leake (Eds.), *Ground-water recharge in the arid and semiarid southwestern United States* (pp. 107–136). U.S. Geological Survey Professional Paper 1703.
Todd, D. K., & Mays, L. W. (2005). *Groundwater hydrology* (3rd ed.). New York: John Wiley & Sons.
Van Genuchten, M. T. (1980). A closed-form equation for predicting the hydraulic conductivity of unsaturated soils. *Soil Science Society of America Journal, 44,* 892–898.
Wang, Z., Feyen, J., van Genuchten, M. T., & Nielsen, D. R. (1998). Air entrapment effects on infiltration rate and flow instability. *Water Resources Research, 34,* 213–222.
Weeks, E. P. (2002). The Lisse effect revisited. *Ground Water, 40*(6), 652–656.
Weir, G. J., & Kissling, W. M. (1992). The influence of airflow on the vertical infiltration of water into soil. *Water Resources Research, 28*(10), 2765–2772.
Wood, W. W. (1999). Use and misuse of the chloride-mass balance method in estimating ground water recharge. *Ground Water, 37,* 2–3.
Wood, W. W., Rainwater, K. A., & Thompson, D. B. (1997). Quantifying macropore recharge: examples from a semi-arid area. *Ground Water, 35,* 1097–1106.

Chapter 4
Groundwater Recharge and Aquifer Water Budgets

4.1 Introduction

Recharge is essentially the processes by which water present on land surface is added to an aquifer. Recharge may be either direct or indirect. Direct recharge, also referred to as diffuse and distributed recharge, is water that enters an aquifer by vertical infiltration and percolation at the site of precipitation or application. Indirect or focused recharge involves the runoff of water from an area of precipitation and its infiltration and percolation to the water table at another location where it is concentrated. The term "direct recharge" is also used with respect to managed aquifer recharge (MAR) to describe the injection of water directly into aquifers as opposed to applying the water to a land surface or introducing it into the vadose zone below land surface using, for example, dry wells or infiltration trenches.

Indirect recharge sites include depressions and channels into which runoff flows and ponds (e.g., ephemeral stream channels, flood plains, alluvial fans, and dry or vadose playas). Direct recharge is known to be of decreasing significance with increasing aridity (Simmers 1990, 1998; De Vries and Simmers 2002). Most of the water that directly infiltrates into the soil in arid lands is retained in the shallow soil to make up soil moisture deficits and is subsequently lost to evapotranspiration (ET). Recharge only occurs where runoff is concentrated or where macropores allow for rapid percolation to the water table.

Anthropogenic aquifer recharge (AAR) is performed by:

- spreading water on to land surfaces
- introducing water into vadose zone below land surface (e.g., using dry wells and infiltration trenches, galleries, or shafts)
- injection into the saturated zone
- inducing recharge from surface water bodies by lowering the water table by pumping
- modifying land surfaces or channels to increase their rates and duration of infiltration.

R. G. Maliva, *Anthropogenic Aquifer Recharge*,
Springer Hydrogeology, https://doi.org/10.1007/978-3-030-11084-0_4

Adding additional freshwater to aquifers that are being depleted (i.e., are experiencing declining water levels or pressures due to excessive pumping) is usually inherently beneficial. However, it is important to have a quantitative understanding of the magnitude, duration, and location of the actual hydrological benefits of proposed and operational MAR systems.

For MAR systems whose purpose is to augment water supplies, anthropogenic recharge has value if it creates an additional water supply that would not otherwise be available, which is referred to as the "useful storage" of water (Maliva and Missimer 2008). In the case of ASR systems that store freshwater in brackish aquifers, usefully stored water is the amount of water that can be recovered when needed at a quality suitable for its intended use. Evaluation of useful storage is more complex where freshwater is stored in freshwater aquifers. Where groundwater pumping is restricted by local impacts (e.g., spring flows, wetlands hydroperiods, and saline-water intrusion), the location and timing of recharge can be important factors in additional to the overall amount of recharge. For example, if the driving issue for an MAR project is maintaining or restoring the flow of a spring or water levels (hydroperiod) of a wetland during dry periods, then the fundamental question is the degree to which recharge of a given volume of water at a given location and time of the year contributes to those specific goals. The actions (e.g., pumping) of other aquifers users can also impact whether an MAR project achieves project-specific objectives.

MAR and unmanaged and unintentional recharge need to be considered in the context of aquifer water budgets, which are essentially accounts of all the water flows into and out of aquifers over specified periods of time. Increased recharge through MAR may not necessarily result in a corresponding increase in the volume of water available in the future. The additional input of water into an aquifer may be offset by increased discharge, decreased natural recharge, or groundwater pumping.

Water budgets can be tabulated through estimation of the values of each of the main components. However, water budgets are now commonly evaluated through numerical groundwater modeling. Groundwater modeling is also used to evaluate the local and temporal hydrological effects of MAR. Numerical modeling has the cachet of being more technically rigorous, but the adage "garbage in, garbage out" is certainly applicable to numerical groundwater modeling, as it is to other areas of computer sciences. Inverse modeling using automated calibration software, in which parameter values are determined from observational data (e.g., water levels in observation wells), is increasingly being used where field data are limited. As is the case for all modeling, the quality assurance/quality control procedures should consider whether the values of the various water budget components and aquifer hydraulic and solute-transport parameter values are reasonable (consistent with available data). An unfortunate reality is that in many areas, including rather heavily regulated areas of developed countries, the values of some basic water budget parameters (e.g., actual ET rates) are still poorly constrained.

4.2 Aquifer Water Budget Concepts

The water budget of an aquifer can be expressed as

$$\sum \text{Inflows} - \sum \text{Outflows} = \text{Increase in storage} \tag{4.1}$$

If more water enters an aquifer in a given year than leaves as a result of pumping, discharge, and other outflows, then the volume of water stored in the aquifer has increased, as manifested by an increase in aquifer water levels or heads (pressure). Similarly, if there is no change in water levels over a given time period, then there has been no change in storage and total inflows and total outflows were balanced. This basic concept is of fundamental importance to MAR systems whose purpose is to increase the volume of water stored in an aquifer but is not always fully appreciated. For example, the author a number of years ago had a discussion with the operator of an ASR system (kept anonymous out of courtesy) that stores freshwater in a freshwater aquifer. The operator reported that they had banked a very large volume of water. When asked how much have water levels risen, he reported that they had not and, in one part of the site, water levels had actually dropped. The unfortunate reality is that no water was actually physically banked as there was no increase in stored water volume. The injected water had been pumped by other aquifer users or perhaps spread out so thinly across the aquifer that any increase in water levels was not detectable. At best the water recharged in the ASR system helped to maintain more stable water levels, which might otherwise had dropped more. A review of historic performance data from ASR systems revealed other examples were a stated purpose of systems was to store water, but monitoring data indicated that injection did not result in a persistent increase in water levels (Maliva and Missimer 2010).

Inflows and outflows each have multiple components, which may be either implicitly or explicitly considered in water-budget analyses depending on the aquifer and purpose of the analysis. With respect to catchment-scale groundwater models, recharge and ET are usually modeled by either (Doble and Crosbie 2017):

- boundary conditions within groundwater flow models
- coupling 1-D unsaturated zone models to groundwater flow models
- using fully coupled saturated-unsaturated (integrated surface water-groundwater) models.

Recharge is most commonly represented in groundwater models using the MODFLOW code as a time-varying value for each cell and ET as linear or piecewise functions of groundwater depth (Doble and Crosbie 2017). Fully coupled saturated-unsaturated zone models allow for the simulation of more complex processes, but have the disadvantages of long computational times and data intensity. The great flexibility of distributed parameter models to capture local hydrological data is of little value if there is no local data for the various parameter values.

Natural recharge is equal to the difference between precipitation and ET minus any net surface-water flow or interflow (i.e., flow within the vadose zone) out of the groundwater basin or study area:

$$R = P - ET - (SW_{out} - SW_{in}) - (IF_{out} - IF_{in}) \tag{4.2}$$

where (in units of volume/time)

R	recharge
P	precipitation
ET	evapotranspiration
SW_{out}	runoff out of study area
SW_{in}	runoff into study area
IF_{out}	interflow out of study area
IF_{in}	inferflow into study area

Evapotranspiration includes crop canopy interception, plant transpiration, and evaporation from the soil and groundwater. Transpiration may consist of water derived from either the soil or groundwater. A distinction is sometimes made between gross and net recharge (Delleur 2006; Doble and Crosbie 2017). Gross recharge is the water that crosses the water table. Net recharge is gross recharge minus transpiration and evaporation of groundwater. Recharge refers to net recharge herein.

For an unconfined aquifer, inputs consist of local recharge (natural and anthropogenic) and any lateral or vertical groundwater flows into the aquifer. The main outflows are groundwater pumping, discharge (e.g., flows to springs, streams, and wetlands), and any lateral or vertical flows out of the aquifer:

$$\Delta S = (NR + AR + GH_{in} + GV_{in}) - (Q + D + GH_{out} + GV_{out}) \tag{4.3}$$

where (in units of volume/time)

ΔS	change in stored water volume
NR	natural recharge
AR	anthropogenic recharge
GH_{in}	horizontal groundwater flow into the aquifer
GV_{in}	vertical groundwater flow into the aquifer
Q	groundwater pumping
D	discharges
GH_{out}	horizontal groundwater flow out of the aquifer
GV_{out}	vertical groundwater flow out of the aquifer

Water budget equations simplify where the value of some parameters is null (or negligible).

4.3 Precipitation (Rainfall)

Precipitation is usually the largest inflow in water budgets and is conceptually the easiest to measure. The two main methods for measuring rainfall are rain gauges and radar-based remote sensing. Simple and inexpensive rain gauges are commercially available or can be constructed. Rainfall over a monitoring period is calculated from the volume of water collected in a container with an opening of known area. Rain gauges provide point measurements of rainfall. However, rainfall often has great spatial and temporal variability, especially in areas in which rainfall occurs mainly as convective storms. For example, during the summer rainy season in Southwest Florida, it is not uncommon for there is to be a large rainfall from a thunderstorm at the author's office, whereas at the same time it is sunny and dry at his home eight km away. The density of rain gauges is typically much too coarse to obtain an accurate record of local short-term (e.g. daily) precipitation within large watersheds or catchment areas. Convective storm cells are often smaller than catchment areas and may or may not reach any given rain gauge within a catchment area during a given storm event.

In humid regions, spatial and temporal variability in rainfall tends to average out over time, but significant (>10%) variations in annual precipitation may still occur in a given year between nearby stations. The effects of spatial and temporal variability in rainfall are greater with increasing aridity as one or several large storms may dominate annual rainfall. Spatial differences in rainfall between stations over a monitoring period may be either random variations or may reflect topographic or geographic controls over local rainfall. Where data on the spatial distribution of rainfall are needed, remote sensing techniques (particularly radar-based) are available with the general limitation of some loss of accuracy. If local (site-specific) rainfall data are important for an MAR project (e.g., for quantification of recharge at a surface-spreading site), then installation of an on-site rain gauge is the preferred solution.

4.3.1 *Rain Gauges*

Rain gauges consist essentially of a collection container with an opening of known area (A) and a means for measuring the volume of rain (V) collected over a given time period. The precipitation rate is the collected volume divided by elapsed time (t) and the opening area (V/tA). Snowfall can be measured as equivalent rainfall by allowing the snow to melt and then measuring the water volume. More commonly, a specially designed board is used to make these measurements. The collected volume of water can be measured manually, using a graduated cylinder or measuring tube, or automatically using either a tipping-bucket system, from the weight of the collected water, or from the height of water in a measuring tube recorded using a pressure transducer.

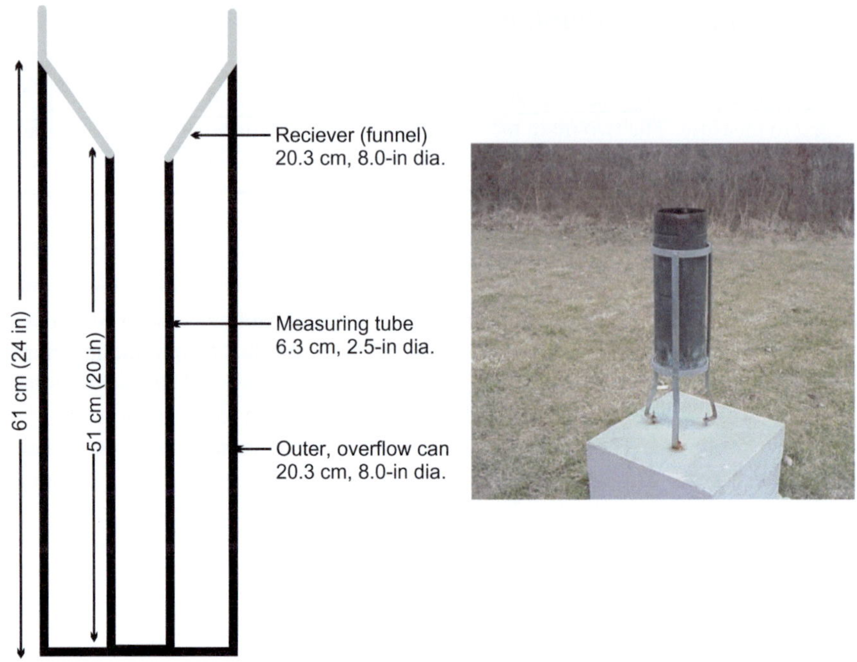

Fig. 4.1 Diagram and photograph of a standard U.S. National Weather Service non-recording rain gauge. Photo source National Weather Service

The U.S. National Weather Service (NWS) standard 8 in. (20.3 cm) non-recording rain gauge is used throughout the world for official rainfall measurements (Fig. 4.1). The NWS gauge has four main components: an 8 in. (20.3 cm) diameter overflow can, an 8 in. (20.3 cm) diameter collector funnel, a 2.5 in. (6.4 cm) diameter measuring tube, and a measuring stick. The collector funnel and measuring tube are sized so that the cross-section area of the collector is ten times the area of the measuring tube and thus the amount of rainfall (in units of length) is one tenth the height of accumulated water in the measuring tube.

A general limitation of rain gauges is that they are not reliable under windy conditions in which there is a strong horizontal component to the rainfall. Non-recording rain gauges are also labor intensive, requiring frequent (commonly daily) manual measurements. Non-recording rain gauges also only provide data on the total amount of rainfall between measurements, not on the intensity of rainfall events. Evaporative losses can impact data, especially if readings are performed infrequently. Powdered cork is added to some gauges to record rainfall events as a cut line between readings so as to reduce the evaporative loss error.

Tipping-bucket systems collect rainfall using a funnel that channels water into a small seesaw-like container (bucket) of known capacity. After the design volume of water fills the bucket, a lever tips, the collected water is dumped, and an electronic signal is sent. Rainfall volume is determined from the number of bucket tips, and the rate of rainfall is determined from the frequency of tips. Tipping-bucket rain gauges tend to slightly underestimate rainfall as small rain events may not trigger the bucket tip and the gauges may experience losses of water during high-intensity rains. Tipping-bucket rain gauges are less labor intensive as only infrequent trips to a station are required to download the collected data. Remote telemetry units (RTUs) are used by some operators to transmit the data into a central office in real time.

4.3.2 Remote Sensing Measurement of Rainfall (Radar and Satellite)

The Next Generation Radar (NEXRAD) weather radar system (technically named the WSR-88D Doppler radar system) is used for rainfall measurements in the United States and employs a 10 cm (3.9 in.) wavelength radar wave. Weather radar involves the transmittance of directional pulses of microwave radiation and recording the return signal from reflections off of droplets of water or ice particles in the atmosphere. The intensity of the return pulse is related to the concentration of droplets or particles in the atmosphere. The radar return, which is expressed as a reflectivity factor (Z), is converted to a radar rainfall estimate (P_r) through an empirical Z-P_r relationship. The Z-P_r relationship varies as a function of many factors (including drop size) and it is not possible to derive a single equation that is accurate for every point in a given radar domain and for every storm type and intensity (Hunter 1996; Hardegree et al. 2008). Rain accumulation is determined by multiplying the average value over a point or area by the time between images. Total accumulations are calculated by adding up the accumulations from all images over the time period of interest. A correction factor is applied to convert radar rainfall (R_r) to gauge rainfall (R_g).

A comparison of NEXRAD and gauge precipitation indicates that radar is more likely to detect high-intensity rainfall events rather than light rains and is, therefore, more appropriate for flood forecasting than for long-term water balance and natural rainfall modeling applications (Hardegree et al. 2008). Smith et al. (1996) observed a systematic underestimation of rainfall by NEXRAD relative to rain gauges for paired gauge-radar rainfall estimates from the southern plains of the United States, A comparison of NEXRAD and rain gauge precipitation data in South Florida also indicated that NEXARD tended to underestimate rainfall, particularly large rainfall amounts, relative to the gauge network (Skinner et al. 2009). NEXRAD may detect high-altitude moisture that does not reach the ground as actual rainfall in arid regions.

Radar rainfall measurements are increasingly being used in surface water and groundwater modeling because they provide spatial and temporal data that can be readily incorporated into models using a GIS framework. However, it is critical that radar rainfall data be ground truthed and calibrated against gauge data. The effects of uncertainty in rainfall rates on groundwater models should be evaluated as part of a sensitivity analysis. Radar rainfall data are commercially available in some areas. For example, the OneRain Company provides gauge-adjusted radar rainfall estimates in the United States based on the NWS NEXARD data.

Satellite rainfall estimates offer the potential of much greater areal coverage compared to land-based radar measurements. The current state of the art is a combination of passive microwave (PMW) data from low-earth orbiting satellites and infrared (IR) data from geostationary satellites (Gebremichael et al. 2010; Pereira Filho et al. 2010). PMW data provide more direct information on rainfall, but has the limitation of a low sampling frequency. IR data has a weak physical relationship to surface rainfall, but has a high spatial and temporal frequency. Combined PMW and IR data can overcome some of the limitations of each individual data source for estimating rainfall (Gebremichael et al. 2010; Pereira Filho et al. 2010). However, Gebremichael et al. (2010) recommended extreme caution with using satellite rainfall estimates because the estimates could be subject to significant errors.

4.4 Evapotranspiration and Lake Evaporation

Evapotranspiration is sum of evaporation and plant transpiration. It is typically expressed in units of length of a water column per time (e.g., mm/d, cm/yr, in/yr). ET is of fundamental importance in water budgets because it is usually by far the largest outflux of water from land surface. Three main types of ET are referred to in the literature: actual ET (ET_a), reference ET (ET_o), and potential ET (PET). ET can also be expressed in terms of the latent heat of vaporization (λ), which is the energy or heat required to vaporize water. Latent heat of vaporization is a function of temperature. The units of latent heat flux are energy divided by the product of area and time (e.g., $Jm^{-2}s^{-1}$ or Wm^{-2}).

Actual ET is the flux of water to the atmosphere from the land surface. The similar parameters ET_o and PET are the ET rates that would occur from a reference surface that is not short of water. ET_o and PET are dependent on local meteorological conditions. Actual ET is also a function of water availability. In arid and semiarid regions, ET_o and PET rates may be very high, while ET_a rates are quite low due to the paucity of available water.

There are two generally recognized reference surfaces: short crop and long crop. The short crop is a hypothetical reference crop with an approximate height of 0.12 m (4.72 in.), a fixed surface resistance of 70 s m^{-1}, and an albedo of 0.23 (Allen et al. 1998). The short crop is similar to an extensive surface of clipped green grass, actively growing, completely shading the ground, and with adequate water (Allen et al. 1998; Task Committee on Standardization of Reference Evapotranspiration 2005). The tall crop has an approximate height of 0.5 m (1.6 ft) and is similar to full cover alfalfa. PET has been largely superseded by ET_o because the former is imprecisely defined. Penman (1948) defined PET in terms of a "short green crop," which could correspond to many types of horticultural and agronomic crops (Irmak and Haman 2003).

Reference ET rate is a climatic parameter that expresses the evaporative power of the atmosphere and is independent of biological and non-weather related variables, such as crop type, crop development, crop roughness, management practices, ground cover, plant density, and soil moisture (Allen et al. 1998). All of the above factors impact ET_a rates. ET_o can be computed from weather data. The recommended standard procedure to calculate ET_o is the FAO Penman-Monteith method, which requires input data on temperature, wind speed, relative humidity, and solar radiation (Allen et al. 1998). Actual ET rates are related to ET_o rates by a crop coefficient (K_c):

$$ET_a = ET_o K_c \tag{4.4}$$

Once the ET_o in an area of interest is determined and the K_c values are known, then ET_a rates can be calculated for the purpose of determining irrigation water requirements. As ET_o rates are based on fully watered conditions, Eq. 4.4 cannot be used to calculate ET_a rates during periods of less than fully watered conditions when ET_a is controlled by water availability.

Evapotranspiration is a major component of the water budget, but measurement of ET_a and surface water evaporation rates is complex and subject to significant errors (uncertainties). Despite its importance, there is usually very limited data on actual ET rates in most, if not all, areas, especially for natural and non-agricultural land covers. In South Florida, for example, ET_a data are available from only several studies performed by the U.S. Geological Survey (USGS) and South Florida Water Management District that were of short duration ($\leq$3 years) and included only a limited number of natural land covers. Essentially no data are available for most of the residential and agriculturally developed areas.

Actual ET rates are measured or estimated in four main manners:

- direct point measurements (lysimeter, sap flow)
- evaporation pans
- micrometeorological point measurements
- remote sensing.

4.4.1 Lysimeters

Lysimeters are essentially cylinders or boxes of soil and vegetation that are installed flush to land surface and are constructed so that changes in water mass and drainage out the bottom (recharge) can be measured. Weighing lysimeters are constructed on balances (or another type of scale) that allow for the measurement of slight changes in the mass of the soil column (Fig. 4.2). Water losses from ET can be estimated from the decrease in the mass of the soil column, corrected for any drainage and precipitation that occurred during the measurement period.

The great advantage of lysimeters is that they can provide the most accurate measurements of ET_a rates. The major limitations of lysimeters are they are expensive to construct and operate, and they provide only a point measurement of ET_a. Lysimeters are also physically limited as to the types of land covers that can be tested, and work

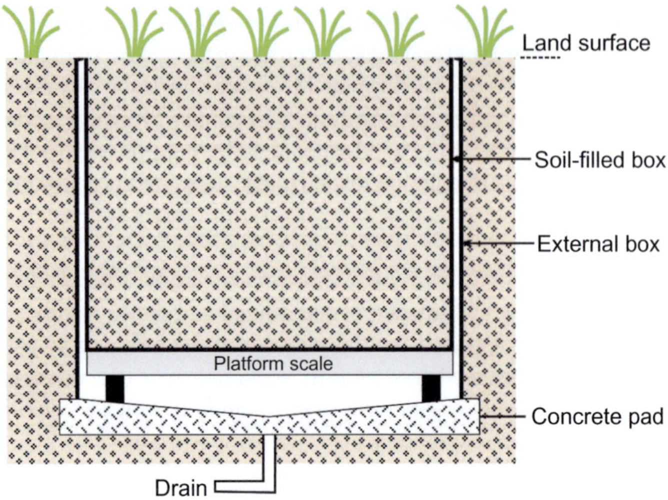

Fig. 4.2 Schematic diagram of a weighing lysimeter. The soil-filled sample box may also be equipped with tensiometers and a neutron probe access tube (after Gee and Jones 1985)

best where vegetation is restricted to grasses and small plant covers. Lysimeters are not practical for measuring ET_a rates in forested areas because mature trees cannot normally be grown within them. Recharge measurements can be impacted by the disturbance of the soil during construction (i.e., the soil structure within a lysimeter may differ from that of adjacent undisturbed soil). Installation of a widespread lysimeter network for evaluation of an aquifer water budget is not economically viable. Relatively high-accuracy lysimeter ET_a data could be used to calibrate ET_a data collected by other means.

4.4.2 Soil Moisture Depletion

Similar to the lysimeter method, ET_a can be estimated from measurements of the change in the water content of a representative sample of undisturbed soil and vegetation, while measuring precipitation and run-on/runoff, and estimating deep drainage from the sample plot (Shuttleworth 2008). Soil moisture changes can be measured using, for example, resistance blocks, tensionmeters, neutron probes, time-domain reflectometers, and capacitance sensors. The soil-moisture depletion method is a point measurement and depends upon accurate measurement of changes in soil moisture throughout the root zone of vegetation, which in some areas can be quite deep.

4.4.3 Sap Flow

The sap flow measurement is a direct, non-destructive means of measuring transpiration. The-heat pulse velocity (HPV) and the thermal-dissipation probe (TDP) methods are commonly used. The velocity of sap flow is typically measured from the transport of heat pulses. The commonly used HPV sap flow sensors consist of three thermocouple needles: a central heater needle and two sampling needles. The sampling needles are inserted upstream and downstream of the heater needle. Short pulses of heat (1–6 s) are released into the sap stream, and the rate of movement of the heated sap is determined from temperature changes monitored in the upstream and downstream needles. The TDP method measures the rate of sap flow using the rate of heat dissipation, which is determined from the measured difference in temperature between a continuously electrically heated needle in the sapwood and second needle installed some distance below the heated needle. Sap flow velocity is inversely related to the temperature difference between the probes with the maximum temperature difference occurring with the velocity is zero.

Measurements for both types of systems are performed automatically using microprocessor/datalogging systems. Sap flow sensors are commercially available. Volumetric sap flow (flux) within a tree can be estimated by multiplying the corrected sap velocity by the cross-sectional area of the water-conducting sapwood.

The main challenge associated with using measurements of sap flow velocity to estimate the ET_a rate of a land cover is upscaling the data from a limited number of trees. Several factors impact sap flow data including the location of the sap-flow sensor on a stem, variability in transpiration among trees and between plots within a catchment, and variability in stand density, composition, sapwood area, and leaf area (Smith and Allen 1996; Köstner et al. 1998; Granier et al. 1996; Ford et al. 2007; Glenn et al. 2011; Uddin et al. 2014). ET_a estimated from sap flow may underestimate total ET_a from a land cover due to understory ET_a (Granier et al. 1996). Each scaling step from leaf, to canopy, to stand, and then to ecosystem involves separate calculations (each with an associated error) that propagate into the cumulative ET_a error (Glenn et al. 2011).

4.4.4 Pan Evaporation

Evaporation pans are widely used to estimate evaporation rates from surface water bodies (e.g., lakes) and reference evapotranspiration rates. The most commonly used pan design in the United States is the Class A evaporation pan (Fig. 4.3), which is a steel cylinder with a diameter of 47.5 in (120.7 cm) and a depth of 10 in (25 cm). The pans are placed in open ground upon a level slatted wooden platform. The less commonly used Colorado sunken pan is a 92 cm (3 ft) square pan that is 46 cm (18 in.) deep. The Colorado sunken pan is placed in the ground with the rim 5 cm (2 in.) above the soil level (Allen et al. 1998).

Fig. 4.3 Class A evaporation pan, Lake Pleasant, Arizona

Evaporation is typically measured every 24 h by either the change in water level or the amount of water that needs to be added to bring the water level back up to a reference level. Pan evaporation readings can be impacted by extraneous factors, such as precipitation and animals (particularly birds) drinking from and splashing the water. Therefore, pans are usually placed inside chain-link or wire mesh cages. Meteorological data such as temperatures (maximum and minimum), humidity, and wind speed are commonly also recorded at evaporation pan stations as these data can assist in the interpretation of the data.

To estimate ET_o and surface water evaporation rates, pan evaporation rates (ET_{pan}) are multiplied by a pan coefficient (K_p):

$$ET_o = K_p(ET_{pan}) \tag{4.5}$$

Pan coefficients are a function of multiple variables, particularly local meteorological conditions and pan design, location, and condition (e.g., Snyder 1992; Snyder et al. 2005). Pan evaporation rates, and thus pan coefficients, may change over time because of aging, deterioration, and repainting of the pan systems (Allen et al. 1998). It is recommended that ET_{pan} values be calibrated against ET_o values computed with the Penman-Monteith method (Allen et al. 1998). Pan evaporation rates (ET_{pan}) are not equivalent to actual surface-water evaporation rates. Evaporation rates from surface-water bodies depend on local conditions, including size (area and shape), depth, vegetation, wind fetch, wind speed, and albedo, which vary between surface water bodies. Pan evaporation coefficient values in the range of 0.7–0.8 are commonly used to estimate free surface-water evaporation rates. Although the 0.1 range in coefficients is small, the coefficient is applied to a large number. Hence, the error in calculated ET_o and surface water evaporation rates may be large relative to recharge rates.

4.4.5 Micrometeorological Techniques—Eddy Covariance Method

The eddy covariance method is the most direct and widely used micrometeorological technique for measuring ET_a. The underlying principle of the method is that ET rates can be calculated from measurements of fast (turbulent) fluctuations in vertical wind speed and water vapor (Tanner and Greene 1989). Vertical components of wind eddies transport pulses of water vapor either upward or downward. The upward components of eddies tend to have higher water vapor content from ET at land surface or in the plant canopy than the downward component. The net sum of the vertical vapor pulses (upward—downward) corresponds to the upward flux of water vapor and thus the ET_a rate. The method assumes that net lateral fluxes of water vapor are negligible.

The eddy covariance method is most commonly performed using stationary towers (flux towers) that are installed above the top of the canopy of the study area. The height of the tower required depends upon the height of the vegetation canopy. Relatively small, tripod-mounted towers are used to measure ET_a over agricultural fields and grass lands, and are commercially available. Very tall towers are needed to measure ET_a rates in forested areas. The towers used by the USGS to measure ET_a rates over pine and cypress forests in the Big Cypress National Preserve in Florida were 38 m high (Fig. 4.4; Shoemaker et al. 2011).

Flux towers are equipped, at a minimum, with a sonic anemometer (i.e., instrument to measure wind speed), infrared gas analyzer, and temperature and humidity probes. The anemometer and infrared gas analyzer perform fast measurements of wind speed and humidity (water vapor). All of the instruments are connected to a datalogging system. Eddy covariance stations are often equipped with other meteorological instruments, such as a pyranometer or radiometer (i.e., devices used measure solar radiation).

4.4.6 Micrometeorological Techniques—Energy Balance Methods

ET_a rates can be estimated from latent heat fluxes determined as the residual of energy balances. Latent heat fluxes are fluxes of heat from the Earth's surface to the atmosphere that are associated with the evaporation of water at the surface. The basic underlying principle is the conservation of energy. Energy arriving at an evaporating surface must be balanced by energy leaving the surface over the same time period (Allen et al. 1998). The conservation of energy equation for a system in which only vertical fluxes are considered is

$$R_n = G + L + H \tag{4.6}$$

where (in units of $Jm^{-2}s^{-1}$ or Wm^{-2})

Fig. 4.4 USGS ET tower in the Big Cypress National Preserve. Note man on tower for scale. *Source* Shoemaker et al. 2011

R_n net radiation to the surface
G soil heat flux
H sensible heat flux (i.e., heat transferred to the air during a change of temperature that is not accompanied by a change of state)
L latent heat flux from the surface

Latent heat flux and, in turn, ET_a rate can be calculated if the values of *Rn*, *G*, and *H* are measured. The Energy Balance Bowen Ratio (EBBR) method utilizes the Bowen Ratio (β) that relates sensible to latent heat flux:

$$\beta = \frac{H}{L} = \gamma\left(\frac{K_h}{K_w}\right)\left(\frac{\frac{\partial T}{\partial z}}{\frac{\partial e}{\partial z}}\right) \tag{4.7}$$

where,

K_h eddy diffusivity for heat (m^2s^{-1})

K_W eddy diffusivity for water vapor (m^2s^{-1})
T temperature (°C)
Z elevation above surface (m)
e vapor pressure (kPa)
γ psychrometric constant (kP °C^{-1})

The psychrometric constant is defined as

$$\gamma = \frac{C_p P}{\lambda_v \epsilon} \tag{4.8}$$

where,

C_p specific heat of air at constant pressure (Jkg^{-1} °C^{-1}),
P atmospheric pressure (kPa),
λ_v latent heat of water vaporization (Wm^{-2}),
ε ratio of the molecular weight of water vapor to dry air (≈0.622)

The eddy diffusivities for heat and water vapor are approximately equal. For two measurements at different vertical elevations, Eq. 4.7 reduces to

$$\beta = \gamma \left(\frac{\Delta T}{\Delta e} \right) \tag{4.9}$$

and

$$L = \frac{R_n - G}{1 + \beta} \tag{4.10}$$

Actual ET rates are obtained using the EBBR method from the difference in temperature (ΔT) and vapor pressure (Δe) between two elevations, atmospheric pressure, net radiation to the surface (R_n), and soil heat flux (G).

4.4.7 Remote Sensing ET Measurements

For water budget analyses, ET_a data are needed on large spatial scales. Large errors may be introduced through the interpolation and extrapolation of scattered point measurements (if available) over heterogeneous landscapes. Numerous studies have been performed over the past several decades on the use of remote sensing data to determine ET_a rates on large spatial scales.

Remote sensing data are used to estimate ET_a rates in two main manners. Vegetation indices derived from canopy reflectance data are used to estimate basal crop coefficients, which are then used to convert reference ET to actual crop ET values (Gonzalez-Dugo et al. 2009). More commonly, evapotranspiration rates are estimated using satellite spectral information and ground meteorological data on the surface energy budget elements (Bastiaanssen et al. 1998; Alberich 2002; Kalma et al. 2008; Glenn et al. 2011; van der Tol and Pardoi 2012; Zhang et al. 2016). Satellite measurements of surface temperature are used to estimate sensible heat flux (H). Soil heat flux (G) cannot be measured remotely. Therefore, the G/Rn is taken as a constant or G is estimated from several other parameters (Kalma et al. 2008).

Land-based ET_a data are needed to calibrate satellite remote sensing data. Kalma et al. (2008) reported, based on a review of published validation data, that the relative error of satellite ET_a measurements compared to land-based flux measurements was 15–30%. A key point is that remote sensing ET_a methods can only be as accurate as the ground method data used for calibration and validation (Glenn et al. 2011). Eddy covariance towers have accuracy errors on the order to 15–30%, which limits the accuracy of the remote sensing methods that depend upon them for validation or calibration (Glenn et al. 2011).

Satellite-based ET_a measurements are also constrained by the satellite return period and pixel size, cloud cover, and the smaller spatial scales of land-based data compared to the pixel-size of satellite data. Kustas and Normal (1996) observed that

> For many of these RS models, differences with ET observations can be as low as 20% from hourly to daily time scales, approaching the level of uncertainty in the measurement of ET and contradicting some pessimistic conclusions concerning the utility of remotely sensed radiometric surface temperature for determining the surface energy balance.

It remains a challenge that data on ET_a rates are needed for evaluation of aquifer water balances, but the error in both land-based and remotely sensed data are large relative to aquifer recharge rates.

4.5 Discharge

4.5.1 Discharge Basics

Groundwater discharge is the flow of groundwater to land surface, surface water bodies (lakes and rivers), or the sea (submarine groundwater discharge). Groundwater discharge may occur as discrete concentrated flows, such as from spring outlets, or diffuse flow. Diffuse discharge includes flows to streams (baseflow) and low-lying land areas (wetlands). In anthropogenically undisturbed aquifers at steady state (i.e., stored water volume does not change over time), recharge is balanced by discharge. Groundwater pumping can induce in changes in recharge and discharge, which are referred to as "capture." Bredehoeft and Durbin (2009) observed that

> Capture is an all-important concept in managing ground water; a ground water system can only be maintained indefinitely if the pumping is equaled by capture – a combined decrease in the undeveloped discharge and increase in the undeveloped recharge. If pumping continually exceeds capture, then water levels in the system can never stabilize, and the system will continue to be depleted.

Anthropogenic aquifer recharge (AAR) can cause increases in discharge and decreases in recharge. Increases in water levels from aquifer recharge can convert losing streams into gaining streams. For example, canal seepage and return flows from irrigation with wastewater in the Tula Valley of Mexico increased local river flows and caused the appearance of several springs (Jiménez and Chávez 2004).

The relationship between groundwater pumping and discharge can be expressed as (Lohman 1972; Bredehoeft et al. 1982; Bredehoeft 2002):

$$Q = \Delta R - \Delta D - \Delta S \tag{4.11}$$

where (units of volume/time)

- Q pumping rate
- ΔR change in recharge (including interaquifer flow)
- ΔD change in discharge
- ΔS change in stored water volume

Under steady-state conditions, $\Delta S = 0$, and

$$Q = \Delta R - \Delta D \tag{4.12}$$

which indicates that the volume of water pumped is balanced by additional (induced) recharge and/or reduced discharge.

Captured discharge can have both beneficial and adverse impacts. Submarine groundwater discharge represents a net loss of freshwater that might otherwise be beneficially used. Groundwater discharge may introduce nutrients and other contaminants into surface water bodies. Capture may adversely impact surface water users through a reduction in stream baseflow and spring flows (Theis 1940). Reductions in discharge can adversely impact groundwater dependent ecosystems. Conversely, MAR can result in increases in discharge, which might be beneficial to local groundwater dependent ecosystems but represents an unintended loss of water that was recharged for planned future use. Hence, understanding and quantifying discharge processes is important for water budget analyses and analyzing the benefits and impacts of MAR systems.

Measurement of discharge is relatively straightforward where there is just one or several gaugeable spring outlets or a stream is entirely fed by springs. Quantifying diffuse discharge is more complex. Point measurements of discharge may be accurate, but up-scaling the measurements to reflect areal, much less aquifer-wide, values is a challenge due to aquifer heterogeneity. Aquifer-wide discharge rates can be estimated as the residual of the aquifer water budget provided that the values of the other elements can be estimated with reasonable accuracy. Follows is a summary of common discharge types and some measurement methods.

4.5.2 Stream and Lake Discharge

Many studies of surface water/groundwater interaction involved intense instrumentation of relatively short stream reaches and are often cost prohibitive for routine application (SKM and CSIRO 2012). The stream flow component sourced from groundwater is referred to as baseflow. Methods used to estimate baseflow are divided between those that focus on quantifying discharge along a stream at one point in time and those that focus on changes in discharge over time at single points in a stream (e.g., at a gauging station). Methods and data used to estimate discharge to streams and lakes include (Zamora 2008; SKM and CSIRO 2012):

- seepage meters
- hydrograph separation
- chemical hydrograph separation
- longitudinal chemistry changes
- horizontal hydraulic gradient
- vertical hydraulic gradient
- flow difference methods
- inverse modeling using thermal gradient data.

4.5.2.1 Seepage Meters

Seepage meters directly measure the seepage across the sediment-water interface. The basic design of a seepage meter is an open cylinder that is driven into the sediment and a means for measuring the flow of water through the cylinder. An early, but still used, design by Lee (1977) to measure water flow in meters employs a plastic bag filled with water to measure seepage volume. The seepage volume is determined from the difference in the volume of water in the bag between the start and end of the test. More sophisticated methods have since been developed to measure seepage rates, such as using flowmeters and dye and thermal pulse-displacement (Taniguchi and Fukuo 1993; Rosenberry 2008; Koopmans and Berg 2011; Zhu et al. 2015).

The advantages of seepage meters are that they are (Sophocleous 2004):

- relatively lightweight and easy to operate
- conceptually simple and rapid measurements can be made
- a direct measurement of seepage flux.

Technical challenges include that small head losses can reduce seepage rates, flowing water can impart hydraulic pressure on the measuring bag, current scouring may breach the hydraulic seal around the seepage meter, seepage rates may be very slow, and disturbance of the sediment during installation can impact measured rates (Zamora 2008). Seepage meters are a point measurements and many measurements may be required to adequately capture the effects of aquifer heterogeneity. Large differences in rates can occur between closely spaced tests.

4.5.2.2 Hydrograph Baseflow Separation Method

The hydrograph separation method distinguishes between streamflow derived from surface runoff and baseflow by separation of hydrographs (plots of water flow versus time) into relatively short periods of high runoff-sourced flow and more uniform lower background baseflow. The hydrograph separation method is popular because it utilizes often existing streamflow data from gauging stations and hence there is minimal additional data acquisition costs. Limitations of the method are that it is sensitive to subjectivity in the data analysis and is most accurate when surface runoff events are well-defined and represent a relatively small proportion of the total flow to a stream (SKM and CSIRO 2012).

4.5.2.3 Chemical Hydrograph Separation Method

Groundwater discharge into rivers is often estimated using water chemistry data. Chemical methods for quantifying baseflow are based on the use of mixing equations to determine the fraction of runoff and groundwater-derived water in a river water sample. The method requires that there is a significant chemical difference between the surface water flow into a river reach and the local groundwater, and that the natural tracer concentrations in the source groundwater and surface water end members are accurately known. End-member concentrations should also not change over time (SKM CSIRO 2012). The binary mixing equation used for the chemical hydrograph method is (Kish et al. 2010; SKM and CSIRO 2012):

$$\frac{Q_g}{Q_t} = \frac{c - c_r}{c_g - c_r} \tag{4.13}$$

Q_g groundwater sourced flow
Q_t total flow
c concentration in river
c_g concentration in groundwater
c_r concentration in runoff (i.e., inflowing surface water)

Additional water sources can be considered by using multiple tracers.

4.5.2.4 Longitudinal Chemistry Method

The longitudinal chemistry methods involve making a series of chemical measurements at multiple points along a river reach at one point in time. Discharge is quantified through the downstream increase in the fraction of groundwater-derived water in the river. This method requires field access and can result in large costs if new bores need to be drilled to obtain data on local groundwater chemistry. The longitudinal chemistry method is based on mixing equations and requires accurate data on end-member compositions (SKM CSIRO 2012). It assumes a uniform groundwater composition for the tracer entering the stream reach. As is the case for the chemical hydrograph separation method, the longitudinal chemistry method works best if the tracer is non-reactive and there is a large difference in tracer concentration between the inflowing river water and groundwater.

4.5.2.5 Horizontal Hydraulic Gradient Method

The hydraulic gradient method is based on Darcy's law using the observed hydraulic gradient between a stream and nearby observation wells. The method requires accurate measurement of aquifer transmissivity (hydraulic conductivity and thickness). The hydraulic gradient method is sensitive to both aquifer hydraulic conductivity and the local hydraulic gradient. Streambed hydraulic conductivity may differ from the aquifer value due to the presence of higher permeability, coarser riverbed sediments or a low permeability clogging layer (SKM and CSIRO 2012). Site-specific discharge values obtained at one or several points may not be representative of a larger river reach (SKM CSIRO 2012).

4.5.2.6 Vertical Hydraulic Gradient

The rate of upwards flow towards the sediment-water interface can be calculated using Darcy's law from vertical hydraulic gradient data collected using nested piezometers. The vertical hydraulic conductivity of the sediment needs to be determined at the test site. The vertical hydraulic gradient method is a point measurement that is sensitive to heterogeneities in sediment composition and thus vertical hydraulic conductivity. Piezometers need to be very carefully installed to obtain representative head measurements. Accurate measurement of the vertical hydraulic conductivity of the streambed sediments is also required.

4.5.2.7 Flow Difference Method

The flow difference method assumes that changes in flow between two gauging stations are due to groundwater inflows or outflows (i.e., transmission losses or gains) over the stream reach. The method assumes that all other inflows and outflows are negligible or can be quantified. The methods works best when there are large differences between downstream and upstream flow (including any tributary flows) because of the significant errors (5–15%) associated with manual flow gauging and gauging station rating curves (SKM and CSIRO 2012). Another source of error is flow within the streambed that is not captured by gauging stations (SKM and CSIRO 2012).

4.5.2.8 Thermal Gradient Methods

Discharge to surface-water bodies can be estimated using the measured thermal gradient from the surface water through the underlying shallow sediments and a one-dimensional advective-conductive heat flow equation. Methods that use temperature as a natural tracer have the great advantage that a large number readings can be economically obtained over a short period of time. Temperature data are collected using multilevel temperature probes (e.g., Meinikmann et al. 2013) or fiber-optic distributed temperature sensors (FO-DTS; Blume et al. 2013). FO-DTS allows for spatially detailed measurements over large distances and at greater resolutions.

Discharge rate estimates using temperature measurements are point measurements and may not be representative of the total groundwater discharge. Local groundwater discharge rates depend on local aquifer hydraulic conductivity, thickness, and hydraulic gradient. Meinikmann et al. (2013) noted that uncertainties in temperature-based groundwater discharge rates are minimized by reducing their roles to weighting factors instead of using them as absolute values.

Ferguson and Bense (2010) demonstrated through stochastic modeling that the validity of the use of the 1D advective-conductive heat flow equation is in question when there is the potential for lateral flow. They specifically cautioned that "using temperature measurements to estimate bulk groundwater discharge may not be practical because of problems with heterogeneities, losing reaches, hyporheic flow, and horizontal flow components" and that "temperature-based estimates of specific discharge has enormous potential but the results should be confirmed through the use of other field techniques or tested by more complex numerical models".

4.5.3 *Submarine Groundwater Discharge*

Submarine groundwater discharge (SGD), in the broadest sense, is any and all flows of water upward across the sea floor into the overlying water (SCOR and LOICZ 2004). It consists of both concentrated spring flow and diffuse seepage, and includes both the discharge of fresh groundwater originating on land and recirculated seawater. Discharged freshwater may undergo mixing with seawater, and thus increase in salinity, before reaching a submarine outlet. There are historical examples where SGD of freshwater was captured offshore for potable use (SCOR and LOICZ 2004). Freshwater SGD may be an important component of water budgets and of nutrient and contaminant fluxes into nearshore marine waters. SGD will occur anywhere that an aquifer with a positive head relative to sea level is hydraulically connected to a seawater body (Burnett et al. 2006). SGD may also be caused by tidal, wave, storm, or current induced pressure gradients, density (salinity) driven convection, geothermal heating, and seasonal fluctuations in the position of the saline-water interface (Burnett et al. 2006).

SGD can be measured using (SCOR and LOICZ 2004, Burnett et al. 2006):

- seepage meters (or other forms of benthic chambers)
- natural tracers (e.g., temperature, salinity, radium and radium) using mixing equations and their concentrations in the receiving water, and seawater and groundwater end members
- natural tracers using vertical compositional gradients evaluated with a vertical one-dimensional advection-diffusion model. Salinity gradients may be obtained either by direct measurement or from bulk groundwater electrical conductivity (Stieglitz et al. 2008)
- Darcy's law with data on the local vertical hydraulic gradient (from nested piezometers) and hydraulic conductivity
- Aquifer water budgets where discharge is the residual, and recharge and storage changes are well constrained.

The limitation of point measurement methods is that they may not be representative of average rates and may include seawater recirculation flows. Large volume-of-investigation methods include inverse numerical groundwater modeling and analytic calculations using shoreward hydraulic gradient and aquifer transmissivity data. However, models used to evaluate potable water resources may over simplify the saline water/freshwater transition (SCOR and LOICZ 2004). Water levels must be converted into equivalent freshwater heads where there are significant differences in groundwater salinity and thus density. Burnett et al. (2006), based on an investigation including multiple sites, recommended that SGD measurement strategies should consider spatial and temporal variations in SGD including the general pattern of decreasing discharge with distance from the shore and the contributions of recirculating seawater.

4.5.4 Wetland Discharge

In arid regions, groundwater-supported wet meadows and playas are present in some basins despite extremely low mean annual precipitation rates (Sanderson and Cooper 2008). Due to orogenic effects, rainfall and associated groundwater recharge can be much greater in surrounding mountains. Where wetlands and playas do not receive surface water flows, discharge rates can be estimated from ET_a measurements. Daily ET_a rates were measured by Sanderson and Cooper (2008) at five studied intermontane wet meadow and playa sites in the San Luis Valley of Colorado. Evapotranspiration from groundwater (ET_g) was calculated as the measured ET_a minus any precipitation and surface water inflows. The wet meadows were supplied mainly by groundwater flow, whereas the two playa sites received surface water flows. In the wet meadows, ET from groundwater (ET_g) was determined to be 75–88% of ET_a.

Micrometeorological (tower) data were analyzed using the Bowen Ratio Energy Balance (BREB) methods. The wet meadows data corroborated the generalization that ET_a from groundwater (ET_g) decreases as the depth to groundwater increases. Declining water levels due to climate variations or pumping would be partially offset by "salvaged" ET. However, in the long term, changes in water levels may trigger changes in vegetation (Sanderson and Cooper 2008). Sanderson and Cooper (2008) also demonstrated that existing ET_g versus water table depth models may substantially under- or over-estimate rates from a shallow aquifer.

ET_a rates from groundwater discharge were evaluated in the Death Valley area of Nevada and California by Laczniak et al. (2001). Annual ET losses were calculated as the product of the acreage of phreatophytes within a discharge area and an annual ET rate representative of the vegetation and soil conditions of the discharge area. Acreages were estimated by delineating vegetation from aerial photographs and field mapping. Annual ET rates were estimated from values reported for similar plant assemblages found throughout the western United States.

An estimate of mean annual groundwater discharge from each discharge area was calculated by summing individual estimates of mean annual groundwater discharge computed for each ET unit within the discharge area. Mean annual groundwater discharge from each ET unit was computed as the product of an adjusted mean annual ET rate and the acreage of the ET unit. Adjustments were made to remove any water from the ET estimates that was contributed by precipitation. The remainder of the water consumed by ET was assumed to have originated from groundwater.

4.6 Storage Change

4.6.1 Water-Level Based Methods

Changes in stored water volume (ΔS) are the product of head changes (Δh), aquifer storativity (storage coefficient; S), and aquifer area (A):

$$\Delta S = \Delta h S A \tag{4.14}$$

Changes in total aquifer stored water volume are the product of head change and storativity integrated over the area of an aquifer. Geoprocessing software packages are available that provide multiple methods for interpolating water level and storativity data onto grid. Gridded data can then be processed to calculate storage changes. Changes in storage are calculated in MODFLOW in each grid cell from the head change, specific storage, and cell dimensions (McDonald and Harbaugh 1988).

Methods for estimate storativity, specific storage, and specific yield were summarized by Maliva (2016). The storativity of confined aquifers is typically determined from multiple-well (pumped and one or more observation wells) aquifer pumping tests.

Determination of the specific yield of unconfined aquifers is much more challenging because of the delayed-yield phenomenon. Specific yield is defined as the volume of water released from storage by gravity drainage per unit surface area of an aquifer per unit decline of the water table. In fine-grained sediments, the gravity-drainage of water can take a very long time (10 s of days to over a year) to reach completion. Specific retention is the volume of water retained against gravity. During aquifer recharge, the volume of water required for a unit increase in water table elevation per unit aquifer area may be greater than the specific yield if the water content of the soil is initially below its specific retention. In completely dry sediments, the volume of water required for a unit rise in the water table is approximately equal to the total porosity (assuming no isolated or closed pores).

Methods used to calculate specific yields include (Neuman 1987; Maliva 2016):

- **laboratory drainage measurements**: volume of water that drains from samples under gravity
- **volume-balance method**: based on the volume of the cone of depression caused by the pumping of a known volume of water
- **aquifer pumping tests**: Boulton (1963) and Neuman (1975) type-curve methods
- **microgravity**: relative gravity surveys are used to determine the change in the mass/volume of water associated with the water table decline measured in a well
- **water table decline associated with ET**: ratio of field measured ET_a and water table decline during a period with no rainfall
- **water table rise associated with recharge**: ratio of field measured precipitation and water table rise, corrected for any runoff and ET.

4.6.2 Relative Microgravity

In unconfined aquifers, relatively microgravity surveys are used to measure changes in the mass, and thus volume, of water underlying a survey point. Relative gravity surveys involve the performance of time series of measurements of the gravitational acceleration (g) at the exact same locations, which eliminates the need for some corrections, such as for terrain and latitude. Relative gravity data still need to be corrected for instrument drift, Earth tides, and environmental effects (Davis et al. 2008). Changes in gravitational acceleration (Δg; μGals) caused by changes in the water table elevation can be calculated, based on the Bouguer slab equation, as (Pool and Eychaner 1995; Pool 2008):

$$\Delta g = 41.9 Sy \cdot b(\mathrm{m}) \tag{4.15}$$

$$\Delta g = 12.77 Sy \cdot b(\mathrm{ft}) \tag{4.16}$$

where

Sy specific yield (dimensionless)
b water table fluctuation (m, ft)

Microgravity measurements provide information on only changes in the total mass of water, not where the changes occurs. Koth and Long (2012) reported that changes in groundwater storage could be quantified with an accuracy of about ±0.5 ft (0.15 m) of water per unit area of aquifer. Gravity changes measured at land surface may also result from aquifer compaction in addition to the amount of water stored in pore spaces. Changes in water storage may occur in the vadose zone, a perched aquifer, or the phreatic zone (or a combination thereof). Relative gravity survey data augment, but do not replace, water level data from wells.

Microgravity data can be used to increase the number of monitoring points and to evaluate changes in vadose zone storage associated, for example, with surface-spreading MAR systems. Maliva and Missimer (2012) summarized several reported successful applications of microgravity to MAR sites including an ephemeral river recharge site near Tucson, Arizona (Pool and Schmidt 1997), an ASR project in Arvada, Colorado, in which water is stored in an abandoned coal mine (Davis et al. 2005, 2008), an ASR system using an alluvial aquifer in Lancaster, California (Howle et al. 2002), and a MAR system using infiltration basins in the Weber River Basin of Utah (Chapman et al. 2008).

Microgravity surveys require meticulous field procedures and accurate corrections for extraneous factors in order to obtain accurate data on changes in water storage. Microgravity measurements are point measurement and require remobilization to a site for each measurement. Microgravity surveys are typically performed to evaluate long-term changes in water storage.

4.6.3 Grace

The Gravity Recovery and Climate Experiment (GRACE) space mission was launched to study processes involving changes in the Earth's mass distribution, including changes in water storage (Wahr et al. 2004). The GRACE mission, which was launched by NASA in 2002, consists of two identical satellites orbiting in the same plane and acting in unison (NASA 2003). The distance between the two satellites is measured using an extremely precise (within 10 μm) microwave ranging system. Global coverage is approximately every 30 days, so data are locally available in approximately monthly time steps. The basic principle behind the GRACE mission is that local changes in the underlying gravitational field of the Earth induce changes in the speed and distance between the twin satellites. As the satellites pass over a gravity anomaly, the speed of the lead satellite either increases or decreases, which results in a change in the distance between the satellites. The difference in distance closes as the trailing satellite passes over the anomaly. Satellite global positioning systems (GPS) determine the exact position of the satellites over the Earth to within a centimeter or less. The GRACE data are processed to provide information on changes in terrestrial water storage (ΔTWS), which are equivalent to the sum of changes in soil moisture (ΔSM) and groundwater storage (ΔGWS).

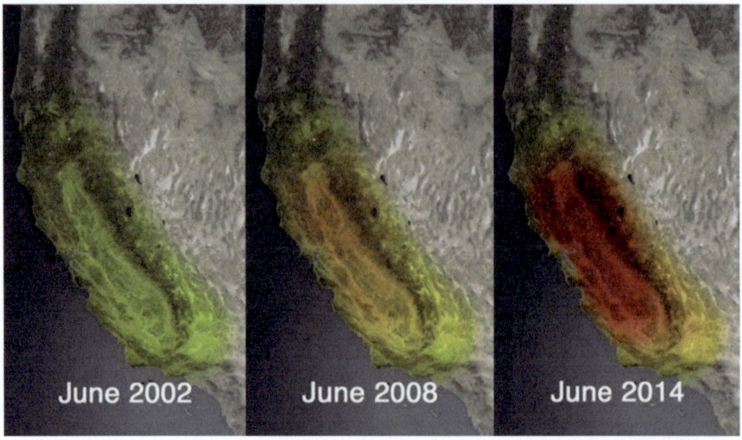

Fig. 4.5 GRACE images showing declining water storage in California in June 2002 (left), June 2008 (center) and June 2014 (right). Colors progressing from green to orange to red represent greater accumulated water loss between April 2002 and June 2014 (NASA/JPL-Caltech/University of California, Irvine)

Satellite gravity measurements have the limitation of a very coarse spatial resolution (Becker 2006). The spatial resolution for water mass variability was initially estimated to be 200,000 km^2 and could be closer to 500,000 km^2 (Rodell et al. 2007). There have been a large number of published studies in which GRACE data have been used to evaluate regional changes in aquifer water levels (e.g., Figure 4.5). GRACE data has too coarse of a spatial resolution to be suitable for local investigations associated with MAR projects.

4.7 Groundwater Pumping

4.7.1 Introduction

Under ideal circumstances, all groundwater pumping would be recorded using calibrated flowmeters with the data promptly sent to a central repository for storage in a readily accessible database (preferably remotely accessible). Regrettably, data on groundwater pumping is typically incomplete, fragmented, often of questionable accuracy, and may not be accessible. Some of the challenges associated with accurately quantifying groundwater pumping for water balance analysis are:

- metering (or other means of recording) of groundwater pumping was not required by regulatory agencies or independently performed
- unpermitted water use

- inaccurate water metering (e.g., pumped volumes in farms are often roughly estimated from pumping times and pump capacity)
- inaccurate or incomplete reporting of pumping.

Public water supply and large industrial water supply wellfields are commonly metered, whereas agricultural irrigation water use may be less rigorously measured and reported. In undeveloped countries, metering of groundwater pumping may not be required or is not enforceable. Small-volume household water use is commonly not required to be metered. For example, residential potable water and irrigation wells (such as used by the author) are not required to be metered in Florida. While the groundwater pumping at one residence may be small, the cumulative pumping of a great number of residences can be a substantial component of a local aquifer water budget.

Where accurate groundwater pumping data are unavailable, then other independent means are needed to estimate rates. For domestic use, groundwater pumping may be estimated from population and household number data and average per capita or per household water use rates.

Permitted allocations may be used to estimate agricultural irrigation water use. However supplemental irrigation water requirements vary between years depending upon weather conditions and irrigators may use more water than necessary. Groundwater pumping rates are commonly estimated for irrigated areas using crop supplemental water requirements, such as approximated using the Blaney-Criddle method. The Blaney-Criddle method is used to estimate ET_o on monthly time steps using mean daily temperature and mean daily percentage of annual daytime hours (Allen and Pruitt 1986; Brouwer and Heibloem 1986). Irrigation requirements are calculated for crop types, growing season, soil types, and efficiency of the irrigation method.

Actual water use for irrigation areas by crop type may be determined from data from representative farms where water use is metered or where farm operators can otherwise provide reliable information. Estimations of groundwater pumping from crop water requirements assumes optimal water use for crop and irrigation system type (i.e., excessive irrigation is not occurring).

4.7.2 Aerial Photography and Satellite Remote Sensing

Meijerink (2007), p. 239 observed that

> If measures are planned to stop groundwater mining, it is obvious that the amount and pattern used for irrigation, if irrigation is the main groundwater use, has to be determined. This task is a common problem in the practice of groundwater management or groundwater studies if a large number of wells with no or poor registration exist. By using remote sensing, the spatial pattern of actual groundwater use can be estimated, as an alternative to doing so by a well inventory in the field.

Fig. 4.6 True color (left) and false color infrared (red) aerial photographs from near Burlington, Vermont (USA). Surface water appears black in the infrared photograph and green healthy vegetation appears red. *Source* U.S. Geological Survey (2001)

This most basic means of estimating irrigated area using remote sensing is through examination of aerial photographs. Calculation of irrigated area is facilitated if the photographs are orthorectified in that the images are geometrically correct and the scales are uniform. Otherwise, aerial photographs can be used in conjunction with detailed maps to establish or confirm land use on mapped parcels. Irrigated crops can be differentiated from undisturbed (uncropped) areas on historical black and white photographs using characteristics such as shape, tone, texture, shadow, association, and pattern (Turney 2012).

Color infrared film allows for the differentiation of live and healthy vegetation, such as the growth phase of irrigated crops (Turney 2012). Healthy vegetation has a high reflectance of infrared radiation, whereas surface water has a high absorbance of infrared radiation (USGS 2001). Colors are assigned to the different infrared signal intensities. By convention, high reflectance areas are shaded red and low reflectance areas are shaded black (Fig. 4.6). As is the case for remote sensing in general, ground truthing is necessary for interpretation of the imagery and validation of interpretations.

Satellite photographic systems, such as the Landsat 7 satellite Enhanced Thematic Mapper Plus (ETM +), collect data in multiple wavelength bands. The ETM + system collects images in seven wavelength bands ranging from blue-green (0.45–0.52 μm) to thermal infrared (10.40–12.50 μm) (NASA n.d.; USGS 2003). Bands may be used separately or combined to create composite false-color images that best differentiate the objects of interest. For example, the false color images generated in Fig. 4.7 highlight areas of healthy vegetation that corresponds to well-irrigated farm fields.

The Normalized Difference Vegetation Index (NDVI) represents the amount of green biomass (i.e., live green plant canopies) with index values corresponding to

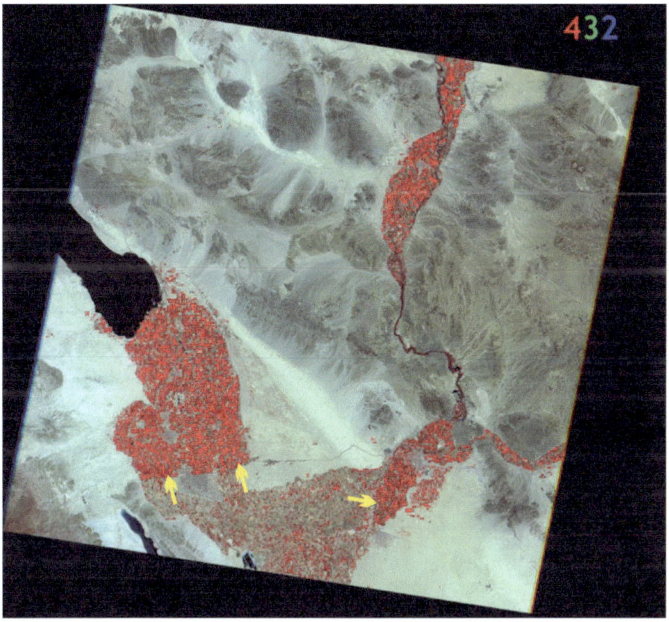

Fig. 4.7 False-color image Landsat 7 ETM + image of the Salton Sea and Imperial Valley of California (USA). Irrigated areas with healthy vegetation in this desert environment appear bright red. The border between the United States and Mexico (arrows) can be located from the difference in vegetation. *Source* NASA, http://landsat.gsfc.nasa.gov/education/compositor/pdfs/Landsat_7_Compositor.pdf

the presence and condition of the vegetation (Rouse et al. 1973; Tucker 1979). The NVDI is calculated as

$$NVDI = \frac{(NIR - VIS)}{(NIR + VIS)} \tag{4.16}$$

where *VIS* and *NIR* are the spectral reflectance measurements acquired in the visible (red) and near-infrared regions, respectively. NDVI values range from −1 to +1 with values greater than 0.1–0.2 generally indicative of vegetation.

The U.S. Geological Survey study of the Lake Altus drainage basin in Oklahoma and Texas provides a good example of the use of remote sensing to estimate irrigated areas and water use (Masoner et al. 2003). Enhanced Thematic Mapper Plus (ETM+) imagery was used to map land use and irrigated croplands during the 2000 growing season. Pixels were categorized into broad land-cover types and three classes of agriculture: row crops, small grains, and hay/pasture. Irrigated crop areas were mapped using a vegetation index consisting of a near infrared band (band 4) divided by a visible red band (band 3). Masoner et al. (2003) reported that "Identification of agricultural crops using satellite imagery requires knowledge of crop phenology, climate for the particular growing season, and ground reference information about specific agricultural practices in the drainage basin." Reference evapotranspiration, crop evapotranspiration, and crop irrigation water requirements were empirically estimated using the solar radiation-based evapotranspiration model of Doorenbos and Pruitt (1977).

Satellite remoting sensing systems have different spatial resolutions. Landsat 7 ETM + has a spatial resolution of about 30 m in the visible bands and 60 m in the infrared band. For mapping irrigated areas, the hypotheses that "the finer the spatial resolution of the sensor used, the greater the irrigated area derived" appears to be true as fragmented areas (i.e., smaller farms) are better detected with finer resolution systems (Velpuri et al. 2009).

Thenkabail et al. (2009) compared remote-sensing derived irrigated areas and census-derived statistics reported in the national system of India. Irrigated area was derived for the nominal year 2000 using time-series data from two satellite remote sensing systems with different spatial resolutions: Advanced Very High Resolution Radiometer (AVHRR) with 10 km resolution and Moderate Resolution Imaging Spectroradiometer (MODIS) with a 500 m resolution. Thenkabail et al. (2009) noted that a number of issues contribute to significant uncertainties in irrigated area estimates, including definitional issues. Remote-sensing derived irrigated areas were consistently greater than irrigated areas reported by national statistics. However, large variations in irrigated areas were also reported between two ministries that are part of the same governmental system.

4.8 Recharge Estimates

Methods used to estimate natural recharge were reviewed by Simmers (1988), Stephens (1996), Scanlon et al. (2002), De Vries and Simmers (2002), Hogan et al. (2004), Sophocleous (2004), Healy (2010), and Maliva and Missimer (2012), from which sources this summary was largely derived. Follows is a summary of some of the more widely used methods to estimate recharge that have practical applications for quantifying anthropogenic aquifer recharge. References are provided to more detailed discussions of each method. Some methods used to evaluate natural recharge rates are either inapplicable or poorly applicable to estimating recharge rates of recent waters (e.g., age dating techniques) or have excessive data collection requirements. As an example of the later, calculation of the recharge rates using the Darcy-Buckingham equation is generally not feasible in applied settings because of the effort and costs involved with obtaining accurate suction head and unsaturated hydraulic conductivity values and, as point measurements, the large number of analyses that may be required to adequately capture aquifer heterogeneity. Similarly, lysimeters can provide accurate point measurements of recharge but are generally too costly to construct and operate for most applications.

4.8.1 Residual of Aquifer Water Budgets

Recharge can be estimated as the residual of aquifer water budgets if the values for all of the other main parameters in the budget can be determined or estimated with a high degree of accuracy. The calculations can either be made directly from measured or estimated values of the main water-budget parameters or, commonly, recharge is estimated by inverse modeling through the calibration of groundwater flow models.

The water-budget method has greater uncertainty in arid and semiarid regions than in humid regions because the precipitation rate is frequently only slightly different than the actual evapotranspiration rate. Small errors in either or both components can cause proportionally large errors in calculated recharge rates (Stephens 1996; De Vries and Simmers 2002; Scanlon et al. 2002; Sophocleous 2004; Maliva and Missimer 2012). Accurate data on local actual ET rates are seldom available.

4.8.2 Water Budgets of Surface Water Bodies

Recharge within surface water bodies can be estimated as the residual of their water budgets (Lerner et al. 1997; Scanlon et al. 2002; Maliva and Missimer 2012):

$$R = P + (SW_{UP} - SW_{DOWN}) + \sum SW_{IN} - ET_a - Q - \frac{\Delta S}{\Delta t} \tag{4.18}$$

where (in units of volume/time)

SW_{UP}, SW_{DOWN}	flow rates at upstream and downstream ends of reach
$\sum SW_{IN}$,	rates of tributary inflows
P	precipitation on surface water body
Q	surface water withdrawals
ET_a	evaporation from surface water or the stream bed (including transpiration from vegetation)
$\Delta S/\Delta t$	rate of change of surface water body and unsaturated zone storage.

Equation 4.18 can be more practically used where multiple elements of the budget are not applicable. For example, in infiltration basins and reservoirs fed by ephemeral streams, the water budget equation after a flow event during a period with no precipitation simplifies to

$$R = -ET_a - \frac{\Delta S}{\Delta t} \tag{4.19}$$

The recharge rate is equal to the change in stored water volume divided by elapsed time (Δt) minus the evaporative loss rate (ET_a). ET_a can be estimated using a nearby evaporation pan. A stage-volume relationship is needed to determine changes in storage from changes in water levels, which requires a detailed topographic survey (digital elevation model) of the reservoir basin. Equation 4.19 assumes that changes in vadose zone storage and interflow are negligible. Similarly, recharge rates from streams are estimated from the transmission loss (SW_{UP}–SW_{DOWN}), corrected for any additional inputs or outputs of water between gauging stations (e.g., tributary flow, evapotranspiration, precipitation, withdrawals, and interflow).

The accuracy of the recharge estimates using surface-water budget methods is constrained by uncertainties in the values of the various budget elements. The greatest uncertainties lie in stream flow rate measurements and tributary flows, which may not be gauged. The uncertainties in flow rates obtained from gauging stations may be greater than recharge rates.

In semiarid and arid environments with thick vadose zones, the volume of water that is actual recharge (i.e., water that reaches the water table) from small rainfall events may be a small fraction of the transmission loss because of large soil-moisture deficits and subsequent losses of infiltrated water to ET.

4.8.3 Water-Table Fluctuation Method

The water-table fluctuation (WTF) method estimates recharge rates from aquifer responses to rainfall events or anthropogenic applications. The WTF method was reviewed by Healy and Cook (2002). For an unconfined aquifer, the recharge from an individual rainfall event is estimated from the increase in water table elevation (Δh) and the aquifer specific yield (Sy):

$$R = Sy\Delta h \tag{4.20}$$

The WTF method is sensitive to the specific yield values of the water table aquifer. Cumulative recharge is the sum of the responses to all rain events over a period of time. Total recharge from a localized application of water (e.g., infiltration basin or channel) can be estimated using geoprocessing software by interpolation of water table elevation change data from observation wells to a grid and using known or estimated specific yield values. Multiple observation wells are required to adequately capture the volume of the mound of recharged water (e.g., around a stream channel or infiltration basin system).

The basic assumption of the WTF method is that water table responses fully reflect only the water added into storage. A considerable delay may occur between a rainfall event (or surface spreading of water) and the response of the water table, which depends on the thickness and hydraulic properties of the vadose zone sediments or rock. Deep aquifers may show muted responses to rainfall events because wetting fronts tend to disperse over long distances (Healy and Cook 2002).

Fundamental requirements of the WTF method are that the water table level responds to rainfall events and that there is no net transport of water from the water table aquifer (e.g., downward leakage). The method is best applied to shallow unconfined aquifers with a single-porosity (as opposed to dual-porosity fractured or karstic aquifers) that exhibit rapid water-level rises and declines in response to rainfall and drainage. The method is most appropriate for the analysis of short-term, pulsed water-level rises in response to individual storms as opposed to long-duration, low-intensity precipitation (Sophocleous 1991; Healy and Cook 2002). The water table elevation response must fully reflect the water added into storage. Barometric pressure, groundwater flow (saturated zone and interflow), diurnal fluctuations in evapotranspiration, entrapped air, and groundwater pumping can also impact water table responses (Sophocleous 1991; Healy and Cook 2002).

The WTF method has the advantages of simplicity, ease of use, low cost (if existing wells and well data are used), and absence of assumptions on the mechanism by which water flows through the vadose zone (Healy and Cook 2002). A disadvantage of the WTF method are that calculated recharge rates are a function of specific yield, for which there is often considerable uncertainty, and water table elevation changes can be impacted by extraneous factors (e.g., local groundwater pumping and irrigation return flows).

4.8.4 Chloride Mass-Balance Method

The chloride mass-balance (CMB) method is based on the concept that rainfall contains low concentrations of chloride (and other conservative ions) that are concentrated by evaporation to a degree inversely proportional to the recharge rate (Ericksson and Khunakasem 1969). Evaporation of precipitation (on the land surface and in the vadose zone) decreases the volume of water and increases the concentrations of dissolved ions. Chloride is the preferred ion used because it usually behaves in a conservative manner and can be accurately and relatively inexpensively measured. Recharge rates (R) can be calculated if the precipitation rate (P) and concentrations of chloride in the rainfall (Cl_p) and shallow groundwater (Cl_{GW}) are known:

$$R = \frac{Cl_P}{Cl_{GW}} P \tag{4.21}$$

The CMB method is based on the following assumptions (Wood and Scanford 1995; Wood 1999; Goni 2002):

- Chloride in groundwater originates only from the precipitation that recharges the aquifer.
- There is no net runoff or run-on of chloride-containing water. The chloride concentration of runoff may be greater than that of precipitation because local runoff may become enriched in chloride from interaction with surficial sediments and rock or through evaporation (Flint et al. 2002).
- Chloride is conservative in the system (i.e., its concentration is not changed by chemical reactions).

- The chloride mass flux (precipitation rate and chloride concentration) has not changed over time.
- One-dimensional downward piston flow of groundwater occurs (i.e., there is minimal lateral flow of water in the vadose zone).
- There is no recycling or concentration of chloride within the aquifer. Recycling of chloride could occur, for example, if groundwater is pumped from the aquifer and the irrigation return flow is evaporatively concentrated.

Soils typically contains some chloride deposited as dry fall, which is dissolved atmospheric chloride or wind-blown particulate salt (in coastal or salt pan areas) that was deposited and retained on the ground surface. Dry fall chloride may be mobilized by infiltrating precipitation. The total chloride (Cl_T) is thus the sum of the amount in rainfall (Cl_p) and the contribution from dry fall (DF) multiplied by precipitation rate:

$$R = \frac{Cl_P + DF}{Cl_{GW}} P = \frac{Cl_T}{Cl_{GW}} \tag{4.22}$$

Dry fall is often not explicitly considered and is either combined in the Cl_p term or a regional mean dry bucket to wet bucket ratio is used (e.g., Krulikas and Giese 1995).

There is a general paucity of local data on rainfall chloride concentrations, which are a function of local geography and climatic conditions such as proximity to the coast and the prevailing wind direction and speed. Long-term records of wet and dry chloride deposition in arid and semiarid regions worldwide are needed (Scanlon et al. 2006). Reported rainfall chloride concentrations are often in the range of 0.4–0.7 mg/L but may be higher in some locations, particularly near the coast. Using assumed or regional values increases the uncertainty of recharge estimates. Chloride concentrations may vary seasonally and between rainfall events. An error would be introduced into recharge estimates if the chloride concentration of major rainfalls, and thus recharge events, is significantly different from the average concentration. CMB analyses can be refined to include a covariation of precipitation and chloride concentrations (Subyani and Şen 2006; Şen 2008).

The CMB method can also be used to estimate the amount of recharge from infiltration ponds and reservoirs. Sukhija (2008) used a variation of the CMB method to estimate the percentage of recharge of water stored in several infiltration ponds ("tanks") in India after monsoon rains. The underlying concept is that infiltration will reduce the total mass of chloride in a pond. The amount infiltration over time is calculated from the change in the total mass of chloride in a pond and the average chloride concentration of the water. This method assumes that chloride leaves a pond only by infiltration and no additional chloride is added to the pond during the calculation period.

References

Alberich, C. (2002). Estimation of the spatial distribution of evapotranspiration using optical remote sensing. In W. Kinzelbach, W. Aeschbach, C. Alberich, I. B. Goni, U. Beyerle, P. Brunner, W.-H. Chiang, J. Rueedi & K. Zoellmann (Eds.), *A survey of methods from groundwater recharge in arid and semiarid regions,* Early Warning and Assessment Report Series UNEP/DEWA/RS.02-2. Nairobi: United Nations Environment Programme.

Allen, R. G., Pereira, L. S., Raes, D., & Smith, M. (1998). *Crop evapotranspiration—Guidelines for computing crop water requirements, FAO Irrigation and Drainage Paper 56*. Rome: Food and Agricultural Organization of the United Nations.

Allen, R. G., & Pruitt, W. O. (1986). Rational use of the FAO Blaney-Criddle formula. *Journal of Irrigation and Drainage Engineering, 112*(2), 139–155.

Bastiaanssen, W. G. M., Menenti, M., Feddes, R. A., & Holtschlag, A. A. (1998). A remote sensing surface energy balance algorithm for land (SEBAL). *Journal of Hydrology, 212–213,* 198–229.

Becker, M. W. (2006). Potential for satellite remote sensing of ground water. *Ground Water, 44,* 306–318.

Blume, T., Krause, S., Meinikmann, K., & Lewandowski, J. (2013). Upscaling lacustrine groundwater discharge rates by fiber-optic distributed temperature sensing. *Water Resources Research, 49,* 7929–7944.

Boulton, N. S. (1963). Analysis of data from non-equilibrium pumping tests allowing for delayed yield from storage. *Institute of Civil Engineers Proceedings (London), 26,* 469–482.

Bredehoeft, J. D. (2002). The water budget myth revisited: why hydrogeologists model. *Ground Water, 40,* 340–345.

Bredehoeft, J. D., & Durbin, T. (2009). Ground water development—the time of capture problem. *Ground Water, 47*(4), 506–514.

Bredehoeft, J. D., Papadopoulos, S. S., & Cooper, H. H., Jr. (1982). The water-budget myth. *Scientific basis of water resources management, Studies in geophysics* (pp. 51–57). Washington, DC: National Academy Press.

Brouwer, C. & Heibloem, M. (1986). *Irrigation water management: irrigation water needs.* Rome: FAO Land and Water Development Division, International Support Programme for Irrigation Water Management.

Burnett, W. C., Aggarwal, P. K., Aureli, A., Bokuniewicz, H., Cable, J. E., Charette, M. A., et al. (2006). Quantifying submarine groundwater discharge in the coastal zone via multiple methods. *Science of the Total Environment, 367*(2), 498–543.

Chapman, D. S., Sahm, E., & Gettings, P. (2008). Monitoring aquifer recharge using repeated high-precision gravity measurements: a pilot study in South Weber, Utah. *Geophysics, 73*(6), WA83–WA93.

Davis, K., Li, Y., Batzle, M., & Raynolds, B. (2005). *Time-lapse gravity monitoring of an aquifer storage recovery project in Leyden, Colorado*. Golden, CO: Colorado School of Mines, Center for Gravity, Electrical & Magnetic Studies.

Davis, K., Li, Y., & Batzle, M. (2008). Time-lapse gravity monitoring: a systematic 4D approach with application to aquifer storage and recovery. *Geophysics, 73*(6), WA61–WA-69.

De Vries, J. J., & Simmers, I. (2002). Groundwater recharge: an overview of processes and challenges. *Hydrogeology Journal, 10,* 5–17.

Delleur, J. W. (2006). *The handbook of groundwater engineering*. Boca Baton: CRC Press.

Doble, R. C., & Crosbie, R. S. (2017). Current and emerging methods for catchment-scale modelling of recharge and evapotranspiration in shallow aquifers. *Hydrogeology Journal, 25,* 3–23.

Doorenbos, J. & Pruitt, W. O. (1977). *Guidelines for predicting crop water requirements,* Irrigation and Drainage Paper no. 24 (rev.). Paris: FAO-ONU.

Ericksson, E., & Khunakasem, V. (1969). Chloride concentration in groundwater, recharge rate and rate of deposition of chloride in Israel Coastal Plain. *Journal of Hydrology, 7,* 178–197.

Ferguson, G., & Bense, V. (2010). Uncertainty in 1D heat flow analysis to estimate groundwater discharge to a stream. *Ground Water, 49*(3), 336–347.

Flint, A. L., Flint, L. E., Kwicklis, E. M., Fabryka-Martin, J. T., & Bodvarson, G. S. (2002). Estimating recharge at Yucca Mountain, Nevada, USA, comparison of methods. *Hydrogeology Journal, 10,* 180–204.

Ford, C. R., Hubbard, R. M., Kloeppel, B. D., & Vose, J. M. (2007). A comparison of sap flux-based evapotranspiration estimates with catchment-scale water balance. *Agricultural and Forest Meteorology, 145*(3), 176–185.

Gebremichael, M., Anagnostou, E. N., & Bitew, M. M. (2010). Critical steps for continuing advancement of satellite rainfall applications for surface hydrology in the Nile River Basin. *Journal American Water Resources Association, 46*(2), 360–366.

Gee, G. W., & Jones, T. L. (1985). *Lysimeters at the Hanford Site: Present use and future needs, PNL-5578*. Richland, WA: Pacific Northwest Laboratory.

Glenn, E. P., Doody, T. M., Guerschman, J. P., Huete, A. R., King, E. A., McVicar, T. R., et al. (2011). Actual evapotranspiration estimation by ground and remote sensing methods: The Australian experience. *Hydrological Processes, 25*(26), 4103–4116.

Goni, I. B. (2002). Chloride method in the unsaturated zone. In W. Kinzelbach, W. Aeschbach, C. Alberich, I. B. Goni, U. Beyerle, P. Brunner, W.-H. Chiang, J. Rueedi & K. Zoellmann (Eds.), *A survey of methods from groundwater recharge in arid and semiarid regions*, Early warning and assessment report series UNEP/DEWA/RS.02-2 (pp. 22-31). Nairobi: United Nations Environment Programme.

Gonzalez-Dugo, M. P., Neale, C. M. U., Mateos, L., Kustas, W. P., Prueger, J. H., Anderson, M. C., et al. (2009). A comparison of operational remote sensing-based models for estimating crop evapotranspiration. *Agricultural and Forest Meteorology, 149*(11), 1843–1853.

Granier, A., Biron, P., Köstner, B., Gay, L. W., & Najjar, G. (1996). Comparisons of xylem sap flow and water vapour flux at the stand level and derivation of canopy conductance for Scots pine. *Theoretical and Applied Climatology, 53*(1), 115–122.

Hardegree, S. P., Van Vactor, S. S., Levinson, D. H., & Winstall, A. H. (2008). Evaluation of NEXRAD radar precipitation products for natural resource applications. *Rangeland Ecology & Management, 61*(3), 346–353.

Healy, R. W. (2010). *Estimating groundwater recharge* (p. 256). Cambridge, UK: Cambridge University Press.

Healy, R. W., & Cook, P. G. (2002). Using ground water levels to estimate recharge. *Hydrogeology Journal, 10*(1), 91–109.

Hogan, J. E., Phillips, F. M., & Scanlon, B. R. (2004). *Groundwater recharge in a desert environment: the Southwestern United States*. Washington, DC: American Geophysical Union.

Howle, J. F., Phillips, S. P., & Ikehara, M. E. (2002). Estimating water-table change using microgravity surveys during an ASR program in Lancaster, California. In P. P. Dillon (Ed.), *Management of aquifer recharge for sustainability* (pp. 269–272). Lisse: Swets & Zeitlinger.

Hunter, S. M. (1996). WSR-88D rainfall estimation: capabilities, limitations and potential improvements. *National Weather Digest, 20*, 26–36.

Irmak, S., & Haman, D. Z. (2003). *Evapotranspiration: potential or reference? Document ABE 343*. Gainesville, FL: University of Florida Institute of Food and Agricultural Sciences.

Jiménez, B., & Chávez, A. (2004). Quality assessment of an aquifer recharged with wastewater for its potential use as drinking water source: "El Mezquital Valley" case. *Water Science and Technology, 50,* 269–276.

Kalma, J. D., McVicar, T. R., & McCabe, M. F. (2008). Estimating land surface evaporation: A review of methods using remotely sensed surface temperature data. *Surveys In Geophysics, 29*(4–5), 421–469.

Kish, G. R., Stringer, C. E., Stewart, M. T., Rains, M. C., & E Torres, A. (2010). *A geochemical mass-balance method for base-flow separation, upper Hillsborough River watershed, west-central Florida, 2003–2005 and 2009*. U.S. Geological Survey Scientific Investigations Report 2020-5092.

Koopmans, D., & Berg, P. (2011). An alternative to traditional seepage meters: Dye displacement. *Water Resources Research, 47*(1), W01506.

Köstner, B., Granier, A., & Cermák, J. (1998). Sapflow measurements in forest stands: Methods and uncertainties. *Annales des Sciences Forestières, 55*(1–2), 13–27.
Koth, K. R., & Long, A. J. (2012). *Microgravity methods for characterization of groundwater storage changes and aquifer properties in the karstic Madison aquifer in the Black Hills of South Dakota, 2009–2012*. U.S. Geological Survey Scientific Investigations Report 2012-5158.
Krulikas, R. K., & Giese, G. L. (1995). *Recharge to the Surficial Aquifer System in Lee and Hendry Counties, Florida*. U.S. Geological Survey Water-Resources Investigations Report 95-4003.
Kustas, W. P., & Norman, J. M. (1996). Use of remote sensing for evapotranspiration monitoring over land surfaces. *Hydrological Sciences Journal, 41*(4), 495–516.
Laczniak, R. J., Smith, J. L., Elliott, P. E., DeMeo, G. A., Chatigny, M. A., & Roemer, G. J. (2001). *Estimates of evapotranspiration, Death Valley regional flow system, Nevada and California*. U.S. Geological Survey Water-Resources Investigations Report 01-4195.
Lee, D. R. (1977). A device for measuring seepage flux in lakes and estuaries. *Limnology and Oceanography, 22*(1), 140–147.
Lerner, D. N., Issar, A. S., & Simmers, I. (1997). Groundwater recharge. In O. M. Saether & P. de Caritat (Eds.), *Geochemical processes, weathering and groundwater recharge in catchments* (pp. 109–150). Rotterdam: AA Balkema.
Lohman, S. W. (1972). *Ground-water hydraulics*. U.S. Geological Survey Professional Paper 708.
Maliva, R. G. (2016). *Aquifer characterization techniques*. Berlin: Springer.
Maliva, R. G., & Missimer, T. M. (2008). ASR, useful storage, and the myth of residual pressure. *Ground Water, 46*(2), 171.
Maliva, R. G., & Missimer, T. M. (2010). *Aquifer storage and recovery and managed aquifer recharge using wells: planning, hydrogeology, design, and operation*. Houston: Schlumberger Corporation.
Maliva, R. G., & Missimer, T. M. (2012). *Arid lands water evaluation and management*. Berlin: Springer.
Masoner, J. R., Mladinich, C. S., Konduris, A. M., & Smith, S. J. (2003). *Comparison of irrigation water use estimates calculated from remotely sensed irrigated acres and state reported irrigated acres in the Lake Altus drainage basin, Oklahoma and Texas, 2000 growing season*. USGS Water-Resources Investigations Report 03–4155.
McDonald, M. G., & Harbaugh, A. W. (1988). *A modular three-dimensional finite-difference ground water flow model*. U.S. Geological Survey Techniques of Water-Resources Investigation Report 06-A1.
Meijerink, A. M. J. (2007). *Remote sensing applications to groundwater*, IHP-VI, Series on Groundwater No. 16. Paris: UNESCO.
Meinikmann, K., Lewandowski, J., & Nützmann, G. (2013). Lacustrine groundwater discharge: Combined determination of volumes and spatial patterns. *Journal of Hydrology, 502,* 202–211.
NASA (n.d.). Landsat science. National Aeronautics and Space Administration. https://landsat.gsfc.nasa.gov/. Accessed March 30, 2018.
NASA (2003). Studying the Earth's gravity from space: The Gravity Recovery and Climate Experiment (GRACE). NASA Facts. The Earth System Sciences Pathfinder Series, FS-2002-1-029-GSFC (rev. December 2003).
Neuman, S. P. (1975). Analysis of pumping test data from anisotropic unconfined aquifers considering delayed gravity response. *Water Resources Research, 11,* 329–342.
Neuman, S. P. (1987). On methods of determining specific yield. *Ground Water, 25,* 679–684.
Penman, H. (1948). Natural evaporation from open water, bare soil and grass. *Proceedings of the Royal Society, A193,* 120–146.
Pereira Filho, A. J., Carbone, R. E., Janowiak, J. E., Arkin, P., Joyce, R., Hallak, R., et al. (2010). Satellite rainfall estimates over South America—possible applicability to the water management of large watersheds. *Journal American Water Resources Association, 46*(2), 344–360.
Pool, D. R. (2008). The utility of gravity and water-level monitoring at alluvial aquifer wells in southern Arizona. *Geophysics*, *73*(6), WA49-WA59.

Pool, D. R., & Eychaner, J. H. (1995). Measurement of aquifer-storage change and specific yield using gravity surveys. *Ground Water, 33,* 425–432.
Pool, D. R., & Schmidt, W. (1997). *Measurement of ground-water storage change and specific yields using the temporal-gravity method near Rillito Creek, Tucson, Arizona*. U.S. Geological Survey Water Resources Investigations Report 97-4125.
Rodell, M., Chen, J., Kato, H., Famiglietti, J.-S., Nigro, J., & Wilson, C. R. (2007). Estimating groundwater storage changes in the Mississippi River Basin (USA) using GRACE. *Hydrogeology Journal, 15,* 159–166.
Rosenberry, D. O. (2008). A seepage meter designed for use in flowing water. *Journal of Hydrology, 359*(1), 118–130.
Rouse, J. W., Haas, R. H., Schell, J. A., & Deering, D. W. (1973). Monitoring vegetation systems in the Great Plains with ERTS. Third ERTS Symposium, National Aeronautics and Space Administration SP-351(1), 309–317.
Sanderson, J. S., & Cooper, D. J. (2008). Ground water discharge by evapotranspiration in wetlands of an arid intermountain basin. *Journal of Hydrology, 351*(3), 344–359.
Scanlon, B. R., Healy, R. W., & Cook, P. G. (2002). Choosing appropriate techniques for quantifying groundwater recharge. *Hydrogeology Journal, 10,* 18–39.
Scanlon, B. R., Keese, K. E., Flint, A. L., Flint, L. E., Gaye, C. B., Edmunds, W. M., et al. (2006). Global synthesis of groundwater recharge in semiarid and arid regions. *Hydrological Processes, 20,* 3335–3379.
SCOR & LOICZ (Scientific Committee on Oceanic Research & Land-Ocean Interactions in the Coastal Zone) (2004). *Submarine groundwater discharge. Management implications, measurements and effects*. Paris: UNESCO.
Şen, Z. (2008). *Wadi hydrology*. Boca Raton: CRC Press.
Shoemaker, W. B., Lopez, C. D., & Duever, M. (2011). *Evapotranspiration over spatially extensive plant communities in the Big Cypress National Preserve, southern Florida, 2007–2010*. U.S. Geological Survey Scientific Investigations Report 2011–5212.
Shuttleworth, W. J. (2008). Evapotranspiration measurement methods. *Southwest Hydrology, 7*(1), 22–23.
Simmers, I. (1988). *Estimation of natural groundwater recharge*. Dordrecht: D. Reidel.
Simmers, I. (1990). Aridity, groundwater recharge and water resources management. In D. L. Lerner, A. S. Issar, & I. Simmers (Eds.), *Groundwater recharge, a guide to understanding and estimating natural recharge, Contributions to Hydrogeology 8* (pp. 1–20). Kenilworth, UK: International Associations of Hydrogeologists.
Simmers, I. (1998). Groundwater recharge: an overview of estimation "problems" and recent developments. In N. S. Robins (Ed.), *Groundwater pollution, aquifer recharge and vulnerability, Special Publication 130* (pp. 107–115). London: Geological Society.
Skinner, C., Bloetscher, F., & Pathak, C. S. (2009). Comparison of NEXRAD and rain gauge precipitation measurements in South Florida. *Journal of Hydrologic Engineering, 14*(3), 248–260.
SKM and CSIRO (2012). *Methods for estimating groundwater discharge to streams—summary of field trials*. Summary Report (January 2012). SKM and CSIRO.
Smith, D. M., & Allen, S. J. (1996). Measurement of sap flow in plant stems. *Journal of Experimental Botany, 47*(12), 1833–1844.
Smith, J. A., Seo, D. J., Baeck, M. L., & Hudlow, M. D. (1996). An intercomparison study of NEXRAD precipitation estimates. *Water Resources Research, 32*(7), 2035–2045.
Sophocleous, M. (1991). Combining the soilwater balance and water-level fluctuation methods to estimate natural ground-water recharge: Practical aspects. *Journal of Hydrology, 124,* 229–241.
Sophocleous, M. (2004). Groundwater recharge. In L. Silveira, S. Wohnlich, and E. J. Usunoff (Eds.), *Encyclopedia of life support systems (EOLSS)*: Oxford, UK: Eolss Publishers (http://www.eolss.net).
Stephens, D. B. (1996). *Vadose zone hydrology*. Boca Raton: CRC Press.

Stieglitz, T., Rapaglia, J., & Bokuniewicz, H. (2008). Estimation of submarine groundwater discharge from bulk ground electrical conductivity measurements. *Journal of Geophysical Research: Oceans, 113,* C08007.

Subyani, A., & Şen, Z. (2006). Refined chloride mass-balance method and its application in Saudi Arabia. *Hydrological Processes, 20,* 4373–4380.

Sukhija, B. S. (2008). Adaptation to climate change: strategies for sustaining groundwater resources during droughts. In W. Dragoni & B. S. Sukhija (Eds.), *Climate change and groundwater, Special Publication 288* (pp. 169–181). London: Geological Society.

Snyder, R. L. (1992). Equation for evaporation pan to evapotranspiration conversions. *Journal of Irrigation and Drainage Engineering, 118,* 977–980.

Snyder, R. L., Orang, M., Matyac, S., & Grisner, M. E. (2005). Simplified estimation of reference evapotranspiration from pan evaporation data in California. *Journal of Irrigation and Drainage Engineering, 131,* 249–253.

Taniguchi, M., & Fukuo, Y. (1993). Continuous measurements of ground water seepage using an automatic seepage meter. *Ground Water, 31*(7), 675–679.

Tanner, B. D., & Greene, J. P. (1989). *Measurements of sensible heat and water vapor fluxes using eddy correlation methods*: Final report prepared for the U.S. Army Dugway Proving Grounds, Dugway, Utah.

Task Committee on Standardization of Reference Evapotranspiration. (2005). *The ASCE standardized reference evapotranspiration equation: Reston.* VA: Environmental and Water Resources Institute, American Society of Civil Engineers.

Theis, C. V. (1940). The source of water derived from wells. Essential factors controlling the response of an aquifer to development. *Civil Engineering, 10,* 277–280.

Thenkabail, P. S., Dheeravath, V., Biradar, C. M., Gangalakunta, O. R. P., Noojipady, P., Gurappa, C., et al. (2009). Irrigated area maps and statistics of India using remote sensing and national statistics. *Remote Sensing, 1*(2), 50–67.

Tucker, C. J. (1979). Red and photographic infrared linear combinations for monitoring vegetation. *Remote Sensing of Environment, 8*(2), 127–150.

Turney, T. C. (2012). Grayscale mapping of historically irrigated lands using GIS. *Journal of Geographic Information Systems, 4,* 33–36.

Uddin, J., Smith, R., Hancock, N., & Foley, J. (2014). Evaluation of sap flow sensors to measure the transpiration rate of plants during canopy wetting and drying. *Journal of Agricultural Studies, 2*(2), 105–119.

USGS (2001). *Understanding color-infrared photographs.* U.S. Geological Survey Fact Sheet 129-01.

USGS (2003) *Landsat: A global land-observing program.* U.S. Geological Survey Fact Sheet 023-03.

van der Tol, C. & Parodi, G. N. (2012). Guidelines for remote Sensing of evapotranspiration. In A. Irmak (Ed.) *Evapotranspiration—remote sensing and modeling*. InTech, https://doi.org/10.5772/18582.

Velpuri, N. M., Thenkabail, P. S., Gumma, M. K., Biradar, C., Dheeravath, V., Noojipady, P., et al. (2009). Influence of resolution in irrigated area mapping and area estimation. *Photogrammetric Engineering & Remote Sensing, 75*(12), 1383–1395.

Wahr, J., Swenson, S., Zlotnicki, V., & Velicogna, I. (2004). Time-variable gravity from GRACE: First results. *Geophysical Research Letters, 31,* L11501.

Wood, W. W. (1999). Use and misuse of the chloride-mass balance method in estimating ground water recharge. *Ground Water, 37,* 2–3.

Wood, W. W., & Sanford, W. E. (1995). Chemical and isotopic methods for quantifying groundwater recharge in a regional, semiarid environment. *Ground Water, 33,* 458–468.

Zamora, C. (2008). *Estimating water fluxes across the sediment-water interface in the lower Merced River, California.* U.S. Geological Survey Scientific Investigations Report 2007-5216.

Zhang, K., Kimball, J. S., & Running, S. W. (2016). A review of remote sensing based actual evapotranspiration estimation. *Wiley Interdisciplinary Reviews: Water, 3*(6), 834–853.
Zhu, T., Fu, D., Jenkinson, B., & Jafvert, C. T. (2015). Calibration and application of an automated seepage meter for monitoring water flow across the sediment-water interface. *Environmental Monitoring and Assessment, 187*(4), 171.

Chapter 5
Geochemistry and Managed Aquifer Recharge Basics

5.1 Introduction

Groundwater over time approaches chemical equilibrium with respect to the reactive minerals present in an aquifer. Aquifers may contain some minerals that are essentially unreactive in that their dissolution and precipitation rates are exceedingly slow under the temperature, pressure, and chemical conditions that occur in most near-surface groundwater environments. Anthropogenic aquifer recharge (AAR) very commonly results in the introduction of water into an aquifer that is compositionally different from the native groundwater and in disequilibrium with aquifer minerals. The introduced water may also have a different temperature. Geochemical reactions and processes important in aquifer recharge include:

- non-reactive mixing of recharged and native groundwater
- precipitation of minerals as cements
- dissolution of aquifer minerals
- oxidation reactions resulting from the introduction of dissolved oxygen (DO) into aquifers containing chemically reducing conditions
- reduction reactions caused by decreases in the oxidation-reduction (redox) potential (ORP, eH, pE) of groundwater as DO and other electron acceptors are consumed by the oxidation of organic matter (dissolved and particulate) in the recharged water
- sorption and cation exchange reactions
- clay swelling and dispersion.

Geochemical processes can impact AAR systems in various manners:

- Mineral precipitation can clog pores, locally reducing the hydraulic conductivity of the aquifer and, in turn, decreasing recharge and production rates.
- Dissolution of minerals can widen pores, resulting in increases in hydraulic conductivity.
- Mineral dissolution, alteration, sorption, and cation exchange reactions can adversely impact recharged water quality by releasing dissolved ions (including arsenic and metals) into the water.

R. G. Maliva, *Anthropogenic Aquifer Recharge*,
Springer Hydrogeology, https://doi.org/10.1007/978-3-030-11084-0_5

- Contaminant attenuation processes during recharge and aquifer storage and transport can improve the quality of recharged water.
- Swelling and dispersion of clay minerals can dramatically reduce the hydraulic conductivity of affected strata, causing severe clogging.

A wide range in effort and sophistication of approaches has been taken to evaluate the geochemistry of AAR systems, ranging from detailed geochemical characterization and modeling, to essentially not considering geochemistry at all. The latter is the norm for small decentralized systems in developing countries where projects are too small to justify the costs of geochemical investigations, and financial and technical resources are unavailable. Lesser-scale geochemical investigations can be appropriate where operational experience with similar systems in the same area can be called upon. At a minimum, it is important to have at least a qualitative understanding of the geochemical processes that might occur in a prospective project. For example, if oxic surface water or potable water is to be recharged into a confined aquifer with chemically reducing conditions, then the possibility that oxidation of reduced mineral species could impact stored water quality requires consideration. An informed decision could then be made as to what further investigation is prudent. This chapter is a primer on geochemical principles. Chap. 6 addresses specific geochemical processes applicable to AAR.

5.2 Chemical Equilibrium Thermodynamics

Chemical equilibrium is the state at which the concentrations of both the reactants and products of a chemical reaction do not tend to change over time. The basic purpose of thermodynamics is to predict the equilibrium composition of a chemical system from the properties (concentrations, temperature, pressure) of its components. Equilibrium thermodynamics will predict, for example, whether a mineral phase will ***tend*** to precipitate or dissolve in water introduced into an aquifer. A separate issue is the kinetics or rates of reactions. Some geochemical reactions occur at exceedingly slow rates in the temperature, pressure and chemical environment of aquifers. For example, both the precipitation and dissolution of quartz (and other silicate minerals) occur at extremely low rates in most groundwater environments and the minerals are essentially nonreactive in the time-scale of operation of aquifer recharge systems. This section provides a basic overview of chemical equilibrium thermodynamics, which is addressed in detail in low-temperature aqueous geochemistry textbooks and reference books (e.g., Garrels and Christ 1965; Hem 1985; Stumm and Morgan 1996; Drever 1997; Langmuir 1996; Ryan 2014), and summarized in some groundwater textbooks (e.g., Freeze and Cherry 1979; Domenico and Schwartz 1998; Fetter 2000; Todd and Mays 2004).

Chemical reactions will tend to proceed so long as they result in a reduction in the free energy of the system. The lowest free energy state for aqueous reactions under given pressure and temperature conditions is the chemical equilibrium state, at which there is no tendency for the concentrations of reactants and products to further change with time. Chemical reactions involve both a forward and reverse reaction, for example, the addition and removal of ions to and from the surface of a crystal (i.e., dissolution and precipitation). At chemical equilibrium, the rates of the forward and reverse reaction are the same, rather than that the reactions do not occur.

The relationship between the products and reactants of a reaction at equilibrium is expressed by the equilibrium constant, which is related to the standard free energy difference between the products and reactants:

$$\mathrm{Log}K_{eq} = \frac{-\Delta G_r^o}{2.303RT} \tag{5.1}$$

where

K_{eq} equilibrium constant
ΔG_r^o Standard Gibbs free energy of the reaction (J/mole, kJ/mol)
T absolute temperature (°K)
R universal gas constant (8.314 J/mol °K)

The standard Gibbs free energy of a reaction is the sum of the standard Gibbs free energies of formation (ΔG_f^o) of the products minus those of the reactants. The standard Gibbs free energy of formation of a compound is the change in Gibbs free energy resulting from the formation of 1 mol of the compound in its standard state from its constituent elements in their standard states. Tables of standard Gibbs free energy of formation for a pressure of 1 atm (or 1 bar) and a temperature of 298.15 K (25 °C) are available in reference books (e.g., Stumm and Morgan 1996).

For example, the mineral anhydrite (calcium sulfate) disassociates into calcium and sulfate ions:

$$CaSO_4(\text{solid}) \leftrightarrow Ca^{2+}(\text{aqueous}) + SO^{-4}(\text{aqueous}) \tag{5.2}$$

The standard Gibbs free energy of the reaction at 1 atm and 25 °C is the ΔG_f^o of Ca^{2+} and SO^{-4} ions minus the ΔG_f^o of the mineral ahydrite (using values from Stumm and Morgan 1996, in KJ mol^{-1}):

$$\Delta G_r^o = -553.54 - 744.6 - (-1321.7) = 23.56 \tag{5.3}$$

which gives an equilibrium constant of $10^{-4.13}$ or 7.46×10^{-5}. In the case of the ionic solids, which include virtually all minerals encountered in aquifers, the equilibrium constant is approximately equivalent to the solubility product (K_{sp}).

The general ion activity product equation for the reaction

$$aA_{(s)} \leftrightarrow bB_{(aq)} + cC_{(aq)} \tag{5.4}$$

is

$$IAP = \frac{(B)^b (C)^c}{(A)^a} \tag{5.5}$$

where (A), (B), and (C) are the activities of the solid 'A' and ions 'B' and 'C'. The activity of a pure solid or liquid phase is 1. Chemical activities are the product of the molal concentration (moles/kg) of each ion (m) and the activity coefficient of the ion (λ). For calcium, as an example,

$$(\mathrm{Ca}^{2+}) = m_{\mathrm{Ca}^{2+}} \lambda_{\mathrm{Ca}^{2+}} \tag{5.6}$$

In relatively dilute water solutions, molal concentrations are essentially equal to molar concentration (moles/L), as the density of water is approximately equal to 1 kg/L. For dilute solutions, the values of activity coefficients also approach one, and activities are approximately equal to the molal concentrations of ions. With increasing salinities (ionic strengths), activity coefficients depart from one. For charged species, the activity coefficients are usually less than one. Commonly used methods to calculate activity coefficients from ionic strength (and other parameter) data are the Debeye-Hückel, Davies, Truesdell-Jones and Pitzer equations.

Returning to anhydrite, the IAP of the dissolution reaction is product of the measured activities of calcium and sulfate ions:

$$\mathrm{IAP} = \left(\mathrm{Ca}^{2+}\right)_{\mathrm{measured}} \left(\mathrm{SO}^{-4}\right)_{\mathrm{measured}} \tag{5.7}$$

If the IAP is less than K_{sp}, then the solution is said to be undersaturated with respect to the mineral and dissolution will tend to occur. If the IAP is greater than K_{sp}, then the solution is said to be supersaturated with respect to the mineral and precipitation will tend to occur. The saturation state of solutions with respect to minerals is commonly expressed in terms of their saturation index (SI), which is the logarithm of the ratio of their IAP and solubility product:

$$\mathrm{SI} = \log\left(\mathrm{IAP}/\mathrm{K}_{\mathrm{sp}}\right) \tag{5.8}$$

Solutions that are supersaturated with respect to a mineral have an SI of greater than zero, whereas undersaturated solutions have SI values of less than zero.

IAPs are calculated using free ion species activities. Metals also occur as complexed ions, which are species formed of a central metal ion bounded to one or more molecules or ions, which are referred to as ligands or complexing agents. Laboratory analyses performed using methods such as inductively coupled plasma-atomic emission spectrometry (ICP-AES) and atomic absorption spectroscopy (AAS) report the total concentration of metals. Measured total metals concentrations include both free (uncomplexed) ions (e.g., Ca^{+2} ions) and complexed ions (e.g., $CaSO_4^o$ and $CaHCO_3^+$). Only the concentration of uncomplexed ions are used to calculate IAP values. Hence, total measured concentrations need to be adjusted to account for complexed ions.

In practice, saturation state calculations are now almost always performed using mineral equilibrium and speciation software packages, such the U.S. Geological Survey MINTEQA2 (Allison et al. 1991), WATEQ 4F (Ball and Nordstrom 1991), and PHREEQC (Parkhurst and Appelo 1999) codes. The codes can be used to evaluate the saturation state of end member solutions and mixed solutions (native groundwater and recharge water), and compositional changes after reaction of recharged water with aquifer minerals. The USGS codes are included in geochemical modeling and data analysis software packages, such as Geochemist Workbench and AquaChem, which have data storage and pre- and post-processing capabilities. As is always the case when using modeling software, it is important that users fully understand how the programs work, their underlying assumptions, and the meaning and appropriateness of various modeling options.

A critical issue for equilibrium thermodynamic modeling of AAR systems is collection of adequate data of sufficient quality. At a minimum, water sampling and analysis program should include all major dissolved species (i.e., species that constitute the bulk of the total dissolved solids) and all minor species that are involved in dissolution and precipitation reactions with reactive minerals present in the aquifer or that could potentially precipitate. Field parameters such as temperature, pH, and ORP also need to be carefully measured. The latter two parameters should be measured using a flow-through device that minimizes the potential for atmospheric exchange.

Review of ASR practices in the United States (Maliva and Missimer 2010) revealed that water chemistry sampling programs are commonly inadequate to evaluate fluid-rock interactions. Parameters lists were often based solely on regulatory requirements, which focus on heath-based drinking water standards and do not include some geochemically important parameters, such as calcium and magnesium, for which there is not a drinking water standard. ORP and pH measurements are commonly unreliable due to sampling procedures that allow for degassing of CO_2 and introduction of DO.

5.3 Carbonate Mineral Reactions

Carbonate mineral reactions are important in AAR because the minerals are commonly present as either primary or secondary constituents of aquifer rocks and are reactive in groundwater physicochemical environments. Many groundwaters and ocean water are close to saturation with respect to calcium carbonate minerals. Changes in temperature, pH, and dissolved CO_2 concentration impact the saturation state of groundwater with respect to carbonate minerals causing either precipitation or dissolution. Precipitation of calcium carbonate is a common cause of clogging of wells and infiltration surfaces. Dissolution of carbonate minerals can increase the hydraulic conductivity of aquifer strata and the hardness of recharged water. Calcium carbonate "scale" can precipitate rapidly (within weeks) under some geochemical circumstances.

Calcium carbonate ($CaCO_3$) occurs in near surface environments as the polymorphs calcite and aragonite. Polymorphs are minerals that have the same composition but different crystal structures. Under near-surface pressure and temperature conditions, calcite has a lower solubility product than aragonite and is thus more stable. However, some organisms (e.g., corals, some groups of mollusks) precipitate aragonite in their shells, and aragonite precipitates directly in some supersaturated surface environments. Magnesium substitutes for calcium in the calcite crystal structure to varying degrees. High-magnesian (magnesium) calcites tend to be more soluble that low-magnesian calcites. Both aragonite and high-magnesian calcite tend to dissolve and reprecipitate as low-magnesian calcite in groundwater environments. Hence, aragonite and high-magnesian calcites are abundant in some recent sediments but decrease in abundance with increasing age and tend to be rare in post-Pleistocene strata.

Dolomite is an ordered calcium magnesium carbonate whose solubility relative to calcite depends upon the aqueous magnesium to calcite ratio. In solutions with high Mg^{+2}/Ca^{+2} ratios, including mean seawater, the replacement of calcite by dolomite is thermodynamically favored. However, dolomite does not readily precipitate in near-surface environments, irrespective of the groundwater saturation state. The abundance of dolomite in some ancient sedimentary rocks and its paucity in recent carbonate environments has long been an enigma in sedimentary geology. The dissolution of dolomite is also more sluggish than calcite in near-surface environments.

Calcium carbonate equilibria is addressed in virtually all low-temperature (aqueous) geochemistry texts. The saturation state of waters with respect to calcite is expressed by four equilibrium equations:

$$K_{CO_s} = \frac{(H_2CO_3)}{P_{CO_2}} \tag{5.9}$$

$$K_1 = \frac{(H^+)(HCO_3^-)}{(H_2CO_3)} \tag{5.10}$$

$$K_2 = \frac{(H^+)(CO_3^{-2})}{(HCO_3^-)} \tag{5.11}$$

$$K_{cal} = (Ca^{2+})(CO_3^{2-}) \tag{5.12}$$

The relationship between calcite solubility (expressed as equilibrium dissolved calcium activity and molal concentration), pH (hydrogen activity), and the partial pressure of carbon dioxide (P_{CO_2}) are obtained by rearranging Eqs. 5.9 through 5.12 (Drever 1997):

$$(Ca^{2+}) = \frac{(H^+)^2 K_{cal}}{P_{CO_2} K_1 K_2 K_{CO_2}} \tag{5.13}$$

$$(Ca^{2+}) = (10^{-pH*2}) \frac{K_{cal}}{P_{CO_2} K_1 K_2 K_{CO_2}} \tag{5.14}$$

$$m^3_{Ca^{2+}} = P_{CO_2}\frac{K_1 K_{cal} K_{CO_2}}{4K_2\lambda_{Ca^{2+}}\lambda^2_{HCO_3^-}} \tag{5.15}$$

The above equations show that calcium carbonate solubility is sensitive to pH and dissolved CO_2 concentration. As pH is the negative of the log of hydrogen ion activity, small changes in pH corresponds to large changes in hydrogen ion activity and thus calcite solubility. Hence, accurate measurement of pH is critical for evaluation of calcium carbonate saturation states. The non-linear relationship between equilibrium calcium concentration and P_{CO_2} is important for carbonate geochemistry as the mixing of two waters that are both at calcite saturation can result in a solution that is either undersaturated or supersaturated with respect to calcite (Runnells 1969; Wigley and Plummer 1976; Drever 1997).

Calcium carbonate precipitation can be induced in AAR systems by:

- degassing of CO_2 from groundwater to the atmosphere, which can decrease the P_{CO_2} and increase the calcium carbonate saturation state (e.g., travertine springs)
- decreases in P_{CO_2} in shallow water bodies by photosynthesis
- increases in temperature, which result in a decrease in CO_2 solubility
- sulfate reduction and methanogenesis.

Calcium carbonate dissolution can be induced in ARR systems by:

- introduction of dilute waters with low calcium concentrations.
- aerobic oxidation of organic matter, which produces CO_2.
- oxidation of reduced (e.g., sulfide) minerals, which can decrease the pH of waters (e.g., acid mine drainage).

5.4 Redox Reactions

5.4.1 Redox Basics

Oxidation-reduction reactions, commonly referred to as "redox" reactions, involve the transfer of electrons. Chemical species that donate electrons are reducing agents and are said to be oxidized during reactions. Species that receive electrons (i.e., electron acceptors) are oxidizing agents and are said to be reduced during reactions. Species that are oxidized have an increase in their oxidation state and species that are reduced have a decrease in their oxidation state. Redox reactions are important in AAR because:

- recharge often introduces water into an aquifer with a different redox state than native groundwater
- the solubility of some metals (e.g., iron and manganese) varies between oxidation states, and redox reactions can mobilize metals and metalloids (e.g., arsenic, Mn^{2+} and Fe^{2+}) to the detriment of groundwater quality

- some natural contaminant attenuation processes (pathogen inactivation and the biodegradation of organic chemicals) are sensitive to redox state and may occur to greater degrees in either reducing or oxidizing groundwater environments
- redox reactions may be specifically taken advantage of to remove specific contaminants (e.g., denitrification)
- minerals precipitation caused by redox reactions can clog pores
- precipitated oxide and hydroxide minerals have high sorptive capacities
- oxidation of reduced minerals can consume DO.

Inasmuch as aqueous solutions do not contain free electrons, redox reactions involve complementary half reactions that release or receive electrons. A common and important redox reaction in groundwater systems is the oxidation of ferrous iron (Fe^{2+}, Fe(II)) to ferric iron (Fe^{3+}, Fe(III))

$$4Fe^{2+} \rightleftharpoons 4Fe^{3+} + 4e^{-} \tag{5.16}$$

where e^{-} is an electron. A complementary half reactions is the reduction of oxygen gas:

$$O_{2(gas)} + 4e^{-} \rightleftharpoons 2O^{2-} \tag{5.17}$$

$$2O^{2-} + 2H_2O \rightleftharpoons 4OH^{-} \tag{5.18}$$

The combined equations for the oxidation of ferrous iron and reduction of oxygen is

$$4Fe^{2+} + O_2 + 4H^{+} \rightleftharpoons 4Fe^{3+} + 2H_2O \tag{5.19}$$

Ferric iron reacts with water to produce low-solubility iron hydroxides, which tend to undergo recrystallization (transformation) to more stable iron (oxy)hydroxides and iron oxides:

$$4Fe^{3+} + 12H_2O \rightleftharpoons 4Fe(OH)_3 + 12H^{+} \tag{5.20}$$

Manganese has a similar redox behavior as iron in groundwater environments. Manganese geochemistry was reviewed by Tebo et al. (2004) and Wu et al. (2016). Mn is more soluble in anoxic conditions where it occur as divalent Mn^{2+}. Mn is commonly present as Mn^{2+} in sulfates and carbonates. Under oxidizing conditions, Mn^{2+} tends to be oxidized to the much less soluble Mn^{+3} and Mn^{+4} states:

$$2Mn^{+2} + O_2 + 2H_2O \rightleftharpoons 2MnO_2 + 4H^{+} \tag{5.21}$$

Mn^{4+} occurs in aquatic environments as insoluble oxides, oxyhydroxides, and hydroxides. Reducing conditions (suboxic and anoxic) tends to result in the reductive dissolution of Mn oxides, releasing Mn^{2+} to solution

Another important redox reaction in groundwater systems is the reduction of sulfate (SO_4^-) to sulfide (S^{2-}) and the reverse reaction, the oxidation of sulfide to sulfate:

$$SO_4^{2-} + 8H^+ + 8e^- \rightleftharpoons S^{2-} + 4H_2O \tag{5.22}$$

During sulfate reduction, sulfur changes from the +6 (S^{+6}) redox state to the −2 (S^{-2}) state.

Redox reactions in groundwater environments are usually biologically mediated, as opposed to being abiotic. Redox reactions that occur during groundwater flow and storage are typically exothermic (i.e., they release energy). Microbes couple electron donors (commonly oxidizable organic carbon) and electron acceptors (DO, nitrate, ferric iron, sulfate) to obtain energy. Coupled reactions are specific to microbial genera. A compound that receives (accepts) an electron during the oxidation of a carbon source (e.g., during cellular respiration) is referred to as a terminal electron acceptor (TEA).

Organic carbon oxidation in groundwater typically proceeds in a sequence from the highest energy yield downward (Table 5.1). The typical TEA order is

(1) oxygen
(2) nitrate
(3) manganese
(4) iron(III)
(5) sulfate
(6) CO2.

If DO is present, then organic carbon will be oxidized by aerobic respiration. Once DO is largely depleted, then the next highest energy TEA (nitrate) will tend to be utilized. However, there may be some overlap in TEA utilization. The impor-

Table 5.1 Organic matter oxidation reactions and energy yields

Reaction	Equation	Energy yield, ΔG (kJ/mole) at pH = 7
Aerobic respiration	$CH_2O + O_2 \rightarrow CO_2 + H_2O$	−502.3
Denitrification	$CH_2O + (4/5)NO_{3-} \rightarrow CO_2 + (2/5)N_2 + (7/5)H_2O$	−476.8
Mn^{+4} reduction	$CH_2O + 2MnO_2 + 4H^+ \rightarrow 2Mn^{2+} + 3H_2O + CO_2$	−340.3
Fe^{+3} reduction	$CH_2O + 4Fe(OH)_3 + 8H^+ \rightarrow 4Fe^{2+} + 11H_2O + CO_2$	−115.0
Sulfate reduction	$CH_2O + (1/2)SO_4{}^{2-} + (1/2)H^+ \rightarrow (1/2)HS^- + H_2O + CO_2$	−104.7
Methanogenesis	$CH_2O \rightarrow (1/4)CH_4 + (1/2)CO_2$	−92.9

Source Champ et al. 1979, and Baveye et al. 1998

tance of the TEAs in organic matter oxidation depends upon their concentration in groundwater. Usually, nitrate, Mn^{+4}, and Fe^{+3} are present at low concentrations in groundwater. On the contrary, the concentration of sulfate is typically so high in brackish and saline aquifers relative to the organic matter concentration that the supply of available organic matter is exhausted long before the sulfate supply is exhausted.

Heterogeneities in aquifer sediment and groundwater composition can impact DO and other TEA consumption. DO removal depends largely upon both the organic content of water and the oxidizable components of sediment and rock. For example, DO may be completely removed and groundwater chemistry progress further down the TEA sequence in organic-rich sediments than in nearby organic-poor sediments.

Sharp spatial variations in redox state may develop in aquifer recharge systems that are used to store reclaimed water or surface water with relatively high organic contents. Vanderzalm et al. (2002), for example, documented that a reactive zone developed adjacent to the ASR well at the Bolivar system in South Australia, which was characterized by high microbial activity related to the relatively large flux of nutrients through the aquifer immediately surrounding the well. Sulfate-reducing conditions developed in the immediate vicinity of the ASR well and were not detected in an observation well located 13 ft (4 m) away.

Aquifers can be subdivided into zones defined by the most active TEA (e.g., sulfate-reducing zone) and identified by the following criteria:

- absence (or greatly reduced concentration) of higher energy-yielding TEAs
- reduction in the TEA concentration (corrected for mixing)
- increased concentrations of the products of TEA reduction (e.g., H_2S gas from sulfate reduction)

The oxidation of iron sulfide minerals (FeS_s) is important in AAR where water containing DO is introduced in a chemically reducing (anoxic) aquifer. The predominant reactions for iron sulfide oxidation involve oxygen, nitrate (NO_3^-), or ferric iron as the electron acceptors (Vanderzalm and Le Gal La Salle, 2005) as follows:

$$4FeS_2 + 15O_2 + 16HCO_3^- \rightarrow 4Fe(OH)_3 + 8SO_4^{2-} + 16CO_2 + 2H_2O \quad (5.23)$$

$$5FeS_2 + NO_3^- + 4H^+ \rightarrow 5Fe^{2+} + 10SO_4^{2-} + 7N_2 + 2H_2O \quad (5.24)$$

$$FeS_2 + 14Fe^{3+} + 8H_2O \rightarrow 15Fe^{2+} + 2SO_4^{2-} + 16H^+ \quad (5.25)$$

The ferrous iron released in the latter two reactions would in turn be oxidized to ferric iron, which would precipitate as an iron (oxy)hydroxide mineral.

5.4.2 *Oxidation-Reduction Potential*

The oxidation-reduction or redox potential (ORP, Eh) of a solution is a measure of the tendency of its constituents to gain or lose electrons and, therefore, to be either reduced or oxidized. Eh is typically measured in millivolts (mV) and expressed relative to a standard hydrogen electrode (SHE). As described by Drever (1997), the SHE consists of a piece of finely divided platinum in contact with a solution containing hydrogen ions at unit activity and hydrogen at a pressure of 1 atm, with the whole system at 25 °C. Depending on the "activity of electrons" in a half cell of a solution relative to the activity in the SHE, electrons will tend to flow via a conductor either to or from the SHE. Eh is positive, by convention, if the activity of electrons in a half-cell containing a sample solution is less than that in the SHE and electrons flow from the SHE to the sample half-cell. A high positive Eh value is indicative of oxidizing conditions

The Nernst equation describes the equilibrium relationship between Eh (volts) and the activities of species in a redox pair:

$$\mathrm{Eh} = \mathrm{E}^{\mathrm{o}} + \frac{2.303\mathrm{RT}}{\mathrm{nF}} \log\left(\frac{\text{activity product of oxidzed species}}{\text{activity product of reduced species}}\right) \tag{5.26}$$

where

R gas constant (8.314 J mol^{-1} K^{-1})
T absolute temperature (°K),
F Faraday constant (96,485 coulombs/mol)
n number of electrons that are involved in the reaction
E^{o} is the standard electron potential

The Nernst equation at 25 °C can be simplified to

$$\mathrm{Eh} = \mathrm{E}^{\mathrm{o}} + \frac{0.059}{\mathrm{n}} \log\left(\frac{\text{activity product of oxidzed species}}{\text{activity product of reduced species}}\right) \tag{5.27}$$

Values of E^{o} can be obtained directly from chemical reference sources or calculated from Gibbs free energy values. With respect to the iron (Fe^{2+} and Fe^{3+}) redox pair, Eh is related to the activities of the species as follows at 25 °C.

$$\mathrm{Eh} = \mathrm{E}^{\mathrm{o}} + 0.059\log\left(\frac{\mathrm{Fe}^{3+}}{\mathrm{Fe}^{2+}}\right) \tag{5.28}$$

Oxidation-reduction potential is commonly expressed in the chemistry literature in terms of electron activity (pE or pε), which is defined in a similar manner as pH as

$$\mathrm{pE} = -\log\left(\mathrm{e}^{-}\right) \tag{5.29}$$

where (e^-) = the activity of electrons. At a temperature of 25 °C

$$pE = 16.9Eh \tag{5.30}$$

Eh is commonly used to describe oxidation-reduction potential in groundwater investigations, especially with respect to field measurements, and is therefore used herein.

The Nernst equation can be used to calculate the activities of redox pair species (e.g., Fe^{2+} and Fe^{3+}) if the Eh and total dissolved concentrations of species are known. Alternatively, if the activity ratio of a redox pair is independently determined, then the Nernst equation can be used to calculate in the Eh of the solution. The Nernst equation expresses the equilibrium relationship between redox pair species. It is important to recognize that equilibrium conditions may not exist in some groundwater systems, particularly if there is a sudden change in water chemistry such as may occur during aquifer recharge.

5.4.3 Redox State Measurement

It has long been appreciated that accurate measurement of the redox potential of most aquatic environments is extremely difficult (Bass Becking et al. 1960; Faust and Vecchioli 1974). Despite the importance of redox state in the geochemistry of MAR systems, it has been observed in a number of investigations that field measurements of DO and oxidation-reduction potential are generally poor (WRRF 2007). DO is commonly introduced during the sampling process, resulting in misinterpretation of anoxic waters as having higher (more positive and oxic) redox states. Indeed, the commonly used procedure of measuring redox state using a field meter is too inaccurate for quantitative geochemical analyses.

Field Eh meters measure potential relative to electrodes other than the SHE. Field readings need to be corrected to Eh values relative to the SHE. Instruction manuals for Eh meters normally provide the necessary conversion information or the instruments are programmed to make the conversion. It is imperative that samples being measured for Eh not be exposed to atmospheric oxygen. Measurements of Eh (and also pH) should be performed using a flow-through system (cell) in which the water sample is kept under positive pressure. It has been the author's observation that the attention to detail in water sample collection varies depending upon whether the purpose of the sampling and analysis is to meet regulatory requirements or to obtain data for a rigorous geochemical investigation under the supervision of a geochemist.

Even when using ideal sampling and measurement procedures, questions may remain as to the meaning and significance of measured Eh values. Unless all redox couples in a sample are in equilibrium, one cannot speak of an Eh of a solution (Drever 1997). For some systems at equilibrium, a reliable Eh value can be calculated from the activities of the dissolved species participating in an equilibrium

(Fe^{2+}/Fe^{3+}, NO_2^{-1}/NO_3^{-1}, SO_4^{2-}-H_2S) using the Nernst equation. However, redox reactions involving S and N do not seem to proceed at a significant rate without biota and are commonly far from equilibrium (Faust and Vecchioli 1974). Care must be taken in the collection and handling of samples to ensure that the concentrations of the components of redox couples do not change before laboratory analysis. The concentration of both components of the redox couples should be well above laboratory detection limits and analysis error. Ideally several redox couples should be measured.

5.4.4 Eh-pH Diagrams

Eh-pH and pε-pH diagrams (also referred to as Pourbaix diagrams) are used to illustrate and graphically analyze mineral redox stability relationships. Procedures used to generate Eh-pH diagrams are included in general aqueous geochemistry texts (e.g., Garrels and Christ 1965; Stumm and Morgan 1996; Drever, 1997) and compilations of diagrams are provided by Brookins (1988) and Takeno (2005). Eh-pH diagrams illustrate the ions and mineral species that are stable in pH and eH fields for a specific combinations of physicochemical conditions, including the pressure, temperature, and total concentrations of the elements considered. Mineral phases are also assumed to be in their standard state. An Eh-pH for iron and sulfur is provided as an example (Fig. 5.1).

Eh-pH diagrams can be manually created using basic thermodynamic data or, more conveniently, by using geochemical analysis software. Mineral precipitation and dissolution in low-temperature groundwater environments may be controlled by kinetics, particularly during the relatively short time periods involved in AAR systems. Some minerals precipitate very slowly and minerals may be metastable in that they persist under Eh-pH conditions outside of their stability fields. The most stable mineral phase under a given Eh-pH condition may either not directly precipitate or may precipitate very slowly. Instead, a less stable mineral phase may form first because it can more readily precipitate. Hence, Eh-pH diagrams should include only mineral phases that may actually form or dissolve in the groundwater system of interest.

Iron precipitation under oxic conditions illustrates the effect of kinetics on mineral precipitation. The most energetically favorable reaction in response to the addition of DO to groundwater containing ferrous (Fe^{2+}) iron in solution is the precipitation of an iron oxide mineral such as hematite (Fe_2O_3). However, in the groundwater environment, a variety of metastable ferric hydroxide ($Fe(OH)_3$) and (oxy)hydroxide ($FeO(OH)$) minerals (e.g., ferrihydrite, goethite) may instead precipitate. Hematite is the predominant mineral phase in older (>2 million years) iron-rich rocks (red beds), whereas hematite in uncommon in younger rocks and sediments, and goethite is the main iron mineral phase (Goss 1987). Over time, ferric (oxy)hydroxides tend to recrystallize into the more stable hematite.

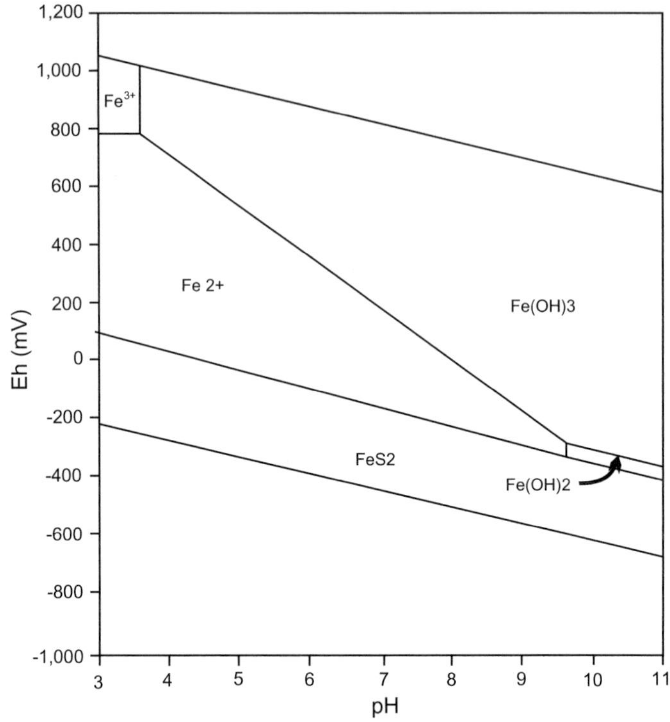

Fig. 5.1 Eh-pH diagram for iron and sulfur at 25 °C. Total iron = 1×10^{-6} M and total sulfur = 1×10^{-3} M

5.5 Kinetics

Reaction kinetics is the study of the rate of chemical reactions. Understanding fluid-rock interactions in aquifers require consideration of both equilibrium thermodynamics and kinetics. The former determines whether a reaction is energetically favorable, and will thus tend to occur, whereas the latter determines the speed of the reaction and thus the degree to which it will occur over the timeframe of interest.

Many minerals react so slowly that groundwaters do not reach chemical equilibrium with respect to the minerals, at least over the time scale of operation of an AAR system. Many of the mineral phases found at land surface and within aquifers are metastable in that they persist in a non-equilibrium state for extended periods of time. For example, quartz (SiO_2), the main mineral of most siliciclastic sediments and rocks, both dissolves and precipitates very slowly in most low-temperature groundwater environments. Most other silicate minerals (e.g., feldspars, micas, amphiboles) are also very poorly reactive in near surface physicochemical environments. The preponderance of metastability near the earth's surface (including in its aquifers) is a direct consequence of the slowness of chemical reactions at low temperatures (Berner

1981). Kinetics plays a more leading role in near surface geochemistry than it does at higher temperatures as reaction rates increase with increasing temperatures (Berner 1981).

Reaction rates are function of a reaction rate constant or coefficient (k), the concentration of reaction species, and the reaction order for each species. The order of a reaction expresses the functional relationship between concentration and rate. Consider the general reaction

$$A + 3B \rightarrow C \tag{5.31}$$

The reaction rate (rate of increase in the concentration of C) in molar or molal units per sec (m/s or M/s) can be expressed as

$$rate = \frac{dC}{dt} = k[A]^x[B]^y \tag{5.32}$$

where t is time, $[A]$ and $[B]$ are the molal concentrations of species A and B, and x and y are the reaction orders of species A and B, respectively. The sum of $x + y$ is the order of the overall reaction. The form of the reaction rate law and reaction orders are empirically determined by the reaction mechanism rather than by the reaction stoichiometry. The units of the reaction rate constant vary with reaction order and can be obtained by dividing the unit of rate (m/s) by the concentration units (m) to the order $(x + y)$ exponent. For example, the reaction rate constant unit for a second order reaction using molal concentration units is (m/s)/m^2 or m^{-1}s^{-1}.

Zero-order reactions (order = 0) have a constant rate that is independent of the concentrations of the reactants. A first-order reaction has a rate proportional to the concentration of one of the reactants. Second order reactions (order = 2) have rates proportional to the concentration of the square of a single reactant or the product of the concentration of two reactants. Common examples of first-order reactions are radioactive decay and the inactivation of pathogens. The first order rate equation is

$$-\frac{dB}{dt} = k[B] \tag{5.33}$$

which can be integrated to

$$B_T = B_0 e^{-kt} \tag{5.34}$$

$$\log \frac{[B_t]}{[B_0]} = \frac{-kt}{2.302} \tag{5.35}$$

where B_0 = the concentration of species B at time $t = 0$, and B_t is the concentration of species of B at time t, and the rate constant (k) has units of 1/time. For a 50% and one log (90%) decrease in concentration, the equations become

$$t_{\frac{1}{2}} = \frac{0.693}{k} \tag{5.36}$$

$$t_{\frac{1}{10}} = \frac{2.303}{k} \tag{5.37}$$

Equation 5.36 is the familiar half-life equation and Eq. 5.37 is the $\log_{10}$ removal equation used for pathogen attenuation.

Reaction rate constants are positively related to temperature. For most reactions, the relationship between the reaction rate constant and absolute temperature (T) can be expressed by the Arrhenius equation

$$k = Ae^{\frac{-E_a}{RT}} \tag{5.38}$$

where A is the pre-exponential factor and E_a is the activation energy. Both A and E_a are usually empirical values.

Modeling of reaction kinetics in near-surface environments is fraught with difficulties because there is considerable uncertainty concerning reaction processes, rate law equations, and rate constants. Reaction rates measured under laboratory conditions or determined through theoretical calculations may differ greatly from actual rates in the field (Berner 1981).

With respect to AAR systems, it is important to differentiate between minerals that are metastable (i.e., have kinetic inhibitions against their precipitation and dissolution under the physicochemical conditions in aquifers) and those that are reactive within the operational time frames of the systems. Reactions that normally should not materially impact aquifer recharge projects for kinetic reasons include (Maliva and Missimer 2010):

- precipitation and dissolution of quartz and other silicate minerals
- dolomite precipitation and, to a lesser degree, dissolution
- heavy mineral precipitation and dissolution
- phosphate mineral (apatite) precipitation and dissolution
- clay mineral dissolution, precipitation, and transformation (recrystallization).

The relatively few geochemical reactions that are of significant concern in AAR systems include (Maliva and Missimer 2010):

- calcite dissolution and precipitation.
- evaporite mineral (e.g., gypsum) dissolution.
- redox reactions involving iron and manganese sulfide, oxide, (oxy)hydroxide, and carbonate minerals.
- sorption reactions (including hydration, and cation and anion exchange) involving clay minerals, iron and manganese oxide and hydroxide minerals, and organic matter.
- biodegradation of organic matter.

Geochemical evaluation of AAR systems should focus largely on the above-listed potentially active geochemical reactions and processes that are both thermodynamically and kinetically favored in low-temperature groundwater environments. Mineralogical investigations should focus on determination of the presence and abundance of reactive minerals. Water quality analyses should include parameters involved in the equilibria of reactive minerals, in addition to major cations and anions.

5.6 Clay Minerals, Cation Exchange and Adsorption

5.6.1 Clay Mineralogy

The term "clay" refers to both clay minerals and sedimentary particles with diameters of less than 0.002 mm. Clay minerals are very finely crystalline aluminosilicates that belong to the phyllosilicate group of minerals. Clay-sized material very commonly contains abundant clay mineral particles. Clay minerals are important in aquifer recharge in the following manners:

- clay-rich sediments and rock tend to have very low permeabilities and often act as confining strata because of their very small particle and pore sizes
- clay minerals tend to have large adsorption capacities because of high specific surface areas and the occurrence of structural and surface charges
- swelling and dispersion of clay minerals during recharge can greatly reduce permeability.

The basic structural configuration of clay minerals is combinations of alternating sheets of tetrahedral silica oxides and octahedral hydroxides. The tetrahedral (T) sheets have the composition of $Si_2O_5OH^{-3}$ and the octahedral (O) sheets have the general compositions of either $Al_2(OH)_6$ (gibbsite) or $Mg_3(OH)_6$ (brucite). The former (gibbsite-type) octahedral sheets, in which each OH ion is coordinated with two trivalent cations, are called dioctahedral sheets, whereas latter type, in which each OH ion is coordinated with three divalent cations, are called trioctahedral sheets.

The most basic clay structure is the 1:1 or T-O clays, which consists of one tetrahedral sheet and one octahedral sheet. The mineral kaolinite ($Al_2Si_2O_5(OH)_4$), is a common 1:1 clay composed of one tetrahedral sheet and one aluminum hydroxide dioctahedral sheet:

$$Si_2O_5OH^{-3} + Al_2(OH)_6 + 3H^+ \rightarrow Al_2Si_2O_5(OH)_4 + 3H_2O \qquad (5.39)$$

The tetrahedral-octahedral layers are held together by weak Van der Waals bonds, which are caused by temporary attractions between electron-rich (negatively charged) regions of one layer and electron-poor (positively charged) regions of another.

Three (2:1) sheet (T-O-T) clay minerals consist of an octahedral sheet sandwiched between two tetrahedral sheets. The addition of another silica tetrahedral sheet to the kaolinite structure gives the mineral pyrophyllite ($Al_2Si_4O_{10}(OH)_2$). The trioctahedral equivalent of pyrophyllite (i.e., brucite octahedral sheets) is the mineral talc

($Mg_3Si_4O_{10}(OH)_2$). The three-sheet pyrophyllite and talc structure is electrically neutral and the layers are very weakly bound together by Van der Waals bonds. Talc is very soft because of the very weak bonding.

The basic clay mineral structure, both two (1:1) or three (2:1) sheets, is subject to ionic substitutions, which affect the composition, surface charge, and behavior of clay minerals. Isomorphous substitution involves the substitution of silica (Si^{4+}) by aluminum (Al^{3+}) in a tetrahedral sheet, or of aluminum by a divalent cation (e.g., Fe^{2+}, Mg^{2+}) in an octahedral sheet. The substitution of lesser charged ions give the clay mineral layer a net negative charge, which is satisfied by bonding with cations between layers. Interlayer cations bind the clay layers together with varying strengths, depending upon their type, abundance, and regularity.

The mineral muscovite (mica) consist of two pyrophyllite layers in which Al^{3+} substitutes for every 4th SiO_4 in the tetrahedral layers and interlayer K^+ ions balance the charge (T-O-T–K^+–T-O-T structure). The composition of muscovite is $KAl_2AlSi_3O_{10}(OH)_2$. The two T-O-T layers that make up the muscovite structure are strongly bound, but adjoining T-O-T–K^+–T-O-T layers are bound by weak Van der Waals bonds. The common clay mineral illite has the structure of muscovite but a less regular composition of $K_{1-1.5}Al_4(Si_{7-6.5})Al_{1-1.5}O_{20})(OH)_4$, which reflects its composition as two pyrophyllite layers in which there has been some isomorphous substitution for Si^{4+} by Al^{3+} and the sharing of interlayer K^+ cations to maintain electrical neutrality.

Another important group of clay minerals for AAR are the smectites (or swelling clays). Montmorillonite, which has the general composition $(0.5Ca,Na)_{0.7}(Al,Mg,Fe)_4(Si,Al)_8O_{20})\cdot 7nH_2O$, is a common smectite clay. Montmorillonite has the muscovite and illite structure of two three-layer (T-O-T) sheets but has undergone lesser isomorphous substitution than illite and its interlayer space is filled with varying amounts of water molecules and exchangeable cations. Smectites are referred to as swelling clays because they tend to swell (take in water and expand) when exposed to freshwater (such as during aquifer recharge). Mixed-layer clays (e.g., illite/smectite) have intercalations of different types of sheets.

5.6.2 *Adsorption and Ion Exchange*

Adsorption is the process whereby solutes accumulate on the surface of solids. Absorption is the process whereby solutes diffuse into the solid. The term "sorption" encompasses both adsorption and absorption, and implicitly recognizes that multiple processes can contribute to the interaction of solutes with solid phases, some of which may not be identified or characterized. Sorbed solutes include cations, anions, and organic compounds. Microorganisms and other particulates can also be adsorbed onto the surfaces of aquifer solid phases. Ion exchange is the exchange of sorbed ions between solids and solutions.

Sorption process are important in AAR because they can remove contaminants from recharged waters. However, sorbed solutes and particles may be subsequently released back into the groundwater. The capacity of an aquifer to sorb given contaminants is finite and can be exhausted over time. Sorption processes also act to retard (slow) the transport of contaminants. The sorption of contaminants provides additional aquifer residence time for their removal by biodegradation processes. Ion-exchange processes can change the chemistry of recharged water, which, depending upon the specific exchange, could be either beneficial or adverse.

Clay mineral sorption processes are driven by surface charges. Positive surface charges will attract negative ions (anions) and particles with negative surface charges. Negative surface charges attract positively charged ions (cations) and particles with positive surface charges. Surface charges on clay minerals are generated by isomorphous substitution, and proton binding and disassociation (Stumm and Morgan 1996). Proton binding and disassociation also result in negative surface charges on hydroxide and oxide minerals and on various functional groups of organic matter (Chang and Page 1985).

Hydroxyl (–OH) groups on the outer surfaces of clay minerals obtain charges by interactions with hydrogen ions (protons). In solutions with low pH values (and thus high concentrations of hydrogen ions in solution), hydrogen ions tend to attach to surface hydroxyl groups creating a positive surface charge (Fig. 5.2). At high pH values (and thus low hydrogen ion concentrations), hydrogen ions tend to detach from the hydroxyl groups creating a negative surface charge.

The relationship between the surface charge of solids and pH values varies between solids. The isoelectric point (IEP) is the pH at which a surface has no net charge (i.e., positive and negative charges are balanced). The similar zero point of charge (ZPC; also referred to as the point of zero charge) is defined as the pH at which the electrical charge density on a surface is zero. The IEP and ZPC are usually similar with the subtle difference that the charge balance on surfaces for the IEP may be influenced by the presence of absorbed ions. The IEP and ZPC determine whether a surface with have cation or anion exchange properties. Clay minerals have low ZPCs (typically <5) and, therefore, have negative charges at the circumneutral pH of groundwater in most aquifer systems and tend to adsorb cations.

Cation exchange capacity (CEC) is a measure of its total capacity of a substance (soil or clay mixture) to hold exchangeable cations. CEC is commonly expressed in units of milli-equivalents (meq) per 100 grams. Each clay mineral has a range of CEC values because of variations in structure and chemical composition (Carroll 1959). The CEC of some common clay mineral types are 1–10 meq/100 g for kaolinite, 10–40 meq/100 g for illites, and 80–150 meq/100 g for smectites (montmorillonite) (Drever 1997). Humic substances present in soils also adsorb and exchange cations and can have CECs greater than those of clay minerals.

The composition of the exchangeable ions depends upon their concentrations in solution and the affinity of ions for an exchangeable site. Ion affinity is a function of the charge and hydrated size of ions. More highly charged (divalent and trivalent) ions are held more strongly onto surfaces than lesser charged (monovalent) ions. Hydrated ion size also impacts replaceability with smaller ions more strongly held. Carroll (1959) reported the following order of replaceability of common cations in clays and other minerals:

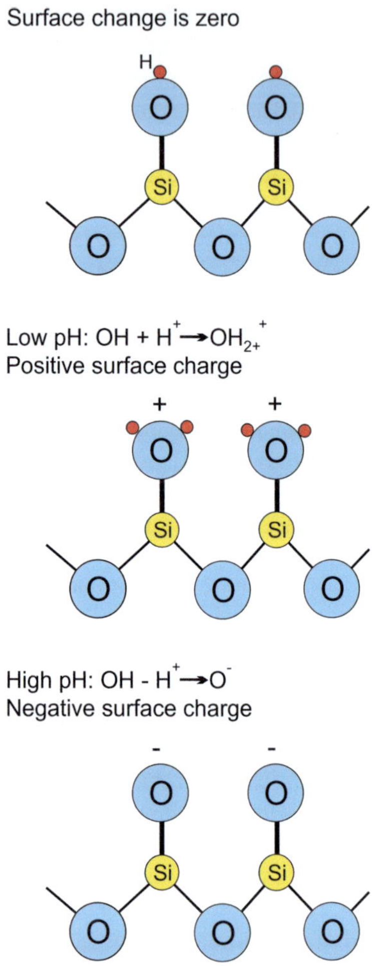

Fig. 5.2 Schematic diagram of the effects of pH on the surface charge of clay minerals. In low pH solutions, hydroxyl groups on surfaces may gain a proton (hydrogen ion) and obtain a positive change. The opposite occurs in high pH solutions where the hydroxyl groups lose a proton to solution and obtain a negative charge

$$Li^+ < Na^+ < K^+ < Rb^+ < Cs^+ \text{ and } Mg^{+2} < Ca^{+2} < Sr^{+2}, < Ba^{+2}$$

With respect of aquifer recharge, cation exchange will commonly tend to cause the concentration of dissolved calcium to decrease and that of sodium to increase. However, the change in composition is usually too small to impact subsequent uses of the recharged water.

5.6.3 Sorption Isotherms

Isotherms are equations or curves that relate the amount of a substance adsorbed (i.e., the adsorbate) on a solid at equilibrium to its concentration in solution. The linear sorption isotherm is the simplest equation, in which the mass of sorbed solute is a linear function of its concentration in solution:

$$C_s = K_d C \tag{5.40}$$

where

C_s mass of solute sorbed per dry weight of solid (mg/kg)
C concentration of solute in equilibrium with the mass of sorbed solute (mg/L)
K_d distribution (adsorption or soil-water partition) coefficient (L/kg).

The distribution coefficient is also the slope of a linear regression of C_s versus C. Fundamental constraints of the linear isotherms are that the relationship between sorbed concentration and solution concentration may not be linear and that the equation has no upper limit on the amount of solute than can be sorbed onto a solid (Fetter 1993). The amount of solute that a given mass of solid can adsorb is finite. Eventually solid surfaces become fully saturated with the solute in question and no further adsorption can occur. Depending upon circumstances, the upper limit may not be approached and a linear isotherm may be appropriate.

The Langmuir isotherm assumes a maximum adsorption capacity for forming a monolayer (C_{max}) of sorbed solute. The amount of sorbed solute asymptotically approaches C_{max}, which has units of mg/kg:

$$C_s = \frac{C_{max} K_l C}{1 + K_l C} \tag{5.41}$$

where K_l is the Langmuir distribution constant

Other isotherms have been developed that may better describe the adsorption behavior of some solutes. The more general Freundlich sorption isotherm has the form:

$$C_s = K_f C^N \tag{5.42}$$

where K_f (Freundlich distribution coefficient) and N are constants. The term N (unitless) is a measure of the intensity of adsorption. When $N = 1$, the Freundlich sorption isotherm becomes the linear isotherm. If N is greater than 1, then the solid has a high affinity for the solute. The Freundlich isotherm also does not have an upper limited on the concentration of solute that may be sorbed onto a given mass of solid.

Distribution coefficients can be determined for either a bulk soil, sediment, or rock, or for a sorptive component of a material, such as organic matter and clay minerals. The values of distribution coefficients for different materials can be determined through batch experiments.

Where sorption is controlled by one component of a solid material whose concentration can be readily measured, evaluating the sorptive properties of the material in terms of the concentration and distribution coefficient of the highly sorptive component is usually the preferred approach. In the case of organic carbon in soil (Bouwer 1991)

$$K_d = f_{oc} K_{oc} \tag{5.43}$$

where

f_{oc} fraction of organic carbon in the soil
K_{oc} organic carbon-water partition coefficient

In contamination investigations, the value of K_{oc} for a chemical is commonly estimated from the octanol-water partition coefficient (K_{ow}), which is a measure of hydrophobicity of a compound. K_{ow} values are obtained by shaking a compound in a mixture of water and n-octanol. The octanol-water coefficient is the ratio of the concentration of the compound in octanol to its concentration in water (Fetter 1993). Fetter (1993) presented a list of equations that relate K_{oc} to K_{ow}, including the Collander relationship (Briggs 1981):

$$\text{Log } K_{oc} = 0.69 \text{ Log } K_{ow} + 0.22 \tag{5.44}$$

K_{OC} can also be estimated from solubility data and the molecular structure of compounds (Fetter 1993).

K_{ow} values represent the tendency of chemicals to partition themselves between organic matter and water. Chemicals with high K_{OW} values are hydrophobic and sorb strongly onto organic matter. Chemicals with low K_{OW} values (<10) tend to hydrophilic and have high solubilities in water.

Adsorption can result in a slowing (retardation) of solute transport in aquifers. As recharged water flows through an aquifer, there may be an initial reduction in solute concentration due to adsorption until the solid surfaces become fully saturated. The breakthrough of the solute can thus be delayed (retarded) relative to the breakthrough of the recharged water in an MAR system. Sorptive processes result in a slower migration of solutes than the average water velocity, which can be measured using a non-sorptive tracer. Solute velocity (v_s) is related to average flow velocity (v) through the retardation factor (R; unitless). In the case of linear equilibrium sorption of a chemical (Freeze and Cherry 1979; Bouwer 1991; Fetter 1993)

$$v_s = \frac{v}{R} \tag{5.45}$$

and

$$R = 1 + \frac{\rho_d K_d}{n} = 1 + \frac{\rho_d f_{oc} K_{oc}}{n} \tag{5.46}$$

where

R retardation factor, which is specific to a particular solute.
ρ_d soil bulk density (kg/L, inverse of K_d unit)
n porosity (fractional)

Retardation factors differ between solutes. If there is no interaction between the solute and soil, then $K_d = 0$ and $R_i = 0$.

In unsaturated flow, n is commonly replaced with θ, the volumetric water content (Bouwer 1991). If there is both mobile and immobile water, then θ is the volume fraction of the mobile water:

$$R_i = 1 + \frac{K_d(1-n)\rho}{\theta} \tag{5.47}$$

where

ρ mass density of the solid phase
$(1-n)\rho$ dry bulk density
θ volume fraction of water that is moving or the "effective" water content

The linear, Langmuir, and Freundlich sorption isotherms are equilibrium models that assume that the rate of change of concentration is much greater than other geochemical processes and groundwater flow (Fetter 1993). Nonequilibrium (kinetic) sorption models have been developed that consider sorption rates (through rate constants) and diffusion through an immobile phase near solid surfaces (Fetter 1993).

5.6.4 Clay Dispersion

The surface chemistry of clays is summarized in many aqueous chemistry textbooks and reference books (e.g., Stumm and Morgan 1996; Drever 1997). Ions interact with clay surfaces to form an electric double layer. Isomorphous substitution and proton binding and disassociation result in a negative charge at clay crystal surfaces. The negative surface charge attracts positive counterions (cations), which form a relatively thin layer that is closely attracted to the surface of the clay particles. The inner, positively charged layer is referred to as the Stern or fixed layer (Fig. 5.3).

Beyond the Stern layer is a second, more loosely bound, layer that is referred to as the Gouy or diffuse layer. The Gouy layer contains free, rather than anchored, ions that move under electric attraction and thermal motion. The Gouy layer electrically screens the inner layer from the aqueous solution and contains the neutralizing charge of counterions. Ions can readily move between the Gouy layer and aqueous solution. The charged Gouy layer results in an electrostatic repulsion between particles. The thickness of the double layer is inversely proportional to solute concentration and is function of the charge and hydrated size of the ions.

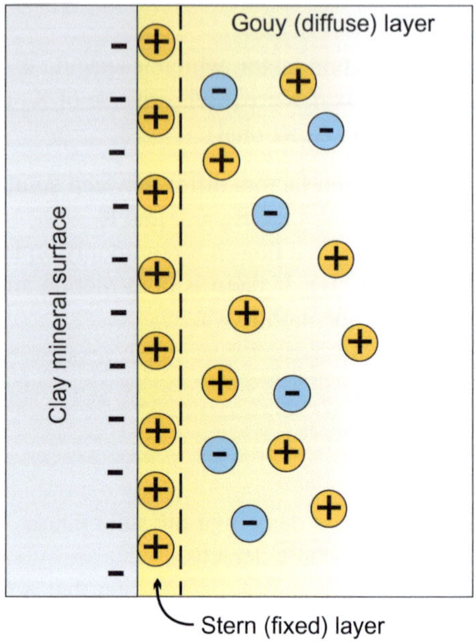

Fig. 5.3 Schematic diagram of the electric double layer

The thickness of the electric double layer shrinks if a clay is immersed in a fluid with a high cation concentration and a high proportion of divalent cations. Thinner electric double layers allow the clay particles to approach each other more closely, ultimately to the point where the physical attraction between particles exceeds the repulsive force of the electric double layer and the clay particles can flocculate. More dilute solutions with mostly monovalent ions tend to have thicker double layers, which increase the dispersion of clay particles.

Dispersed clay particles can migrate and clog pore throats, reducing hydraulic conductivity. Clogging by clay dispersion can be particularly problematic because particle movement and accumulation at pore throats is largely irreversible (Frenkel et al. 1978, 1992). On the contrary, clogging by clay swelling is reversible.

The effects of changes in the salinity and composition of water on clay dispersion and hydraulic conductivity have long been the subject of much interest both in the soil sciences and oil and gas industry. Deflocculation results in the destruction (collapse) of soil structure (aggregates) and a loss of permeability and infiltration capacity. The impacted soil becomes hard and compact when dried. In the oil and gas industry, water flooding in water sensitive formations can cause a rapid reduction in permeability and well capacity. Clay dispersion can be a major cause of clogging in MAR systems where freshwater is recharged into brackish or saline aquifers that contain clays. These situations have the undesirable combination of both dilution and a prevalence of sodium cations. Clogging due to clay dispersion is further addressed in Sects. 11.2.6 and 11.3.6.

Clay dispersion in soils depends largely on the clay mineral content and composition of the soil, the amount of exchangeable sodium in soil, and the composition of the applied water. Columns studies indicate that soils composed predominantly of either kaolinite or illite with a small amount of smectite have high dispersive behaviors (Stern et al. 1991). The amount of exchangeable sodium held by a soil is referred to as it sodicity. Sodic soils are characterized by a disproportionately high concentration of exchangeable sodium. The sodicity of soils is commonly quantified using the exchangable sodium percentage (ESP) parameter:

$$\mathrm{ESP} = 100 * \mathrm{Na}/(\mathrm{Ca} + \mathrm{Mg} + \mathrm{Na} + \mathrm{K} + \mathrm{Al}) \tag{5.48}$$

ESP concentration units are equivalents (milliequivalents) of exchangeable cations per 100 g of soil. Soils with an ESP of greater than 15% are commonly considered sodic (Abrol et al. 1988), but a lower (6%) cutoff is used in Australia. Soils may still be adversely impacted by the addition of sodium at lesser ESPs. Very small increases in ESP from low levels can cause a large increase in clay dispersion (Oster et al. 1980; Frenkel et al. 1992).

Water with high sodium to calcium ratios can cause the exchange of sodium ions in solution with calcium ions sorbed on clay minerals with associated greater swelling of clay colloids and clay dispersion. A commonly used water quality parameter in the soil sciences is the sodium adsorption ratio (SAR)

$$SAR = \frac{\mathrm{Na}^{+}}{\sqrt{\frac{1}{2}\left[\mathrm{Ca}^{2+} + \mathrm{Mg}^{2+}\right]}} \tag{5.49}$$

where Na^{+}, Ca^{2+}, and Mg^{2+} are concentrations expressed in milliequivalents per liter (meq/L). Halliwell et al. (2001) differentiated between practical and effective SAR. Practical SAR is calculated from total concentrations and includes Ca and Mg that have their activity reduced through organic complexing. Effective SAR is calculated from the active (i.e., uncomplexed) concentrations of Ca and Mg, which may be significantly less than total concentrations.

There is no single threshold for an acceptable SAR value as far as avoiding clay dispersion, although values above 10 may be problematic, especially in fresh water (Ayers and Westcott 1985). SAR hazard varies with crop sensitivity, soil type, and salinity. At higher salinities, higher SAR values can be accommodated without causing infiltration problems. However, the introduction of fresher water (e.g., rainfall, irrigation water) can increase the risk of infiltration problems. If water applied to a soil has a high SAR, then the soil may be treated using amendments containing calcium (or magnesium), such as gypsum.

5.7 Geochemical Evaluation

Evaluation of potential geochemical reactions that can occur (or are occurring) in an MAR system should start with a qualitative evaluation of the processes that might occur. Basic issues to be considered include

- the redox state of recharged water relative to the native groundwater chemistry
- changes in redox state that may occur after recharge (e.g., removal of DO by organic matter biodegradation)
- reactive minerals present in the recharge zone
- potential for various sorption processes (e.g., cation exchange capacity of aquifer materials)
- potential for clay dispersion and swelling.

The results of the qualitative analysis should guide subsequent geochemical investigations. For example, if oxygenated water is to be injected into a confined carbonate aquifer with chemically reducing conditions, then it can be deduced that the oxidative dissolution of chemically reduced minerals (with associated metals and metalloid release) and some carbonate mineral dissolution might occur. This knowledge should then be used to evaluate whether these processes are likely to significantly impact the quality of recharged water. As another example, if freshwater is to be injected into a brackish siliciclastic aquifer, then it should be recognized that clay dispersion could cause severe aquifer clogging, and thus the water sensitivity of the formation should be investigated.

Geochemical evaluations of MAR systems (when performed) are most commonly conducted by geochemical modeling using widely available modeling software, such as PHREEEQC. Given accurate data on the chemistry of the recharge water and native groundwater, and recharge zone mineralogy, it is possible to calculate the saturation state of the recharge water and estimate its composition after equilibration with aquifer minerals. Modeling software also allows for the simulation of non-equilibrium conditions (i.e., reaction kinetics), but the accurate values of the various rate constants are typically not available. Hence, large errors must be assumed in modeling of reaction kinetics.

Geochemical evaluations of MAR systems also require complete data on water chemistry. The author has observed that the water quality analyses performed for many ASR projects in Florida were restricted to parameters specifically required by the regulatory agency (Florida Department of Environmental Protection) in its permits, rather than all the parameters needed to evaluate the geochemistry of the systems. The permit requirements focus of environmental considerations (drinking water standards) rather the geochemical evaluation requirements, and some important parameters (e.g., calcium, magnesium, and bicarbonate) for which there is not a drinking water standard are not tested for. Recommended parameters for water chemistry sampling for geochemical evaluation of major MAR systems are listed in Table 5.2.

Table 5.2 Parameters for geochemical compatibility water analyses

Parameters	Parameter types
Sodium Chloride Sulfate Total dissolved solids	Salinity-correlated parameters
Calcium Magnesium Bicarbonate (alkalinity) pH	Carbonate mineral equilibrium parameters
Iron Manganese ORP/Eh Dissolved oxygen Redox couples (e.g., Fe^{2+}/Fe^{3+}, NO_2^{-1}/NO_3^{-1})	Redox mineral reactions
Arsenic Uranium Molybdenum Nickel Zinc Cobalt	Leachable metals and metaloids
Dissolved silica (H_2SiO_4) Potassium Fluoride Barium	Silicate minerals Miscellaneous parameters

Source Maliva and Missimer (2010)

Sampling procedures and analytical techniques should follow accepted industrial standards. However, the unfortunate reality is that sampling and analytical procedures used for regulatory purposes are commonly less than "research grade." Accurate data on pH and eH (Pe), in particular, are critical, and should be carefully obtained using flow-through sampling techniques and the latter from concentration data for redox pairs.

Laboratory-scale column and/or batch testing can provide more direct information on the geochemistry of MAR systems. The preferred testing procedure would be flow-through testing using cores or columns of aquifer materials (ideally undisturbed) and actual recharge water and native groundwater (as a baseline). Such testing can also be used to evaluate pretreatment options. However, project budgets for most MAR systems preclude laboratory geochemical testing. Aquifers are hydraulically and geochemically heterogeneous, and an important issue is scaling laboratory results to the formation/aquifer scale (National Research Council 2008).

References

Abrol, I. P., Yadav, J. S. P., & Massoud, F. I. (1988). *Salt-affected soils and their management*. FAO Soils Bulletin 39. Rome: Food and Agriculture Organization of the United Nations.

Allison, J. D., Brown, D. S., & Novo-Gradac, K. J. (1991). *MINTEQA2/PRODEFA2, A geochemical assessment model for environmental systems: Version 3.0 user's manual,* EPA/600/3-91/021. Athens, GA: U.S. Environmental Protection Agency Environmental Research Laboratory.

Ayers, R. S., & Wescott, D. W. (1985). *Water quality for agriculture*. FAO Irrigation and Drainage Paper 29. Rome: Food and Agriculture Organization of the United Nations.

Ball, J. W., & Nordstrom, D. K. (1991) *User's manual for WATEQ4F, with revised thermodynamic data base and test cases for calculating speciation of major, trace, and redox elements in natural waters.* U.S. Geological Survey Open-File Report 91-183.

Bass Becking, L. G. M., Kaplan, I. R., & Moore, D. (1960). Limits of the natural environment in terms of pH and oxidation-reduction potentials. *Journal of Geology, 68*(3), 243–284.

Baveye, P., Vandevivere, P., Hoyle, B. L., DeLeo, P. C., & de Lozada, D. S. (1998). Environmental impact and mechanisms of the biological clogging of saturated soils and aquifer materials. *Critical Reviews in Environmental Science and Technology, 28*(2), 123–191.

Berner, R. A. (1981). Kinetics of weathering and diagenesis. In A. C. Lasaga & R. J. Kirkpatrick (Eds.*), Kinetics of geochemical processes, Reviews in Mineralogy* (vol 8, pp. 111–134). Washington DC: Mineralogical Society of America.

Briggs, G. G. (1981). Theoretical and experimental relationships between soil adsorption, octanol-water partition coefficients, water solubilities, bioconcentration factors, and the parachor. *Journal of Agricultural and Food Chemistry, 29*(5), 1050–1059.

Brookins, D. G. (1988). *Eh-pH diagrams for geochemistry*. Berlin: Springer Science & Business Media.

Bouwer, H. (1991). Simple derivation of retardation equation and application to preferential flow and macrodispersion. *Ground Water, 29*(1), 41–46.

Carrol, D. (1959). Ion exchange in clays and other minerals. *Geological Society of America Bulletin, 70,* 749–780.

Champ, D. R., Gulens, J., & Jackson, R. E. (1979). Oxidation-reduction sequences in ground water flow systems. *Canadian Journal of Earth Sciences, 16*(1), 12–23.

Chang, A. C., & Page, A. L. (1985). Soil deposition of trace metals during groundwater recharge using surface spreading. In T. Asano (Ed.), *Artificial recharge of groundwater* (pp. 609–626). Stoneham MA: Butterworth.

Domenico, P. A., & Schwartz, F. W. (1998). *Physical and chemical hydrogeology* (2nd ed.). New York: Wiley.

Drever, J. I. (1997). *The geochemistry of natural waters: Surface and groundwater environments* (3rd ed.). Upper Saddle River, NJ: Prentice Hall.

Faust, S. D., & Vecchioli, J. (1974). Injecting highly treated sewage into a deep-sand aquifer. *Journal American Water Works Association, 66*(6), 371–377.

Fetter, C. W. (1993). *Contaminant hydrogeology*. New York: Macmillan Publishing Company.

Fetter, C. W. (2000). *Applied hydrogeology* (4th ed.). Englewood Cliffs, NJ: Prentice Hall.

Freeze, R. A., & Cherry, J. A. (1979). *Groundwater*. Englewood Cliffs, NJ: Prentice Hall.

Frenkel, H., Goertzen, J. O., & Rhoades, J. D. (1978). Effects of clay type and content, exchangeable sodium percentage, and electrolyte concentration on clay dispersion and soil hydraulic conductivity. *Soil Science Society of America Journal, 42*(1), 32–39.

Frenkel, H., Levy, G. J., & Fey, M. V. (1992). Clay dispersion and hydraulic conductivity of clay-sand mixtures as affected by the addition of various anions. *Clays and Clay Minerals, 40,* 515–551.

Garrels, R. M., & Christ, C. L. (1965). *Solutions, minerals, and equilibria*. San Francisco: Freeman Cooper.

Goss, C. J. (1987). The kinetics and reaction mechanism of the goethite to hematite transformation. *Mineralogical Magazine, 51*(361), 437–451.

Halliwell, D. J., Barlow, K. M., & Nach, D. M. (2001). A review of the effects of wastewater sodium on soil physical properties and their implications for irrigation systems. *Australian Journal of Soil Research, 39*(6), 1259–1267.
Hem, J. D. (1985). *Study and interpretation of the chemical characteristics of natural water* (3rd ed.). U.S. Geological Survey Water Supply Paper 2254.
Langmuir, D. (1996). *Aqueous environmental geochemistry*. Englewood Cliffs, NJ: Prentice Hall.
Maliva, R. G., & Missimer, T. M. (2010). *Aquifer storage and recovery and managed aquifer recharge using wells: Planning, hydrogeology, design, and operation*. Houston: Schlumberger Corporation.
National Research Council. (2008). *Prospects for managed underground storage of recoverable water*. Washington, DC: National Academy Press.
Oster, J. D., Shainberg, I., & Wood, J. D. (1980). Flocculation value and gel structure of sodium/calcium montmorillonite and illite suspensions. *Soil Science Society of America Journal, 44*(5), 955–959.
Parkhurst, D. L., & Appelo, C. A. J. (1999). *PHREEQC (Version 2)—a computer program for speciation, batch reaction, one-dimensional transport, and inverse geochemical calculations*. U.S. Geological Survey Water-Resources Investigations Report 99-42549.
Runnells, D. D. (1969). Diagenesis, chemical sediments, and the mixing of natural waters. *Journal of Sedimentary Petrology, 39*, 1188–1201.
Ryan, P. (2014). *Environmental and low temperature geochemistry*. Chichester: Wiley.
Stumm, W., & Morgan, J. J. (1996). *Aquatic chemistry* (3rd ed.). New York: Wiley-Interscience.
Stern, R., Ben-Hur, M., & Shainberg, I. (1991). Clay mineralogy effect on rain infiltration, seal formation and soil losses. *Soil Science, 152*(6), 455–462.
Takeno, N. (2005). *Atlas of Eh-pH diagrams*. Geological Survey of Japan Open File Report 419.
Tebo, B. M., Bargar, J. R., Clement, B. G., Dick, G. J., Murray, K. J., Parker, D., et al. (2004). Biogenic manganese oxides: Properties and mechanisms of formation. *Annual Review Earth and Planetary Sciences, 32*, 287–328.
Todd, D. K., & Mays, L. W. (2004). *Groundwater hydrology* (3rd ed.). New York: Wiley.
Vanderzalm, J. L., Le Gal La Salle, C., Hutson, J. L., & Dillon, P. J. (2002). Water quality changes during aquifer storage and recovery at Bolivar, South Australia. In P. J. Dillon (Ed.), *Management of aquifer recharge for sustainability* (pp. 82–88). Lisse: A.A. Balkema.
Vanderzalm, J., & Le Gal La Salle, C. (2005) Impacts of redox cycling on inorganic chemistry during ASR. In P. Dillon & S. Toze (Eds.), *Water quality improvements during aquifer storage and recovery. Volume 1: water quality improvement processes* (pp. 25–38). Denver: AWWA Research Foundation.
Wigley, T. M. L. & Plummer, L. N. (1976). Mixing of carbonate waters. *Geochimica et Cosmochimica Acta*, 40, 989–995.
WRRF. (2007). *Reclaimed water aquifer storage and recovery: Potential changes in water quality*. Alexandria, VA: WateReuse Research Foundation.
Wu, X., Zhang, W., Du, S., Shi, X., Yu, X., Huan, Y., et al. (2016). Migration and transformation of manganese during the artificial recharging of a deep confined aquifer. *Arabian Journal of Geosciences, 9*(2), 1–12.

Chapter 6
Anthropogenic Aquifer Recharge and Water Quality

6.1 Introduction

The composition of recharge water evolves as it passes through the unsaturated zone and enters and flows through an aquifer. Infiltrated and injected waters interact with aquifer minerals and organic matter, and mix and react with native groundwater. Geochemical processes during and after aquifer recharge can either improve or cause a deterioration of water quality. The concentrations of pathogens and some chemical contaminants are reduced during recharge, transport, and residence in groundwater environments. The storage of reclaimed and surface waters in groundwater environments increases their recycling (residence) time, thereby allowing more time for the biodegradation of contaminants that degrade slowly (Dillon et al. 2006). The intentional use of aquifer recharge to improve water quality is referred to as "natural aquifer treatment" (NAT). Conversely, fluid-rock interactions may release metalloids (arsenic) and metals into recharged water, causing an unacceptable deterioration in quality.

Although a great number of geochemical processes are theoretically possible in anthropogenic aquifer recharge (AAR) systems, the number of processes that can significantly impact water quality is rather small. The envelope of potentially significant geochemical processes is constrained by the limited number of mineral species that either are present and reactive in aquifers or could precipitate out of solution under the temperature, pressure, and chemical conditions found in most shallow groundwater environments. Processes that could occur during and after recharge in a shallow groundwater environment can be categorized based on whether the processes could significantly impact the quality of stored water. Significance is based on impacts to the suitability of recovered water for intended uses or environmental protection criteria (e.g., drinking water and groundwater quality standards).

Geochemical evaluations of AAR systems should start with the system-specific evaluation of the potential water-quality impacts of the main geochemical processes known to be active in AAR systems, which include

R. G. Maliva, *Anthropogenic Aquifer Recharge*,
Springer Hydrogeology, https://doi.org/10.1007/978-3-030-11084-0_6

- mixing
- chemical disequilibrium (dissolution and precipitation)
- redox reactions
- arsenic and metals leaching
- sorption and ion exchange
- pathogen attenuation
- dissolved organic carbon attenuation and transformations
- trace organic compounds attenuation.

Recharged waters frequency pass through different vertical and horizontal geochemical zones over time. Most basic are what have been referred to as the recharge-proximal aquifer zone and the distal aquifer zone (Stuyfzand 2011). The recharge-proximal aquifer zone consists of the first few meters of transport and is the geochemically most active zone, characterized by (Stuyfzand 2011):

- high biological activity
- relatively short residence times
- exponential decline of reaction rates with distance.

The distal aquifer zone is characterized by (Stuyfzand 2011)

- relatively low biological activity
- high residence times
- reaction rates remaining relatively constant.

The distal aquifer zone may have a vertical redox zonation with depth from suboxic (O_2 and NO_3^- reducing), to anoxic (Fe^{3+} reducing), to deep anoxic (SO_4^{2-} reducing; Stuyfzand 2015). Large reductions in pathogen concentrations often occur in the proximal aquifer zone. Processes active in the distal aquifer zone include cation exchange and sorption, oxidation of organic matter and sulfide minerals (pyrite), and reductive dissolution of iron (oxy)hydroxides.

6.2 Mixing Equations and Curves

The most basic water quality change that occurs during AAR is mixing of recharged water and native groundwater. If mixing is conservative, in that the concentrations of the various solutes do not change because of chemical reactions, then the composition of mixtures can be expressed using a simple binary mixing equation:

$$C_{i,w} = C_{i,r} X_i + C_{i,gw}(1 - X_i) \tag{6.1}$$

where

$C_{i,w}$ concentration of species "i" in the water sample (mixture)
$C_{i,r}$ concentration of species "i" in recharged water

$C_{i,gw}$ concentration of species "i" in native groundwater
X_i fraction of recharged water

A basic requirement for mixing analysis is that the concentrations of end-member waters be accurately known. Consideration also needs to be given to variations in end member concentrations over time. For example, the salinity and nutrient concentrations of some surface water bodies may have significant temporal variation.

The general equation for multiple end-member mixing analysis with "j" end members is

$$\sum_{i}^{m}\sum_{j}^{n} C_{i,j}X_j = C_{i,w} \tag{6.2}$$

where

m number of chemical species ("i")
n number of end members ("j")
X_j fraction of end member "j"
$C_{i,j}$ concentration of species "i" in end member "j".

Equation 6.2 can be solved by simple algebraic manipulation if $m = n-1$. Mixing equation analysis requires that chemical species are conservative, significant compositional differences occur between end-member waters, and that the concentrations of the chemical species are not closely correlated. A correlation of concentrations might occur by evaporative concentration or a relationship to seawater salinity. Chloride is a very commonly used tracer species because it tends to be conservative, its concentration tends to vary, and its concentration can be inexpensively and accurately measured. Halide ratios are useful traces, especially when an end member is wastewater from an area that fluoridates potable water (Vengosh and Pankratov 1998; Panno et al. 2006). Other potential tracers include boron, oxygen and hydrogen stable isotopes, and nitrate in oxic aquifers (Buszka et al. 1994).

Mixing analyses can be used to interpret water chemistry changes and as a predictive tool. If data are available on end-member water chemistries and mixing ratios (e.g., determined using chloride or total dissolved solids concentration data), then the composition of other species in water mixtures can be estimated.

In an early investigation of the recharge of reclaimed water by injection into the siliciclastic Magothy Aquifer at Bay Park (Nassau County, New York), Faust and Vecchioli (1974) determined the fraction of reclaimed water in samples using a binary mixing equation with chloride as a conservative tracer. Cation exchange was evaluated by comparing the measured concentrations of cations in samples versus their theoretical concentrations calculated assuming conservative mixing. The analysis indicated losses of Ca^{2+}, Mg^{2+}, K^+, Na^+, and $NH_4{}^+$ and the release of H^+.

Three-component mixing equations were used to analyze recovered water quality data from the Hillsborough County (Florida) Northwest Dechlorination Facility ASR system using chloride and fluoride as tracers (Maliva and Missimer 2010). The injected water was reclaimed water that had an elevated fluoride to chloride ratio

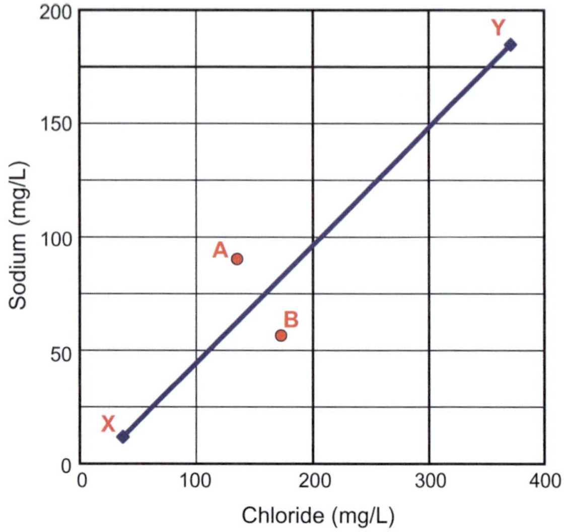

Fig. 6.1 Two-component linear mixing curve for sodium using chloride as a conservative tracer. Mixtures of the source water (X) and the native groundwater water (Y) would plot on the mixing curve. If a water sample (A) plots above the curve, then some process other than mixing of the two end-member waters was responsible for a sodium enrichment. If a water sample plots below the mixing curve (B), then some geochemical process has resulted in the removal of sodium

because the local potable water supply is fluoridated. The mixing analysis indicated that the recovered water was predominantly the injected reclaimed water, which was contaminated by a small component of high-salinity water with a fluoride to chloride ratio similar to seawater.

Mixing analysis using mixing curves is a useful screening tool to identify the occurrence of geochemical reactions by non-conservative mixing behavior. Mixing curves are cross plots of the concentrations of two dissolved constituents. One of the constituents should be a conservative tracer (e.g., chloride). The composition of the recharged water and ambient (native) groundwater are plotted as end-member water compositions and a conservative mixing curve is drawn as a line connecting the two end-member compositions (Fig. 6.1). Water samples whose compositions reflect the simple mixing of the two end members should plot on or very close to the conservative mixing curve. If a water sample plots off the curve, then it indicates that processes other than mixing impacted its composition, assuming that the end-member concentrations are representative and sample water chemistry has been accurately measured.

If geochemical processes caused an increase in the concentration of the parameter being evaluated, then the sample should plot above the conservative mixing curve. Samples that plot below the mixing curve may have been impacted by processes that removed the constituent from solution. Mixing curves were used, for example, to analyze water quality data from a managed aquifer recharge (MAR) project near Palo Alto, California (Hamlin 1985, 1987).

An advantage of mixing curve analysis is that it allows for a quick screening of the data. Once the water quality data from a system are entered into a spreadsheet, it is a minor task to generate a series of mixing curve plots. Mixing-curve analysis, combined with a knowledge of the potential geochemical reactions that may occur in the system, can provide insights into the geochemical processes that might be impacting the chemistry of the recharged water. For example, if a water sample plots above the calcium-chloride mixing curve, then possibility of calcium release by calcite dissolution should be considered, particularly if the aquifer contains abundant calcite. Similarly, the release of sodium by desorption from clays may be revealed by water samples that plot above the sodium-chloride mixing curve.

6.3 Dissolution, Precipitation, and Replacement

Dissolution, precipitation, and replacement reactions can modify porosity and permeability, and add or remove solutes from groundwater. Whether a reaction can occur depends on both equilibrium thermodynamics and reaction kinetics (Chap. 5). Thermodynamics determines whether a reaction is energetically favorable under aquifer temperature, pressure, and chemical conditions. Kinetics determines the rates of reactions and thus whether they will occur to a significant degree under the time frame of concern for AAR systems. For dissolution reactions, mineral phases should be both present in an aquifer and exposed to groundwater. Some trace mineral phases occur as very small crystals included in (i.e., encapsulated within) larger crystals and thus are not exposed to circulating groundwater.

Aquifer heterogeneity also impacts the type and degree of modification of water quality by fluid-rock interactions, as some aquifer zones may receive a disproportionate amount of recharged water, whilst other less permeable intervals (e.g., shale beds) may receive very minor, if any, flow. Less permeable beds may be in only diffusional contact with recharged water flowing through more permeable beds. Hence consideration needs to be given to where in an aquifer mineral phases are located.

Mineral phases present in carbonate, siliciclastic, and crystalline (igneous and metamorphic) aquifers are summarized in Table 6.1. The abundance of mineral phases in the three main aquifer rock types are categorized as being either common (C), moderate (M), trace (T), or rare or absent (R), as defined below:

Common: Main constituent—usually present at greater than 5% by weight
Moderate: Minor constituent (0.5–5.0%) or common in <25% of aquifer material
Trace: Often present, but in very small amounts (<0.5%)
Rare or absent: Not present or present in a very small percentage of samples

Mineral phases in bold text in Table 6.1 are normally reactive under typical aquifer temperature, pressure, and chemical conditions. Clay minerals are considered poorly reactive in that they tend not to dissolve, precipitate, or be recrystallized into other minerals during the time frame of operation of AAR systems, but they may still impact recharged water through cation exchange and sorption processes.

Table 6.1 indicates that there are relatively few reactive minerals that could impact recharged water quality. The reactive minerals and their potential water qual-

Table 6.1 Mineral phases found in aquifers

Mineral		Composition	Rock types		
			Carbonate	Siliciclastic	Crystalline
Calcite		$CaCO_3$	C	M	R
Aragonite		CaCO3	M-T	R	R
Dolomite		$CaMg(CO_3)_2$	C	M-T	R
Siderite		$FeCO_3$	T	T	R
Quartz		SiO_2	M—C	A	A
Opal		$SiO_2 \cdot H_2O$	R	R	R
Feldspar	Orthoclase Microcline	$KAlSi_3O_8$	R	C	C
	Albite	$NaAlSi_3O_8$	R	M-T	C
	Anorthite	CaAl2Si2O8	R	T	C
Amphiboles		$(Ca,Mg,Fe)Si_8O_{22}(OH)_2$	R	T-T	C
Pyroxenes		$(Ca,Mg,Fe)SiO_3$	R	R	M-R
Olivenes		$(Mg,Fe)_2SiO_4$	R	R	M-R
Muscovite		$KAl_2(AlSi_3O_{10})(OH)$	R	T	C-M
Biotite		$K(Mg,Fe)_3(AlSi_3O_{10})(OH)$	R	T	M
Kaolinite		$Al_2Si_2O_5(OH)_4$	T	C-T	R
Illite		$K_{1\text{-}1.5}Al_4(Si_{7\text{-}6.5})Al_{1\text{-}1.5}O_{20})(OH)_4$	T	C-T	R
Smectite (montmorillonite)		$(0.5Ca,Na)_{0.7}(Al,Mg,Fe)_4 (Si,Al)_8O_{20})\ 7nH_2O$	T	C-T	R

(continued)

Table 6.1 (continued)

Mineral	Composition	Rock types		
		Carbonate	Siliciclastic	Crystalline
Chlorite	$(Mg,Al,Fe)_{12}(Si,Al)_8O_{20})(OH)_{16}$	T	T-C	R
Glauconite	$(K,Na,Ca)_{1.2\text{-}1.0}$ $(Mg,Al,Fe)_4$ $(Si,Al)_8O_{20}(OH)_4 \cdot 7nH_2O$	T	T-M	R
Pyrite, marcasite	FeS_s	T	T	R
Pyrrhotite	$(Fe_{0.8\text{-}1}S)$	R	R	R
Hematite	Fe_2O_3	R	T-M	R
Goethite	$FeO \cdot OH$	R	T-M	R
Limonite	$FeO \cdot OH \cdot nH_2O$	R	T-M	R
Bernalite	$Fe(OH)_3$	R	R	R
Pyrolusite	MnO_2	R	T	R
Manganite	$MnO \cdot OH$	R	R	R
Gypsum	$CaSO_4 \cdot 2H_2O$	T	T	R
Anhydrite	$CaSO_4$	T	T	R
Halite	$NaCl$	R	R	R
Barite	$BaSO_4$	T	T	R
Celestite	$SrSO_4$	R	R	R
Carbonate fluorapatite (francolite)	$Ca(PO_4CO_3)_3F$	T	R	R
Fluorite	CaF_2	R	R	R

Notes C = Common, M = Moderate abundance, T = Tracer, R = Rare or absent

Table 6.2 Main mineral reactions in aquifers and water chemistry impacts

Minerals		Main reaction types	Water chemistry effects
Group	Common species		
Carbonates	Calcite	pH-controlled dissolution and precipitation	Changes in Ca^{2+}, pH, and alkalinity
Evaporites	Gypsum, anhydrite, halite	Dissolution	Increases in Ca^{2+}, SO_4^{2-}, Na^+, Cl^-, TDS
Low solubility ionic crystals	Fluorite, barite, celestite	Precipitation and dissolution	Changes in the concentrations of minor elements (e.g., Ba^{2+}, F^-)
Sulfide minerals	Pyrite, marcasite	Oxidation	Increases in Fe^{2+}, Mn^{2+}, and trace metalloids (As) and metals concentrations, decrease in pH
Oxide and hydroxide minerals	Iron oxy(hydroxides)	Reductive dissolution, recrystallization, precipitation, and sorption	Changes in Fe^{2+}, Mn^{2+}, and trace metalloids (As) and metals concentrations

ity impacts are summarized in Table 6.2. Redox reactions during and after aquifer recharge are further discussed in Sect. 6.4.

Carbonate minerals The solubility of carbonate minerals is pH controlled. Calcite readily dissolves and precipitates in groundwater environments with associated changes in Ca^{2+} concentration and pH. Dolomite is much less reactive under the pH conditions typically present in groundwater.

Evaporite minerals Evaporite minerals, such as gypsum, anhydrite, and halite, have high solubilities and readily dissolve is fresh water. Dissolution increases the concentrations of Ca^{2+}, SO_4^{2-}, Na^+. Cl^-, and TDS (total dissolved solids). Evaporite minerals are rarely present in aquifers used for AAR because of their high solubility and association with high-salinity waters. In arid regions, evaporite minerals may accumulate in the vadose zone and be mobilized during surface spreading, adversely impacting the quality of recharged water.

Low-solubility ionic minerals Low-solubility ionic minerals include a variety of minerals that have significantly lesser solubilities than evaporite minerals, such as fluorite, barite, and celestite. Low-solubility ionic crystalline minerals are uncommon and, when present, are typically found in trace quantities. The dissolution and precipitation of these minerals can impact the concentrations of minor constituents, such as Ba^{2+}. Sr^{2+}, and F^-, but the effects on water quality tend to be insignificant.

Sulfide minerals Introduction of dissolved oxygen (DO), and associated increase in Eh and pE, can result in the oxidation of chemically reduced minerals, such as sulfides (e.g., pyrite), and the release of iron, manganese, and trace metalloids and metals to solution. Oxidation of sulfide minerals also results in a decrease in pH. Dissolved metals concentration may not increase if the metals are reprecipitated as, or sorbed onto, neoformed (oxy)hydroxides.

Oxide and hydroxide minerals A transition to chemically reducing conditions can result in the reductive dissolution of iron and manganese oxide and hydroxide minerals. Reducing conditions can result in the change of iron and manganese ions to the more soluble Fe^{2+} and Mn^{2+} redox states. Reductive dissolution of oxide and hydroxide minerals may also result in the release of sorbed metals and metalloids into solution. Introduction of DO into chemically reducing groundwater containing dissolved Fe^{2+} can cause the precipitation of iron (oxy)hydroxides.

6.4 Redox Reactions

Changes in oxidation-reduction potential control the dissolution, precipitation, and alteration of redox-reactive minerals, of which iron and manganese sulfides, oxides, and hydroxides are most important in groundwater systems. Redox reactions involving iron and manganese minerals can cause elevated concentrations of iron and manganese in recharged water. Trace metals and metalloids (e.g., arsenic, molybdenum) that are either incorporated into, or sorbed onto, redox-reactive minerals may be released by redox reactions. For example, the oxidative reduction of iron sulfide minerals containing arsenic has been identified as the source of elevated arsenic concentrations in waters stored in some aquifer storage and recovery (ASR) systems in Florida and elsewhere (Sect. 6.5). Redox reactions involving iron minerals can cause well and aquifer clogging, and contribute to further changes in aquifer redox state. Oxidation of sulfide minerals, for example, may contribute to a decrease in DO concentration during storage.

Hydrous iron and manganese oxides and hydroxides have extremely high sorption capacities and affinities for heavy metals (Drever 1997). The exceedingly fine crystal sizes of oxides and hydroxides results in very large specific surface areas and thus adsorptive capacities. In oxidizing conditions, iron and manganese oxides and hydroxides are effective in scavenging metals (and metalloids) out of solution, including Co, Ni, Cu, Pb, Ag, Cd, and As. Reductive dissolution of oxides and hydroxides may release sorbed ions back into solution. The surface charge (negative) of manganese and iron oxides and hydroxides increases with increasing pH. The more negative surface charge at high pH values results in the stronger adsorption of the positively changed metal ions (Drever 1997). Eh and pH changes in AAR systems can either increase or decrease the concentrations of metals in groundwater.

Iron and manganese mineral precipitation involves a variety of metastable mineral phases. Oxidized iron mineral phases that form in groundwater environments include a variety of ferric hydroxides and oxyhydroxides that are hydrated to varying degrees.

Initially formed oxides and hydroxides may be poorly crystalline or amorphous, which makes their characterization difficult. Houben (2003) reported that ferric iron initially precipitates as an amorphous iron oxide (ferrihydrite, $Fe_2^{3+}O_3 \cdot 0.5(H_2O)$) that is thermodynamically unstable and recrystallizes ("ages") with time to form more stable phases, mainly goethite ($Fe^{3+}O\,(OH)$). Iron sulfide (FeS_2) polymorphs (pyrite and marcasite) are the main reduced iron minerals typically encountered in groundwater environments. Iron sulfide precipitation in some instances may involve the initial precipitation of an iron monosulfide (FeS) phase, which later recrystallizes into pyrite (Berner 1970).

With respect to AAR, waters often pass through zones with distinctly different redox states. In the case of surface spreading, recharged waters are commonly initially oxic and experience oxic conditions in the vadose zone where they are still in contact with the atmosphere. After entering the phreatic zone, DO may be consumed by the biological oxidation of organic matter and the oxidation of reduced minerals, eventually resulting in anoxic conditions. The transition from oxic to anoxic conditions is taken advantage of in soil-aquifer treatment systems to remove nitrogen from recharged water (Bouwer 1973, 1974). Aerobic (oxic) conditions are required for conversion of ammonium to nitrate, and subsequent anaerobic conditions allow for the denitrification of nitrate to nitrogen gas.

Similarly, groundwater in confined aquifers and deep unconfined aquifers is usually anoxic as waters are isolated from the atmosphere and DO present at the time of recharge has long been consumed. The transition to anoxic conditions may be quite rapid in injected waters with sufficiently high concentrations of organic matter.

Field and modeling results demonstrate that redox reactions tend to be most active in the immediate vicinity of ASR injection and recovery wells (Greskowiak et al. 2005a, b; 2006). Aerobic, sulfate-reducing, and methanogenic conditions, with their associated mineralogical reactions, may co-exist over short distances. Vanderzalm et al. (2002), for example, documented a reaction zone that formed adjacent to the ASR well at the Bolivar ASR system in South Australia. High microbial activity was related to a relatively large flux of nutrients through the aquifer immediately surrounding the ASR well. Sulfate-reducing conditions, for example, developed only in the immediate vicinity of the ASR well and were not detected in an observation well located 4 m (13 ft) away.

The passage of recharged water through different redox zone is important for the attenuation of pathogens and chemical contaminants because the removal rates of some species varies with redox state.

6.4.1 *Recharge of Oxic Water into Reduced (Anoxic) Aquifers*

Recharge of oxic water into an anoxic aquifer creates a redox front across which large changes in the saturation state of redox-reactive minerals may occur. On the native groundwater side of the front, sulfide minerals may be in chemical equilibrium with anoxic groundwater. Groundwater on the oxic recharged water side of the front is

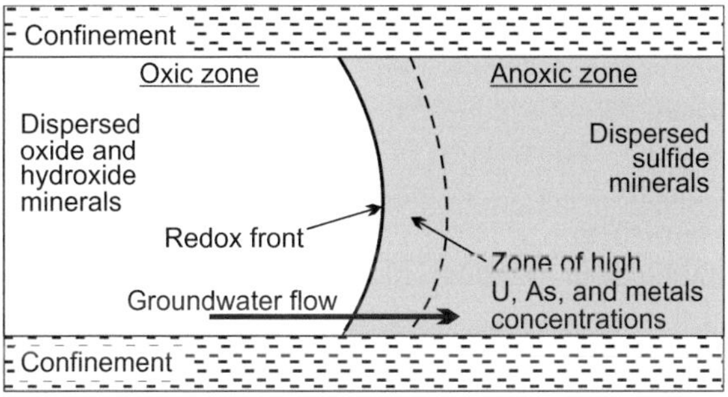

Fig. 6.2 Conceptual diagram of a uranium roll-front system. The rate of groundwater flow is greater than the rate of migration of the redox front, and uranium, arsenic, and metals mobilized by oxidative dissolution of sulfide minerals at the redox front are transported across the front into an enriched zone

undersaturated with respect to sulfide minerals, but saturated or supersaturated with respect to oxide and hydroxide minerals.

Water quality and mineralogical changes associated with recharge of oxic water into an anoxic sedimentary aquifer is well studied in the economic geology realm with respect to uranium roll front deposits, as summarized by de Vito (1978), Granger and Warren (1979), Drever (1997) and Langmuir (1997). Iron sulfides with associated uranium and other trace elements are dispersed in the chemically reducing formation (aquifer). The flow of oxic water into the formation causes a redox front to develop between the oxidizing and reducing waters (Fig. 6.2). The change to oxic conditions as the redox front passes results in the dissolution or alteration of sulfides and other chemically reduced minerals and the release to the groundwater of some trace elements, such as uranium, selenium, arsenic, and molybdenum. Iron (oxy) hydroxides form on the oxic side of the front.

The redox front moves in the direction of groundwater flow but at a slower rate. The rate of movement of the redox front depends on the capacity of the aquifer to consume DO. Released metals are transported by the more rapidly flowing groundwater across the redox front into the reducing environment where they tend to reprecipitate. Depending on the rate of movement of the front, water chemistry, and the kinetics of the precipitation reaction, the concentrations of some elements (e.g., arsenic) may, at least temporarily, remain elevated in the groundwater. The mobilization of uranium by the introduction of oxic fluids is taken advantage of in the in-situ leach mining process.

The movement of a recharge-induced roll front in some AAR systems mobilizes arsenic and metals to the extent that applicable water quality (e.g., drinking water) standards are exceeded in the recharged (stored) water. Where recovery is performed using the same well as used for recharge, recovery will tend to pull both the redox front and released arsenic and metals back toward the pumped well.

Redox reactions may be limited by amount of redox-reactive material present in an aquifer. For example, in the St Andrews Stormwater ASR system (South Australia), sulfate increased over the first 3 annual injection cycles, which was thought to be due to the oxidation of sulfide minerals (pyrite; Herczeg et al. 2004). A lack of sulfide oxidation after the third year was suggested as being due to sulfide minerals no longer being accessible or the labile fraction of sulfide minerals having been exhausted.

A key water quality issue is the fate of the metalloids and metals released as the result of the introduction of DO. Depending upon geochemical conditions, they may be either immobilized by immediate sorption onto neoformed (oxy)hydroxide minerals, stay in solution, or coprecipitate with or be sorbed onto sulfide minerals that form on the reduced side of the redox front. Reversal of the flow direction during recovery may remobilize the metalloids and metals. Arsenic release during AAR is further addressed in Sect. 6.5.

6.4.2 Recharge of Organic-Rich Water

Organic carbon is usually the most important electron donor in groundwater systems. Recharge of reclaimed water and surface water with high biodegradable organic carbon concentrations can stimulate high levels of microbial activity with an associated local depletion of DO and development of anoxic conditions. Oxidation of pyrite also contributes to the removal of DO. In the U.S. Geological Survey study of reclaimed water injection at Bay Park, New York, it was determined that DO persisted in the injected water only to about 12 feet (3.7 m) into the formation (Ragone et al. 1975; Ehrlich et al. 1979). The decrease in DO was associated with an increase in dissolved iron concentrations to values greater than those present in both the recharged and native groundwater, which indicates oxidation of an iron sulfide (pyrite or marcasite) source (Faust and Vecchioli 1974; Ragone et al. 1975; Ragone 1977). Vanderzalm et al. (2002) similarly documented DO removal and the development of anoxic conditions near a well used for the recharge of reclaimed water. Pyne et al. (1996) presented data from several ASR systems that indicate that injected water becomes anoxic several days or weeks after injection.

Experimental studies of reclaimed water recharge have also shown rapid decreases in DO concentrations. Rinck-Pfeiffer et al. (1998) injected treated wastewater in a 6.3 in. (16 cm) long column containing crushed aquifer material. The DO concentration of the effluent decreased rapidly for first 10 days, and then levelled off at 1–2 mg/L, while more oxic conditions were maintained near the inlet.

The main water quality change associated with the transition to reducing conditions is the reductive dissolution of (oxy)hydroxide minerals and increases in the concentrations of metals that have greater solubility in their reduced form (particularly Fe^{2+}, Mn^{2+}). In many areas, high concentrations of iron and manganese occur in confined aquifers with reducing conditions. Hydrogen sulfide gas (H_2S) released by sulfate reduction can give groundwater an objectionable rotten egg smell.

6.5 Arsenic

Arsenic is a known carcinogen of the skin, lung, bladder, liver, and kidneys. Chronic exposure to arsenic through water consumption continues to be a major public health problem worldwide, affecting hundreds of millions of people (Naujokas et al. 2013). A large amount of research has been performed on the geochemistry of arsenic in response to the tragic arsenic poisoning from groundwater that has occurred in parts of South Asia (western India and Bangladesh). Leaching of arsenic in water stored in some MAR systems continues to be a regulatory and operational challenge. Arsenic leaching in ASR systems in the United States became a more serious concern when the federal primary drinking water standard (i.e. maximum contaminant level) was reduced in 2001 from 50 to 10 μg/L. Some ASR systems in Florida that met the old 50 μg/L drinking water standard became in violation of the new lowered drinking water standard, which is the applicable groundwater quality standard. Arsenic leaching in MAR systems in the United States has not posed a public health risk since it has not resulted in exceedances of the drinking water standard in actual potable water supplies. Nevertheless, arsenic leaching in ASR systems has adversely impacted the permitting of systems and reduced support for the technology.

6.5.1 Sources of Arsenic in Groundwater

Arsenic is a widespread trace constituent in sedimentary and igneous rocks. The key issue concerning arsenic leaching during aquifer recharge is the amount of arsenic present in a reactive or labile form, as opposed to being incorporated into the crystal structure of non-reactive mineral phases. To impact water quality, arsenic-bearing mineral phases must also be in contact with the bulk (flowing) pore waters. Minerals that contain labile arsenic (e.g., pyrite crystals), for example, may be entirely contained (included) within dolomite or calcite crystals and thus be isolated from bulk pore waters. Reactive phases within very low permeability shales or clays may also have a limited (diffusional only) contact with bulk groundwaters.

The is a limited envelope of geochemical conditions and settings in which there is a combination of a high concentration of labile arsenic in vadose zone or aquifer rock and sediment and geochemical conditions favorable for its mobilization in concentrations of potential health concern. Recognized soil or hydrogeological situations where natural (non-anthropogenic) elevated arsenic concentrations may be encountered include (e.g., Schreiber et al. 2000; Welch et al. 2000; Ravenscroft et al. 2001; Smedley and Kinniburgh 2001, 2002; Arthur et al. 2002; Stollenwerk 2002; Alaerts and Khouri 2004; Bhattacharya et al. 2004; McArthur et al. 2004; Maliva and Missimer 2010; Fakhreddine et al. 2015):

- peaty or peaty clayey soils with high humic concentrations and a high water table in which arsenopyrite (arsenic-bearing iron sulfide) crystals are oxidized when soils are drained
- young volcanic deposits or thermal water sources
- loamy or clayey deposits with dissolved or absorbed arsenic
- ancient (non-recent) sedimentary deposits containing iron sulfide minerals (e.g., pyrite, marcasite) in which arsenic in solid solution or sorbed is released upon their oxidation
- oxic groundwater with iron and manganese oxides and hydroxides containing sorbed arsenic.

Smedley and Kinniburgh (2002) provided a very detailed discussion of the natural occurrence of arsenic in groundwater and the factors responsible for its mobilization. In most cases of elevated arsenic concentrations in groundwater, the aquifer sediments had near average arsenic concentrations (1–20 mg/kg range). High dissolved arsenic concentrations on a regional scale require both (1) a geochemical trigger that releases arsenic from a solid phase to groundwater, and (2) conditions that allow arsenic to remain in solution in groundwater (Smedley and Kinniburgh 2002). Smedley and Kinniburgh (2002) described two distinct triggers that can lead to the release of arsenic on a large scale:

1) Development of high pH (>8.5) under oxidizing conditions in semiarid and arid environments, which leads to the desorption of adsorbed arsenic from metal oxides or prevents oxides from being formed.
2) Development of strongly reducing conditions at near neutral pH conditions, which leads to the desorption of arsenic from metal oxides and the reductive dissolution of Fe and Mn oxides and associated release of sorbed As. The onset of reducing conditions tends to occur in sediments with high organic carbon contents and confining layers that retard the diffusion and convection of DO.

On a more local scale, the introduction of DO into sediments can result in the oxidation of iron sulfides, and the release of sulfate, acidity, and trace constituents including arsenic (Smedley and Kinniburgh 2002). Arsenic and trace metals (e.g., Cd, Pb, Au, Sb, W, and Mo) are scavenged out of solution by either solid solution into, or sorption onto, iron sulfide minerals (Huerta-Diaz and Morse 1992; Chappaz et al 2014; Gregory et al. 2015). Organic-rich environments with high sulfate concentrations, such as coastal marshes and swamps, may be intensely reducing and favor the formation of trace element-rich iron sulfide minerals (e.g., Harbison 1986; Helz et al. 2004). On a finer-scale, iron sulfides may form within sedimentary rocks at dispersed locations that are enriched in organic matter, such as within fossil shells and organic-rich laminae and burrows. Subsequent oxidation of iron sulfide minerals will release the trace elements into solution.

6.5.2 Arsenic in ASR Systems in Florida

Florida Geological Survey (FGS) investigated fluid-rock interaction processes during ASR in the state, particularly the mobilization of arsenic and trace metals. Several general patterns were evident in three studied ASR facilities in southwestern Florida (Arthur et al. 2001, 2002). Elevated arsenic concentrations were associated with elevated concentrations of metals (Co, Fe, Mn, Mo, Ni, and V). Arsenic concentrations decreased during subsequent operational cycles if the injected volumes were approximately equal. However, if the injected volume in a subsequent cycle was increased, relatively high arsenic concentrations may occur, which was attributed to a "new" aquifer volume being exposed to injected water (Arthur et al. 2005a). Arsenic concentrations were also relatively low in storage-zone monitoring wells compared to that in water recovered from ASR wells. The operational cycle testing data were interpreting as having two possible explanations: (1) flushing of a finite amount of As present in a labile form in the aquifer and (2) an armoring effect whereby FeOOH-coatings form on crystal surfaces, reducing their rate of oxidation.

FGS laboratory experiments, which included a sequential extraction procedure, indicated a strong association of arsenic with the insoluble (non-carbonate) fraction of the carbonate aquifer rock (Arthur et al. 2005a, 2007). Oxidation of arsenian pyrite was identified as the likely source of arsenic, but it was noted that other phases, such as organic matter, also contain arsenic and metals. The metals detected at elevated concentrations were often associated with framboidal and euhedral pyrites.

Price and Pichler (2006) subsequently examined the distribution of arsenic within the Suwannee Limestone (Oligocene), which is being used as an ASR storage zone in southwest-central Florida. High arsenic concentrations were associated with abundant non-carbonate minerals. Arsenic was detected in pyrite using SEM-EDX analyses, but not all pyrites had high arsenic concentrations. Electron microprobe analyses of pyrites gave a mean arsenic concentration of 2,300 ppm and a range of 100–11,200 ppm, with the highest measured concentrations occurring in pseudo-framboidal pyrite. Lazareva and Pichler (2007) obtained similar results in an investigation of the occurrence of arsenic in the Arcadia Formation (Oligocene to Miocene) in southwestern Florida. The results of both investigations demonstrate the importance of identification and characterization of the trace mineral components of ASR storage zone strata.

Modeling results indicated that pyrite is stable in contact with the native groundwater of the Suwannee Limestone of southwest-central Florida (Jones and Pichler 2007). The sulfate-sulfide redox couple was used as an indicator of redox conditions because of the difficulty of obtaining accurate field measurements of Eh. The stability of pyrite in native groundwater indicates that arsenic concentrations in well waters should be low, which is supported by the measured As concentrations in all tested well water samples being $\leq 0.036\ \mu g/L$. Stability diagrams indicate that hydrous ferric oxides (HFOs) are not stable in the full range of waters from native groundwater to nearly pure injected water, which indicates that arsenic could not be removed from solution by HFOs. HFOs are rare in the studied limestones. Jones and Pichler (2007) suggested that at some distance from injection wells, low DO and the presence of nutrients could stimulate microbes that are capable of precipitating pyrite.

The Jones and Pichler (2007) modeling results a support a model in which arsenic is released by pyrite oxidation and largely stays in solution. Arsenic was suggested to remain in solution when it is released into groundwaters with suboxic conditions or in the presence of nitrate (a competing electron acceptor with ferric iron; Vanderzalm et al. 2007; Mirecki et al. 2013). Operational data from the Destin Waters Users (Florida) reclaimed ASR system (Maliva et al. 2013, 2018) supports a model of the arsenic released by the oxidative dissolution of a limited labile source largely remaining in solution.

Mirecki (2006) performed inverse geochemical modeling using the PHREEQC code of existing data from three ASR sites in South Florida: the Olga and North Reservoir systems in Lee County and the East Hillsboro system in Palm Beach County. The ASR systems use the upper Floridan Aquifer System (a limestone aquifer) as a storage zone. The modeling was hampered by incomplete and, in some instances, poor quality water chemistry data, which were collected for regulatory rather the scientific research purposes. A goal of the modeling was to elucidate the processes responsible for arsenic mobility observed in the three systems. Key operational observations are that:

- arsenic concentrations were below detection limits in the recharged water and native groundwater
- elevated arsenic concentrations occurred in the recovered water with the concentration increasing over each recovery cycle
- elevated arsenic was not detected in storage-zone monitoring wells located 250–400 ft (76–122 m) from the ASR wells, even though the recharged water reached the wells
- the recovered water became geochemically reducing (sulfate reduction occurred).

Mirecki (2006) proposed a model in which:

- arsenic was released by oxidative dissolution of arsenic-bearing pyrite caused by DO in the recharged water
- iron (oxy)hydroxides precipitates were stable so long as the Eh was in the +50 to +150 Mv range
- arsenic released by pyrite dissolution was sorbed onto neoformed iron (oxy)hydroxides
- reducing conditions subsequently became established in the recharged water (ORP <-200 mV)
- reductive dissolution of iron (oxy)hydroxides occurred as low Eh water was pulled toward ASR wells, which resulted in the release of sorbed arsenic.

An unresolved issue is whether the small modeled mass of neoformed iron (oxy)hydroxides was sufficient to sequester and release the arsenic concentrations (up to 68 μg/L) measured in the recovered water.

The working hypothesis of arsenic release by oxidative dissolution of pyrite and subsequent sorption and desorption of arsenic on neoformed hydrous ferric oxides

(HFOs) is supported by modeling studies by Stuyfzand and Pyne (2010) and Wallis et al (2010, 2011), bench-top experimental studies by Arthur et al. (2005b, c, 2007) and Lazareva et al. (2015), and field studies by Vanderzalm et al. (2007).

Mirecki et al. (2013) performed geochemical modeling of the Kissimmee River ASR Project in which treated surface water is stored. Reduction in arsenic concentrations over three cycle tests was attributed to the oxidative dissolution of arsenic-bearing iron sulfide minerals and formation of iron (oxy)hydroxides under oxic conditions. With return of reducing conditions, iron (oxy)hydroxides underwent reductive dissolution, releasing arsenic. Where sulfate-conditions are re-established and there is sufficient dissolved iron, arsenic is proposed to be sequestered by coprecipitation with iron sulfide. The pyrite resequestration model for arsenic is appropriate for ASR systems having the following characteristics: (1) the recharged water has sufficient organic carbon to stimulate aquifer microbes, (2) recharged water has negligible concentrations of other electron acceptors (manganese and nitrate) that inhibit sulfate reduction, and (3) a native aquifer sulfate-reducing redox environment (Mirecki et al. 2013). In potable water ASR systems with low dissolved organic carbon, iron, and sulfate concentrations, sulfate-reduction may be insufficient to sequester arsenic (Mirecki et al. 2013).

6.5.3 Arsenic in the Bolivar, South Australia Reclaimed Water ASR System

The Bolivar reclaimed water ASR systems in South Australia is perhaps the most exhaustively investigated system to date. The native (ambient) brackish groundwater was reported to have an average arsenic concentration of 5 μg/L and the recharged reclaimed water had an average arsenic concentration of 3 μg/L (Vanderzalm et al. 2007). Elevated arsenic concentrations were reported in water recovered from the ASR well and a monitoring well located 4 m (13 ft) away. However, elevated arsenic concentrations were not detected in a monitoring well located 50 m (164 ft) from the ASR well, at the edge of the injected water plume. The maximum reported arsenic concentration of 186 μg/L occurred at the start of recovery.

Vanderzalm et al. (2007) proposed that the source of the arsenic is the oxidative dissolution iron sulfide minerals. Under oxic conditions, the arsenic occurs primarily as the charged arsenate ($HAsO_4^{2-}$) species, which tends to be sorbed onto iron (oxy)hydroxides. As reducing conditions became reestablished during storage by the oxidation of organic carbon present in the injected reclaimed water or by the reversal of flow during recovery, iron (oxy)hydroxides were proposed to have been dissolved and the adsorbed arsenic released in the more soluble, uncharged arsenite (H_3AsO_3) species form. Arsenic concentrations decreased rapidly during recovery, but the average arsenic concentration of approximately 14 μg/L exceeded the WHO drinking water guideline of 10 μg/L. Nevertheless, all but the initially recovered water was suitable for its intended agricultural use for which the applicable guideline value is 100 μg/L (Vanderzalm et al. 2007).

6.5.4 Arsenic in Recharge Systems in the Netherlands

Stuyfzand (1998a) reviewed geochemical data from 11 deep recharge experiments in the Netherlands. The introduction of oxic water into deep reduced aquifers caused the dissolution of pyrite and associated mobilization of trace metalloids and metals (As, Co, Ni, Zn). Arsenic stayed in solution longer than the released metals because it was released in the reduced form, which has a low sorption affinity for iron hydroxides. The reduced arsenic was subsequently converted to the oxidized arsenate state, which is much less mobile due to preferential sorption onto neoformed iron hydroxides.

Wallis et al. (2010) performed a geochemical investigation of the Langerak ASTR trial site in the Netherlands. Freshwater was injected into a deep anoxic (methanogenic) freshwater fluvial sand aquifer. The geochemical data indicate that the oxidation of pyrite increased SO_4^{2-} and Fe^{2+} concentrations and mobilized As (up to 90 μg/L), Ni, and Zn. DO and nitrate concentrations decreased. Wallis et al. (2010) proposed the following model:

1) release of Fe^{2+} by the oxidative dissolution of pyrite, which precipitates as ferrihydrite $(FeOH)_3$—hydrous ferric oxides (HFOs)
2) released arsenic (As^{3+}) stays in solution until its kinetically controlled oxidation to arsenate by abiotic and/or biologically mediated processes.
3) As^{3+} that is transformed over time to As^{5+} is sorbed onto neo-precipitated HFOs.

6.5.5 Management of Arsenic Leaching

Considerable attention has been paid to developing strategies to prevent or manage arsenic leaching where it occurs to the degree that it impacts the use of stored water and/or violates environmental protection regulations. Regulatory issues include groundwater quality requirements and water quality standards for the various uses of recovered water. Underground injection control requirements in the United States prohibit the endangerment of underground sources of drinking water (USDWs), which are defined as non-exempt aquifers containing less than 10,000 mg/L of total dissolved solids. With respect to arsenic, endangerment is considered causing a violation of the primary drinking water standard, the 10 μg/L maximum contaminant level. The U.S. Environmental Protection Agency rules are highly conservative in that drinking water standards are applied to brackish aquifers that contain water that is not directly potable and consideration is not given to the treatment brackish groundwater requires (commonly reverse osmosis desalination) for potable use. The main strategies proposed or implemented to manage arsenic in ASR and other MAR projects are summarized below.

6.5.5.1 Attenuation Over Operational Cycles

The amount of leachable arsenic present in most aquifers is quite small. During each recharge and recovery cycle, some of the leachable arsenic is removed resulting in decreased leachable arsenic concentrations over time. The supply of leachable arsenic in a formation will eventually be exhausted. This strategy is being successfully employed in the Destin Water User (Florida) reclaimed water ASR system (Maliva et al. 2013, 2018) in which arsenic concentrations in recovered water have progressively decreased over time (system operation) and in most wells are now below the 10 μg/L MCL.

Orange County Utilities (Florida) similarly addressed arsenic leaching in its ASR system by conditioning the storage zone by the introduction of DO-rich potable water (Thomas et al. 2017). Over several operational cycles, arsenic concentrations in the recovered water steadily decreased until concentrations were consistently below the 10 μg/L MCL for two consecutive operational cycles. The storage zone was progressively "scrubbed" of leachable arsenic. Stored water with high arsenic concentrations was removed from the aquifer. Injected water reached a storage zone monitoring well located 510 ft (155 m) from the ASR well, as indicated by the presence of fluoride (the potable water supply is fluoridated whereas fluoride is not detected in the native groundwater), but at no point did the arsenic concentration exceed the 10 μg/L MCL.

Over recovery (recovery of greater than 100% of injected volume) may accelerate the arsenic removal process by removing dissolved arsenic from the storage zone. The advantages of attenuation over a series of operational cycles are its low cost and finality. Once the leachable arsenic supply has been exhausted, arsenic leaching will no longer occur.

6.5.5.2 Target Storage Volume Approach

The target storage volume (TSV) approach is based on the observation that arsenic tends to accumulate within a buffer zone around an ASR well. Pyne (2007) proposed that if ASR operations are conducted to avoid recovery of the buffer zone, then arsenic concentrations should be acceptable after a few operational cycles. The emplacement of a TSV (i.e., initially injecting a large volume of water) was proposed as key to achieving this goal. The initial injection of a large volume of water would push water bearing leached arsenic away from the ASR well. A regulatory issue is that water with elevated arsenic concentrations may remain in the storage zone and eventually reach monitoring wells.

6.5.5.3 Dissolved Oxygen Removal

Where arsenic leaching is caused by the oxidative reductive of iron sulfide (pyrite) minerals, leaching can be prevented by pretreating recharged water so that its pH and eH are in the stability field of the sulfide minerals. At a minimum, DO must be removed. Other oxidants (e.g., residual chlorine, nitrate) may also require removal if present in the recharged water at high concentrations.

DO removal technologies potentially appropriate for ASR systems were reviewed and evaluated by ASR Systems (2006) and CH2M Hill (2007) in studies prepared for the Southwest Florida Water Management District. Three main types of pre-treatment processes for DO removal were identified as being potentially economically viable:

- uncatalyzed chemical reduction
- microbially-catalyzed reduction
- volatilization.

Uncatalyzed chemical reduction involves removal of DO by reaction with a reduced sulfide compound, such as a sulfide (S^{2-}), sulfite (SO_3^{-}) or thiosulfite ($S_2O_3^{2-}$). The main disadvantages of uncatalyzed chemical reduction are that it adds dissolved solids to the water, may reduce the pH of the water, and the DO removal reaction may be incomplete at land surface and aquifer temperatures (ASR Systems 2006). Several times more reactant may be required for complete DO removal than is indicated by reaction stoichiometry with associated increases in cost. Safety issues associated with the storage of reactive agents are also a consideration. Uncatalyzed chemical reduction may be used as a polishing step for further reduction of eH.

Microbially-catalyzed reduction uses microbial aerobic respiration to remove DO. DO is consumed by the oxidation of organic compounds. Microbially-catalyzed reduction may also remove other compounds of concern, such as nitrate and phosphate (ASR Systems 2006). The main disadvantages of microbially-catalyzed reduction are that there is no operational history with respect to ASR, a relatively large footprint may be required, and the system may not be effective if operated in a discontinuous mode (ASR Systems 2006).

Volatilization uses either a carrier gas or negative pressure to strip DO out of solution. It has the important advantages that it does not involve chemical addition and has a relatively small footprint. CH2M Hill (2007) evaluated both gas membranes (Liqui-Cel) and the GDT^{TM} centrifugal separator method for DO stripping. Gas transfer membranes allow for the passage of gases but not liquids. Operational testing at the Bradenton (Florida) ASR system demonstrated that DO removal using a membrane degasification system and dechlorimination (i.e., removal of chloramines) using sodium bisulfite can successfully prevent arsenic leaching from exceeding the applicable 10 μg/L standard (Norton et al. 2012).

For all deoxygenation methods, bench top and pilot testing, including batch testing with aquifer material, is needed to determine whether the DO removal is sufficient to prevent the oxidation of sulfide minerals. The main disadvantages of DO removal are its costs and that it will be a perpetual process. As the iron sulfide minerals are not removed, all recharged water must be treated for the operational life of the ASR system.

6.5.5.4 Enhanced Oxidation of Sulfide Minerals

An alternative strategy to prevent sulfide oxidation is to accelerate the processes to either remove the minerals or coat the crystals with a protective layer of iron (oxy)hydroxide. The enhanced oxidation method for arsenic control is essentially an extension of the in-situ iron removal (ISIR) process.

ISIR involves the cyclic injection of oxygenated water into an aquifer and the subsequent withdrawal of injected water and groundwater in which iron (and manganese) concentrations are lower than in the native groundwater (Appelo et al. 1999; Appelo and de Vet 2003). In an early study, Hallberg and Martinell (1976) describe the "Vyredox" method, which is used to reduce the iron concentration of produced groundwater by injecting oxygenated water to oxidize Fe^{2+} to Fe^{3+} in an aquifer. Oxygenated water is injected into both the production well and a ring of aeration wells installed around the production well. The efficiency ratio, defined as the ratio of iron-free water recovered to the amount of oxygenated water injected, increased over successive operational cycles and then plateaued. The example given achieved a final efficiency ratio of about 9.

Appelo et al. (1999) proposed that the operational success of ISIR may be explained by the oxidation of exchangeable and sorbed Fe^{2+}. The process involves displacement of Fe^{2+} from exchange and sorption sites during injection of oxygenated water, and subsequent oxidation to form iron (oxy)hydroxides. During withdrawal, the exchange sites, including the neoformed iron oxyhydroxides, sorb Fe^{2+} from the groundwater as it flows past. Appelo et al. (1999) and Appelo and de Vet (2003) suggested that the efficiency of ISIR is limited by the amount of oxidant in the injected water, the exchange capacity of the aquifer, and the amount of exchangeable Fe^{2+} that can consume the oxidant during the injection stage. Mettler et al. (2001) examined the iron precipitate from an ISIR system in La Neuveville, Switzerland, using chemical extraction techniques, XRD, and Mössbauer spectroscopy. The ferric oxides consisted mainly of goethite (50–100%), minor ferrihydrite (<=12%) and small amounts of Fe-oxides associated with phyllosilicates.

Field data from a site in the Netherlands and modeling results indicate that ISIR can reduce the concentration of arsenic in recovered water to values below natural background levels (Appelo and de Vet 2003). Injection of oxidized water results in the oxidation of As^{3+} to the more strongly sorbed As^{5+}. The arsenic concentration of the initially recovered water was higher than expected (but still below 10 μg/L), possibly due to arsenic sorbed on colloidal iron (oxy)hydroxide particles. As recovery proceeded during the initial cycles, spikes in arsenic concentration occurred above natural background concentrations, which were related to a redox transition and the displacement of sorbed arsenic by phosphate as reduced water approached the well (Appelo and de Vet 2003). The effectiveness of ISIR to manage the concentrations of iron, manganese, and arsenic in ASR and other MAR systems would depend upon the maintenance of oxic conditions in the aquifer so that iron (oxy)hydroxides remain stable and sorbed iron and arsenic are not released.

A similar strategy for managing the leaching of metals was proposed for siliciclastic aquifers in the southern Netherlands based on testing for the Herten pilot ASR system (Stuyfzand and Doomen 2004; Stuyfzand et al. 2006). A recommendation to speed up inactivation is to add an oxidant to the water with a buffer (sodium hydroxide) to prevent the pH decrease from the oxidation of the sulfides.

Antoniou et al. (2014) documented laboratory column experiments of the pre-oxidation of originally anoxic fluvial siliciclastic sediments with a 0.02 M solution of potassium permanganate ($KMnO_4$), a strong oxidant. The results of multiple-cycle simulation of the operation of an ASR system using oxygenated tap water and the permanganate solution indicated that the permanganate pre-treatment can reduce pyrite oxidation through the competition for DO by sorbed Mn(II) and Fe(II) on newly formed Mn-oxides and the deactivation of pyrite by the dissolution of the most reactive crystals. If reducing conditions become reestablished during storage periods, substantial amounts of Mn may be released, adversely impacting ASR system performance in the subsequent cycle. Repeated permanganate treatments might be required (Antoniou et al. 2014), which would be expected to restrict application of the technique.

Van Halem et al. (2010) documented field testing of in situ iron removal to reduce naturally high levels of arsenic in groundwater in Bangladesh. The source of iron is believed to be the reductive dissolution of iron hydroxides. ISIR successfully reduced dissolved iron concentrations, but the system did not prove to be effective yet for the removal of arsenic. Potential reasons suggested were a short contact time and competition with other anions for adsorption sites.

The effectiveness of ISIR-type processes as a means to manage the concentrations of iron, manganese, and arsenic in ASR systems will depend upon the maintenance of oxic conditions in the aquifer during storage and thus the stability of iron (oxy)hydroxides. If reducing conditions become reestablished, by either the consumption of DO by microorganisms or the oxidation of reduced minerals, sorbed metals may be released back into solution. The ISIR method may give poor results for surface water or reclaimed water ASR systems in which the injected water has a high biodegradable organic carbon concentration (Maliva and Missimer 2010).

6.5.5.5 Non-Redox-Related Arsenic Leaching Management

Release of arsenic into recharged water was reported in surface basins in Orange County (California; Fakhreddine et al. 2015). Both the recharged and native groundwater have similar oxidation-reduction potentials, which excludes a redox reaction source. Solid phase analyses suggest that the released arsenic resides primarily in the clay fraction. The results of batch and column tests indicate that divalent cations (Ca^{2+} and Mg^{2+}) were most influential in decreasing arsenic release, with the most substantial decrease in arsenic elution occurring in columns treated with dolomitic lime ($CaO{\cdot}MgO$). Ca^{2+} and Mg^{2+} were proposed to increase the positive charge of clay surfaces, which subsequently increases adsorption of negatively charges arsenate species. Fakhreddine et al. (2015) concluded that "Commonly used water amendments, including quicklime and dolomitic lime, can provide a source of divalent cations for recharge water and represent a potential mitigation strategy for large-scale aquifer storage and recovery projects."

6.6 Sorption and Cation Exchange

6.6.1 Introduction

Sorption and desorption are processes that transfer metals and chemical to and from a solid phase. Sorption and cation exchange processes are summarized in Sect. 5.6. Low polarity and nonpolar organic compounds are sorbed primarily by natural carbonaceous matter, whereas ionic molecules and polar organic compounds may interact with mineral surfaces that have a surface charge (National Research Council 2008). Sorption processes can be important in some AAR systems through the removal and release of trace elements, organic chemical contaminants, and nutrients. For example, sorption is an important mechanism for the removal of phosphate by soils (Tunesi et al. 1999; Leader et al. 2008).

Sorption can attenuate contaminants and dampen variability, but it may not be a sustainable contaminant sink because contaminants are not removed and sorption capacity is finite (National Research Council 2008). Reversible sorption and desorption reactions act as a temporary storage reservoir for contaminants in an aquifer. Once aquifer solids equilibrate with a particular contaminant concentration, sorption/desorption reactions will not have any further net effect on dissolved contaminant concentrations (National Research Council 2008). If sorption is the primary removal mechanism, then breakthrough of contaminants could eventually occur. However, with respect to AAR, aquifer sorption capacities may be so large relative to the contaminant load that the sorption capacity is not exhausted over the operational life of a system. For example, Idelovitch et al. (2003) reported that the Dan Region Reclamation Project (Shafdan) SAT system still had excellent and stable removal of trace elements and phosphorous by "non-sustainable" process (adsorption and chemical precipitation) after 25 years of system operation.

Sorptive processes are also important for the removal of fine particles, such as viruses. Viral attachment is a function of pH, the isoelectric point of the porous medium, and the isoelectric point of viruses (Guan et al. 2003). Particles suspended in water are negatively charged if the pH is above their isoelectric point. Guan et al. (2003) examined the transport of two viruses with different isoelectric points (bacteriophages MS2 (pH = 3.9) and øX174 (pH = 6.6) in a model aquifer. The experimental results demonstrated that particles with different isoelectric points exhibit dramatic changes in attachment behavior when a specific water pH (i.e., the critical pH) is reached that causes them to be oppositely charged. A key observation was that as pH increases, a critical value is reached, called the supercritical pH condition, at which there is a dramatic decrease in viral removal due to the porous medium and viruses both having negative or weakly positive changes (Guan et al. 2003). Viruses and other pathogens that are temporarily or permanently immobilized by adsorption and other processes (e.g., filtration) have longer aquifer residence times for inactivation to occur.

Adsorption processes in ASR were summarized by Dillon and Pavelic (1996). The amount of solute (contaminant) adsorption onto a solid surface of a porous medium depends on the characteristics of the solute and the composition and character of the porous medium. Adsorption is affected by the total organic carbon, mineralogy, and surface area of the porous medium and the pH and ionic strength of the groundwater. Dillon and Pavelic (1996) noted that the adsorption of many contaminants may be sufficiently small so that it is unlikely that adsorption can make a significant difference in the maximum contaminant concentrations in aquifers. Sorption may either allow for more time for biodegradation or other natural attenuation processes to occur (Dillon and Pavelic 1996) or, conversely, it may reduce the bioavailability of a compound so that it is less degraded (Storck et al. 2012).

The sorption behavior of soils and aquifer materials with respect to organic compounds can be predicted with limited accuracy from the organic carbon content of the materials and the octanol-water partition coefficient of the compounds (Sect. 5.6.3). More accurate, system-specific data on the sorption behavior can be obtained from batch-type procedures using actual soil, sediment, or rock samples from a project site. Batch-type procedures and guidelines for estimating soil adsorption of chemicals were published by the USEPA (Roy et al. 1992). The basic procedure is to mix a solution containing a known concentration of the chemical(s) in question with a given mass of adsorbent (soil or aquifer material) for a given period of time. The solution is then separated from the adsorbent and analyzed for the chemical(s) tested. The mass of sorbed chemical can be calculated from the initial and final concentrations. Sorption isotherms are obtained by performing a series of batch runs using different initial chemical concentrations. As discussed by Roy et al. (1992), multiple experimental parameters can affect the sorption of a given constituent including contact time, mixing ratio, temperature, and pH. Hence, experimental conditions should match field conditions as closely as practicable.

The cation exchange capacity (CEC) of soils and others materials are commonly measured by soil laboratories by first exchanging the cations in a sample with a single cation (index ion; Hendershot and Duquette 1986; ASTM 2010.). Ammonium acetate or barium chloride solutions of known concentration are often used. The index ion is then displaced using another salt solution and the amount of the displaced index ion is measured. In the ASTM (2010) standard method for measuring the CEC of fine sediments, ammonium acetate is used as the index ion and KCl is used as the salt solution. Total CEC is calculated from the final nitrogen concentration of the salt solution and the soil mass.

Less satisfactory methods for estimating the CEC of soil or aquifer materials is through predictions from soil physical and chemical properties (e.g., Manrique et al. 1991; Yukselen and Kaya 2006). The CEC of aquifer material can be estimated from clay mineral concentrations and an average or presumed representative CEC value for each clay mineral type. This method has a very high degree of uncertainty in the estimated CEC values resulting from errors in both estimated clay mineral concentrations and the assumed CEC value for each clay mineral type.

Sorption and cation exchange processes contributing to the final chemistry of water samples have been evaluated through inverse modeling procedures of varying sophistication. Most basic is mixing equation and mixing curve evaluation, in which sorption and cation exchange processes are indicated in recharged water samples by metal (or other constituent) concentrations that differ significantly from values that would occur by conservative mixing.

General geochemical modeling software (e.g., PHREEQC; Parkhurst and Appelo 1999) and mass-balance programs, such as NETPATH (Plummer et al. 1991, 1994) and Easy Leacher (Stuyfzand 1998b, 2001, 2005), can interpret water chemistry changes between initial and final waters along a hydrological flow path by considering the potential contributions of a suite of geochemical processes that can release or remove chemical constituents. More sophisticated tools are reactive solute transport models, which combine groundwater flow and solute transport codes with geochemical modeling software (e.g., PHREEQC) to simulate various biogeochemical reactions. The PHAST (Parkhurst and Kipp 2002) and PHT3D (Prommer et al. 2003) codes have been used to interpret data from ASR systems.

6.6.2 Ion Exchange and MAR Water Quality

The exchangeable ions present in a soil or aquifer rock depend upon their concentrations in solution and the affinity of ions for exchangeable sites. Ion affinity is a function of the charge and the hydrated size of ions, with small higher charged (divalent and trivalent) ions having a greater affinity than lesser charged (monovalent) ions. Divalent cations, such as Ca^{+2} and Mg^{2+}, tend to readily exchange with (replace) monovalent cations, such as N^{+} and K^{+}. In aquifer recharge systems were the native groundwater is brackish or saline, the groundwater cations are dominated by sodium and a large fraction of the exchangeable sites will contain sodium ions, which are susceptible to exchange when waters with low sodium concentrations and proportionally higher calcium and magnesium concentrations are introduced. Indeed, cation exchange in aquifer recharge systems often results in decreases in the concentration of calcium in recharged water and increases in the sodium concentration (e.g., Smith and Hanor 1975; Hamlin 1985, 1987; Castro and Gardner 1997; Mirecki et al. 1998; Petkewich et al. 2002). Desorption of sodium, and associated increases in sodium concentration in stored water, may also be caused by the displacement of native brackish groundwater with very fresh water (harvested rainwater; Zuurbier 2015). The sorption of calcium might result in a slight degree of calcite undersaturation and associated calcite dissolution.

Zuurbier (2015) documented field evidence and reactive-transport modeling results from Nootdorp, The Netherlands, that indicate that cation exchange played a dominant role in controlling the Na^+ concentration of recovered water. In the lower part of the ASR storage zone, Na^+ was adsorbed by upward migrating brackish water. Desorption of Na^+ occurred when the injected freshwater (harvested rainwater) replaced the brackish water.

Castro and Gardner (1997) suggested, based on inverse modeling results, that anion exchange with phosphate minerals may release fluoride into solution. Cation and anion exchange reactions need to be considered in interpretations of the evolution of water chemistry in ASR and other AAR systems. However, the effects of cation and anion exchange on the quality of recharged water tend to not be significant because they usually result in only small changes in the concentrations of parameters present at relatively high concentrations and/or with water quality (e.g., drinking water) standards that are relatively high. For example, the amount of sodium released by cation exchange is usually small, especially relative to the drinking water standard for sodium.

6.6.3 Sorption and MAR Water Quality

Sorption reactions can greatly impact the quality of recharged water because they can reduce the concentrations of trace elements and organic compounds that pose a potential health risk at very low concentrations. Very notably, sorption and desorption of arsenic can affect whether its concentration exceeds the very low WHO guideline and USEPA primary drinking water standard of 10 μg/L. Sorption of phosphorus onto soils affects both its availability to plants and its concentration in runoff and stormwater. Various soil amendments are used to increase phosphorus sorption and improve the quality of recharged water in infiltration basins, bioretention systems, and other MAR systems involving land application (Sect. 23.13).

The sorption capacity of soil and rock depends upon its composition. Hydrous iron and manganese oxides and hydroxides have extremely high adsorption capacities and affinities for heavy metals (Drever 1997). In oxidizing conditions, the iron and manganese oxides and hydroxides are very effective in scavenging metals (and metalloids) out of solution, including Co, Ni, Cu, Pb, Ag, Cd, and As. The negative surface charge of manganese and iron oxides and hydroxides increases with increasing pH, which results in stronger adsorption of positively changed metal ions (Drever 1997). Subsequent reducing conditions can result in the reductive dissolution or alteration of oxide and hydroxide minerals and the release of sorbed metals back into solution. Changes in Eh and pH in AAR systems can result in either the removal of metals from solution or their release back into solution.

Stuyfzand (2015) investigated trace element patterns in Rhine river water recharged into a dune aquifer system in the Netherlands. As, Mo and U showed a redox-dependent behavior. The following trace elements show (or also show) a clear sorptive behavior: B, Ba, Co, Cs, Cu, Li, Mo, Rb, Sb, Sr, U, W. Field and laboratory studies of metals removal from recharged water and their retention in the soils of soil-aquifer treatment systems indicate the sorption of metals including Cd, Cr,

Cu, Cr, Ni, Pb, and Zn (Lee at el. 2004; Lin et al. 2008). Trace metals occur in water both as dissolved species and included within or sorbed onto fine particles. Particle-bound metals are removed with varying efficiencies by filtration during infiltration, percolation, and saturated-zone flow. To understand trace metal behavior in aquifer recharge systems, it is important to determine whether the metals are present in the dissolved or particulate form. Dissolved metals concentrations are determined by first filtering water samples (typically using a 0.45 μm filter) before adding an acid (HNO_3) preservative.

Sorption process are also important for the retardation and removal of trace organic compounds, which is addressed with respect to natural aquifer treatment in Chap. 7. Sorption and other processes that impact water quality are further addressed in sections on the various AAR methods.

References

Alaerts, G. J., & Khouri, N. (2004). Arsenic contamination of groundwater: mitigation strategies and policies. *Hydrogeology Journal, 12,* 103–144.

Antoniou, E. A., Hartog, N., van Breukelen, B. M., & Stuyfzand, P. J. (2014). Aquifer pre-oxidation using permanganate to mitigate water quality deterioration during aquifer storage and recovery. *Applied Geochemistry, 50,* 25–36.

Appelo, C. A. J., & de Vet, W. W. J. M. (2003). Modeling of in situ iron removal from groundwater with trace elements such as arsenic. In A. H. Welch & K. G. Stollenwerk (Eds.), *Arsenic in groundwater* (pp. 381–401). Boston: Kluwer Academic.

Appelo, C. A. J., Drijver, B., Hekkenberg, R., & de Jonge, M. (1999). Modeling in situ iron removal from ground water. *Ground Water, 37*(6), 811–817.

Arthur, J. D., Cowart, J. B., & Dabous, A. A. (2001). *Florida aquifer storage and recovery geochemical study: Year three progress report.* Florida Geological Survey Open-File Report No. 83.

Arthur, J. D., Dabous, A. A., & Cowart, J. B. (2002). Mobilization of arsenic and other trace elements during aquifer storage and recovery, southwest Florida. In G. R. Aiken & E. K. Kuniansky (Eds.), *U.S. Geological Survey Artificial Recharge Workshop Proceedings, April 2–4, 2002, Sacramento, California,* (pp. 47–50). U.S. Geological Survey Open-File Report 02-89.

Arthur, J. D., Dabous, A. A., & Cowart, J. B. (2005a). Water-rock geochemical considerations for aquifer storage and recovery: Florida case studies. In C.-F. Tsang & J. A. Apps (Eds.), *Underground injection science and technology, Developments in Water Science 52* (pp. 65–77). Amsterdam: Elsevier.

Arthur, J. D., Dabous, A. A., & Fischler, C. (2005b). *Bench-scale geochemical assessment of water-rock interactions: Sanford aquifer storage and recovery facility*, Draft report submitted the Camp Dresser and McKee, Inc. (September 21, 2005). Tallahassee: Florida Geological Survey.

Arthur, J. D., Dabous, A. A., & Fischler, C. (2005c). *Bench-scale geochemical assessment of water-rock interactions: Seminole County ASR core samples, Draft report submitted the Camp Dresser and McKee, Inc.* Florida Geological Survey: Tallahassee.

Arthur, J. D., Dabous, A. A., & Fischler, C. (2007). Aquifer storage and recovery in Florida: Geochemical assessment of potential storage zones. In P. Fox (Ed.), *Management of aquifer recharge for sustainability, proceedings of the 6th International Symposium on managed aquifer recharge of Groundwater* (pp. 185–197). Phoenix: Acacia Publishing.

ASR Systems, LLC (2006). *Evaluation of arsenic mobilization processes occurring during aquifer storage recovery activities. Task 2—Technical memorandum, literature review, arsenic mobi-*

lization processes during ASR operations. Report prepared for the Southwest Florida Water Management District. Gainesville, FL: ASR Systems LLC.

ASTM. (2010). *Standard test method for measuring the exchange complex and cation exchange capacity of inorganic fine-grained soils (Standard D7503-10)*. West Conshohocken, PA: ASTM International.

Berner, R. A. (1970). Sedimentary pyrite formation. *American Journal of Science, 268,* 1–23.

Bhattacharya, P., Welch, A. H., Ahmed, K. M., Jacks, G., & Naidu, R. (2004). Arsenic in groundwater of sedimentary aquifers. *Applied Geochemistry, 19*(2), 163–260.

Bouwer, H. (1973). Design and operation of land treatment systems for minimum contamination of groundwater. In J. Braunstein (Ed.), *Underground water management and artificial recharge, Publication 110* (pp. 273–290). London: International Association of Hydrological Sciences.

Bouwer, H. (1974). Renovating municipal wastewater by high-rate infiltration for groundwater recharge. *Journal American Water Works Association, 66*(3), 159–163.

Buszka, P. M., Brock, R. D., & Hooper, R. P. (1994). *Hydrogeology and selected water-quality aspects of the Hueco Bolson Aquifer at the Hueco Bolson Recharge Project area, El Paso, Texas*. U.S. Geological Survey Water-Resources Investigations Report 91-4092.

Castro, J. E., & Gardner, L. R. (1997). A geochemical model for aquifer storage and recovery project at Myrtle Beach, SC. In D. R. Kendall (Ed.), *Conjunctive use of water resources: Aquifer storage and recovery, Proceedings AWRA Symposium*, Long Beach, California (pp. 201–210). Middleburg, VA: American Water Resources Association.

Chappaz, A., Lyons, T. W., Gregory, D. D., Reinhard, C. T., Gill, B. C., Li, C., et al. (2014). Does pyrite act as an important host for molybdenum in modern and ancient euxinic sediments? *Geochimica et Cosmochimica Acta, 126,* 112–122.

CH2 M Hill (2007). *Arsenic mobilization in two Suwanee Limestone aquifer storage recovery systems*. Final Technical Report submitted to the Southwest Florida Water Management District (August 2007). CH2 M Hill.

De Vito, R. H. (1978). *Uranium geology and exploration*., Lecture notes and references Golden, CO: Colorado School of Mines.

Dillon, P. J. & Pavelic, P. (1996). *Guidelines on the quality of stormwater and treated wastewater for injection into aquifers for storage and reuse*, Report No. 63A. Adelaide: Centre for Groundwater Studies.

Dillon, R., Toze, S., Pavelic, P., Vanderzalm, J., Barry, K., Ying, G.-L., Kookana, R., Skjemstad, J., Nicholson, B., Miller, R., Correll, R., Prommer, H., Greskowiak, J., & Stuyfzand, P. (2006). Water quality improvements during aquifer storage and recovery at ten sites. In *Recharge systems for protecting and enhancing groundwater resources, Proceedings of the 5th International Symposium on Management of Aquifer Recharge*, Berlin, Germany, 11–16 June 2005 (pp. 85–94). Paris: UNESCO.

Drever, J. I. (1997). *The geochemistry of natural waters: Surface and groundwater environments* (3rd ed.). Upper Saddle River, NJ: Prentice Hall.

Ehrlich, G. G., Ku, H. F. H., Vecchoili, J., & Ehlke, T. A. (1979). *Microbiological effects of recharging the Magothy aquifer, Bay Park, New York with tertiary-treated sewage*. U.S. Geological Survey Professional Paper 751-E.

Fakhreddine, S., Dittmar, J., Phipps, D., Dadakis, J., & Fendorf, S. (2015). Geochemical triggers of arsenic mobilization during managed aquifer recharge. *Environmental Science and Technology, 49,* 7802–7809.

Faust, S. D., & Vecchioli, J. (1974). Injecting highly treated sewage into a deep-sand aquifer. *Journal American Water Works Association, 66*(6), 371–377.

Granger, H. C. & Warren, C. G. (1979). *The importance of dissolved free oxygen during formation of sandstone-type uranium deposits*. U.S. Geological Survey Open-File Report 79-1603.

Gregory, D. D., Large, R. R., Halpin, J. A., Baturina, E. L., Lyons, T. W., Wu, S., et al. (2015). Trace element content of sedimentary pyrite in black shales. *Economic Geology, 110*(6), 1389–1410.

Greskowiak, J., Prommer, H., Vanderzalm, J., Pavelic, P., & Dillon, P. (2006). Quantifying biogeochemical changes during ASR of reclaimed water at Bolivar, South Australia. In *Recharge systems for protecting and enhancing groundwater resources, Proceedings of the 5th International Symposium on Management of Aquifer Recharge*, Berlin, Germany, 11–16 June 2005 (pp. 360–365). Paris: UNESCO.

Greskowiak, J., Prommer, H., Vanderzalm, J., Le Gal La Salle, C., Pavelic, P., & Dillon, P. (2005a). PHT3D modeling of water quality changes during ASR at Bolivar. In P. Dillon & S. Toze (Eds.), *Water quality improvements during aquifer storage and recovery. Volume 1: water quality improvement processes*, Report 91056F (245–277). Denver: AWWA Research Foundation.

Greskowiak, J., Prommer, H., Vanderzalm, J., Pavelic, P., & Dillon, P. (2005b). Modeling of carbon cycles and biogeochemical changes during injection and recovery of reclaimed water at Bolivar, South Australia. *Water Resources Research, 41,* W10418.

Guan, H., Schulze-Makuch, D., Schaffer, S., & Pillai, S. D. (2003). The effect of critical pH on virus fate and transport in saturated porous medium. *Ground Water, 41*(5), 701–708.

Hallberg, R. O., & Martinell, R. (1976). Vyredox—In situ purification of ground water. *Ground Water, 14*(2), 88–93.

Hamlin, S. N. (1985). *An investigation of ground-water recharge by injection in the Palp Alto Baylands, California; hydraulic and chemical interactions—final report.* U.S. Geological Survey Water-Resources Investigations Report 84-4152.

Hamlin, S. N. (1987). Hydraulic/chemical changes during ground-water recharge by injection. *Ground Water, 25,* 267–274.

Harbison, P. A. T. (1986). Mangrove muds—A sink and a source for trace metals. *Marine Pollution Bulletin, 17*(6), 246–250.

Helz, G. R., Vorlicek, T. P., & Kahn, M. D. (2004). Molybdenum scavenging by iron monosulfide. *Environmental Science and Technology, 38*(16), 4263–4268.

Hendershot, W. H., & Duquette, M. (1986). A simple barium chloride method for determining cation exchange capacity and exchangeable cations. *Soil Science Society of America Journal, 50*(3), 605–608.

Herczeg, A. L., Rattray, K. J., Dillon, P. J., Pavelic, P., & Barry, K. E. (2004). Geochemical processes during five years of aquifer storage recovery. *Ground Water, 42,* 438–445.

Houben, G. J. (2003). Iron oxide incrustations in wells. Part 1: Genesis, mineralogy and geochemistry. *Applied Geochemistry, 18*(6), 927–939.

Huerta-Diaz, M. A., & Morse, J. W. (1992). Pyritization of trace metals in anoxic marine sediments. *Geochimica et Cosmochimica Acta, 56*(7), 2681–2702.

Idelovitch, E., Icekson-Tal, N., Avraham, O., & Michail, M. (2003). The long-term performance of soil aquifer treatment (SAT) for effluent reuse. *Water Science & Technology: Water Supply, 3*(4), 239–246.

Jones, G., & Pichler, T. (2007). Relationship between pyrite stability and arsenic mobility during aquifer storage and recovery in southwest central Florida. *Environmental Science and Technology, 41*(2), 723–730.

Langmuir, D. (1997). *Aqueous environmental geochemistry*. Upper Saddle River, NJ: Prentice Hall.

Lazareva, O., Druschel, G., & Pichler, T. (2015). Understanding arsenic behavior in carbonate aquifers: Implications for aquifer storage and recovery (ASR). *Applied Geochemistry, 52,* 57–66.

Lazareva, O., & Pichler, T. (2007). Naturally occurring arsenic in the Miocene Hawthorn Group, southwestern Florida: Potential implication for phosphate mining. *Applied Geochemistry, 22,* 953–973.

Leader, J. W., Dunne, E. J., & Reddy, K. R. (2008). Phosphorus sorbing materials: Sorption dynamics and physicochemical characteristics. *Journal of Environmental Quality, 37*(1), 174–181.

Lee, S. Y., Lee, J. U., Choi, H., & Kim, K. W. (2004). Sorption behaviors of heavy metals in SAT (soil aquifer treatment) system. *Water Science and Technology, 50*(2), 263–268.

Lin, C., Negev, I., Eshel, G., & Banin, A. (2008). In situ accumulation of copper, chromium, nickel, and zinc in soils used for long-term waste water reclamation. *Journal of Environmental Quality, 37*(4), 1477–1487.

Maliva, R. G., Griswold, R. F., & Autrey, M. M. (2013). Prototype for a reclaimed water aquifer storage recovery system benefits and operational experiences. *Florida Water Resources Journal, 65*(3), 54–59.

Maliva, R. G., Autrey, M. M., Law, L., Manahan, E. S., & Missimer, T. M. (2018). Reclaimed water aquifer storage and recovery system: Update on a groundbreaking system in Florida. *Florida Water Resource Journal, 69*(2), 52–59.

Maliva, R. G., & Missimer, T. M. (2010). *Aquifer storage and recovery and managed aquifer recharge using wells: Planning, hydrogeology, design, and operation*. Houston: Schlumberger.

Manrique, L. A., Jones, C. A., & Dyke, P. T. (1991). Predicting cation-exchange capacity from soil physical and chemical properties. *Soil Science Society of America Journal, 55*(3), 787–794.

McArthur, J. M., Banerjee, D. M., Hudson-Edwards, K. A., Mishra, R., Purohit, R., Ravenscroft, P., et al. (2004). Natural organic matter in sedimentary basins and its relation to arsenic in anoxic ground water; the example of West Bengal and its worldwide implications. *Applied Geochemistry, 19,* 1255–1293.

Mettler, S., Abdelmoula, M., Hoehn, E., Schoenenberger, R., Weidler, P., & Von Gunten, U. (2001). Characterization of iron and manganese precipitates from an in situ ground water treatment plant. *Ground Water, 39*(6), 921–930.

Mirecki, J. E., Campbell, B. G., Conlon, K. J., & Petkewich, M. D. (1998). Solute changes during aquifer storage recovery testing in a limestone/clastic aquifer. *Ground Water, 36,* 394–403.

Mirecki, J. E. (2006). *Geochemical models of water-quality changes during aquifer storage recovery (ASR) cycle tests, Phase 1: Geochemical models using existing dat*a, Final Report ERDC/EL TR-06-8. Jacksonville: U.S. Army Corps of Engineers.

Mirecki, J. E., Bennett, M. W., & López-Baláez, M. C. (2013). Arsenic control during aquifer storage recovery cycle tests in the Floridan Aquifer. *Groundwater, 51*(4), 539–549.

Natural Research Council (2008). *Prospects for managed underground storage of recoverable water*. Washington, DC: National Academies Press.

Naujokas, M. F., Anderson, B., Ahsan, H., Aposhian, H. V., Graziano, J. H., Thompson, C., et al. (2013). The broad scope of health effects from chronic arsenic exposure: update on a worldwide public health problem. *Environmental Health Perspectives, 121*(3), 295–302.

Norton, S., Ellison, D., & Kohn, S. (2012). Minimizing arsenic mobilization during aquifer storage and recovery by source water degasification. Paper presented at 2012 NGWA Ground Water Summit. May 6–10, 2012, Garden Grove, CA.

Panno, S. V., Hackley, K. C., Kelly, W. R., & Hwang, H. H. (2006). Isotopic evidence of nitrate sources and denitrification in the Mississippi river, Illinois. *Journal of Environmental Quality, 35,* 495–504.

Parkhurst, D. L., & Kipp, K. L. (2002). Parallel processing for PHAST—A three-dimensional reactive-transport simulator. In S. M. Hassanizadeh, R. J. Schotting, W. G. Gray, & G. G. Pinder (Eds.), *Computational methods in water resources, Developments in water science 47* (pp. 711–718). Amsterdam: Elsevier.

Parkhurst, D. L., & Appelo, C. A. J. (1999). *PHREEQC (Version 2)—A computer program for speciation, batch reaction, one-dimensional transport, and inverse geochemical calculations*. U.S. Geological Survey, Water-Resources Investigations Report 99-42549.

Petkewich, M. D., Parkhurst, D. L., Conlon, K. J., & Mirecki, J. E. (2002). *Hydrologic and geochemical evaluation of aquifer storage recovery in the Santee Limestone/Black Mingo aquifer, Charleston, South Carolina, 1998–2002*. U.S. Geological Survey Scientific Investigations Report 2004-5046.

Plummer, L. N., Prestemon, E. C., & Parkhurst, D. L. (1994). *An interactive code (NETPATH) for modeling NET geochemical reactions along a flow PATH–version 2.0*. U.S. Geological Survey Water-Resources Investigations Report 94-4169.

Plummer, L. N., Prestemon, E. C., & Parkhurst, D. L. (1991). *An interactive code (NETPATH) for modeling NET geochemical reactions along a flow PATH*. U.S. Geological Survey Water-Resources Investigations Report 91-4078.

Price, R. E., & Pichler, T. (2006). Abundance and mineralogical association of arsenic in the Suwannee Limestone (Florida): Implications for arsenic release during water-rock interaction. *Chemical Geology, 228,* 44–56.

Prommer, H., Barry, D. A., & Zheng, C. (2003). MODFLOW/MT3DMS based reactive multi-component transport modeling. *Ground Water, 41,* 347–357.

Pyne, R. D. G. (2007). *Evaluation of arsenic mobilization processes occurring during aquifer storage recovery activities. Task 3—Technical Memorandum: field data analysis*, Report prepared for the Southwest Florida Water Management District (October 16, 2007). Gainesville, FL: ASR Systems.

Pyne, R. D. G., Singer, P. C., & Miller, C. T. (1996). *Aquifer storage recovery of treated drinking water*. Denver: American Water Works Association Research Foundation.

Ragone, S. E. (1977). *Geochemical effects of recharging the Magothy Aquifer, Bay Park, with tertiary-treated sewage*. U.S. Geological Survey Professional Paper 751-D.

Ragone, S. E., Ku, H. F. H., & Vecchioli, J. (1975). Mobilization of iron in water in the Magothy Aquifer during long-term recharge with tertiary-treated sewage, Bay Park, New York. *U.S. Geological Survey Journal of Research*, 3, 93–98.

Ravenscroft, P., McArthur, J. M., & Hoque, B. A. (2001). Geochemical and palaeohydrological controls on pollution of groundwater by arsenic. In W. R. Chappel, C. O. Abernathy, & R. Calderon (Eds.), *Arsenic exposure and health effects IV* (pp. 53–78). Oxford: Elsevier.

Rinck-Pfeiffer, S. M., Ragusa, S. R., & Vandevelde, T. (1998). Column experiments to evaluate clogging and biochemical reactions in the vicinity of an effluent injection well. In J. H. Peters, et al. (Eds.), *Artificial recharge of groundwater* (pp. 93–97). Rotterdam: Balkema.

Roy, W. R., Krapac, I. G., Chou, S. F. J., & Griffin, R. A. (1992). *Batch-type procedures for estimating soil adsorption of chemicals*. Cincinnati: USEPA Risk Reduction Engineering Laboratory.

Schreiber, M. E., Simo, J. A., & Freiberg, P. G. (2000). Stratigraphic and geochemical controls on naturally occurring arsenic in groundwater, eastern Wisconsin, USA. *Hydrogeology Journal, 8,* 161–176.

Smedley, P. L., & Kinniburgh, D. G. (2001). Source and behaviour of arsenic in natural waters. In *United Nations synthesis report on arsenic in drinking water*. World Health Organization.

Smedley, P. L., & Kinniburgh, D. G. (2002). A review of the source, behaviour and distribution of arsenic in natural waters. *Applied Geochemistry, 17*(5), 517–568.

Smith, C. G., & Hanor, J. S. (1975). Underground storage of treated water: a field test. *Ground Water, 13,* 410–417.

Stollenwerk, K. G. (2002). Geochemical processes controlling transport of arsenic in groundwater: a review of adsorption. In A. H. Welch & K. G. Stollenwerk, (Eds.), *Arsenic in ground water, geochemistry and occurrence* (pp. 67–100). New York: Springer.

Storck, F. R., Schmidt, C. K., Lange, F. T., Henson, J. W., & Hahn, K. (2012). Factors controlling micropollutant removal during riverbank filtration (PDF). *Journal American Water Works Association, 104*(12), E643–E652.

Stuyfzand, P. J. (1998a). Quality changes upon injection into anoxic aquifers in the Netherlands: Evaluations of 11 experiments. In J. H. Peters, et al. (Eds.), *Artificial recharge of groundwater* (pp. 283–291). Rotterdam: Balkema.

Stuyfzand, P. J. (1998b). Simple models for reactive transport of pollutants and main constituents during artificial recharge and bank filtration. In J. H. Peters, et al. (Eds.), *Artificial recharge of groundwater* (pp. 427–434). Rotterdam: Balkema.

Stuyfzand, P. J. (2001). *Modeling of quality changes upon artificial recharge and bank infiltration: Principles and user's guide of EASY-LEACHER*, Report SWI-99.199. Rijswijk: Kiwa.

Stuyfzand, P. J. (2005). East-Leacher modeling of water during deep well injection at Someren. In P. Dillon & S. Toze (Eds.), *Water quality improvements during aquifer storage and recovery* (Vol. 1, pp. 197–213). Report 91056F Denver: AWWA Research Foundation.

Stuyfzand, P. J. (2011). Hydrogeochemical processes during riverbank filtration and artificial recharge of polluted waters. In C. Ray & M. Shamrukh (Eds.), *Riverbank filtration for water security in desert countries* (pp. 97–128). Dordrecht: Springer.

Stuyfzand, P. J. (2015). Trace element patterns in Dutch coastal dunes after 50 years of artificial recharge with Rhine River water. *Environmental Earth Sciences, 73*(12), 7833–7849.

Stuyfzand, P. J., & Doomen, A. (2004). *The Dutch experience with MARS (Managed Aquifer Recharge and Storage), a review of facilities, techniques and tools.* Water Research Publication 05.001. Rijswijk: KIWA.

Stuyfzand, P. J., & Pyne, R. D. G. (2010). Arsenic behavior in SW Florida ASR and its export modeling. Proceedings ISMAR-7, Abu Dhabi, 9–13 October 2010.

Stuyfzand, P. J., Wakker, J. C., & Putters, B. (2006). Water quality changes during aquifer storage and recovery (ASR): Results from the pilot Herten (Netherlands) and their implications for modeling. In *Recharge systems for protecting and enhancing groundwater resources, Proceedings of the 5th International Symposium on Management of Aquifer Recharge*, Berlin, Germany, 11–16 June 2005 (pp. 164–173). Paris: UNESCO.

Thomas, M. F., Kuihiro, K., Traexler, K., & Johnston, M. (2017). Who needs pretreatment? Not Orange County Utilities' operational aquifer storage and recovery. *Florida Water Resources Journal, 68*(2), 34–36.

Tunesi, S., Poggi, V., & Gessa, C. (1999). Phosphate adsorption and precipitation in calcareous soils: The role of calcium ions in solution and carbonate minerals. *Nutrient Cycling in Agroecosystems, 53*(3), 219–227.

Vanderzalm, J. L., Le Gal La Salle, C., Hutson, J. L., & Dillon, P. J. (2002). Water quality changes during aquifer storage and recovery at Bolivar, South Australia. In P. J. Dillon (Ed.), *Management of aquifer recharge for sustainability* (pp. 82–88). Lisse: A.A. Balkema.

Vanderzalm, J. L., Dillon, P. J., & Le Gal La Salle, C. (2007). Arsenic mobility under variable redox conditions induced during ASR. In P. Fox (Ed.), *Management of aquifer recharge for sustainability, Proceedings of the 6th International Symposium on Managed Aquifer Recharge of Groundwater* (pp. 209–219). Phoenix: Acacia Publishing.

Van Halem, D., Heijman, S. G. J., Johnston, R., Huq, I. M., Ghosh, S. K., Verberk, J. Q. J. C., et al. (2010). Subsurface iron and arsenic removal: Low-cost technology for community-based water supply in Bangladesh. *Water Science and Technology, 62*(11), 2702–2709.

Vengosh, A., & Pankratov, I. (1998). Chloride/bromide and chloride/fluoride ratios of domestic sewage effluents and associated contaminated ground water. *Ground Water, 36*(5), 815–824.

Wallis, I., Prommer, H., Simmons, C. T., Post, V., & Stuyfzand, P. J. (2010). Evaluation of conceptual and numerical models for arsenic mobilization and attenuation during managed aquifer recharge. *Environmental Science and Technology, 44,* 5035–5041.

Wallis, I., Prommer, H., Pichler, T., Norton, S. B., Annable, M. D., & Simmons, C. T. (2011). Process-based reactive transport model to quantify arsenic mobility during aquifer storage and recovery of potable water. *Environmental Science and Technology, 45,* 6924–6931.

Welch, A. H., Westjohn, D. B., Helsel, D. R., & Wanty, R. B. (2000). Arsenic in ground water of the United States: Occurrence and geochemistry. *Ground Water, 38,* 589–604.

Yukselen, Y., & Kaya, A. (2006). Prediction of cation exchange capacity from soil index properties. *Clay Minerals, 41*(4), 827–837.

Zuurbier, K. G. (2015). *Increasing freshwater recovery upon aquifer storage. A field and modelling study of dedicated aquifer storage and recovery configurations in brackish-saline aquifers*. Ph.D. Dissertation, Technische Universiteit Delft.

Chapter 7
Contaminant Attenuation and Natural Aquifer Treatment

7.1 Introduction

Both operational and experimental results demonstrate that the concentrations of some contaminants decrease during infiltration, percolation, and saturated-zone flow and residence. The improvement in water quality may be either unplanned (incidental) or intentional. The intentional use of natural contaminant attenuation process to improve water quality is referred to as "natural aquifer treatment" (NAT). Recharge of "impaired" water (i.e., water that does not meet drinking water standards) potentially introduces pathogens and chemical contaminants into aquifers. Impaired waters include wastewater, stormwater, and surface waters treated to various degrees. The recharge of impaired waters into aquifers can result in improvement in water quality through filtration, fluid-rock interactions, and microbiological processes, and by providing residence (retention) time to allow for the biodegradation of slowly degradable contaminants. Natural contaminant attenuation processes can be also beneficial for systems recharging potable water by reducing the concentration of disinfection byproducts.

Anthropogenic aquifer recharge (AAR) systems vary greatly in the degree to which there is intention, planning, and management of the attenuation of contaminants by natural processes. At one end of the spectrum are unplanned and unmanaged aquifer recharge systems for which contaminant attenuation is a fortuitous benefit. For example, natural attenuation of pathogens present in raw wastewater occurs in aquifers locally recharged by sewer main leakage. There is no planning or management of the pathogen attenuation processes. At the other end of the spectrum are managed aquifer recharge (MAR) systems, such as soil-aquifer treatment (SAT), riverbank filtration (RBF), and aquifer storage transport and recovery (ASTR), that are specifically designed to take advantage of natural contaminant attenuation processes to improve water quality. NAT may be either an initial treatment element (such as is often the case for RBF systems) or may serve an additional "polishing" function to further treat water that has already received some treatment. Aquifer recharge and subsequent recovery can be important for obtaining public acceptance of indirect

R. G. Maliva, *Anthropogenic Aquifer Recharge*,
Springer Hydrogeology, https://doi.org/10.1007/978-3-030-11084-0_7

potable reuse of treated wastewater because it can result in the loss of identity (or perceived "taint") of the water as being wastewater. NAT can also provide a buffer to variations in source-water quality prior to engineered treatment (Dillon et al. 2008).

NAT has been used to improve water quality for over two hundred years in RBF systems in Europe (Chap. 20). Surface water is drawn indirectly from a river or other surface water body using wells or galleries located on land near the surface-water body rather than by direct pumping of the water source. RBF systems take advantage of the natural filtration that occurs as water passes through riverbed sediments and underlying aquifer sediment and rock. SAT systems further treat wastewater by controlled infiltration through soil in constructed basins (Chap. 19). SAT systems take advantage of physical, chemical, and biological processes within the vadose zone and shallow aquifer to remove pathogens and reduce the concentrations of nutrients, metals, and organic chemicals. ASTR systems, such as the Hueco Bolson system near El Paso, Texas, recharge treated wastewater using wells to take advantage of aquifer contaminant attenuation processes to further remove viruses and other residual contaminants before the water is recovered using wells.

NAT includes both sustainable and non-sustainable processes. Sustainable processes include biodegradation of chemicals and pathogen inactivation. Non-sustainable processes include sorption because the sorption capacity of soils and aquifer materials is finite. Eventually all sorption sites become fully occupied and contaminant breakthrough will occur, although the time required for this to occur could be very long, beyond the operational life of an MAR system. Criteria for NAT to be effective and sustainable are (Dillon et al. 2008):

- turbidity is constrained so that physical clogging is manageable
- labile organic matter is constrained so that biological clogging is manageable
- dissolution of aquifer material is not excessive so that the integrity of an aquifer and aquitards is maintained and unacceptable increases in dissolved solids does not occur
- beyond an attenuation zone, existing beneficial uses of an aquifer are not compromised by MAR activities
- beneficial use of an aquifer would not be compromised if in the future recharge operations are to cease.

The effectiveness of NAT varies between compounds and microorganisms. Organic compounds with greater halogen contents are more readily degraded under reducing conditions, whereas non-halogenated compounds tend to be more readily transformed under more oxidizing conditions (National Research Council 2008). Experimental and field studies have shown that many trace organic compounds (i.e., compounds of emerging concern) are attenuated in the soil and groundwater environment (Sect. 7.4). However, some trace organic compounds, such as the antiepileptic drugs carbamazepine and primidone, are recalcitrant to degradation in most groundwater environments. Similarly, *Cryptosporidium* and *Giardia* oocysts tend to have very slow inactivation rates in groundwater, whereas *E. coli* bacteria are much more rapidly attenuated. The attenuation rates of various pathogens and chemical contaminants also depends on the physical, chemical, and biological conditions of an aquifer.

Evaluation of contaminant attenuation during aquifer recharge and the design of NAT systems are based evaluations of

- the pathogens and chemical contaminants present in the recharge water and their concentrations
- contaminant attenuation rates under system-specific groundwater physical and chemical conditions (e.g., temperature, redox state, salinity)
- transport distance and aquifer residence time
- mixing with native groundwater
- water quality targets for recovered water.

Accurate evaluations of transport distance, aquifer residence time, and mixing with native groundwater need to consider the effects of aquifer heterogeneity. The concentration of flow into thin flow zones can result in much greater flow rates (and lesser residence times) between a recharge well or basin and a recovery well than would occur under more homogeneous aquifer conditions. Infiltration and percolation through secondary porosity features (macropores) can result in more rapid flow through soil and bypassing of vadose zone contaminant attenuation processes.

The passage of water through different redox environments (oxic and anoxic) maximizes the opportunity for contaminant attenuation (Dillon et al. 2006; Stuyfzand et al. 2007). Water that infiltrates or is recharged by wells in MAR systems often have high dissolved oxygen (DO) concentrations. Reducing conditions may later become established in the aquifer, especially if the recharged water has a high organic carbon concentration.

Water quality targets for various contaminants that might be present in water, in general, are based on some type of risk assessment, which is broadly defined as the processes of estimating the probability of occurrence of an event and the probable magnitude of adverse effects on safety, health, and ecology over a specified time period (Asano et al. 2007). Risk assessments involve four main components (Asano et al. 2007):

(1) **Hazard identification**: recognition of the microorganisms and chemicals potentially present that can increase the incidence of an adverse health condition
(2) **Exposure assessment**: evaluation of exposure scenarios and the probability (and frequency) of exposure of an individual to a given chemical or pathogen dose over a specified time
(3) **Dose-response assessment**: quantification of the risk of disease or infection of an individual from a given chemical or pathogen dose
(4) **Risk characterization**: combination of the exposure and dose-response assessments to estimate the incidence of a given adverse impact on a population.

Risks from pathogens also depend on the host's immune status. Limited data are available on the pathogen doses that are necessary to cause infection for most microorganisms. There is also limited understanding of the relationship between infection and the various forms of illness (Macler and Merkle 2002). Sensitive populations, including children, the elderly, and people with compromised immune systems, stand a greater risk of severe outcomes (National Research Council 1998). The acceptability of a given risk for a given activity depends upon (Asano et al. 2007):

- costs (human, social, and economic) associated with the risk
- costs of implementing measures to reduce the risk
- societal resources available that could be mobilized to reduce the risk
- other risks that could be reduced with available resources.

With respect to the use of wastewater for irrigation, the World Health Organization (2006) recommends that the tolerable impact from treated water consumption and water reuse in agriculture is $\leq 1 \times 10^{-6}$ disability-adjusted life year (DALY) per person per year. A DALY is equal to one year of healthy life lost and includes both time lost as the result of premature death and time spent disabled by disease. Health impacts are weighted in terms of severity within a range of 0 for no impact to 1 for death (NRMMC-EPHC-AHMC 2006). DALYs per person per year are obtained by multiplying the frequency of disease (number of cases per year) by DALYs per case.

With respect to natural contaminant attenuation and AAR, risks associated with the recharge of impaired waters is commonly managed by either requiring recharge water to meet promulgated water quality standards or guidelines, giving treatment credits for NAT (based on system type and travel or residence time), or water quality standards for finished water. NAT is best used as an element in a multiple-barrier approach to protecting public health. As noted by the NRMMC-EPHC–NHMRC (2009), "The potential for the inactivation of pathogens in aquifers highlights the potential use of aquifers as robust treatment barriers in the multibarrier approach to pathogens."

Depending upon local environmental regulations, some types of NAT may not be allowed. For example, in the United States, injection of water that contains contaminants at concentrations exceeding primary (health-based) drinking water standards (or ambient groundwater concentration) is not allowed (without a formal exemption) into aquifers considered underground sources of drinking water (USDWs). A USDW is defined as a non-exempt aquifer that contains less than 10,000 mg/L of total dissolved solids. Elsewhere (e.g., Australia), a treatment zone or zone of discharge is permitted whereby compliance with groundwater standards is at the boundary of the zone. A regulatory issue is the credit that NAT systems may receive toward meeting potable water treatment requirements. For example, riverbank filtration systems in the United States are eligible for 0.5 or 1.0 log *Cryptosporidium* removal credits if certain conditions are met (USEPA 2010).

Natural attenuation processes for pathogens, disinfection byproducts, trace organic compounds, dissolved organic carbon (DOC) and metals are reviewed in this chapter. The performance of various AAR systems in improving recharged water quality is addressed in subsequent chapters and sections on system types.

7.2 Pathogen NAT

7.2.1 Pathogen Retention and Inactivation

The fate and transport of pathogens in subsurface environments has long been of great interest because pathogens represent the greatest potential health risk associated with water. Microorganisms associated with waterborne disease are primarily enteric pathogens (i.e., they originate within the intestines of humans or other animals), have a fecal-oral or fecal-dermal route of infection (either human-to-human or animal-to-human), and survive in water (Gerba and Goyal 1985; National Research Council 1998; Bos et al. 2010). A one-time exposure to enteric pathogens may be sufficient to cause disease. The level of exposure required for human infection varies between pathogens. Ingestion of as few as 10 to 30 oocysts of *Cryptosporidium parvum* may cause infection in a healthy person (Hlavsa et al. 2005). Chemical contaminants are rarely present in drinking waters at concentrations sufficient to cause health impacts from a single exposure. Instead, the health risks associated with chemical contaminants arise from long-term, chronic exposure at low concentrations. The U.S. Environmental Protection Agency (USEPA) maximum contaminant levels are based on lifetime ingestion of 2 L of water per day for a 70-kg adult.

The concentrations of pathogens are reduced in groundwater environments by physical retention (e.g., filtration, straining, sedimentation, adhesion), inactivation (dying off), and dilution. Straining is a purely physical removal process governed by the size of pore throats and microbial particles. The effectiveness of straining depends on the ratio of the diameter of media (i.e., sand grain size) and particle diameters. Straining is insignificant when the ratio is greater than 20, significant when the ratio is between 10 and 20, and no particle penetration occurs when the ratio is less than 10 (McDowell-Boyer et al. 1986). For *Cryptosporidium* oocysts, which are about 5 μm in diameter, straining is significant for sediment grain sizes below 100 μm.

Filtration removes particles smaller than pore sizes through transport and attachment to particles surfaces. Per the colloidal filtration theory, the rate of attachment of microorganisms to sand grains can be expressed in terms of the collector efficiency and collision efficiency (Harvey and Garabedian 1991; Schijven et al. 2002; Gupta et al. 2009). These terms quantify the rate of collisions of particles with collectors (e.g., sand grains) and the fraction of colliding particles that remain attached to the collector. Collision efficiency depends upon the net repulsive and attractive forces of the surfaces of particles and collectors, which vary depending on water chemistry.

The primary factors that affect pathogen retention are (Gerba and Goyal 1985; Jørgensen 2001; Jørgensen and Peters 2001; National Research Council 2008):

- **Soil type**: The straining and filtering efficiency of soil and aquifer material is dependent upon pore-throat size distribution, porosity, pore sizes and shapes, mineralogy, and grain and crystal size, shape, and surface characteristics. Fine-textured soils containing clay minerals retain microorganisms more effectively that sandy soils. Clay minerals, iron (oxy)hydroxides, and organic matter display high sorption capacities for viruses.

- **Solute characteristics**: Ionic strength, pH, temperature, and the concentrations of DOC, surfactants, and nutrients influence particle and microorganism transport and adhesion. Low pH values (<5) favor viral retention as the viruses become less electronegative or even electropositive, and thus tend to be attracted to (or repulsed less by) negatively charged soil surfaces. Divalent cations favor retention and very low salinities favor transport.
- **Dissolved organic matter**: Humic and fulvic acids (which color water) appear to cause increased viral transport through soils by interfering with viral adsorption or by causing desorption.
- **Infiltration rate**: Slower infiltration rates tend to result in greater viral retention.
- **Rainfall events**: Rainfall may mobilize previously retained bacteria and viruses, greatly promoting their transport.
- **Type of organism**: Particle size, shape, surface charge, and hydrophobicity affect adsorption. Differences in adsorption between virus types are probably related to physicochemical differences in virus capsid surfaces.

Most enteric microorganisms have a low survivability outside of human intestines. Pathogen survival in soils and aquifers is controlled by factors including (Gerba and Goyal 1985; John and Rose 2005; Toze 2005)

- **Pathogen type**: Differences in survivability occur between bacteria, viruses, protozoa, and helminths, and between species.
- **Indigenous microorganism populations**: Greater inactivation rates have been observed in nonsterile versus sterile soil and groundwater samples. Indigenous microorganism may compete with or prey upon introduced enteric pathogens, which are not adapted to the groundwater environment.
- **Soil moisture**: Bacteria and viruses tends survive for only a very short time under dry conditions.
- **Temperature**: Higher temperatures favor more rapid inactivation.
- **Groundwater chemistry**: Pathogen survival is impacted by salinity, redox state, and nutrient concentrations.

Wastewater and other impaired waters may contain a great variety of different pathogens with different concentrations and virulences. Whereas virtually all pathogens are attenuated in soil and groundwater environments, their removal rates are highly variable. Routine monitoring for all pathogens that could be present in wastewater is not feasible, would be highly expensive, and cannot be performed in real time (Asano and Cotruvo 2004). Evaluations of pathogen attenuation in AAR systems need to consider organisms indicative of both health risk (particularly fecal contamination indicators), the range of microorganisms that might be present (e.g., bacteria, viruses, and protozoan parasites), and represent the range of inactivation rates that occur in groundwater environments.

Pathogen removal rates are commonly expressed in $\log_{10}$ removals:

$$\log_{10} \text{removals} = -\log\left(\frac{C_t}{C_0}\right) \tag{7.1}$$

where,

C_0 initial number of organisms (at $t = 0$)
C_t number of organisms at time "t".

A one-$\log_{10}$ removal is equal to a 90% reduction in concentration. Log_{10} removal time (or just "log removal time"; τ) is defined as:

$$\tau = t / \log_{10}(C_0/C_t) \tag{7.2}$$

where t = time (days).

Pathogen removal rates and efficiencies during aquifer recharge have been the subject of considerable investigation. Pathogen attenuation investigations fall into several categories:

- field evaluation of changes in pathogen concentration along flow paths from recharge sites
- laboratory "bench top" batch and column studies
- in situ diffusion chamber studies.

7.2.2 Field Evaluations of Pathogen Attenuation During Aquifer Recharge

Field investigations of pathogen removal during MAR commonly involve analyses of the recharge water and samples from production wells and monitoring wells located between the recharge location and production wells. Pathogen attenuation has been most thoroughly investigated in riverbank filtration systems in which removal of pathogens is a primary objective.

As a field study example, Betancourt et al. (2014) evaluated virus removal at three operational reclaimed MAR system sites in the United States:

- **Prairie Water Project, Brighton, Colorado**: riverbank filtration and soil aquifer treatment
- **Sweetwater Recharge Facility, Tucson, Arizona**: spreading basins for groundwater recharge
- **San Gabriel River Coastal Basin Spreading Center, Pico Rivera, California**: spreading basins for groundwater recharge.

Source water before recharge and recovered water samples were tested for adenoviruses, enteroviruses, Aichi viruses, and pepper mild mottle virus (PMMoV) by quantitative polymerase chain reaction. Samples of groundwater that tested positive for any virus were further tested for the presence of infectious viruses by cell culture.

PMMov and Aichi viruses were detected in some groundwater samples with a short travel time. A key result of this investigation is that all viruses were removed to below detection limits in groundwater samples with a residence time of 14 days or greater. The MAR systems had a 1 to greater than 5 log removal of viruses.

The Betancourt et al. (2014) study highlights a main limitation of using operational data to evaluate pathogen removal; quantification of pathogen removal efficiency is constrained by their concentrations in the infiltrated waters. Initial (surface water) concentrations may be too variable and, for some parameters (e.g., *Cryptosporidium* and *Giardia* oocycsts), too low for accurate quantification of removal efficiency. Field data may demonstrate, for example, that an RBF system is effective at removing the low concentrations of *Cryptosporidium* oocycsts present in a surface water source. However, the system may have been able to remove much higher concentrations of oocycsts if they were present. The calculated removal efficiencies are thus a minimum value, and the system may have a higher potential removal efficiency than can be calculated from the field data.

Few studies have taken the more rigorous approach of seeding recharged water with high concentrations of microorganisms, or their surrogates, to better quantify removal efficiencies. Schijven et al. (1999) investigated the removal of viruses in the Castricum artificial recharge system in the Netherlands. The Castricum system recharges a shallow dune aquifer with treated water from the River Rhine and Lake Ijssel. The water is recovered for potable supply using about 600 shallow wells. High concentration of the bacteriophages MS2 and PRD1 were added to the recharge water as virus surrogates and sampled for in down-gradient monitoring wells. Based on the quality of the treated surface water, it was determined that a virus reduction of at least 8 $\log_{10}$ may be required to comply with the maximum allowable concentration of 1.8×10^{-7} plaque-forming units per liter. The experiment results indicated that at least an 8 $\log_{10}$ reduction is achieved with a 30-m aquifer passage, which corresponds to a travel time of about 25 days. An approximately 3 $\log_{10}$ removal occurred within the first 2.4 m of passage and the remaining 5 $\log_{10}$ removal occurred in a linear fashion in the following 27 m.

The rate of reductions during transport were much greater than the field measured inactivation rates of free bacteriophages of 0.12 day^{-1} and 0.030 day^{-1} for PRD-1 and MS2, respectively, which correspond to $\log_{10}$ removal times of 8.3 and 33.3 days (Schijven et al. 1999). The main removal process was thus found to be attachment. The high initial viral removal rates were suggested to be due to the presence of more favorable attachment sites within the first few meters of soil passage (Schijven et al. 2002).

Schijven et al. (2000) examined the removal of microorganisms by deep well injection into a sandy aquifer near the village of Someren in southeastern Netherlands. The Someren system was essentially a pilot ASTR system in which water seeded with high concentrations of bacteriophages (MS2 and PRD1), bacterial spores (*Clostridium bifermentans R5*) and bacteria (*Escherichia coli* WR1), were injected into one well while a production well located 98 m away was pumped. The injected water was treated surface water from a nearby canal (Zuid—Willemsvaart). Monitoring wells located 8, 12, 22, and 38 m from the injection well, in line with the production well, were sampled for the injected microorganisms. High levels of microorganism removal occurred within the first 8 m of transport. Concentrations of MS2 and PRD1 were reduced by 6 $\log_{10}$, R5 by 5 $\log_{10}$, and WR1 by 7.5 $\log_{10}$ over the first 8 m of transport.

The large retention of microorganisms within 8 m of the injection well was attributed to their attachment onto ferric (oxy)hydroxides that formed as the result of injection of oxygenated water into the anoxic deep aquifer (Schijven et al. 2000). Schijven et al. (2000) cautioned that under anoxic conditions, further from the injection well, ferric oxyhydroxides may not be present and attachment of microorganisms may be "very low or negligible."

Where seeding recharge water with actual microorganisms is not practical or permitted, surrogates may be used. Microspheres can be selected that mimic the size and surface properties (charge) of microorganisms of concern (Pang et al. 2009). For example, fluorescent microspheres were used in Miami-Dade County, Florida, to investigate the potential for *Cryptosporidium parvum* oocysts to be transported from deep mine lakes into water production wells (Harvey et al. 2008, 2011; Mohanram et al. 2010). Oocyst-sized (1.6, 2.9, and 4.9 mm diameter) carboxylated polystyrene microspheres were injected into a karstic flow zone that was isolated using packers while a production well located 97 m away, open to the same flow zone, was pumped (Harvey et al. 2008; Mohanram et al. 2010). The test demonstrated particle transport through the flow zone, but not whether *Cryptosporidium parvum*-sized particles could actually infiltrate from a lake and be transported into a production well. A constraint on the use of fluorescent microspheres for tracer testing is their high cost at the quantities needed for tests (Harvey et al. 2011), especially if they are to be introduced in large quantities into surface-water bodies to evaluate removal during bank filtration. Use of surrogates would be particularly challenging and expensive when the source water is flowing and surrogates need to be continuously introduced.

7.2.3 Laboratory "Bench Top" Batch and Column Studies

Bench-top microorganism survival experiments basically involve preparing bottles of test solutions whose composition can be varied in terms of water source, salinity, nutrients, and physical and chemical parameters (e.g., temperature, pH, Eh). The bottles are then seeded with the microorganisms to be tested and placed in a climate-controlled, dark setting to replicate aquifer conditions. Microorganism concentrations are periodically measured to quantify the decay in concentrations over time.

John et al. (2004) performed a literature review of the factors affecting the survival of microorganisms in groundwater and performed bench-top survival studies of microorganisms in representative surface and groundwater samples in Florida. The objective of the investigation was to obtain data on microorganism survivability that could assist with informed decision-making concerning the potential impacts of the injection of impaired surface waters into aquifers in Florida. The bench-scale

Table 7.1 Ranges of $2\log_{10}$ (99%) decline rates (days) in natural water samples

Microorganism	Groundwater				Surface water			
	Avon park well		Lake lytal well		Bill evers res.		Clear lake res.	
	22 °C	30 °C	22 °C	30 °C	22 °C	30 °C	22 °C	30 °C
Fecal coliform	17–22	8–12	35–45	11–12	4–6	1–2	6–10	4–5
Enterococci	5–17	2–7	32–35	13–16	3–4	1–2	6–7	4
F+ RNA coliphage	1–3	1	4	1	7	2–3	2	1
DNA coliphage	28–45	10–16	24–41	10–14	11–19	5–11	19–28	6–13
PRD-1 bacteriophage	89–186	109–192	52–132	39–53	10–31	8–15	23–25	14–23
Cryptosporidium	>200	18	48	17	45	10	30	11
Giardia	51	20	66	19	>200	25	>200	26

Source John et al. (2004) Table 5

studies were designed to evaluate the relative importance of temperature and TDS under controlled conditions. Raw and pasteurized samples of representative surface and ground waters that may potentially (or actually) be utilized in ASR systems were used. The studied microorganisms were

- fecal coliform bacteria
- enterococci bacteria
- DNA coliphage
- F+ RNA coliphage
- PRD-1 bacteriophage
- *Cryptosporidium parvum*
- *Giardia lamblia*

Testing was performed using two groundwater samples (from Avon Park and Lake Lytal Park sites) and two surface-water samples (Bill Evers Reservoir and Clear Lake Reservoir) from central Florida at temperatures of 22 °C (72 °F) and 30 °C (86 °F). The results of the bench-top studies, expressed as 2 $\log_{10}$ (99%) inactivation times, are provided in Table 7.1. The testing results indicate that the protozoan parasites (*Cryptosporidium parvum* and *Giardia lamblia*) and PRD-1 bacteriophage are more persistent than the fecal coliform bacteria and enterococci. If the parasite inactivation rates are extrapolated to 99% inactivation, then between 10 and over 200 days would be needed for *Cryptosporidium* and 24 to over 200 days would be needed for *Giardia*. John et al. (2004) cautioned that the parasite values may not be a very accurate determination of actual longevity due to the small sample size and extrapolation from an only 25 day-long experiment.

Toze (2005) compiled published data on the decay rates of bacteria and viruses in groundwater, which are summarized in Table 7.2. The Toze (2005) study was part of an American Water Works Association Research Foundation investigation of water quality improvement during ASR. The wide variation in reported decay or

Table 7.2 Summary of published decay rates of bacteria and viruses in groundwater

Microorganism	1-$\log_{10}$ removal time (days) (90% removal)	2-$\log_{10}$ removal time (days) (99% removal)
Bacteria		
Escherichia coli	2.8–33	5.8–66
Fecal streptococci	4.2–33	8.4–66
Streptococcus faecalis	3.2	6.4
Salmonella typhimurium	4.5–7.7	9.0–15.4
Clostridium perfringens	>83	>166
Viruses		
Poliovirus 1	0–38	0–76
Coxsackievirus	9.1–20	18.2–40
Echovirus 6	9.1	18.2
Echovirus 11	10	20
Echovirus 24	20	40
Hepatitis A virus	>3.3	>6.6
Rotavirus SA-11	2.8	5.6
Coliphage f2	0.7–2.6	1.4–5.2
Coliphage T2	5.9	11.8
Coliphage T7	6.7	15.4
FRNA phage MS2	2.6–36	5.2–72

Source Toze (2005)

inactivation rates between different organisms, as well as between different studies of the same organism, indicate that pathogen survival rates are likely highly site specific (Toze 2005). Nevertheless, the data indicate that most organisms will undergo at least a 4-$\log_{10}$ decline over six to 12 months with *Cryptosporidium*, *Giardia*, and the PRD-1 bacteriophage being most resistant (having the slowest removal rates).

Inactivation rates may be influenced by multiple factors, and some studies attempted to isolate the effects of specific groundwater conditions. For example, Sidhu et al. (2006) examined the effects of redox conditions on the decay rates of coxsackievirus B2, adenovirus 41, bacteriophage MS2, and *Cryptosporidium*. Laboratory bioreactor experiments were performed under nitrate-reducing, sulfate-reducing, and control anoxic (no additional electron acceptors added) conditions. Decay rates were slower than previously observed under aerobic conditions. Different enteric pathogens behave differently under similar redox conditions (Sidhu et al. 2006). A greater removal of *Cryptosporidium* occurred in the presence of active indigenous groundwater microorganisms than under sterile conditions. The viruses showed differing responses to redox conditions. More rapid decay of adenovirus occurred under nitrate-reducing conditions than under sulfate-reducing conditions, whereas the opposite relationship occurred with the MS2 bacteriophage.

Several studies have demonstrated the importance of indigenous groundwater microorganisms on pathogen removal. Indigenous microorganisms are adapted to the groundwater environment and may out-compete or prey upon introduced enteric pathogens (Gordon et al. 2002; Gordon and Toze 2003; Toze and Hanna 2002; Wall et al. 2006, 2007). Other variables, such as nutrient concentrations and DO, are thought to be secondary influences on viral survival through their effect on the activity of native groundwater organisms and increased production of compounds instrumental in viral inactivation.

Gordon et al. (2002), Toze and Hanna (2002), Sidhu et al. (2006), Wall et al. (2006, 2007) performed experimental studies on the inactivation of pathogens and indicator organisms that might be present in reclaimed water used for artificial recharge. The Gordon et al. (2002) study included poliovirus and coxsackievirus. Toze and Hanna (2002) studied additional microbial pathogens, including *Salmonella typhimurium*, *Aeromonas hydrophila*, *Escherichia coli*, and coliphage MS2. Both studies showed that the presence of indigenous groundwater microorganisms was the most important factor influencing the inactivation of the studied microorganisms.

Wall et al. (2006, 2007) investigated the specific effects of indigenous groundwater organisms on the survival of the enteroviruses poliovirus type 1, coxsackievirus B3, and adenovirus 41. Only 27% of the isolated indigenous bacteria strains studied showed an ability to reduce mean viral copy numbers of poliovirus and coxsackievirus B3 over time compared to sterile controls. Viral decay was observed to be faster in the groundwater microcosm as whole than within individual isolates. The results suggest that there may be a number of compounds, such as protease enzymes, produced by a small number of the active indigenous groundwater microorganisms that work in conjunction to produce the consistently faster decay times observed in groundwater microcosms. Antiviral activity by groundwater microorganisms is virus specific and, in some instances, may not require the direct presence of groundwater microorganism cells (i.e., antiviral activity can be caused by the production of virucidal compounds that can act at some distance). That viral decay is not the result of one individual groundwater bacteria suggests a more robust removal process and lesser health risk (Wall et al. 2006).

7.2.4 *Diffusion Chamber Studies*

Diffusion chambers are essentially small chambers equipped with porous membranes through which groundwater can flow, but particles (e.g., microorganisms) larger than the membrane pore size cannot pass. In situ inactivation of microorganisms can be evaluated by seeding the chambers with a known concentration of microorganisms and suspending the chamber below water in a well (Pavelic et al. 1998). In-situ diffusion chambers more closely replicate ambient aquifer conditions and hence more reliable and representative pathogen inactivation rates can be obtained. However, inactivation rates from diffusion chambers are impacted by system design, including membrane pore size and composition.

Table 7.3 Water quality and pathogen inactivation rates from an ASTR system in Salisbury, South Australia

Parameter	Reed bed	Groundwater
Temperature (°C)	10.4 (±1.8)	20 (±0.5)
Electrical conductivity (μS/cm)	219 (±29.9)	2,000 (±484)
Dissolved organic carbon (mg/L)	4.7 (±0.39)	<0.005
Dissolved oxygen (mg/L)	5.29 (±1.97)	0.48 (±0.44)
Pathogen	Log_{10} reduction time (days)	
Escherichia coli	4	0.1
Enterococcus faecalis	6	2.5
Salmonella typhimurium	5	0.7
Campylobacter jejuni	No data	0.2
Coxsackievirus	No data	>34
Adenovirus	33	>34
Rotavirus	No data	>34
Cryptosporidium oocysts	>48	>34

Source Sidhu et al. (2010)

Sidhu et al (2010) evaluated pathogen attenuation using diffusion chambers in a pretreatment constructed reed bed and ASTR system in Salisbury, Adelaide, South Australia (Table 7.3). The attenuation data did not consider adsorption, sedimentation, and resuspension. The data show the bacteria have more rapid attenuation rates than viruses and *Cryptosporidium* oocysts. Sidhu et al. (2010) concluded that post-recovery treatment would be required to bring the water to potable quality.

Sidhu and Toze (2012) compared pathogen inactivation rates obtained using in-situ diffusion chambers versus laboratory microcosms. Teflon diffusion chambers with 0.010 and 0.025 μm pore sizes were used, with the later, larger pore size allowing for a greater flow rate through the chamber. Pathogens evaluated were:

- *Salmonella enterica*
- *Enterococcus faecalis*
- *Escherichia coli*
- bacteriophage MS2
- adenovirus
- *Cryptosporidium parvum*

Laboratory microcosm studies tended to give slower inactivation rates than diffusion chambers, which would result in an underestimation of removal rates. For some microorganisms (*Salmonella enterica* and adenovirus), a slower inactivation rates

Table 7.4 Pathogen inactivation time from diffusion chamber tests in Australia

Type	Pathogen	Log_{10} (T90) removal times (days)
Bacteria	*Salmonella enterica*	0.7–2
	Escherichia coli	0.1–1.5
	Enterococcus faecalis	1–3
Viruses	Coxsackievirus B3	18.5–120
	Adenovirus strain 41	27.5 to no reduction (>200)
	Rotavirus	34 to no reduction (>200)
Protozoa oocysts	*Cryptosporidium parvum*	38–120

Source Sidhu et al. (2015)

occurred in 0.010 μm versus 0.025 μm pore-size diffusion chamber, which suggests that flow rate through the chamber affects the results.

Sidhu et al. (2015) evaluated pathogen decay at four MAR sites in Australia. Multiple diffusion chambers were each seeded with either bacteria, viruses, or *Cryptosporidium* oocysts and then suspended in monitoring wells. The diffusion chambers contained either recharge water or native groundwater. The study sites and type of MAR systems were:

- **Parafield, Adelaide, South Australia**: stormwater ASTR
- **Floreat, Perth, Western Australia**: reclaimed water infiltration galleries
- **Bolivar, South Australia**: reclaimed water ASR
- **Alice Springs, Northern Territory**: reclaimed water SAT

Viruses were quantified using polymerase chain reaction (PCR). The infectious status of the viruses was not determined. Bacteria showed rapid decay (log_{10} removal times $\leq$3 days) for all species across all sites (Table 7.4). Viruses had much slower decay rates, with adenovirus being the most resistant to decay. As a generalization, much slower decay rates were observed under anoxic and anaerobic conditions compared to aquifers under aerobic conditions.

Lisle (2016) investigated the inactivation of *E. coli* bacteria in groundwater samples collected from two zones (Upper Floridan Aquifer and Avon Park Permeable Zone) of the Floridan Aquifer System (Florida) each at three monitoring wells in south-central Florida. Inactivation rates were evaluated using diffusion chambers held within an above-ground mesocosm system, which insulated the chambers from atmospheric oxygen and temperature, and minimized geochemical alteration of the samples. Native groundwater was allowed to flow through the mesocosm under artesian pressure at rates similar to those in the aquifer. Groundwater within the aquifer zones is anaerobic with ORPs in the range of −309 to −365 mV.

Inactivation data was explained by a biphasic model with an initial rapid decline (K1) that was followed by a greatly reduced rate of decline (K2). The biphasic behavior was explained by the presence of two subpopulations, with one subpopulation being more physiologically susceptible to inactivation than the other (Lisle 2016). The measured inactivation rates are

$$K1 = 0.172 - 0.684\,h^{-1}(\text{mean}\,0.380\,h^{-1})$$
$$K2 \leq 0.0001 - 0.0182\,h^{-1}(\text{mean}\,0.009\,h^{-1})$$

K1 was approximately 42-fold greater than K2. Ninety five percent of the *E. coli* population was inactivated in <11 h. The study results demonstrated that anaerobic and reduced groundwater can enhance the natural inactivation of *E. coli* at significantly greater rates than previously published rates from other groundwater systems (Lisle 2016).

7.2.5 Prediction of Pathogen Inactivation by MAR

It has been well established that pathogens are inactivated in groundwater environments and that additional removal occurs during transport by straining and filtration processes. A question remains as to how NAT can be best utilized as a water treatment step in MAR systems. Microbiological risks associated with MAR requires consideration of (Maliva and Missimer 2012)

- pathogens present in the recharge wastewater and their concentrations
- attenuation processes active in the storage aquifer between the times of recharge and recovery
- residence time
- pathogen decay rates
- the degree to which natural attenuation processes are being relied upon for pathogen removal
- the uses of the reclaimed water and associated water quality requirements, particularly health-based water quality criteria
- post-recovery treatment processes.

The greatest microbiological risks occur where recharged water contains high concentrations of multiple pathogens (e.g., untreated wastewater) and the residence time is low (Page et al. 2010; Toze et al. 2010). The simplest approach for addressing the health risks associated with pathogens relates groundwater residence time to the inactivation (removal) rate of each pathogen within the aquifer. The target residence time for each pathogen is equal to the product of the target number of $\log_{10}$ removals and the $\log_{10}$ removal rate of the pathogen.

The target $\log_{10}$ removals can be estimated as (NRMMC-EPHC-AHMC 2006; Page et al. 2015)

$$\log(R) = \log \frac{nVF}{D} \tag{7.3}$$

where

$\log(R)$ required number of $\log_{10}$ removals
n concentration of organisms in the recharge water (n/L)
V exposure volume (L)
F exposure frequency per year
D dose equivalent to 10^{-6} disability-adjusted life year (DALY)

The dose equivalent to 10^{-6} DALY recommended by the NRMMC-EPHC-AHMC (2006) are

$$\begin{aligned} &Cryptosporidium\text{: } 1.6 \times 10^{-2} \\ &\text{rotavirus: } 2.5 \times 10^{-3} \\ &\text{campylobacter: } 3.8 \times 10^{-2} \end{aligned}$$

The NRMMC-EPHC–NHMRC (2009) opined that inactivation is the only factor that should be used to assess aquifer treatment, and other processes, such as retardation, should only be considered an added benefit, which is reasonable approach for most MAR systems. However, methods have been proposed for incorporating other processes into removal rate estimates. Credits for $\log_{10}$ removals in an ASTR systems (one-directional flow) can be calculated based on measured inactivation rates and estimated attachment rates (Schijven and Hassanizadeh 2000; Page et al. 2015):

$$\log_{10}\left(\frac{C}{C_0}\right) = -\frac{(\mu_i + k_{att})L}{2.3v} \tag{7.4}$$

C concentration at time "t"
C_0 concentration at time "0"
μ_i inactivation rate coefficient for pathogens measured using in situ diffusion chambers (1/T)
k_{att} attachment coefficient (1/T)
L travel distance (L)
v Darcy velocity (L/T)

The attachment coefficient depends on the water flow and diffusion characteristics of viruses, as well as the surface properties of viruses and aquifer sediments. Schijven and Hassanizadeh (2000) presented a detailed methodology for predicting the removal of viruses during soil passage that incorporates the greater removal rates observed during the first few meters of soil passage. The greater initial removal rates were attributed to the presence of favorable attachment sites. However, the main limitation of methods to predict removal under field conditions remains the need for detailed information on soil properties and inactivation rates under project site-specific conditions. Hence, the question becomes whether the effort to obtain the required information needed to estimate attachment rates can be justified in terms of the still considerable uncertainty in the predictions.

MAR systems should be conservatively designed to achieve the highest required target residence time for all the pathogens of concern. For example, if pathogen "A" has an estimated $\log_{10}$ removal rate of 6–10 days (based on field or published data), then the World Health Organization (2006) target of 6–7 $\log_{10}$ pathogen removal could be achieved by a 70-day residence time. If pathogen "B" has a $\log_{10}$ removal rate of 20 days, then 140 days would be required for a 7 $\log_{10}$ removal. If pathogen "B" has the longest $\log_{10}$ removal time of the considered pathogens, then the MAR system might be designed to achieve a 140-day residence time or other risk-reducing elements may be added (post-treatment of recovered water). However, a limitation of this approach is uncertainty in $\log_{10}$ removal times. In some instances, decay curves (rates) are not linear (Page et al. 2010). Instead of using one estimated $\log_{10}$ removal rate for each pathogen, risks associated with pathogens may be better evaluated by quantifying the risks associated with minimum, maximum, and most likely removal rates (Page et al. 2010).

Indicator pathogens are used to assess pathogen risk because it is not technically and economically practical to routinely test for all known pathogens and data on the inactivation rates for many pathogens will not be available under project groundwater conditions. For example, Page et al. (2010) used rotavirus, *Cryptosporidium*, and *Campylobacter* as representative pathogens in a comparative study of subsurface pathogen treatment. Conservative pathogens (i.e., species with slow inactivation rates) should be chosen. The NRMMC-EPHC–NHMRC (2009) proposed that enteric viruses and protozoa should be included in pathogen survival studies because they are most resilient. Consideration should also be given to the likely abundance and relative health risk of pathogens present in a recharge water source.

The NRMMC-EPHC–NHMRC (2009) cautioned that comparison of the numbers of pathogens in recovered and recharged water is rarely adequate for an appropriate microbial risk assessment. Direct testing of pathogen decay rates has been strongly recommended. In situ diffusion chambers have been used to determine $\log_{10}$ reductions in pathogen numbers for MAR research sites (e.g., Page et al. 2010; Toze et al. 2010) but are too involved to be a routine tool for most MAR projects. A large-scale reclaimed water recharge project in a developed country may have a sufficiently large budget and technical resources to performed detailed pathogen removal studies, but it is important to be cognizant that MAR projects encompasses a broad range of project types, sizes, and budgets. Testing that is technically and economically feasible for large projects in the United States, Australia, and Europe will not be viable for small (domestic or village scale) projects in, for example, South Asia or Africa.

Conservatively long inactivation rate data for various pathogens obtained from the scientific literature could be used where site-specific testing is not practical, particularly if data are available from geochemically similar systems. Bekele et al. (2006) developed a predictive tool for pathogen die off during storage based on laboratory survival experiment data that considers the effects of temperature, redox state, and nutrient levels (degradable organic carbon) on die off rates. It was recognized that the activities of indigenous microorganisms are very important for pathogen attenuation and that physicochemical parameters may impact pathogen attenuation through impacts on indigenous microorganism populations. A computer program was developed in which information on groundwater environment is input and pathogen-specific inactivation rates are provided. Such an approach has clear value if the database reflects the range of environmental impacts on inactivation rates and potential synergistic and antagonistic effects.

Toze (2006) cautioned that current understanding of the influence of environmental factors on pathogen survival is incomplete and when combined with the complexity of the environment receiving wastewater, it is still difficult to accurately predict the stability of various pathogens in different environments. Additional safety factors are therefore prudent to provide greater assurance. Multiple preventative measurements (including MAR) can and should be applied to achieve target reductions in pathogen exposure (NRMMC-EPHC-AHMC 2006; World Health Organization 2006).

7.3 Disinfection Byproducts

7.3.1 Introduction

Disinfection byproducts (DBPs) are compounds that form by the reaction of disinfectants, such as chlorine and ozone, with organic and inorganic compounds in water. Hundreds of DBPs may potentially form during disinfection, the vast majority of which are either not regulated (i.e., there is no numerical drinking standard), are not detected during standard water quality testing, and/or are present at very low concentrations, well below thresholds for causing illness. The two main classes of DBPs, trihalomethanes (THMs) and haloacetic acids (HAAs), form during disinfection with chlorine and to a lesser degree with chloramines (Table 7.5). THMs and HAAs are believed to raise the risk of various cancers. THMs and HAAs can also form after aquifer recharge by the reaction of residual free chlorine with organic carbon in the recharged water or native groundwater. Chloramination (disinfection with compounds of chlorine and ammonia) results in lesser production of THMs and HAAs, but produces *N*-nitrosodimethylamine (NDMA), a known carcinogen.

The USEPA maximum contaminant level (MCL; primary drinking water standard) for total THMs is 0.080 mg/L. The USEPA MCL for total HAAs is 0.060 mg/L. The World Health Organization (2004) has much higher guideline values for THMs, which are 0.3 mg/L for chloroform, 0.1 mg/L for bromoform, 0.1 mg/L for dibromochloromethane, and 0.060 mg/L for bromodichloromethane. NDMA is listed as a priority toxic pollutant in the United States Code of Federal Regulations (40 CFR 131.36), but a MCL has not yet been established for drinking water. The World Health

Table 7.5 Disinfection byproducts

Treatment	Compound	Formula
Chlorination and to a lesser extent chloramination	Trihalomethanes (THMs)	
	Trichloromethane (chloroform)	$CHCl_3$
	Bromodichloromethane	$CHBrCl_2$
	Dibromochloromethane	$CHBr_2Cl$
	Tribromomethane (bromoform)	$CHBr_3$
	Haloacetic Acids (HAAs)	
	Monochloroacetic acid	$CH_2ClCOOH$
	Dichloroacetic acid	$CHCl_2COOH$
	Trichloroacetic acid	CCl_3COOH
	Monobromoacetic acid	$CH_2BrCOOH$
	Dibromoacetic acid	$CHBr_2COOH$
Chloramination	*N*-Nitrosodimethylamine (NDMA)	$C_2H_6N_2O$
Ozonation	Bromate	BrO_3^-
Chlorine dioxide	Chlorite	ClO_2^-

Organization drinking water guideline for NDMA is 0.1 μg/L (10 ng/L). The current California Department of Health notification level for NDMA in drinking water is 10 ng/L.

Bromate (BrO_3^-) is an inorganic ion that forms when ozone reacts with naturally-occurring bromide in water. Bromate is a suspected carcinogen for which the USEPA established an MCL of 0.010 mg/L. Chlorite (ClO_2^-) is an inorganic ion that forms when chlorine dioxide is used as a disinfectant. The USEPA MCL for chlorite is 1.0 mg/L.

The water recharged in ASR and other MAR systems (particularly those using wells) is commonly disinfected before recharge to prevent the introduction of pathogens into the aquifer and to prevent biological fouling by microbial growth. DBPs may impact the operation of MAR systems in four main ways (Maliva and Missimer 2010):

(1) Injection of water containing DBPs at concentration exceeding applicable groundwater standards may not be permitted.
(2) Formation of DBPs within a storage aquifer may be a regulatory violation if underground injection causes an exceedance of one or more groundwater standards.
(3) Formation of DBPs within the storage aquifer may limit future uses of the stored water if the concentrations of DBPs increase to values above applicable drinking water (or other use) standards. If recovered water exceeds drinking water MCLs, then it might not be useable for potable water supply without blending or post-treatment.

(4) Insufficient attenuation of DBPs during storage may result in violations of MCLs upon re-disinfection of the recovered water prior to its being sent into a distribution system.

An important challenge for the recharge of reclaimed and other impaired waters in ASR systems (and other MAR systems using wells) is optimizing the disinfection system so that both pathogen and DBPs standards are met. If too high a chlorine dose is used, then pathogen standards may be met but DBP standards exceeded. Conversely, if too low a chlorine dose in used, then DBP standards may be met but pathogen standards exceeded. The WHO (2004) emphasized that adequate disinfection should never be compromised in attempting to meet guidelines for THMs and other DBPs because of the much greater health risk from pathogens than DBPs. A great attraction of NAT is that it can used to reduce both pathogen and DBP concentrations. A lower chlorine dose may be acceptable if NAT can be relied upon to remove residual pathogens.

7.3.2 Formation of THMs and HAAs in MAR Systems

The formation of THMs and HAAs in ASR and other MAR projects was reviewed by Miller et al. (1993), Singer et al. (1993), Pyne et al. (1996), Fram et al. (2003), Nicholson and Ying (2005), and Pavelic et al. (2005a, b, 2006). After recharge of water containing free chlorine, an initial increase in DBPs may occur, followed by a reduction in concentrations by natural attenuation processes. Data from multiple ASR systems suggest that early DBP formation after injection may be more common than previously realized because of the low frequency of sampling during early stages of storage (Pavelic et al. 2006).

THMs and HAAs form by the reaction of free chlorine with water to form the weak hypochlorous acid (HOCl), which is the disinfecting compound. Hypochlorous acid reacts with dissolved organic carbon (DOC) to form DBPs. The formation of brominated DBPs involves an additional step in which the chlorine oxidizes bromide to form hypobromous acid (HOBr), which then reacts with DOC to form DBPs. DOC contains a wide variety of compounds whose composition varies between sources and hydrologic histories.

The formation of THMs and HAAs after injection is controlled by the concentrations of free chlorine and DOC that is susceptible to THM formation. Fram et al. (2003) referred to the latter as the “quality” of the DOC. The humic fraction (humic and fulvic acids) of DOC serves as the precursor material for the formation of THMs and HAAs (Miller et al. 1993). The susceptibility of water samples to THM formation is quantified as their trihalomethane formation potential (THMFP). THMFP is defined as the difference between the initial THM concentration of a sample and the THM concentration after the sample has been spiked with a sufficiently large chlorine dose so that residual free chlorine remains after the test period. HAA formation potential is similarly defined. THMFP potentials are a function of both sample DOC concentration and composition, and its prior disinfection history. A standard method for measuring THMFP is provided as Method 5710B of the Standard Methods for the Examination of Water and Wastewater (Rice et al. 2012).

7.3.3 Attenuation of THMs and HAAs in MAR

Field and laboratory studies of the natural attenuation of THMs and HAAs in MAR systems were reviewed by Maliva and Missimer (2010). The concentration of THMs and HAAs decrease over time in aquifers by three main processes:

(1) dilution with native groundwater in the aquifer
(2) sorption onto aquifer solids, particularly organic matter
(3) biotransformation (also referred to as biodegradation).

Abiological transformations (e.g., hydrolysis) of THMs and HAAs occur too slowly relative to other processes to be significant in ASR systems (Nicholson et al. 2002; Pavelic et al. 2005a, b; Nicholson and Ying 2005). Attenuation by sorption processes depends upon the organic content and composition of the aquifer. Sorption of THMs and HAAs will be minimal in aquifers with low organic contents (Nicholson et al. 2002; Nicholson and Ying 2005).

Experimental and field studies have shown that the biotransformation rate of THMs, and other low molecular weight haloaliphatic compounds, depends upon redox environment (Bouwer and Wright 1988). In general, as the environment (electron acceptor condition) becomes more oxidizing, more compounds tend to persist. THMs are resistant to biodegradation under oxic conditions (Singer et al. 1993; Landmeyer et al. 2000; Nicholson et al. 2002; Fram et al. 2003), whereas HAAs undergo biodegradation under both oxic and anoxic conditions (Singer et al. 1993; Landmeyer et al. 2000; Nicholson et al. 2002).

THMs undergo microbial biotransformation under anoxic conditions by reductive dehalogenation. The stability of the carbon-halogen bond influences reaction rates and stability. The ease (and thus rate) of reductive dehalogenation in the same redox environment generally follows the order of I > Br > Cl > F. Brominated compounds should undergo reductive dehalogenation more readily than chlorinated compounds (Bouwer and Wright 1988). Experimental studies indicate that highly brominated THMs ($CHBr^3$ and $CHBr_2Cl$) do indeed undergo the fastest degradation rates with chloroform degradation being the slowest. Significant biodegradation of chloroform is initiated either under sulfate-reducing (Gutpta et al. 1996a, b) or methanogenic (Bouwer and Wright 1988) conditions. The rate of biodegradation of THMs increases as the groundwater environment becomes progressively more reducing, from denitrifying to sulfate-reducing to methanogenic (Bouwer and Wright 1988).

Biological attenuation of compounds present in trace concentrations, such as NDMA, may be limited in highly treated waters in which almost all nutrients and organic carbon are removed. Compounds present in very low concentrations may be incapable of supporting the microbial metabolism necessary for the removal of the compounds during subsurface transport (National Research Council 2008). Microbial biodegradation of NDMA may occur under aerobic and anaerobic conditions, but field data of its removal in MAR systems is not available (Nicholson and Ying 2005).

7.3.4 Field Studies of THM and HAAs in ASR Systems

DBP formation and attenuation were investigated in detail in the Las Vegas Valley (Nevada) Water District (LVVWD) and Bolivar (South Australia) ASR systems. Singer et al. (1993) and Pyne et al. (1996) evaluated DBP attenuation in six ASR systems in the United States and United Kingdom. Field data from operational ASR system supports the general model of some additional THMs formation after recharge, followed by rapid early biodegradation of HAAs and preferential removal of brominated THMs in anoxic aquifers (Maliva and Missimer 2010). Chloroform tends to be refractory. A large spread in attenuation rates occur between sites. In general, warm, highly-reducing environments with high biodegradable DOC concentrations favor THMs attenuation.

The fate of HAAs and THMs in the Las Vegas Valley Water District (LVVWD) ASR system (Nevada) was examined in an early series of studies (Miller et al. 1993; Bernholtz et al. 1994; Thomas et al. 2000; Landmeyer et al. 2000). Chlorination of the recharged surface water produces approximately 15 μg/L of total HAAs and 50 μg/L of total THMs. Water samples collected during the first days of recovery had reported total HAAs and THMs concentrations as high as 29 and 90 μg/L, respectively. An initial increase in concentration was observed in some wells, followed by a decrease in the concentrations of total HAAs and THMs over time and recovered volume. The decrease in the concentration of total HAAs was relatively rapid and none were detected after several days of water storage. The decline in the concentration of brominated THMs (particularly $CHBr_3$ and $CHBr_2$) was greater than predicted by dilution and was attributed as likely being due biotransformation. Chloroform is more refractory in the LVVWD ASR system and declines in concentration can be explained almost entirely by dilution (Miller et al. 1993; Bernholtz et al. 1994; Thomas et al. 2000).

The Bolivar ASR system was used to store treated wastewater in an anoxic brackish aquifer. Operational data show both the formation and biodegradation of THMs (Nicholson et al. 2002; Pavelic et al. 2005a, b, 2006). The concentration of the more refractory THMs ($CHCl_3$ and $CHBrCl_2$) initially increased as the residual chlorine reacted with organic matter, whereas the concentration of the more labile THMs ($CHBr_2Cl$ and $CHBr_3$) continuously decreased because they were degraded more quickly than they were being formed. The total THMs concentration in the ASR well decreased from 145 μg/L at the last day of recharge to <4 μg/L after 109 days of storage. Breakthrough of THMs was not observed at an observation well located 164 ft (50 m) from the ASR well, even though the injected water had reached the reached the well (Pavelic et al. 2005b).

More rapid degradation rates (shorter half-lives) of THMs, particularly $CHCl_3$, near the ASR well was attributed to greater microbial activity caused by the accumulation of injected particulate organic matter (Pavelic et al. 2005b). Methanogenic conditions in the immediate vicinity of the ASR well appear to be responsible for the relatively rapid THM attenuation, compared to the nitrate-reducing environment present at the observation well located 13 ft (4 m) away from the ASR well (Pavelic et al. 2005b).

The Lancaster, Antelope Valley, California ASR system stores treated surface water in two storage zones that both contain aerobic freshwater (Fram et al. 2003). The general pattern observed during cycle testing was an increase in THMs after injection (above source water concentrations) followed by a decrease over time. Dilution by mixing of the injected and natural waters was the primary cause of the measured decreases in THMs concentrations. Natural attenuation mechanisms (biodegradation and sorption) did not occur to a significant degree.

7.4 Trace Organic Compounds

7.4.1 Introduction

Trace organic compounds (TrOCs) are a very broad group of chemical compounds that have been detected in surface waters, groundwaters, and drinking water supplies at very low concentrations. TrOCs are also referred to as "compounds of emerging concern" (CECs), "emerging contaminants", "emerging pollutants", "emerging pollutants of concern", and "emerging substances of concern". TrOCs have been suggested to have potentially deleterious human health and ecotoxicological effects, but many questions remain as to whether TrOCs pose a real risk to human health at the concentrations at which they have been detected.

Virtually all compounds that are used in society have pathways through which they may be discharged into the environment and enter water supplies (Ongerth and Khan 2004). TrOCs include both the chemicals themselves and their various metabolic breakdown products. TrOCs include:

- pharmaceuticals (prescription and non-prescription drugs and their breakdown products)
- antibiotics
- synthetic and natural hormones
- personal care products (PCPs)
- detergent metabolites
- antimicrobial agents (disinfectants)
- brominated flame retardants
- perfluorooctane surfactants
- fragrance and flavoring compounds
- insect repellants (e.g., DEET)

- X-ray contrast agents (e.g., iopromide)
- plasticizers
- caffeine.

The potential for direct release of TrOCs into the environment exists anywhere humans live or visit (Daughton and Ternes 1999). Pharmaceutical compounds may enter the wastewater stream, and eventually the environment, through the disposal of unused pills and solutions in toilets, excretion of either partially or wholly unmetabolized compounds after human ingestion, and land application of biosolids. The origin, fate, analytical methods, risks, and treatment technologies of pharmaceuticals and PCPs were reviewed in detail by Ternes and Joss (2006) and Richardson (2006). TrOCs are now being identified more frequently in groundwater, surface water, and reclaimed water because analytical technologies have improved so that these compounds are now detectable at extremely low concentrations, often on the order of nanograms (10^{-9} g) per liter (Sedlak et al. 2000). TrOCs were present in the environment in the past, but they could not be detected using then existing analytical technologies.

According to the Chemical Abstracts Services, over 65 million chemical products are available and about 15,000 new chemicals are added to the registry daily (Snyder 2014). The existing paradigm of evaluating the health-risks of chemicals is far too slow and vastly incapable of coping with the rapid discovery of new chemical contaminants in water (Snyder 2014). Treatment process such as oxidation and natural biodegradation process may significantly reduce the concentrations of some TrOCs, but if the total organic carbon concentration of a water is not correspondingly reduced, it means that these chemicals are being transformed into other chemicals of mostly unknown toxicity (Snyder 2014).

The health risks associated with TrOCs are still the subject of debate. Although the presence of TrOCs in the environment and water supplies is undesirable, there is not a clear picture as to the seriousness of the problem (i.e., whether TrOCs pose a material health risk at the detected concentrations if ingested) or the best course of action (Daughton 2009; Stanford et al. 2010). The National Research Council (2012) concluded with respect to TrOCs in wastewater reuse that

> collectively while these findings are insufficient to ensure complete safety, toxicological and epidemiological studies provide supporting evidence that if there are any health risks associated with exposure to low levels of chemical substances in reclaimed water, they are likely to be small.

TrOCs removal during MAR has been investigated with respect to systems with a primary water treatment objective (e.g., SAT and RBF) and systems where reclaimed water is used for aquifer recharge. A limitation of field studies of TrOC removal (which are summarized in sections on each system type) is that relatively few compounds may be present in the source waters and concentrations in source waters tend to vary over time. Laboratory (column) studies allow for the investigation of potential TrOC removal in MAR systems under controlled conditions but have the limitation that laboratory conditions will differ in some degree from field conditions. Follows is a review of some investigations of the natural attenuation of TrOCs from laboratory experiments, RBF systems, surface spreading (SAT) systems, and in the groundwater (phreatic) environment.

7.4.2 Laboratory Studies of TrOCs Removal During MAR

Laboratory columns experiments have been performed to evaluate the chemical controls of TrOCs removal during aquifer recharge. The studies typically evaluated the impacts of one or more parameters on the removal efficiency of a series of TrOCs. Key variables investigated were redox state and the amount and type of biodegradable organic carbon present. The National Research Council (2008) noted that reverse-osmosis treatment removes almost all organic carbon, except for some low molecular-weight, non-polar compounds (e.g., NDMA and 1,4 dioxane). By removing almost all nutrients and organic carbon, it has been suggested that biological alteration mechanisms in aquifers may be limited. Compounds present at very low concentrations (ng/L) may be incapable of supporting a microbial metabolism and thus are unlikely to be removed during subsurface transport.

Rauch et al. (2006) investigated how differences in source-water chemistry promote the microbial breakdown of TrOCs. The question investigated was whether the composition of biodegradable organic carbon in recharge water has a major effect on soil microbial communities and soil biomass activity. Columns filled with silica sand were fed water with a total organic carbon (TOC) concentration of 3 mg/L and the following organic carbon matrices:

- hydrophilic carbon (HPI)
- hydrophobic acids (HPO-A)
- colloidal organic carbon
- effluent organic matter (efOM).

Reverse-osmosis treated effluent was used as a control. The feed water was spiked with the following TrOCs: ketoprofen, naproxen, phenacetin, and gemfibrozil. Columns fed HPO-A and RO-treated effluent exhibited the most efficient removal of all TrOCs. The results support the conclusion that efficient TrOCs removal is possible in advanced treated effluents with very low biodegradable DOC (BDOC) concentrations (Rauch et al. 2006).

Rauch-Williams et al. (2010) performed column studies of the role of organic matter in the removal of TrOCs. The columns contained aquifer material that were each acclimated by long-term (>2 years) feed of either secondary treated wastewater or specific compound fractions of secondary treated wastewater. Ten TrOCs were evaluated:

- carbamazepine (anticonvulsant)
- diclofenac (analgesic/anti-inflammatory)
- gemfibrozil (blood lipid regulator)
- ibuprofen (analgesic/anti-inflammatory)

- ketoprofen (analgesic/anti-inflammatory)
- naxaprofen (analgesic/anti-inflammatory)
- phenacetin (antipyretic)
- primidone (anticonvulsant)
- propyphenazone (analgesic)
- TCEP—tris(2-chloroethyl)phosphate; flame retardant).

The chemical conditions evaluated were:

- anoxic
- aerobic—hydrophobic acids
- aerobic—hydrophilic carbon
- aerobic—organic colloids
- aerobic—effluent organic matter.

The results of experiments included the following:

- Carbamazepine and primidone were refractory under all conditions.
- Gemfibrozil, ibuprofen, ketoprofen, naxaprofen, and propyphenazone were significantly degraded under anoxic conditions.
- The most degradable TrOCs (gemfibrozil, ibuprofen, ketoprofen, and naxaprofen) underwent significantly greater transformation under aerobic conditions.
- Removal efficiencies were generally better under aerobic conditions except for diclofenac, which underwent faster transformation under anoxic conditions.
- Unexpected high removal efficiencies for all degradable TrOCs occurred under low BDOC conditions.

Rauch et al. (2006) and Rauch-Williams et al. (2010) concluded that naturally present organic matter in the soil can promote TrOCs removal by serving as a substrate for their removal. BDOC in the form of colloidal and hydrophilic carbon stimulates soil biomass growth and induces secondary utilization of TrOCs. Microbial adaptation to TrOCs took place, as removal rates increased with increased exposure of the columns to trace pollutants. Under low BDOC (oligotrophic) conditions, it was postulated that a specialized slow-growing, diverse microbial community develops that grows on refractory carbon substances (hydrophobic acids) and is capable of degrading TrOCs.

A series of laboratory column investigations were performed at King Abdullah University of Science and Technology (KAUST) in Saudi Arabia on the impacts of primary organic carbon substrate and microbial communities on the removal of TrOCs. Li et al. (2013) evaluated the impacts of BDOC on microbial communities. Two sets of columns receiving different DOC and BDOC concentrations were employed to explore the response of microbial communities to DOC availability. Levels of BDOC were varied by adjusting the concentrations of peptone, yeast extract, and humic acid. Microbial community structure was examined using DNA extraction and pyrosequencing and quantitative real-time polymerase chain reaction (qPCR).

BDOC was quantified as the difference in DOC between the column influent and effluent. For both moderate and low BDOC feeds, greater than 80% DOC removal occurred during the first 30 cm of transport (~5 to 6 h of residence time). The greatest community diversity occurred in the samples collected from the top 1 cm of the 120 cm columns and in the column with the lowest BDOC. There was a strong similarity in microbial diversity in the moderate and low BDOC columns between 30 and 120 cm, with only a slight increase with depth.

Increasing microbial diversity occurred with decreasing BDOC concentration, which was suggested to be due to the diverse refractory carbon source (humic acids) selecting for coexistence of numerous microbial groups with different metabolic functions (Li et al. 2013). BDOC rather than DOC was suggested as being the dominant factor influencing microbial community composition in MAR systems. Community evolution took 3–4 months to reach steady state with an associated improved DOC removal.

In a subsequent study (Li et al. 2014), sand columns were fed solutions with different total BDOC concentrations and ratios of peptone, yeast extract, and humic acid. The columns were spiked with the TrOCs gemfibrozil, caffeine, trimethoprim, diclofenac, and atenolol. The test results demonstrated that the concentration and composition of the primary substrate shape the makeup of the microbial community, which appears to be correlated with increased removal of TrOCs. Lower BDOC and high ratios of humic acids (i.e., more refractory BDOC) promote a more diverse microbial community and greater TrOCs removal.

Alidina et al. (2014a) hypothesized that the primary substrate represented by the biodegradable portion of the bulk organic carbon shapes the microbial community in MAR systems and thus influences TrOCs removal. The "primary substrate" provides the energy source for microbes as TrOCs are an inadequate source due to their parts per trillion concentration. Column experiments were performed on the degradation of six TrOCs representing different degrees of biodegradability (atenolol, caffeine, diclofenac, primidone, and trimethoprim). Feed solutions differed in their total organic carbon concentration and concentrations of peptone, yeast, and humic acids, with the latter representing the more refractory component of wastewater.

Primidone was not attenuated and caffeine was easily degradable with greater than 75% removal regardless of the nature of the primary substrate. The concentrations of atenolol, gemfibrozil, and diclofenac were affected by the concentration and composition of the primary substrate. Greater removal of TrOCs occurred when the primary substrate consisted mostly of refractory compounds. The column experiment results suggest that microbial populations vary in the presence of specific enzymes capable of degrading different TrOCs. It was concluded that the microbial groups responsible for the TrOC removal are likely slow growing and unable to compete against other microorganisms when the primary substrate consists largely of easily degradable carbon but appear to be able to flourish under more refractory carbon conditions. The presence of a more refractory carbon substrate results in a more diverse biocommunity (Alidina et al. 2014a).

Alidina et al. (2014b) investigated the effects of pre-exposure of soil to TrOCs on their subsequent removal. The issue is whether an adaptation period is required for microbial populations to acquire the ability to transform TrOCs. Column experiments were performed using paired pre-exposed and non-exposed columns with different primary substrates (synthetic treated wastewater). Microbial community structure was evaluated by DNA extraction and pyrosequencing, and qPCR.

No systematic difference in TrOCs attenuation was observed between pre-exposed and non-exposed columns. The presence of TrOCs at ng/L concentrations had no obvious influence on the microbial community phylogenetic structure. The experiment results indicate that an adaptation period is not required and that TrOCs removal is the result of co-metabolism rather than secondary substrate utilization. In simpler terms, microorganisms do not use TrOCs as a primary energy source, but rather that enzymes needed for TrOCs degradation are produced for the purpose of microbial utilization of the DOC energy source. The enzymes were not produced for TrOCs degradation but happen to result in their degradation. The practical conclusion is that MAR systems are robust in removing new TrOCs as there is not an adaptation period with low removal efficiencies.

Scheytt et al. (2004) evaluated the transport of three pharmaceutically active compounds (clofibric acid, diclofenac, and propyphenazone) that were detected in water recovered from the Lake Tegel bank filtration system near Berlin using saturated column experiments. Clofibric acid had the same transport behaviour as a non-reactive tracer and showed no degradation. Diclofenac and propyphenazone showed retardation. The experiment results suggest that the concentrations observed at the Lake Tegel bank filtration site are controlled by sorption, desorption, and input variation rather than degradation.

Ying et al. (2003) conducted laboratory experiments of sorption and degradation of five endocrine disrupting compounds (EDCs) that have been detected in wastewater: the estrogenic steroids E2 and EE2, bisphenol A (BPA, an industrial chemical), 4-tert-octylohenol (4-t-OP, a surfactant degradation product), and 4-n-nonylphenol (4-n-NP, a surfactant degradation product). Degradation experiments under aerobic conditions gave calculated half-lives of 2-days for E2, 7 days for 4-n-NP, and 81 days for EE2 (Ying et al. 2003). The concentrations of BPA and 4-t-OP were unchanged. Under anaerobic conditions, the concentrations of the five EDCs remained almost unchanged over 70 days. Only E2 underwent very slow degradation with an estimated half-life of 107 days.

7.4.3 TrOCs Removal During Riverbank Filtration

Field studies have shown that riverbank filtration is effective in attenuating some TrOCs (Heberer 2002; Heberer et al. 2004, 2006; Verstraeten et al. 2002; Zuehlke et al. 2004; Massmann et al. 2006, 2008; Hoppe-Jones et al. 2010). TrOCs concentrations in surface waters depend on a variety of factors including water use patterns in the watershed, wastewater treatment processes, and physical and biological removal processes. Hoppe-Jones et al. (2010) documented that TrOCs found in European rivers were not detected in the Ohio River and Cedar River RBF sites in the United States. It was suggested that their absence may be due to a greater per capita water use in the United States and thus greater dilution. TrOC concentrations are reduced in RBF systems by dilution, sorption, and biodegradation processes. The most rapid reduction in TROCs concentrations occurs during the initial infiltration period when most TOC removal also occurs (Hoppe-Jones et al. 2010). TrOC attenuation rates tend to be lower at colder temperatures (<10 °C). Some of the more refractory TrOCs (e.g., carbamazepine and primidone) are not significantly attenuated during RBF and have been detected at very low concentrations in groundwater samples from nearby production wells.

Heberer et al. (2004) studied the fate and transport of TrOCs at two bank filtration sites in Berlin, Germany. Antibiotics, estrogens, bezafibrate, and indomethacin were completely removed, whereas the following compounds were either persistent or underwent only partial removal: carbamazepine, primidone, clofibric acid (metabolite of blood lipid regulating drugs), AMDOPH (drug metabolite), propyphenazone (drug metabolite), bentazone (pesticide residue), mecoprop (pesticide residue), tris(2-chloroethyl)-phosphate (TCEP, a flame retardant), and tris-(chloroisopropyl)-phosphate (TCPP; flame retardant). AMDOPH and propyphenazone residues were believed to have originated mainly from a spill at a former upstream production facility rather than from treated wastewater. Except for AMDOPH, the concentrations of the TrOCs detected in production wells were less than 500 ng/L.

Massmann et al. (2006, 2008) and Greskowiak et al. (2006) reported on the effects of redox state on the attenuation of some TrOCs at a bank filtration site in Berlin, Germany. The study area, located near Lake Tegel, consists of one of three ponds surrounded by over 40 production wells. The ponds received water from Lake Tegel, which, in turn, receives treated wastewater. The studied pond contains DO year round. In the summer, DO in the groundwater decreases and is eventually totally removed by microbial activity before reaching the first well. Redox state in the groundwater becomes denitrifying and manganese reducing. Due to lesser biological activity and greater gas solubility with low temperatures, DO is not totally removed in the winter, except in a deep monitoring well located the greatest distance from the pond.

The pond water contained phenazone and its metabolites (propyphenazone, AMDOPH, DP, AAA, and FAA) and carbamazepine. A tendency was observed for elevated groundwater concentrations of phenazone and some of its metabolites (propyphenazone, AAA, and FAA) in the summer, whereas complete removal occurred in the winter when DO was present. Only marginal removal of AMDOPH occurred in the winter, and carbamazepine was persistent in both oxic and anoxic conditions. Greskowiak et al. (2006) developed a calibrated reactive solute-transport model using the PHT3D code that simulated both the evolution of redox conditions and removal of phenazone. The Massmann et al (2006, 2008) study demonstrates a variable temperature-dependent redox sensitivity of TrOC removal.

7.4.4 TrOCs Removal During Soil-Aquifer Treatment

Drewes et al. (2002) evaluated TrOCs removal in treated effluent (recharge water) and monitoring well samples from five SAT systems in the United States. The main observations were

- Anti-inflammatory drugs (e.g., diclofenac, ibuprofen, fenoprofen, ketoprofen, naproxen) are present in secondary and tertiary treated effluents. Their concentrations are significantly lower in facilities employing nitrification and denitrification.
- TrOCs were not detected in reverse-osmosis (RO) permeate from Scottsdale Water Campus (Arizona).
- Antiepileptic drugs carbamazepine and primidone were detected in all downgradient monitoring wells. Significant removal was not evident.
- SAT has a high potential for removing acidic drugs (analgesics/anti-inflammatory and lipid regulators).

Field data from SAT systems in Mesa and Tucson, Arizona, documented the reduction of 17β-estradiol, estriol, and testosterone concentrations to below detection limits (Amy and Drewes 2007). The only pharmaceutically active compounds (PHACs) detected after SAT were carbamazepine and primidone. The chlorinated flame retardant TCIPP was also detected after SAT in the most down-gradient monitoring well but at a significantly reduced concentration.

Alkylphenol polyethoxylates (APEOs) are surfactants and, with their degradation metabolites, are among the most frequently detected aquatic contaminants in the environment (Montgomery-Brown et al. 2003). A field study performed at the Mesa Northwest Water Reclamation Plant found that SAT was effective in removing most APEOs metabolites (Montgomery-Brown et al. 2003). The alternating flooding and drying cycles of SAT system (and thus alternating aerobic and anaerobic conditions) appear to enhance overall APEO removal efficiencies (Montgomery-Brown et al. 2003). Drewes et al. (2003) documented that long-term SAT reduced the concentrations of EDTA, APECs (alkylphenol polyethoxycarboxylates) and NTA (nitrilotriacetic acid) by 89%, 99%, and 100% respectively.

In field studies at two SAT sites, the hormones estriol and testosterone were not detected in lysimeter samples taken 1.5 m (4.9 ft) below land surface (Mansell and Drewes 2004). Substantially reduced concentrations of 17β-estradiol were detected at 1.5 m (4.9 ft) below land surface in the vadose zone, but the hormone was not detected in down-gradient monitoring wells.

Cordy et al. (2004) performed laboratory experiments in which secondary treated wastewater samples were passed through a 2.4 m (7.9 ft) sediment column and the concentrations of TrOCs in the influent and effluent waters were compared. Only eight TrOCs were detected in the influent and effluent samples: carbamazepine, sulfamethoxazole, benzophenone, 5-methyl-1H-benzotriazole, *N,N*-diethyl-toluamide, tributylphosphate, tri(2-chloroethyl) phosphate, and cholesterol. Cordy et al. (2004) concluded that under recharge conditions similar to those in arid and semiarid climates, some TrOCs and pathogens may persist in treated effluent after SAT and reach the groundwater. However, the Cordy et al. (2004) study did not address the additional attenuation that will occur to some degree in the groundwater environment.

The mechanism of removal of steroids (17β-estradiol, estriol, and testosterone) during SAT was investigated in bench-top experiments by Mansell et al. (2004) and Mansell and Drewes (2004). Estriol and testosterone were below detection limits after 1 m of travel through the porous media in the column experiments, whereas the concentration of 17β-estradiol decreased from about 200 ng/L to 1.1 ng/L. Physical sorption was the primary method of removal and was most efficient in soils containing high concentrations of silt, clay, and organic matter. Microbial degradation resulted in further removal of the hormones. Testosterone showed the greatest retardation, followed by estriol and 17β-estradiol. All compounds were also subject to removal by biodegradation regardless of the type of organic carbon present in the sample. The biodegradation occurred under both aerobic and anoxic conditions.

The performance of SAT systems in terms of the removal of more refractory TrOCs may improve over time as the biological activity of the systems increases with ripening. Column testing results showed that the removal of gemfibrozil, diclofenac, and bezafibrate increased from less than 20% in reactors ripened for five days to over 90% when the reactors were ripened for 240 days (Abel 2014). Phenacetin, paracetamol, ibuprofen and caffeine were easily removed under various operating conditions. Reduced removal of some TrOCs may thus occur after drying and scrapping of SAT basins.

7.4.5 TrOCs Removal During Surface Spreading

Laws et al. (2011) examined the attenuation of TrOCs in reclaimed water at a fully instrumented research basin located at the Montebello Forebay in California. The percentage of reclaimed water in samples was estimated using temperature and the refractory TrOC primidone as tracers. The general geochemical background is that ammonia is rapidly removed in the vadose zone (aerobic nitrification) followed by nitrate reduction. Nitrate is stable down to about 7.6 m, which suggests that there is insufficient DOC to support efficient heterotrophic denitrification. TOC concentration was reduced by 55% in the upper aquifer (travel time <3 days) and by 79% in the deep aquifer (travel time = 60 days). The reclaimed water had a BDOC concentration of 4.6 mg/L. BDOC concentration was reduced to 0.8 ± 0.2 mg/L in the upper aquifer (above 9.5 m) and to <0.1 mg/L in the lower aquifer (deeper than 10 m). Easily assimilable organic carbon is thus being consumed under aerobic conditions. TrOCs showed a wide range of removals and can be categorized in four main groups based on removal percentages after 3 and 60 days (Table 7.6).

Quanrud (2003) and Quanrud et al. (2004) examined the fate of estrogenic activity during surface water discharge and incidental recharge in the Santa Cruz River, Tucson, Arizona. Boron concentration and isotope data were used to determine the volumetric contribution of wastewater to the water samples. The boron data confirm that the Santa Cruz River water samples are entirely of wastewater origin. Estrogenic

Table 7.6 Removal of trace organics at Montebello Forebay research basin (Laws et al. 2011)

Group	Compounds	% removal at 3 days	% removal at 60 days without dilution
Ready rapid removal	Atenol	96	>99.9
	Iopromide	98	95
	Fluoxetine	≥97	≥97
	Gemfibrozil	93	94
	Naproxen	63	88
	Diclofenac	55	≥99
Slow removal (>70% at 60 days)	Trimethoprin	Negligible	90
	Triclosan	Negligible	≥86
	TCPP	Negligible	82
	Ibuprofen	47	74
	DEET	24	75
Partial removal (35–50% at 60 days)	Meprobamate	Negligible	50
	TCEP	Negligible	49
	Sulfamethoxazole	Negligible	26
Negligible removal	Phenytoin	36	Negligible
	Carbamazepine	Negligible	Negligible
	Primidone	Negligible	Negligible

Source Laws et al. (2011)

activity was reported to decrease in surface water by about 60% over a 35 km reach downstream of the discharge point, whereas DOC concentration decreased by only about 20%. Little attenuation of estogenic activity appeared to occur during percolation through the vadose zone. The decrease in estrogenic activity during recharge was significantly less than occurs at the nearby Sweetwater Recharge Facility (SRF) SAT system.

The Santa Cruz River and SRF data suggest that MAR of wastewater effluent (SAT) can provide water quality benefits that are not realized, at least not immediately, during unmanaged incidental recharge of reclaimed water in a riverbed. SAT systems have the near continuous maintenance of a biochemically active layer through which most of the water infiltrates, which does not occur in a natural riverbed (Quanrud 2003; Quanrud et al. 2004).

Drewes et al. (2002, 2003) documented that caffeine, analgesic/anti-inflammatory drugs (e.g., diclofenac, ibuprofen, ketoprofen, naproxen, and fenoprofen) and blood lipid regulators (e.g., gemfibrozil) were quickly reduced to concentrations near or below detection limits during groundwater recharge at facilities in the Western United States. Antiepileptic drugs (e.g., carbamazepine, primidone) were not removed during recharge under either oxic or anoxic conditions. Other refractory CECs include clofibric acid (a blood-lipid regulator), some antibiotics (e.g., sulfamethoxazole),

X-ray contrast media, and chlorinated flame retardants (TCEP, TCPP). Snyder et al. (2004) similarly documented the variable removal of TrOCs during artificial recharge by land application in an arid environment (Las Vegas, Nevada, area). TROCs were not detected in down-gradient monitoring wells, which suggests continued biodegradation in the saturated zone.

7.4.6 TrOCs Attenuation in Groundwater (Recharge by Injection)

Stuyfzand et al. (2007) reported on TrOCs removal in basin and deep well recharge systems in the Netherlands. Recharge and observation well water samples were analyzed for 140 organic micropollutants. Both young (1 year) suboxic and older anoxic infiltrates were analyzed. Significant biodegradation (>90%) of phenazone, iohexol, iomeprol, and iopamidol occurred in the suboxic environment and of sulfamethoxazole and amidotrizoic acid in the anoxic environment. The following compounds were found to be very persistent in young suboxic and old anoxic infiltrates: carbamazepine, MCPP, bentazone, tertiary octlphenol, iso-nonylphenol, PFOA, PFOS, TCEP, 1,4-dioxane, and diglyme.

The attenuation of TrOCs was investigated in a salinity barrier system located southwest of Barcelona, Spain (Candela et al. 2016). Tertiary-treated wastewater from the Depurbaix wastewater facility was injected into a Quaternary unconsolidated siliciclastic aquifer (LLobregat Delta) to create a hydraulic barrier against saline-water intrusion. The testing was performed during Phase I of the salinity barrier project, which had three active injection wells. The injected water and samples from a monitoring system were analyzed. The travel distance to the monitoring wells or piezometers (with the exception of one well located 4 m from an injection well) ranged between 250 and 800 m, and the sampling was performed for over a three-year period (2007–2010).

The water sampled were tested for 81 trace organic compounds. Only 11 compounds were detected in both the treated wastewater and in monitoring wells:

- carbamazepine
- ciprofloxacin
- diazepam
- diclofenac
- hydrochlorothiazide
- mepivacaine
- sulfamethazine
- sulfamethoxazole
- sulfapyridine
- 4-AAA
- 4-FAA.

Candela et al. (2016) concluded that dilution was much more important than attenuation or degradation processes toward reducing TrOC concentration.

7.4.7 TrOCs Removal by NAT Summary

Field and experimental studies have demonstrated that the concentrations of some TrOCs decrease rapidly in soil and groundwater environments through sorption and biodegradation processes. TrOCs have a wide range of structural and functional groups that give them markedly different physiochemical properties (Benotti and Snyder 2009). In general, hydrophilic compounds tend to be more persistent in MAR systems, as hydrophobic compounds tend to be adsorbed onto solids (particularly organic matter). Adsorption of TrOCs is strongly related to the organic content of the media (Khan and Rorije 2002). Aquifer redox conditions also strongly influence the rate of attenuation of TrOCs.

Maeng et al. (2011a, b) summarized removal efficiencies of TROCs during MAR:

- antibiotics, NSAIDS, beta blockers and steroid hormones exhibit good removal efficiencies, especially for compounds showing hydrophobic-neutral characteristics
- anticonvulsant (antiepileptic) drugs (e.g., carbamazepine) generally show poor removal
- phenazone-type pharmaceuticals exhibit better removal efficiencies under oxic conditions
- some compounds, such as x-ray contrast agents and sulfamethoxazole, show no significant removal under oxic conditions but are removed under anoxic conditions.

TrOC removal may be optimized in MAR systems by having recharged water experience both oxic and anoxic conditions. Nevertheless, some compounds are persistent, which is often due to structural features that prevent enzymatic attack (National Research Council 2008). The antiepileptic drugs carbamazepine and primidone, in particular, are resistant to degradation in water and wastewater treatment processes and within soil and groundwater environments, and therefore, may be used as indicators of anthropogenic input (Clara et al. 2004). MAR is effective in removing many, but not all, TrOCs, and additional post-treatment may be required to achieve non-detection in potable water. The question still remains as to whether non-detection is a reasonable goal in consideration of the lack of evidence for actual health risks from TrOCs at their detected concentrations after MAR.

7.5 Dissolved Organic Carbon

Dissolved organic carbon (DOC) from both natural and anthropogenically impacted waters contains a great diversity of compounds that vary in their biodegradability.

The bulk organic matter (OM) of wastewater effluent is a mixture of natural organic matter (NOM) and effluent organic matter (efOM). EfOM contains a wide variety of refractory and biodegradable compounds including TrOCs. NOM is dominated by humic and fluvic compounds. Organic carbon is removed by a combination of filtration, sorption, oxidation-reduction, and biodegradation (National Research Council 2008). Biodegradation is the primary removal method. High molecular weight components tend to be hydrolyzed to lower molecular weight compounds, which serve as substrates for microorganisms.

Maeng et al. (2011a, b) reviewed methods for characterizing bulk organic matter and the changes in bulk organic matter content that occur during aquifer recharge and bank filtration. Several parameters are used to characterize the DOC present in waters. Specific ultraviolet absorbance (SUVA), which is the ratio of DOC to UV absorbance at 254 nm, represents the relative aromaticity of organic matter. XAD resin fractionation separates NOM into hydrophobic and hydrophilic fractions. Size extraction chromatography describes the molecular weight distribution of DOC and categorizes it into biopolymers, humic substances, building blocks, neutrals, and low molecular weight fractions.

Field and laboratory studies by Drewes and Fox (1999) demonstrated that short-term SAT results in the preferential removal of ultra-hydrophilic compounds, which consist primarily of amino acids, proteins, and polysaccharides, all of which are potentially biodegradable. Short-term SAT was found to increase SUVA by the preferential removal of non-aromatic compounds. Sustained biodegradation of more poorly degradable organic compounds occurred during long-term SAT. Long-term SAT also resulted in a change in the structure of the DOC with a loss of aromatic and carboxylic character, an increase in aliphatic compounds, and associated decrease in SUVA. Further investigation of the fate of wastewater effluent organic matter (EfOM) and TrOCs at operational SAT sites located in Mesa (Northwest Water Reclamation Plant) and Tucson (Sweetwater Underground Storage and Recovery Facility), Arizona, indicated that 50–70% of the DOC is removed after accounting for dilution (Fox et al. 2001; Drewes et al. 2003; Amy and Drewes 2007).

DOC changes at the Mesa Northwest Water Reclamation Facility SAT system were examined by Fox et al. (2001). A naturalization process occurs whereby major differences between reclaimed water OM and NOM are eliminated by short-term transformations. After several years of travel, DOC concentration approached background concentration in the aquifer and the bulk of the OM could not be distinguished from NOM.

Compositional changes in the character of efOM was evaluated at the Mesa, Tucson, and Rio Honda Spreading Grounds (Los Angeles, California). A reduction in DOC occurred at all three sites, with the following compositional changes (Sattler et al. 2006)

- almost complete elimination of dissolved organic nitrogen (DON)
- removal of polysaccharides and proteins during early SAT
- increased SUVA reflecting preferential biodegradation of non-humic (non-aromatic) efOM

- lesser, but still significant, removals of humic substance under long-term anoxic conditions.

Fox et al. (2005) analyzed soil samples from year-long bench-top studies and an operational SAT system (Mesa Northwest Water Reclamation Plant, Arizona) to determine whether reductions in DOC measured in SAT systems are caused by sorption or sustainable biological processes. Organic carbon accumulated mostly in the upper 2 cm (0.8 in.) of the soil and organic carbon did not accumulate below 8 cm (3.1 in.). The TOC accumulation in the surficial soil was less than 20% of the TOC load that was applied. The results of the Fox et al. (2005) investigation indicate that organic carbon removal in SAT systems is sustainable because it occurs by biodegradation processes rather than by sorption (Fox et al. 2005).

Organic carbon removal in an operational SAT system in Tucson, Arizona, showed the typical pattern of rapid DOC removal in the upper 3 m (10 ft) of the soil (Quanrud et al. 2003). Comparison of recharge basins that had been in operation for 10 years with new (less than 2-year-old) basins demonstrated that DOC removal efficiency does not decrease over time and that the systems are sustainable. SAT systems were found to be most effective in the removal of hydrophilic biodegradable organic compounds. THM formation potential was reduced by an average of 91% across the vadose zone. Drewes (2009) also observed that, in general, the majority of EfOM removal occurs in the initial phase of infiltration.

Data on the fate of DOC injected in ASR and other MAR systems that use wells for recharge is more limited. Skjemstad et al. (2002, 2005) investigated the fate of DOC injected in the Bolivar reclaimed water ASR system in South Australia. The composition of the DOC in the injectate and well waters at the Bolivar site are typical of NOM present in other surface waters and groundwaters. Filtration of particulate organic matter occurred near the ASR well. Microbial assimilation of the particulate organic matter released lower molecular weight molecules as DOC. The injected water and water samples collected at monitoring wells located 4, 50, and 75 m (13, 164, and 246 ft) from the injection well were analyzed. A net loss of DOC occurred during passage through the aquifer with the high molecular weight and more acidic materials having been preferentially lost. The hydrophilic fraction of the DOC was suggested to have been preferentially sorbed by the mineral matrix, whereas the hydrophobic fraction is less strongly sorbed on the mineral matrix and is transported more readily through the aquifer (Skjemstad et al. 2002). The continued increase in DOC content observed after breakthrough in a monitoring well was suggested as indicating that a limited sorption capacity of the aquifer material had been gradually overcome.

Schoenheinz (2011) proposed that DOC can be a useful indicator parameter to evaluate groundwater flow and transport processes during aquifer passage. DOC behavior is considered to reflect the collective behavior of the dissolved organic compounds in water. To be useable as an indicator parameter, the DOC in the recharged water should be measurable (mg/L level) and occur at a greater concentration than in the native (non-impacted) groundwater.

It is assumed that DOC decreases by first order kinetics. Dividing DOC into four compound groups: (1) easily, (2) moderately, (3) poorly, (0) non-degradable (as a function of system time scale and flow path), total concentration (CT) is expressed as (Schoenheinz 2011)

$$C_T = C_1 \cdot \left(e^{-\lambda_1 t} - 1\right) + C_2 \cdot \left(e^{-\lambda_2 t} - 1\right) + C_3 \cdot \left(e^{-\lambda_3 t} - 1\right) + C_0$$

where

C_0 is the concentration of non-degradable compounds and C_n and λ_n are the initial concentrations and degradation rate constant (1/day) of compound group "n". The ratio of λ_1 for easily degradable compounds to λ_2 for moderately degradable compounds was found to be about 10:1. Similarly the ratio of λ_2 to λ_3 was also taken to be about an order of magnitude. Rate constant values can be obtained from laboratory experiments and monitoring well profiles (Schoenheinz 2011). The method provides a conceptual model for the changes in organic carbon concentrations in MAR flow systems and may be used as a tool to find necessary residence times and flow path lengths to achieve organic carbon degradation objectives (Schoenheinz 2011).

7.6 Metals

Metals removal during surface spreading was reviewed by Chang and Page (1985). Metals occur in water in the dissolved form (free or complexed ions) and in the particulate form. Where metals are present as finely divided suspended solids (or sorbed onto such particles), their removal occurs mainly through straining and filtration. Dissolved metals are removed by ion exchange reactions, precipitation, and surface adsorption.

When evaluating metals removal processes, it is important to determine whether metals are present in the dissolved or particulate form by sampling for both total and dissolved metals. Dissolved metals sampling involves first filtering the sample (typically using a 0.45 μm filter) before the addition of an acid preservative (usually HNO_3). Metals removal during MAR is dependent on oxidation-reduction (redox) potential as the solubility of some metal species varies greatly with redox state (Sect. 4.4). Changes in oxidation-reduction potential can substantially impact the solubility of some metals and impact sorption and desorption through the precipitation and dissolution of iron and manganese minerals that have high sorption capacities.

Trace elements are not subject to decomposition and are thus retained in the soil essentially permanently. Long-term, continuous high-rate application could, therefore, result in the accumulation of trace elements to the extent that it could render the application area unsuitable for some subsequent uses. The sorption capacity of soils with respect to trace elements is finite, but appears to be large enough so that MAR systems could operate for long periods of times (decades) before a soil's capacity to attenuate trace elements is exhausted.

References

Abel, C. D. T. (2014). *Soil aquifer treatment. Assessment and applicability of primary effluent reuse in developing countries*. Ph.D. Dissertation, Delft University of Technology, The Netherlands.

Alidina, M., Li, D., & Drewes, J. E. (2014a). Investigating the role for adaptation of the microbial community to transform trace organic chemicals during managed aquifer recharge. *Water Research, 56,* 172–180.

Alidina, M., Li, D., Ouf, M., & Drewes, J. E. (2014b). Role of primary substrate composition and concentration on attenuation of trace organic chemicals in managed aquifer recharge systems. *Journal of Environmental Management, 144,* 58–66.

Amy, G., & Drewes, J. (2007). Soil aquifer treatment (SAT) as a natural and sustainable wastewater reclamation/reuse technology: Fate of wastewater effluent organic matter (EfOM) and trace organic compounds. *Environmental Monitoring and Assessment, 129*(1), 19–26.

Asano, T., Burton, F. L., Leverenz, H. L., Tsuchihashi, R., & Tchobanoglous, G. (2007). *Wastewater reuse. Issues, technologies and applications*. New York: McGraw-Hill.

Asano, T., & Cotruvo, J. A. (2004). Groundwater recharge with reclaimed wastewater: Health and regulatory considerations. *Water Research, 38,* 1941–1951.

Bekele, E. B., Toze, S., Sidhu, J., & Hanna, J. (2006). Development of a microbial pathogen attenuation model to aid management of aquifer recharge systems. In *Recharge systems for protecting and enhancing groundwater resources*, Proceedings of the 5th International Symposium in Management of Aquifer Recharge, ISMAR5, Berlin, Germany, June 11–16, 2005 (pp. 337–341). Paris: UNESCO.

Benotti, M. J., & Snyder, S. A. (2009). Pharmaceuticals and endocrine disrupting compounds: Implications for ground water replenishment with recycled water. *Ground Water, 47*(4), 499–502.

Bernholtz, A., Brothers, K., & Katzer, T. (1994). Artificial ground-water recharge in Las Vegas, Clark County, Nevada: Storing today, treating tomorrow? In A. I. Johnson & R. D. G. Pyne (Eds.), *Artificial recharge of ground water, II: Proceedings of the Second International Symposium on Artificial Recharge of Ground Water* (pp. 548–557). Reston, VA: American Society of Civil Engineers.

Betancourt, W. Q., Kitajima, M., Wing, A. D., Regnery, J., Drewes, J. E., Pepper, I. L., et al. (2014). Assessment of viruses removal by managed aquifer recharge at three full-scale operations. *Journal of Environmental Sciences and Health, Part A: Toxic/Hazardous Substances and Environmental Engineering, 49*(14), 1685–1692.

Bos, R., Carr, R., & Keraita, B. (2010). Assessing and mitigating wastewater-related health risks in low-income countries. In P. Dreschel, C. A. Scott, L. Raschid-Sally, M. Redwood & A. Bahri (Eds.), *Wastewater irrigation and health. Assessing and mitigating risk in low income countries* (pp. 29–47). Ottawa: International Development Research Centre.

Bouwer, E. J., & Wright, J. P. (1988). Transformations of trace aliphatics in anoxic biofilm columns. *Journal of Contaminant Hydrology, 2,* 155–169.

Candela, L., Tamoh, K., Vadillo, I., & Valdes-Abellan, J. (2016). Monitoring of selected pharmaceuticals over 3 years in a detrital aquifer during artificial groundwater recharge. *Environmental Earth Sciences, 75*(3), 1–13.

Chang, A. C., & Page, A. L. (1985). Soil deposition of trace metals during groundwater recharge using surface spreading. In T. Asano (Ed.), *Artificial recharge of groundwater* (pp. 609–626). Stoneham MA: Butterworth.

Clara, M., Strenn, B., & Kreuzinger, N. (2004). Carbamazepine as a possible marker in aquatic environments: Investigations on the behaviour of carbamazepine in wastewater treatment and during groundwater infiltration. *Water Research, 38,* 947–954.

Cordy, G. E., Duran, N. L., Bouwer, H., Rice, R. C., Furlong, E. T., Zaugg, S. D., et al. (2004). Do pharmaceuticals, pathogens, and other organic waste water compounds persist when waste water is used for recharge? *Groundwater Monitoring & Remediation, 24*(2), 58–69.

Daughton, C. G. (2009). Chemicals from the practice of healthcare: Challenges and unknowns posed by residues in the environment. *Environmental Toxicology and Chemistry, 28,* 2490–2494.

Daughton, C., & Ternes, T. (1999). Pharmaceuticals and personal care products in the environment: Agents to subtle change? *Environmental Health Perspectives, 107*(supplement 6), 907–938.

Dillon, P., Page, D., Vanderzalm, J., Pavelic, P., Toze, S., Bekele, E., et al. (2008). A critical evaluation of combined engineered and aquifer treatment systems in water recycling. *Water Science and Technology, 57*(5), 753–762.

Dillon, P., Toze, S., Pavelic, P., Vanderzalm, J., Barry, K., Ying, G.-L., et al. (2006). Water quality improvements during aquifer storage and recovery at ten sites. In *Recharge systems for protecting and enhancing groundwater resources*, Proceedings of the 5th International Symposium on Management of Aquifer Recharge, Berlin, Germany, June 11–16, 2005 (pp. 85–94). Paris: UNESCO.

Drewes, J. E. (2009). Ground water replenishment with recycled water—Water quality improvements during managed aquifer recharge. *Ground Water, 47*(4), 502–505.

Drewes, J. E., & Fox, P. (1999). Fate of natural organic matter (NOM) during groundwater recharge using reclaimed water. *Water Science and Technology, 40*(9), 241–248.

Drewes, J. E., Heberer, T., Rauch, T., & Reddersen, K. (2003). Fate of pharmaceuticals during groundwater recharge. *Ground Water Monitoring and Remediation, 23,* 64–72.

Drewes, J. E., Heberer, T., & Redderson, K. (2002). Fate of pharmaceuticals during indirect potable reuse. *Water Science and Technology, 46*(3), 73–80.

Fox, P., Aboshanp, W., & Alsamadi, B. (2005). Analysis of soils to demonstrate sustained organic carbon removal during soil aquifer treatment. *Journal of Environmental Quality, 34*(1), 156–163.

Fox, P., Houston, S., & Westerhoff, P. (2001). *Soil aquifer treatment for sustainable water supply.* Denver: American Water Works Association Research Foundation.

Fram, M. S., Bergamaschi, B., Goodwin, K. D., Fujii, R., & Clark, J. F. (2003). *Process affecting the trihalomethane concentrations associated with the third injection, storage, and recovery test at Lancaster, Antelope Valley, California, March 1998 through April 1999.* U.S. Geological Survey Water-Resources Investigations Report 03-4062.

Gerba, C. P., & Goyal, S. M. (1985). Pathogen removal from wastewater during groundwater recharge. In T. Asano (Ed.), *Artificial recharge of groundwater* (pp. 283–328). Stoneham, MA: Butterworth.

Gordon, C., & Toze, S. (2003). Influence of groundwater characteristics on the survival of enteric viruses. *Journal of Applied Microbiology, 95,* 536–544.

Gordon, C., Wall, K., Toze, S., & O'Hara, G. (2002). Influence of conditions on the survival of enteric viruses and indicator organisms in groundwater. In P. J. Dillon (Ed.), *Management of aquifer recharge for sustainability* (pp. 133–138). Lisse: A.A. Balkema.

Greskowiak, J., Prommer, H., Massmann, G., & Nützmann, G. (2006). Modeling seasonal redox dynamics and the corresponding fate of the pharmaceutical residue phenazone during artificial recharge of groundwater. *Environmental Science and Technology, 40*(21), 6615–6621.

Gupta, M., Gupta, A., Suidan, M. T., & Sayles, G. D. (1996a). Biotransformation rates of chloroform under aerobic conditions-II, sulfate reduction. *Water Research, 30,* 1387–1394.

Gupta, V., Johnson, W. P., Shafieian, P., Ryu, H., Alum, A., Abbaszadegan, M., et al. (2009). Riverbank filtration: Comparison of pilot scale transport with theory. *Environmental Science and Technology, 43*(3), 669–676.

Gupta, M., Sharma, D., Suidan, M. T., & Sayles, G. D. (1996b). Biotransformation rates of chloroform under aerobic conditions-I. Methanogenesis. *Water Research, 30,* 1377–1385.

Harvey, R. W., & Garabedian, S. P. (1991). Use of colloid filtration theory in modeling movement of bacteria through a contaminated sandy aquifer. *Environmental Science and Technology, 25*(1), 178–185.

Harvey, R. W., Metge, D. W., Shapiro, A. M., Renken, R. A., Osborn, C. L., Ryan, J. N., et al. (2008). Pathogen and chemical transport in the karst limestone of the Biscayne aquifer: 3. Use of microspheres to estimate the transport potential of Cryptosporidium parvum oocysts. *Water Resources Research, 44*(8), W08431.

Harvey, R., Metge, D., Sheets, R., & Jasperse, J. (2011). Fluorescent microspheres as surrogates in evaluating the efficacy of riverbank filtration for removing *Cryptosporidium parvum* oocysts and other pathogens. In C. Ray & M. Shamrukh (Eds.), *Riverbank filtration for water security in desert countries* (pp. 81–96). Amsterdam: Springer.

Heberer, T. (2002). Tracking persistent pharmaceutical residues from municipal sewage to drinking water. *Journal of Hydrology, 266*(3–4), 175–189.

Heberer, T. (2006). Occurrence, transport, attenuation and removal of pharmaceutical residues in the aquatic environment and their relevance to drinking water supply in urban areas. In *Recharge systems for protecting and enhancing groundwater resources*, Proceedings of the 5th International Symposium on Management of Aquifer Recharge, Berlin, Germany, June 11–16, 2005 (pp. 529–534). Paris: UNESCO.

Heberer, T., Mechlinski, A., Franck, B., Knappe, A., Massmann, G., Pekdeger, A., et al. (2004). Field studies on the fate and transport of pharmaceutical residues in bank filtration. *Ground Water Monitoring and Remediation, 24,* 70–77.

Hlavsa, M. C., Watson, J. C., & Beach, M. J. (2005). Cryptosporidiosis surveillance—United States 1999–2002. *Morbidity and Mortality Weekly Report (MMWR) Surveillance Summaries, 54*(1), 1–8.

Hoppe-Jones, C., Oldham, G., & Drewes, J. E. (2010). Attenuation of total organic carbon and unregulated trace organic chemicals in U.S. riverbank filtration systems. *Water Research, 44,* 4643–4659.

John, D. E., & Rose, J. B. (2005). Review of factors affecting microbial survival in groundwater. *Environmental Science and Technology, 39,* 7345–7356.

John, D. E., Rose, J. B., & Karmarainen, A. (2004). *Survival of fecal indicator bacteria, bacteriophage and protozoa in Florida's surface and ground water*, Final Report of the Fate of Microorganisms in Aquifer Study. Brooksville: Southwest Florida Water District, and West Palm Beach: South Florida Water Management District.

Jørgensen, C. (2001). Migration and survival of bacteria during artificial recharge. In *Artificial recharge of groundwater*, Project ENV4-CT95-0071 Final Report (pp. 231–234). Luxembourg: European Commission.

Jørgensen, C., & Peters, J. (2001). Migration and survival of viruses. In *Artificial recharge of groundwater*, Project ENV4-CT95-0071 Final Report (pp. 142–146). Luxembourg: European Commission.

Khan, S. J., & Rorije, E. (2002). Pharmaceutically active compounds in aquifer storage and recovery. In P. J. Dillon (Ed.), *Management of aquifer recharge for sustainability* (pp. 169–174). Lisse: A. A. Balkema.

Landmeyer, J. E., Bradley, P. N., & Thomas, J. M. (2000). Biodegradation of disinfection byproducts as a potential removal process during aquifer storage recovery. *Journal of the American Water Resources Association, 26,* 861–867.

Laws, B. V., Dickenson, E. R. V., Johnson, T. A., Snyder, S. A., & Drewes, J. E. (2011). Attenuation of contaminants of emerging concern during surface spreading aquifer recharge. *Science of the Total Environment, 409,* 1087–1094.

Li, D., Alidina, M., & Drewes, J. E. (2014). Role of primary substrate composition on microbial community structure and function and trace organic chemical attenuation in managed aquifer recharge systems. *Applied Microbiology and Biotechnology, 98*(12), 5747–5756.

Li, D., Alidina, M., Ouf, M., Sharp, J. O., Saikaly, P., & Drewes, J. E. (2013). Microbial community evolution during simulated managed aquifer recharge in response to different biodegradable dissolved organic carbon (BDOC) concentrations. *Water Research, 47*(7), 2421–2430.

Lisle, J. T. (2016). Natural inactivation of Escherichia coli in anaerobic and reduced groundwater. *Journal of Applied Microbiology, 120,* 1739–1750.

Macler, B., & Merkle, J. C. (2002). Current knowledge of groundwater microbial pathogens and their control. *Hydrogeology Journal, 8,* 29–40.

Maeng, S. K., Sharma, S. K., & Amy, G. (2011a). A framework for assessment of organic micropol lutant removals during managed aquifer recharge and recovery. In C. Ray & M. Shamrukh (Eds.), *Riverbank filtration for water security in desert countries* (pp. 137–149). Dordrecht: Springer.

Maeng, S. K., Sharma, S. K., Lekkerkerker-Teunissen, K., & Amy, G. L. (2011b). Occurrence and fate of bulk organic matter and pharmaceutically active compounds in managed aquifer recharge: A review. *Water Research, 45*(10), 3015–3033.

Maliva, R. G., & Missimer, T. M. (2010). *Aquifer storage and recovery and managed aquifer recharge using wells: Planning, hydrogeology, design, and operation.* Houston: Schlumberger Corporation.

Maliva, R. G., & Missimer, T. M. (2012). *Arid lands water evaluation and management.* Berlin: Springer.

Mansell, J., & Drewes, J. E. (2004). Fate of steroidal hormones during soil-aquifer treatment. *Groundwater Monitoring & Remediation, 24*(2), 94–101.

Mansell, J., Drewes, J. E., & Rauch, T. (2004). Removal mechanisms of endocrine disrupting compounds (steroids) during soil aquifer treatment. *Water Science and Technology, 50*(2), 229–237.

Massmann, G., Dünnbier, U., Heberer, T., & Taute, T. (2008). Behaviour and redox sensitivity of pharmaceutical residues during bank filtration—Investigation of residues of phenazone-type analgesics. *Chemosphere, 71*(8), 1476–1485.

Massmann, G., Greskowiak, J., Dünnbier, U., Zuehlke, S., Knappe, A., & Pekdeger, A. (2006). The impact of variable temperatures on the redox conditions and the behaviour of pharmaceutical residues during artificial recharge. *Journal of Hydrology, 328*(1), 141–156.

McDowell-Boyer, L. M., Hunt, J. R., & Sitar, N. (1986). Particle transport through porous media. *Water Resources Research, 22,* 1901–1921.

Miller, C. J., Wilson, L. G., Amy, G. L., & Brothers, K. (1993). Fate of organochlorine compounds during aquifer storage and recovery: The Las Vegas experience. *Ground Water, 31,* 410–416.

Mohanram, A., Ray, C., Harvey, R. W., Metge, D. W., Ryan, J. N., Chorover, J., et al. (2010). Comparison of transport and attachment behaviors of *Cryptosporidium parvum* oocysts and oocyst-sized microspheres being advected through three mineralogically different granular porous media. *Water Research, 44,* 5334–5344.

Montgomery-Brown, J., Drewes, J. E., Fox, P., & Reinhard, M. (2003). Behavior of alkylphenol polyethoxylate metabolites during soil aquifer treatment. *Water Research, 37*(15), 3672–3681.

National Research Council. (1998). *Issues in potable reuse: The viability of augmenting drinking water supplies with reclaimed water.* Washington, DC: National Academy Press.

National Research Council. (2008). *Prospects for managed underground storage of recoverable water.* Washington, DC: National Academy Press.

National Research Council. (2012). *Potential for expanding the nation's water supply through the reuse of municipal wastewater.* Washington, DC: National Academy Press.

Nicholson, B. C., Dillion, P. J., & Pavelic, P. (2002). Fate of disinfection by-products during aquifer storage and recovery. In P. J. Dillon (Ed.), *Management of aquifer recharge for sustainability* (pp. 155–160). Lisse: A. A. Balkema.

Nicholson, B., & Ying, G.-G. (2005). Attenuation of disinfection by-products during aquifer storage and recovery In P. Dillon & S. Toze (Eds.), *Water quality improvements during aquifer storage and recovery. Volume 1: Water quality improvement processes*, Report 91056F (pp. 39–64). Denver: AWWA Research Foundation.

NRMMC-EPHC-AHMC. (2006). *National guideline for water recycling: Managing health and environmental risks, Phase 1*. Natural Resource Management Ministerial Council, Environment Protection and Heritage Council, Australian Health Ministers Conference.

NRMMC-EPHC–NHMRC. (2009). *Australian guidelines for water recycling: Managing health and environmental risks. Phase 2: Managed aquifer recharge* (July 2009). Natural Resource Management Ministerial Council, Environment Protection and Heritage Council and National Health and Medical Research Council.

Ongerth, J. E., & Khan, S. (2004). Drug residuals: How xenobiotics can affect water supply sources. *Journal American Water Works Association, 96*(5), 94–101.

Page, D., Dillon, P., Toze, S., Bixio, D., Genthe, B., Jiménez Cisneros, B. E., et al. (2010). Valuing the subsurface pathogen treatment barriers in water recycling via aquifers for drinking supplies. *Water Research, 44,* 1841–1852.

Page, D., Gonzalez, D., Torkzaban, S., Toze, S., Sidhu, J., Miotliński, K., et al. (2015). Microbiological risks of recycling urban stormwater via aquifers for various uses in Adelaide. *Australia. Environmental Earth Sciences, 73*(12), 7733–7737.

Pang, L., Nowostawska, U., Ryan, J. N., Williamson, W. M., Walshe, G., & Hunter, K. A. (2009). Modifying the surface charge of pathogen-sized microspheres for studying pathogen transport in groundwater. *Journal of Environmental Quality, 38,* 2210–2217.

Pavelic, P., Dillon, P., & Nicholson, B. (2005a). Analysis of the fate of DBPs at eight ASR sites. In P. Dillon & S. Toze (Eds.), *Water quality improvements during aquifer storage and recovery. Volume 1: Water quality improvement processes*, Report 91056F (pp. 39–64). Denver: AWWA Research Foundation.

Pavelic, P., Dillon, P. J., & Nicholson, B. C. (2006). Comparative evaluation of the fate of disinfection byproducts at eight aquifer storage and recovery sites. *Environmental Science and Technology, 40,* 501–508.

Pavelic, P., Nicholson, B. C., Dillon, P. J., & Barry, K. E. (2005b). Fate of disinfection by-products in groundwater aquifer storage and recovery with reclaimed water. *Journal of Contaminant Hydrology, 77,* 351–373.

Pavelic, P., Ragusa, S. R., Flower, R. L., Rinck-Pfeiffer, S. M., & Dillon, P. J. (1998). Diffusion chamber method for in situ measurement of pathogen inactivation in groundwater. *Water Research, 32*(4), 1144–1150.

Pyne, R. D. G., Singer, P. C., & Miller, C. T. (1996). *Aquifer storage recovery of treated drinking water*. Denver: American Water Works Association Research Foundation.

Quanrud, D. (2003). Estrogenic activity and volume of fraction of wastewater origin in monitoring wells along the Santa Cruz River, Arizona. In *Proceedings 11th Biennial Symposium on Groundwater Recharge*, June 5–7, 2003, Phoenix, AZ.

Quanrud, D. M., Hafer, J., Karpiscak, M. M., Zhang, J., Lansey, K. E., & Arnold, R. G. (2003). Fate of organics during soil-aquifer treatment: Sustainability of removals in the field. *Water Research, 37*(14), 3401–3411.

Quanrud, D. M., Quast, K., Conroy, O., Karpiscak, M. M., Gerba, C. P., Lansey, K. E., et al. (2004). Estrogenic activity and volume fraction of waste water origin in monitoring wells along the Santa Cruz River. *Arizona. Groundwater Monitoring & Remediation, 24*(2), 86–93.

Rauch, T., Munoz, J., Drewes, J. E., Amy, G., & Choi, H. (2006). Fate of pharmaceuticals during soil infiltration loading to artificial groundwater recharge. In *Recharge systems for protecting and enhancing groundwater resources,* Proceedings of the 5th International Symposium in Management of Aquifer Recharge, ISMAR5, Berlin, Germany, June 11–16, 2005 (pp. 562–567). Paris: UNESCO.

Rauch-Williams, T., Hoppe-Jones, C., & Drewes, J. E. (2010). The role of organic matter in the removal of emerging trace organic chemicals during managed aquifer recharge. *Water Research, 44*(2), 449–460.

Rice, E. W., Baird, R. B., Easton, A. D., & Clesceri, L. S. (Eds.). (2012). *Standard methods for examination of water and wastewater* (22nd ed.). Washington, DC: American Public Health Association Publications.

Richardson, S. D. (2006). Environmental mass spectrometry: Emerging contaminants and current issues. *Analytical Chemistry, 78,* 4021–4046.

Sattler, A., May, G., & Drewes, J. (2006). Fate of wastewater effluent organic matter (EfOM) through soil aquifer treatment (SAT). In *Recharge systems for protecting and enhancing groundwater resources*, Proceedings of the 5th International Symposium in Management of Aquifer Recharge, ISMAR5, Berlin, Germany, June 11–16, 2005 (pp. 573–582). Paris: UNESCO.

Scheytt, T., Mersmann, P., Leidig, M., Pekdeger, A., & Heberer, T. (2004). Transport of pharmaceutically active compounds in saturated laboratory columns. *Groundwater, 42*(5), 767–773.

Schijven, J., Berger, P., & Miettinen, I. (2002). Removal of pathogens, surrogates, indicators, and toxins using riverbed filtration. In C. Ray, G. Melin, & R. B. Linksy (Eds.), *Riverbank filtration. Improving source-water quality* (pp. 73–116). Dordrecht: Kluwer Academic Press.

Schijven, J. F., & Hassanizadeh, S. M. (2000). Removal of viruses by soil passage: Overview of modeling, processes, and parameters. *Critical Reviews in Environmental Science and Technology, 30*(1), 49–127.

Schijven, J. F., Hoogenboezem, W., Hassanizadeh, S. M., & Peters, J. H. (1999). Modeling removal of bacteriophages MS2 and PRD1 by dune recharge at Castricum, Netherlands. *Water Resources Research, 35,* 1011–1111.

Schijven, J. F., Medema, G., Vogelaar, A. J., & Hassanizadeh, S. M. (2000). Removal of microorganisms by deep well injection. *Journal of Contaminant Hydrology, 44*(3), 301–327.

Schoenheinz, D. (2011). Dissolved organic carbon as an indicator parameter for groundwater flow and transport. In C. Ray & M. Shamrukh (Eds.), *Riverbank filtration for water security in desert countries* (pp. 151–168). Dordrecht: Springer.

Sedlak, D. L., Gray, J. L., & Pinkston, K. E. (2000). Understanding microcontaminants in recycled water. *Environmental Science and Technology, 35,* 508A–515A.

Sidhu, J., Hanna, J., & Toze, S. (2006). Influence of groundwater redox conditions on decay of enteric viruses and *Cryptosporidium.* In *Recharge systems for protecting and enhancing groundwater resources,* Proceedings of the 5th International Symposium in Management of Aquifer Recharge, ISMAR5, Berlin, Germany, June 11–16, 2005 (pp. 511–517). Paris: UNESCO.

Sidhu, J. P. S., & Toze, S. (2012). Assessment of pathogen survival potential during managed aquifer recharge with diffusion chambers. *Journal of Applied Microbiology, 113*(3), 693–700.

Sidhu, J. P. S., Toze, S., Hodgers, L., Barry, K., Page, D., Li, Y., et al. (2015). Pathogen decay during managed aquifer recharge at four sites with different geochemical characteristics and recharge water sources. *Journal of Environmental Quality, 44,* 1402–1412.

Sidhu, J. P. S., Toze, S., Hodgers, L., Shackelton, M., Barry, K., Page, D., et al. (2010). Pathogen inactivation during passage of stormwater through a constructed reedbed and aquifer transfer, storage and recovery. *Water Science and Technology, 62*(5), 1190.

Singer, P. C., Pyne, R. D. G., Mallikarjum, A. V. S., Miller, C. T., & Majonnier, C. (1993). Examining the impact of aquifer storage and recovery on DBPs. *Journal American Water Works Association, 85*(11), 85–94.

Skjemstad, J. O., Hayes, M. H. B., & Swift, R. S. (2002). Changes in natural organic matter during aquifer storage. In P. J. Dillon (Ed.), *Management of aquifer recharge for sustainability* (pp. 149–154). Lisse: A.A. Balkema.

Skjemstad, J., Swift, R., & Hayes, M. (2005). Evaluation of changes in NOM during ASR at the Bolivar site. In P. Dillon & S. Toze (Eds.), *Water quality improvements during aquifer storage and recovery. Volume 1: Water quality improvement processes*, Report 91056F (pp. 99–110). Denver: AWWA Research Foundation.

Snyder, S. A., Leising, J., Westerhoff, P., Yoon, Y., Mash, H., & Vanderford, B. (2004). Biological and physical attenuation of endocrine disruptors and pharmaceuticals: Implications for water reuse. *Groundwater Monitoring & Remediation*, *24*(2), 108–118.

Snyder, S. A. (2014). Emerging chemical contaminants: Look for greater harmony. *Journal American Water Works Association, 106*(5), 38–52.

Stanford, B. D., Snyder, S. A., Trenholm, R. A., Holandy, J. C., & Vaderford, B. J. (2010). Estrogenic activity of US drinking waters: A relative exposure comparison. *Journal American Water Works Association, 120*(11), 55–65.

Stuyfzand, P. J., Segers, W., & van Rooijen, N. (2007). Behavior of pharmaceuticals and other emerging pollutants in various artificial recharge systems in the Netherlands. In P. Fox (Ed.), *Management of aquifer recharge for sustainability*, Proceedings of the 6th International Symposium on Managed Aquifer Recharge of Groundwater (pp. 231–245). Phoenix: Acacia Publishing.

Ternes, T. A., & Joss, A. (2006). *Human pharmaceuticals, hormones and fragrances—The challenge of micropollutants in urban water management*. London: IWA Publishing.

Thomas, J. M., McKay, W. A., Cole, E., Landmeyer, J. E., & Bradley, P. M. (2000). The fate of haloacetic acids and trihalomethanes in an aquifer storage and recovery program, Las Vegas, Nevada. *Ground Water, 38,* 605–614.

Toze, S. (2005). Pathogen survival in groundwater: A review of the literature. In P. Dillon & S. Toze (Eds.), *Water quality improvements during aquifer storage and recovery. Volume 1: Water quality improvement processes*, Report 91056F (pp. 123–140). Denver: AWWA Research Foundation.

Toze, S. (2006). Water reuse and health risks—Real vs perceived. *Desalination, 187,* 41–51.

Toze, S., Bekele, E., Page, D., Sidhu, J., & Schackleton, M. (2010). Use of static quantitative microbial risk assessment to determine pathogen risks in an unconfined aquifer used for managed aquifer recharge. *Water Research, 44,* 1038–1049.

Toze, S., & Hanna, J. (2002). The survival potential of enteric microbial pathogens in a reclaimed water ASR project. In P. J. Dillon (Ed.), *Management of aquifer recharge for sustainability* (pp. 139–146). Lisse: A.A. Balkema.

USEPA. (2010). *Long term 2 enhanced surface water treatment rule toolbox guidance manual.* Washington, D.C.: U.S. Environmental Protection Agency Office of Water.

Verstraeten, I. M., Heberer, T., & Scheytt, T. (2002). Occurrence, characteristics, and transport and fate of pesticides, pharmaceutically active compounds, and industrial and personal care products at bank filtration sites. In C. Ray, G. Melin, & R. B. Linksy (Eds.), *Riverbank filtration. Improving source-water quality* (pp. 175–227). Dordrecht: Kluwer Academic Press.

Wall, K., Toze, S., & O'Hara, G. (2006). Interactions of indigenous bacteria with enteric viruses during water quality improvement by aquifer storage and recovery (ASR). In *Recharge systems for protecting and enhancing groundwater resources*, Proceedings of the 5th International Symposium in Management of Aquifer Recharge, ISMAR5, Berlin, Germany, June 11–16, 2005 (pp. 518–524). Paris: UNESCO.

Wall, K., Toze, S., & O'Hara, G. (2007). Indigenous groundwater microorganism processes that influence the decay of enteric viruses. In P. Fox (Ed.), *Management of aquifer recharge for sustainability*, Proceedings of the 6th International Symposium on Managed Aquifer Recharge of Groundwater (pp. 246–256). Phoenix: Acacia Publishing.

World Health Organization. (2004). *Trihalomethanes in drinking water summary statement*, Document WHO/SDE/WSH/04.08/64s (September 2004). Geneva: World Health Organization.

World Health Organization. (2006). *WHO guidelines for the safe use of wastewater, excreta, and greywater, Volume 1, Policy and regulatory aspects*. Geneva: World Health Organization.

Ying, G.-G., Kookana, R. S., & Dillon, P. (2003). Sorption and degradation of selected endocrine disrupting chemicals in aquifer material. *Water Research, 37,* 3785–3791.

Zuehlke, S., Heberer, T., Duennbier, U., & Fritz, B. (2004). Analysis of endocrine disrupting steroids Investigation of their release into the environment and their behavior during bank filtration. *Ground Water Monitoring and Remediation, 24*(2), 78–85.

Chapter 8
MAR Project Implementation and Regulatory Issues

8.1 Project Plan

The proverb "he who falls to plan is planning to fail" (Winston Churchill, paraphrased from Benjamin Franklin) certainly applies to managed aquifer recharge (MAR) projects. Planning was defined by the Business Dictionary (n.d.) as

> A basic management function involving formulation of one or more detailed plans to achieve optimum balance of needs or demands with the available resources. The planning process (1) identifies the goals or objectives to be achieved, (2) formulates strategies to achieve them, (3) arranges or creates the means required, and (4) implements, directs, and monitors all steps in their proper sequence.

Maliva and Missimer (2010) noted with respect to aquifer storage and recovery (ASR) systems that

> It is not an over statement that the single most important process for the successful implementation of an ASR project is planning. Proper planning of ASR projects increases the probability of success and can reduce project costs by potentially eliminating unexpected surprises during system construction, testing, and operation. Planning of ASR projects involves giving due consideration to the various factors that can impact the cost and success of an ASR project and proactively developing strategies for dealing with them. The essence of proper project planning is being in control of the project rather than leaving things to chance.

The importance of planning applies to MAR projects in general. The planning process also includes assessment of project risks and developing strategies to mitigate identified risks. Risk is broadly defined herein as the probability of an adverse outcome, which for an MAR project would be the system not meeting expectations. Depending upon the system type, adverse outcomes include:

- target recharge rate could not be achieved due to excessive clogging
- poor recovery of water stored in an ASR system
- unacceptable deterioration of stored water quality
- failure to achieve water quality improvement (natural contaminant attenuation) goals

R. G. Maliva, *Anthropogenic Aquifer Recharge*,
Springer Hydrogeology, https://doi.org/10.1007/978-3-030-11084-0_8

- greater than expected construction and operational costs
- adverse environmental impacts
- inability to obtain all required regulatory approvals for a system.

Various aspects of the planning processes involved in ASR and MAR projects have been addressed by the American Society of Civil Engineers (2001), Brown (2005), Pyne (2005), Dillon and Molloy (2006), National Research Council (2008), NRMMC-EPHC–NHMRC (2009) and Maliva and Missimer (2010). The basic MAR planning and implementation process starts with the identification of project goals (success criteria). A multi-phase feasibility investigation is then performed to determine whether MAR is technically and economically feasible in the proposed project area. Feasibility assessments involve both a review of existing data (i.e., a desktop investigation) and field testing. If an MAR project is determined to be feasible, then the system is designed and required regulatory approvals obtained. The final project phases are the construction and operational testing of a pilot or full-scale system. Project planning should incorporate flexibility in the project design, construction, and operation to allow for adaptive management ("learning by doing").

8.2 Project Success Criteria

Based on a review of global ASR implementation, Maliva and Missimer (2010) observed that there has been a great reluctance to publically categorize any ASR system as being "failed" regardless of obvious poor operational results. This is attributed to basic human nature and a desire not to apply a negative light on the technology. As an extreme example, the author attended a technical presentation in which an MAR system that clearly failed to even approach project goals was described as not being a failure, but rather that the system objectives just needed to be redefined. The quote by an unknown author that "success has many fathers, but failure is an orphan" is applicable to historic ASR implementation (Maliva and Missimer 2010). While the concern that negative news from a small number of failed projects could be disingenuously exploited by system or technology opponents is valid, the downside is that valuable lessons may be lost that could provide guidance to improve future implementation.

Project success criteria should be realistic and projects not oversold. For example, it has been stated that successful ASR systems should approach a 100% recovery efficiency (i.e., nearly all of the recharged water is recovered). However, it has been the experience in Florida (and elsewhere) that recovery efficiencies of 70–80% is a more realistic goal for systems that store freshwater for potable use in brackish aquifers. A system operator in Southwest Florida once commented to the author that their ASR system was a real lifesaver as far as meeting water demands during a previous dry season, but he wanted to know why the system was not working well because its recovery efficiency was only about 70%. The answer is that their system was indeed working very well in that it cost effectively provided the utility with the design capacity flow when needed. A 100% recovery efficiency goal is simply not achievable based on system hydrogeology.

Meaningful evaluation of MAR system performance requires that objective criteria for project success be established at the start of a project. The success criteria should be quantitative and objective so that there is no ambiguity as to whether they are being achieved (Maliva and Missimer 2010) and memorialized in writing. Memorialization is important to avoid "moving the goalposts" whereby system expectations and success criteria are readjusted downward so that an under-performing system is presented as meeting expectations.

It is important that all parties involved in an MAR project (e.g., owner, operator, consultants, and other stakeholders) come to an agreement as to specific project goals and success criteria. Success criteria are minimum performance standards. Hypothetical examples of project success criteria are:

- An ASR system shall be capable of providing each year 1.5 million gallons per day (Mgd; 5680 m^3/d) of water on demand that meets primary drinking water standards for 100 consecutive days during the seasonal dry or high-demand period.
- A riverbank filtration system shall be capable of continuously providing 5.0 Mgd (18,930 m^3/d) of raw water with a minimum 2-log removal of all pathogens.
- A rapid infiltration basin system shall achieve an annual hydraulic loading rate of 50 m/yr (164 ft/yr) without causing adverse impacts, such as waterlogging of surroundings areas.

8.3 MAR Feasibility Assessment

The word feasible is commonly defined broadly as "capable of being done." MAR feasibility assessments thus evaluate whether a proposed project can be implemented that meets project success criteria. Feasibility assessments necessarily involve economic considerations inasmuch as there is little that cannot be done with unlimited resources. Hence the germane question is whether MAR is capable of being performed at a cost commensurate with project benefits or competitive with other water supply, storage, and treatment options.

A related issue is site selection, which addresses the preferred location to construct a proposed MAR system. Some sites can have a combination of features or attributes that make them more favorable for MAR than other sites within a general project area.

Feasibility issues for MAR projects can be subdivided into hydrogeological, infrastructure and logistical, regulatory, and economic factors. Hydrogeological factors, such as vadose zone and aquifer hydraulic properties and geochemistry, determine whether an MAR system can be successfully operated at a site. Infrastructure and logistical factors include issues related to the integration of MAR systems into existing water or wastewater treatment and transmission systems, and the ability to physically construct the system. Regulatory factors determine whether authorization can be obtained to construct and operate a system and the operational and monitoring requirements mandated by regulatory agencies. Economic factors address the costs to construct and operate an MAR system. Infrastructure, logistical and regulatory factors tie into economics because they can be important cost items. For example, regulatory-driven pretreatment and monitoring requirements can be major cost items that impact the economic viability of some MAR projects. Similarly, land availability (and cost if land has to be purchased) and proximity to existing water and wastewater treatment and transmission infrastructure (and thus pipe line and pumping costs) can have large impacts on the economics of MAR projects.

MAR feasibility is not a simple "yes/no" or "pass/fail" question. Each feasibility factor must be evaluated as to the degree to which local conditions are favorable or unfavorable for successful project implementation. Site conditions for a given feasibility factor may be characterized as being optimal, acceptable, poor, or a fatal flaw:

- **Optimal**: Site conditions are within the range of values most favorable for implementation of MAR. Infrastructure is readily available on site (e.g., system is to be located on a water treatment plant site). Hydrogeological conditions are highly favorable for achieving system capacity, recovery efficiency, and water quality targets. No significant regulatory obstacles will be encountered.
- **Acceptable**: Site conditions are favorable for meeting minimum performance requirements but are less than ideal. For example, needed infrastructure is not present at a system site but could be provided at a cost that is in line with the anticipated project budget and system benefits. Less than optimal aquifer or vadose zone hydraulic conductivity values may necessitate additional recharge wells or infiltration basin area (with associated greater costs) to achieve system capacity targets.
- **Poor**: Site conditions are outside of the range of values normally considered acceptable for an MAR system, but a project may still be viable. Major infrastructure costs can substantially impact the cost-benefit ratio of a system. However, cost advantages of MAR compared to other potential water supply, storage, or treatment option makes MAR still worth pursuing.
- **Fatal flaw**: One or more site conditions make ASR either technically or economically unviable. A fatal flaw is a condition that is so unfavorable as to render an MAR project infeasible at a site, irrespective of all other factors, which could be highly favorable.

Many fatal flaws are ultimately economic issues in that they raise the cost of an MAR system to the point where costs exceed benefits (Maliva and Missimer 2010). Some examples of fatal flaws are:

- Inadequate land is available at an acceptable cost to construct a planned MAR system.
- Regulatory water quality standards for recharged water would necessitate such a high-level of pretreatment that it would render a project economically unviable.
- Storage zone transmissivity for an ASR system is inadequate for achieving minimum target injection and recovery rates.

- Soil contamination exists at a proposed surface-spreading site, which would likely be mobilized and contaminant local groundwater.
- Local hydrogeological conditions are present (e.g., well-developed karst conduits system or high salinities) in an ASR storage zone that would result in unacceptably poor recovery efficiencies.

Failure to recognize a fatal flaw will result in considerable time, effort, and money spent on a project that is not viable. Hence, a fatal flow analysis should be performed early in the development of a project (Maliva and Missimer 2010). A fatal flow analysis consists of the identification of possible fatal flaws for a project and then evaluating the potential for such conditions to exist at a site.

A site proposed for an MAR system will seldom have optimal conditions for all feasibility factors and tradeoffs are often necessary. The germane issue is whether the combination of site conditions would allow a system to be constructed and operated that achieves project goals (success criteria) at an acceptable cost. The site selection process involves evaluating prospective system sites to identify the location that has the most favorable combination of feasibility factors. Numerical scoring systems are often used to compare sites (Sect. 8.5).

8.4 MAR Feasibility Factors

8.4.1 Water Needs and Sources

Perhaps the most fundamental feasibility requirement for MAR systems is the availability of a suitable supply of water to meet project needs. The potential usefulness of an ASR system, for example, depends upon the amount and duration of excess demand that cannot be met with current water supplies and the amount of excess water that can be captured and stored. Evaluation of future water supplies and demands requires consideration of historic water supply and demand trends and extrapolation of these trends into the future. If sufficient excess water will not be available that can be captured and recharged, then MAR will not be viable (or may be able to meet only part of future demands).

MAR systems in water scarce regions often experience a recharge bottleneck (Fig. 8.1) in which large volumes of water may be available for short time periods (e.g., after major rainfall events) and there is ample underground storage space, but aquifer recharge is limited by a relatively slow rate of recharge. System injection rates are the product of well capacities and number of wells. Recharge rates in surface-spreading facilities are the product of average infiltration rate and infiltration area. To increase total recharge, either temporary storage is needed to increase the duration of recharge, or the number of wells or the infiltration area need to be increased.

While it may be superficially attractive to construct an ASR or other MAR system with a capacity adequate to ensure that all water-supply demands can be met under all circumstances, this is usually not a cost-effective strategy (Maliva and Missimer

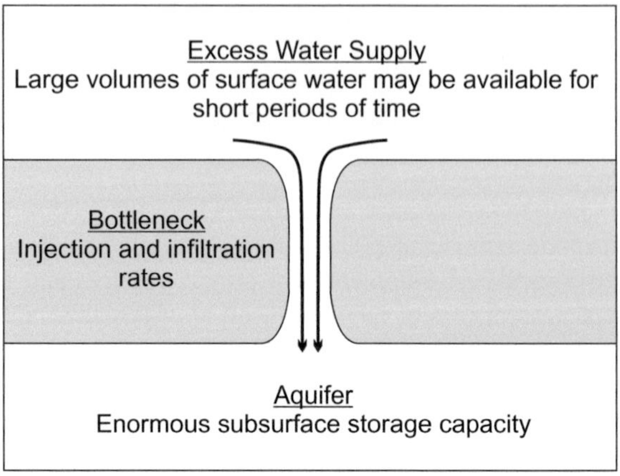

Fig. 8.1 Injection and infiltration rates often act as a bottleneck that constrains managed recharge of water supplies with a limited temporal availability

2010). Economic analyses need to consider the expected frequency of water shortages of different magnitudes, the impacts of the shortages, and the costs to mitigate the impacts of the shortages.

As an extreme example, it would not make economic sense to build an ASR system to store water that is available only after a 100-year rainfall period or to supply water in preparation for a 100-year drought (at least for non-potable uses) because of the very low frequency of times in which the systems would provide benefits. Potable water ASR systems do not need to be designed for meeting entire normal (average) demands during severe droughts, when it can be more cost effective to implement emergency conservation measures to reduce demands (Maliva and Missimer 2010). Due to the discount factor, cost-benefit analyses for water projects that provide infrequent benefits in the distant future are seldom favorable (Maliva 2014).

8.4.2 Hydrogeological Factors

Hydrogeological factors consist of a variety of hydraulic and hydrogeochemical parameters that control infiltration rates, pumping and injection rates, the movement and mixing of anthropogenically recharged water, and fluid-rock interactions. Hydrogeological feasibility factors include (but are not limited to):

- aquifer transmissivity and hydraulic conductivity
- aquifer heterogeneity
- aquifer dispersivities
- aquifer thickness, storativity, and effective porosity
- storage zone native groundwater salinity and water chemistry
- vadose zone unsaturated and saturated hydraulic conductivities
- presence of confining strata and their properties (leakance)

- vadose zone sediment composition (e.g., grain size and organic matter concentration)
- mineralogy.

The importance and optimal values of the hydrogeological factors controlling MAR system performance varies between system types. The hydrogeological controls on MAR system performance are addressed in the chapters and sections on specific system types.

8.4.3 Infrastructure and Logistical Issues

Infrastructure and logistical issues affect the ability to physically construct, operate, and maintain MAR systems. Some of the main infrastructure and logistical issues impacting MAR systems are summarized below.

8.4.3.1 Land Availability

Sufficient land must be available to construct a proposed MAR system and any planned or anticipated future system expansions. The preferred situation is to take advantage of land already owned by an MAR system owner. If land must be purchased, then acquisition costs can be an important factor in the economic feasibility of projects. Where available land is limited or too expensive, MAR systems that use wells for recharge (which have a small surface footprint), rather than surface spreading, may be the preferred option. Land availability tends to be a greater challenge for systems constructed in urban or suburban settings. For public utility systems, easements for system construction may be obtained either voluntarily or involuntary through the eminent domain process or, depending upon country, some other form of compulsory purchase or acquisition. The time required to purchase land can be a major project scheduling issue.

8.4.3.2 Site Accessibility

MAR project sites need to be accessible for equipment (e.g., drilling rigs) and storage of supplies during construction and periodic visits by the system operator for routine and emergency maintenance work, inspections, and the collection of water samples. MAR system sites should be readily accessible to the operator at all times. Site accessibility involves consideration of legal access, driving time, and road conditions.

8.4.3.3 Site Security

Water and wastewater utility infrastructure, including MAR systems, may be targets for terrorism, malicious vandalism, and theft. Equipment installed in remote, isolated areas is particularly vulnerable to theft (e.g., for scrap metal) and vandalism. In parts of the United States, aboveground objects are potential firearms targets. From a site-security perspective, the preferred location for an MAR system is within the secured perimeters of existing utility infrastructure. If locating a system within a secured perimeter is not practical, then high-value items (e.g., wells, pumps, electrical equipment) should be located within a fenced and locked enclosure. Additional security measures may be necessary including barbed wire, alarm systems, and security cameras. Aboveground features (e.g., wellheads) located near roads should be protected using bollards (or other means) against vehicular impacts.

8.4.3.4 Proximity to Water and Wastewater Distribution Infrastructure

MAR system may recharge water from water and wastewater treatment and distribution systems or produce water to be sent to such systems. Transmission of water long distances can be very expensive in terms of pipeline and pumping station construction costs and operational costs (e.g., energy). It is therefore preferable that MAR systems be located close to the recharge water source or to the point of use of recovered water. Preferred locations are at or near treatment plants, surface water sources, or strategic points in the water distribution or transmission system.

8.4.3.5 Proximity to Electrical Power Infrastructure

Many MAR systems need to be supplied with electrical power for the operation of pumps and surface facilities (e.g., pre- and post-treatment systems, monitoring equipment). Most pumps with a capacity of 0.5 MGD (1,890 m^3/day) or greater, using more than 5 or 10 hp, will require 480 V, 3-phase electrical power (or equivalent). The optimal situation is for MAR systems to be located near existing electrical power lines located at water treatment plant sites, at pumping stations in the distribution system, and near most above-ground storage facilities. Providing electrical power to remote locations without distribution lines can be a significant additional expense.

8.4.3.6 Source Water Quality, and Pretreatment Requirements

MAR systems require an adequate source of water of suitable quality. Water recharged in MAR systems includes potable water, reclaimed water (treated sewage effluent), and surface and stormwater treated to varying degrees. Recharge water must be of suitable quality to allow for the operation of MAR systems with an acceptable and manageable degree of clogging. The water also needs to meet applicable regulatory standards for groundwater recharge, which varies between jurisdictions and system types. Systems that recharge by injection usually require a higher quality water to minimize well clogging and meet often more rigorous regulatory requirements than exist for surface-spreading systems. Source water quality dictates pretreatment requirements. Pretreatment costs can exceed the costs of the actual recharge system and are thus an important economic feasibility issue. Pretreatment requirements can be particularly onerous where reclaimed water is being recharged and indirect potable reuse is a possibility or reality (Chap. 12).

8.4.4 Regulatory and Political Issues

Depending upon project location, a very wide range of regulations may apply to the construction and operation of MAR projects. Regulations may pertain to well and surface construction activities, land development, and the extraction, recharge, and recovery of water. Peripheral issues related to land development and environmental protection can arise that can delay projects and add to project costs. The following types of regulations might affect the permitting, construction, and operation of ASR and other MAR projects (Maliva and Missimer 2010):

- underground injection control regulations, which control the quantity and quality of injected fluids, well construction, monitoring, systems operations, and reporting
- water management regulations that confer the right to extract and use surface waters
- groundwater use regulations that control the recovery and use of produced water
- environmental protection regulations that impact the disposal of produced brackish and saline waters during well drilling and testing
- groundwater use regulations that protect recharged water from other aquifer users
- rules regarding the ownership of underground storage space
- drinking water quality regulations for potable water systems, which address post-treatment and disinfection, monitoring, and construction requirements specific to drinking water facilities
- wastewater facilities regulations, which can include permitting requirements for the recharge of treated wastewater, and specific treatment and monitoring requirements prior to recharge and for recovered water before it is sent to a reuse system
- water well construction and permitting regulations
- local zoning, land use, and planning regulations, which may control the location of water utility infrastructure and include specific land development requirements
- local building department regulations, which may require permits and the meeting of building codes for on-site construction (e.g., well houses)
- environmental protection regulations from various levels of government having jurisdiction over site activities, including rules related to endangered species and wetlands protection and construction in stream channels and flood plains

- historical preservation regulations, if an MAR site is discovered to be of historic significance (e.g., contains archeological sites)
- occupation health and safety regulations, which may affect well drilling and on-site construction activities, and the storage and handling of chemicals used during construction, well development, and system operation.

Follows is an abbreviated summary of the types of regulations that could impact MAR projects with a focus on the United States. Some countries and states have specific regulations related to aquifer recharge, whereas projects in other areas are authorized under a general environmental impact assessment processes. It therefore behooves professionals involved in MAR projects to become intimately familiar with all the regulations that could apply to a given project.

8.4.4.1 U.S. Underground Injection Control Regulations

The Safe Drinking Water Act (SDWA) of 1974 (and subsequent amendments) required the U.S. Environmental Protection Agency (USEPA) to establish a system for the regulation of underground injection activities, which resulted in the formation of the USEPA Underground Injection Control (UIC) program. Individual states and Native American tribes can obtain primary enforcement authority (primacy) for all or some types of injection wells. If a state or tribe receives primacy, it must still meet the USEPA UIC regulations. States or tribes can establish more restrictive regulations.

The USEPA regulations concerning injection wells are promulgated in the Code of Federal Regulations (CFR), Title 40 (Protection of Environment), Parts 144 through 148. The overriding objective of the USEPA UIC program is to prevent endangerment of "underground sources of drinking water" (USDWs). A USDW is defined as a non-exempted aquifer that contains water with less than 10,000 mg/L of total dissolved solids (TDS). Endangerment is considered any activity that results in the presence of any contaminant such that it results in non-compliance with any national or state primary drinking water regulation. The USEPA UIC regulations apply only to injection wells. A "well" is defined in the UIC regulations (40 CFR 144.3) as a "bored, drilled, or driven shaft whose depth is greater than the largest surface dimension; or, a dug hole whose depth is greater than the largest surface dimension; or, an improved sinkhole." ASR and most other MAR injection wells are categorized as Class V injection wells because they typically discharge into a USDW.

Water injected in MAR systems must meet health-based primary drinking water standards (i.e., maximum contaminant levels; MCLs). If the concentration of a parameter in the native groundwater in the injection (storage) zone is greater than its MCL, then the native groundwater concentration becomes the applicable standard for the injected water. Underground injection may not <u>cause</u> a violation of a primary drinking water standard. For example, the leaching of arsenic into water recharged in an ASR system is a violation of UIC rules if the concentration exceeds the MCL of 10 μg/L even if arsenic in the injected water was below the MCL. Depending upon jurisdiction, the compliance point for meeting primary drinking water standards may be either at the wellhead or at the boundary of a zone of discharge. The latter option allows for the natural attenuation of contaminants.

The UIC regulations (40 CFR 146.4) allow for an exemption from non-endangerment requirements, which would allow injected water to exceed some primary drinking water standards if the injection zone locally does not currently serve as a source of drinking water and it cannot now and will not in the future serve as a source of drinking water. The USEPA rules have the flexibility to allow for less stringent aquifer protection requirements where it is recognized that an aquifer will not be used as a drinking water source. In practice, this flexibility is illusory (Maliva and Missimer 2010). It has been very difficult in most circumstances to obtain an aquifer exemption for political reasons because granting of an aquifer exemption is viewed as being tantamount to allowing contamination of a drinking water supply, irrespective of the actual present or future uses of an aquifer.

The Code of Federal Regulations contains very little in terms of specific construction, testing, and operational requirements for ASR and other Class V injection wells, other than the fundamental UIC program-wide requirement that underground injection activities shall not endanger USDWs. There is considerable variation between primacy states on the regulations adopted for the different types of Class V injection wells. All injection must be authorized under either general rules or specific permits. Wells authorized by rule do not require a permit if they comply with basic specified requirements. For example, stormwater injection well systems are authorized by rule in some states where the practice is not expected to impair groundwater quality.

8.4.4.2 U.S Surface Water Law

MAR system owners must have a legal "right" to the water used for recharge. Regulation of surface water withdrawals in the United States is under the purview of each state. Surface water is allocated either under the riparian doctrine, prior appropriation doctrine, or a hybrid combination of the two doctrines. The riparian doctrine limits the use of surface water to landowners that own a parcel of land adjacent to the watercourse. The water must be put to a reasonable use and the abstraction of water should not interfere with the reasonable use of downstream riparian landowners. Riparian use is typically regulated by a permitting system that evaluates whether a proposed use is reasonable and allocates the quantity of water that may be used and time of use. The permitting process also allows for consideration of environmental impacts of withdrawals. In Florida, for example, the amount and timing of surface water and groundwater withdrawals is limited by minimum flows and levels (MFLs) established for some watercourses. No new allocations may be issued and existing allocations may be reduced if an MFL is not being met. If an MFL is being met, then excess water may be available for at least part of some years for recharge in MAR systems.

The prior appropriation doctrine, which has been adopted by many states in the western United States, allocates water based on the concept "first in time, first in right." Water rights are not tied to land ownership and can be sold. The basic concept is that the earliest, most senior appropriator has the highest priority for water relative to other less senior appropriators. In times of shortage, the senior appropriator may request a "call" on the river, which would force more junior appropriators to curtail their use so that more senior appropriators can take their full appropriative rights. Appropriators are required to put the water to a beneficial use. Prior appropriation rights can be lost if the water has not been used for a number of years. However loss of rights due to nonuse seldom occurs due to the value of water rights. A criticism of the prior appropriation doctrine is that the "use it or lose it" doctrine encourages wasteful, unnecessary use of water to preserve rights.

8.4.4.3 Groundwater Use Law

Groundwater law issues related to MAR include authorization for use of an aquifer, ownership of recharged and stored water, the relationship between groundwater and surface water, and the protection of stored water from other aquifer users. Groundwater use law in the United States is fragmented with differences in legal doctrines, principles, rules, and regulations existing between states, within districts in a state, and between aquifers. States and jurisdictions vary on how they address the impacts of groundwater withdrawals on surface waters and what legally constitutes groundwater. A distinction is often made, particularly in the western states, between shallow groundwater that is closely connected to a stream or river (i.e., "tributary" water) and deeper groundwater with a lesser connection to surface water bodies.

Local groundwater law often does not specifically address some key issues related to MAR systems, particularly the ownership of recharged water and the relationship between the right of system owners to recover stored water versus the rights of existing aquifer users. Four main groundwater law doctrines have been adopted in the United States; (1) absolute ownership, (2) reasonable use, (3) prior appropriation, and (4) correlative rights. The four doctrines are end-members. Some states have adopted combinations of the doctrines and there are variations in how the doctrines are applied. Professionals involved in MAR projects must have a clear understanding of local groundwater law.

The doctrine of absolute ownership (also referred to as the English Rule) holds that a landowner owns the groundwater underlying his or her property. A landowner can withdraw as much water as desired without any liability for impacts to adjacent and more distant landowners. Water, in essence, belongs to the user with the biggest pump. The doctrine of absolute ownership provides no incentives for the implementation of MAR as it offers no protection of stored water. A neighboring landowner could install a well on his property, near a neighboring MAR system, and "steal" the stored water. The doctrine of absolute ownership is becoming obsolete as groundwater resources are becoming over exploited. For example, groundwater use in Texas has historically not been regulated because a court ruled in 1899 that the movement of groundwater was unknowable and in the realm of the "occult." Texas groundwater law is evolving with establishment of groundwater management areas and groundwater conservation districts, which have some authority to regulate the spacing of water wells, production from water wells, or both.

The reasonable use doctrine (also referred to as the American Rule) allows a landowner beneficial use of groundwater as long as the use does not harm other aquifer users, the aquifer system, or the environment. Groundwater use is permitted by a state or a more local regulatory agency. Groundwater may be considered to be owned by either the state or landowner. The reasonable use doctrine has been adopted by the majority of eastern states, and also by Arizona and Nebraska. For example, groundwater (and surface water) use in Florida is regulated by five water management districts that manage the consumptive use of water, managed aquifer recharge, well construction, and surface-water management under Chapter 373 of the Florida Statutes. The consumptive (water) use allocation is not "owned" by the applicant. Permits have a finite duration and there is not an inherent right to a renewal (although most permits are routinely renewed). An applicant for a water use permit must demonstrate that the water would be put to a beneficial use and that the proposed withdrawals will not adversely impact existing permitted and legal users, the aquifer, and the environment. The beneficial use requirement also leads to permit conditions requiring the implementation of conservation measurements to minimize the waste of water. The issuance of water use permits in Florida requires that a use be reasonable, beneficial, and in the public interest. The permitting process can recognize MAR and provide protection to MAR system owners and operators. The permitting process could prevent theft of water stored in an ASR system by considering new proposed withdrawals near an existing ASR system to be an unacceptable impact to an existing aquifer user.

The prior appropriation doctrine (also referred to as Western Water Law) has been adopted by a majority of the western states to also regulate groundwater law. Under the prior appropriation doctrine, each water user has a water right that has an annual quantity and an appropriation date. The water rights may control pumping rates, well spacing, and well construction. Water rights are issued until all the available water is allocated with the general requirement that the proposed water use be reasonable and beneficial. Senior users (i.e., holders of water rights with the oldest appropriation date) have first rights on groundwater use. During drought periods, or other circumstances in which water use must be reduced, pumping by more junior users must be curtailed first.

Under the correlative rights doctrine, groundwater rights are proportional to a landowner's share of the land overlying an aquifer. The water is considered to be owned by the landowner, but its use is controlled by the state. The use of groundwater must be reasonable and beneficial, and is correlated with the rights of other groundwater users. In the event of shortages, due to droughts or excessive aquifer drawdowns, all water users share in a pro-rated reduction in groundwater use related to their land owned; no distinction is made between senior and junior water users or to the use of the water (e.g., public supply versus agricultural).

The restatement of torts doctrine combines elements of the absolute ownership and reasonable use doctrines. A landowner is entitled to withdraw groundwater for a beneficial purpose and is not liable for interference with the water uses of others unless either (1) the withdrawal causes unreasonable harm by lowering water levels or pressures, (2) exceeds the owners reasonable share of the annual supply of groundwater, or (3) causes a direct and substantial adverse effect on a water course or lake and causes harm to a person entitled to use its water.

8.4.4.4 Environmental Protection Regulations

Environmental protection regulations apply to MAR projects with respect to impacts during the construction and operation of systems and the impacts of operation of systems on local surface hydrology. The construction of MAR systems is similar to other construction projects in that the potential exists for impacts to endangered or threatened species, particularly where construction is occurring on undeveloped land. In Florida, for example, a federal permit may be required for proposed land uses within 660 ft (200 m) of an active bald eagle nest, with "active" defined as any activity within the past five years. Maliva and Missimer (2010) reported on how the presence of three gopher tortoises, a species of special concern in Florida, on an ASR site in Florida delayed the permitting of the system expansion by over a year because the tortoises had to be captured, blood testing performed for a gopher tortoise upper respiratory tract disease (no longer required), and a suitable relocation site identified and approved. An endangered species survey, for both fauna and flora, may be required for projects involving construction on undeveloped land.

Surface water withdrawals and the operation of MAR systems using shallow aquifers for storage zones can impact surface hydrology and groundwater-dependent ecosystems during both recharge and recovery. MAR can have beneficial environment impacts. For example, MAR is being implemented in the western United States as a tool to support instream flows to maintain salmonid populations. Conversely, groundwater pumping during recovery may locally lower the water table, contributing to the dehydration of wetlands, land subsidence, and reduction in stream flow. The hydrological impacts of the operation of MAR systems can be evaluated through groundwater flow or integrated groundwater/surface water modeling.

8.4.4.5 Land Development and Construction Codes

A wide variety of local land use and development regulations and building codes can apply to MAR projects. These regulatory requirements primarily affect the above-ground parts of MAR systems (e.g., infiltration basins, wellheads, fences, well houses, access roads) and site development work. Local zoning, land use, and planning regulations usually have greatest impacts on construction on undeveloped land or in residential areas, and the least impact on construction within existing developed parts of utility properties. Some design and permitting items that have arisen in MAR and production well projects the author has worked on in Florida include:

- indigenous vegetation removal permits (for clearing native vegetation at well sites)
- fire-fighting equipment access road requirements to ASR and monitoring well sites (including the vehicle turn radius)
- decorative-concrete wall design for the perimeter of a site
- landscaping and irrigation plans
- stormwater management plans
- conservation easements (e.g., part of a small, approximately 30 m by 30 m, well site had to be set aside as a conservation area).

MAR and production well projects have been temporarily shut down during construction because it was determined that a necessary permit or approval was not obtained. Compliance with land-use and planning regulations can become a major cost item for some MAR projects. All local land development and building regulations that may impact the construction of an MAR system should be identified and evaluated as part of feasibility assessments.

8.4.4.6 Political and Public Support (Public Involvement)

MAR projects may elicit opposition from parts of local populations for a number of reasons. Some groups have opposed ASR projects as a matter of principle because of objections to injecting anything into what are perceived to be pristine aquifers. MAR systems using reclaimed water may arouse concerns over the water eventually entering the drinking water supply (i.e., that indirect potable reuse will occur). Some objections can be expected for virtually any utility project from the "not in my backyard" (NIMBY) and "people against virtually everything" (PAVE) members of the public. Utility expansion projects have also been opposed under the belief that any project that provides more water is bad because it allows for greater local population growth and is a disincentive for conservation.

The potential for significant public opposition, which can be translated into political opposition, should be evaluated as part of an initial feasibility assessment. A proactive approach involving public outreach is recommended to maximize public acceptance and preempt opposition. Education and involvement of key stakeholders early in the project development process can be of great value. Over time, public opinion on MAR and wastewater reuse has become more favorable as the public has developed a greater understanding of the water challenges their communities face.

8.5 Economic Analysis and MAR Feasibility

The basic principles of the economics of MAR were reviewed by Maliva (2014). MAR projects should be both economically and financially feasible (Cowdin and Peters 1988; National Research Council 2008). Economic feasibility means that the benefits of a project are greater than its costs. Financial feasibility addresses whether financial resources are available to construct and operate a system. Projects may be economically feasible but not financially viable. In developing countries, for example, MAR projects may be able to yield benefits exceeding their costs over time, but the projects may not be implemented because up-front money for their construction is not available.

Cost–benefit analysis (CBA) principles and methods are addressed in microeconomics textbooks and some dedicated books (e.g., Layard and Glaister 1994; Boardman et al. 1996). The net present value (NPV) method is commonly used in which both the initial investment in a project and the benefits and costs expected to be achieved or incurred over the life of a project are considered. Future benefits and costs are discounted at an appropriate rate. The basic NPV equation is

$$\mathrm{NPV} = -\mathrm{C}_0 + \sum \mathrm{B_i}/(1+\mathrm{r})^{\mathrm{i}} - \sum \mathrm{C_i}/(1+\mathrm{r})^{\mathrm{i}} \tag{8.1}$$

where C_0 is the initial (capital) costs in year 0, B_i and C_i are the benefits and costs in year "i", and r is the discount rate. Cost-effectiveness analysis (least-cost analysis) and lifecycle costs analysis consider only the costs to achieve a pre-set objective or criterion (benefits are considered to be a constant).

The costs of MAR projects include both capital, operation and maintenance costs, and finance costs (debt service). Capital costs are fixed, one-time expenses incurred during the design and construction of an MAR system. Capital costs include (Maliva 2014):

- land
- testing costs
- feasibility analyses
- consulting services for the design, permitting, and supervision of construction
- construction costs (e.g., wells, basins, pumps, land development, roads, piping, instrumentation, controls, and pretreatment systems)
- regulatory testing requirements during construction and operational testing.

Operation and maintenance costs include (Maliva 2014)

- labor (system operation, regulatory compliance, administration)
- electricity
- consulting services
- regulatory testing requirements (e.g., water quality testing)
- maintenance costs (e.g., parts replacement, periodic well and basin rehabilitation)
- pre-treatment costs (additional treatment prior to recharge)
- post-treatment costs (e.g., chlorination)
- raw water costs.

A key issue in CBAs is that marginal (i.e., additional), rather than average costs, should be used and sunk costs (i.e., costs that have already been incurred or will be incurred whether or not a project proceeds) should not be considered. For example, the marginal operational labor cost is zero if existing plant staff can operate the system (i.e., there is no increase in total labor costs). Labor costs are included in a CBA if additional staff (or contracted labor) are needed to operate and maintain the system. Sunk costs that should not be included are, for example, previously performed hydrogeological investigations, an existing well that is no longer used, and existing intakes and piping. CBAs should consider opportunity costs associated with land, which include money that could have been obtained if a property was sold or leased and the value of goods and services that might have been obtained if the land were put to an alternative use (Maliva 2014).

Estimation of MAR project costs is relatively straight forward compared to estimations of project benefits. Economic analysis of MAR projects is complicated by the difficulty of applying an economic value to water. Todd (1965) in a pioneering paper on the economics of artificial recharge observed that

> It is clear the analysis of the benefits of artificial recharging is dependent on what value can be assigned to a unit volume of water

and

> in assessing the benefits of artificial recharge, consideration must be given to the importance of water to the total economy, to the value of water for various uses, as well as to the direct and intangible benefits that may accrue.

Quantification of the value of water is most straightforward where water is sold in a free market, which is seldom the case. A fundamental challenge in quantifying the economic benefits of water projects is that there is seldom a free market with respect to water and observed prices often do not reflect the social value of a good or service (Gibbons 1986; Colby 1989). Water has social and environmental values, which are difficult to quantify. MAR can also provide benefits by increasing the volume of water in storage and thus stabilizing or increasing aquifer water levels. The total economic value of the recharged water includes its extraction value plus in situ (non-use) values derived from groundwater being in place. In situ benefits include avoiding the adverse impacts of land subsidence and reduced pumping costs.

Methods used to monetize the benefits of MAR systems were reviewed by Maliva (2014). Economic value is ultimately measured based on substitutability, which can be expressed in terms of willingness to pay (WTP) and willingness to accept compensation (WAC; Freeman 1993). With respect to water, WTP is the amount of money that someone would be willing to pay for a given amount of water rather than do without. WAC is the minimum amount of money someone would require to voluntarily forgo the use of a given amount of water. The economic value to society of a good or service is the aggregate of the WTPs of all individuals. The value of water used for irrigation (and other purposes) can also be quantified in terms of the marginal productivity of water, which the extra value of output that can be obtained from additional applications of water.

CBAs of water projects have been greatly abused to justify government investment in politically favored but economically inefficient projects. Reinser (1999) in his book "Cadillac Desert: The American West and its Disappearing Water" described how false economic analyses were widely used to give the perception that major water supply projects in the western United States made economic sense, when in fact they could never be economically justified because the farmers (primary beneficiaries) could never afford the true cost of the delivered water.

In consideration of the complexity of assigning a meaningful value to project benefits, cost-effectiveness analysis and lifecycle costs analysis are more commonly employed in practice. MAR projects are compared against other water supply, storage, and treatment options that provide the same water resources or environmental benefits. For example, if it is determined that there is a need for an additional amount of potable water to meet the demands of a community during peak demand periods, then cost-effectiveness analysis could be used to evaluate ASR versus other supply and storage options. A limitation of cost-effectiveness analysis is that an entire list of projects could be ranked without any assurance that any of them are actually worth doing (Pearce et al. 2006).

CBAs need to consider risks and uncertainties (Boardman et al. 1996). It would clearly be improper to assume in a CBA of an MAR project that a 100% favorable result will be obtained when there is a real potential for poorer results. Not considering risk and uncertainty biases CBAs by increasing expected benefits. Risk and uncertainty can be incorporated into CBA through an expected value analysis (Boardman et al. 1996). The future is characterized in terms of a series of distinct, mutually exclusive contingencies. To evaluate risks, a probability is assigned to each possible contingency. Expected net benefits (ENB) are calculated as

$$\mathrm{ENB} = \sum \mathrm{P_i}(\mathrm{B_i}-\mathrm{C_i}) \quad (8.2)$$

where P_i = probability of contingency "i", and B_i and C_i are the present value of the benefits and costs of contingency "i". The sum of the probabilities for all of the contingencies is equal to one. The probability of each contingency can be based on historic experience or the subjective opinions of experts. For example, possible contingencies for an ASR project could theoretically be

- **Optimal performance**: system provides target 1.5 million gallons per day (Mgd; 5,680 m^3/d) of water that meets primary drinking water standards for 100 consecutive days during the summer peak demand period.
- **Suboptimal**: system can provide only 1.0 Mgd (4,790 M^3/d) of potable quality water for 100 days or 1.5 Mgd (5,680 m^3/d) for 66 days.
- **Poor**: system has a very poor recovery efficiency and can provide 1.5 Mgd (5,680 m^3/d) of water for less than 30 days.

Individual CBAs would be performed for each contingency and the ENB would be the cumulative value weighted for the estimated probability of each contingency.

8.6 Project Implementation Strategies

MAR projects are typically implemented in a phased manner. Several quite similar phasing schemes have been proposed for the implementation of ASR and other MAR projects, which mainly differ in whether some phases are further subdivided. The

National Research Council (2008) and Maliva and Missimer (2010) presented the following implementation sequence, which represents the actual project implementation strategy employed for many ASR projects:

Phase I: Desktop feasibility evaluation
Phase II: Field investigations and testing of potential system sites
Phase III: Design
Phase IV: Pilot system or full-scale construction and testing
Phase V: Project review and adaptive management
Phase VI: System expansion

Phase IV involves the construction of an entire small-scale MAR system (e.g., single injection well ASR system or a small area infiltration basin system). For large-scale projects, Phase IV might involve pilot testing of a single well ASR system or a single infiltration basin. The completion of each phase is a Go/No Go decision point, at which the results of the phase are evaluated and the feasibility of the project is reassessed. MAR should be approached with goal of avoiding the financial commitment to construct a full-scale system until there is a high degree of certainty that the project will meet performance expectations set at the start of the project.

The Australian Managed Aquifer Recharge Guidelines (NRMMC-EPHC–NHMRC 2009) proposed a similar project phasing approach in which project risks are identified and progressively evaluated. The Australian MAR guidelines assess 12 hazards common to MAR projects (Table 8.1) in a four-step process. Increasing costs of acquiring information are incurred as confidence in the viability of a project increases. The NRMMC-EPHC–NHMRC (2009) MAR guidelines include a maximal and residual risk assessment. A maximal risk assessment identifies inherent risks in the absence of preventative measures. A residual risk assessment evaluates residual risks after consideration of potential preventative measures. For example, a maximal risk assessment of a reclaimed water ASR system would likely identify pathogens in the recovered water as a significant risk element. The residual risk assessment might consider risks remaining after natural attenuation during the planned storage period and post-treatment (e.g., disinfection) of the recovered water.

The recommended project phasing and risk assessment procedures under the NRMMC-EPHC–NHMRC (2009) MAR guidelines are as follows:

Stage 1—Desktop study Evaluation if a project is likely to be viable and the likely difficulty in its implementation. Stage 1 Involves consideration of both technical and regulatory issues and identifies data that will need to be obtained in Stage 2.

Stage 2—Investigations and Assessment Collection of additional information needed to assess risks, such as aquifer residence time, source (recharge) and native groundwater chemistry and quality, and characterization of reactive aquifer minerals. A "maximal risk assessment" is undertaken to estimate the risks in the absence of any controls or preventive measures. A pre-commissioning residual risk assessment is performed to evaluate potential preventative measurements and operational procedures that could be implemented to ensure acceptably low residual risks to human

Table 8.1 Risks or hazards associated with MAR projects

Risk type	Examples
Operational	(1) System not meeting performance objectives (a) Low recovery efficiency (b) Low infiltration/recharge rates (c) Does not achieve target hydraulic response (increase in water levels or heads) (2) Excessive well or aquifer clogging (3) Loss of well mechanical integrity
Recharge water quality	(1) Recharge water does not meet applicable water quality standards and impairs the aquifer or adversely impacts other aquifer users (a) Microbiological (parameters) (b) Salinity (c) Nutrients (d) Chemicals (e) Radionuclides (2) Anticipated water quality improvements (natural aquifer treatment) not realized; recovered water does not meet water quality standards (3) Excessive migration of recharged water
Fluid-rock interaction	(1) Deterioration of recharge water quality (e.g., arsenic and metals leaching)
Hydraulic impacts	(1) Pressure-induced fracturing (2) Land subsidence (3) Aquifer dissolution (loss of integrity caused by low pH or other conditions)
Environmental impacts	(1) Impacts on groundwater dependent ecosystems (e.g., reductions in stream or spring flows and wetland hydroperiods during recovery) (2) Water logging
Greenhouse gases	(1) Excess energy consumption relative to other options

Sources NRMMC-EPHC–NHMRC (2009), Maliva and Missimer (2010)

health and the environment. The residual risk assessment may be repeated until the residual risk is acceptable by the addition of extra preventive measures.

Stage 3—Commission Trials A pilot or full-scale system is tested. The MAR system is trialed to validate the effectiveness of preventive measures and operational controls, and to assess the suitability of the recovered water for its intended uses. Stage 3 includes monitoring of processes that could not be measured prior to the start of recharge. The commission trials include an operational risk assessment, which assesses whether on-going operation of the project has acceptably low human health and environmental risks. The aim is to identify unforeseen risks and required preventive measures.

Stage 4—Operation The final stage involves the actual operation of the MAR system with a management plan and regular operational monitoring. Validation monitoring is performed to assess whether the quality of the recovered water is acceptable and to verify that the environmental value of the affected aquifer is protected.

The phasing of the Australian MAR guidelines (NRMMC-EPHC–NHMRC 2009) follows the normally employed MAR project phasing of a desktop study, followed by field testing, and then construction and testing of a pilot system. The Australian MAR guidelines provide a technically sound and flexible approach for assessing the risks, and thus the feasibility, of projects by prompting consideration of many of the factors that impact project feasibility and risks. The guidelines have no formal legal status but were designed to provide guidance for individual governments in the development of their MAR policies and regulations (NRMMC-EPHC–NHMRC 2009). The application of the guidelines to a trial ASTR project is discussed by Page et al. (2010). It is important to recognize that the degree of effort and technical sophistication applied in the risk assessment tasks will necessarily vary between locations and projects depending upon local financial and technical resources, project complexity, and inherent risks.

8.7 Desktop Feasibility Assessment

MAR desktop feasibility evaluations consist of a review of available information on water supply and demands, hydrogeology, utility infrastructure, and regulatory requirements. It is referred to as "desktop" in that it does not involve significant field work other than site inspections. Maliva and Missimer (2010) recommended that desktop feasibility assessments be divided into an initial Conceptual Feasibility Assessment and a subsequent much more-detailed Preliminary Feasibility Assessment and Design Study. The Conceptual Feasibility Assessment is, in essence, an initial screening of basic feasibility issues including:

- water needs, sources, and potential storage requirements evaluation
- potential MAR system types
- potential system locations
- potential storage zones
- key regulatory issues
- system goals and expectations
- fundamental economics.

It has been the author's experience that sufficient data are commonly available early in the desktop feasibility assessment process to allow for an informed initial Go/No Go decision to be made as to whether MAR is locally feasible, and to establish project goals, directions, and priorities.

The Preliminary Feasibility Assessment and Design Study is a more detailed investigation of the key infrastructure, logistical, hydrogeological, and regulatory issues that affect the feasibility, design, and operation of an MAR system. If MAR is determined to be feasible, then the study should also include a preliminary system design and cost estimates. The desktop study should also identify data deficiencies that need to be addressed in subsequent project phases.

A checklist or scoring system approach is the preferred method for evaluating MAR feasibility because it forces consideration and evaluation of all recognized feasibility factors as to whether they are optimal, acceptable, poor, or a fatal flaw. The list of factors considered should be comprehensive and will vary between system types, objectives, and locations.

CH2M Hill (1997) developed an ASR feasibility screening tool for the St. Johns River Water Management District (SJRWMD) in Florida (Sect. 13.5.1). The feasibility screening tool consists of evaluations of

(1) facility planning factors that determine whether there is a storage need that could be met by ASR
(2) hydrogeological, design, and operational factors that determine whether ASR is likely to be technically feasible
(3) cost
(4) regulatory issues.

8.8 Site Selection

The choice between several MAR options (e.g., potential system types, designs, and locations) usually depends upon a series of variables that impact system performance, feasibility, and costs. Numerical scoring should be employed to increase the objectivity of the system type and site selection process. Multiple criteria decision analysis (MCDA) techniques are widely used to assist decision-makers to select between numerous and sometimes conflicting objectives. MCDA techniques have been employed using GIS to identify sites most favorable for MAR. Decision support systems (DSSs) allow for even more sophisticated evaluation of MAR implementation options.

8.8.1 Multiple Criteria Decision Analysis

MCDA methods were reviewed by Stewart and Scott (1995) and Belton and Stewart (2002). MCDAs differ in the mathematical algorithms utilized, means by which options for each element or objective are scored, methods of application of weights, and the manner in which stakeholder preferences are considered. Weighted scoring systems are a simple type of MCDA that are widely used to evaluate water and wastewater infrastructure options. MCDA has been used to compare MAR to other water supply and storage options, to determine the preferred type of MAR project, and to select sites for MAR systems.

The weighted-sum method is perhaps the simplest and most understandable MCDA. For site selection, the weighted-sum method consists of the following basic procedures

- identification of the criteria that are considered important for the planned activity or system
- development of a scoring system for each criterion based upon the degree to which site conditions meet performance and economic goals
- assignment of a weighting factor for each criterion, which reflects its relative importance for the satisfactory performance of the activity or system
- assignment of scores for each of the criteria for each of the potential sites
- calculation of a cumulative score for each potential site.

The performance score for scenario "i" (Z_i) is calculated as

$$Z_i = \sum_{j=1}^{n} w_j z_{ij} \tag{8.3}$$

where,

W_j weight factor for criterion "j"
Z_{ij} performance value criterion "j" in scenario "i"

Performance values are assigned with the optimal values or conditions receiving the greatest scores. Decreasing scores are assigned with increasing departure from the optimal value or condition. Assignment of performance values and weight factors inherently involve considerable subjectivity, particularly where performance cannot be readily quantified. For example, it can be difficult to objectively assign a numerical performance value to the preservation of an ecosystem function criterion. A professional judgement has to be made as to relative importance of very different parameters. For the site selection of ASR systems, for example, weights need to be assigned to each of the different hydrogeological parameters (e.g., storage zone salinity, transmissivity, aquifer heterogeneity) and between different feasibility categories (hydrogeological, infrastructure and logistics, regulatory, and economics) to obtain overall site scores.

There have been numerous applications of MCDA to MAR and other water development projects. For example, Close et al. (2001) evaluated different recharge types and locations for the City of Phoenix North Gateway Water Reclamation Plant. Options were scored (1–4) for the following criteria:

- permitting requirements
- life expectancy (rehabilitation potential)
- land area requirements
- habitat modification (impacts)

- flood control impacts
- proximity to surrounding land owners
- public education effort required
- cost

Total score was the cumulative unweighted score for all eight parameters.

In a similar study, Cross et al. (2001) investigated potential locations for large-scale underground storage facilities in western Arizona for water from the CAP aqueduct. The screening procedure consisted of:

(1) Initial screening to identify regions in which large-scale recharge operations would not be technically or economically feasible or would cause unacceptable impacts.
(2) The remaining area was divided into 22 regions each with similar conditions that differ from those of other regions.
(3) Regions were evaluated based on eight criteria with the first two hydrogeological characteristics used for an initial screening.

    - long-term potential recharge capacity
    - chemical quality of the groundwater
    - proximity to the CAP aqueduct
    - land ownership
    - land use
    - feasibility for recovery of stored water
    - potential for land subsidence
    - potential for regulatory constraints.

Long-term potential recharge capacity was evaluated in terms of

- potential infiltration capacity of near-surface sediment
- predominance of coarse-grained strata in the upper vadose zone
- depth to the top of a regional poorly-permeable sequence
- thickness of the regional poorly-permeable sequence
- aquifer boundary constraints
- average depth to groundwater.

MCDA can be a valuable approach for evaluation of MAR options and site selection. However, the author has observed that frequently the answer is obvious without MCDA. MCDA not uncommonly serves to give a more technical appearance to the selection process (i.e., make conclusions appear more objective). The MCDA process can be subject to abuse in that the scoring system can be adjusted to give a desired result through the assignment of weight factors and performance values to the various criteria considered. Hence, to reduce the potential for the introduction of personal biases, it is recommended that the MCDA scoring system to be used in a project be determined at the start of a project and that the process (e.g., assigning weight factors to be used) involve key stakeholders.

8.8.2 Geographic Information Systems

A geographic information system (GIS) is defined as any system that captures, stores, analyzes, and displays data that are linked to location (i.e., are georeferenced). GIS broadly refers to all aspects of managing, manipulating, and using digital georeferenced data. Rather than applying the MCDA process to discrete sites, GIS can be used for regional site selection analyses using a weighted overlay function.

The underlying feature of GIS is the association of geographic data (coordinates) with non-geographic data (attributes). Vector data are stored as a series of x, y and optionally z coordinates with respect to a coordinate reference system (e.g., latitude and longitude, Universal Transverse Mercator system). A vector can be a point, line, polyline, or polygon. Polygons are defined by multiple (4 or more) vertices in which the first and last are the same (i.e., there is closure). Examples of polygons relevant to MAR are wetlands, land use zones, soil types, vegetation types, surficial geology zones, municipal boundaries, and property ownership (Maliva and Missimer 2012).

Raster data consists of a grid of columns and rows of cells (pixels) each of which stores a single value that represents a condition in the cell. Common types of raster data are digitized georeferenced aerial and satellite photographs and remote sensing data. Digital elevation models (DEMs) a commonly used type of raster data in which each pixel represents elevation above a datum (e.g., sea level).

Vector and raster layers can be overlain for spatial analysis. Each vector layer may contain only one type of geographic data and commonly they are set up to represent one type of feature (theme). Thematic layers are developed for each criterion being considered. For example, a thematic layer may contain surface geology data with each polygon representing a specific geological formation or rock type. Each layer is commonly stored as a shape-file, which consists of a minimum of three subfiles: a main file that stores geometric information (.shp), a shape index file (.shx), and a file that stores attribute data (.dbf). Various remote sensing data are well suited for the development of thematic layers. Many remote sensing data are obtained in a digital form (e.g., satellite images). Non-digital data can be digitized and processed to construct thematic layers.

MCDA is performed by overlaying thematic layers. For example, Ghayoumian et al. (2007) performed a GIS site-selection analysis for recharge basins in the Gavbandi River Basin in southern Iran. The site selection criteria used were (1) slope, (2) infiltration rate based on soil texture-permeability relationships, (3) depth to groundwater, (4) groundwater salinity from electrical conductivity, and (5) land use and geomorphology. Thematic maps were created for each of the criteria and each criterion was divided into 4 or 5 classes. The thematic layers were classified, weighted and integrated in a GIS environment using either Boolean or fuzzy logic (Ghayoumian et al. 2007). In Boolean logic, only satisfactory or unsatisfactory conditions are considered (i.e., areas as assigned a value of either 0 or 1). In fuzzy logic, membership is expressed as a continuous scale from 0 to 1. Each class is assigned a membership function value, which reflects its degree of acceptability (1 being the optimal condition). A map was generated showing the location of areas suitable for MAR based on weighted cumulative score.

Other GIS MCDA studies vary in the thematic layers used. Chowdury et al. (2010) integrated remote sensing, GIS, and MCDA techniques to map favorable areas for MAR in the West Medinipur district of West Bengal, India. The generated thematic layers were geomorphology, geology, drainage density, slope, and aquifer transmissivity. Rahman et al (2012), in an MAR site selection investigation in the Algarve Region, Portugal, used layers for slope (topography), infiltration rate (soil type), sub-surface impermeable layer thickness, groundwater depth, distance to groundwater pollution sources, aquifer thickness, transmissivity, aquifer storage volume, groundwater quality (chloride and nitrate), and residence time. The Rahman et al. (2012) investigation included defined threshold values (discarding conditions) of selected criteria for MAR constraint mapping (i.e., fatal flaw values). Mahmoud et al. (2014) screened sites for groundwater recharge structures in the Jazan Region of Saudi Arabia based on soil types, land cover and land use, slope (topography), runoff coefficient, and rainfall surplus precipitation.

GIS and groundwater modeling were used to investigate ASR feasibility in the Pajaro Valley Groundwater Basin of central coastal California (Russo et al. 2015). Eleven datasets were considered in the GIS MAR suitability analysis:

- surficial geology (G)
- soil infiltration capacity
- land use
- elevation (slope)
- verified (measured) infiltration rate
- aquifer thickness
- aquifer hydraulic conductivity
- confining layer thickness
- aquifer storativity
- vadose zone thickness
- historical changes in water table elevation (D).

Datasets were integrated into effective properties:

- effective infiltration capacity (I_E)
- effective transmissivity (T_E)
- available storage space (V)

The MAR suitability index is the weighted sum of the classified values (from 1 to 5) of I_E, T_E, V, D, and G. The impacts of hypothetical MAR scenarios were then evaluated in terms of their impact on increasing aquifer heads and managing saline-water intrusion. Potential ASR systems were simulated at areas determined to have a high MAR suitability.

Multicriteria GIS analysis was used to map the optimal locations for MAR of reclaimed water in the northwestern part of the Region of Murcia, Spain (Pedrero et al. 2016). The following criteria were used in the analysis:

Economic criteria

- transport distance from a wastewater treatment plant (not to exceed 8 km)
- pumping costs—elevation change should not exceed 15 m

Environmental criteria

- distances from water supply sources, urban agglomerations, and natural ecological reserves

Technical criteria

- slope (0–12%)
- soil texture (sandy loams, loamy sand, and fine sand soils)
- soil clay content (<10%)
- groundwater depth (≥5 m)
- road access (<50 m from road)
- land use.

Of the total study area (237,960 ha), only 2.7% (6,442 ha) was identified as being optimal for groundwater recharge.

8.8.3 Decision Support Systems

Decision support systems (DSSs) are computer programs that combine data management systems and analytical methods to formulate and evaluate multiple solutions to problems. The basic components of DSSs are a graphic user interface (GUI), databases, and simulation and optimization models. Simulation models are used to evaluate water resources behavior and may include groundwater and surface water flow models of varying degrees of sophistication. Water management scenarios, such as various MAR implementation options, can be evaluated in terms of their impacts on water resources and the environment, social and economic benefits, and costs.

Models are composed of objects and processes (Jordan 2006). Objects are defined by their properties and behaviors, and include various water supply components (e.g., reservoirs, MAR systems, conveyances, treatment systems, and groundwater) and demand components. Objects are linked by processes that quantitatively describe how objects communicate (interact) with other objects. Objects are related to other objects by inputs and outputs, such as water flow into and out of a reservoir (Jordan 2006). Analytical solutions or look-up tables are used to describe the relationship between objects, such as the relationship between well pumping and aquifer recharge on groundwater levels, the costs of pumping water, and supply and demand components.

The great attraction of DSSs for water resources management is that they allow for the examination of numerous factors involved in the design and management of more sustainable water resources systems (Loucks and Gladwell 1999). The basic modeling procedure is to first develop a reference ("business as usual") scenario and then develop one or more "what if" policy scenarios with alternative assumptions about future developments. A "what if" scenario might be construction of an MAR system with a given capacity at a given location to recharge an overdrafted aquifer. A DSS could be developed would evaluate the hydrological impacts of the recharge and associated costs and benefits.

Water management DSSs were reviewed by Chemonics International (2004). DSSs vary greatly in their sophistication in terms of their scales, issues considered, and modeling procedures. Some DSSs (e.g., WEAP21) can be linked to groundwater flow models, such as MODFLOW. Generic dynamic system simulation software (e.g., STELLA) may also be employed to evaluate water management options. The main challenges associated with DSSs are gathering the correct information, formulating proper questions, and interpreting the output (Purkey and Huber-Lee 2006). As is generally the case with any sort of predictive modeling, the accuracy of simulation outputs depends upon the quality of the data and the assumptions used in the simulation.

An important question is whether the effort (and cost) to develop DSS models is commensurate with their benefits. Does a DSS provide new insights or is it used to validate what is already known or could be determined using simpler methods (Maliva and Missimer 2012)? There is always the danger with computer models with impressive graphics that their results may have a greater acceptance (particularly by lay people) than is warranted by the accuracy of the raw data used and the underlying assumptions.

8.9 Phase II: Field Investigations and Testing of Potential System Sites

The objective of field testing programs is to obtain site-specific hydrogeological data to be used to further evaluate MAR feasibility at a site and for system design and permitting. Field investigation programs should obtain data identified in the desktop investigation as being necessary for a more informed evaluation of potential system performance and for system design. The testing performed as part of field investigation programs is system specific and depends upon the system type, system location, and the availability of existing data. Project budget is also an important consideration as, in practice, there is seldom sufficient funding available to perform all the testing that might be found useful. Hence, prioritization is an important part of the design of field testing programs.

For ASR and other MAR system using wells for recharge, field investigations commonly involve the drilling and testing of exploratory wells. In some locations, sufficient data are available from existing wells in the project site vicinity so that the characteristics of the storage or recharge zone are known and its approximate depth, transmissivity, well yield, and water quality can be reasonably ascertained. Existing data on aquifer mineralogy and water chemistry (e.g. major and minor cations and anions, mineral saturation states) are usually much more limited. MAR systems utilizing freshwater aquifers tend have the most existing data available because aquifers in overdraft inherently tend to have numerous existing wells, which are the ultimate cause of the overdraft.

ASR systems that use brackish or saline aquifers as storage zones tend to have the greatest hydrogeological data requirements because of the need to consider solute-transport in addition to the hydraulic response to recharge and recovery. However, aquifers containing brackish, or otherwise poor quality, water tend to have relatively little available hydrogeological information because few wells have been completed in the aquifers. Where groundwater is not directly useable, there was often no economic reason for aquifer testing.

Exploratory well programs involve both formation and hydraulic testing and the collection of water chemistry data. Some specific elements of testing programs include:

- evaluation of geology through description of well cuttings
- coring, and description and analysis of core samples
- borehole geophysical logging
- collection and analysis of water samples while drilling and after completion of wells
- aquifer pumping tests (single well or preferable multiple well tests)
- packer tests.

MAR systems that utilize surface spreading require data on the properties of the vadose zone, particularly its vertical hydraulic conductivity, which is usually measured by some form of infiltration testing. Key issues are obtaining sufficient data that are representative of the entire infiltration surface and identifying and characterizing any low permeability strata within the vadose zone that could impede the percolation of applied water to the water table. Data are often also needed on the hydraulic properties of the water table aquifer, which controls the rate of lateral flow away from infiltration sites and thus mounding potential. Test wells and surface geophysics are often used to obtain data on the hydrogeology of shallow strata. Aquifer hydrogeological testing programs are addressed in Chap. 9 and vadose zone testing techniques are summarized in Chap.10.

8.10 Phase III: MAR System Design

MAR system design involves the selection of system type and location, as well as the actual design of the chosen system. System design should also incorporate operational considerations. MAR systems are prone to clogging and, therefore, designs should attempt to reduce the potential for (or severity of) clogging, and allow for ready system rehabilitation once clogging occurs. As a general principle, MAR systems should be designed to incorporate as much flexibility as possible to accommodate futures conditions. Initial system design and location should also consider the potential for future system expansion, even if the need may not occur until well into the future. For example, potential project sites should ideally have room for additional recharge wells or infiltration basins, which might be needed at some time in the future. Potential future well or basin locations should be identified and conceptual-level plans developed on how to connect the new wells or basins to the initial project infrastructure. Pretreatment systems should also be readily expandable.

The successful design of MAR systems depends greatly on local conditions. It is therefore not possible to provide generic standards, guidelines and blueprints (Bouwer 1999). The design of MAR systems requires knowledge of system types, their processes, controls and limitations, and a thorough assessment of local water quality, aquifer hydraulics, and hydrogeological conditions. Bouwer (2002, p. 140) succinctly noted that

> Design and management of artificial recharge systems involves geological, geochemical, hydrological, biological, and engineering aspects. Because soils and underground formations are inherently heterogeneous, planning, design, and construction of groundwater recharge schemes must be piecemeal, first testing for fatal flaws and general feasibility and then proceeding with pilot and small-scale systems until the complete system can be designed and constructed.

The American Society of Civil Engineers (2001) "Standard Guidelines for the Artificial Recharge of Groundwater" provides a general overview of the planning, design, and operation of some types of MAR systems. Pyne (2005) and Maliva and Missimer (2010) provide an overview of the design of ASR systems and other MAR systems that use wells.

8.11 Phase IV: Pilot System Construction

Large MAR systems should be designed and constructed in a phased manner, which initially includes a pilot test. For a multiple-well ASR system, pilot testing normally involves the construction and operational testing of a single-well ASR system. For a large infiltration basin or SAT system, pilot testing might include the operation of a single cell (basin), which may be temporarily subdivided to allow alternating wetting and drying cycles. The primary objectives of pilot testing are to determine if the system will operate as designed and to obtain additional data that could be used to optimize the design of the full-scale system.

Pilot testing is also important for evaluating water chemistry issues associated with MAR systems, including the occurrence of adverse geochemical reactions (e.g., arsenic and metals leaching) and beneficial processes (e.g., attenuation of contaminant concentrations). Pilot testing also allows for the evaluation of clogging rates and field testing of pretreatment and remedial options to manage clogging.

Operational testing should, to the extent possible, mimic actual anticipated long-term operational conditions. For example, the water used during operational testing should be essentially the same water that will be recharged during the operation of full-scale systems. Also, the duration of the recharge, storage, and recovery cycles should be similar to the anticipated mode of system operation.

8.12 Phases V and VI: Project Review, Adaptive Management, and System Expansion

The performance of MAR systems should be periodically and objectively reviewed. Upon completion of pilot testing or initial operation of an MAR system, an objective review of project performance should be performed. System performance should be compared to performance objectives set at the initiation of a project, which could be a target recovery efficiency, injection rates (specific injectivity), basin infiltration rate, or water quality improvement. The causes of any system under performance should be investigated. Operational procedures may need to be adjusted or the system modified to improve system performance.

Adaptive management is a structured iterative decision-making process in the face of uncertainty. Uncertainty is reduced and management improved by system monitoring and adapting to new and different information. Adaptive management is often referred to as simply "learning by doing." Adaptive management approaches are employed when there is uncertainty over the performance and impacts of a proposed activity. For example, an adaptive management approach is being taken in the multi-billion (U.S.) dollar Comprehensive Everglades Restoration Plan (CERP) in South Florida because it is not possible to accurately predict the performance and interaction of each the numerous elements of the project. The key feature of adaptive management plans is that they encourage and incorporate flexibility in the design and operation of systems to allow for adaptation to unexpected results.

Adaptive management should be considered during system design. For example, MAR systems should have a robust design that can accommodate some unexpected conditions. There should also be design and operational flexibility, which allows for opportunities to adjust the design or operation of a system to adapt to information obtained from monitoring and project experiences. Adaptive management also involves learning (continuous on-going system evaluation) in which, for example, monitoring data are compared to initial predictive modeling results and the model is recalibrated or refined as necessary based on new information.

Injection wells typically experience a loss of performance over time. An adaptive management approach should be taken in order to identify operational changes and rehabilitation strategies that allow for the wells to maintain performance at the lowest cost. An adaptive management approach may also be fruitful for wadi recharge and infiltration basin systems in which different operational procedures are tested in order to find the operational protocols that most cost-effectively result in the greatest recharge rates.

Most MAR systems are expandable in that system capacity may be increased by the installation of additional wells or infiltration basins. The operational data and experiences from pilot testing and subsequent project phases are used to improve the design of a future system expansion. A groundwater model calibrated against operational data can be a valuable tool for optimizing the design of an MAR system expansion. Different design options can be simulated to find the option that most efficiently achieves project objectives.

References

American Society of Civil Engineers. (2001). *Standard guidelines for artificial recharge of ground water, Report EWRI/ASCE 34-01*. Reston, VA: American Society of Civil Engineers.

Belton, V., & Stewart, T. J. (2002). *Multiple criteria decision analysis: An integrated approach*. Norwell, MA: Kluwer Academic Publishers.

Boardman, A., Greenberg, D., Vining, A., & Weimer, D. (1996). *Cost-benefit analysis: Concepts and practice*. Upper Saddle River, NJ: Prentice Hall.

Bouwer, H. (1999). Artificial recharge of groundwater: Systems, design, and management. In Mays, L. E. (Ed.) *Hydraulic design handook* (p. 24.1–24.44). New York: McGraw Hill.

Bouwer, H. (2002). Artificial recharge of groundwater: Hydrogeology and engineering. *Hydrogeology Journal, 10*(1), 121–142.

Brown, C. J. (2005). Planning decision framework for brackish water aquifer, storage and recovery (ASR) projects: Ph.D. thesis, University of Florida, Gainesville.

Business Dictionary (n.d.). Planning. http://www.businessdictionary.com/definition/planning.html. Accessed June, 18, 2017.

CH2M Hill. (1997). *Water supply needs and sources assessment, alternative water supply strategies investigation. A tool for assessing the feasibility of aquifer storage recovery*. Special Publication SJ97-SP4. Palatka: St. Johns River Water Management.

Chemonics International Inc. (2004). *International survey of decision support systems for integrated water management*. Report prepared for U.S. Agency for International Development.

Chowdhury, A., Jha, M. K., & Chowdary, V. M. (2010). Delineation of groundwater recharge zones and identification of artificial recharge sites in West Medinipur district, West Bengal, using RS, GIS and MCDM techniques. *Environmental Earth Sciences, 59*(6), 1209–1222.

Close, C., Conroy, A., Shannon, J., & Williams, D. (2001). Recharge feasibility assessment for the City of Phoenix North Gateway Water Reclamation Plant. In *Proceedings Tenth Biennial Symposium. Artificial Recharge of Groundwater*, Tucson, AZ, June 2001, pp. 51–60.

Colby, B. G. (1989). Estimating the value of water in alternative uses. *Natural Resources Journal, 29,* 511–527.

Cowdin, S. W., & Peters, H. J. (1988). The economics of ground water recharge projects. In Johnson, A. I. & D. J. Finlayson (Eds.) *Artificial recharge of groundwater. Proceedings of the international symposium on artificial recharge of groundwater*, August 23–27, 1988, Anaheim, California (pp. 14–22). New York: American Society of Civil Engineers.

Cross, M., Lyons., E., Harbour, T., & Cullom, C. (2001). Reconnaissance investigations for sitings of underground storage facilities in western Arizona. In *Proceedings Tenth Biennial Symposium. Artificial Recharge of Groundwater*, Tucson, AZ, June 2001, pp. 41–50.

Dillon, P., & Molloy, R. (2006). *Developing aquifer storage and recovery (ASR) opportunities in Melbourne, technical guidance for ASR*. Canberra: CSIRO Land Water.

Freeman III., A. M. (1993). *The measurement of environmental and resource values*. Washington, DC: Resources for the Future.

Ghayoumian, J., Mohseni Saravi, M., Feiznia, S., Nouri, B., & Malekian, A. (2007). Application of GIS techniques to determine areas most suitable for artificial groundwater recharge in a coastal aquifer in southern Iran. *Journal of Asian Earth Sciences, 30,* 364–374.

Gibbons, D. C. (1986). *The economic value of water*. Washington, D.C.: Resources for the Future.

Jordan, D. (2006). Understanding decision support systems: A tool for analyzing complex systems. *Southwest Hydrology, 5*(4), 16–34.

Layard, R., & Glaister, S. (1994). *Cost-benefit analysis* (2nd ed.). Cambridge, UK: Cambridge University Press.

Loucks, D. P., & Gladwell, J. S. (1999). *Sustainability criteria for water resources systems*. Cambridge, UK: Cambridge University Press.

Mahmoud, S. H., Alazba, A. A., & Amin, M. T. (2014). Identification of potential sites for groundwater recharge using a GIS-based decision support system in Jazan region-Saudi Arabia. *Water Resources Management, 28*(10), 3319–3340.

Maliva, R. G. (2014). Economics of managed aquifer recharge. *Water, 6*(5), 1251–1279.
Maliva, R. G., & Missimer, T. M. (2010). *Aquifer storage and recovery and managed aquifer recharge using wells: Planning, hydrogeology, design, and operation*. Houston: Schlumberger Corporation.
Maliva, R. G., & Missimer, T. M. (2012). *Arid lands water evaluation and management*. Berlin: Springer.
National Research Council. (2008). *Prospects for managed underground storage of recoverable water*. Washington, DC: National Academy Press.
NRMMC-EPHC–NHMRC. (2009). *Australian guidelines for water recycling: Managing health and environmental risks. Phase 2: Managed aquifer recharge natural resource.* Management Ministerial Council, Environment Protection and Heritage Council and National Health and Medical Research Council.
Page, D., Dillon, P., Vanderzalm, J., Toze, S., Sidhu, J., Barry, K., et al. (2010). Risk assessment for aquifer storage transfer and recovery with urban stormwater for producing water for a potable quality. *Journal of Environmental Quality, 39*(6), 2029–2039.
Pearce, D., Atkinson, G., & Mourato, S. (2006). *Cost-benefit analysis and the environment: Recent developments*. Paris: Organization for Economic Co-operation and Development.
Pedrero, F., Alarcón, J. J., Abellán, M., & Perez-Cutillas, P. (2016). Optimization of the use of reclaimed water through groundwater recharge, using a geographic information system. *Desalination and Water Treatment, 57*(11), 4864–4877.
Purkey, D. R., & Huber-Lee, A. (2006). A DSS for long-term water utility planning. *Southwest Hydrology, 5*(4), 18–31.
Pyne, R. D. G. (2005). *Aquifer storage recovery, a guide to groundwater recharge through wells*. Gainesville, FL: ASR Systems.
Rahman, M. A., Rusteberg, B., Gogu, R. C., Ferreira, J. L., & Sauter, M. (2012). A new spatial multicriteria decision support tool for site selection for implementation of managed aquifer recharge. *Journal of Environmental Management, 99,* 61–75.
Reisner, M. (1999). *Cadillac desert: The American West and its disappearing water* (2nd ed.). New York: Penguin.
Russo, T., Fisher, A. T., & Lockwood, B. S. (2015). Assessment of managed aquifer recharge site suitability using a GIS and modeling. *Groundwater, 53*(3), 389–400.
Stewart, T. J., & Scott, L. (1995). A scenario-based framework for multicriteria decision analysis in water resources planning. *Water Resources Research, 31*(11), 2835–2843.
Todd, D. K. (1965). Economics of ground water recharge. *Journal of The Hydraulic Division, Proceeding of The American Society of Civil Engineers, 91,* 249–270.

Chapter 9
MAR Hydrogeological and Hydrochemistry Evaluation Techniques

9.1 Information Needs

Hydrogeological and hydrochemical data collection programs for managed aquifer recharge (MAR) systems need to be designed to meet project-specific information requirements. Project-specific data requirements are determined during desktop feasibility evaluations, which have varying effort levels and degrees of sophistication. The scope of work for field testing programs depends upon the data requirements for further feasibility assessment, system design, and environmental and human health impact assessments, and how much of the required data that can be obtained of sufficient quality from existing sources.

Projects vary greatly in their scale, sophistication, and budgets. Cleary the data requirements for the design of multiple-well salinity barrier and aquifer storage and recovery (ASR) systems and major riverbank filtration systems are much greater than that required for a small stormwater infiltration basin. If numerical groundwater modeling is an integral part of the assessment and design of a project, then the field investigation program should be tailored to obtaining the specific data needed to construct and populate the model. The amount and type of data required for projects where solute transport and water quality are of concern (e.g., salinity barriers and ASR systems in which freshwater is stored in a brackish aquifer) tend to be considerably greater than the requirements for projects concerned only with hydraulic impacts (e.g., MAR systems in which freshwater is recharged into freshwater aquifers). More detailed and higher quality water chemistry data are needed for geochemical modeling.

Inasmuch as projects have finite budgets, prioritization is an important part of development of testing programs. Program elements should be selected to preferentially obtain the most critically needed data. Consideration also needs to be given to the relative data quality from the different techniques and costs. Basic MAR technical issues and information requirements are summarized in Table 9.1.

R. G. Maliva, *Anthropogenic Aquifer Recharge*,
Springer Hydrogeology, https://doi.org/10.1007/978-3-030-11084-0_9

Table 9.1 Summary of MAR data requirements

Technical issue	Primary information requirements
System unit capacity (e.g., recharge rate per injection well or infiltration rate per unit surface area)	Aquifer transmissivity, well efficiency and specific capacity, infiltration rates, unsaturated and saturated vertical and horizontal hydraulic conductivities
Potential cumulative system capacity—overall recharge rate possible at a project site	Aquifer transmissivity and unit capacities
Fate of injected water (i.e., where recharged water flows and its hydrologic impacts)	Aquifer heterogeneity (with respect to hydraulic conductivity), effective porosity
Clogging potential—maintainability of recharge rates over time (system operation)	Recharged water quality, aquifer porosity, pore types, pore-size distribution, and aquifer heterogeneity
Water quality changes during and after recharge (beneficial and adverse)	Aquifer mineralogy and native groundwater and recharge water quality (chemistry)
Recharged water recoverability (recovery efficiency of ASR systems)	Native groundwater quality (salinity), aquifer heterogeneity (dispersivity), local hydraulic gradient, confining strata properties

9.2 Testing Methods Overview

Aquifer and vadose zone characterization techniques are discussed to varying degrees in most general hydrogeology and groundwater textbooks and in dedicated texts. Hydrogeology field (aquifer characterization) techniques were reviewed by Assad et al. (2004), Weight (2008) and Maliva (2016). Vadose (unsaturated) zone characterization techniques were reviewed by Wilson et al. (1995) and Stephens (1996). Comprehensive reference books on water well construction and testing were prepared by Driscoll (1986), Roscoe Moss Company (1990), Misstear et al. (2006), and Sterrett (2007). Maliva and Missimer (2010, 2012) provided an overview of aquifer characterization techniques applicable to ASR, MAR (in general), and arid lands investigations. The U.S. Geological Survey published a series of Techniques of Water-Resources Investigations Reports, which are available online (http://pubs.usgs.gov/twri/). ASTM International (www.astm.org) has published over 12,000 technical standards, some of which on soil and aquifer characterization are referenced herein with respect to specific testing methods. ASTM standards are developed by voluntary committees and are considered recommendations, unless they are specifically referenced as a mandatory requirement in a contract (i.e., project specifications) or governmental regulation.

Aquifer characterization techniques commonly applied in MAR investigations are summarized in Tables 9.2 and 9.3 by information types. The application, advantages, and limitations of the techniques are addressed and references are provided to more detailed discussions.

Table 9.2 Techniques to determine aquifer hydraulic and transport parameters and infiltration rates

Information	Field technique
Aquifer transmissivity, hydraulic conductivity, storativity, and well capacity	• Single- and multiple-well pumping tests • Slug tests
Aquifer heterogeneity (variations in hydraulic properties with depth)	• Packer (pumping and slug) tests • Pumping tests while drilling • Borehole geophysical logging (flowmeter and advanced logs) • Direct-push technology (unlithified strata) • Coring and permeability testing
Infiltration rates	• Infiltrometer tests (single and double ring) • Pilot (basin) infiltration tests
Unsaturated zone hydraulic properties	• In situ permeameter tests
Porosity	• Core analyses • Borehole geophysical logs
Pore types (primary versus secondary)	• Core descriptions • Borehole imaging logs • Optical and SEM petrography
Dispersivity	• Tracer tests
Depth, orientation, and continuity of strata	• Surface geophysics

Table 9.3 Techniques used for geochemical evaluations

Information	Techniques
Discrete zone water sample collection	• Packer test • Direct-push technology • Monitoring wells
Water quality (salinity) profiles	• Borehole geophysical logging • Surface geophysical surveys • Direct-push technology • Packer testing
Lithology	• Well cuttings • Cores • Borehole geophysical logging
Mineralogy	• Thin section petrography • X-ray diffraction analysis
Fluid-rock interactions (geochemical compatibility)	• Laboratory bench-top testing • Push-pull tracer tests • Geochemical modeling
Clogging potential	• Core flow-through testing • Source water quality analyses

9.3 Exploratory Wells

Exploratory well programs should be designed to most cost effectively obtain required data. Although exploratory wells are drilled using some of the same methods used for water production wells, they differ in that their main purpose is aquifer testing. Hence, the drilling and construction methods of exploratory wells should be designed specifically to accommodate planned testing elements. For example, some borehole geophysical logs require an open (non-cased) hole with a diameter between tool-specific minimum and maximum values (Maliva and Missimer 2010). Exploratory well drilling and construction programs should be designed to provide the required borehole conditions, rather than pre-determined well drilling plans constraining the testing program. Exploratory wells should be ideally constructed so that the wells could serve a subsequent required function for the MAR system. For example, exploratory wells for ASR systems are often completed as storage-zone monitoring wells.

Exploratory wells can also be designed to be used as one of the recharge wells for an MAR system, in which case any specific design requirements for recharge wells need to be considered. Recharge wells should be constructed of materials that will not significantly corrode in the site-specific water chemistry conditions. Recharge wells (and injection wells in general) are vulnerable to clogging. Therefore, recharge wells should be designed, to the extent possible, to minimize clogging potential and accommodate some clogging. Wells should be constructed and developed to maximize well efficiency. In screened wells, greater slot sizes and open areas than typically used for production wells will allow for retention of well capacity after some clogging (Maliva and Missimer 2010). Well and wellhead design should also readily allow for well rehabilitation methods.

The main well drilling methods used for MAR exploratory wells, and their advantages and limitations are summarized in Table 9.4. Drilling techniques used for aquifer characterization were reviewed by Maliva (2016).

9.3.1 Mud-Rotary Method

The direct mud-rotary method is widely used for drilling groundwater wells in both consolidated and unconsolidated strata. The borehole is drilled using a rotating bit that is attached to the bottom of a string of drill pipe. The most commonly used drill bit is the tricone roller-type bit, which consists of three conically shaped rollers mounted with hardened steel or tungsten carbide teeth (Fig. 9.1). Drilling fluid is circulated down the drill pipe, out the bit, and up to land surface through the annulus between the borehole wall and drill pipe (Fig. 9.2). Drill cuttings are removed from the circulating drilling fluid by either settling in mud pits or using desanders and a shale shaker.

Table 9.4 Summary of exploratory well drilling methods

Method	Advantages	Limitations
Mud-rotary	• Flexibility, can be used in most lithologies	• Aquifer testing and water sampling during drilling is difficult • Drilling mud must be completely developed from the formation
Direct air-rotary	• Drilling mud is not used, which reduces development requirements • Loss of circulation is usually not a problem	• Greater costs • Drilled strata should produce adequate water for transport of cuttings to the surface • Formation must be stable (lithified/consolidated)
Reverse-air rotary	• Drilling mud is not used and development is minimized • Superior cuttings recovery • Ready monitoring of water quality trends while drilling	• Requires a consolidated formation • Not suitable for unsaturated and poorly productive strata
Dual-tube methods	• Flexibility, can be used in most lithologies • Drilling mud is not required, reducing development • Superior cuttings recovery (core-sized pieces) • Superior water sampling while drilling • Well completion and some testing can be performed in unconsolidated formations	• Equipment is not widely available • Small diameters borehole may needed to reamed to install a casing and screen
Dual-rotary	• Allows setting of casings and screens in unconsolidated and unstable formations • Slant wells can be drilled	• Limited equipment availability in some areas
Cable-tool	• Suitable for drilling through coarse glacial till, boulder deposits, and highly fractured or cavernous strata • Formation is cased off during drilling, which allows for the collection of discrete cuttings and water samples	• Slow drilling rates and thus higher costs than other methods • Limited contractor availability in some areas
Rotary-sonic	• Rapidity • Good-quality continuous core samples can be recovered • Minimal generation of solid waste (well cuttings)	• Some sample deformation (compaction and core growth may occur)
Hollow-stem auger	• Rapidity • Low cost • Suitable for unconsolidated strata	• Drilling and well installation can be difficult in hard consolidated rock and in glacial deposits with coarse cobbles and boulders
Wireline coring	• Recovery of high-quality cores of lithified strata • Speed	• Small diameter corehole may require reaming for testing, geophysical logging, and completion as a monitoring well

Fig. 9.1 Photos of bits used in water well drilling. **a** Tri-cone bit. **b** Tiered reamer bit with lead tri-cone bit. **c** Wing or drag bit

Bentonite mud is the most commonly used drilling fluid, but a variety of other drilling fluids (e.g., biodegradable organic polymers) and additives are used depending upon site geological conditions. The drilling fluid cools, lubricates, and cleans the drill bit and transports cuttings to the surface. The viscosity and density of drilling fluids should be monitored during drilling. The viscosity of the drilling fluid should be high enough to allow cuttings to be carried to the surface while not being so high as to impede pumping. In unconsolidated formations, the weight of the drilling fluid (determined by its density and the drilling depth) should results in a hydrostatic

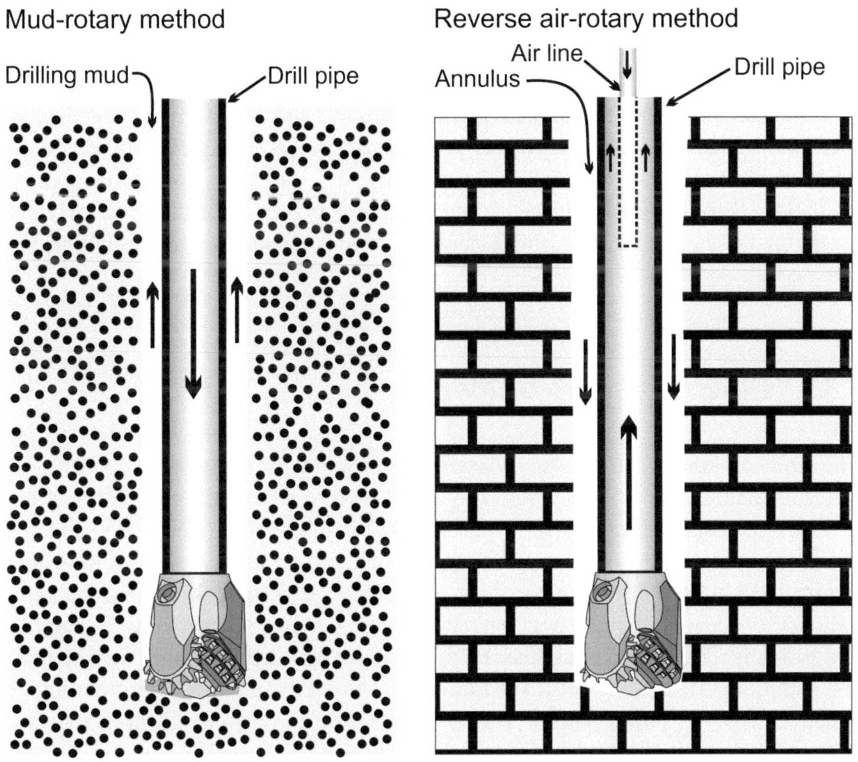

Fig. 9.2 Mud-rotary and reverse-air drilling diagrams. In mud-rotary drilling, mud is pumped down the drill pipe and returns to the surface with entrained cuttings through the annulus between the drill pipe and the formation. In reverse-air rotary drilling, water is pumped up the drill pipe using an air line

pressure sufficient to prevent the borehole from collapsing and fluid flow into the borehole. Barite is often added to drilling mud to increase its density.

The greater hydrostatic pressure in a borehole than in the adjoining formation results in the flow (invasion) of drilling fluid into the formation, which will preferentially occur in more permeable strata. In extreme situations, drilling fluid will flow into high-permeability intervals rather than upward through the annulus, which is referred to as a loss of circulation. Suspended material within the drilling fluid (mud and fine cuttings) are filtered out to varying degrees at the borehole wall forming a "mudcake," which tends to be thickest opposite the more permeable strata. Wells drilled using the mud-rotary methods must be thoroughly developed to remove mudcake and drilling fluids that entered into (i.e., invaded) injection and production zones. Well development (i.e., removal of drilling fluids and formation damage) is especially critical for recharge wells because injection will tend to force residual drilling fluids into the formation causing clogging.

A main limitation of the mud-rotary drilling method is that collection of water-quality samples and aquifer hydraulic testing during drilling are time-consuming, and thus expensive, as the drilling fluid needs to be removed from the tested interval. In consolidated formations, water-quality sampling and hydraulic testing may be performed using straddle or single (off-bottom) packers. A temporary well screen and artificial filter pack are usually installed for sampling and testing of unconsolidated formations. Profiles of transmissivity and water quality-versus-depth can be obtained by the installation of a series of temporary screens during drilling, although this process is time consuming.

9.3.2 *Direct Air-Rotary Drilling*

The direct air-rotary method uses compressed air and commonly small quantities of water or foam as the drilling fluid. The method is used in consolidated formations in which the drilling fluid is not required to stabilize the borehole. The advantages of air-rotary drilling are the rapid transport of cuttings to land surface (and thus minimal mixing of cutting samples from different depths) and a cleaner borehole that requires less development. Direct air-rotary drilling is used where loss of circulation is a problem. Well yields can also be estimated while drilling using this method, and sampling and analysis of discharge water samples can provide information on changes in water quality with depth. Air and water have a much lower viscosity than drilling mud and thus a lesser capacity to transport cuttings to land surface. However, the addition of organic polymers (foaming agents) can greatly increase the viscosity of the drilling fluids. The main limitations of direct air-rotary drilling are greater costs associated with the required large-capacity compressor and the penetrated strata need to produce adequate water for the transport of cuttings. The method is not recommended for locations with a deep water table.

9.3.3 *Reverse-Air Rotary Method*

The reverse-air rotary method reverses the drilling fluid circulation employed in direct-rotary methods by using an air line within the drill pipe to pump water. Water and cuttings travel upward through the drill pipe from the drill bit to land surface (rather than upward through the annulus between the drill pipe and formation; Fig. 9.2). The produced water may be either discharged to waste (open-circulation method), or after the cuttings are removed using desanders and a shale shaker, returned to the well through the annulus between the drill pipe and borehole wall (closed-circulation method).

The reverse-air rotary method is the preferred method for drilling through consolidated, water-producing strata. The potential for drilling fluid-induced clogging of the formation is greatly reduced by not using drilling muds. The flow of water from the formation into the drill pipe acts to develop the well. Cuttings tend to be clean (not covered with drilling mud) and are transported rapidly up the small-diameter drill pipe to land surface, which reduces depth errors and the mixing of materials. Well yield and water quality can also be estimated while drilling using this method.

Water circulates downward through the annulus to the drill bit. Hence, the composition of produced water reflects a mixture of water produced at the drill bit and the circulating water. Changes in the chemistry of the produced water monitored at land surface usually reflect changes in the chemistry of the groundwater at the bottom of the hole. An increase in the salinity of the water produced at a given depth may indicate that a productive horizon of higher salinity water has been encountered, but the salinity of the produced water may not reflect the actual salinity of the native groundwater in the productive horizon due to mixing.

The reverse-air rotary method is only suitable for consolidated or lithified formations with stable boreholes because there is no drilling mud to stabilize the borehole. The method also requires that the formation produce sufficient water to allow for an upward flow in the drill pipe to land surface. The reverse-air rotary method is not appropriate for drilling through the unsaturated zone or in very poorly productive strata.

9.3.4 Dual-Tube Methods

Dual-tube methods utilize a second concentric tube or pipe to stabilize the borehole during drilling. A reverse-circulation flow of drilling fluids is used whereby the flow is downward between the two drill pipes and upward through the inner pipe. Dual-tube (or dual-wall) drilling is performed using either the rotary or percussion method (Strauss et al. 1989). In the rotary method, a roller bit is attached to the inner drill pipe. The bit is usually only one nominal size larger than the diameter of the outer casing to minimize the width of the annulus between the outer pipe and formation (Fig. 9.3). For percussion drilling, an open-face bit is used, which is attached to the outer pipe and driven by an aboveground pile hammer.

Dual-tube methods are attractive for MAR exploratory well programs because:

- they can be used in both consolidated and unconsolidated formations
- they do not require the use of drilling muds
- the small annulus between the outer drill pipe and formation minimizes fluid flow in the annulus, which allows for more representative water sample to be obtained by pumping (air lifting) through the drill string without a downward return flow
- superior cutting samples can be obtained.

Large core pieces can be recovered using the dual-tube method if a hollow bit is used, although the recovery tends to be less than usually obtained using standard coring methods. In unconsolidated formations, the borehole will likely cave in as the drill string is removed. Geophysical logging and screen installation can be performed in unconsolidated strata by introducing drilling mud to stabilize the borehole.

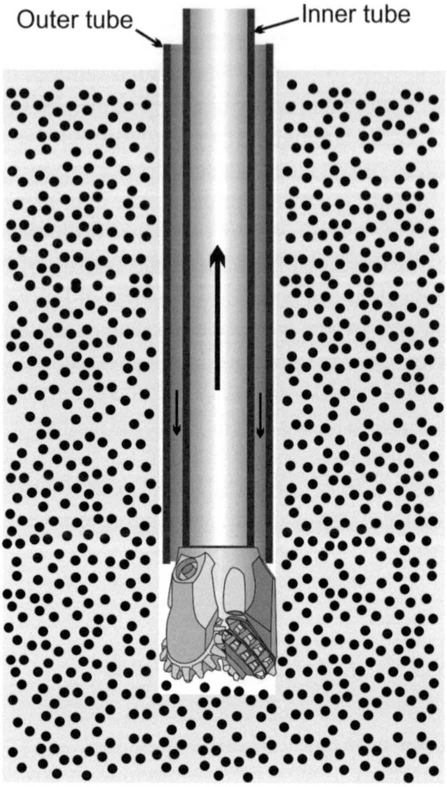

Fig. 9.3 Schematic diagram of the dual-tube reverse-circulation rotary drilling method. The very small annulus between the outer tube and the formation minimizes production of formation water from above the bit during drilling

Depending upon the type of drill bit used, small diameter (1–2 in.; 2.5–5.1 cm) screens may be installed through the drill pipe (Driscoll 1986). In aquifer characterization programs, dual-tube methods can be first used to obtain higher quality lithological samples and water quality data (e.g., salinity-versus-depth profiles). The borehole may later be reamed to obtain a sufficient diameter for well completion. Such an approach was successfully used by the author for an ASR exploratory at the Seminole Tribe of Florida, Brighton Reservation.

9.3.5 Dual-Rotary Drilling

The dual-rotary drilling method was developed for efficient drilling through unconsolidated formations. Dual-rotary drilling rigs are commonly referred to as "Barber" drilling rigs as the drilling technology was developed in 1979 by Barber Industries (now Foremost Industries; Herrick 1994; Henahan 1999; Foremost Industries 2003). The dual-rotary drilling rig has two independent drive units. The lower drive unit advances an outer casing to which a carbide studded shoe is welded to the bottom joint (pipe segment). A top-drive rotary head handles an inner drill string equipped with either a down-hole hammer, drag bit, or roller cone bit, for drilling inside or ahead of the casing. The top and lower drive units are operated independently, which allows the drill bit to be positioned either ahead of or behind the casing shoe. Depth-specific water and formation samples can be obtained because the outer casing largely seals off the overlying strata, minimizing the potential for cross contamination.

The major advantages of the dual-rotary method are that the outer casing keeps the borehole open and drilling fluids are not required to stabilize the hole. The dual-rotary method is commonly used is central Florida to set a casing through epikarst, which is prone to borehole collapse, sand production, and loss of circulation. Once the target depth is reached, the inner drill string is removed and the casing and/or screen are installed. The outer casing is simultaneously pulled out as the filter pack and sealing material (grout) are added. A limitation of dual-rotary drilling is that it requires specialized drilling rigs, which may not be locally available.

9.3.6 Cable-Tool Drilling

The basic cable-tool drilling method consists of repeatedly raising and dropping a heavy string of drilling tools with a chisel-shaped bit into a borehole, which breaks, crushes and loosens the formation. The crushed or loosened material forms a slurry in the well, which is removed by either bailing or using a sand pump. The drill bit is slightly rotated each stroke to form a circular borehole. The well is kept open during drilling (especially when drilling through unconsolidated or unstable formations) by lowering a steel casing, which may be either the permanent casing or a temporary casing. Where a temporary casing is used, a permanent casing and well screen are installed by lowering them inside the temporary casing. A filter pack is then added and the well grouted as the temporary casing is lifted out of the well.

The cable-tool method may be the best, and in some cases the only method available, for drilling in coarse glacial till, boulder deposits, and aquifers that are highly fractured, disturbed, or cavernous (Driscoll 1986). Loss of circulation of drilling fluids is not a problem because the cable-tool system does not require circulation of drilling fluids to remove cuttings. Cable-tool drilling allows for accurate water and lithologic sampling during drilling as the overlying strata are cased off during drilling. The impacts from the drill bit may also induce fracturing that enhances permeability near the borehole.

Cable-tool rigs are relatively reliable and inexpensive. The major disadvantages of the cable-tool method include that the penetration rates are relatively slow and that heavier walled steel casings may be required (Driscoll 1986). Cable-tool-drilled wells are often slower to construct and, therefore, more expensive. Cable-tool drillers may not be locally available.

9.3.7 Rotary-Sonic Drilling

The rotary sonic-drilling method, also referred to as the sonic and rotasonic method, utilizes high-frequency vibrational energy with downward pressure and rotation to advance the drilling tool. Sonic drilling is a dual-tube technique in which an inner drill string and core barrel are vibrated into the formation. An outer override casing is next advanced to seal off the upper strata and prevent collapse of the borehole. A one-piece core barrel is commonly used, but split barrels are also available, which allows for the recovery of less disturbed cores. The core barrel is then recovered and the core removed. Drilling proceeds by reinstalling the core barrel inside the override casing and driving it to the next depth. The inner drill string can be fitted with a screen for formation water sampling and hydraulic testing during drilling (ASTM 2004). Permanent monitoring wells are constructed by installing a screened casing and filter pack within the override casing. Grout is emplaced as the override casing is removed.

The rotary sonic method is typically restricted to depths of 150 m or less, although greater depths are possible by drilling in stages. The depth limitation is due to the dampening of vibrational energy, which is transmitted to the borehole wall (Stephan 1995). The principal advantages of rotary-sonic drilling are that it is rapid, good-quality continuous core samples can be recovered, a drilling fluid is not required, drilling can be performed through both lithified and unlithified strata, and there is minimal generation of solid wastes (well cuttings).

9.3.8 Hollow-Stem Auger Method

Hollow-stem augers are widely used for the installation of shallow (<50 m) monitoring wells because the method is relatively fast and inexpensive and, for most applications, does not involve the use of drilling fluids. The basic method is to drill to the target well depth with the bottom of the auger string sealed using either a pilot assembly consisting of a bottom plug, center bit, and center rod assembly or a disposable (commonly wooden) knock-out plate. The pilot assembly and auger string are connected to the spindle of the drilling rig using a double-adapter drive cap that ensures that the center rod and pilot assembly rotate along with the auger column. The augers act as a temporary casing to stabilize the borehole during drilling. The ability to withdraw the center plug and bit for sampling, while the augers are still in place, is a principal advantage of the hollow-stem auger method (Davis et al. 1991). Sediment samples can be collected during drilling using split spoons, Shelby (thin-walled) tubes, or core barrels.

Upon reaching total depth, the pilot assembly is removed or knockout plate dislodged and the well casing and screen are installed. The casing and screen are usually installed within the augers and the filter pack, a sealant (bentonite pellets or chips), and then Portland cement or bentonite slurry are added as the augers are withdrawn from the well.

A main disadvantage of the hollow-stem auger method is disturbance of the formation, particularly the smearing of clays and silts on more permeable sand and gravel intervals (Keely and Boateng 1987a, b). Mixing of sediments also occurs to varying degrees during drilling. Hollow-stem augering is most effective in unconsolidated muds, silts, and sands. Drilling and well installation can be difficult in hard consolidated rock (in which penetration can be very slow) and in glacial deposits with coarse cobbles and boulders.

9.3.9 Wireline Coring

Continuous wireline coring is a simple and economical method for obtaining long cores of lithified materials. The wireline coring system is a dual-wall system in which the bit is attached to the outer core barrel, which is attached to the drill string. A short-length inner core barrel is positioned at the base of the drill string. As drilling proceeds, the core is pushed into the inner core barrel. To remove the core, an overshot is lowered on the end of a wireline, which attaches to the top of the inner core barrel. When the wireline is pulled back, the inner core barrel is disengaged and retrieved to land surface. After the core is removed, the inner core barrel is lowered back to the bottom of the drill string where it re-engages to the base of the outer core barrel.

The major advantage of the wireline coring system is that the outer core barrel and drill string do not have to be tripped out of the hole to recover the core, which allows for rapid coring. Coring bits are available for drilling through different types of rock. High recovery of high-quality cores is often obtainable from well-lithified, unfractured strata.

Three standard wireline core sizes are used in groundwater investigations, which are designated NQ, HQ, and PQ (trademarks of the Boart Longyear Corporation). The NQ, HQ, and PQ cores have diameters of 4.78, 6.35, and 8.51 cm (1.88, 2.50, and 3.35 in.), respectively. The core holes are relatively small; the PQ hole is only 12.27 cm (4.83 in.) in diameter. Core holes may require reaming if they are to be geophysically logged or converted to a well. A pilot is recommended for reaming, which has a protruding rod in front of the bit to reduce the risk of deviation of the reamed hole from the core hole (i.e., a double-hole condition).

9.4 Aquifer Pumping Tests

9.4.1 Introduction

Conceptually, the performance of aquifer pumping or performance tests (APTs) is simple; one or more wells are pumped and water levels are recorded in the pumped and, ideally, a number of observation wells. Dedicated books have been written on the performance of APTs and interpretation of test data (e.g., Walton 1962, 1987; 1997; Stallman 1969; Kruseman and de Ridder 1991; Dawson and Istok 1991; Kasenow 1997, 2006). Basic pumping test methods and data analysis techniques are addressed in most general groundwater texts and reference books as the subject is so fundamental to groundwater investigations. Several software packages are available for the interpretation of pumping test data, including the Aquifer Test Pro and AQTESOLV commercial packages and the U.S. Geological Survey AQTESTSS series of spreadsheet programs (Halford and Kuniansky 2002). Excel-based programs for basic aquifer test analyses are available as freeware.

The basic technical and operational challenges for APTs lie in obtaining accurate data that are readily interpretable and representative of local aquifer conditions. Attention to detail is essential for a successful APT. Interpretation of APTs is typically based on time-versus-drawdown data at a constant pumping rate. Time-drawdown data are usually collected during both the initial pumping phase of a test and the subsequent recovery period after the pump is turned off. APTs may also be performed by injecting water at a constant rate.

Maliva and Missimer (2010) and Maliva (2016) provided some practical recommendations for performing APTs. Critical issues for a successful test and data interpretation include:

- achieving a constant, uninterrupted pumping rate throughout the test
- accurate recording of the time since the actual moment that pumping started and the corresponding water level changes (i.e., drawdown) in the pumped and observation wells
- accurate recording of the time since pumping was terminated and corresponding water levels changes during recovery
- correcting time-drawdown data for any changes in water levels that are not due to the pumping performed as part of the APT (e.g., from tides, rainfall, and pumping by other aquifer users)
- obtaining time-drawdown data at a sufficient frequency (particularly at the start of the test) to capture the shape of the time-versus-drawdown plot
- if practicable, tests should have sufficient duration to detect leakage effects and, in the case of unconfined aquifers, delayed-yield effects.

Produced water should be discharged in a manner so that it does not recharge the pumped aquifer during the test. APTs can be performed by recording water level changes in a single pumped well (single-well test) or by pumping one well (or much less commonly multiple wells) and monitoring water levels in one or more observation wells (piezometers). APTs using observation wells are preferred because (Maliva 2016):

- more accurate measurement of storativity values can be obtained
- time-drawdown data obtained from observation wells are less sensitive to well construction, well and formation clogging (skin damage), well-bore storage, and well development than data from pumped wells

- data from multiple observation wells can be interpreted using distance-drawdown methods
- observation well time-drawdown data are less sensitive to variations in pumping rate
- data from multiple observation wells can potentially detect directional aquifer anisotropy.

Single-well tests have the advantage of lower costs, especially if existing wells can be used and dedicated observation wells would need to be installed for a multiple-well test. In pumped wells, drawdowns from well losses may be greater than the formation response, which can make interpretation of the data difficult and introduce errors in the data analysis.

Water level changes are now usually measured using electronic pressure transducers and dataloggers, which have the great advantage of allowing for high-frequency, automatic measurement and recording. Older-style units have down-hole pressure transducers that are connected by a cable to a datalogger at land surface. Downhole water-level probes that contain both a pressure sensor and internal datalogger are now more commonly used. Probes may also have sensors that measure temperature, conductivity, and other parameters. Datalogging systems need to be programmed to start at the exact moment pumping begins or the start of pumping must be recognizable in the data. Where all the dataloggers are synchronized, the start of pumping can be estimated with sufficient accuracy from the initial drawdown in the pumped well, provided that water level measurements are recorded at a high frequency.

9.4.2 Pumping Test Data Analysis

Aquifer performance test data are usually interpreted using the Theis (1935) non-equilibrium equation, which is based on confined aquifer conditions, or variations thereof developed to interpret data from tests performed under conditions other than those assumed for the equation. For example, Hantush and Jacob (1955) and Walton (1960, 1962) modified the Theis non-equilibrium equation to provide a solution for leaky confined aquifers with no storage in confining layers. All solutions have underlying assumptions. It important to consider whether test conditions meet the assumptions of a solution and, if not, the potential impacts of the uncertainty or deviations on calculated aquifer parameters. Too frequently, test data are input into aquifer test analysis software with little evaluation of the quality of data, the appropriateness of the methodology, and any limitations of the interpretations.

Elementary interpretation of APT data using the Theis equation is covered in virtually all introductory groundwater textbooks. The Theis (1935) method is a curve-matching technique based on the equations

$$u = \frac{r^2 S}{4Tt} \tag{9.1}$$

$$s = \frac{Q}{4\pi T} W(u) \tag{9.2}$$

where

s	drawdown (m)
Q	pumping/discharge rate (m^3/d)
T	transmissivity (m^2/d)
u	an empirically derived function
r	distance between pumped and observation wells (m)
t	time since pumping/discharge began (m)
S	storage coefficient (unitless)
$W(u)$	the "well function" or "Theis well function"

The logarithmic plot of $W(u)$ versus *1/u* values (included in most groundwater textbooks) is referred to as the Theis curve (Fig. 9.4). Pumping test data are interpreted using the Theis method by plotting the time-drawdown data from an APT on a square, logarithmic grid of the same grid size as a plot of $W(u)$ versus *1/u*. The graphs are shifted until the time-drawdown data are superimposed on the Theis curve with the axes of the two graphs parallel (Fig. 9.4). A match point is selected, at which a set of values of s, t, $W(u)$, and u are obtained. Commonly $W(u) = 1$, and $1/u = 1$ are used (to simplify calculations) and the s and t values are recorded for that match point. Transmissivity and storativity values are calculated using Eqs. 9.1 and 9.2 from the s, t, $W(u)$, and u values.

The Theis non-equilibrium equation is based on the following assumptions:

- discharge from the pumped well is constant
- the pumped well fully penetrates and is open through the entire thickness of the aquifer
- flow into the well is radial, horizontal, and laminar
- the aquifer is homogenous and isotropic
- the aquifer thickness is uniform
- the aquifer is confined and remains saturated throughout the entire test
- the aquifer is of infinite areal extent
- the potentiometric surface is flat
- the radius of the well is very small so that casing storage in negligible
- instantaneous removal of water from storage with a decline in head.

Examination of the shape of the Theis curve reveals the importance of collection of accurate early test data to capture the bend of the curve and obtain the correct, unique curve-match. If the late test data plot off the Theis curve, then aquifer conditions have departed from the assumption of the method. In the case of leaky aquifers, late test data will plot below the Theis curve. In a bounded aquifer, late test data may plot above the Theis curve.

Cooper and Jacob (1946) published a modification of the Theis non-equilibrium equation (also known as the "straight-line" method), which uses a semi-logarithmic plot of drawdown versus time, with drawdown on the linear scale. The Cooper and

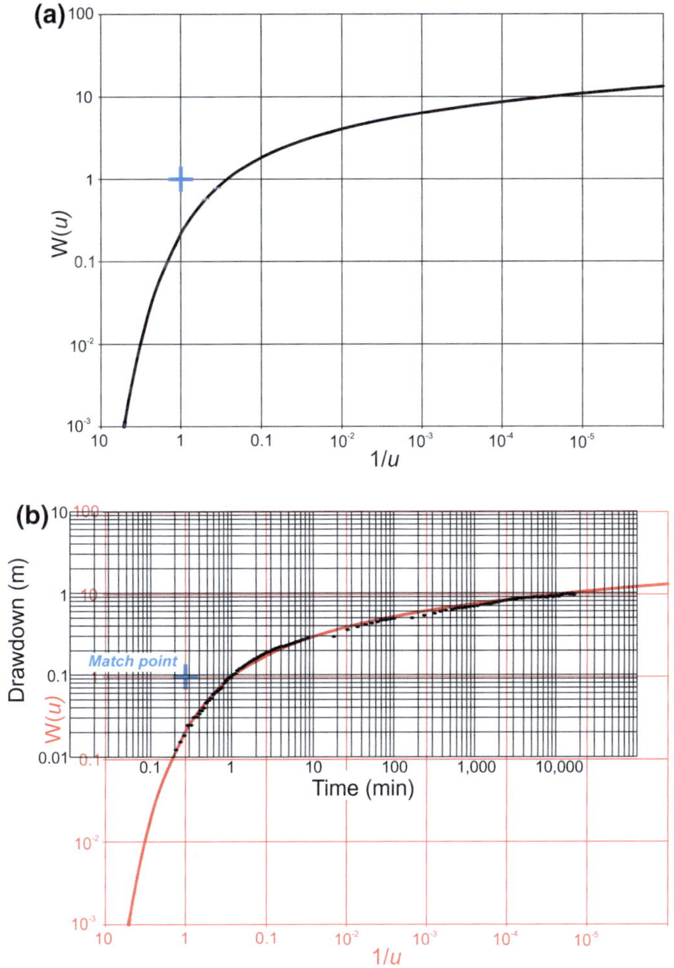

Fig. 9.4 **a** Theis curve. **b** Application of Theis method to observation well data from an APT in Lakeland, Florida. The match points are W(u) = 1, 1/u = 1, s = 0.09 m and t = 0.27 min, and the calculated transmissivity is 13,700 m^2/d for a pumping rate of 16,400 m^3/d

Jacob method is perhaps the most widely used method to analyze pumping test data because of its simplicity. Transmissivity and storativity are estimated from the slope of the plot (Δs, change in drawdown over 1 log cycle) as follows:

$$T = \frac{2.3Q}{4\pi\,\Delta s} \tag{9.3}$$

$$S = \frac{2.25T\,t_o}{r^2} \tag{9.4}$$

where t_o = time at drawdown $s = 0$. The units of the parameters must be consistent (e.g., $T = m^2/d$, $Q = m^3/d$, $t = d$, $r = m$). The Cooper-Jacob method has all the assumptions of the Theis method, in addition to the requirement that the value of μ be very small (less than about 0.05 or 0.01). The Cooper and Jacob method was designed for confined aquifers but can be used with caution for unconfined aquifers (Walton 1962).

The Cooper and Jacob method is valid only for data that plot on the Theis curve on log-log plots (i.e., data that meet the assumptions of the Theis equation). Departures of data from a straight line on semi-log plots indicate that test conditions have departed from the Theis assumptions. For example, a flattening of the time-drawdown curve (deflection to the right), may be evidence that water is being added to the aquifer and, as a result, there is less drawdown at given times during the latter part of the test. Water may be added to an aquifer by leakage through adjoining aquitards (semiconfining units) or recharge (e.g., to an unconfined aquifer from a surface water body or precipitation). If the time-drawdown curve becomes horizontal (i.e., there is no change in water levels over time), groundwater pumping is being balanced by water added to the aquifer from leakage or recharge. It is good practice to always prepare a log-log plot to determine what part of the time-drawdown data plot on the Theis curve.

9.4.3 Water Quality Testing

The discharge water should be periodically analyzed for field parameters (e.g., specific conductance, temperature, pH) and basic salinity parameters (e.g., chloride), especially if the aquifer and adjoining strata contain waters of markedly different quality. Particular attention should be placed to the stability of water chemistry where the tested aquifer is underlain by more saline groundwater and/or the site is located near the coast (i.e., in horizontal proximity to the saline-water interface). Significant changes in water quality during pumping may be indicative of enhanced hydraulic connection with poorer quality water (e.g., through fractured zone or conduits), which can adversely impact the performance of an MAR system. A pronounced increase in salinity during an APT may be a prognosticator of poor recovery efficiency for an ASR system.

9.5 Slug Testing

Slug tests are used to determine the hydraulic conductivity of strata from rate of change of water level in a well after the instantaneous addition or withdrawal of a volume of water (i.e., a "slug"). The tests are now commonly performed by either the insertion or withdrawal of a solid pipe or by inducing water level changes in a well using air pressure or a vacuum. Water level data are usually recorded using pressure transducer and datalogging systems because they allow for accurate measurement and recording of water levels at small time intervals. Aquifer hydraulic conductivity is calculated from the water level-versus-time response and equations appropriate for the specific test conditions. Slug testing procedures were reviewed by Butler (1998), Cunningham and Schalk (2001), Weight (2008), Chen et al. (2012), and Maliva (2016).

Slug tests have the advantages that they are quick and inexpensive, can be performed on small-diameter wells, and do not produce water requiring disposal. The latter can be an important consideration for contaminated sites. Slug tests have a relatively small volume of investigation compared to aquifer pumping tests, which needs to be considered when utilizing the data. Slug test data can provided information on local aquifer heterogeneity, but the hydraulic conductivity values may not be representative of average aquifer conditions. A major disadvantage of slug tests is that they are susceptible to well skins and formation damage affecting test results. The rate of flow into a well can be impacted by clogging of the borehole wall. Hence, well development is a critical issue for slug testing programs.

Simple, inexpensive tools for slug tests can be readily fabricated by filling lengths of PVC or steel pipe of various diameters and lengths with concrete. An eyelet anchored in the concrete is used to attach a rope or monofilament line to the slug. A common methodology is to first perform a falling level (slug-in) test, whereby water level in a well is raised by rapidly lowering the tool into the well. After water levels have recovered to background levels, a rising level (slug-out) test is performed by quickly removing the tool.

Pneumatic slug tests are performed by lowering water levels in a well by pressurizing the well casing using either air or another gas (e.g., nitrogen). Pneumatic slug-out tests are performed by increasing the pressure in a casing to lower the water level and then quickly releasing the pressure, allowing water level to recover. Slug-in tests uses a vacuum pump to increase the water level in a well. A near instantaneous decrease in water level is achieved by quickly releasing the vacuum. Pneumatic tests have the advantage of allowing for a near instantaneous change in water level, which is critical for tests performed in formations with very high hydraulic conductivities.

Although slug tests are conceptually simple, attention to detail is required to obtain data that are representative of the formation. The most important issues are collection of accurate, high-frequency water level-versus-time data and ensuring that the tested wells are adequately developed. Butler et al. (1996) and Butler (1998) proposed guidelines for improving the performance and analysis of slug tests. Performance of multiple tests for each well (or well interval) using different displacements was recommended to allow the viability of the theory underlying the analysis model to be assessed and to evaluate well condition. Skin effects may be indicated by changes in calculated hydraulic conductivity values between successive tests on the same well or between slug-in and slug-out tests (Butler and Healey 1998).

Profiles of hydraulic conductivity with depth can be obtained by multilevel slug tests, which involves performing a series of slug tests on aquifer intervals isolated using single or straddle packers. Multilevel slug tests can be performed on wells with open-hole or screened completions. However, data from screened wells may

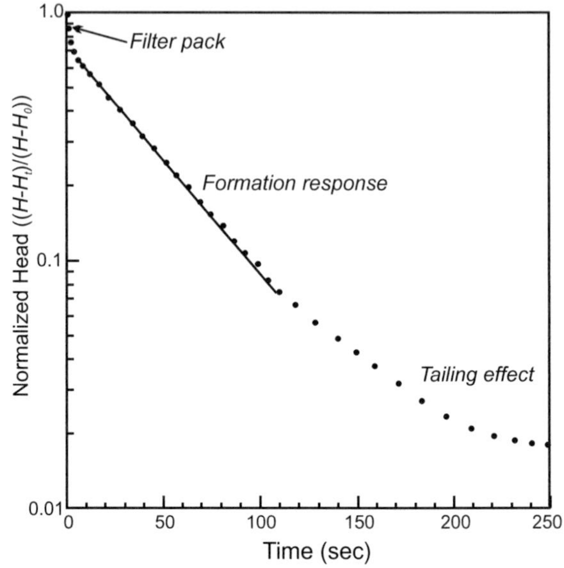

Fig. 9.5 Hypothetical time-normalized head data for a slug test showing the triple-line effect. The first steep segment represents drainage from the filter packer. The second segment, which includes most of the change in normalized heads, is the formation response, which is used for the test analysis. The third segment is a tailing effect and represents a small fraction of the total recovery

be compromised by bypass flow around packers through high-permeability backfill material or if the formation does not collapse against the screen (Meville et al. 1991).

A variety of different methods (and refinements thereof) have been developed to interpret the water level-versus-time data from slug tests. The most commonly used methods are those of Hvorslev (1951), Bouwer and Rice (1976), and Cooper et al. (1967), with the former two by far most popular. Slug tests data were originally interpreted using manual graphic methods. Commercial and freeware software packages are now available and commonly used for more automated interpretation of slug test data. The author has frequently observed incorrect test interpretations using software because either the wrong method was used or the wrong part of the test data was used. Slug test data may show a double or triple-line effect, in which early, steeper segment reflects drainage from the gravel pack (Fig. 9.5). The later, less-steep segment should be used to estimate the hydraulic conductivity of the formation.

A question that arises in slug interpretation is which of the different interpretation methods provides hydraulic conductivity values that most accurately reflect actual values. There is no consensus on the subject. The most appropriate methods depend upon the extent to which conditions at a tested well meet the underlying assumptions of the analytical methods. The values of hydraulic conditions (e.g., anisotropy ratios) at test sites are commonly not well constrained (Maliva 2016). A recommended approach is to interpret slug-test data using multiple methods and compare their corresponding hydraulic conductivity values. If the different methods give similar results, then the mean or median value might be utilized in an aquifer characterization program for at least an initial estimate of local aquifer hydraulic conductivity. However, agreement of values is not assurance that the calculated values are correct (i.e., are representative of the formation), if the rate of flow into the well is strongly influenced by skin effects. Field conditions, such as skin effects, can be a greater source of error than differences between the conventional interpretation techniques used for data interpretation (Hyder et al. 1994).

9.6 Packer Tests

Packer testing involves the isolation of part of a borehole for hydraulic testing and/or water quality sampling using inflatable packers. It is a particularly important technique for aquifer characterization for MAR programs because it allows for zonal isolation and thus greater vertical resolution of aquifer heterogeneity. A basic limitation of packer testing is that it requires a stable borehole, which is needed to both seat the packers and to allow for pumping of (or injection into) the tested interval. Packer testing is most commonly performed in groundwater investigation on open boreholes in lithified strata. Testing may also be performed on screened intervals, but the potential exists for leakage around the packers through the annulus (gravel pack) between the screen and formation. Tested intervals must also be developed to remove drilling fluids and minimize or eliminate skin effects (e.g., mudcake and formation damage).

The commonly used inflatable retrievable packers are constructed with rubber elements that are reinforced with a high-strength material (commonly Kevlar) and a pipe mandrel upon which the rubber elements are attached. The pipe mandrel also provides a conduit for pumping or injection of water and room for instrumentation cables. Packer systems can be installed using either drill pipe or a wire line. Inflatable packer systems are available for a wide range of bore-hole sizes, including, at the low-end of the spectrum, systems that can be run in Boart Longyear NQ (75.7-mm, 3-in. borehole diameter) core holes.

Aquifer pumping tests using packers are performed in a similar manner as aquifer pumping tests run on entire wells. The tested intervals are pumped at a constant rate and the change in water levels versus time during pumping and subsequent recovery are recorded. Packer test intervals may also be performed by injecting water. In addition to evaluating aquifer hydraulic parameters, injection tests can be used to evaluate clogging potential and through push-pull tracer tests, aquifer transport properties and fluid-rock interactions (Sect. 9.9).

Two basic types of packer tests are commonly performed in groundwater investigations: straddle (dual-packer) tests and single (off-bottom) tests (Fig. 9.6). Straddle-packer tests utilize two separate packers to isolate an intervening depth interval of a borehole. The length of the test interval can be adjusted by changing the spacing between the packers. Single-packer tests employ one packer that is used to isolate the bottom or top of a borehole. Straddle-packer tests can be run after drilling and geophysical logging of the entire borehole. The depth intervals for tests can be determined from review of lithological and geophysical logs. Multiple tests can also be efficiently performed by lowering or raising the packer assembly without having

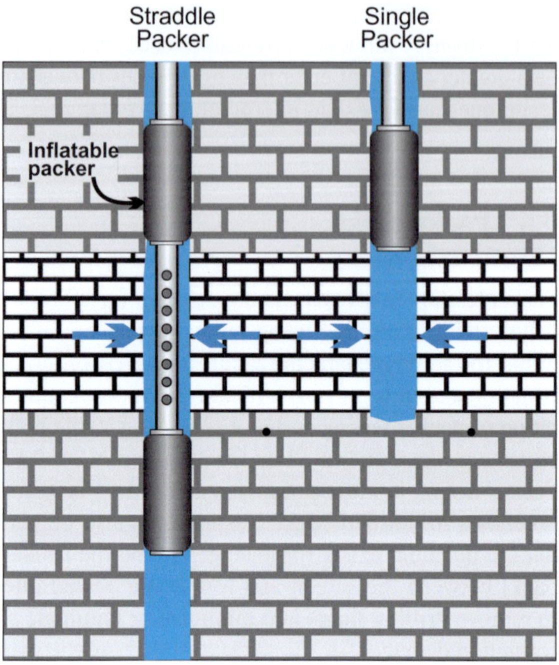

Fig. 9.6 Schematic diagram of straddle-packer and single-packer tests of a target unit (white)

to trip in and out of the hole for each tests. Single-packer tests have the advantage or being less prone to leakage as only one packer is set. Off-bottom testing can be performed immediately after drilling to a target depth, which may reduce the development time.

The main source of error associated with packer testing is short-circuiting of water around the packer elements, which can be caused by a poor packer seal (e.g., due to borehole roughness or irregularity), use of an improperly sized packer, or fracturing at the packer seat depth. A caliper log and/or borehole video should be run on the borehole to identify areas with borehole conditions that are suitable for the tight seating of a packer (a round, smooth, constant diameter interval).

Measured drawdowns may be impacted by wellbore effects such as skin effects and frictional head losses. Head losses due to friction will be much greater in packer tests because water is being pumped through a small-diameter drill pipe or tubing. Hence, drawdown measured in the drill pipe near land surface are often significantly greater than the drawdown (pressure decrease) measured using a transducer set in the packer zone.

Packer test data are analyzed using the same methods as applied to well testing (i.e., the Theis non-equilibrium method and modifications thereof). An average hydraulic conductivity value of the tested interval is obtained by dividing the transmissivity value obtained for the tested interval by the thickness of the interval. Packer tests usually do not follow the assumptions of the Theis non-equilibrium equation, particularly that partially penetrating conditions and, therefore, vertical flow occurs. The occurrence of vertical flow in packer tests tends to result in an over estimation of hydraulic conductivity (Johnson and Frederick 1997).

9.7 Testing and Sampling While Drilling

The properties of strata penetrated during drilling can be evaluated by performing a series of pumping tests while drilling to the total depth of a borehole. For each test, the calculated transmissivity value will increase as a greater thickness of aquifer strata is penetrated by the borehole. The transmissivity of the depth interval between two tests is approximately equal to the difference in the transmissivity between the tests. As a hypothetical example, if the transmissivity from a test performed on a borehole from 0 to 50 m below a casing is 100 m^2/d and the transmissivity from a successive test performed from 0 to 75 m below a casing is 300 m^2/d, then it can be estimated that the transmissivity of the 50–75 m depth interval is 200 m^2/d. The salinity of depth intervals can be estimated from the salinity of the water produced from each test and the fraction of water produced from each tested interval (assumed to be proportional to their transmissivity) by simple mixing equations.

Testing while drilling is particularly cost-effective when drilling using the reverse-air rotary method, in which case a test can be readily performed by tripping out the drill string and installing a temporary pump. This method has been employed in multiple ASR exploratory wells in Florida. Testing while drilling allows for the evaluation of the hydraulic properties and water quality of potential storage, monitoring, and confining zones. Short-term pumping tests can be performed while the drill string is still in well, but the drawdowns are impacted by head losses from the drill bit and pipe. A series of specific capacity (i.e., pumping rate divided by drawdown) measurements during reverse-air drilling (e.g., after each drill pipe is down) can provide semi-quantitative information on the location of transmissive intervals.

Testing while drilling is more involved when drilling using the mud-rotary method in unconsolidated strata. Hydraulic testing and water sampling can be performed by the installation of a temporary screen in the depth interval to be tested. A basic procedure is to drill a borehole to the bottom of each target test interval and then installing a temporary screen in the depth interval to be tested. The screen is attached either to the bottom of the drill pipe or a temporary casing. Depending upon borehole conditions, a gravel filter pack may have to be installed, which should be sealed with bentonite or another low permeability material. The screened interval is then developed to remove the drilling fluids and allow for collection of representative water samples. After testing is completed, the screen is removed and drilling resumed.

An alternative strategy is to drill the borehole to total depth, and then based on analysis of the cuttings and geophysical logs, determine the depth interval(s) to be screened. Testing would be performed from the bottom of the borehole upward.

9.8 Direct-Push Technology

Direct-push technology (DPT) includes a series of relatively low-cost methods used for the characterization of shallow unconsolidated or semi-consolidated strata. DPT involves the pushing or vibrating of a drive point (bit), screen, and drill rod into sediment or soft rock. The DPT tool string is usually advanced with a hydraulic ram supplemented with vehicle weight or, more commonly, high-frequency percussion hammers. The advantages of DPT over standard drilled test wells are greater speed and site accessibility, minimal generation of cuttings, and often lower costs (Butler et al. 1999; 2002). Applications of DPT include (Maliva 2016):

- discrete, one-tine, water sample collection
- water quality profiling (one-time collection of multiple samples at different depths at a given location
- installation of permanent small-diameter monitor wells
- collection of small diameter core samples
- slug testing (single or multiple)
- continuous electrical conductivity profiling (to map salinity and lithology changes)
- hydraulic conductivity profiling.

The implementation and development of new uses for direct-push technologies accelerated in the early 1990s and the development of new tools for aquifer characterization uses is expected to continue. DPT was initially used in aquifer characterization for water sampling. For example, DPT is as cost effective means to obtain numerous water samples used to delineate the extent of a contaminant plume. The water quality data might be used to guide the location of permanent monitoring and remediation recovery wells. More recently developed DPT tools are increasingly being used for stratigraphic, lithologic, and hydraulic conductivity profiling.

Direct-push electric conductivity (EC) logs are essentially small-diameter versions of borehole resistivity logs. The probe contains a series of closely spaced electrodes. Schulmeister et al. (2003) utilized an electrical conductivity tool that had four electrodes in a Wenner-type array (i.e., a collinear, evenly spaced configuration) with a spacing of 2 cm. DPT EC logging using direct-push systems can provide high-resolution data on changes in formation resistivity. Lithologic information can be obtained if the tested strata have significant differences in resistivity (the reciprocal of conductivity) and variations in the salinity (specific conductance) of the groundwater are small. EC logs may distinguish between clay-rich (low resistivity) confining strata and clean (clay-poor) aquifer sands and gravels that have a high resistivity. Where little variation occurs in the resistivity of sediment and rock types, DPT EC can be used to evaluate vertical changes in salinity.

The direct-push permeameter (DPP; Lowry et al. 1999, Butler et al. 2007), direct-push injection logger (DPIL; Dietrich et al. 2008) and hydraulic profiling tool (HPT; Kober et al. 2009) allow for evaluation of changes in hydraulic conductivity with depth (i.e., aquifer heterogeneity). These methods are based on the principal that changes in specific injectivity (ratio of injection rate and pressure) are related to changes in hydraulic conductivity. Water is injected at a constant rate and changes in pressure are measured using either one or more transducers located on the tool (either above or below the injection port or screen; DPP), within the tool, or at land surface (DPIL and HPT). The DPP and DPIL methods provide point measurements, whereas the HPT provides a continuous profile. The DPP provides a measure of actual hydraulic conductivity, whereas the DPIL and HST measure relative hydraulic conductivity. Actual hydraulic conductivity values are obtained by calibration of the injection rate/pressure profiles against hydraulic conductivity values obtained by DPP, direct-push slug tests, or sieve analysis data (Dietrich et al. 2008).

Greater information can be obtained from DPT by combining methods. Liu et al. (2009, 2012) described a prototype tool (high-resolution K Tool) that couples the DPP and DPIL into a single tool. Hydraulic conductivity values obtained from the DPP measurements are used to transform high-resolution DPIL data into actual hydraulic conductivity profiles. Kober et al. (2009) combined data from direct-push slug tests (DPST), DPIL, and HPT to generate an aquifer model that was evaluated against data from a natural-gradient tracer test. The combined DPIL and DPST data allowed for mapping of high and low-permeability zones in the relatively homogenous test aquifer. Model simulations using the DPT data showed a good reproduction of the measured tracer breakthrough.

A main limitation of DPT is that it can be used in only unlithified or poorly lithified strata that can be penetrated by the tools. DPT has considerable value for characterization of shallow aquifers as part of the initial field investigations for surface-spreading-type MAR systems.

9.9 Single-Well (Push-Pull) Tracer Tests

Tracer tests are an important tool in groundwater investigation as they provide information on the transport properties of aquifers, particularly flow paths and dispersivity. Single-well tracer tests (also referred to as push-pull and single-well pulse tests) involve injection of a known volume of water with a known concentration of a tracer into a well and then pumping the well to recover the tracer. A series of measurements of recovered volumes and tracer concentrations are recorded.

Single-well tracer tests have the advantages of requiring only a single well or well zone (isolated by packers) and that tests can be performed to recover all of the tracer, which may be a regulatory concern. Single-well tracer tests can be used to obtain an initial estimate of dispersivity values, but have the limitation of a relatively small volume of investigation (Gelhar and Collins 1971; Pickens and Grisak 1981; Fetter 1998).

The operational (cycle) testing of an ASR system is in essence a large-scale single-well tracer tests when the injected water has distinct differences in composition compared to the native groundwater. For example, longitudinal dispersivity and effective porosity values were estimated for the storage zone of an ASR system in Destin, Florida, through the calibration of a solute-transport model for the initial operational cycles (Maliva et al. 2013). Operational testing data from an ASR system were similarly used to evaluate the significance of secondary porosity (fractures) on groundwater flow at a site in Melbourne, Australia (Miotiński et al. 2011).

There is considerable potential for the use of single-well (push-pull) tracer tests to evaluate in-situ geochemical processes in aquifers to be used for MAR systems. The injected water should contain at least one non-reactive parameter that can be used as a tracer for mixing (Istok et al. 1997; Haggerty et al. 1998). Chloride is a preferred tracer if there is a difference in concentration between the injected water and native groundwater. For example, single-well tracer tests could be an effective means to evaluate the potential for adverse geochemical reactions in ASR systems, such as arsenic leaching (Norton 2007), as well as obtaining data on aquifer transport properties. Water chemically similar to the water to be stored would be injected and the recovered water analyzed for arsenic, metals, and other parameters of concern. The single-well tests could be performed on an exploratory well, monitoring well, or existing well to determine if adverse fluid-rock interactions will be a challenge at a site, in advance of the construction of a pilot or full-scale system (Maliva 2016).

9.10 Borehole Geophysical Logging

Borehole geophysical logging is a fundamental tool for aquifer characterization because it can provide essentially continuous in situ measurements of the properties of the logged strata, such as

- porosity
- aquifer heterogeneity (variations in hydraulic conductivity with depth), including the location of preferential flow zones
- location, size, and orientation of fractures and other secondary porosity features
- groundwater electrical conductivity (salinity)
- groundwater temperature
- aquifer mineralogy
- borehole conditions (diameter and shape)
- permeability (hydraulic conductivity).

The basic logging procedure involves slowly lowering into a borehole (or raising from the bottom) a tool (sonde) that is designed to measure one or more properties of the logged strata. The raw data are measurements of the properties versus depth, which are then processed to obtain the parameters of interest. Borehole geophysical logging methods and applications to groundwater investigations were reviewed by Keys (1989, 1990, 1997), Collier (1993), Wempe (2000), Kobr et al. (2005), Maliva et al. (2009), Maliva and Missimer (2010), and Maliva (2016).

Borehole geophysical logs can be roughly divided into basic and advanced logs based on their sophistication, information provided, availability, and cost to run. The basic geophysical log suite typically run for groundwater investigations includes some, or all, of the following logs: caliper, natural gamma ray, resistivity (long-short normal and dual induction), spontaneous potential, temperature, fluid resistivity, flowmeter, and sonic (acoustic). Borehole logs that use a radioactive source (neutron and density logs) are much less commonly run in groundwater investigations because of concerns over (or local prohibitions against) the use of radioactive sources in fresh-water aquifers. Advanced geophysical logs include imaging logs, nuclear magnetic resonance, and gamma ray spectroscopy logs (Maliva et al. 2009).

Borehole geophysical logging has long been a critical tool in the oil and gas industry, and most of the logs now used in groundwater investigations were originally developed for the oil and gas industry. Some of the basic geophysical logs that are still widely used for groundwater investigations have been supplanted in the oil and gas industry by more advanced versions. However, the older generation of logs still provide useful information and are typically inexpensive to run. Local geophysical loggers or well-drillers with logging equipment that can run at least some of the basic geophysical logs are present in most areas.

There is great variation in the sophistication of borehole geophysical log interpretation. Geophysical logs are commonly only qualitatively interpreted in groundwater investigations. Quantitative analyses allows for greater information to be extracted from logs, but requires greater technical expertise and rigorous quality assurance and control procedures to ensure that high-quality data are obtained. The greatest value is obtained from borehole geophysical logging when multiple logs are run in conjunction with other aquifer testing methods, such as aquifer pumping tests, packer tests, and cuttings and core analyses. For example, salinity-versus-depth profiles may be obtained using porosities obtained from a sonic log and formation resistivity data obtained from a deep resistivity (or dual induction log). Similarly, a hydraulic conductivity-versus-depth profile may be obtained from a core porosity-versus-hydraulic conductivity transform and a porosity log.

It is critical when developing (specifying) a logging program to understand the data provided by each log, the required borehole conditions, and the limitations of each log. Borehole geophysical logs are typically run on open (uncased) holes. However, there are some logs that can be run on cased wells, although the quality of the data tends to be less due to the attenuation of signals by the casing. Some logs can be run in borehole filled with drilling muds, whereas others require a water-filled and competent borehole. The main geophysical logs used in groundwater investigations and the information they directly or indirectly provide are summarized in Table 9.5.

All exploratory wells should be geophysically logged because of the valuable information provided. Exploratory well construction programs should accommodate the planned logging programs. When drilling using the mud-rotary method, a basic suite of caliper, natural gamma ray, resistivity (including spontaneous potential), and sonic log can provide information on:

- basic lithology
- depth control for well cuttings
- porosity
- salinity-versus-depth profile
- location of potential high permeability zones.

Table 9.5 Geophysical logs and information provided and uses

Log	Information
Caliper	• Borehole diameter and shape • Hardness of strata • Presence of large secondary pores (fractures or conduits)
Natural gamma ray	• Natural radioactivity of strata • Lithology (e.g., location of clay beds and clean sands) • Lithological correlation between wells.
Spontaneous potential	• Location of permeable beds • Location of shale or clay beds (confining units) • Inter-well correlation • Determination of formation water resistivity.
Electrical resistivity, Dual induction	• Groundwater resistivity (and thus salinity) • Identification of permeable zones from drilling-fluid invasion • Porosity estimation • Lithology • Inter-well correlation
Sonic (acoustic)	• Porosity • Presence of fractured zones
Fluid conductivity	• Conductivity (and thus salinity) of borehole fluids • Location of flow zones
Temperature	• Temperature • Location of flow zones
Flowmeter	• Relative or absolute transmissivity of strata • Location of flow and relatively tight (low permeability) zones
Borehole imaging (optical, acoustic and micro resistivity)	• Structural interpretations, such as the detection and orientation of faults, fractures, and structural dip • Differentiation between open and healed (closed) fractures • Measurement of fracture apertures, which is used to estimate fracture permeability • Identification and characterization of sedimentary structures, such as bedding (scale and orientation) • Stratigraphic and structural interpretations, including bedding (scale and orientation) and biogenic structures (e.g., bioturbation) • Detection and quantification of diagenetic structures, such as secondary porosity and nodules
Gamma ray spectroscopy (elemental spectroscopy)	• Mineralogy • Elemental composition
Nuclear magnetic resonance	• Porosity and pore size distribution • Permeability

Logging opportunities are greater when drilling is performed using water (e.g., reverse-air rotary method) in stable strata (i.e., borehole is not prone to collapse). Flowmeter, fluid conductivity and temperature logs can be used to locate flow zones in wells. Flowmeter logs, which measure the velocity of vertical flow within a well, can be used to apportion the total transmissivity of the logged interval (as determined from a pumping tests) between different depth intervals.

Caliper logs measure borehole diameter. Changes in diameter may indicate differences in hardness. Softer material tends to be eroded ("washed out") during drilling, whereas borehole diameter will tend to be close to bit size ("gauge") in hard, well-lithified strata. The response of some geophysical logs depends upon borehole diameter. For a given upwards volumetric flow rate during pumping, the velocity measured with a flowmeter log will be greater in a smaller borehole diameter. Hence, caliper logs are routinely run if flowmeter logging is to be performed.

Imaging log are also valuable tools for characterizing penetrated strata. Optical methods (e.g., borehole video and the more advanced optical borehole imaging logs) require that the borehole be filled with clear (non-turbid) water. Acoustic and microresistivity imaging logs can be run on both water and mud-filled boreholes.

Some application of advanced borehole geophysical logging to MAR investigations were reviewed by Maliva et al. (2009). Elemental spectroscopy logs have a neutron source and determine elemental composition, and in turn mineralogy, from emitted gamma rays (Barson et al. 2005). Nuclear magnetic resonance (NMR) logs provide data on pore size distribution, which can be further processed to obtain permeability. NMR logs can capture fine-scale aquifer heterogeneity.

Technical teams for major MAR project should include an expert in borehole geophysical logging who is familiar with all of the available logs, the information they provide, their limitations and borehole condition requirements, and costs. Another consideration is the local availability of loggers capable of running the various logs. Local loggers are available in many areas that can run some or all of the basic logs. Local contractors capable of running advanced geophysical logs may not be available outside of active oil and gas provinces. The costs of mobilizing a geophysical logger over a long distance may make running advanced geophysical logs cost prohibitive, unless the provided data are critical for a specific project.

9.11 Surface and Airborne Geophysics

Surface and airborne geophysical methods can be a valuable element of aquifer characterization programs for some MAR projects because they typically are less expensive and can be performed quicker than methods that require the drilling of boreholes. The lesser unit costs allow for a larger number of measurements to be performed and thus greater spatial coverage. The greater spatial coverage of surface and airborne geophysical techniques comes at the expense of lesser vertical resolution. Inversion methods used to interpret the data also provide non-unique solutions.

Table 9.6 Surface geophysical methods and provided information

Log	Information
Resistivity and electromagnetic methods	• Interface between saline and fresh groundwater • Contact between porous rock and impermeable (very high resistivity) bedrock • Location of relatively high resistivity freshwater-bearing coarse sediments (e.g., channel deposits) amidst less resistive clayey deposits • Top of the water table
Ground-penetrating radar (GPR)	• Depth to bedrock • Depth to the water table • Thickness of soil • Vadose zone structures (e.g., buried channels, layering of strata)
Surface nuclear magnetic resonance	• Saturated porosity volume • Pore size and thus hydraulic conductivity
Seismic reflection and refraction	• Stratigraphy—depth to strata with marked differences in acoustic impedance • Subsurface structures (e.g., folds, faults, collapse features)
Relative gravity surveys	• Changes in stored water volume • Storage and movement of recharged water in the unsaturated zone • Estimation of specific yield

Specific applications of surface geophysical methods for aquifer characterization include determination of (ASTM 1999):

- depth, thickness, and areal extent of soil and unconsolidated sediments
- depth to bedrock
- depth, thickness, and lateral continuity of rock layers
- depth to the water table
- salinity and salinity changes, such as the position of vertical and horizontal freshwater/saline-water interfaces
- location of fractures and fault zones
- location of voids (caverns) and sinkholes
- soil and rock properties.

Basic references on surface geophysics and applied geophysics include Zohdy et al. (1974), Telford et al. (1990), Eastern Research Group (1993), USACOE (1995); Reynolds (1997), Sharma (1997), American Society of Civil Engineers (1998), Milson (2003), Burger et al. (2006), Kirsch (2008), and Dentith and Mudge (2014). The main surface geophysical methods and the information they provide are summarized in Table 9.6.

The greatest applications of resistivity (or conductivity) based techniques, such as DC resistivity, frequency-domain electromagnetic (FEM) and time-domain electromagnetic (TDEM) methods, are for detecting and mapping the contacts between materials with significant differences in resistivity. Therefore, a critical consideration is that the target should have a significant resistivity contrast in order for it to be detected. For example, resistivity-based methods may not be able to detect the water table if it is not associated with a significant change in resistivity. Resistivity methods are well suited for detecting changes in salinity and have been demonstrated to be particularly useful for determining the location and shape of coastal freshwater/saline-water interfaces.

Ground-penetrating radar (GPR) uses pulses of electromagnetic radiation in the radio and microwave bands (10–3,000 MHz) to detect subsurface structures. The velocity of the pulses is primarily a function of the permittivity (dielectric constant) of the material. When the GPR pulse hits an object or layer with a different permittivity, part of the signal is reflected back to the surface and is detected by a receiving antenna. Another part of the wave energy continues to travel downward and may be reflected back by deeper reflectors. Low-conductivity materials, such as unsaturated sediment and solid rock, cause little signal attenuation and have relatively great penetration depths. Saturated sediments, on the contrary, have relatively high electrical conductivities and permittivities, and thus high signal attenuation. GPR can be performed rapidly, is relatively inexpensive to run, has high vertical and horizontal resolution, and usually close to real-time initial interpretation. Its main applications for MAR lies in characterization of the vadose zone for surface-spreading systems. GPR may be used, for example, to identify and map potential confining strata within the vadose zone and the top of bedrock.

Surface nuclear magnetic resonance (SNMR), which is also referred to as magnetic resonance soundings (MRS), is the only method that can directly detect freshwater in the subsurface and can provide information on pore-size distribution, which can be used to estimate hydraulic conductivity. Applications of SNMR to groundwater investigations were reviewed by Maliva (2016). The results of some earlier published studies demonstrated the value of SNMR as a screening tool, which combined with other surface geophysical methods, may allow for the identification of areas that are more likely to have higher transmissivities and greater well yields. Although it has been claimed that SNMR has already passed the experimental stage and is evolving into a useful tool for applied hydrogeophysics (Yaramanci and Müller-Petke 2009), the question still remain as to whether the informational benefits of the technology are commensurate with its costs.

Seismic reflection and refraction techniques are used to obtain information on subsurface structure from the reflection or refraction of sound (seismic) waves off the boundary between materials that have different acoustic impedances, which is the product of the seismic wave velocity and density of the rock. The equipment consists of a source, one or more receivers called geophones (hydrophones in waterborne surveys), and seismographs, which are equipment that record the geophone outputs. Seismic surveys may be performed both on land and on marine and freshwater bodies. Seismic surveys can identify subsurface structures, but do not provide information on whether identified features are hydraulically active. For example, a fault detected in a seismic survey could be either a flow conduit, an impermeable feature, or could have no impact on vertical and horizontal groundwater flow. For MAR systems, seismic reflection surveys can be used to map the boundary between unlithified aquifer strata and underlying bedrock, formation boundaries (if associated with a major lithological change), and large-scale structures within an aquifer. The question arises as to whether either a resistivity-based or a seismic method can most cost-effectively provide the required structural information.

Temporal changes in the mass of subsurface strata caused by changes in the volume of water may be detected by a series of high-resolution measurements of gravitation acceleration at land surface. Relative gravity surveys are typically performed for groundwater investigations in which a time series of measurements is performed of the gravity difference between survey points and a base station located in a stable area not effected by aquifer water level changes (e.g., bedrock area adjacent to a groundwater basin). Changes in water levels in unconfined aquifers caused by pumping or recharge (natural or artificial) can be quantified from the gravity change, if sufficiently large.

Microgravity data has particular value for monitoring of MAR in unconfined aquifers in which changes in well water levels are due to changes in water volume rather than mainly pressure. It has been used to increase the density of monitored points as a less expensive alternative to the installation of additional monitoring wells (e.g., Pool and Schmidt 1997; Howle et al. 2002; Davis et al. 2005, 2008; Chapman et al. 2008). Microgravity-determined water levels are less accurate than standard well measurements and continuous monitoring is not practical. Microgravity monitoring data augments, not replaces, water level data from monitoring wells. Relative microgravity data reflect changes in the total mass of water, which includes both the unsaturated zone and saturated zone. Where the vadose zone is thick and changes in the mass of water in zone are sufficiently large to be detectable by microgravity, relative microgravity surveys can be used to map the storage and movement of water in the zone.

It has not been uncommon for the results of surface geophysical investigations to fail to meet expectations. A thorough understanding of the theory, field procedures, and interpretation technique of each method, along with an understanding of local geology are necessary for successful completion of surface geophysical surveys (ASTM 1999). The most important question for surface geophysical investigations is whether the required data can be obtained using methods under consideration. For example, if the objective is to map a confining unit, then the basic question is whether there is a sufficient contrast in physical properties between the unit and adjoining strata to be detectable using the geophysical methods under consideration. Key requirements for successful geophysical investigations include (Maliva 2016):

- a conceptual understanding of the problem (target) and the physical contrast that is likely to exist
- selection of a method that is appropriate to the target
- using available control to reduce non-uniqueness and equivalence in the processing and interpretation of the data.

9.12 Core Analyses

Cores are collected as part of aquifer characterization programs to obtain high-quality samples for lithological, petrophysical, and chemical analyses. Core analyses for groundwater investigations of lithified strata are usually performed on either whole core samples or core plugs drilled from cores. Core plugs have standard diameters of 1 in. (2.5 cm) and 1.5 in. (3.8 cm) and are taken in either the horizontal or vertical direction. Core analyses can also be performed on unconsolidated strata using artificial cores, which are packed cylinders of sediment samples or actual cores collected using thin-walled samplers (e.g., Shelby tubes). Core sample analysis is commonly performed by commercial core laboratories. Some geotechnical laboratories, as well as some universities and geological surveys, are also equipped to perform routine core analyses, which include porosity, permeability (hydraulic conductivity), and grain density measurements.

Core analyses can provide accurate data on the properties of the tested samples, but have the limitation of a very small volume of investigation. Hydraulic conductivity values obtained from core samples cannot be directly scaled up to the aquifer scale. Cross-plots (transforms) of core porosity versus hydraulic conductivity can be used to establish the numerical relationship between the two parameters, which can then be used to estimate hydraulic conductivity values from a porosity geophysical log (e.g., sonic log).

The minipermeameter has emerged as a cost-effective tool for performing rapid and inexpensive permeability measurements on cores and outcrops (Goggin 1993; Sutherland et al. 1993; Sharp et al. 1994; Hurst and Goggin 1995). The minipermeameter has a hollow probe tip that is fitted with an O-ring to allow for a tight pressure seal with a slabbed core or outcrop surface. Air (or nitrogen gas) is injected into the sample through the probe. The rate of flow is a function of the injection pressure and permeability of the sample. Steady-state air flow rate and injection pressure data are converted to permeability values using empirical relationships based on standard core analyses and minipermeameter readings of the same samples. The great advantage of the minipermeameter is that it avoids the need to drill core plugs and each analysis is quick and inexpensive. Hence, a large number (100 s) of measurements can be performed on a slabbed core to obtain data on the statistical variation of permeability.

Coring is useful in MAR investigations for obtaining high-quality samples for mineralogical and geochemical investigations. Core samples are used for thin section petrography and x-ray diffraction analyses. Core flow-through tests can be used to evaluate potential fluid-rock interactions and clogging during recharge. Flow-through tests involve passing the proposed recharge water through a core sample (usually at a constant pressure) and monitoring changes in the flow rate and effluent water quality over time. Changes in flow rate over time would reflect either clogging processes, which decrease hydraulic conductivity, or dissolution processes, which increase hydraulic conductivity. Core-flow tests have particular value for evaluating the potential for clogging from clay dispersion and swelling because hydraulic conductivity decreases from these properties are usually rapid. Change in water chemistry between influent and effluent can provide insights into fluid-rock interactions that may be active in an MAR system. However, some fluid-rock interactions may occur at too slow of rate to be detectable after the short-duration passage of water through core samples.

9.13 Mineralogical Analyses

Data on the mineralogy are needed to evaluate the potential for fluid-rock interactions in AAR systems. As discussed in Sect. 6.3, many of the mineral phases present in sediments and rocks are not reactive under the temperature, pressure, and water chemistry conditions that occur in near surface groundwater environments. Hence, mineralogical investigations for MAR systems should focus on identifying the presence of reactive mineral phases, which may be present in only trace quantities. Bulk mineralogy can often be determined from visual observation of cuttings and core samples, perhaps with the assistance of a hand lenses (magnifying glass) or stereomicroscope. More detailed evaluations are performed using thin-section petrography and x-ray diffraction analysis.

Thin-section petrography is the most useful technique for characterizing the composition and textures of rock. A thin section is a thin (typically 30 μm thick) sliver of rock or sediment that is glued onto a glass slide. Thin sections are examined using a petrographic microscope, which is essentially a transmitted-light microscope equipped with two polarizing filters (Nicol prisms) oriented at 90° to each other. Some petrographic microscopes also allow for viewing of samples under reflected light. A well-trained petrographer can identify the minerals present in thin sections based on their form and optical characteristics if the crystals are large and common enough to be viewed. Thin-section petrography is also used to identify cement and porosity types. Often samples to be thin sectioned are vacuum impregnated with colored epoxy to facilitate identification of pores. Thin-section petrography for groundwater investigations is now usually performed by commercial laboratories (or a subcontracted expert) as most groundwater consulting and engineering firms do not have the equipment and trained personnel in house.

Powder x-ray diffraction (XRD) analysis allows for the identification of very finely crystalline material, including clays. XRD is based on the diffraction of an x-ray beam off the planes of atoms in crystal structures. The diffraction angles and relatively intensity of the diffracted beams are a function of the crystal structure of the mineral phase. XRD provides the most definitive mineral identification. Diffraction patterns can be conceptualized as unique fingerprints of minerals. The detection limit for XRD is approximately 2%, so the technique may not be able to detect minerals present in trace quantities.

XRD analyses are performed by commercial and university laboratories. Only a small sample of powdered rock is needed for routine analyses. Special sample preparation techniques are required for clay mineralogical analysis. The clay-sized fraction is first separated and then oriented slides prepared. Multiple XRD analyses are performed following various sample treatments (e.g., heating and glycolation) to facilitate clay mineral identification.

Scanning electron microscopy (SEM) uses an electron beam to provide a high-magnification and high-resolution image of the surface of a material. The interaction of the electron beam with material surfaces generates secondary electrons, backscattered electrons, and X-rays. Secondary electrons are emitted from the surface of the sample and are used primarily for imaging. Backscattered electrons (BSE) are electrons from the primary beam that are reflected back from the material surface. Backscattered imaging is commonly performed on polished thin sections and is used to differentiate between minerals. The interaction of the electron beam with samples also causes the emission of X-rays, which have energy levels characteristic of each element. SEMs, particularly newer models, are commonly equipped with an Energy Dispersive X-ray Spectroscopy (EDS) unit that is used for elemental analyses. Mineralogical determinations are made from the shape of crystals and their ratios of elements.

The electron microprobe (EM) is similar to the SEM in that it emits a beam of electrons and obtains elemental data from the emitted x-rays but has a very fine resolution and can provide more accurate elemental analyses. EMs are sophisticated tools and require training to properly operate and obtain accurate data. Most major universities have an EM and some allow outside use of the equipment under supervision for a fee. The SEM and EM have some specialized uses for MAR investigation. For example, both SEM and EM were used to confirm that arsenic-bearing pyrite was the likely source of the arsenic that leached into water stored in some ASR systems in Florida (Price and Pichler 2006; Lazareva and Pichler 2007).

9.14 Geochemical Investigations

Evaluation of potential geochemical reactions that may occur (or are occurring) in an MAR system should start with a qualitative evaluation of the processes that could occur. Basic issues to be considered include:

- redox state of recharge water relative to the native groundwater
- changes in redox state that may occur upon recharge (e.g., removal of dissolved oxygen by organic matter biodegradation)
- reactive minerals present in the recharge zone
- potential for various sorption processes.

The results of the qualitative analysis should guide subsequent geochemical investigations. For example, if oxygenated water is to be injected into a confined aquifer with chemically reducing conditions, then it can be deduced that oxidative dissolution or alteration of chemically reduced minerals (with associated metals and metalloid release) may occur. This knowledge should then be used to evaluate whether these processes are likely to significantly impact the quality of recharged water and/or cause clogging.

Geochemical evaluations of MAR systems, where performed, are most commonly carried out by geochemical modeling using widely available modeling software, such as PHREEEQC (Parkhurst and Appelo 1999). Given accurate data on recharge zone mineralogy and the chemistries of the water to be recharged and native groundwater, it is possible to calculate the saturation state of the waters and different mixtures of recharged water and native groundwater with respect to various minerals. The composition of the recharged water after equilibration with aquifer minerals can also be predicted. Modeling software also allows for the simulation of non-equilibrium conditions (i.e., reaction kinetics). Recommended water chemistry parameters for sampling for geochemical evaluation of major MAR systems are listed in Table 5.2.

Sampling procedures and analytical techniques should follow accepted industrial standards. Accurate data of pH and eH (Pe), in particular, are critical, and should be carefully obtained using flow-through sampling techniques and concentration data from redox pairs.

Laboratory-scale column and/or batch testing can provide more direct information on the geochemistry of MAR systems. The preferred testing procedure would be flow-through testing using cores or columns of aquifer materials (ideally undisturbed) and actual recharge water and native groundwater (as a baseline). Such testing can also be used to evaluate pretreatment options. An important consideration is scaling limited laboratory results to the formation/aquifer scale (National Research Council 2008).

The Florida Geological Survey (FGS) commenced in 1995 a research program to investigate fluid-rock interaction processes that occur during ASR. Bench-top leaching and sequential extraction testing was performed (Arthur et al. 2005a, 2007). For the leaching studies, samples of crushed aquifer rock were placed in sealed glass reaction vessels partially filled with either distilled deionized (DDI) water or natural groundwater (Fig. 9.6). The water could have either high or low dissolved oxygen concentrations. Samples of leachate were periodically collected for cation and anion analyses. The sequential extraction technique involved sequentially immersing samples in different fluids to isolate mineral-metal associations. Fluids used were DDI water (to extract water soluble metals and metalloids), 0.1 M acetic acid (to extract acid-soluble elements), 0.1 hydroxylamine hydrochloride (to extracts metals fixed on Fe and/or Mn-oxides), 0.1 M sodium pyrophosphate (to decomposed organics), and HNO_3 and $HClO_4$ solution (to extract metals fixed in sulfides, silicates, and heavy minerals) (Fig. 9.7).

The FGS was contracted to performed bench-top leaching tests of core samples from an ASR exploratory well in Sanford, Florida (Arthur et al. 2005b). Arsenic concentrations in the leachate exceeded the primary drinking water and groundwater standard of 10 μg/L. High levels of molybdenum leaching also occurred. The testing results were a prognosticator of the arsenic leaching that subsequently occurred during operational testing of the ASR system.

Fig. 9.7 Reaction vessel used in Florida Geological Survey leachability analyses (*Source* Florida Geological Survey)

9.15 Modeling

Applications of modeling to ASR and MAR projects were reviewed by Maliva and Missimer (2010), Kloppmann et al. (2012), and Ringleb et al. (2016). These reviews summarized the different model codes that have been, or potentially could be, used for MAR projects. Well-established modeling tools that are widely used for groundwater investigation (e.g., MODFLOW family of codes) are suitable to MAR investigations. More advanced codes may be needed for some specific applications, such as simulation of solute-transport through dual-porosity aquifers (Guo et al. 2015). Biogeochemical and aquifer ecosystem modeling may be needed to simulate pathogen decay process and the biological mediation of many aquifer geochemical processes (Kloppmann et al. 2012). With respect to ASR projects, Maliva and Missimer (2010) identified the following basic types of modeling:

- **Theoretical (conceptual)**: Simulations of a real aquifer or hypothetical aquifer to develop a better basic understanding of processes that control the performance of systems. The results are used as a general guide for implementation.
- **Uncalibrated predictive**: Models of a proposed MAR system are developed in advance of system construction and operational testing using best available local

and regional hydrogeological data. The objective is typically to evaluate potential system performance as part of a feasibility assessment and for preliminary design.
- **Calibrated groundwater and solute transport**: Models are calibrated against historical operational (monitoring) data. The calibration process (i.e., inverse modeling) is used to obtain a better understanding of local hydrogeological and solute-transport conditions. The calibrated model can be used as predictive tool to evaluate various operational and system expansion options.
- **Geochemical and reactive solute transport**: Models are used to evaluate geochemical processes that could impact water quality and system performance (e.g., mineral precipitation causing clogging). The greatest value of geochemical modeling has been as an interpretative tool through the calibration process rather than as a predictive tool.

Other basic types of modeling applied to MAR include (Maliva and Missimer 2012; Ringleb et al. 2016):

- **Unsaturated zone flow**: Models are used to simulate surface spreading-type systems (e.g., mounding evaluations).
- **Surface water (watershed or water balance)**: Various applications include evaluation of runoff and infiltration rates.
- **Land subsidence**: Simulation of the effects of operation of an MAR system toward mitigation of compaction and associated land subsidence.

Basic applications of modeling to MAR projects include (Maliva and Missimer 2010; Ringleb et al. 2016):

- **Feasibility assessment**: Evaluation of whether a proposed system at a given site will likely meet project objectives.
- **Data evaluation:** Model calibration (i.e., inverse modeling) is used obtain insights on site hydrogeological and geochemical conditions.
- **Performance assessment**: Extrapolation of initial operational results into the future (e.g., prediction of long-term recovery efficiencies of ASR systems from early operational testing results).
- **Design and optimization**: Evaluation of various system design and operational options to determine optimal design and operational protocols.
- **Impact analysis/risk assessments**: Evaluation of potential for adverse impacts (e.g., mounding).
- **Migration and residence time**: Estimation of how far recharged water might migrate and the travel time to recovery or production wells (and other sensitive receptors). These are key issues where indirect potable reuse is possible or a reality.
- **Water quality and clogging prediction**: Evaluation of geochemical process that may impact recharged water quality or impact system operations (e.g., cause clogging). Geochemical modeling is commonly used to determine potential reactions between recharged and native groundwater and between recharged water and aquifer minerals. Fluid-mixing and fluid-rock interactions may either improve water quality (i.e., natural aquifer treatment) or could adversely impact groundwater quality (e.g., arsenic and metals leaching).

- **Saline-water intrusion**: Density-dependent solution transport modeling is used to evaluate the impacts of recharge on the position of the saline-water interface and the potential effectiveness of salinity-barrier system options.

Discussion of modeling procedures and the various codes is well beyond the scope of this summary. For any MAR project in which modeling is being considered, the project team should include an experienced modeling who is familiar with the various modeling codes and their strengths and weakness for various applications. It is critical to appreciate the limitations of models, which usually stems from inadequate hydrogeological data, rather than intrinsic limitations of the modeling software. A sensitivity analysis is a critical part of the modeling process in which the impacts of uncertainty on model results are evaluated. A sensitivity analysis involves performing multiple simulations in which the values of model parameters are systematically varied. The sensitivity analysis process gives a range of predicted values (e.g., recovery efficiencies) rather than a single value.

References

American Society of Civil Engineers. (1998). *Geophysical exploration for engineering and environmental investigations*. Reston, VA: ASCE Press.

Arthur, J. D., Dabous, A. A., & Cowart, J. B. (2005a). Water-rock geochemical considerations for aquifer storage and recovery: Florida case studies. In C.-F. Tsang & J. A. Apps (Eds.), *Underground injection science and technology*, Developments in Water Science 52 (pp. 65–77). Amsterdam: Elsevier.

Arthur, J. D., Dabous, A. A., & Fischler, C. (2005b). *Bench-scale geochemical assessment of water-rock interactions: Sanford aquifer storage and recovery facility*, Draft report submitted the Camp Dresser and McKee, Inc. (September 21, 2005). Tallahassee: Florida Geological Survey.

Arthur, J. D., Dabous, A. A., & Fischler, C. (2007). Aquifer storage and recovery in Florida: Geochemical assessment of potential storage zones. In P. Fox (Ed.), *Management of aquifer recharge for sustainability. Proceedings of the 6th International Symposium on Managed Aquifer Recharge of Groundwater* (pp. 185–197). Phoenix: Acacia Publishing.

Assaad, F. A., LaMoreaux, J. E., & Hughes, T. (Eds.). (2004). *Field techniques for geologists and hydrogeologists*. Berlin: Springer.

ASTM (1999). *Standard guide for selecting surface geophysical methods*. D6429-99 (reapproved 2011). West Conshohocken, PA: ASTM International.

ASTM (2004). *Standard practice for sonic drilling for site characterization and the installation of subsurface monitoring devices*. D6914-04. West Conshohocken, PA: ASTM International.

Barson, D., Christensen, R., Decoster, E., Grau, J., Herron, M., Herron, S., et al. (2005). Spectroscopy: The key to rapid, reliable petrophysical answers. *Oilfield Review, 17*(2), 14–33.

Bouwer, H., & Rice, R. C. (1976). A slug test for determining hydraulic conductivity of unconfined aquifers with completely or partially penetrating wells. *Water Resources Research, 12,* 423–428.

Burger, H. R., Sheehan, A. F., & Jones, G. H. (2006). *Introduction to applied geophysics: Exploring the shallow subsurface*. New York: W. W. Norton.

Butler, J. J., Jr. (1998). *The design, performance, and analysis of slug tests*. Boca Raton: Lewis Publishers.

Butler, J. J., Jr., McElwee, C. D., & Liu, W. (1996). Improving the quality of parameter estimates obtained from slug tests. *Ground Water, 34,* 480–490.

Butler, J. J., Jr., & Healey, J. M. (1998). Relationship between pump-test and slug-test parameters: Scale effect or artifact? *Ground Water, 36,* 305–313.

Butler, J. J., Jr., Dietrich, P., Wittig, V., & Christy, T. (2007). Characterizing hydraulic conductivity with direct push permeameter. *Ground Water, 45,* 409–419.

Butler, J. J., Jr., Healey, J. M., Zheng, L., McCall, W., & Schulmeister, M. K. (1999). *Hydrostratigraphic characterization of unconsolidated alluvial deposits with direct-push sensor technology*. Kansas Geological Survey Open-File Report 99-40.

Butler, J. J., Jr., Healey, J. M., McCall, G. W., Garnett, E. J., & Loehide, S. P., II. (2002). Hydraulic tests with direct push equipment. *Ground Water, 40,* 25–36.

Chapman, D. S., Sahm, E., & Gettings, P. (2008). Monitoring aquifer recharge using repeated high-precision gravity measurements: a pilot study in South Weber, Utah. *Geophysics, 73*(6), WA83–WA93.

Chen, C.-S., Sie, Y.-C., & Lin, Y.-T. (2012). A review of the multilevel slug test for characterizing aquifer heterogeneity. *Terrestrial Atmospheric and Oceanic Sciences, 23*(2), 131–143.

Collier, H. A. (1993). *Borehole geophysical techniques for determining the water quality and reservoir parameters of fresh and saline water aquifers in Texas*, Report 343. Austin: Texas Water Development Board.

Cooper, H. H., Jr., Bredehoeft, J. D., & Papadopoulos, I. S. (1967). Response of a finite-diameter well to an instantaneous charge of water. *Water Resources Research, 3,* 263–269.

Cooper, H. H., Jr., & Jacob, C. E. (1946). A generalized graphical method for evaluating formation constants and summarizing well-field history. *Transactions. American Geophysical Union, 27,* 526–534.

Cunningham, W. L., & Schalk, C. W. (Comps.). (2001). *Groundwater technical procedures for the U.S. Geological Survey*. U.S. Geological Survey Techniques and Methods 1-A1.

Davis, H. E., Jehn, J., & Smith, S. (1991). Monitoring well drilling, soil sampling, rock coring, and borehole logging. In D. M. Nielsen (Ed.), *Practical handbook of ground-water monitoring* (pp. 195–237). Chelsea. MI: Lewis Publishing.

Davis, K., Li, Y., Batzle, M., & Raynolds, B. (2005). *Time-lapse gravity monitoring of an aquifer storage recovery project in Leyden, Colorado*. Boulder: Colorado School of Mines, Center for Gravity, Electrical & Magnetic Studies.

Davis, K., Li, Y., & Batzle, M. (2008). Time-lapse gravity monitoring: a systematic 4D approach with application to aquifer storage and recovery. *Geophysics, 73*(6), WA61–WA69.

Dawson, K. J., & Isotak, J. D. (1991). *Aquifer testing: Design and analysis of pumping and slug tests*. Boca Raton: Lewis Publishers.

Dentith, M., & Mudge, S. T. (2014). *Geophysics for the mineral exploration geoscientist*. Cambridge, UK: Cambridge University Press.

Dietrich, P., Butler, J. J., Jr., & Faiβ, K. (2008). A rapid method for hydraulic profiling in unconsolidated formations. *Ground Water, 46,* 323–328.

Driscoll, F. G. (1986). *Groundwater and wells* (2nd ed.). St. Paul, MN: Johnson Filtration Systems.

Eastern Research Group (1993). *Use of airborne, surface, and borehole geophysical techniques at contaminated sites. A reference guide*, Document EPA/625/R-92-007. Washington, D.C.: U.S. Environmental Protection Agency, Office of Research and Development.

Fetter, C. W. (1998). *Contaminant hydrogeology* (2nd ed.). Englewood Cliff, NJ: Prentice-Hall.

Gelhar, L. W., & Collins, M. A. (1971). General analysis of longitudinal dispersion in nonuniform flow. *Water Resources Research, 7,* 1511–1521.

Goggin, D. J. (1993). Probe permeametry: Is it worth the effort? *Marine and Petroleum Geology, 10,* 299–308.

Guo, W., Coulibaly, K., & Maliva, R. G. (2015). Simulated effects of aquifer heterogeneity on ASR system performance. *Environmental Earth Sciences, 73*(12), 7803–7809.

Halford, K. J., & Kuniansky, E. L. (2002). *Spreadsheets for the analysis of aquifer-test and slug-test data, version 1.2*. U.S. Geological Survey Open-File Report 02-197.

Haggerty, R., Schroth, M. H., & Istok, J. D. (1998). Simplified method of "push-pull" test data analysis for determining in situ reaction rate coefficients. *Ground Water, 36,* 314–324.

Hantush, M. S., & Jacob, C. E. (1955). Non-steady radial flow in an infinite leaky aquifer. *American Geophysical Union Transactions, 36,* 95–100.

Henahan, M. (1999). Dual rotary drilling in unconsolidated overburden. *Water Well Journal*, July 1999, (pp. 66–68).

Herrick, D. (1994). Dual-rotary drilling—is it for you. *Water Well Journal, 1994,* 50–54.

Howle, J. F., Phillips, S. P., & Ikehara, M. E. (2002). Estimating water-table change using micro-gravity surveys during an ASR program in Lancaster, California. In P. Dillon (Ed.), *Management of aquifer recharge for sustainability* (pp. 269–272). Lisse: Swets & Zeitlinger.

Hurst, A., & Goggin, D. J. (1995). Probe permeametry: An overview and bibliography. *American Association of Petroleum Geologists Bulletin, 79,* 463–473.

Hvorslev, J. (1951). *Time lag and soil permeability in ground-water observations*. Vicksburg, MS: Waterways Experimental Station, U.S. Army Corps of Engineers.

Hyder, Z., Butler, J. J., Jr., McElwee, C. D., & Liu, W. (1994). Slug tests in partially penetrating wells. *Water Resources Research, 30,* 2945–2957.

Industries, Foremost. (2003). *Benefits of dual rotary drilling in unstable overburden formations*. Calgary: Foremost Industries.

Istok, J. D., Humphrey, M. D., Schroth, M. H., Hyman, M. R., & O'Reilly, K. T. (1997). Single well, "push-pull" test for in situ determination of microbial activities. *Ground Water, 35,* 619–631.

Johnson, G. S., & Frederick, D. B. (1997). *Straddle-packer aquifer test analyses of the Snake River Plain aquifer of the Idaho National Engineering Laboratory*. Moscow, ID: Idaho Water Resources Institute.

Kasenow, M. (1997). *Applied groundwater hydrology and well hydraulics*. Highlands Ranch, CO: Water Resources Publications.

Kasenow, M. (2006). *Aquifer test data: Evaluation and analysis*. Highlands Ranch, CO: Water Resources Publications.

Keely, J. F., & Boateng, K. (1987a). Monitoring well installation, purging, and sampling techniques—Part 1: Conceptualizations. *Ground Water, 25,* 300–313.

Keely, J. F., & Boateng, K. (1987b). Monitoring well installation, purging, and sampling techniques—Part 2: Case histories. *Ground Water, 25,* 427–439.

Keys, W. S. (1989). *Borehole geophysics applied to ground water investigations*. Dublin, OH: National Water Well Association.

Keys, W. S. (1990). *Borehole geophysics applied to ground-water investigations*. Techniques of Water-Resources Investigations of the United States Geological Survey, Book 2, Chapter E2.

Keys, W. S. (1997). *A practical guide to borehole geophysics in environmental investigations*. Boca Raton: Lewis Publishers.

Kirsch, R. (Ed.). (2008). *Groundwater geophysics: A tool for hydrogeology*. Berlin: Springer.

Kloppmann, W., Aharoni, A., Chikurel, H., Dillon, P., Gaus, I., Guttman, J., et al. (2012). Use of groundwater models for prediction and optimisation of the behaviour of MAR sites. In C. Kazner, T. Wintgens, & P. Dillon (Eds.), *Water reclamation technologies for safe managed aquifer recharge* (pp. 311–349). London: IWA Publishing.

Kober, R., Hornbruch, G., Oeven, C., Tischer, L., Großmann, J., Dietrich, P., et al. (2009). Evaluation of combined direct-push methods used for aquifer model generation. *Ground Water, 47,* 536–546.

Kobr, M., Mareš, S., & Paillet, F. (2005). Borehole geophysics for hydrogeological studies: Principles and applications. In Y. Rubin & S. S. Hubbard (Eds.), *Hydrogeophysics* (pp. 291–331). Springer: Dordrecht.

Kruseman, G. P., & de Ridder, N. A. (1991). *Analysis and evaluation of pumping test data*. Wageningen, Netherlands: International Institute for Land Reclamation and Improvements Publications.

Lazareva, O., & Pichler, T. (2007). Naturally occurring arsenic in the Miocene Hawthorn Group, southwestern Florida: Potential implication for phosphate mining. *Applied Geochemistry, 22,* 953–973.

Liu, G., Butler, J. J., Jr., Bohling, G. C., Reboulet, E., Knobbe, S., & Hyndman, D. W. (2009). A new method for high-resolution characterization of hydraulic conductivity. *Water Resources Research, 45,* W08202.

Liu, G., Butler, J. J., Jr., Reboulet, E., & Knobbe, S. (2012). Hydraulic conductivity profiling with direct push methods. *Grundwasser—Zeitschrift der Fachsektion Hydrogeologie, 17,* 19–29.

Lowry, W., Mason, N., Chipman, V., Kisiel, K., & Stockton, J. (1999). *In-situ permeability measurements using direct push techniques: Phase II topical report,* report prepared for the DOE Federal Energy Technology Center. Sante Fe, NM: Science and Engineering Associates.

Maliva, R. G. (2016). *Aquifer characterization techniques*. Berlin: Springer.

Maliva, R. G., & Missimer, T. M. (2010). *Aquifer storage and recovery and managed aquifer recharge using wells: Planning, hydrogeology, design, and operation*. Houston: Schlumberger Corporation.

Maliva, R. G., & Missimer, T. M. (2012). *Arid lands water evaluation and management*. Berlin: Springer.

Maliva, R. G., Clayton, E. A., & Missimer, T. M. (2009). Application of advanced borehole geophysical logging to managed aquifer recharge investigations. *Hydrogeology Journal, 17,* 1547–1556.

Maliva, R. G., Griswold, R. F., & Autrey, M. M. (2013). Prototype for a reclaimed water aquifer storage recovery system benefits and operational experiences. *Florida Water Resources Journal, 65*(3), 54–59.

Melville, J. G., Molz, F. J., Güvem, O., & Widdowson, M. A. (1991). Multilevel slug tests with comparisons to tracer data. *Ground Water, 29,* 897–907.

Milson, J. (2003). *Field geophysics* (3rd ed.). Chichester: John Wiley & Sons.

Miotiński, K., Dillon, P. J., Pavelic, P., Cook, P. G., Page, D. W., & Levett, K. (2011). Recovery of injected freshwater to differentiate fracture flow in low-permeability brackish aquifer. *Journal of Hydrology, 409,* 273–282.

Misstear, B., Banks, D., & Clark, L. (2006). *Water wells and boreholes*. Chichester: Wiley.

National Research Council. (2008). *Prospects for managed underground storage of recoverable water*. Washington, DC: National Academy Press.

Norton, S. B. (2007). Quantifying the near-borehole geochemical response during aquifer storage and recovery: Application of "push-pull" analytical techniques to ASR cycle testing. Master's thesis), University of Florida, Gainesville.

Parkhurst, D. L., & Appelo, C. A. J. (1999). *PHREEQC (Version 2)—A computer program for speciation, batch reaction, one-dimensional transport, and inverse geochemical calculations.* U.S. Geological Survey Water-Resources Investigations Report 99-42549.

Pickens, J. F., & Grisak, G. E. (1981). Scale-dependent dispersion in a stratified granular aquifer. *Water Resources Research, 17,* 1191–1211.

Pool, D. R., & Schmidt, W. (1997). *Measurement of ground-water storage change and specific yields using the temporal-gravity method near Rillito Creek, Tucson, Arizona*. U.S. Geological Survey Water Resources Investigations Report 97-4125.

Price, R. E., & Pichler, T. (2006). Abundance and mineralogical association of arsenic in the Suwannee Limestone (Florida): Implications for arsenic release during water–rock interaction. *Chemical Geology, 228,* 44–56.

Reynolds, J. M. (1997). *An Introduction to applied and environmental geophysics*. Chichester: Wiley.

Ringleb, J., Sallwey, J., & Stefan, C. (2016). Assessment of managed aquifer recharge through modeling—a review. *Water, 8,* 579.

Roscoe Moss Company. (1990). *Handbook of ground water development*. New York: Wiley.

Schulmeister, M. K., Butler, J. J., Jr., Healey, J. M., Zheng, L., Wysocki, D. A., & McCall, G. W. (2003). Direct-push electrical conductivity logging for high-resolution hydrostratigraphic correlation. *Ground Water Monitoring and Remediation, 23*(3), 52–62.

Sharma, P. V. (1997). *Environmental and engineering geophysics*. Cambridge, UK: Cambridge University Press.

Sharp, J. M., Jr., Fu, L., Cortez, P., & Wheeler, E. (1994). An electronic minipermeameter for use in the field and laboratory. *Ground Water, 32,* 41–46.

Stallman, R. W. (1969). *Aquifer testing design, observation, and data analysis*. U. S. Geological Survey Techniques of Water Resources Investigations, Chapter B1, Book 3.

Stephan, R. (1995). Rotasonic drilling: an alternative to conventional technology in environmental characterization studies. *Proceedings or the National Outdoor Action Conference and Exposition, Aquifer Remediation/Ground Water Monitoring/Geophysical Methods, 9,* 325–337.

Stephens, D. B. (1996). *Vadose zone hydrology*. Boca Raton, FL: CRC Press.

Sterrett, R. J. (2007). *Groundwater and wells* (3rd ed.). St. Paul, MN: Johnson Screens.

Strauss, M. F., Story, S. L., & Mehlhorn, N. E. (1989). Application of dual-wall reverse-circulation drilling in ground water exploration and monitoring. *Ground Water Monitoring and Remediation, 9*(2), 63–71.

Sutherland, W. J., Halvorsen, C., Hurst, A., McPhee, C. A., Robertson, G., Whattler, P. R., et al. (1993). Recommended practice for probe permeametry. *Marine and Petroleum Geology, 10,* 309–317.

Telford, W. M., Geldart, L. P., & Sheriff, R. E. (1990). *Applied geophysics*. Cambridge, UK: Cambridge University Press.

Theis, C. V. (1935). The relation between the lowering of the piezometric surface and the rate and duration of discharge of a well using groundwater storage. *Transactions. American Geophysical Union, 16,* 519–524.

USACOE (1995). *Geophysical exploration for engineering and environmental investigations: Engineer manual*. Manual No. 1110-1802. Washington, DC: U.S. Army Corps of Engineers.

Walton, W. C. (1960). *Leaky artesian aquifer conditions in Illinois*. Illinois State Water Survey Report of Investigations.

Walton, W. C. (1962). *Selected analytical methods for well and aquifer evaluation*. Illinois State Water Survey Bulletin 49.

Walton, W. C. (1987). *Groundwater pumping tests: Design and analysis*. Chelsea, MI: Lewis Publishers.

Walton, W. C. (1997). *Practical aspects of ground water modeling* (3rd ed.). Westerville, OH: National Ground Water Association.

Weight, W. D. (2008). *Hydrogeology field manual* (2nd ed.). New York: McGraw Hill Professional.

Wempe, W. L. (2000). Predicting flow properties using geophysical data: Improving aquifer characterization. Dissertation, Stanford University.

Wilson, G. J., Everett, L. G., & Cullen, S. L. (Eds.). (1995). *Handbook of vadose zone characterization and monitoring*. Boca Raton: CRC Press.

Yaramanci, U., & Müller-Petke, M. (2009). Surface nuclear magnetic resonance—a unique tool for hydrogeophysics. *The Leading Edge*, October 2009, 1240–1247.

Zohdy, A. A. R., Eaton, G. P. & Mabey, D. R. (Eds.) (1974). *Application of surface geophysics to ground-water investigations*. Techniques of Water-Resources Investigations of the United States Geological Survey, Book 2, Chapter D1.

Chapter 10
Vadose Zone Testing Techniques

10.1 Introduction

Vadose zone hydraulic properties impact the performance of surface-spreading MAR systems by controlling the rate of flow of water from land surface to the water table. The presence of a low permeability layer within the vadose zone can impede vertical flow and result in perched aquifer conditions. Vadose zone hydraulic properties obtained from infiltration tests at land surface may differ greatly from those at depth, especially where the vadose zone is very thick. Evaluation of hydraulic properties of the vadose zone is complicated by the zone being a two-phase (water and air) system, as opposed to the single-phase system of the saturated (phreatic) zone.

An additional consideration is the volume and area of investigation of tests relative to spatial heterogeneity. For example, the results of one or several small-area infiltration tests may not be representative of the properties of an entire infiltration basin area. Project budgets may greatly limit the amount of testing that can be performed. Where testing is limited, the use of safety factors in design can compensate for uncertainty.

In the United States, state, city and other governmental stormwater management, land development, and transportation department manuals include detailed vadose zone testing procedures, which are either regulatory requirements or recommendations. The testing procedures tend to be quite similar reflecting common sources. Methods presented herein are given as examples of standard procedures.

10.2 Air Entrainment

The term "entrapped" air is commonly used to describe a discontinuous gas phase that is present within otherwise saturated soil or sediment and is no longer connected to the atmosphere (Faybishenko 1995; Marinas et al. 2013). The gas phase may be either air or it may compositionally differ from air. Air may be entrapped as the result

R. G. Maliva, *Anthropogenic Aquifer Recharge*,
Springer Hydrogeology, https://doi.org/10.1007/978-3-030-11084-0_10

of natural fluctuations in the water table and during anthropogenic infiltration. Air entrapment occurs as the result of non-uniform, preferential flow during infiltration. Water will preferentially infiltrate (flow) through larger pores and pore throats, while nearby smaller pores remain not fully saturated (Faybishenko 1995). Increases in entrapped air (gas) may also be caused by increases in temperature, decreases in pressure, and the decomposition of organic matter (Faybishenko 1995). Entrapped air in the vadose zone affects measured hydraulic conductivity values during infiltration by both reducing water-saturated porosity and obstructing pores and pore throats.

The term "quasi-saturated hydraulic conductivity" is used to describe the hydraulic conductivity of soils or sediments beneath the water table that contain entrapped air (Faybishenko 1995). The term "unsaturated hydraulic conductivity" is reserved for soils above the water table. The related term "field-saturated hydraulic conductivity" (K_{fs}) is used to describe conditions during the generally short-term occurrence of entrapped air during an infiltration event (Marinas et al. 2013). The term "effective hydraulic conductivity" is also used to denote the hydraulic conductivity of soils and sediments that contain entrapped air.

The effects of entrapped air on groundwater flow in soils have received considered study. In an early study, Christiansen (1944) reported that during week to month-long laboratory soil infiltration tests, three distinct phases are evident:

- an initial period in which permeability decreases, sometimes by only a minor degree
- a period in which permeability increases, in some instances to more than 30 times the previous minimum rate
- a final period of steady gradual decrease in permeability.

Faybishenko (1995) proposed that the initial decrease in permeability is due to the mobilization of some entrapped air and its accumulation in, and blocking of, larger pores. Christiansen (1944) demonstrated that the increase in permeability during the second phases is due to the displacement and dissolution of the entrapped air. The third phase is due to surface sealing by microbial and/or physical processes (Faybishenko 1995).

A number of investigations have evaluated the impacts of entrapped air on measured soil properties by flushing soils with carbon dioxide gas. CO_2 is much more soluble than air and thus more quickly and completely dissolves in soil water. The importance of entrapped air in field permeameter (air-entry and borehole methods) measurements was evaluated by Stephens et al. (1984) using CO_2 replacement at a field site in the Sevilleta National Wildlife Refuge, located north of Socorro, New Mexico. In the case of borehole infiltration tests, long-term infiltration rates were about the same with and without CO_2 treatment. However, the final infiltration rate was achieved in the untreated test in about twice as much time.

The results of field and laboratory infiltration tests performed by the U.S. Geological Survey in which the amount of air encapsulation was reduced by flushing samples with CO_2 gas suggest that a significant portion of the entrapped air resides within the transmission zone of the soil (Constantz et al. 1988). The transmission zone is located above the wetting front (zone) and is characterized by a relatively uniform water content and a hydraulic gradient primarily driven by gravitational forces. Over the duration of the tests, residual air results in the effective hydraulic conductivity of the transmission zone being no greater than 20% of the saturated hydraulic conductivity of the soil (Constantz et al. 1988).

As summarized by Wang et al. (1998), as water infiltrates into the vadose zone, soil air is displaced and may become compressed in front of the wetting front, which can result in a substantial decrease in the rate of infiltration. When the pressure become sufficiently high, air will escape from the soil, resulting in a sharp decrease in air pressure and major increase in the rate of infiltration. Wang et al. (1998) performed laboratory experiments to evaluate the effects of air entrapment on infiltration in dry sands. The experimental results demonstrated that infiltration rates were controlled by the rate of air flow out of the sand. Under air-confining conditions, in which air ahead of the wetting front cannot readily escape to the atmosphere, compression of the air was found to cause flow fingering, which may lead to accelerated, but heterogeneous, transport of water.

From an applied perspective, field hydraulic conductivity values obtained from short-duration infiltration tests may by substantially less than saturated hydraulic conductivity values that may be achieved during long-term continuous operation of infiltration systems. Bouwer (1966) reported that experience indicates that field saturated hydraulic conductivity is approximately 50–75% less than the actual saturated hydraulic conductivity (K_s) and, therefore, calculated hydraulic conductivity values should be multiplied by at least two to account for entrapped air. ASTM (2016) similarly reported that hydraulic conductivity measured in the field (i.e., field-saturated hydraulic conductivity) may be reduced by as much as a factor of two compared to a trapped air-free condition.

10.3 Soil Infiltration Rates and Hydraulic Conductivity Measurements

Methods to measure saturated hydraulic conductivity in soils were reviewed by Klute and Dirksen (1986), Stephens (1996), ASTM (2016), and Angulo-Jaramillo et al. (2016). Vadose zone hydraulic testing is performed using infiltrometers and permeameters. Infiltrometers typically measure hydraulic conductivity at the soil surface, whereas permeameters are used to measure conductivity at different depths. Commonly used methods include:

- single-ring and double-ring infiltrometer
- pilot (basin) infiltration test
- borehole permeameter
- velocity permeameter
- double-tube method
- air-entry permeameter
- core testing.

Measured infiltration rates are influenced by both the permeability of the soil and capillary effects, which depend upon soil properties (grain and pore sizes and textures) and initial soil moisture content. Tests should have sufficient duration so that steady-state conditions are achieved and capillary effects are minimized. Infiltration rates can be impacted by soil air or gas pressure, clay dispersion and swelling, and particle rearrangement. Borehole permeameter test data are impacted by the configuration of the flow field around a borehole, which is highly dependent on the geometry of the borehole, hydraulic properties of the soil, the capillary suction of the soil, and borehole conditions (formation damage).

Tests can be performed under constant-head or falling-head conditions. Falling-head tests have the advantages of being quicker and requiring less water. However, long-duration constant-head tests may more closely approach conditions that may be encountered during the long-term operation of an MAR system.

10.4 Single- and Double-Ring Infiltrometers

10.4.1 Methods

Infiltration testing essentially involves measuring the rate of infiltration under controlled conditions. The objective of infiltration testing is to determine the rate of steady-state vertical infiltration, which is taken as being essentially equal to the vertical saturated hydraulic conductivity. Infiltration rate test data from small area tests are scaled up to design full-scale managed infiltration systems. Two basic technical issues with infiltrometer and permeameter tests are (1) divergence of flow and (2) air entrapment. Lateral divergence from vertical infiltration is due to the capillary action of soils and variations in hydraulic conductivity that may favor lateral flow, such as a low permeability layer below the infiltrometer. The effects of divergence increase as infiltration area decreases and as permeability decreases with depth (Johnson 1963). A partial solution to divergence is a double-ring test in which a constant or near constant water level is maintained between the inner and outer rings to obtain more vertical infiltration below the inner ring.

The lateral component of flow in small-scale infiltration tests results in greater infiltration rate values (i.e., overestimation of vertical infiltration rates). Even double-ring tests, designed to reduce the impacts of divergence, can still significantly overestimate infiltration rates. Divergence can be addressed by either increasing the test area or correcting for divergence in the data analysis. Large-area tests are generally more expensive, take longer to perform, and require greater volumes of water but would be expected to provide data more representative of overall site conditions than small-area tests.

Johnson (1963), in the U.S. Geological Survey's "Field method for measurement of infiltration," noted that infiltration rates are affected by:

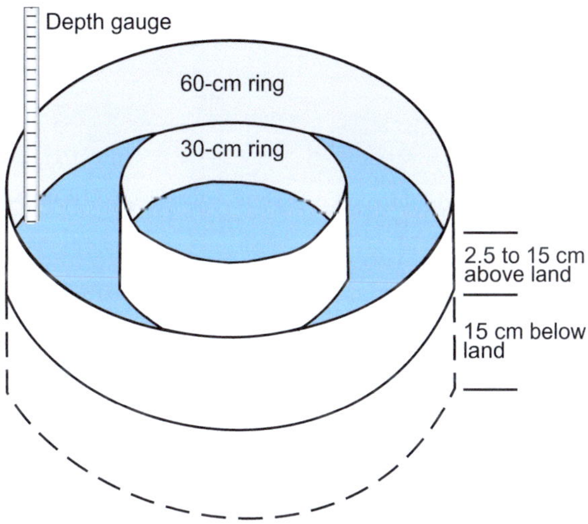

Fig. 10.1 Double-ring infiltrometer

- sediment soil texture and structure, which control permeability
- amount and distribution of soil moisture
- chemical and physical state of water including temperature, turbidity, and salinity
- head of applied water
- depth to groundwater
- temperature of sediments
- atmospheric pressure
- amount of entrapped air in sediments
- type of equipment and method used
- variation in permeability with depth.

Tests performed by ponding a large area are considered most reliable but their high costs usually dictate that infiltrometer rings be used. Large-area methods are necessary in coarse-grained materials where particle size is large relative to ring size (Johnson 1963).

Johnson (1963) observed that infiltration rates are most rapid early in tests when sediments are still unsaturated and that infiltration rates gradually decrease as the uppermost sediments become saturated and infiltration becomes controlled by less permeable sediments at depth. The critical soil zones controlling infiltration are the least permeable zones. Infiltration rates can be considerably decreased by disturbance of the sediment surface during tests (Johnson 1963). Test results may be impacted by clay dispersion if the chemistry of the water used in the test is significantly different from that of the soil-zone water (Bouwer 2002).

Johnson (1963) provided a series of recommendations for the design and performance of infiltration tests (Figs. 10.1 and 10.2) as follows:

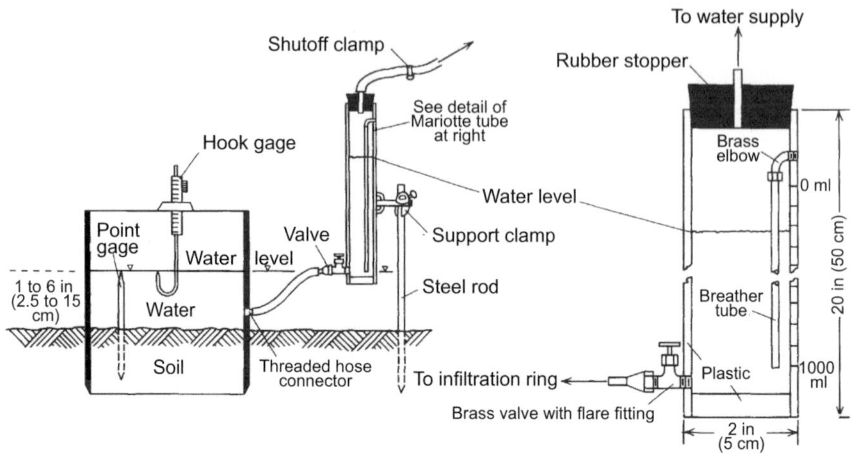

Fig. 10.2 Infiltrometer ring installation and Mariotte tube details (*Source* Johnson 1963)

- cylinders should be 20-in. (50.8 cm) high, 1/8-in. (3.2 mm) thick and be constructed of a hardened aluminum alloy with bottom edge bevels from the outside inward
- for double-ring tests, 12 and 24-in. (30.5 and 61-cm) diameter rings are recommended
- rings should be driven 6–8 in. (15–20 cm) into the soil
- rings should be driven using driving caps, which are ½-in. (12.7 mm) thick aluminum alloy disks with centering pins around the edge and a diameter slightly larger than the ring, and centered wooden blocks
- rings may be driven with a heavy sledge or jacked into the surface using a jack beneath a heavy truck
- constant water levels should be maintained either manually or using a Mariotte bottle (flask, siphon) or a float valve for high rate and long duration tests
- water levels should be maintained 1–6 in. (2.5–15 cm) above the soil
- the amount of fluid added and water levels in the ring(s) should be recorded every 15 min for the first hour of the test, every 30 min for second hour, hourly thereafter
- trenching (excavation) of the test site after the completion a test is recommended to evaluate spreading
- adding dyes to the water can facilitate identification of newly moistened areas.

The ASTM (2003) standard method for double-ring infiltrometer testing is largely based on the Johnson (1963) method. Infiltration velocity (v_{ir}) is calculated as

$$v_{ir} = \Delta V_{ir}/A_{ir}\Delta t \tag{10.1}$$

where

ΔV_{ir} volume of water added over timing period
A_{ir} cross-sectional area of the inner ring

Δt duration of time interval

Water flow rate versus time for the inner ring is monitored by means of a Mariotte bottle or other constant-head water supply until it stabilizes at a constant value. The method assumption is that the soil layer immediately below the ponded area (rings) is fully saturated and thus the matric potential is essentially zero. Under these saturated conditions, the hydraulic gradient is unity, and the final infiltration rate (final flux) is presumed to be equal to the saturated vertical hydraulic conductivity of the soil (ASTM 2003).

Standard (ASTM) sized double-ring infiltrometers are commercially available. Alternative systems have also been developed and marketed. For example, the Turf-Tec® infiltrometer is a commercial semi-automated double-ring infiltrometer that uses an inner ring 2-3/8 in. (6.03 cm) in diameter and an outer ring that is 4-1/4 in. (10.79 cm) in diameter (Turf Tech n.d.). The double-ring cutting blades are pushed into ground about 5 cm (2 in.) until a depth limiting ring is against the land surface. Once the soil is saturated, a test is performed by filling both rings with clean water and a 15 min timer is started. Fifteen minutes later, when a timer beeps, the position of a pointer on a scale gives the water infiltration rate for 15 min in inches or millimeters.

10.4.2 Single-Ring Infiltration Screening

Small diameter (±30 cm) single-ring tests can be performed relatively quickly and inexpensively, and thus may be preferred for site screening as they allow for more tests to be performed over a greater area. If initial testing results are favorable (infiltration rates are sufficiently large), then the next step would be to conduct infiltration tests using one or more larger test basins (Bouwer 2002). Bouwer (2002) presented a single-ring infiltrometer method based on sequentially performing a series of falling-head tests in which water levels are allowed to drop by about 5–10 cm, the elapsed time recorded, and them the cylinder is refiled and the next test step performed. This procedure is repeated for about 6 h or until the accumulated infiltration has reached about 50 cm, whichever comes first.

Bouwer (2002) corrected for divergence using field measurements of the lateral and vertical extent of infiltrated water for the last time interval of the test. Downward infiltration (I_W) is calculated from the rates measured inside the infiltrometer cylinder (I_n) as follows (Bouwer 2002):

$$I_w = \frac{I_n \pi r^2}{\pi (r + x)^2} \tag{10.2}$$

where (using consistent units)

r radius of infiltrometer

x horizontal distance of the edge of the wetted zone from the edge of the infiltrometer

The measured infiltration rate ($Y_n/\Delta t_n$) is the change in water level inside the infiltrometer (Y_n) over the time Δt_n for the last step (water-decline period).

Hydraulic conductivity (K) is calculated as (Bouwer 2002)

$$K = I_w L(z + L - h_{we}) \tag{10.3}$$

where

- L depth of the wetting front after the last test water-level decline period (test step; calculated or measured)
- I_W infiltration rate during last water level decline period
- z average depth of water in the cylinder during the last water-level decline period
- h_{we} capillary pressure or negative suction at the wetting front (units of length; estimated from soil type)

The calculated K value is used as the long-term infiltration rate in inundated areas without surface clogging and low-permeability restricted layers deeper down (Bouwer 2002).

Bargarello et al. (2004) and Nimmo et al. (2009) proposed single-ring falling-head infiltration test procedures as a simple, quick, and inexpensive option for performing multiple tests. Because of the spatial variability of field saturated hydraulic conductivity (K_{fs}) over broad areas, it was proposed that it is generally better to perform a large number of measurements in as many locations as possible, even at the expense of a somewhat increased uncertainty in individual measurements (Nimmo et al. 2009). In the case of a single-ring infiltrometer, Nimmo et al. (2009) proposed that it can be preferable to mathematically correct for subsurface radial spreading effects associated with small-diameter, single-ring tests, then to instrumentally minimize the effects by using very large rings or double-ring apparatuses. The U.S. Geological Survey developed a system that involves a bare minimum instrumentation consisting of a (Nimmo et al. 2009):

- steel bucket with its bottom removed (instrument used: 22 cm high, 21-cm diameter at base, 26-cm diameter at top)
- small hovel
- stop watch
- few liters of water
- soil-bentonite mixture to seal the outside bucket edge (if needed)
- rubber mat (placed on soil during water addition to minimize disturbance).

A preselected volume of water is added to achieve of ponding depth of 0.03–0.1 m, the rubber mat removed, and the time to complete infiltration is recorded. Saturated hydraulic conductivity (K_{fs}) is calculated using the equations

$$K_{fs} = \frac{L_G}{t_f} \ln\left(1 + \frac{D_o}{L_G + \lambda}\right) \tag{10.4}$$

$$L_G = C_i d + C_2 b \tag{10.5}$$

where

L_G	ring-installation scaling length (m)
t_f	time to head (ponding depth) of zero
D_o	initial ponded depth (m)
λ	macroscopic capillary length of soil (0.08 m for most soils with significant structural development, 0.03 m for extremely coarse and gravelly soils, 0.25 m for fine-textured soils without macropores)
d	ring infiltration depth (m)
b	ring radius (m)
C_1, C_2	empirical constants (0.993 and 0.578 respectively, from Reynolds and Elrick 1990)

Nimmo et al. (2009) evaluated the effects of the accumulation of infiltrated water on K_{fs} values. Water movement into the initially dry soil was found to be sufficiently limited so that the first few falling-head measurements yielded progressively lesser K_{fs} values. The tests were rapid enough to allow for multiple repetitions and K_{fs} values stabilized after about 5–10 cm of water had been applied. The method will have a greater uncertainty in areas where biotic crusts and other adsorptive materials take up much of the applied water and it does not account for heterogeneity (Nimmo et al. 2009).

10.5 Pilot (Basin) Infiltration Tests

Pilot infiltration tests (PITs) are essentially constant-head infiltration tests performed on excavated basins. The important advantage of PITs is a larger test area than that of a ring infiltrometer and thus a greater likelihood that test results are representative of a planned infiltration system. Basic recommendations for performing PITs include (Philips and Kitch 2011; City of Tacoma 2012; WADOE 2013):

- the bottom of the pit should be the infiltration surface, which is the planned base of infiltration ponds or the elevation of the top of native soil for bioretention systems
- the horizontal area of the bottom of the test pit should be approximately 100 ft^2 (9.3 m^2)
- the size and geometry of the test pit should be accurately documented
- slide slopes of the depth interval of ponding during tests should be vertical (if possible) to facilitate calculation of infiltrated volume
- a vertical measuring rod or pressure transducer (or both) is used to measure water levels
- a splash plate at the bottom of the test pit is recommended to reduce side wall erosion and excessive disturbance of the pit bottom
- presoak pit for a least 6 h at the maximum ponding depth
- add water to the pit and adjust flow rate to maintain a constant water level

- start with a constant-head test with the water level maintained at the maximum design ponding depth
- perform tests until the flow stabilizes (variation in flow rate is less than 5%) and record the cumulative infiltrated volume and instantaneous flow rate every 15 min
- add water until one hour after the flow rate has stabilized while maintaining the same pond water level
- calculate the hourly flow rate for the stabilization period, which should be the lowest recorded rate
- after the constant-head test (flow rate and water level stabilized), turn off water and the record infiltration rate (change in water level times pit area) until the pit is empty
- at the conclusion of testing, over-excavate the pit to examine the distribution of the infiltrated water (e.g., observe whether it mounded on a shallow restrictive layer).

Philips and Kitch (2011) additionally recommended using a very highly permeable geotextile at base of the pit to reduce disturbance of the soil.

The basic PIT and basin infiltration test procedures are straightforward. Modifications to the procedure have been developed to facilitate and automate testing procedures. Castanos-Vegas and Lansey (2001) documented a prototype automated infiltrometer for measuring infiltration rates in recharge basins that used float switches connected to pumps to equalize water levels within the infiltrometer to the level in the recharge basin(s). Waters levels and pumping times are recorded with a datalogger. Infiltration rates are calculated from the water balance:

$$\Delta V_{\text{within infiltrometer}} = V_{\text{pumped in}} - V_{\text{pumped out}} - V_{\text{infiltrated volume}} \tag{10.6}$$

Pumped volume was calculated from pumping running times and pumping rates. It was noted that a planned design modification would operate pumps using water depth sensors, eliminating the need for float switches.

Gain et al. (2003) reported on an automated infiltrometer that was field tested at the Sweetwater Recharge Facility in Tucson, Arizona. The system used pressure transducers to measure water levels, two pumps (inside and outside of the infiltrometer), and a datalogger to control the pumps and store and transfer data. When the difference in water level exceeds a set value of 0.05 ft (1.5 cm), the program turns on one of the two pumps to either pump water into or out of the infiltrometer. Additional transducers were installed to measure surface saturation and water level in a perched groundwater mound.

10.6 Air-Entry Permeameter

Downward flow of infiltrated water in the unsaturated zone is driven by the hydraulic head of ponded water (positive water pressure) and the downward pull of water from negative pressure in the soil zone (capillary-driven flow). Bouwer (1966) introduced the air-entry permeameter as a means of including negative-pressure flow in saturated hydraulic conductivity measurements of initially unsaturated soils. The basic procedures are as follows (Bouwer 1966; ASTM 2016):

(1) A metal cylinder approximately 25-cm in diameter is driven approximately 10 cm into the ground. The cylinder is connected to a water-supply reservoir with a supply valve and a vacuum gauge.
(2) Water is applied to the cylinder at a relatively high head until the wetted front is expected to have reached a depth approximately equal to the cylinder penetration. The rate of fall of water level in the reservoir is recorded. The approach of the wetting front to the bottom of the ring may be determined by a change in soil tension measured using a tensiometer, a change in electrical conductivity measured using a resistivity probe, and/or from the volume of water infiltrated using an assumed porosity (Stephens 1996).
(3) The water supply valve is closed and the vacuum developed inside the cylinder by the suction exerted by the finer pores is recorded.
(4) The minimum pressure (maximum vacuum) above the standing water is recorded, which occurs at the incipient entry of air into the cylinder at the bottom of the wetted zone.
(5) Depth of the wetted zone is determined by direct observation (excavation), use of dyes, an electrical conductivity probe, or other means.

The air-entry pressure is used as an approximation of the wetting-front pressure head for determination of the hydraulic gradient, and, in turn, the field-saturated hydraulic conductivity (ASTM 2016).

Air entry pressure (P_a) is calculated as

$$P_a = P_{min} + G + L \tag{10.7}$$

P_{min} maximum reading on vacuum gauge (cm)
G height of vacuum gauge above soil surface (cm)
L depth of wetted front (cm)

K_s is calculated as (cm/sec)

$$K_s = 2\frac{dh}{dt}L\frac{r_R^2}{r_C^2}/(H_i + L - 0.5P_a) \tag{10.8}$$

where

dh/dt rate of fall of water level in the reservoir just before closing the valve (cm/sec)
H_i height above the soil surface of the water level in the reservoir at the time the supply valve is closed (cm)
r_R radius of the reservoir (cm)
r_C radius of the cylinder (cm)

The coefficient of 2 is an empirical value to correct for the effects of entrapped air. It has been observed that the 2 multiplier may result in an over estimation of K_s (Lee et al. 1985). The air-entry permeameter allows for rapid measurements of the vertical hydraulic conductivity of initially unsaturated soils. A disadvantage of the method is that it tests only a small area and may not reflect the effects of macropores on hydraulic conductivity (Sai and Anderson 1991).

10.7 Borehole Permeameters

Borehole permeameter tests have the advantages of being in situ measurements, operational simplicity, and allowing for field-saturated hydraulic conductivity measurements to be made at different depths. Hence, a series of tests can be performed at increasing depths to obtain profiles of saturated hydraulic conductivity versus depth. Several variations of the tests have been proposed including both falling-head and constant-head designs. A general constraint of borehole permeameters is that the rate of flow of water from a borehole into the adjoining formation depends upon borehole conditions (e.g., formation damage or smearing of the borehole wall) in addition to soil properties. Methods used to interpret test data need to account for borehole and test conditions (capillary effects).

Borehole permeameters are commonly performed on augered boreholes, which are often not stable in non-cohesive sediments (e.g., gravels). Miller et al. (2011) presented a direct-push vadose-zone permeameter for use in coarse alluvial gravels, in which tests are performed by driving a slotted pipe to the target sampling depth.

Constant-head tests are most commonly performed using borehole permeameters. The tests involve the measurement of the steady-state rate of flow into a borehole required to maintain a constant water level. The basic procedures are:

(1) Install a screen to the desired depth by either augering (hand or machine-driven), driving, or other drilling methods. An open hole may be used if the borehole is stable (not subject to collapse).
(2) Pre-wet the borehole to obtain saturated conditions near the screen.
(3) Add water to the borehole, adjusting the flow rate so that a constant water level is achieved. A constant head may be achieved using a float-switch type device, Mariotte (bottle/tube) device, or manual adjustments.
(4) Record the flow rates required to achieve a constant head until the rate stabilizes (i.e., rate does not significantly change between successive readings).
(5) Excavate area.

The basic equations for shallow water table conditions, defined by the distance between the water level in the borehole and water table being less than three times the depth of water in the well, are (Zangar 1953; USBR 1974, Stephens and Neuman 1980; Stephens 1996):

$$K_s = \frac{2Q}{C_u r(T_u + H - A)} \quad (10.9)$$

$$C_u = \frac{2\pi\left(\frac{H}{r}\right)}{\sinh^{-1}\left(\frac{H}{r}\right) - 1} \quad (10.10)$$

where (using consistent units)

K_s saturated hydraulic conductivity
Q steady state flow rate
r borehole radius
T_u distance from water in the borehole to the water table
H depth of water in the borehole
A length of tested interval ($H = A$ for fully screened wells)
sinh hyperbolic sine function

For deeper water table conditions, the corresponding equations are (Zangar 1953; Stephens and Neuman 1980):

$$K_s = \frac{Q}{C_u r H} \quad (10.11)$$

$$C_u = \frac{2\pi\left(2AH - A^2\right)}{rH\left[sinh^{-1}\left(\frac{A}{r}\right) - \left(\frac{A}{H}\right)\right]} \quad (10.12)$$

10.8 Guelph Permeameter

The Guelph permeameter is a commercially available small-diameter, constant-head borehole permeameter system. The Guelph permeameter is relatively easy to use and software are available to process test data. The equipment can be transported, assembled, and operated easily by one person. Measurements can normally be made in ½–2 h, depending on soil type, and require only about 2.5 L of water. The basic limitation of the method is that it is a small-volume technique.

The Guelph permeameter is essentially a Mariotte siphon system that measures the steady-state rate of infiltration into saturated soil from a cylindrical borehole in which a constant head (depth) of water is maintained. Measurements are obtained at a given depth using either a single constant head or two different constant heads. The one-head method is simpler and quicker, but less accurate. The one-head procedure is used for applications where the saturated hydraulic conductivity (K_s) needs to be known only within a factor of 2, which for many applied engineering applications is probably sufficient (Eijkelkamp Agrisearch Equipment 2011).

An interpretation method of Guelph permeameter data was presented by Glover (1953) and updated by Elrick et al. (1989). The rate of steady-state inflow (Q) can be expressed as (Elrick et al. 1989):

$$Q = f(H, a)K_{fs} + g(H, a)\phi_m \tag{10.13}$$

where

f, g functions of the constant height of the ponded water (H) and radius of the well (a)
K_{fs} fluid-filled hydraulic conductivity
ϕ_m matric fluid potential

The first term is the saturated component of flow and the second the unsaturated or capillary component of flow. The matric fluid potential is a measure of the capillarity of the soil.

In the Glover (1953) relationship, the effects of capillarity are not considered, and

$$Q = 2\pi H^2 K_{fs}/C_G \tag{10.14}$$

where C_G is a dimensionless shape factor that depends on the (H/a) ratio.

Field saturated hydraulic conductivity can be calculated as (Stephens et al. 1988)

$$K_{fs} = \frac{Q\left(sinh^{-1}[H/a] - 1\right)}{2\pi H^2} \tag{10.15}$$

The Glover (1953) methods tends to overestimate K_{fs} values as it does not account for capillary effects and unsaturated flow (Stephens et al. 1988). To incorporate the effects of capillarity, an additional term needs to be added. The Reynolds and Elrick (1986) method is based on simultaneously solving two equations (Elrick et al. 1989):

$$Q_1 = A_1 K_{fs} + B_1 \phi_m \tag{10.16}$$

$$Q_2 = A_2 K_{fs} + B_2 \phi_m \tag{10.17}$$

where

$$\begin{aligned} A_1 &= \left[\left(2\pi H_1^2/C_1\right) + \pi a^2\right] \\ A_2 &= \left[\left(2\pi H_2^2/C_2\right) + \pi a^2\right] \\ B_1 &= (2\pi H_1/C_1) \\ B_2 &= (2\pi H_1/C_2) \end{aligned}$$

C_1 and C_2 are dimensionless shape factors that depend primarily on the (H/a) ratio and also somewhat on the K_{fs}/ϕ_m ratio (Reynolds and Elrick (1986). Once values of C_1 and C_2 are estimated, then the above equations can be solved as there are two equations and two unknowns.

Elrick et al. (1989) proposed a simplified method, based on the fact that the parameter α^* (K_{fs}/ϕ_m) has a limited range of values in soils:

$$K_{fs} = Q/\left[A + (B/\alpha^*\right] = CQ/\left(2\pi H^2 + \pi a^2 C + 2\pi H/\alpha^*\right) \quad (10.18)$$

$$\phi_m = Q/\left[A\alpha^* + B\right] = CQ/\left(2\pi H^2 + \pi a^2 C)\alpha^* + 2\pi H\right) \quad (10.19)$$

The value of α* ranges between 1 and 36 m^{-1} (Elrick et al. 1989), with 12 m^{-1} recommended for structured soils and clays through clay loams, and unstructured medium and fine sands and sandy loams (first choice for most soils), 4 m^{-1} for unstructured fine-textured soils, and 36 m^{-1} coarse and gravelly sands.

Guelph permeameters are widely available. The Eijkelkamp Agrisearch Equipment (2011) instrument has a maximum practical operating depth of 315 cm and is recommended for 6 cm diameter augered boreholes. Profiles are obtained by augering deeper after each reading. For each reading (depth), the rate of fall of water is recorded (cm/min) until the rate of fall does not significantly change in three consecutive time intervals (usually of 2 min duration each). Software is available to process data ("Guelph-permeameter-calculator.xls") using either one-head or two-head methods. Inputs are head, borehole radius, soil texture category, and steady-state recharge rate (cm/s). C (shape) factors are calculated based on soil texture category after Zhang et al. (1998).

The advantages of the Guelph permeameter are that it is a fairly simple procedure, commercial availability, the tripod-mounted system can be operated by one person, and a moderate cost (less than $5,000 USD). The main disadvantage are a small volume of investigation and susceptibility to borehole skin effects impacting results.

10.9 Velocity Permeameter

The velocity permeameter, developed by Merva (1987, 1995), is an in situ falling-head permeameter that is used on cores isolated by driving a core barrel (cylinder) into the soil. Both vertical and horizontal saturated hydraulic conductivity can be measured. Horizontal hydraulic conductivity is measured by driving a core barrel into the side-wall of a trench or hole. Hydraulic conductivity is measured from the rate of fall of water in a head tube as a function of time and head based on the equation

$$K_s = \frac{dv}{dh} s \quad (10.20)$$

where

(dv/dh) change in the rate (velocity) of fall of water with change in head
s distance through which head is dissipated (length of core)

In practice, the velocity of fall is calculated from a series of measurements of the time required for head to fall through successive intervals of Δh. The rate of water flow through the core is obtained from the measured velocity by correcting for the diameters of the head tube and cores. An advantage of the method is that the effects of preexisting soil-water potentials are eliminated in the calculations as they cancel out.

10.10 Comparisons of Infiltrometer and Permeameter Systems

Gregory et al. (2005) compared the results of constant-rate and falling-head double-rings tests performed using 15- and 30-cm rings and constant-head tests performed using ASTM-standard 30- and 60-cm diameter rings. The constant-rate tests with the smaller rings provided infiltration rates that were statistically greater than rates from ASTM standard tests and falling-head tests with smaller rings. Nevertheless, Gregory et al. (2005) concluded that constant-rate tests using the smaller rings was adequate to represent the sandy soils tested, and they proposed a Mariotte siphon (bottle) device that allows for one person to easily perform the test.

Philips and Kitch (2011) performed a comparative study of in situ hydraulic conductivity testing and indirect measurements at three sites in southern California. Infiltration rate measurements considered included:

- direct measurements (infiltration tests: PIT, double-ring, and borehole infiltrometer)
- indirect measurements of soil properties (grain size analyses, CPT tests).

The results of the Philips and Kitch (2011) investigation confirmed the long-reported observation that direct measurements with a larger area of investigation are most accurate and representative. They recommended the PIT as the preferred method. Small-scale and indirect methods are appropriate for preliminary studies (e.g., site screening) because they avoid doing a more expensive PIT test on what may turn out to be an inappropriate site. Philips and Kitch (2011) noted that substantial correction factors need to be applied for design and that larger factors are needed for borehole and double-ring infiltrometer data than PIT data.

A number of studies have compared different methods for measuring saturated hydraulic conductivity in soils, several of which are summarized below. Lee et al. (1985) compared the following three methods relative to soils types at four different locations in southern Ontario

- air-entry permeameter (AEP)
- Guelph permeameter (GP)
- laboratory falling-head permeameter applied to small cores (SC).

In general, the AEP tended to provide higher values of mean K_s than the GP, which, in turn, yielded equivalent or higher values than the SC method. Lee et al. (1985) suggested that the difference between methods may reflect a decreasing importance of macropores on K_s from AEP to GP to SC. They also suggested that the AEP may work best for macropore-dominated K_s, whereas SC may be most appropriate for matrix-dominated K_s. The GP may be most appropriate where an average of matrix and macropore-dominated K_s is required. All three methods gave approximately the same K_s values for structureless soils (i.e., soils in which particles are not aggregated).

Kanwar et al. (1990) compared saturated hydraulic conductivity measurements obtained using the Guelph permeameter, velocity permeameter and a laboratory constant-head permeameter on silt-loam soils at eight sites at an agricultural research center near Ames, Iowa. The mean K_s values obtained by the Guelph permeameter tended to lie between the vertical and horizontal K_s values obtained using the velocity permeameter, with a better agreement to horizontal K_s values. It was noted that the Guelph permeameter measures some combination of vertical and horizontal K_s, whereas the velocity permeameter measures either. Laboratory K_s values were about 10–800 times greater than the field measurements, which was suggested to be due to macropores that connected one end of the core to the other.

Mohanty et al. (1994) compared the following methods for measuring saturated hydraulic conductivity in a Wisconsin-age glacial-till soil at a research farm near Boone, central Iowa:

- Guelph permeameter
- velocity permeameter
- disk permeameter (constant head measurements using an emplaced sand layer insider a 25.4-cm ring)
- double-tube method (Bouwer 1964)
- laboratory constant-head permeameter method.

The double-tube method was found to be impractical for most attempted measurements due to soil conditions at the site. The main results of the comparative study are (Mohanty et al. 1994):

- The Guelph permeameter gave the lowest K_s values, possibly due to a small sample size, wall smearing, and air entrapment.
- Disk permeameter and double-tube methods gave the highest K_s values, probably due to a large sample size.
- The velocity permeameter gave values closer to those from detached cores measured in the laboratory.
- Laboratory analyses had the greatest variation (standard deviation), possibly due to the presence or absence of macropores.

The various methods developed to measure vertical infiltration rates and saturated hydraulic conductivity each have specific advantages and disadvantages. Choice of methods need to consider accuracy, site (soil conditions), speed, equipment and man-power requirements, and cost (Lee et al. 1985), project data requirements, and how the data will be used. With respect to MAR system design, method errors and aquifer heterogeneity are usually addressed by applying safety factors rather than conducting more detailed and expensive site investigations. Inasmuch as infiltration rates are of primary concern for MAR, then techniques that directly measure infiltration rates (e.g., infiltrometer tests and PITs) are preferred.

References

Angulo-Jaramillo, R., Bagarello, V., Iovino, M., & Lassabatere, L. (2016). *Infiltration measurements for soil hydraulic characterization*. Berlin: Springer.

ASTM (2003). *Standard test method for infiltration rate of soils in field using double-ring infiltrometer*. ASTM Standard D3385-03. West Conshohocken, PA: ASTM International.

ASTM. (2016). *Standard guide for comparison of field methods for determining hydraulic conductivity in vadose zone (ASTM D5126–16e1)*. West Conshohocken, PA: ASTM International.

Bagarello, V., Iovino, M., & Elrick, D. (2004). A simplified falling-head technique for rapid determination of field-saturated hydraulic conductivity. *Soil Science Society of America Journal, 68*(1), 66–73.

Bouwer, H. (1964). Measuring horizontal and vertical hydraulic conductivity of soil with the double-tube method. *Soil Science Society of America Journal, 28,* 19–23.

Bouwer, H. (1966). Rapid field measurement of air entry value and hydraulic conductivity of soil as significant parameters in flow system analysis. *Water Resources Research, 2*(4), 729–738.

Bouwer, H. (2002). Artificial recharge of groundwater: Hydrogeology and engineering. *Hydrogeology Journal, 10*(1), 121–142.

Castanos-Vegas, A., & Lansey, K. (2001). Measuring infiltration rates in soil-aquifer treatment recharge basins using an automated infiltrometer. In *Proceedings Tenth Biennial Symposium Artificial Recharge of Gr*oundwater, Tucson, AZ, June 2001, pp. 117–123.

Christiansen, J. E. (1944). Effect of entrapped air upon the permeability of soils. *Soil Science, 58*(5), 355–366.

City of Tacoma (2012). Appendix B procedure for conducting a pilot infiltration test. In *Stormwater management manual* (2012 ed.). City of Tacoma, Washington Environmental Services Department, Science and Engineering Division. Retrieved May 20, 2018 from http://cms.cityoftacoma.org/surfacewater/swm2012/V5-AppB.pdf.

Constantz, J., Herkelrath, W. N., & Murphy, F. (1988). Air encapsulation during infiltration. *Soil Science Society of America Journal, 52*(1), 10–16.

Eijkelkamp Agrisearch Equipment (2011). Operating Instructions. 09.07 Guelph Permeameter. Retrieved May 20, 2018 from https://www.eijkelkamp.com/download.php?file=b2082939.

Elrick, D. E., Reynolds, W. D., & Tan, K. A. (1989). Hydraulic conductivity measurements in the unsaturated zone using improved well analyses. *Groundwater Monitoring and Remediation, 9*(3), 184–193.

Faybishenko, B. A. (1995). Hydraulic behavior of quasi-saturated soils in the presence of entrapped air: laboratory experiments. *Water Resources Research, 31*(10), 2421–2435.

Gain, J., Ela, W., Arnold, R., & Lansey, K. (2003). Measuring localized surface infiltration rates in secondary effluent basins. In *Proceedings 11th Biennial Symposium on Groundwater Recharge*, June 5–7, 2003, Phoenix, AZ.

Glover, R. E. (1953). Flow from a test hole located above groundwater level. In C. N. Zangar (Ed.), *Theory and problems of water percolation, Engineering Monograph 8*. U.S. Bureau of Reclamation: Denver.

Gregory, J. H., Dukes, M. D., Miller, G. L., & Jones, P. H. (2005). Analysis of double-ring infiltration techniques and development of a simple automatic water delivery system. *Applied Turfgrass Science*, (1st ed., Vol. 2). https://doi.org/10.1094/ats-2005-0531-01-mg.

Johnson, A. I. (1963). *A field method for measurement of infiltration*. U.S. Geological Survey Water-Supply Paper 1544-F.

Kanwar, R. S., Rizvi, H. A., Ahmed, M., Horton, R., & Marley, S. J. (1990). Measurement of field-saturated hydraulic conductivity by using Guelph and velocity permeameters. *Transactions of the ASAE, 32*(6), 1885–1890.

Klute, A., & Dirksen, C. (1986). Hydraulic conductivity and diffusivity: Laboratory methods. In A. Klute, A. (Ed.), *Methods of soil analysis, Part 1* (2nd ed.), Monograph 9 (pp. 687–734). Madison, WI: American Society Agronomy—Soil Science Society of America.

Lee, D. M., Elrick, D. E., Reynolds, W. D., & Clothier, B. E. (1985). A comparison of three field methods for measuring saturated hydraulic conductivity. *Canadian Journal of Soil Science, 65*(3), 563–573.

Marinas, M., Roy, J. W., & Smith, J. E. (2013). Changes in entrapped gas content and hydraulic conductivity with pressure. *Ground Water, 51*(1), 41–50.

Merva, G. E. (1987). The Velocity Permeameter technique for rapid determination of hydraulic conductivity in-site. In *Proceedings of the Third International Workshop on Land Drainage*, pp G-55–G-66. Columbus: Ohio State University.

Merva, G. E. (1995). Horizontal hydraulic conductivity determination using the Velocity Permeameter. In H. W. Beclcher & F. M. D'Itri (Eds.), *Subirrigation and drainage control* (pp. 151–168). Boca Raton: CRC Press.

Miller, R. B., Heeren, D. M., Fox, G. A., Storm, D. E., & Halihan, T. (2011). Design and application of a direct-push vadose zone gravel permeameter. *Groundwater, 49*(6), 920–925.

Mohanty, B. P., Kanwar, R. S., & Everts, C. J. (1994). Comparison of saturated hydraulic conductivity measurement methods for a glacial-till soil. *Soil Science Society of America Journal, 58*(3), 672–677.

Nimmo, J. R., Schmidt, K. M., Perkins, K. S., & Stock, J. D. (2009). Rapid measurement of field-saturated hydraulic conductivity for areal characterization. *Vadose Zone Journal, 8*(1), 142–149.

Philips, C. E., & Kitch, A. A. (2011). A review of methods for characterization of site infiltration with design recommendations. In *Proceedings 43rd Symposium of Engineering Geology and Geotechnical Engineering*. University of Las Vegas, Nevada, March 2011.

Reynolds, W. D., & Elrick, D. E. (1986). A method for simultaneous in situ measurement in the vadose zone of field-saturated hydraulic conductivity, sorptivity and the conductivity-pressure head relationship. *Groundwater Monitoring and Remediation, 6*(1), 84–95.

Reynolds, W. D., & Elrick, D. E. (1990). Ponded infiltration from a single ring: I. Analysis of steady flow. *Soil Science Society of America Journal, 54*(5), 1233–1241.

Sai, J. O., & Anderson, D. C. (1991). *State-of-the art field hydraulic conductivity measurements of compacted soils*. Cincinnati: U.S. Environmental Protection Agency Risk Reduction Engineering Laboratory.

Stephens, D. B. (1996). *Vadose zone hydrology*. Boca Raton: CRC Press.

Stephens, D. B., Lambert, K., & Watson, D. (1984) Influence of entrapped air on field determinations of hydraulic properties in the vadose zone. In *Proceedings of Conference on Characterization and Monitoring in the Vadose Zone* (pp. 57–76).

Stephens, D. B., & Neuman, S. P. (1980). *Analysis of borehole infiltration tests above the water table*, Technical report #35). Tucson: Natural Resource Systems, University of Arizona.

Stephens, D. B., Unruh, M., Havlena, J., Knowlton, R. G., Jr., Mattson, E., & Cox, W. (1988). Vadose zone characterization of low-permeability sediments using field permeameters. *Ground Water Monitoring and Remediation, 8*(2), 59–66.

Turf Tech (n.d.). Turf-Tec Infiltrometer Instructions. Tallahassee: Turf-Tec International. Retrieved 20 May, 2018 from http://www.turf-tec.com/in/IN2-W-Instructions.pdf.

WADOE. (2013). *Eastern Washington low impact development guidance manual*. Lacey, WA: State of Washington Department of Ecology.

Wang, Z., Feyen, J., Genuchten, M. T., & Nielsen, D. R. (1998). Air entrapment effects on infiltration rate and flow instability. *Water Resources Research, 34*(2), 213–222.

USBR. (1974). *Earth manual* (2nd ed.). Washington DC: United States Bureau of Reclamation.

Zangar, C. N. (1953). *Theory and problems of water percolation, Monograph 8*. Denver: United States. Department of the Interior, Bureau of Reclamation.

Zhang, Z. F., Groenevelt, P. H., & Parkin, G. W. (1998). The well-shape factor for the measurement of soil hydraulic properties using the Guelph Permeameter. *Soil and Tillage Research, 49*(3), 219–221.

Chapter 11
Clogging

11.1 Introduction

Clogging is a local reduction in permeability caused by the filling or obstruction of pores. Clogging can be the result of physical, chemical, or biological processes, or some combination of the three. The principle loci of clogging in managed aquifer recharge (MAR) systems is at or just below the infiltration surface (e.g., basin floor) in land application systems and at the borehole wall in systems that use wells for recharge. It is through these interfaces that the recharge water flows must pass with any entrained suspended solids, air bubbles, nutrients, and other ions and uncharged species. The infiltration surface and borehole wall are thus the preferred site of the filtration of suspended solids, biological growth, and biogeochemical reactions (mineral precipitation and alteration). Clogging results in a reduction in recharge rates and is often the primary operational challenge for MAR systems. Hence, an important element in the design and operation of MAR systems is minimizing the potential for clogging and developing cost-effective strategies for rehabilitating systems to restore their performance.

Clogging tends to occur most rapidly and severely in MAR systems that use wells for recharge. Huisman and Olsthoorn (1983, p. 237) noted that

> Without any doubt, the most important drawback to the use of injection wells is the danger of clogging, primarily caused by an entrance rate into the aquifer which is one to two orders of magnitude higher than that with spreading ditches.

Injection wells force a large volume of water through a relatively small surface area, the borehole wall. The suspended solids load of the injected water is either filtered out at the borehole wall or passes through the borehole wall and is filtered out a short distance into the adjoining aquifer. In the case of infiltration basins and other surface-spreading MAR systems, the applied water passed through a much larger surface area. In addition to physical clogging by filtration of suspended solids, clogging of injection wells may also occur as the result of gas (air) binding, biological growth, and chemical precipitation (Sniegocki 1963; Huisman and Olsthoorn 1983). Injection wells can seriously clog in days if the injected water is of poor quality, not

R. G. Maliva, *Anthropogenic Aquifer Recharge*,
Springer Hydrogeology, https://doi.org/10.1007/978-3-030-11084-0_11

properly pretreated (e.g., disinfected to prevent microbial growth), or is geochemically incompatible with the aquifer rock or native groundwater. MAR systems that utilize trenches and pits have similar clogging causes as occur wells and may also clog rapidly, especially when they receive poor quality water.

In addition to gross impacts to recharge rates, clogging can impact the spatial pattern of recharge. The concentration of flow into the most permeable strata can result in the preferential clogging of such strata. For example, data from an early aquifer storage and recovery (ASR) test program in Norfolk, Virginia, demonstrated how aquifer plugging can impact of the recovery of freshwater recharged into a brackish aquifer (Brown and Silvey 1977). Flowmeter log data indicated that zones that clog during injection may become unplugged, and thus productive, during recovery. The clogged zones contain mostly formation water because they accepted relatively little freshwater during injection. During recovery, production from the formerly clogged zones tended to increase the salinity of the recovered water.

As reviewed herein, the causes of clogging are now well understood, as are means for their management, which includes measures to both slow or prevent clogging from occurring and for restoring system capacity. Nevertheless, the rate of clogging is a complex, system-specific function of recharge water quality, formation lithological properties, and geochemical environment, and generally cannot be predicted in advance with great confidence. Clogging may still occur even when the recharge water is of very high quality (e.g., potable water and reverse-osmosis treated reclaimed water). The term "recharge water" is used herein in refer to water sent to a recharge system, as opposed to "recharge<u>d</u> water," which is water that actually enters an aquifer.

The main technical challenge with respect to management of clogging now is developing the most cost-effective design (including pretreatment) and operational and rehabilitation protocols to maintain long-term system capacity. A trade off occurs between treatment to prevent clogging and rehabilitation to reverse it. In simple terms, is it more economical to treat recharge water so that clogging will not occur (or will occur at a lower rate) or to accept that clogging will occur and focus on efficiently rehabilitating the recharge system? Dillon et al. (2016) observed with respect to ASR systems that:

> Trading off treatment and well rehabilitation has, in the past, biased outcomes to low-cost treatments and some acceptance of higher maintenance costs of ASR operations, simply as a means of deferring costs. The risk of economic failure of such systems is high, and the inconvenience of being forced to schedule well remediation at short notice would likely give ASR a bad name among operators.

Clearly, there is a need to be able to accurately predict clogging probabilities and rates in advance of system construction and to accurately incorporate risks (i.e., potential costs) associated with clogging into economic analyses of MAR systems. Economic analyses of MAR systems will be biased in the favorable direction if only a favorable outcome is assumed (Maliva 2014). In practice, management of clogging necessarily still involves some informed trial and error, in which options are tested and monitored for both initial improvement (i.e., restoration of performance) and the duration of the improvement. Perhaps the resources management term "adaptive management" (i.e., learning by doing) is more appropriate. The various clogging mechanisms and potential responses (and associated costs) should be identified at the start of a project along with an initial estimate of the potential for their occurrence based on site hydrogeology, source water quality, and experiences with similar systems.

Advancing beyond a trial and error approach requires research and investigations (Dillon et al. 2016). Uncertainty associated with clogging may be reduced through laboratory (column) and field testing, but the up-scaling of results from short-term, small-scale testing to the long-term operation of a full-scale system is far from perfect. A basic issue for MAR projects, in general, is how much testing should be performed prior to the construction of a project, which ties into the economics of information (Maliva 2016). Aquifer characterization and source water quality and treatment evaluation efforts are economically justifiable if the value of the information obtained exceeds the costs of the information (Ma 2011; Maliva 2016). Dillon et al. (2016) proposed a Bayesian analysis approach to clogging investigations, which considers the probability that the results of an experiment (testing) will provide a good indication of the success of a future ASR system.

11.2 Causes of Well Clogging

The processes that cause clogging in recharge and other injection wells have long been understood and have been summarized in a number of books and technical papers (e.g., Sniegocki 1963; Olsthoorn 1982; Huisman and Olsthoorn 1983; Pyne 1995, 2005; Maliva and Missimer 2010; Martin 2013). Nine potential causes of clogging of MAR wells were listed by Sniegocki (1963) over fifty years ago (Table 11.1). Clogging processes can categorized as being primarily either physical, chemical, or biological.

A key element in the design and operation of MAR systems using wells is evaluating the potential for each of the clogging processes to occur, and then developing and implementing strategies to minimize their effects on system performance.

11.2.1 Entrapment and Filtration of Suspended Solids

Suspended solids in recharge water are filtered out either at the well screen, filter pack, or borehole wall, or enter the aquifer where they fill and obstruct pores. The permeability of media depends largely upon the diameters of pore throats along the fluid flow path. Pore throats are the constrictions that connect adjacent pores through which water flows. In general, fine-grained materials tend to have smaller pores and pore throats, and thus both lower permeabilities and greater abilities to filter out particles than coarser-grained sediments. Physical clogging by entrapment,

Table 11.1 Principal causes of clogging in recharge systems using wells

Category	Cause of clogging
Physical	Entrapment and filtration of suspended solids present in recharge water
	Mechanical jamming of an aquifer caused by particle rearrangement when the direction of water movement is reversed
	Gas binding; entrapment of air entrained in recharge water or gas generated in a formation
Chemical	Chemical reactions between groundwater and recharge water causing the precipitation of insoluble products
	Precipitation of iron in injected water and native groundwater as a result of aeration (introduction of dissolved oxygen)
	Clay swelling
	Clay dispersion; ion exchange reactions mobilize clays that physically clog an aquifer
Biological	Bacterial clogging by bacterial biomass and extracellular polymers (biofilms) caused by either bacterial contamination of an aquifer by the recharge water or increased growth of indigenous bacteria
	Bacterial clogging by iron bacteria

straining, and filtration occurs by the deposition of a layer of very fine-grained, low permeability material through which recharge water must flow or by fine particles entering an aquifer and clogging pore throats.

Clogging of injection wells has been compared to the clogging of a filter (Huisman and Olsthoorn 1983; Pyne 1995), which was described by Schippers and Verdouw (1980) as occurring in a three-stage process. The first stage is blocking filtration in which suspended particles physically block pore spaces. The second stage is cake or gel filtration, in which suspended particles form a progressively thickening layer of filtrate on the screen or borehole surface. Injection pressure during the cake or gel filtration stage linearly increases with time as the thickness of the filter cake increases. The final, and most severe stage of clogging, is referred to as cake filtration with compression, which results in the greatest increase in required injection pressure at a given injection rate. Once cake filtration with compression is reached, continued operation of a well may not be practical because of either low injection rates or required high injection pressures (Pyne 1995).

Torkzaban et al. (2015) after Ryan and Elimelech (1996) divided particles into colloidal particles, with diameters between 0.01 and 10 μm, and suspended particles, with diameters between 10 and 100 μm. Colloids are defined as mixtures in which particles do not settle and cannot be separated out by ordinary filtering or centrifuging. Milk, for example, is a colloid. Potential clogging material is referred to herein as "ultrafine particles," which includes material in the very-fine silt through colloid size range.

The ultrafine particles are typically not mobile under prevailing aquifer conditions (Torkzaban et al. 2015). Disturbances resulting from changes in physicochemical and hydrodynamic conditions mobilize them. The mobilized particles may be transported downstream by flowing groundwater, where they may be deposited, particularly at pore constrictions, with an associated reduction in permeability.

Aquifers contain a wide range of ultrafine particles that differ in composition, size, shape, and surface properties, making it difficult to characterize and predict their detachment, transport and depositional behaviors in response to various physicochemical processes (Torkzaban et al. 2015). Ultrafine particles include particles that are already present in a formation and, less abundantly, particles that form by the in situ precipitation of new mineral phases or degradation (alteration) of existing phases. The main types of ultrafine particles are (Torkzaban et al. 2015):

- silicate particles (clay minerals)
- carbonates
- humic substances
- iron and manganese (oxy)hydroxides
- aluminum hydroxides
- sulfides and polysulfides
- microorganisms.

Ultrafine particle deposition occurs by McDowell-Boyer et al. (1986), Torkzaban et al. (2015):

- surface or cake filtration
- straining (size exclusion)
- bridging (several particles arrive at pore constriction at the same time and wedge together)
- filtration (attachment to surfaces).

Filtration and attachment involve interception, diffusion, and sedimentation of particles. Models have been developed to predict filtration, which involve quantification of collision probability, attachment (sticking) efficiency, porosity, and grain diameter (Yao et al. 1971). However, such modeling efforts have no practical applied value for MAR systems.

Clearly, recharge of water with high colloidal and suspended solids concentrations should be avoided. However, the behavior of aquifers to colloidal and suspended solids is specific to both the formation and recharge water type and is not predictable with confidence (Huisman and Olsthoorn 1983). Formations composed of fine-grained sediment with small pores and pore throats, in general, have the greatest susceptibility to physical clogging. On the contrary, carbonate aquifers in which flow is dominated by large secondary pores generally have a low susceptibility to physical clogging.

11.2.2 Mechanical Jamming

Periodic reversal of flow direction (e.g., between injection and recovery in ASR wells) may lead to a rearrangement of particles into a lower porosity and permeability configuration. However, mechanical jamming appears to be uncommon and generally does not have a significant impact on injection well pressure (Huisman and Olsthoorn 1983). Particle rearrangement, and thus clogging, by mechanical jamming is irreversible.

11.2.3 Gas Binding

Bubbles of air or other gases (e.g., CO_2) decrease permeability by obstructing pore throats. Air bubbles may be entrained within the injected water, liberated from solution when cold water is injected into a warmer aquifer (Sniegocki 1958, 1963), or form when negative (below atmospheric) pressure conditions develop (Huisman and Olsthoorn 1983). Air bubbles may be already entrained within recharge water as it reaches a wellhead, be "sucked" into the water at or near the wellhead through leaks in piping under negative pressure conditions, or be mixed into the recharge water as the result of water being allowed to cascade down a well.

Carbon dioxide, and perhaps also nitrogen, bubbles may form by microbial activities. As is the case for solid particles, gas bubbles may become trapped, and reduce permeability at well screens, within filter packs, or within an aquifer. Air bubbles that enter an aquifer flow outward (away from the recharge well) with the injected water until they reach a pore throat through which they cannot pass. Sniegocki (1958, 1963) noted that the forces that prevent further movement of air bubbles may result from one or more of three causes:

- simple blocking of air bubbles within pores by small pore throats
- the so-called "Jamin" effect
- distortion of gas bubbles when they are forced through capillary openings (pore throats).

The Jamin effect, as described by Sniegocki (1963), is that a capillary tube containing restrictions and filled with a chain of alternating air and water bubbles is capable of sustaining a finite pressure. Aquifers may behave as series of capillary tubes with the Jamin effect and bubble distortion acting to prevent the further movement of entrained air.

Air bubbles in a well screen or filter pack may eventually escape and rise to the surface after injection is terminated. Air binding differs significantly from clogging with solids in that air bubbles dissolve over time. After an initial rapid increase in flow resistance (injection pressure), a dynamic equilibrium is reached as the rate of bubble formation or migration into a formation or filter pack equilibrates with the rate of bubble dissolution (Huisman and Olsthoorn 1983).

Gas binding can cause a rapid clogging of wells. In an early study, gas binding was identified as the major contributor to clogging in recharge tests performed in a Quaternary-aged sand and gravel aquifer in the Grand Prairie area of southeastern Arkansas (Sniegocki 1963). Air entrainment was estimated to have caused at least a 50% reduction in permeability.

Air has a much lower density than water and, therefore, air bubbles will tend to rise in a well. A key issue determining whether air bubbles that enter a well will reach the aquifer is the downhole flow velocity versus the bubble-rise velocity (Huisman and Olsthoorn 1983; Pyne 1995). The velocity at which bubbles rise is a function of their size. If the rate of downward flow in a recharge well is less than the bubble-rise velocity, then air bubbles entrained in the injected water tend to rise rather than travel downward. For air bubbles with diameters of 0.1 to 10 μm, the bubble rise velocity is approximately 1.0 ft/s (0.3 m/s; Huisman and Olsthoorn 1983; Pyne 1995).

Clogging due to air entrainment is largely a design and operational issue rather than a source water quality issue. Injection wells and wellhead should be designed to maintain positive pressures during injection.

11.2.4 Chemical Clogging—Mineral Scaling

Mineral precipitation in a well screen and/or filter pack, on a borehole wall, and within an aquifer can result in dramatic reductions in injection well capacity. Chemical clogging is caused by either the injected water being supersaturated with respect to a clogging mineral or mixing or fluid-rock interactions causing supersaturation. In general, chemical clogging tends not to be a major problem in recharge wells because the injected freshwaters usually have low ion concentrations (i.e., are greatly undersaturated with respect to minerals that could potentially precipitate in a groundwater environment). Chemical clogging is of much greater concern in disposal injection wells that discharge waters that are hypersaline or otherwise have high ion concentrations. The author was once retained to investigate the cause of the rapid clogging of an industrial injection well (within 3 weeks of the start of injection). It was found that the injected water was supersaturated with respect to calcium carbonate minerals (calcite and aragonite) by a factor of about 20, which was remedied by a pH adjustment of the injectate.

Calcium carbonate scaling has occurred in some recharge wells. Calcium carbonate is strongly controlled by pH, which can vary considerably due to changes in the partial pressure of CO_2. Degassing of CO_2 can result in calcium carbonate supersaturation. Pearce and Eckmann (1999) documented that injection rates in the Marco Lakes ASR system in Collier County, Florida, declined from 800 to 100 gpm in a matter of days due to calcium carbonate scaling. Calcium carbonate scaling can be prevented by decreasing the pH of injected water with an acid feed or the introduction of carbon dioxide to form carbonic acid. Carbonate scale can also be readily removed by acid treatment.

Calcite carbonate dissolution is actually more likely than precipitation as the result of the recharge of freshwater. Enhanced hydraulic conductivity caused by calcite dissolution was detected in laboratory experiments performed to simulate clogging in wastewater ASR systems (Rinck-Pfeiffer et al. 1998, 2000, 2002). Calcite dissolution near the inlet resulted in an increase in hydraulic conductivity, but not to pre-wastewater injection values.

11.2.5 Chemical Clogging—Redox Reactions

Iron (oxy)hydroxide precipitation may occur when water containing dissolved oxygen (DO) is recharged into an aquifer containing chemically reducing groundwater and dissolved ferrous iron (Fe^{2+}). Ferric (Fe^{3+}) iron, the oxidized form of iron, is much less soluble than ferrous iron. Recharged water usually contains high DO concentrations and confined aquifers commonly have chemically reducing conditions. Well clogging with neoformed iron (oxy)hydroxide minerals can occur at the injection well borehole and in the immediately adjoining aquifer.

The dissolved iron concentration in groundwater, even under reducing conditions, is seldom so high that its local oxidation during recharge alone could cause aquifer clogging that would materially impact the performance of a recharge system. Dissolved iron concentrations are usually <10 mg/L, so even if all of iron present in the groundwater near a recharge well were oxidized before the native groundwater is displaced by iron-free recharged water, the mass of iron (oxy)hydroxide that precipitates would very minor. For significant clogging to occur, a flow into a well of water containing dissolved iron is required.

Clogging caused by iron bacteria can be a serious problem in production wells that are part of MAR systems as there can be a continuous flux into the wells of iron-rich water during pumping (Sect. 11.2.8).

A more important source of iron is chemically reduced minerals, such as iron sulfides (e.g., pyrite and marcasite). Oxidation of iron sulfide minerals can produce much greater amounts of hydrated iron (oxy)hydroxides minerals, which due to their porous texture can obstruct pores. Released iron (oxy)hydroxide particles may also become mobilized and obstruct pores. Moorman et al. (2002) documented an unusual situation where clogging from iron (oxy)hydroxides occurred in a dedicated recovery well. The possibility was suggested that iron (oxy)hydroxides that form around injection wells may be transported in the aquifer in a colloidal state (Moorman et al. 2002).

11.2.6 Clay Swelling and Dispersion

Clay swelling and dispersion are the result of the expansion of the electrostatic double layer, and associated increase in repulsive forces, caused by a reduction in ion strength (salinity) and a predominance of monovalent cations versus multivalent cations (Sect. 5.6.4). It has long been understood that the swelling and mobilization (dispersion) of clays depend upon the type of clay minerals present, and are induced by changes in pH, ion types, and ion strength (salinity) that affect the properties (thickness) of the electrostatic double layer. Strata that are susceptible to salinity-induced permeability changes are described as being "water sensitive."

Clay minerals tend to have negative surface charges, which are balanced by adsorbed cations. Source of negative surface charges, and thus cation exchange capacity, are (Grim 1968):

- isomorphic substitution (Al^{3+} for Si^{+4})
- lattice imperfections and broken bonds
- exposed hydroxyl groups.

Swelling clays, such as montmorillonites, expand when water penetrates and is adsorbed in interlayer molecular spaces. The amount of expansion depends upon the exchangeable cations contained in the clay. Where sodium is the predominant exchangeable cation, clays can swell to several times their original volume, effectively clogging pore spaces. Clay swelling is a reversible process. Swelled clay minerals may subsequently contract when recharged freshwater is pumped out and the original salinity conditions are restored.

Dispersion is the mobilization and transport of clay particles. Mobilized clay particles are transported in the flowing recharged water until they become lodged in pore throats, locally reducing permeability. Unlike swelling, dispersion results in largely irreversible reductions in permeability (Brown and Silvey 1977) as trapped particles tend to remain lodged in pore throats or continue to fill small pores after groundwater chemistry (salinity) returns to natural conditions.

Clay-water interactions can result in clogging both near injection wells and deeper in the aquifer. The latter results in a gradual loss of injectivity and is very difficult to remediate. The critical salinity is the cation concentration above which the double-layer thickness is sufficiently thin that Van de Waals attractive forces cause clay particles to attach to each other and to pore walls (20 nM Nacl at neutral pH; Torkzaban et al. 2015). Critical salinity values depend upon the mole percent calcium and ionic strength of a solution (Torkzaban et al. 2015). The greatest potential for a loss of permeability by clay dispersion occurs when freshwater is recharged into clay-rich sand or sandstone (siliciclastic) aquifers that contain brackish or saltwater. Montmorillonite and mixed layer clays that are small in size and have large surface charges are usually the most water sensitive (Brown and Silvey 1977).

Decreases in well performance (specific injectivity; injection rate divided by injection pressure) from clay dispersion can be dramatic. In an early ASR test program in Norfolk, Virginia, documented by Brown and Silvey (1973, 1977), clay dispersion caused specific injectivity to decrease from 3.7 gpm/ft (46.0 L/min) to 0.93 gpm/ft (11.5 L/min) after the injection of only 132,700 gallons (502 m^3) of treated water.

During the following injection test (no. 4), a 0.2 N calcium chloride preflush was used to reduce swelling of the interstitial clay. A total of 20.146 Mg (76,250 m^3) of water was injected over a 95-day period. Injection was performed in 39 injection phases, separated by well redevelopment pumping. Even with the frequent well redevelopments, there still was an overall trend of decreasing well specific injectivity over time (Brown and Silvey 1977).

Konikow et al. (2001) performed benchtop column experiments to simulate the effects of clay dispersion on ASR system performance, as observed in the Norfolk

ASR system. The columns were packed with fine-grained sand with 0–5% clay minerals. The greatest reduction in permeability occurred when freshwater was introduced into the sample with the highest initial salinity and a montmorillonite clay content. Columns that contained kaolinite did not experience a loss of permeability with the introduction of freshwater. The effluent from the columns containing montmorillonite showed a large decrease in calcium concentration and increase in sodium concentration, which is consistent with the clay dispersion. The Konikow et al. (2001) experiments demonstrated that clay contents as low as 1% can result in a significant reduction in permeability with the introduction of freshwater.

Barry et al. (2013) discussed clogging issues in a small-scale domestic rainwater harvesting ASR system in Kingswood, South Australia. An approximately 50% decline is specific capacity occurred over the 39 month duration of the test. The cause of the clogging was believed to be swelling and dispersion of montmorillonite clay caused by the lower salinity of the injected water compared to the brackish native groundwater. XRD analysis of material recovered during an ASR pump-out indicates that it is composed of about 33% smectite/montmorillonite. Backflushing of the well removed accumulated suspended solids, as indicated by the high turbidity of the produced water, but was not sufficient to remove all accumulated clogging agents.

Reported instances of clogging of wells in which clay dispersion is the primary cause are uncommon. However, its potential impacts are severe and often largely irreversible. Hence, it important to determine whether swelling clays are present before that start of injection to avoid the risk of formation damage. Clay from drilling fluid may also jeopardize the life of a well (Brown and Silvey 1977). Therefore, swelling clays should not be introduced into an aquifer through the use of bentonite-based drilling mud. If bentonite-based mud is used, then well development should completely remove the mud from the formation prior to the start of injection. Preflush treatments are available to manage clay dispersion (Sect. 12.9.4).

11.2.7 Biological Clogging

Biological activity in wells can cause clogging by combinations of the accumulation of cell biomass, the secretion of extracellular polymers (polysaccharides), and the trapping of sediment and other microorganisms. The resulting biological film or layer is referred to as biofilm. Under favorable conditions, biofilms can develop rapidly (within days) and clog recharge wells.

Biological clogging results from either the introduction of foreign microorganisms into a well, filter pack, or aquifer, or by the stimulation of indigenous organism growth by the introduction of nutrients (Vecchoili et al. 1980). Bacteria preferentially grow where their food is most abundant, which is at well-screen openings and in the filter pack (Huisman and Olsthoorn 1983). The introduction of DO has been reported to be by far the most significant stimulator of microbial growth in wells and aquifers (Mansuy 1999). Aquifers contain diverse and abundant populations of indigenous microorganisms. Introduced DO and nutrients can stimulate the rapid growth of native populations at well screens, in the filter pack, and in the adjoining aquifer.

Biological clogging processes have been investigated in laboratory (column) studies, which have shown that bacteria passing through a porous media tend to adhere as bacterial microcolonies on available surfaces. The attached bacteria produce large amounts of polysaccharides and coalesce to form a plugging biofilm (Shaw et al. 1985; Taylor and Jaffe 1990). Laboratory studies have shown that the biomass within a porous medium can cause a substantial reduction in permeability, up to 3 orders of magnitude. The presence of particles in the recharge water rapidly decreases core permeability because the particles become trapped in the developing biofilm, accelerating clogging (Shaw et al. 1985).

11.2.8 Biological Clogging—Iron Bacteria

Most mineral precipitation in wells is the result of biological activity (Mansuy 1999), which may either create local chemical microenvironments that have thermodynamic conditions favorable for mineral precipitation or act as catalysts for precipitation. Iron bacteria have been defined as "that group of aerobic bacteria, which appear to utilize the oxidation of ferrous and/or manganous ions as an essential component of their metabolic functioning" (Cullimore and McCann 1978). Cullimore and McCann (1978) observed that

> iron bacteria have caused problems in water supplies since the dawn of civilization and there are many references in history to 'red' water, undrinkable water covered in slime, and plugged wells.

Cullimore and McCann (1978) reported on 15 genera of iron bacteria and discussed their identification. The two most recognized iron-oxidizing bacteria (FeOB) are the sheath-forming *Leptothrix ochracea* and stalking-forming *Gallionella ferruginea* (Emerson and De Vet 2015). Iron bacteria can cause problems in wells by (Cullimore and McCann 1978):

- clogging screens
- forming coatings on pipes, impellers, and pump motors, reducing flow rates
- water quality deterioration (reduced potability)
- total plugging/clogging of wells.

FeOB can be important in MAR systems as they can be both a nuisance, through the clogging of wells and biocorrosion of iron and steel pipe, and beneficial through the removal of iron, arsenic, and other species.

The role of FeOB in engineered water ecosystems was reviewed by Emerson and De Vet (2015). FeOB are capable of carrying out the oxidation of Fe^{2+} to Fe^{3+} at circumneutral pH with the precipitation of iron oxyhydroxides. Chemical oxidation of ferrous iron (Fe^{2+}) readily occurs in oxic environments by the reaction

$$4Fe^{2+} + 10H_2O + O_2 \rightarrow 4Fe(OH)_3 + 8H^+ \qquad (11.1)$$

Most active biological oxidation takes place at the interface between oxic and anoxic conditions as FeOB are adapted to growing under low DO conditions (Emerson and De Vet 2015). Pumping and recharge wells can provide a redox interface setting favorable for both the chemical and biological oxidation of iron (Smith and Touvinen 1985). Chemical iron oxidation will occur in high DO conditions, competing with biological oxidation. FeOB are generally found where anoxic iron-rich waters come into contact with air or mix with oxygenated water, such as a result of excessive drawdown in a well or mixing with aerobic water. High levels of soluble iron and manganese in native groundwater and the introduction of DO in recharge wells is a setting favorable for the stimulation of FeOB to form precipitates that cause well clogging (Dillon and Pavelic 1996).

FeOB produce biopolymers (polysaccharides) that result in the adsorption of iron (oxy)hydroxides and their removal from cell surfaces, and help cells maintain a spatial position in the optimal flux of Fe and DO for growth (Emerson and De Vet 2015). Biological growth and accumulation of biomass and biogenic precipitates are major contributors to well clogging. Non-biogenic oxidation of ferrous iron may otherwise occur. However, the biofilms developed by FeOB result in a greater volume of clogging material and more rapid clogging than might have occurred by abiotic precipitation. Hence, it is important to avoid introducing iron bacteria into a well, which can be achieved by disinfecting equipment that enters a well with a strong (250 ppm) chlorine solution and disinfecting the well after the completion of drilling. However, iron bacteria are naturally present in groundwater and, therefore, good housekeeping practices alone during well construction and pump installation may not be sufficient to prevent infestation in wells that have geochemical conditions favorable for their growth. Smith and Touvinen (1985) reported that "some aquifers are considered by water well contractors to be 'bad iron' areas, generally contaminated with iron-precipitating bacteria."

Once FeOB become established in a well, their complete removal can be very difficult because the biopolymers provide some protection of the FeOB cells from disinfecting agents (Mansuy 1999). Accumulations of iron precipitates and particles, and biofilm development may clog distribution systems, cause a discoloration of water, and compromise the biostability of water through increased resistance to secondary (distribution) disinfection (Emerson and De Vet 2015).

As reviewed by Emerson and De Vet (2015), FeOB can have a beneficial use through in situ iron removal. The injection of aerated water into an anaerobic ferruginous aquifer can promote iron removal via oxidation, which appears to be due at least partially to biological processes. Biological iron oxidation may also be used to remove dissolved iron in sand filters. The high porosity and large specific surface area of organic iron oxide and hydroxide microstructures give them excellent sorption characteristics, which might be used for the removal of arsenic (Emerson and De Vet 2015).

11.3 Clogging Prediction and Management

Clogging is a major operational challenge for recharge systems using wells and its prevention and management through pretreatment and remediation can represent large capital and operational costs that impact the economics of MAR systems. Hence, considerable attention has been paid to evaluating the clogging potential of various recharge waters. An objective has been to identify water quality indices that quantify the tendency for and/or rate of well and aquifer clogging. Such criteria could be used as thresholds for both feasibility assessments and as pretreatment targets. Specific indices introduced to estimate clogging potential include (Pérez-Paricio and Carrera 2001; Page and Dillon 2007):

- modified fouling index (MFI; Schippers and Verdouw 1980)
- parallel filtration device or index (PFD or PFI; Schippers et al. 1995)
- assimilable organic carbon (AOC; Van der Kooij et al. 1982)
- biofilm formation potential (BFP; Van der Kooij and Veenendaal 1992)
- biodegradable organic carbon (BDOC).

The objective is to determine a maximum acceptable value for the indices below which significant clogging should not occur. However, strong limitations exist when clogging susceptibility indices are applied to field conditions because clogging rates are also controlled by site-specific factors other than recharge water quality. Heterogeneity plays a primary role in clogging as it is may involve blocking of the most transmissive paths. Huisman and Olsthoorn (1983) cautioned that even the best quality drinking water may contain some clogging substances. It is recognized that as there are multiple causes of well clogging, more than one water quality criterion is needed to quantify clogging potential. Clogging rates also depend upon well construction and development, and formation characteristics.

11.3.1 Suspended Solids Criteria

Total suspended solids (TSS) and turbidity are poor indicators of the clogging potential for potable water ASR and recharge systems because the range of values are small and the detection limits of TSS are too high (Huisman and Olsthoorn 1983; Pyne 1995). A plot of normalized clogging rates versus hydraulic conductivity and TSS for some ASR systems showed no clear overall pattern (Pyne 1995). Pyne (1995) concluded that the failure of some wells to follow the intuitive pattern that clogging rates should increase with increasing TSS concentrations and decreasing hydraulic conductivity might be due to either an inability to measure the controlling factors accurately or to other unconsidered factors impacting clogging rates. Aquifer heterogeneity may impact clogging rates. Clogging may be more rapid in heterogeneous aquifers where most of the injected water, and entrained suspended solids load, flows into a thin flow zone.

The modified fouling index or membrane filtration index (MFI) of Schippers and Verdouw (1980) has been proposed to be a better indicator of clogging potential from suspended solids because it incorporates the effects of suspended solids concentration, particle size, and composition (Huisman and Olsthoorn 1983; Peters 1988; Hutchinson and Randall 1994; Pyne 1995; Hutchinson 1997). MFI is measured from the rate of water flow through a membrane filter, having 0.45 μm pores, at a constant pressure of 30 psi (210 kPa). Once a constant pressure is reached, the volume of filtrate is measured using a graduated cylinder for a maximum of 20 min. For each test, t/v is plotted against v, where t = time in seconds and v = volume in liters. The slope (tan α) of the straight line part of the curve (cake filtration with compaction interval) is determined and the MFI is calculated as follows (Schippers and Verdouw 1980):

$$\mathrm{MFI} = (\eta_{20}/\eta) \cdot (\mathrm{P}/210) \cdot \tan\alpha \tag{11.2}$$

where

η_{20} viscosity at 20 °C
η viscosity of water at the test temperature
P applied pressure in kPa

Alternatively, the (P/210) term can be replaced by (P/30), where P is pressure in psi.

Buik and Willemsen (2002) evaluated the potential for using the MFI to predict the rate of physical clogging of recharge wells with suspended particles. Buik and Willemsen based their analysis on the infiltration theory of Olsthoorn (1982), whereby clogging rates are related to the concentration of suspended solids, the infiltration rate into the borehole wall, and the permeability of the filter cake:

$$\Delta h_v = \left(\frac{1}{\rho_w g}\right)\left(\frac{c\mu_d}{k_c}\right)v^2 t \tag{11.3}$$

Δh_v increase in pressure caused by clogging (m)
ρ_w density of infiltrated water (kg/m^3)
g gravitational acceleration (m/s^2)
C concentration of suspended matter in the injected water (kg/m^3)
μ_d dynamic viscosity (Ns/m^2)
k_c intrinsic permeability of the filter cake (m^2)
v infiltration rate on the borehole wall (m/s)
t infiltration time (s)

Olsthoorn (1982) further related measured MFI values to the concentration suspended sediments and the intrinsic permeability of the filter cake using the equation

$$\mathrm{MFI} = \frac{\mu_d c}{2PA_f^2 k_c} \tag{11.4}$$

where

MFI measured MFI rate (s/L^2)
A_f area of filter (m^2)
P pressure loss (N/m^2, Pa)

Using standard conditions for MFI measurements (0.45 μm filter, $A_f = 1.38 \times 10^{-3}$ m^2, $P = 2 \times 10^5$ Pa), clogging rate was related to MFI value as follows (Buik and Willemsen 2002).

$$V_v = 2 \times 10^{-6} MFI \cdot \mu_{eg} \frac{v_b^2}{\left(\frac{k}{150}\right)^{1.2}} \tag{11.5}$$

where

$$K = 150(D_{50} \cdot 10^3)^{1.65} \tag{11.6}$$

and

V_v clogging rate (m/yr)
v_b infiltration rate on borehole wall (m/h)
μ_{eq} equivalent full loads per year (m^3 infiltrated per year divided by max flow rate (m^3/h)
k hydraulic conductivity (m/d)
D_{50} median grain size of aquifer (m)

An important observation of Buik and Willemsen (2002) is that in heterogeneous aquifers, clogging rates are related to hydraulic conductivity. The beds with the highest hydraulic conductivity will received more water and thus a greater load of suspended solids. A consequence of this relationship is that the differential clogging rates will continue until all layers are receiving the same amount of water. A limitation of the Buik and Willemsen (2002) method is that it can predict the rate of clogging when clogging occurs, but it cannot predict whether or when clogging will occur. The method describes the clogging rate between backflushing events, not the long-term clogging rate and the recovery of hydraulic conductivity by backflushing.

Dillon et al. (2001) reported on an upgraded MFI apparatus that can be used for waters with high particulate concentrations. A pressurized feed tank is used to pass water through a membrane at a constant pressure. Both pressure above the filter and flow volume are recorded automatically. The upgraded MFI apparatus was found to give repeatable results for MFIs up to 900 s/L^2. Dillon et al. (2001) showed that MFI cannot be reliably predicted from water quality parameters, such as TSS, TOC, and turbidity.

Silt density index (SDI) is similar to the MFI and is used to evaluate the clogging potential of waters for membrane water treatment systems. SDI is measured using a similar apparatus as for MFI and is performed by measuring the time required to filter a fixed volume of water through a standard 0.45 μm filter at a constant given pressure of 30 psi (210 kPa). Typically, the sample volume is 500 mL and the difference between the initial (t_i) and final (t_f) measurements is 15 min (T_e). SDI is calculated as

$$\mathrm{SDI} = (100 * (1 - (t_i/t_f)))/T_e \tag{11.7}$$

An SDI of 3 or less is considered suitable for membrane treatment facilities and injection in wells (Pitt and Magenheimer 1997).

11.3.2 Organic Carbon Indices

Organic carbon in recharged water can promote biological clogging. The term "biostability" is used in the water industry to describe the ability of water to support bacterial growth and biofilm development. LeChevallier et al. (2015) noted that "Biologically stable water is produced when all nutrients that might support significant bacterial growth in finished water have been sufficiently removed."

Commonly used methods to measure the organic matter content of water, such as dissolved organic carbon (DOC) and total organic carbon (TOC), include a wide variety of compounds, many of which are refractory (i.e., not readily utilized by microorganisms) and thus do not promote biological activity. Various types of organic matter and organic matter analyses are summarized in Table 11.2. Biodegradable organic matter (BOM) is considered the limiting nutrient in determining the rate of biologically mediated clogging in injection wells (Page and Dillon 2007). BOM is the fraction of organic matter that can be used by bacteria for anabolic (cell growth and reproduction) and catabolic (chemical energy) purposes (Page and Dillon 2007). Biological stability is a function of the BOM flux plus additional factors including temperature, hydraulic conditions, mineralogy, geochemistry, and influent bacteria population (Page and Dillon 2007).

Hijnen and Van der Kooij (1992) studied the effects of easily assimilable organic carbon (AOC) on the rate of clogging of sand filters. AOC concentration is determined by inoculating water samples with two bacteria cultures (*Pseudomonas fluorescens* strain P17 and *Spirillum* sp. strain NOX) and multiplying the maximum colony counts to give the AOC content in units of mg acetate-C equivalent/L. AOC compounds typically constitute only a small (<1%) fraction of the total DOC. Hijnen and Van der Kooij (1992) proposed that AOC contents should be 0.01 mg (10 μg) acetate-C equivalent/L or less to prevent biological clogging of a recharge well for a period of more than a year.

AOC values are less than BDOC, as the former is based on two bacteria, whereas BDOC measurements utilizes a large number of types of bacteria. Different results are obtained depending on whether bacteria are suspended or attached (Page and Dillon 2007). A weakness of AOC is that there is no known link between reference organisms and clogging in water reclamation projects (Page and Dillon 2007). Page

Table 11.2 Water organic matter parameters and analyses (after Page and Dillon 2007, and USEPA methods)

Organic matter parameter or analysis	Description
Total organic carbon (TOC)	Organic carbon in a water sample is oxidized to form carbon dioxide (CO_2) by combustion in an oxidizing gas and UV-promoted or heat-catalyzed chemical oxidation
Dissolved organic carbon (DOC)	Fraction of TOC that passes through a 0.45 μm filter
Chemical oxygen demand (COD)	Amount of oxygen that can be consumed by reaction with a strong oxidizing agent under high temperature (150 °C)
Biochemical oxygen demand (BOD)	Measure of the uptake of DO by microorganisms in a sample at a fixed temperature (20 °C) and given period of time (typically 5 days) in the dark
Specific ultraviolet absorbance (SUVA)	UV-254 value (in cm^{-1}) of a sample divided by the DOC concentration (mg/L) and then multiplied by 100. Measure of aromaticity of sample
Natural organic matter (NOM)	Natural organic material present in surface water or groundwater, including both humic and non-humic fractions. Measured by UV adsorption
Biodegradable organic matter (BOM)	Organic matter that can be broken down (utilized for food) by naturally occurring microorganisms within a reasonable length of time
Biodegradable organic carbon (BDOC)	Filtered water is inoculated with indigenous microorganisms and a series of DOC measurements are made until no further reductions in DOC are observed. Incubation is generally between 10 and 30 days
Assimilable organic carbon (AOC)	Sample is incubated with bioassay organisms, and population growth curves and maximum cell densities are determined and converted into carbon (usually acetate) concentration equivalent using yield coefficients

and Dillon (2007) concluded that a new measure of BOM is needed and that the best candidate is a modified BDOC method employing sand for determining the largest portion of the BOM pool. However, AOC is a standard method that is performed by some commercial laboratories.

11.3.3 Laboratory Studies of Physical and Biological Clogging

Column testing has the advantage of allowing clogging and chemical process to be studied under controlled conditions. Rinck-Pfeiffer et al. (2000) performed column testing to evaluate clogging process that are thought to be active in ASR systems. The testing was performed in advance of the initiation of operational testing of the Bolivar reclaimed water ASR system (South Australia). The columns were 16 cm length, had an inner diameter of 2.5 cm, and were filled with crushed and sieved core material from a sandy limestone, the T2 aquifer from the Northern Adelaide Plains. The T2 aquifer was used as the storage zone for the Bolivar ASR system. The experiments were performed using reclaimed water from the Bolivar wastewater treatment plant. The effluent used was not chlorinated so as to obtain more severe clogging as a worst case scenario. Influent and effluent samples were analyzed for Eh, pH, DO, and electrical conductivity (EC), cations, anions, metals, *E. coli* and total bacteria count. Flow rates through the columns were recorded, along with pressure at multiple points in the column. After the 22-day tests, core samples were analyzed for polysaccharides, total biomass, and calcium.

Hydraulic conductivity measurements showed three main stages (Rinck-Pfeiffer et al. 2000):

- an initial steep decline in hydraulic conductivity over the first seven days, with the greatest decrease occurring near the inlet (0–3 cm).
- a stable period from days 7 to 15
- a gradual increase in hydraulic conductivity after 15 days.

Rinck-Pfeiffer et al. (2000) interpreted the data as indicating an initial clogging process followed by later calcite dissolution that increased the hydraulic conductivity. The initial clogging was attributed first to physical clogging from suspended solids, which was followed by biological clogging. Biological clogging is indicated by a decrease in DO from about 5.5 to 7.5 mg/L in the influent to 1–2 mg/L in the effluent over the first 10 days of the experiment. The highest concentrations of total biomass and polysaccharides occurred in the core sample from the inlet end.

An increase in calcium concentration from day 8 onward is indicative of calcite dissolution. Rinck-Pfeiffer et al. (2000) also documented that the calcium concentration in the column material had decreased (from initial values) near the inlet and had increased near the outlet, suggesting reprecipitation of the dissolved calcite. SEM observations also support calcite precipitation near the outlet. Approximately 10% of the calcite present in the aquifer material was dissolved at the inlet end of the columns, which raised the concern that calcite dissolution could impact the integrity of the formation (Rinck-Pfeiffer et al. 2000).

Subsequently reported column tests compared clogging in columns that received reclaimed water from the Bolivar wastewater treatment plant that were either (a) treated with formalin to remove bacteria (restricting clogging to physical processes), (b) microfiltered to remove suspended solids (restricting clogging to biological processes), (c) received no further treatment, and (d) received no further treatment but had a lower flow rate (2 m/d vs. 10/d; Rinck-Pfeiffer et al. 2002). After 22 days, hydraulic conductivity at the inlet end of the columns (0–3 cm) was reduced to 19% of the initial value for the physical clogging test, to 27% of the initial value for the biological clogging test, and to 5% of the initial value for physical and biological clogging combined. Greater polysaccharide accumulation occurred during the low flow rate test. Calcite dissolution was the dominant geochemical reaction in all of the columns.

Page et al. (2014) performed column testing of three different treatment options for turbid water to reduce clogging. The source water was from the River Darling and the tested aquifer material was medium to coarse-grained sand. The tested waters were:

- coagulated and flocculated water with chlorine disinfection (town water)
- coagulated and flocculated water with granular activated carbon (GAC) treatment and chlorine disinfection (GAC water)
- bank-filtrated water (BF water)

The tests were performed using 16-cm long, 2-cm inner diameter packed columns with a hydraulic conductivity of 2.17 m/d and a flow rate of 4.3 l/d. The GAC test had an 8% decline in hydraulic conductivity after 37 days, whereas the other waters had declines of 26–29%. Most of the clogging occurred in the upper 3 cm of the columns. All three waters had similar declines in hydraulic conductivity over the first 21 days, which were attributed to similar rates of physical clogging. The greater later declines in hydraulic conductivity was due mainly to biological clogging as indicated by the GAC water column having lesser concentrations of polysaccharides and total DNA. The GAC water had the lowest mean DOC concentration (4.5 mg/L) compared to 8.4 mg/L in the town water and 9.7 mg/L in the BF water.

Page et al. (2014) proposed the following treatment targets for water to be recharged in siliceous alluvium using wells:

- turbidity: <0.6 NTU
- MFI: <2 s/L^2
- biodegradable DOC: <0.2 mg/L
- total nitrogen: <0.3 mg/L
- residual chlorine: >0.2 mg/L.

11.3.4 Field Studies of Clogging

The Bay Park, New York, Aquifer Recharge Test Project performed from 1968 to 1973 was one of the earliest and most intensively studied MAR projects of its kind, including 18 observation wells (Ehrlich et al. 1979; Vecchoili et al. 1980). An observation well was installed inside the gravel pack of the injection well, which allowed for the determination of the primary location of clogging. The pilot testing program consisted of the injection of tertiary-treated wastewater into the Late Cretaceous Lower Magothy Formation, which consists predominantly of fine to medium-grained quartz sands. The usual injection rate was 350 gpm (22.1 L/s).

Clogging was a problem during injection, which was caused mainly by filtration of suspended solids in the injectate. No chronological trends in aquifer heads were observed in the observation wells, indicating that injection had not impacted the hydraulic properties of the aquifer. The clogging had occurred at the interface between the gravel pack and the aquifer. The fine to medium-grained sands of the injection zone rendered it very effective in filtering out particulates. Even when suspended solids concentrations were low (<1 mg/L), accumulation of injected solids on or immediately adjacent to the borehole wall (aquifer face) was considerable. The rate of head increase became greater as the specific capacity of the wells decreased, indicating that the material filtered out at the borehole wall contributed to increased filtration and clogging. The average rate of clogging was approximately 3 ft (0.9 m) of excessive head build-up per 1 MG (3785 m^3) of injected water. Redevelopment by pumping and surging was only partially effective in restoring the specific capacity lost during injection.

Clogging due to microbial growth was insignificant so long as a 2 mg/L total residual chlorine was maintained. An injection test performed using unchlorinated reclaimed water experienced much more rapid clogging than a test performed under similar condition using chlorinated water (Ehrlich et al. 1979). A shock dose of total residual chlorine of 200 mg/L was sufficient to restore injectivity by destroying the bacterial slime around the injection well that formed during the injection test with unchlorinated water. Compounds of iron, aluminum and phosphate also contributed to the clogging, but the extent of the precipitation was unresolved. Treatment with 32% commercial grade hydrochloric acid was reported to have resulted in an almost 50% improvement in specific capacity by apparently dissolving some precipitated compounds.

Hijnen et al. (1998) evaluated the clogging potential of water used for the recharge of sandy aquifers in the Netherlands. MFI and AOC were found to be useful to estimate the potential for clogging but could not predict the actual clogging rate of a recharge well. Guideline water quality parameters were proposed (MFI < 3 s/L^2, AOC < 10 μg/L) for recharge in deep sandy aquifers to limit clogging (Hijnen et al. 1998).

Recharge tests were performed at the Langerak aquifer recharge site in the Netherlands using groundwater pretreated by aeration, rapid-sand filtration, and $NaNO_3$ addition (Timmer et al. 2001). The water was reported to have had a low clogging potential with a MFI < 3 s/L^2 and AOC < 10 μg/L (Timmer et al. 2001). Despite the high quality of the recharge water, clogging still occurred with a resulting head buildup of 3.3 m after 4 months. Juttering (intermittent pumping) restored the well capacity, but rapid clogging occurred once recharge resumed. Treatment with a strong oxidant (H_2O_2) had little effect, suggesting that bacterial growth was not the cause of the clogging. The recovered water at the Langerak site had higher concentrations of Fe, Mn, turbidity, and suspended solids compared to the WRK site in Nieuwegein, which had otherwise similar conditions. Clogging was determined to be the result of a low concentration ($\approx$40 μg/L) of ferric iron flocs (between 15 and 40 μm in diameter) that was produced by the oxidation of ferrous iron present in the native groundwater used for the recharge testing. The clogging is believed to have occurred at the interface between the gravel pack and the formation sands.

A recovery well at the Nieuwegein site was reported to have suddenly clogged after 728 days of operation without noticeable advance build-up of hydraulic resistance

(Moorman et al. 2002). Analysis of the material clogging at the well screen indicated that it consisted mainly of an iron (oxy)hydroxide or ferric hydroxiphosphate. The most probable cause of the clogging was thought to be mixing of waters with different redox states around and in the well.

Pavelic et al. (2007a, b) examined the effects of water quality on clogging at the Bolivar reclaimed water ASR site (South Australia) over the initial four-year period of operational testing. The storage zone is a confined sandy limestone aquifer and the source water was secondary-treated reclaimed water that was further treated by detention in a stabilization pond, dissolved air floatation and filtration (DAFF), and chlorine disinfection. Clogging was quantified by temporal changes in relative effective intrinsic permeability calculated from hydraulic conductivity values obtained using Thiem equation and temperature-corrected viscosity values. Permeability values were reported relative to the permeability prior to the start of injection (k/k_0).

Relative permeability-versus-time plots showed periods of declining (25% of the time), stable (55%), and then rising relative permeability. A rapid decline occurred at the beginning of the first cycle with no subsequent long-term decline. For each long-term operational cycle, k/k_0 fell from 0.3–0.6 to 0.1–0.2 within 17–50 days, with most of the decline occurring within the first 7–14 days. Short-term oscillations in relative permeability occurred in response to periodic backwashing events.

Turbidity was largely associated with short-term clogging due to particle retention. Its significance is reduced over the long term as the predominantly organic particles were degraded in the presence of DO and nitrate. The degraded particles provided a substrate for biomass production. The level of total nitrogen in the source water appears to be the limiting factor for microbial growth in the long-term and possibly short-term. Clogging rates were found to be highly dependent on the quality of the injected water. To achieve an acceptably low rate of short-term clogging and no clogging or unclogging in the long term, Pavelic et al. (2007b) recommended:

- turbidity <3 NTU
- total nitrogen <10 mg/L
- pH <7.2

Pavelic et al. (2008) documented the unsuccessful efforts to rehabilitate a clogged ASR well completed in a Quaternary sand aquifer at the Urrbrae wetland site, Adelaide, South Australia. The ASR system was used to store stormwater that was pre-treated by rapid-sand filtration. The well was completed with three segments of wire-wrapped stainless steel screen with 0.5 and 1.0 mm slots. The reported water quality of the injected water was

- turbidity: 0.8–55 NTU
- MFI: 90–389 s/L^2
- bacterial regrowth potential: 39–331 acetate carbon equivalent (μg/L)

After the injection of 4×10^3 m^3 over a 6-week period in spring 1999, the injection rate decreased from 3 L/s to a final value of 0.5 L/s. Remedial efforts that consisted of intermittent backwashing, shock chlorination (300 mg/L), and bailing and surging to recover sand that entered through a perforation of a well screen joint failed to significantly improve specific capacity. Clogging was attributed to high levels of suspended solids and bacterial growth fed by labile organic carbon and other nutrients in the wetland water. The level of pretreatment was inadequate for the low transmissivity (6 m^2/d) of the storage-zone aquifer. The low permeability of the aquifer may have limited chlorine from accessing small pore spaces in the aquifer. The Urrbrae ASR system was cited as an example where ASR did not work (Pavelic et al. 2008).

De La Lomo González (2013) reviewed clogging history at three aquifer storage transfer and recovery (ASTR) systems in the Netherlands. The investigated systems inject into a siliciclastic sand aquifer using screened wells. The injected water is surface water that was pretreated to varying degrees including coagulation and filtration and, in some cases, activated carbon. The wells at all three systems have operated successfully for a least 20 years but have experienced clogging over time. The primary cause of the clogging was attributed to iron precipitates and iron-reducing bacteria. Potential sources of iron and manganese identified are:

- colloidal iron and manganese in the recharge water
- Fe and $Fe(OH)_3$ from the coagulation pretreatment process
- iron and manganese in the ambient groundwater
- iron and manganese mobilized by reductive dissolution of iron and manganese hydroxides in the aquifer during shut-in periods.

Recommendations to reduce clogging include the avoidance of using drilling mud in screened intervals or carefully removing (reaming) the borehole wall to remove residual mud (formation damage), and having adequate pretreatment to achieve target injected water quality (MFI <3 s/L^2 and AOC of <10 μg/L-acetate-C/L). Understanding and controlling the source of iron is also important (De La Lomo González 2013).

Johnston et al. (2013) discussed clogging experiences during injection into a deep confined sandstone aquifer (Leederville aquifer) in Perth, Western Australia. Data were presented from a pilot (Jandakot), a research and development (Mirrabooka), and an operational (Beenyup) recharge system. The recharged aquifer consists of interbedded sandstones, siltstones, and shales that are anoxic and chemically reducing. Clogging due to suspended solids occurred in the Jandakot and Mirrabooka systems. The clogging at the latter facility may have been due to mobilization of aquifer fines allowed by the absence of a gravel filter pack (a natural filter pack was attempted). Backwashing decreased clogging and allowed for recharge of a greater volume of water but did not completely restore wells and prevent further clogging. Clogging did not occur at the Beenyup system, which is used to inject highly treated wastewater. The secondary-treated wastewater is further treated by ultrafiltration, reverse osmosis, and ultraviolet disinfection.

Page and Dillon (2013) reported the results of an investigation of the treatment requirements for stormwater ASR in a Siluro-Devonian siliciclastic fractured aquifer at the Rossdale Golf Club, Aspendale, South Australia. The aquifer has a very low transmissivity (1.2–1.8 m^2/d) and the sandstone aquifer units are mineralogically composed of quartz (48–73%), kaolinite (12–23%) and muscovite (7–18%). Injection and recovery cycles were performed first with potable water and then with treated stormwater. Clogging occurred as indicated by a reduction in specific capacity. Potential clogging mechanisms identified were:

- physical clogging due to suspended solids
- clay dispersion caused by the lower salinity of injected water
- chemical clogging caused by precipitation of iron oxyhydroxides from FeS_2 and $FeCO_3$ in the aquifer
- biological clogging

It was concluded that is difficult to differentiate between the active clogging mechanism. Page and Dillon (2013) recommended that stormwater be treated to potable water quality prior to injection, specifically targets of $\leq$0.6 NTU, $\leq$1.7 mg/L and $\leq$0.2 mg/L for turbidity, DOC, and BDOC, respectively. However, some clogging occurred in the last two of four injection/recovery cycles using potable water. Ultrafiltration with granular activated carbon was eventually selected to treat the raw stormwater and meet the above water quality targets.

11.3.5 Clay Dispersion

The tendency for clays to expand in freshwater may be decreased by replacing their exchangeable cations with cations less inclined to attract water to interlayer sites (Reed 1972; Reed and Coppel 1972). More firmly attached cations tend to decrease the double-layer thickness and thus decrease the tendency for particles to repel each other (Reed 1972). Reed (1972) documented the results of tests of hydoxy-aluminum treatment on core samples of the water-sensitive Berea Sandstone and field testing at a steam injection site in California. Sandstones treated with OH-AL solutions with OH/Al ratios of either 2.0 or 2.4 and aged in freshwater showed little, if any, sensitivity to freshwater. After treatment, it was found to be beneficial to overflush the formation with freshwater.

Torkzaban et al. (2015) performed core testing of clay dispersion. Clogging potential was found to depend on the amount of colloid release and the initial permeability of the cores, which is related to the distribution of pore sizes. A 50–90% reduction in permeability occurred in selected samples of the Precipice aquifer when RO treated water displaced a 0.5 mM NaCl solution. Flow reversal mobilized some clay particles captured at pore constrictions resulting in a sudden permeability improvement. However, flow reversal restored the permeability of cores by only a factor of about 20%. The released particles had a hydraulic radius of about 1.5 μm, and the majority were kaolinite or quartz. Final permeability was always less than the initial permeability. Some key specific results of the core testing are (Torkzaban et al. 2015):

- ultrafine particle release may occur without a permeability reduction, particularly in samples with low clay contents and large pores
- declines in permeability occurred in samples with high clay contents and low permeabilities due to particle release and entrapment
- permeability reduction was negligible in core samples with low (2–3%) clay contents
- the extent of clogging and clay release decreased when calcium cations were added to the water prior to injection
- high pH (>8) promotes clay release and disaggregation; low pH (<6) minimizes clay release
- when samples were initially saturated with high concentrations of divalent cations, clay release was negligible upon RO water injection
- divalent cations are preferentially adsorbed on clay particles when present in a mixture of monovalent and divalent cations
- $CaCl_2$ preflush resulted in very little subsequent reduction in permeability, but clogging may occur if the sample is subsequently saturated with a solution with a high concentration of NaCl
- key variables that controlled permeability loss are the total salt concentration (ionic strength of injected water relative to native groundwater) and cation composition.

Torkzaban et al. (2015) cautioned that "Prior to injecting any freshwater into an aquifer, consideration should be given to the possibility of chemical reaction which might interfere with the injection process." The possibility of permeability reduction as the result of clay dispersion should be thoroughly investigated for any recharge system involving the injection of freshwater into a siliciclastic aquifer containing brackish or saline waters because of the potential for rapid and permanent formation damage.

Standard methods used to evaluate the dispersive characteristics of soils, such as crumb and double hydrometer methods (ASTM 2013, 2017) are too imprecise to evaluate the low levels of clay dispersion that could cause significant reductions in permeability in recharge wells. The preferred option is laboratory flow-through tests using samples of the actual water to be recharged (or chemically similar water) and core samples initially saturated with formation waters. If cores are not available, push-pull (single well) tracer tests on an interval isolated by packers could also be used to evaluate the potential for adverse fluid-rock interactions. The testing should ideally be performed on an exploratory/test well, rather than a recharge well to avoid permanent damage to the latter.

11.3.6 Prediction of Physical and Biological Clogging from Source Water Quality

A major unresolved question remains the relationships between AOC, MFI, SDI values, turbidity, TSS concentrations, and other variables on actual clogging rates in injection/storage zones with different hydraulic conductivities, porosities, pore sizes, and pore types. In general, no universally applicable water quality guidelines for water recharged using wells has been identified. Bouwer (2002) observed that MFI, AOC, and PFI are useful parameters for comparing relative clogging potentials but cannot be used to predict clogging and injection rates for actual recharge wells.

Dillon and Pavelic (1996) are correct in that while these parameters are useful indicators of the potential occurrence of clogging by suspended solids, they are inadequate as predictors of the rate of decline where complex processes are responsible for clogging.

Two independent numerical model codes were developed as part of the European Union Artificial Recharge project to simulate clogging phenomena: CLOG and MIKE-SHE SC (Pérez-Paricio et al. 2001). The CLOG code can simulate attachment and detachment of suspended particles, bacterial growth and die-off, kinetics of precipitation of minerals, and multi-phase flow includes gas. The MIKE-SHE SC code is limited to describing biogrowth and decay processes and the sedimentation and detachment of organic matter. The models were successfully calibrated to field and column test data. The basic limitation of clogging modeling is that numerous variables affect clogging, including kinetic parameters, whose values are poorly known. Hence, there are many uncertainties associated with clogging modeling (Pérez-Paricio et al. 2001). It is therefore not feasible to predict clogging with a satisfactory degree of confidence at the present time (Pérez-Paricio et al. 2001; Pérez-Paricio and Carrera 2001). Modeling of ASR system performance, in general, has been successful in simulating past system performance, but there is still poor predictive ability in the absence of initial data for calibration.

11.3.7 Evaluation of Chemical Clogging Potential

Chemical clogging potential is evaluated by geochemical modeling of the saturation state of the recharge water and mixtures of recharge water and native groundwater. The U.S. Geological Survey MINTEQA2 (Allison et al. 1991), WATEQ4F (Ball and Nordstrom 1991), and PHREEQC (Parkhurst and Appelo 1999) codes are all suitable for equilibrium modelling under the low-temperature conditions of groundwater environments.

The accuracy of modeling results depends upon the completeness and quality of the input data. Accurate measurements of pH is critical for evaluating the saturation state of carbonate minerals. Accurate data on oxidation-reduction potential (Eh or pE) is required for modeling the saturation state of redox minerals. As discussed in Sect. 5.4.3, field (meter) measurements of oxidation-reduction potential tend to be unreliable and Eh is more accurately determined from redox pair concentrations.

Basic field observations are important for a qualitative understanding of the redox state of aquifers. For example, the presence of unaltered iron sulfide minerals and a hydrogen sulfide (rotten egg) odor are indicative of chemically reducing conditions. Cuttings with iron (oxy)hydroxide (rust) staining may be indicative of oxic conditions. Cuttings from aquifers with chemically reducing conditions often develop rust staining after exposure to the atmosphere. Commonly, redox reactions associated with the recharge of oxygenated water into aquifers containing chemically reducing conditions have a greater impact on water quality than on aquifer hydraulic properties.

11.4 Clogging of Surface-Spreading MAR Systems

11.4.1 Causes of Clogging Overview

Infiltration and percolation rates of MAR systems that involve spreading of water on the land surface (as opposed to using wells) depend upon the permeability of the vadose zone sediments and rocks, depth to the water table, water-table aquifer transmissivity (which influence groundwater mounding), and clogging. Clogging is typically the major operational and maintenance issue affecting the long-term of performance of surface-spreading systems. Clogging reduces system infiltration rates and creates maintenance requirements (costs). Maintaining or increasing infiltration rates is of great importance in urban areas where the cost of additional land for system expansion can be prohibitive.

Surface-spreading systems are less prone to clogging than wells for waters of a given quality because of the orders of magnitude greater area through which water infiltrates (land surface area versus borehole wall surface). However, surface-spreading systems are often used for the recharge of less treated, poorer quality water. Land application systems also involve exposure to the atmosphere and sunlight, which allows for the growth of photosynthetic organisms. Infiltration systems may be impacted either positively or negatively by the actions of macrofauna.

Similar to wells, the main caused of clogging of infiltration basins and other types of surface spreading systems are:

- deposition of low permeability, fine-grained materials (silts and clays) on the infiltration surface
- chemical precipitation (e.g., $CaCO_3$)
- biological processes
- clay swelling and dispersion
- air and gas clogging.

Although not clogging, temperature decreases act to increase the viscosity of water and thus decrease hydraulic conductivities and infiltration rates.

Physical clogging is caused by fine particles present in the recharge (influent) water, material mobilized by erosion in the spreading facility, windborne dust, and particulate organic matter. Biological clogging processes include (Bouwer 2002):

- accumulation of algal and bacteria flocs on the infiltration surface
- growth of microorganisms to form biofilms that block pores and/or reduce pore size
- inducement of chemical precipitation through changes in water chemistry (pH)

Entrapped gases include air entrained in the recharge water, gases released as the result of decreases in solubility caused by temperature increases and pressure decreases, and biogenic gases (CO_2 and N_2) generated in the soil.

The main cause of clogging in surface-spreading systems is deposition of suspended sediments and the associated formation of an organic matter-rich (schmutzdecke) layer. Earlier studies of the clogging of surface-spreading basins provided basic insights into clogging processes. Clogging is essentially a surface sealing process, as evidenced by tension developing a short distance below the infiltration surface (Behnke 1969). Clogging occurs initially by gravitational settling, which produces a graded layer as coarser sediment settles first (Behnke 1969). Infiltration rates subsequently decline as finer material is deposited and the layer increases in thickness. Interstitial straining occurs when the uppermost pores become small enough to strain out the remaining particles.

Laboratory studies have shown that infiltration rate-versus-time curves have three segments (Behnke 1969; Fig. 11.1). The initial segment shows little change in infiltration rate with time as not enough material is deposited to form a flow-limiting surface layer. The second segment exhibits a rapid decrease in infiltration rate with time as the clogging layer becomes horizontally continuous and progressively increases in thickness. The last segment shows little change over time as the rate of flow through the clogging layer is so slow that little new material is being added to the layer. The trapping of particles at the front of the filter results in the retention of similar-sized particles and thus a self-filtration process. Self-filtration contributes to clogging by forming a progressively thickening filter cake (Benamar 2013).

Large water (ponding) depths can increase infiltration rates by increasing the hydraulic gradient across clogging layers but can adversely impact infiltration by causing compression, and thus a reduction in the hydraulic conductivity, of clogging layers (Bouwer 1989; Bouwer and Rice 1989). Increased water depths also result in slower turnover rates and the potential for greater algal growth, which can lead to a higher pH and calcium carbonate precipitation (Bouwer 1989; Bouwer and Rice 1989).

Two types of clogging layers occur in infiltration basin and similar surface-spreading systems (Hutchison 2013):

(1) an upper layer of particulate and organic matter located above the original sediment surface (basin floor)
(2) a lower layer in the native sediment below the original surface in which porosity and permeability have been reduced by the intermixture of organic and inorganic matter.

Clogging layers that develop atop the original sediment surface can cause a large and rapid reduction in infiltration rates. However, surficial clogging layers can be

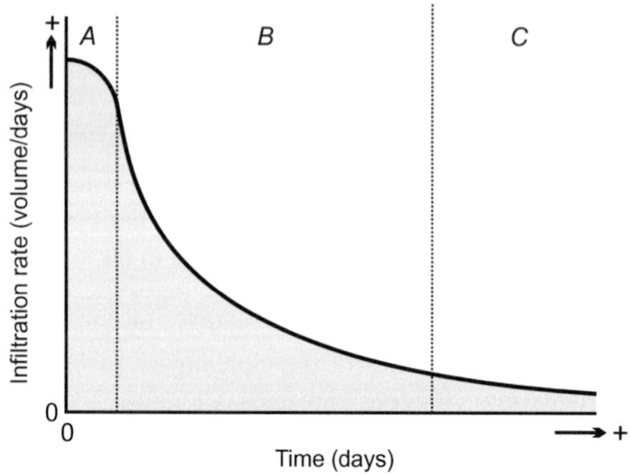

Fig. 11.1 Conceptual infiltration rate versus time curve for a basin in which water level is kept constant. The curve can be divided into three parts (Behnke 1969). Initially there is little change in infiltration rate as not enough material has been deposited to form a limiting layer (A). Part B reflects a rapid decrease in infiltration rate as a clogging layer develops. Finally, the infiltration rate stabilizes as the rate of flow through the layer is so slow that little new material is being added. The area under the curve (shaded gray) represents the total infiltrated volume

more readily removed by scrapping or be disrupted by harrowing and other means. The transport and deposition of suspended sediments deeper below land surface may result in a slower rate of reduction of infiltration rates, but the deeper clogging tends to be much more difficult to remediate and can seriously affect the long-term performance of surface-spreading systems.

Goss et al. (1973) investigated the downward movement of fine particles into the soil by performing infiltration tests using suspensions of sediments in which cesium-237 was added as a radioactive tracer. Cesium is readily sorbed onto clay minerals. As would be suspected, the cesium-tagged sediments traveled deeper below land surface (>45.7 cm; >18 in.) in the test in which large pores were exposed at land surface. In the test where the large pores were destroyed by cultivation, most (>90%) of the cesium-tagged sediment was filtered out in the upper 2.5 cm (1-in.) of sediment, which can be readily removed by scrapping or other means. Very coarse sediments or macropore recharge can thus be unfavorable for the long-term performance of infiltration basins even though they can result in very high initial infiltration rates.

Clogging rate is clearly is a function of the concentration and type of suspended solids in the water being recharged, which varies greatly between systems. Benamar (2013) observed that recent researches indicate that the rate of particle straining in saturated porous media is sensitive to the ratio of particle diameter to sand-grain diameter, shape and surface roughness of the solid matrix, particle size non-uniformity, pore-scale hydrodynamics, and pore water chemistry. Large particles are retained by mechanical filtration and small particles by physiochemical filtration. Clogging rates are also a function of the nutrient content of recharge water (and associated induced biological activity), site hydrogeology, climatic conditions, and the mode of operation of the systems (e.g., wetting and drying cycles and ponding depth).

Systems in Europe commonly recharge treated water of very high quality and thus clogging rates are low. In contrast in the western United States, untreated surface water or reclaimed water treated to different degrees is most commonly recharged by surface spreading and, as a result, management of clogging is a major operational concern.

11.4.2 Laboratory Investigations of Clogging of Surface-Spreading MAR Systems

Pavelic et al. (2006) performed a laboratory column study on the effects of water depth on rates of infiltration. Two sediment types were used in the experiments, a sand and a loam (sand with 17% silt and 13% clay). The sand had an approximately 200-fold greater hydraulic conductivity than the loam but experienced much greater clogging, and as a result, the difference in hydraulic conductivity between the two sediment types decreased over time. Greater ponding depth was found to increase infiltration rates in the short term, but did not translate into higher rates over the longer term. It was suggested that increases in clogging may be due to a greater downward movement of clogging agents (particulates) into the soil matrix, as opposed to compression of a surficial filter cake. Based on the experiment results, a ponding depth of the 30 cm of less was recommended for a proposed soil-aquifer treatment (SAT) system in Alice Springs, Australia.

Phillips et al. (2007) performed column testing on the controls of percolation rates over time using actual foulant (material recovered from the bottom of a basin) and sediment samples from the Orange County Water District (OCWD; California) recharge basins. The change in percolation rates was found to follow a log-decay expression with respect to accumulated foulant as follows:

$$Q_L = Q_0 e^{-rL} \tag{11.8}$$

where,

Q_0 initial percolation rate (m/d) (ft/d)
Q_L percolation rate at end load L (m/d) (ft/d)
L total foulant accumulated per unit area (mg/m^2) (lb/ft^2)
r sediment/foulant interaction coefficient (m^2/mg) (ft^2/lb)

The sediment/foulant coefficient was found to be a function of both foulant composition and total suspended solids (TSS) concentration. For a given total foulant accumulation, a greater reduction in percolation rate occurred when the load was applied more rapidly (i.e., the TSS concentration was greater). Reduction in TSS concentration was, therefore, identified as means of improving system performance (Phillips et al. 2007).

Benamar (2013) performed column experiments to evaluate the effects of grain size and shape and flow rates on permeability reduction. A kaolinite slurry and columns packed with either glass beads or silica sand were used. The experiment results indicate:

- for a given flow volume, greater reduction in permeability occurs at low flow rates
- an angular shape of sand grains provides a greater likelihood of clogging than grains of rounded shape
- maximum clogging occurs in the upstream layer of the bed and is dependent on the suspension concentration, pore structure, and flow rate.

Coulon et al. (2015) performed column studies of clogging based on an infiltration/retention basin (Cheviré basin) in Nantes, France. Columns were filled with composite sediment from the basin and the volume, wetting and drying cycles, and suspended sediment concentration (400 mg/l) of the applied water corresponded to site conditions. The sediment accumulation rate was about 1 cm year^{-1}. Samples of the deposited sediment layer were dehydrated with a water-acetone exchange system, encapsulated in polyester resin containing a UV-sensitive pigment, and photographed using a high-resolution camera. After 6 and 12 months, the sediment retained its particulate structure. Voids between sediment aggregates strongly decreased over time. At 24 months, visible interstitial space was rare and at 36 months the sediment was homogenized (lost its aggregate structure). The evolution of the pore network resulted in a reduction in infiltration rate. As the sediment layer increased in thickness over time, a water layer developed over the sediment surface because of a reduction in hydraulic conductivity (7.5 fold decrease within 30 months).

11.4.3 Field Investigations of Clogging of Surface-Spreading MAR Systems

Reports of field investigations of infiltration basins and other surface-spreading systems tend to focus on water quality changes and hydraulic performance (i.e., maintenance of infiltration). These type of studies are discussed in chapters on the design and operation of each system types. Follows is a summary of some studies that specifically addressed the causes of clogging.

Schuh (1990) evaluated the causes and effects of clogging in a northern temperate climate. The test site was a 15 m by 15 m recharge basin located near Oakes, North Dakota. Infiltration tests were performed in the fall and spring using turbid river water. Large decreases in infiltration rates caused by air entrapment were observed at early times in both the fall and spring, which were followed by a partial recovery. The data from the fall test support the hypothesis of deeper penetration of clay during early operational periods (19–75 h) followed by later interception of all sediment on the basin soil surface as a filter cakes develops (Schuh 1990).

Infiltration rates during the spring test decreased at a faster rate than during the fall. The greater decrease in infiltration rates in the spring were attributed to calcium carbonate precipitation caused by algae photosynthesis-induced increases in pH. Water chemistry data, scanning electron microscopy, and HCl effervescence tests support carbonate precipitation having occurred at grain contacts (Schuh 1990).

Heilweil et al. (2004) performed detailed investigations of the effects of air and gas entrapment at a 60-m diameter infiltration pond at Sand Hollow, southwestern Utah. Helium-4 was used as a gas-partitioning tracer and bromide as a non-partitioning tracer. Helium, and other low solubility gases, are preferentially partitioned into the gas phases and, as a result, their transport is retarded compared to non-partitioning phases. The retardation of helium relative to bromide indicates the presence of substantial amounts of gas-filled porosity beneath the pond. Key observations were:

- some of the air entrapped in porous media beneath newly added surface water will migrate upward and escape to the atmosphere, while much will remain trapped within the porous media until it dissolves into infiltrating water
- trapped air will typically occur as air bubbles within the largest pore space, reducing the permeability of the media
- the effect of trapped gas on permeability depends on the quantity and size of the gas bubbles and the uniformity of the pore throats
- trapped gas should eventually dissolve because of the increase in ponded water depth and hydrostatic pressure
- temperature affects the solubility of gases; gases dissolve more rapidly in cooler water
- clogging can be increased by augmentation of entrapped air bubbles with biogenic gases (CO_2, methane).

Heilweil et al. (2004) noted that the practice of drying and tilling will reintroduce trapped gas, partially offsetting permeability gains. Wet tilling was recommended instead. It was also noted that infiltration rates based on laboratory permeability analyses may greatly over estimate actual infiltration rates under natural conditions because of the presence of gas bubbles.

Heilweil et al. (2008) subsequently investigated clogging within Sand Hollow Reservoir, a 50 km^2 basin underlain by the Navajo Sandstone, which is locally covered with a thin veneer of soil. Seasonal variations in infiltration rates were measured. The minimum infiltration rate occurs during the summer followed by rapidly rising rates during autumn. The effects of temperature on the dynamic viscosity of water and thus intrinsic permeability were removed from the data. The corrected data still shows a similar seasonal trend in recharge rates and apparent intrinsic permeability, including a seven-fold increase between June and October 2007. Core samples were taken of recently deposited silts and measured for vertical hydraulic conductivity, which was found to be equal to or greater than that of the Navajo Sandstone and overlying soils. Silt deposition was, therefore, generally not creating a rate-limiting permeability layer for aquifer recharge.

Trapped gas in the sediment beneath the reservoir was evaluated by measuring total dissolved gas (TDG) pressure in temporary drive-point piezometers. Positive or near-zero excess TDG pressures in some samples from the shallower parts of the reservoir when the water was warmer indicate the presence of gas bubbles. The TDG was depleted in oxygen and enriched in CO_2 and methane, indicating a biogenic respiration and decay origin. Heilweil et al. (2008) proposed that the seasonal variability in recharge rates was due to the generation of biogenic gas bubbles in the warmer summer months and their dissipation during the cool winter months. The decrease in bubble size in the winter is explained by Henry's Law, in that gas solubility increases with decreasing temperature.

An increase in recharge rates was observed from mid-2008 through 2010, which may be explained by the dissolution of trapped air bubbles (Heilweil and Marston 2013). Gas bubble dissolution is suggested by elevated TDG and DO concentrations in four monitoring wells located within 300 m of the reservoir (Heilweil and Marston 2013).

The influence of algal biofilm growth and sediment deposition on clogging was investigated in two infiltration basins located in the Lyons, France, metropolitan area (Gette-Bouvarot et al. 2014). The basins are fed surface water from the Vieux Rhone channel when the turbidity of the water is low (<50 NTU). Three cores were taken from each basin and analyzed for particle size distribution, TOC and total nitrogen concentrations, and biological parameters (protein content, algal biomass, bacterial abundance, and hydrolytic activity of the biofilm). Infiltration tests were performed on the basins after they were completely drained. The results of the investigation indicate that the algal biomass appears to have antagonistic impacts. At low degrees of development, organic matter associated with biofilms may trap silt particles and form micro-aggregates, which result in macropores that enhance flow. As the algal biofilm growth proceeds, pores are occluded with an associated drastic negative impact on hydraulic parameters.

References

Allison, J. D., Brown, D. S., & Novo-Gradac, K. J. (1991). *MINTEQA2/PRODEFA2, A geochemical assessment model for environmental systems: Version 3.0 user's manual.* Athens, GA: U.S. Environmental Protection Agency Environmental Research Laboratory.

ASTM. (2013). *Standard test methods for determining dispersive characteristics of clayey soils by the crumb test*, Method D6572-13e2. West Conshohocken, PA: ASTM International.

ASTM. (2017). *Standard test method for dispersive characteristics of clay soil by double hydrometer*, Method D4221-17. West Conshohocken, PA: ASTM International.

Ball, J. W., & Nordstrom, D. K. (1991). *User's manual for WATEQ4F, with revised thermodynamic data base and test cases for calculating speciation of major, trace, and redox elements in natural waters.* U.S. Geological Survey Open-File Report 91-183.

Barry, K., Dillon, P., & Pavelic, P. (2013). Domestic scale rainwater ASR observations of clogging and effectiveness of its management. In R. Martin (Ed.), *Clogging issues associated with managed aquifer recharge methods* (pp. 184–192). Australia: IAH Commission on Managing Aquifer Recharge.

Behnke, J. J. (1969). Clogging of surface spreading operations for artificial ground-water recharge. *Water Resources Research, 4,* 870–876.

Benamar, A. (2013). Soil clogging phenomena in vertical flow. In R. Martin (Ed.), *Clogging issues associated with managed aquifer recharge methods* (pp. 77–83). Australia: IAH Commission on Managing Aquifer Recharge.

Bouwer, H. (1989). Estimating and enhancing groundwater recharge. In M. L. Sharma (Ed.), *groundwater recharge* (pp. 1–10). Rotterdam: A.A. Balkema.

Bouwer, H. (2002). Artificial recharge of groundwater: Hydrogeology and engineering. *Hydrogeology Journal, 10,* 121–142.

Bouwer, H., & Rice, R. C. (1989). Effect of water depth in groundwater recharge basins on infiltration. *Journal of Irrigation and Drainage Engineering, 115*(4), 556–567.

Brown, D. L., & Silvey, W. D. (1973). Underground storage and retrieval of fresh water from a brackish-water aquifer. In. J. Braunstein (Ed.), *Underground waste management and artificial recharge*, Proceedings of a Symposium held at New Orleans, September 1973 (pp. 379–419). Menasha, WI: G. Banta Company.

Brown, D. L., & Silvey, W. D. (1977). *Artificial recharge to a freshwater-sensitive brackish-water sand aquifer, Norfolk, Virginia*. U.S. Geological Survey Professional Paper 939.

Buik, N. A., & Willemsen, A. (2002). Clogging rate of recharge wells in porous media. In P. J. Dillon (Ed.), *Management of aquifer recharge for sustainability* (pp. 195–198). Lisse: A. A. Balkema.

Coulon, A., Cannavo, P., Charpentier, S., & Vidal-Beaudet, L. (2015). Clogging-up of a stormwater infiltration basin: A laboratory approach using image analysis. *Journal of Soils and Sediments, 15*(8), 1738–1752.

Cullimore, D. R., & McCann, A. E. (1978). The identification, cultivation and control of iron bacteria in ground water. In E. A. Skinner & J. M. Shewan (Eds.), *Aquatic microbiology* (pp. 219–261). London: Academic Press.

De La Lomo González, B. (2013). Clogging of deep well infiltration recharge systems in the Netherlands. In R. Martin (Ed.), *Clogging issues associated with managed aquifer recharge methods* (pp. 163–173). Australia: IAH Commission on Managing Aquifer Recharge.

Dillon, P., & Pavelic, P. (1996). *Guidelines on the quality of stormwater and treated wastewater for injection into aquifers for storage and reuse*. Urban Water Resource Association of Australia Research Report No. 109, 48 p.

Dillon, P. J., Pavelic, P., Massmann, G., Barry, K., & Correll, R. (2001). Enhancement if the membrane filtration index (MFI) method for determining the clogging potential of turbid urban stormwater and reclaimed water used for aquifer storage and recovery. *Desalination, 140,* 153–165.

Dillon, P., Vanderzalm, J., Page, D., Barry, K., Gonzalez, D., Muthukaruppan, M., et al. (2016). Analysis of ASR clogging investigations at Three Australian ASR sites in a bayesian context. *Water, 8,* 442. https://doi.org/10.3990/w8100442.

Ehrlich, G. G., Ku, H. F. H., Vecchoili, J., & Ehlke, T. A. (1979). *Microbiological effects of recharging the Magothy aquifer, Bay Park, New York with tertiary-treated sewage*. U.S. Geological Survey Professional Paper 751-E.

Emerson, D., & De Vet, W. (2015). The role of FeOB in engineered water ecosystems: A review. *Journal American Water Works Associations, 107*(1), E47–E57.

Gette-Bouvarot, M., Mermillod-Blondin, F., Angulo-Jaramillo, R., Delolme, C., Lemoine, D., Lassabatere, L., et al. (2014). Coupling hydraulic and biological measurements highlights the key influence of algal biofilm on infiltration basin performance. *Ecohydrology, 7*(3), 950–964.

Goss, D. W., Smith, S. J., Stewart, B. A., & Jones, O. R. (1973). Fate of suspended sediment during basin recharge. *Water Resources Research, 9,* 668–675.

Grim, R. E. (1968). *Clay mineralogy*. New York: McGraw-Hill.

Heilweil, V. M., & Marston, T. M. (2013). Evaluation of potential gas clogging associated with managed aquifer recharge from a spreading basin, southwestern Utah, USA. In R. Martin (Ed.), *Clogging issues associated with managed aquifer recharge methods* (pp. 84–94). Australia: IAH Commission on Managing Aquifer Recharge.

Heilweil, V. M., Solomon, D. K., & Ortiz, G. (2008). Silt and gas accumulation beneath an artificial recharge spreading basin, Southwestern Utah, U.S.A. *Boletín Geológico y Minero, 120*(2), 1851–1896.

Heilweil, V. M., Solomon, D. K., Perkins, K. S., & Ellet, K. M. (2004). Gas-partitioning tracer test to quantify trapped gas during recharge. *Ground Water, 42*(4), 589–600.

Hijnen, W. A. M., Bunnik, J., Schippers, J. C., Straatman, R., & Fulmer, H. C. (1998). Determining the clogging potential of water used for artificial recharge in deep sandy aquifers. In J. H. Peters (Ed.), *Artificial recharge of groundwater* (pp. 437–440). Rotterdam: Balkema.

Hijnen, W. A. M., & Van der Kooij, D. (1992). The effect of low concentrations of assimilable organic carbon (AOC) in water on biological clogging of sand beds. *Water Research, 26*(7), 963–972.

Huisman, L., & Olsthoorn, T. N. (1983). *Artificial groundwater recharge*. London: Pittman.

Hutchinson, A. (1997). A graphical method for estimating injection well clogging using the modified fouling index. In D. R. Kendall (Ed.), *Conjunctive use of water resources: Aquifer storage and recovery, Proceedings AWRA Symposium, Long Beach, California* (pp. 241–249). Middleburg, VA: American Water Resources Association.

Hutchinson, A. (2013). Surface spreading recharge facility clogging—The Orange County Water District experience. In R. Martin (Ed.), *Clogging issues associated with managed aquifer recharge methods* (pp. 107–118). Australia: IAH Commission on Managing Aquifer Recharge.

Hutchinson, A., & Randall, R. (1994). Evaluation of injection well clogging with the Modified Fouling Index (MFI). In A. I. Johnson & R. D. G. Pyne (Eds.), *Artificial recharge of ground water II*, Proceedings of the Second International Symposium on Artificial Recharge of Ground Water (pp. 710–719). Reston, VA: American Society of Civil Engineers.

Johnston, K., Martin, M., & Higginson, S. (2013). Case study: Recharge of potable and tertiary-treated wastewater into a deep confined sandstone aquifer in Perth, Western Australia. In R. Martin (Ed.), *Clogging issues associated with managed aquifer recharge methods* (pp. 174–183). Australia: IAH Commission on Managing Aquifer Recharge.

Konikow, L. F., August, L. L., & Voss, C. I. (2001). Effects of clay dispersion on aquifer storage and recovery in coastal aquifers. *Transport in Porous Media, 43,* 45–64.

LeChevallier, M. W., Schneider, O. D., Weinrich, L. A., Jjemba, P. K., Evans, P. J., & Hooper, J. L. (2015). *An operational definition of biostability in drinking water*. Denver: Water Research Foundation.

Ma, Y. Z. (2011). Uncertainty analysis in reservoir characterization and management: How much should we know about what we don't know. In Y. Z. Ma & P. R. La Pointe (Eds.), *Uncertainty analysis and reservoir modeling, Memoir 96* (pp. 1–15). Tulsa: American Association of Petroleum Geologists.

Maliva, R. G. (2014). Economics of managed aquifer recharge. *Water, 6*(5), 1251–1279.

Maliva, R. G. (2016). *Aquifer characterization techniques*. Berlin: Springer.

Maliva, R. G., & Missimer, T. M. (2010). *Aquifer storage and recovery and managed aquifer recharge using wells: Planning, hydrogeology, design, and operation*. Houston: Schlumberger Corporation.

Mansuy, N. (1999). *Water well rehabilitation, a practical guide to understanding well problems and solutions*. Boca Raton: Lewis Publishers.

Martin, R. (Ed.). (2013). *Clogging issues associated with managed aquifer recharge methods*. Australia: IAH Commission on Managing Aquifer Recharge.

McDowell-Boyer, L. M., Hunt, J. R., & Sitar, N. (1986). Particle transport through porous media. *Water Resources Research, 22*(13), 1901–1921.

Moorman, J. H. N., Colin, M. G., & Stuyfzand, P. J. (2002). Iron precipitation clogging of a recovery well following nearby deep well injection. In P. J. Dillon (Ed.), *Management of aquifer recharge for sustainability* (pp. 209–214). Lisse: A.A. Balkema.

Olsthoorn, T. N. (1982). *The clogging of recharge wells*, Communications No. 72. Rijwijk, Netherlands: Kiwa.

Page, D., & Dillon, P. (2007). *Measurement of the biodegradable fraction of dissolved organic matter relevant to water reclamation via aquifers*. Canberra: CSIRO Land and Water.

Page, D., & Dillon, P. (2013). Characterisation of clogging during urban stormwater aquifer storage and recovery operations on a low permeability fractured aquifer. In R. Martin (Ed.), *Clogging issues associated with managed aquifer recharge methods* (pp. 193–199). Australia: IAH Commission on Managing Aquifer Recharge.

Page, D., Vanderzalm, J., Miotliński, K., Barry, K., Dillon, P., Lawrie, K., et al. (2014). Determining treatment requirements for turbid river water to avoid clogging of aquifer storage and recovery wells in siliceous alluvium. *Water Research, 66,* 99–110.

Parkhurst, D. L., & Appelo, C. A. J. (1999). *PHREEQC (Version 2)—A computer program for speciation, batch reaction, one-dimensional transport, and inverse geochemical calculations.* U.S. Geological Survey Water-Resources Investigations Report 99-42549.

Pavelic, P., Dillon, P. J., & Barry, K. E. (2007a). Management of clogging for reclaimed water in a carbonate aquifer. In P. Fox (Ed.), *Management of aquifer recharge for sustainability*, Proceedings of the 6th International Symposium on Managed Aquifer Recharge of Groundwater (pp. 148–161). Phoenix: Acacia Publishing.

Pavelic, P., Dillon, P., Barry, K., Armstrong, D., Hodson, A., Callaghan, J., et al. (2008). *Lessons drawn from attempts to unclog an ASR well in unconsolidated sand aquifer.* Canberra: CSIRO Land and Water.

Pavelic, P., Dillon, P. J., Barry, K. E., Vanderzalm, J. L., Correl, R. L., & Rinck-Pfeiffer, S. M. (2007b). Water quality effects on clogging rates during reclaimed water ASR in a carbonate aquifer. *Journal of Hydrology, 334,* 1–16.

Pavelic, P., Mucha, M., Dillon, P., & Barry, K. (2006). Laboratory column study on the effect of ponding depth of infiltration rates during SAT. In *Recharge systems for protecting and enhancing groundwater resources*, Proceedings of the 5th International Symposium on Management of Aquifer Recharge, Berlin, Germany, June 11–16, 2005 (pp. 624–629). Paris: UNESCO.

Pearce, M. S., & Eckmann, D. M. (1999). Designing ASR systems to maximize recovery efficiency and reduce construction and operating costs. In *Proceedings 1999 Florida Section/American Water Works Association Annual Conference*, Orlando, Florida.

Pérez-Paricio, A., & Carrera, J. (2001). Clogging and heterogeneity. In *Artificial recharge of groundwater*, European Commission Project ENV4-CT95-0071 Final Report (pp. 51–58). Luxembourg: European Communities—European Commission Directorate-General for Research.

Pérez-Paricio, A., Carrera, J., & Von Christierson, B. (2001). Numerical modelling of clogging. In *Artificial recharge of groundwater*, European Commission Project ENV4-CT95-0071 Final Report (pp. 100–107). Luxembourg: European Communities—European Commission Directorate-General for Research.

Peters, J. H. (1988). Artificial recharge and water supply in The Netherlands. In A. I. Johnson & R. D. G. Pyne (Eds.), *Artificial recharge of ground water II*, Proceedings of the Second International Symposium on Artificial Recharge of Ground Water (pp. 132–144). Reston, VA: American Society of Civil Engineers.

Phillips, D. W., Jr., Lyon, S., & Hutchinson, A. S. (2007). Development of a percolation decay model to guide future optimization of surface water recharge basins. In P. Fox (Ed.), *Management of aquifer recharge for sustainability*, Proceedings of the 6th International Symposium on Managed Aquifer Recharge of Groundwater (pp. 433–446). Phoenix: Acacia Publishing.

Pitt, W. A. J., & Magenheimer, S. (1997). ASR technology: Avoidance and solutions to aquifer clogging problems. In D. R. Kendall (Ed.), *Conjunctive use of water resources: Aquifer storage and recovery*, Proceedings AWRA Symposium, Long Beach, California (pp. 251–260). Middleburg, VA: American Water Resources Association.

Pyne, R. D. G. (1995). *Groundwater recharge and wells.* Boca Raton: Lewis.

Pyne, R. D. G. (2005). *Aquifer storage recovery, a guide to groundwater recharge through wells.* Gainesville, FL: ASR Systems.

Reed, M. G. (1972). Stabilization of formation clays with hydroxy-aluminum solutions. *Journal of Petroleum Technology, 24*(7), 860–864.

Reed, M. G., & Coppel, C. P. (1972). Sand stabilization with hydroxyl-aluminum solutions, SPE Paper No. 4186. In SPE 43rd California Regional Meeting. Society of Petroleum Engineers.

Rinck-Pfeiffer, S., Dillon, P., Ragusa, S., & Hutson, J. (2002). Injection well clogging processes during aquifer storage and recovery (ASR) with reclaimed water. In P. J. Dillon (Ed.), *Management of aquifer recharge for sustainability* (pp. 189–194). Lisse: A.A. Balkema.

Rinck-Pfeiffer, S., Ragusa, S., Sztajnbok, P., & Vandervelde, T. (2000). Interrelationships between biological, chemical and physical processes as an analog to clogging in aquifer storage and recovery (ASR) wells. *Water Research, 34,* 2110–2118.

Rinck-Pfeiffer, S. M., Ragusa, S. R., & Vandevelde, T. (1998). Column experiments to evaluate clogging and biochemical reactions in the vicinity of an effluent injection well. In J. H. Peters, et al. (Eds.), *Artificial recharge of Groundwater* (pp. 93–97). Rotterdam: Balkema.

Ryan, J. N., & Elimelech, M. (1996). Colloid mobilization and transport in groundwater. *Colloids and Surfaces A: Physicochemical and Engineering Aspects, 107,* 1–56.

Schippers, J. C., & Verdouw, J. (1980). The modified fouling index for determining characteristics of water. *Desalination, 32,* 137–148.

Schippers, J. C., Verdouw, J., & Zweere, G. J. (1995). Predicting the clogging rate of artificial recharge. *Journal of Water Supply: Research and Technology - Aqua, 44*(1), 18–28.

Schuh, W. M. (1990). Seasonal variation of clogging of an artificial recharge basin in a northern climate. *Journal of Hydrology, 121*(1), 193–215.

Shaw, J. C., Bramhill, B., Wardlaw, N. C., & Costerton, J. W. (1985). Bacterial fouling in a model core system. *Applied and Environmental Microbiology, 49*(3), 693–701.

Smith, S. A., & Touvinen, O. H. (1985). Environmental analysis of iron-precipitating bacteria in groundwater and wells. *Ground Water Monitoring & Remediation, 5,* 45–52.

Sniegocki, R. T. (1958). *Plugging by air entrainment in artificial-recharge tests.* U. S. Geological Survey Open File Report 58-97.

Sniegocki, R. T. (1963). *Problems in artificial recharge through wells in the Grand Prairie Region, Arkansas.* U.S. Geological Survey Water-Supply Paper 1615-F, 25 p.

Taylor, S. W., & Jaffe, P. R. (1990). Biofilm growth and the related changes in the physical properties of a porous medium I. Experimental investigation. *Water Resources Research, 26*(9), 2153–2159.

Timmer, H., Verdel, J. D., & Stuyfzand, P. J. (2001). Deep recharge clogging in Langerak (The Netherlands). In *Artificial recharge of groundwater*, European Commission Project ENV4-CT95-0071 Final Report (pp. 89–99). Luxembourg: European Communities—European Commission Directorate-General for Research.

Torkzaban, S., Vanderzalm, J., Treumann, S., & Amirianshoja, T. (2015). *Understanding and quantifying clogging and its management during injection of CSG water permeates, brines and blends.* Australia: CSIRO.

Van der Kooij, D., & Veenendaal, H. R. (1992). Assessment of the biofilm formation characteristics of drinking water. In *American Water Works Association: Water Quality Technology Conference Proceedings*, Toronto, ON (pp. 1099–1110).

Van Der Kooij, D., Visser, A., & Hijnen, W. A. M. (1982). Determining the concentration of easily assimilable organic carbon in drinking water. *Journal American Water Works Association, 74*(10), 540–545.

Vecchoili, J., Ku, H. F., & Sulam, D. J. (1980). *Hydraulic effects of recharging the Magothy Aquifer, Bay Park, New York with tertiary-treated sewage.* U. S. Geological Survey Professional Paper 751-F.

Yao, K. M., Habibian, M. T., & O'Melia, C. R. (1971). Water and waste water filtration. Concepts and applications. *Environmental Science & Technology, 5*(11), 1105–1112.

Chapter 12
MAR Pretreatment

12.1 Introduction

Pretreatment of water prior to managed aquifer recharge (MAR) is often necessary to reduce clogging rates, prevent adverse geochemical reactions, and meet health-based regulatory requirements. The term "pretreatment" is used herein for MAR-specific treatment elements employed prior to recharge. Where MAR is being used as a natural water treatment technology, pretreatment of recharge water can be an important element in a multiple barrier approach for protecting public health and the environment. The combination of pretreatment and MAR can have additive and, in some instances, synergistic effects. MAR may in itself be used as pretreatment. For example, riverbank filtration can be used to both extract and treat surface water for subsequent aquifer recharge.

Key design and operational issues for MAR systems are the degree to which water is treated prior to recharge and the reliance upon natural aquifer treatment processes during MAR to meet water quality goals. Treatment of recharge water to potable standards may reduce (but not eliminate) clogging potential and health risks but may be overkill for projects that do not involve indirect potable reuse. An additional consideration is the treatment that water will receive after recovery. With respect to the management of clogging in both well and surface-spreading systems, an important design and operational consideration is the effort (and associated costs) required to treat water to minimize clogging potential versus the costs, operational impacts, and effectiveness of periodical rehabilitation actions.

Economics also dictate the type of pretreatment implemented. Passive solutions that require no or minimal on-going human intervention are preferred for systems with low value (e.g., water is stored for non-potable uses) and in developing countries where financial and technical resources are limited. Stormwater management systems commonly include passive sedimentation features (settling basins, vegetated swales and filter strips) before the entry of water into infiltration structures.

R. G. Maliva, *Anthropogenic Aquifer Recharge*,
Springer Hydrogeology, https://doi.org/10.1007/978-3-030-11084-0_12

Some of specific MAR pretreatment objectives are:

- reduction in suspended solids concentrations and thus clogging potential
- reduction in particulate organic matter, dissolved organic carbon, and nutrients to reduce the potential for biological clogging
- pathogen removal
- meeting regulatory water quality standards for aquifer recharge
- adjustment of water chemistry to avoid adverse fluid-rock interactions
- removal of trace organic compounds and other contaminants of concern for indirect potable reuse.

This chapter reviews the main types of pretreatment systems that have applications for MAR systems. Where available, examples are provided of their application and effectiveness in achieving MAR treatment goals and their limitations. Most MAR pretreatment systems are standard, or at least widely used, technologies for wastewater treatment and potable water supply facilities, and engineering texts on the design and operation of these facilities include much more detailed information. Pretreatment systems discussed herein are:

- roughing filters
- granular-media filters
- screen filters
- membrane filtration systems
- MIEX® resin ion exchange
- constructed wetlands
- disinfection
- chemical treatments
- multiple-element pretreatment systems

12.2 Roughing Filters

A roughing filter is a coarse-media (typically rock or gravel) filter that is used primarily to reduce turbidity as a pretreatment step. It can be conceptualized as sedimentation tank that is a filled with gravel. Roughing filters can reduce suspended solids concentrations and thus clogging potential in MAR systems. Inasmuch as pathogens and some chemical contaminants may be attached to particles, pretreatment with roughing filters can also contribute toward reducing health risks.

Roughing filters can be an attractive technology in some situations because they can make use of local resources and require minimal mechanical equipment. Hence, they are generally an appropriate pretreatment technology for rural and small urban water supply systems and developing countries (Wegelin 1996). Roughing filter systems may be used instead of conventional pretreatment systems involving coagulation and flocculation, sedimentation, and filtration. The effluent from roughing filters may be further treated by slow-sand filtration.

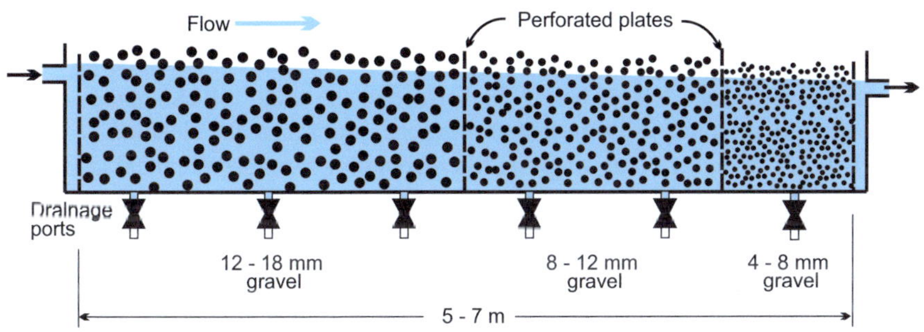

Fig. 12.1 Conceptual design of a horizontal flow roughing filter, based on Wegelin (1996)

Roughing filter design and operation were reviewed by Wegelin (1996). The main process active in roughing filters is sedimentation (gravity settling), although other processes such as screening, interception, adsorption (onto biofilms) and biological oxidation may also contribute to water quality improvements. Roughing filters drastically reduce settling distances as the fines settle onto gravel (a distance of a few mm) rather than the 1 to 3 m to the bottom of a sedimentation tank or basin.

The basic roughing filter design is gravel organized in a series of chambers or layers within a single chamber, with a decrease in grain size in the down flow direction (Wegelin 1996). Laminar flow through the gravel is required, which usually limits velocity to between 0.3 and 1.5 m/hr (Wegelin 1996). The chambers can be designed for upward, downward, or horizontal flow. The horizontal design (Fig. 12.1) has the advantage that the downward drift of sediment regenerates the upper part of the filter (Wegelin 1996). Filters are also regenerated by periodic drainage.

Roughing filters may be incorporated into a surface water treatment train as follows (Wegelin 1996):

(1) removal of coarse material using a sedimentation tank (grit chamber)
(2) aeration to allow for the oxidation of biomass and provision of dissolved oxygen (DO) needed for nitrification (e.g., using a cascade aerator)
(3) roughing filtration to reduce turbidity and improve biological characteristics (1–2 log removal of pathogens)
(4) slow-sand filtration
(5) disinfection.

The basic treatment goal of roughing filtration is to produce water that is suitable for slow-sand filtration (10–20 NTU, total suspended solids between 2 and 5 mg/L).

Wegelin (1996) provided guidance on the design of roughing filters, a few key aspects of which are summarized below. System design will depend on required capacity, and average and peak turbidity and suspended solids concentrations. Horizontal systems normally have lengths of between 5 m and 7 m and grain sizes between 20 and 4 mm (Wegelin 1996). For a system with three chambers, their lengths are recommended to be 3:2:1 with the bulk of removal occurring in the first coarsest and largest chamber. The chambers should be separated by perforated walls. Wegelin

(1996) recommended a maximum depth of 1 m for easy removal of the filter material and that the width should not exceed 4–5 m. Flow control should be provided so that the water level is kept below the surface of the filter. Filters should also be designed with a drainage system that allows for fast drainage for cleaning.

Wegelin (1996) noted that roughing filter technology has been revived in Europe through its use in artificial groundwater recharge plants. For example, in a system described in Dortmund, Germany, raw water falls over an aeration cascade, crosses a sedimentation trough, and then enters roughing filters with lengths of 50–70 m. After passing over a second aeration cascade, the water passes through a sand filter bed and then into the aquifer.

Lin et al. (2006) evaluated roughing filters specifically for the pretreatment of stormwater prior to aquifer storage and recovery (ASR). Roughing filtration would be applied prior to slow-sand filtration to remove suspended solids. A key limitation of slow-sand filtration is strict requirements placed on source water quality to prevent premature filter clogging. Lin et al. (2006) documented an experimental study using clay (kaolinite) and water from the Urrbrae wetlands site. The tested roughing filter had a 2.4 m column with 2.18, 5.18 and 7.55 mm average diameter media. The results of the study were a 75–96% reduction in turbidity and a 78–96% reduction in total suspended solids.

Cumulative removal efficiency was found to improve with longer filter lengths, smaller media size, and slower hydraulic loading rates, and a steady-state model (equation) was developed that relates kaolinite removal to these three factors (Lin et al. 2006). However, the Urrbrae wetlands water had poorer reduction in turbidity and suspended solids than for kaolinite with a preferential removal of larger particles. Organic-rich water has a high membrane fouling index (MFI) due to the organic natures of particles. The organic particles are more easily compressed and thus have a greater tendency to clog the filter paper. Roughing filters appear to be less efficient at reducing MFI.

12.3 Granular-Media Filters

Granular-media filters include rapid-sand filters (RSFs; also known as rapid-gravity filters, RGFs), rapid-pressure filters (RPFs) and slow-sand filters (SSFs). Both RSFs and SSFs operate under gravity, whereas water is forced through the filter under pressure in RPFs. RSFs and RPFs use relatively coarse media and, as their names imply, have more rapid flow rates than slow-sand filters. RSFs and RPFs are used primarily for suspended solids removal with subsequent disinfection to remove microorganisms. SSFs are designed and operated to also provide pathogen removal, although the effluent may still require disinfection.

The design and operation of granular-media filters are addressed in water treatment plant design texts (e.g., Edzwald 2011; Crittenden et al. 2012; Howe et al. 2012; Randtke and Horsley 2012). For potable water treatment plants, granular filters are one element of a multiple component treatment process, which usually involves some combination of pretreatment including straining at the inlet, clarification, sedimentation, flotation processes, and preliminary filtration. In rapid filtration methods, particle removal occurs throughout the filter bed, whereas in slow-sand filtration, particles are not driven far into the bed and most of the filtering occurs at a biologically active surface layer, commonly referred to as the "schmutzdecke," which is German for "dirty skin" or "dirty cover."

A common misconception is that particles are removed in granular media filters by straining, whereby the transport of a particle is stopped where it encounters a smaller-sized pore or pore throat. The actual main mechanism of particle removal is filtration, which involves the collision of suspended particles with media particles (and associated coatings), which is called transport, and attachment. Particles are transported by (Hendricks et al. 1991):

- **interception:** streamlines carry particles to sand grain surfaces so that a brushing effect will occur.
- **sedimentation:** particles settle under the force of gravity combined with convection.
- **diffusion:** random motion causes a collision.

12.3.1 Rapid-Sand Filtration and Rapid-Pressure Filtration

Rapid-sand filtration is the most commonly used type of sand filtration. RSFs have the advantage of requiring a much small area than SSFs to treat a given flow of water, but the produced water is of lesser quality than that produced by SSFs and disinfection is relied upon to much greater degree to inactivate pathogens. Due to high flow rates, and thus suspended particle loads, RSFs experience more rapid clogging than SSFs with a resulting decrease in flow rates. RSFs are restored by frequent backwashing in which water is forced upwards through the filter. Backwashing expands (fluidizes) the filter bed (i.e., dilates the pores), scours and erodes the accumulated fine solids, and transports the solids to the top of the filter. The main design parameters are:

- grain size(s) of the filter material
- composition of the filter material
- height (thickness) of the filter bed
- height of the supernatant water
- filtration rate.

The granular media usually has grain sizes in the 0.5–1.2 mm range and filtration rates of 5–20 m/h, compared to flow rates of 0.1–0.3 m/h in SSFs. RSFs may be composed of a single or multiple layers of media. In downward flowing filters, media particle size decreases downward. Particle size and composition are also chosen so that there is a downward increase in settling velocity, which is necessary to maintain the layering during backwashing. Silica sand is the most common media used in sand filters.

A common design for dual-media filters utilizes an upper layer of coarse-grained, angular anthracite, which has a low density. The angularity of anthracite makes it more effective in trapping particles and its low density results in its tendency to remain at the top of filters during backwashing. A basal layer of fine-grained garnet sand, which has a high density (specific gravity of 4.2), is used in some filters.

The main advantage of rapid-pressure filters is that the higher applied pressure allows for more rapid flow rates and thus smaller system footprints and longer-run times between backwashing. The higher pressure overcomes resistance from clogging and allows filtration to continue for longer periods of time. A disadvantage of pressure filters is that they are enclosed within a steel chamber and the media cannot be directly observed.

12.3.2 Slow-Sand Filters

Slow-sand filters are relevant to MAR both as a pretreatment method and because some MAR systems (infiltration basins) essentially act as slow-sand filters. Slow-sand filters are addressed in water treatment books and some specialty texts (e.g., Hendricks et al. 1991) from which most of this summary was derived. Slow-sand filtration (referred to a "biofiltration") as a pretreatment for MAR was reviewed by Page et al. (2006). The first installed SSF was constructed in 1829 by the Chelsea Water Company, London (Hendricks et al. 1991). SSF accomplishes particle removal through a combination of physical straining, filtration, adsorption, and biodegradation within the schmutzdecke layer. Biological particles attached to the schmutzdecke layer are most likely metabolized by microorganisms (Page et al. 2006).

SSFs are appropriate for small communities because of their low operating costs and passive operation that requires minimal operator intervention (they do not involve a full-time operator on site). A biologically mature SSF can achieve a 2–4-log total coliform removal (Hendricks et al. 1991). The primary negative aspect of SSFs is a relatively large surface area. SSFs are attractive for small communities in developing countries because of their simplicity. If a plant has been properly designed and constructed, filter performance will depend largely upon the conscientiousness of an operator carrying out a daily routine, which for most days will involve just a quick inspection (Visscher 1990).

The main operational issue for SSFs is clogging. The technology is not suitable without pretreatment (e.g., roughing filtration) for waters with high turbidities and suspended solids loads. High turbidities tend to result in rapid clogging (Visscher 1990; Hendricks et al. 1991), but very low turbidities do not necessarily result in very low clogging rates. The preferred raw water turbidity is <10 NTU with a suggested upper limit of 30 to 50 NTU (Hendricks et al. 1991). The total biomass plays a role in clogging (Page et al. 2006). Algae were reported to be the dominant form of biomass present in uncovered SSFs. Protozoa feed on bacteria and detritus attached to sand grains. Macroinvertebrates (e.g., midge larvae) may have a beneficial effect in reducing clogging through burrowing and pelletalization (Page et al. 2006).

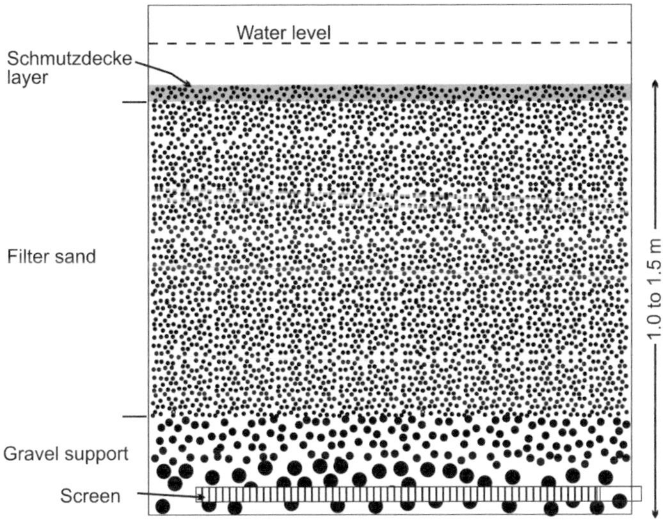

Fig. 12.2 Conceptual diagram of slow-sand filter

The basic design of SSF is illustrated in Fig. 12.2. Some basic design and operational issues are (Huisman and Wood 1974; Hendricks et al. 1991):

- A head water of ≥0.3 m (1 ft) above the bed minimizes sand erosion and short circuiting. The inlet should spread out flow. A uniform hydraulic loading rate over the sand bed is critical.
- The sand filter bed provides both filtration and retention time. The sand bed thickness should be ≥1 m (1.3 m preferred) with a 0.5 m minimum. Thicker beds increase the number of scrapings that can be performed, and extend bed life, but at a cost of greater head losses in accordance with Darcy's law. A thicker bed (>0.5 m) may also be appropriate where the filter is the only water treatment (Visscher 1990).
- The recommended sand properties are a d10 values (i.e., sieve size that will pass 10% of the grains) of 0.2–0.3 mm, a uniformity coefficient (UC; d60/d10) of 1.5–2.0, and a UC upper limit of <3.
- The preferred peak daily flow is 0.1–0.3 m/h. The traditional range of 0.04–0.4 m/h should be acceptable from the standpoint of filter efficiency.
- Lower filtration rates will occur with decreasing temperatures. The temperature effect on the viscosity of water and hydraulic conductivity is almost a factor of 2 between 0 and 25 °C.
- The ripening process tends to be more rapid at higher temperatures and greater nutrient concentrations. Most systems achieve peak performance after a two-day ripening period (Cleasby 1990).
- Filtered water is collected using an underdrain consisting of slotted or perforated pipe (a 1 m spacing between laterals is preferred; 2 m should be satisfactory)

covered with layers of graded gravel. The gravel support must be graded with the coarsest material at the bottom. The grain size of each layer should not allow for the downward movement of fines. Huisman and Wood (1974) provided rules for the design of gravel supports in terms of layer thicknesses and grain sizes.

The efficiency of filtration is greater for sands with a smaller grain size because of their greater straining capacity and higher surface area for biomass attachment. However, a finer grain size results in greater hydraulic resistance (Seelaus et al. 1986). Sand with higher a UC may be used out of necessity based on the local availability of materials. Ripening of the schmutzdecke layer is more important than grain size (Hendricks et al. 1991).

Development of the schmutzdecke layer is critical for filtration, but once it becomes too thick it impedes vertical flow. Filtration rates in SSFs are maintained by periodically scrapping off the schmutzdecke layer. Good run times (i.e., times between scrapings) are considered 30 days to several months. The sand usually does not need to be replaced until its thickness is reduced to 0.5–0.7 m (Hendricks et al. 1991). Removed sand can be washed and reused. Hendricks et al. (1991) suggested a flume system for cleaning sand, whereby particles are suspended and allowed to settle. Where replacement sand is used, Huisman and Wood (1974) suggested placing new sand at bottom and biologically active older sand at the top to increase the ripening rate.

12.4 Screen Filters

Screening is commonly the first step in surface water and wastewater treatment. Coarse traveling or stationary screens are used at intakes to prevent large items (including fish and other aquatic life) from entering a treatment facility. There is a great variety of types and design modifications for screens used for water and wastewater treatment. Screen types vary in the type and size of openings, configuration (e.g., plates, drums, disks), construction material, and cleaning method. Screens will become clogged with the filtered-out material (screenings) and must have some method for periodic cleaning.

Key design parameters for screens are open area and opening size (aperture). Open area is the percentage of effective filtration area (i.e., total area exposed to fluid flow). Greater open areas results in a lesser pressure loss (which is directly proportional to the square of fluid velocity) and a greater "dirt" holding capacity before a preset maximum pressure loss is reached (Allhands 2005). Aperture size controls the size of particles that are excluded (i.e., cannot pass through the screen). Small screen openings result in a greater retention of particles, and thus lesser particle concentrations in the filtrate (water that passes through a filter). However, greater particle retention results in more rapid clogging. Hence, water treatment systems commonly employ a series of filters with progressively decreasing aperture sizes.

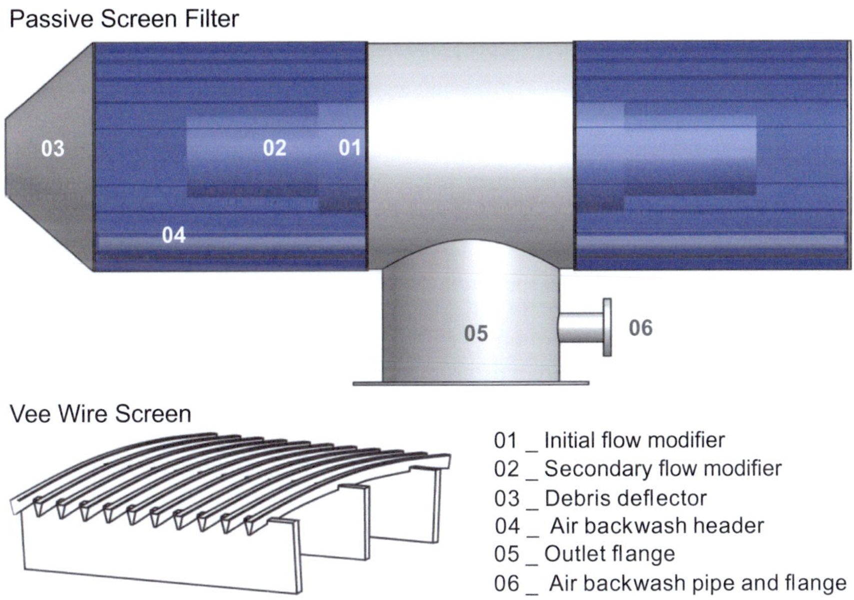

Fig. 12.3 Passive screen filter and Vee Wire screen diagrams. Courtesy of Johnson Screens

Bar screens, which provide initial coarse screening, consist of closely spaced vertical bars, usually between 0.6 and 7.6 cm (1/4 to 3 in.) apart. Bar screen systems may employ either an automatic rake (or other type of automatic cleaning mechanism) or be manually cleaned.

Passive screen intakes are cylindrical wedge-wire screens that are used for marine intakes for desalination plants (Missimer et al. 2015; Fig. 12.3) but also have applications for the surface-water supply for MAR systems. The screens are called "passive" because they have no moving parts. The screens are cleaned using bursts of compressed air, which result in a rapid reversal of flow direction. Most passive screens are "T"-shaped with screened sections located on either side of a central intake to maximize flow capacity (Missimer et al. 2015). Passive screens are used mainly to prevent the impingement and entrainment of marine organisms. The slot aperture is commonly less than 3 mm and the design intake velocities through the screen ranges from 10 to 15 cm/s (Missimer et al. 2015). Passive screen systems should be ideally located and oriented so that there are cross currents parallel to the screen, which allows debris (i.e., suspended sediments, plankton) in the water column to pass by the screen without impinging upon it. Ambient velocities should be equal to or greater than through velocities (Missimer et al. 2015, and references therein). Passive screens are effective at removing larger suspended particles, but are ineffective in reducing turbidity and the concentration of particles finer than the slot size.

Rotating-drum filters consists of a rotating drum that utilizes either perforated plates, a mesh screen, or a filter cloth to remove solids. In the commonly used inside-out configuration, raw water enters the upstream side of the filter and flows outward through the screen. Suspended solids become trapped on the inside surface of the screen. Drum filters are self-cleaning using an internal rake and/or rinsing system in which the solids are collected in a trough and discharged.

Rotating-disk filters consists of steel or concrete filter tanks in which a series of rotating filter disks are mounted. Each disk is composed a stainless-steel framework with a number of segments or panels that usually have a polypropylene (plastic) frame upon which filter cloth media are mounted. In the common outside-inside mode of operation, water enters the filter tank, in which the filters are partially (60–65%) or completely submerged. The water flows though the cloth media and into the center of the disks, from which the filtrate is collected. Solid particles are retained on the surface of the filter cloth. As clogging progresses, the water level at the inlet increases and when a preset maximum level is reached an automatic back-washing procedure is triggered using high-pressure jets of water or vacuum heads. The reject water with suspended solids is sent to waste. In the inside-out configuration, water flows into the central drum conduits and then into the interior of the filter disks. Filtrate flows to the outside of the disks and into the collection tank.

Rotating-disk filters can use filter cloths with opening as small as 10 μm (or even smaller) and are thus capable of removing very fine suspended particles. Because of the ability of cloth media to removal very fine particles, disk filters are used as pretreatment before membrane filtration systems. Disk filters are also used to treat reclaimed water to unrestricted reuse standards (e.g., Knapp and Tucker 2006; Beecher et al. 2013). A great advantage of rotary disk filters over drum filters is a greater filter area, which allows for smaller, more compact systems. Disk filters also have the advantage of being relatively inexpensive and simple, requiring low backwash volumes, and backwashing being performed while the filter remains on line.

Allhands (2005) described a self-cleaning screen filter (manufactured by Amiad) consisting of a cylindrical filter in which the water inflow is upward from base into the interior of the filter. Cleaning commence once the pressure differential reaches a preset value (7 psi). A suction scanner containing six tubular nozzles is rotated and moved linearly to give nozzles a spiral motion that covers the entire screen. There is no interruption of filtration during cleaning. The filter system was reported to be able to remove suspended solids down to 10 μm depending on selected screen size (Allhands 2005).

12.5 Membrane Filtration

Membrane filtration includes a series of technologies used to treat water. A great amount of research is on-going on membrane technologies because of their numerous important water treatment and industrial applications. New membrane technologies are in various stages of development and there are continuous refinements of existing technologies. Membrane technologies now relevant to MAR are pressure-driven

Table 12.1 Types of membrane filtration

Membrane	Pore size (μm)	Applications (excluded material)
Microfiltration (MF)	0.1 or 5.0 (or 10.0)	Large suspended solids, large microorganisms (bacteria, oocysts)
Ultrafiltration (UF)	0.01–0.1	Very fine particles, small microorganisms (viruses), large organic compounds (e.g., large proteins)
Nanofiltration (NF)	0.001–0.01	Some salts (multivalent), heavy metals, and organic compounds, hardness of water
Reverse osmosis (RO)	0.0001–0.001	Salts, metals, and most organic compounds

membrane filtration in which liquid is forced through a filter membrane with a high surface area. The semipermeable membranes allow water to flow through while capturing suspended particles and dissolved substances. Membranes are designed to allow only a certain size range and types of particles and substances to pass through. Four types of membranes are used in wastewater and water treatment, which are differentiated based on their pore size and thus the size of particles and dissolved constituents that are excluded (Table 12.1).

In general, with decreasing pore size membranes can exclude progressively smaller particles and compounds, but greater pressures and associated energy are required to force water through the membranes. Reverse osmosis is widely used for the desalination and purification of water as it can remove salts, metallic ions, and most organic constituents. Some organic chemicals of concern, such as NDMA (N-Nitrosodimethylamine) and trihalomethanes (THMs), are only partially removed by RO. RO membranes can remove the greatest range of particles and compounds but are prone to fouling. Hence, where the source water is not already of high quality, RO is commonly preceded by MF or UF.

For water supply systems using surface water, MF is effective in removing *Cryptosporidium parvum* and *Giardia lamblia* oocycsts, coliform bacteria, and some viruses. MF is used in wastewater treatment as an alternative to conventional granular media filtration to reduce turbidity and chemical usage for disinfection. Membrane treatment systems are increasingly being used to treat wastewater to meet standards for unrestricted public access reuse.

Noble et al. (2003) reported on pilot testing of MF for ASR pretreatment. The reported advantages of MF included:

- 100% removal of coliform bacteria and related pathogens
- capital and O&M costs are competitive with conventional water-treatment techniques
- low residual solids generation and the solids that are generated are not hazardous
- automated operation.

The main disadvantage of MF and other membrane techniques is membrane fouling and the need for pretreatment to reduce the rate of fouling and/or to periodically clean the membranes. RO is an integral component of the full advanced treatment (FAT) train that is used to treat wastewater for indirect potable reuse (Sect. 22.6.1).

12.6 MIEX Process

The MIEX (trademark of Orica Australia) magnetic ionic-exchange resin process was developed for the removal of dissolved organic carbon (DOC) from wastewater. As summarized by Slunjski et al. (2000), the MIEX resin has a strong base functionality and is, therefore, capable of exchanging with weak organic acid ions at the usually neutral pH of most raw waters. The magnetic component of the resin facilitates the agglomeration of $\pm 180\ \mu m$ resin beads into larger, heavier particles, facilitating their settling. The MIEX system is a continuous (rather than batch process) that has three main components: one or more contactors (10–20 min detention time), a settler, and a resin regeneration system. The settled resin is pumped back to the contactors as a slurry with a small fraction sent to the regeneration system.

Zhang et al. (2012) documented pilot testing of a MIEX MAR pretreatment system using secondary effluent from the Gaobeidian Wastewater Treatment Plant (north China). Subsequent steps in the treatment train were flocculation with polyaluminum chloride, sedimentation, and ozonation. The MIEX system provided a 44% reduction in average DOC (from 6.4 to 3.6 mg/L), a 50% reduction in UV_{254} (from 13.1 to 6.5 m^{-1}), an 18% reduction in SUVA (from 2.2 to 1.8 $L(mg\ m)^{-1}$), a 69% reduction in color (from 35 to 11 Pt-Co units) and a 32% reduction in nitrate concentration (from 24.3 to 16.5 mg/L).

In a subsequent study, jar tests were performed on the combination of MIEX and ozonation as pretreatment for soil aquifer treatment (SAT; Zhang et al. 2015). Aromatic DOC with apparent molecular weights of 2–5 kDa were found to be preferentially removed by MIEX, whereas the subsequent ozonation preferentially removed large molecular weight (>10 KDa) compounds, including fulvic and humic acid-like substances. The results of this investigation demonstrated the complimentary nature of the different treatment methods, including SAT, in removing dissolved organic matter (Zhang et al. 2015).

The main applications of MIEX for MAR is for pretreatment of organic-rich surface and reclaimed water both to meet water quality standards and reduce DBP (THMs) formation if chemical disinfection is to be performed. For example, the proposed surface water treatment for the Moore Haven ASR System, which is part of the Central and South Florida Comprehensive Everglades Restoration Plan, is planned to be MIEX for DOC removal coupled with chloramine disinfection (USACOE and SFWMD 2014).

12.7 Constructed Wetlands

Constructed wetlands use natural physical and biological processes active in wetlands to improve water quality. Kadlec and Wallace (2009) provided a very comprehensive review of the history, hydraulics, performance, and design of treatment wetlands, from which most of the following discussion was derived. The USEPA (2010) "Constructed Wetlands Treatment of Municipal Wastewaters" manual also provides a good overview of constructed wetlands design issues. Four underlying concepts for constructed wetlands are (Kadlec and Wallace 2009):

> Wetlands are land areas that are wet during part or all of the year because of their location in the landscape
>
> Wetlands are wet long enough to exclude plant species that cannot grow in saturated soils and to alter soil properties because of the chemical, physical, and biological changes that occur during flooding
>
> Because wetlands have a higher rate of biological activity than most ecosystems, they can transform many of the common pollutants that occur in conventional wastewater into healthier byproducts or essential nutrients that can be used for additional biological productivity
>
> Modern treatment wetlands are man-made systems that have been designed to emphasize specific characteristics of wetland ecosystems for improved treatment capacity.

Three types of treatment wetlands are in widespread use (Kadlec and Wallace 2009):

(1) **Free water surface (FWS):** wetlands with areas of open water and a similar appearance to natural marshes.
(2) **Horizontal subsurface flow (HSSF):** wetlands that typically employ a gravel bed planted with wetland vegetation. Water flows horizontally through the gravel, below land surface.
(3) **Vertical flow (VF):** water is distributed across the surface of a sand or gravel bed planted with wetland vegetation and percolates downward through the root zone.

Free water surface (FWS) system treat water by the processes of sedimentation, filtration, oxidation, sorption, precipitation, photodegradation, and plant uptake. FWS systems are most commonly used in the United States for advanced treatment of secondary-treated wastewater. Constructed wetlands are effective in reducing suspended solids, nutrients, and organic carbon concentrations. The main disadvantage of constructed wetlands is that they require large and appropriately located land areas. They have the great environmental benefit of providing habitats for a wide variety of wildlife because they mimic natural wetlands.

HSSF systems are designed and operated so that the water stays in the root zones of plants. Water is not exposed during the treatment process, which minimizes the exposure of humans and wildlife to pathogens. HSSF systems are commonly used to treated primary effluent from single family homes and small communities. VF systems most commonly employ pulsed surface loading. The operational scheme is comparable to that of a slow-sand filter. They can be operated to oxidize ammonia (nitrification).

FWS treatment wetlands are most appropriate for large-scale MAR systems. With respect to MAR, treatment wetlands can be used to treat urban stormwater and additionally treat reclaimed water prior to recharge. Infiltration directly (leakage) from treatment wetlands may also result in groundwater recharge. Treatment wetlands constructed above the water table can be constructed with a clay (or other liner) to prevent infiltration and recharge of an underlying aquifer. Leakage is a particular concern where the underlying aquifer is used as a local water supply. Where the underlying aquifer contains non-potable groundwater, downward leakage may not be an issue or may be desirable. Constructed wetlands designed for local aquifer recharge are referred to as "leaky wetlands."

Design options depend on the quality of influent water and treatment goals. If nitrogen removal is a goal, then a system design should have water passing through both an oxic and anoxic area. The sequential design model (Gearheart and Finney 1999) considers the dominant physical and biological processes responsible for determining effluent quality in each distinctive area (zone) of a constructed wetland and specific areas of the wetlands are designed for each of the target functions. The compartments or areas may not be specific discrete physical compartments and may overlap in time and space. However, the USEPA (2000) noted that design of a FWS as a sequential series of single-function zones (cells) with individual outlets is not an unattractive concept.

The USEPA (2000) presented a three-zone design for FWS systems to achieve progressive, sequential water quality improvement (Fig. 12.4):

- **Zone 1:** fully vegetated and anaerobic throughout its depth during the growing season. TSS and associated constituents are removed by sedimentation and flocculation. Remaining volatile and semivolatile constituents are also removed.
- **Zone 2:** open water that is regenerated by atmospheric contact and dissolved oxygen (DO) released by submerged macrophytes. Oxidation of carbonaceous compounds and nitrification of ammonia occurs.
- **Zone 3:** similar to zone 3. Denitrification.

Temporary nutrient removal by plant uptake may occur in Zones 1 and 3 at certain times of the year, while the release of these nutrients can occur at other times.

The detailed design of FWS constructed wetlands is well beyond the scope of this section. Basic design variables include:

- total area
- number, size, depth and shape of wetland cells
- hydraulic retention time
- vegetation types and coverage
- inlet and outlet type and location
- internal flow patterns.

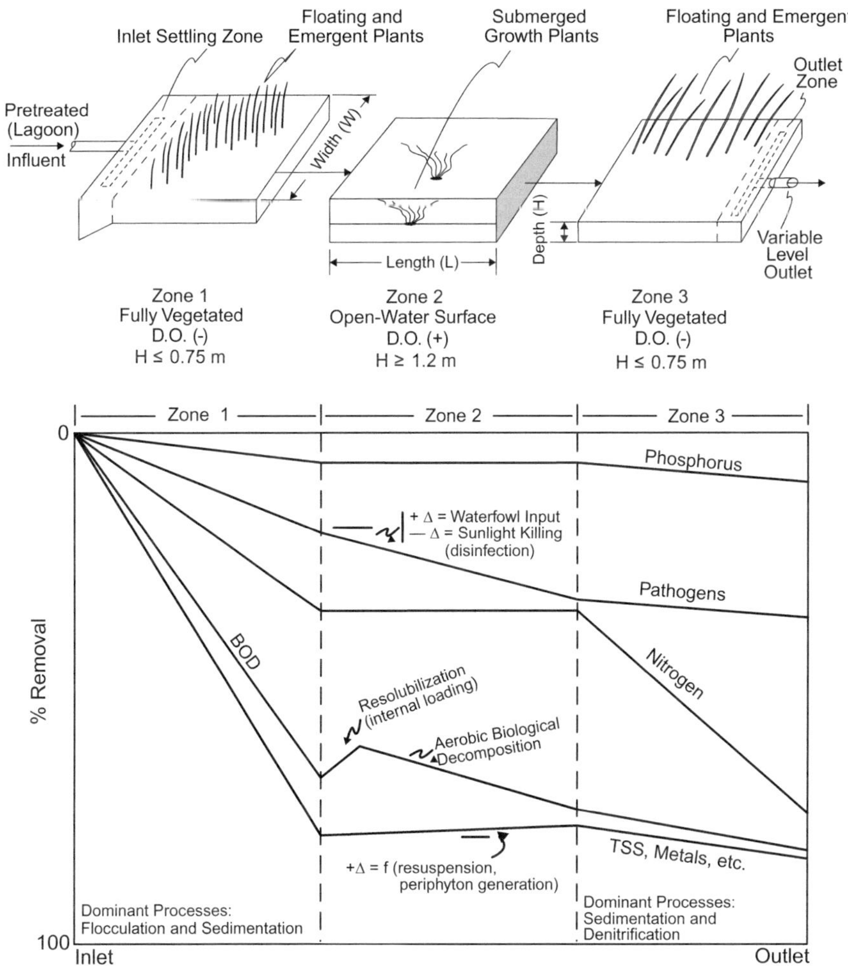

Fig. 12.4 Elements of a three-zone free water surface (FWS) constructed wetland and generic removal of pollutants. *Source* USEPA 2000

The hydrology and hydraulic of FWS treatment wetlands is quite complex due to variation in water depths and the presence of vegetation, as reviewed by Kadlec and Wallace (2009). The ideal conditions for pollutant removal is slow flow through shallow water and dense vegetation to allow sufficient time and contact for various pollutant attenuation process to occur. A key design and operational issue is avoiding channeling of flow through deeper water areas and thus increasing the amount of stagnant water. Some basic design concepts are (Kadlec and Wallace 2009; USEPA 2010; ITRC 2003):

- residence time must be equal to or greater than the reaction time needed to achieve desired effluent concentrations
- required residence times, which are usually in the 4 to 15 days range, depend upon on the contaminant type, their concentrations, their degradation and removal rates, and treatment goals (effluent concentration targets)
- a key variable for pollutant attenuation is the actual wetland detention time (τ), which the wetland volume divided by volumetric flow rate (Q); the nominal detention time (τ_n) is defined as

$$\tau_n = \frac{V_n}{Q} \tag{12.1}$$

where the nominal volume (V_n) is calculated from the geometry of the wetland and water depth

- the actual wetland hydraulic detention time is the product of the nominal detention time and the wetland volumetric efficiency (e_v), which accounts for the volume occupied by plants and pockets of stagnant water that do not contribute to wetland flow:

$$\tau = e_v \tau_n \tag{12.2}$$

- the design objective is slow shallow flow without channelization (preferred flow paths)
- flows should be <0.15 m/s (0.5 ft/s) and laminar with typical depths of 0.09 to 0.6 m (0.3–2.0 ft)
- internal flow patterns should promote mixing and avoid short-circuiting and associated reduced pollutant removal efficiency
- open-water areas should extend across the width of treatment cells to reduce channelization.

Kadlec and Wallace (2009) noted that the Manning equation is widely used to model flow in FWS wetlands but is actually not appropriate for this setting. Under the laminar or transitional flow regime of treatment wetlands, Manning's "n" coefficient is not a constant, but is instead strongly dependent on both flow velocity and plant stem density. The frictional effects that retard fluid flow are dominated by the drag exerted by the stems and litter of the vegetation, which increases over time as the vegetation become denser. The value of "n" is also dependent of local water depth. Values of "n" for open-channel turbulent flow may greatly underestimate the value for a wetland. Published "n" values from similar FWS systems may provide general guidelines for site-specific values.

A key objective of treatment wetlands is nutrient (nitrogen and phosphorous) removal. Phosphorous is retained in wetlands by plant uptake, microbial immobilization, and adsorption onto minerals. Phosphorus incorporated into plant tissue may be either released as the tissue decays or become incorporated as organic matter in the soil profile of the wetland (White et al. 2004). White et al. (2004) examined the retention of phosphorous in wetlands in twelve experimental mesocosms established in an Everglades Nutrient Removal Project wetland in South Florida. Mesocosms varied depending on whether they were continuously or intermittently flooded and contained macrophytes (*Thypha* plants). Samples of the inflow and outflow were analyzed for soluble reactive phosphorous (SRP; orthophosphate), total dissolved phosphorous (TDP), dissolved organic phosphorous (DOP; TDP SRP) and total phosphorous (TP). Soil samples were analyzed for total and extractable phosphorous.

A mean annual reduction of SRP of 85% for all treatments was reported with reductions ranging from 91% for continuously flood treatment to an average of 80% for drawdown treatment. Approximately half of all phosphorous was found in the organic pool of the peat soils. A key result of the study was that drawdown of surface water introduced oxygen into the soils, which increases the rate of mineralization of organic matter and the associated net release of DOP from the mesocosms. It was concluded that the greatest challenge to managing some large, stormwater treatment wetlands will be to maintain soils in a flooded condition during dry months to prevent the release of SRP and DOP from organic sediments.

Treatment wetlands have abundant water and growing emergent plants, which makes them attractive to wildlife. The presence of birds and other wildlife, in turn, makes treatment wetlands attractive to humans interested in the environment for recreation or environmental study (Knight 1997). Treatment wetlands can be designed to provide passive recreational opportunities (e.g., hiking, jogging, biking, photography, and wildlife study) and facilities (e.g., trails, boardwalks, and observation towers) can be provided to enhance the public's ability to observe the diversity of wetland habitats and associated wildlife populations (Knight 1997). For example, the 50-acre (20.2 ha) Wakodahatchee Wetlands in Palm Beach County, Florida, which receives recycled water from the Southern Region Water Reclamation Facility, provides a wildlife habitat that has attracted more than 150 different species of birds, as well as alligators, rabbits, turtles, otters, and foxes (Fig. 12.5a). The Wakodahatchee Wetlands has become an important rookery for the threatened Wood Stork (Fig. 12.5b). Both the Wakodahatchee Wetlands and nearby Green Cay Wetlands, have become two of the most popular bird watching and photography spots in the County and are important environmental assets of the community.

Orlando Wetlands Park (Orlando, Florida) consists of a 1,220 acre (494 ha) man-made wetland treatment system that was completed in July 1987 from pasture areas. The wetlands were designed to treat up to 35 million gallons a day (159,000 m^3/d) of reclaimed water from the Iron Bridge Regional WRF before it is discharged to the St Johns River (City of Orlando n.d.). The Orlando Wetlands Park is home to over 30 species of wildlife that are listed on the Florida Wildlife Conservation Commission's Threatened and Endangered Wildlife list. The Orlando Wetlands Park has more than 20 miles (32 km) of roads and woodland trails leading through marshes, hardwood hammocks and along scenic lakes. The most popular activities are bird-watching, nature photography, jogging, and bicycling (City of Orlando n.d.).

Fig. 12.5 Palm Beach County Utilities (Florida) Wakodahatchee Wetlands. (Top) View from boardwalk. (Bottom) Threatened wood storks nesting in the constructed wetlands

Similarly, the Sweetwater Wetlands in Tucson Arizona, has more than 2.5 miles (4.0 km) of pathways accessible to visitors and has become a popular location to view native wildlife in an urban setting. The streamside riparian zone supports a great variety of wildlife, including dragonflies, raccoons, hawks, bobcats and dozens of other species that make the wetlands their full- or part-time home (City of Tucson n.d.).

Treatment wetlands do pose some ecological risks. Wildlife could be exposed to non-degradable hazardous substances, such as toxic metals and synthetic organic compounds. Generally, the most effective way to avoid toxicity problems in a treatment wetland is through an appropriate level of pretreatment. Hazards to humans may also occur from exposure to pathogens, which can be addressed through pretreatment and avoiding direct contact (Knight 1997). Mosquitos and other biting insects can be a nuisance or vectors for disease transmission, which can be controlled through maintaining an ecosystem containing organisms that prey on mosquito larvae (Knight 1997). For example, in South Florida, mosquito fish (genus *Gambusia*) can be highly effective in reducing the mosquito population of wetlands. Dangerous reptiles, such as venous snakes, alligators, and crocodiles, may be present in treatment wetlands in some regions, the threat from which can be limited by controlling public access (Knight 1997).

Design considerations to improve the ecological benefits of constructed wetlands are (Knight 1997):

- pretreat to avoid excessive loading of toxic metals, organic compounds, and ammonia-N concentrations
- prevent excessive loading with mineral and organic sediments
- maintain a non-zero DO concentration
- design flexibility to control water levels
- incorporate deep-water zones without creating hydraulic short circuits (include islands in open water areas)
- utilize a diversity of plant species including those with known wildlife benefits
- incorporate vertical structure in plant communities by planting herbaceous vegetation, trees, and shrubs.
- incorporate horizontal structure through littoral shelves, benches, and deep zones.
- include structural density by use of an irregular shoreline
- install dead snags and nesting platforms.

Elevated nutrient concentrations can increase biological productivity but excessive concentrations can result in eutrophication (i.e., anaerobic conditions caused by the excessive stimulation and subsequent decay of algal communities).

12.8 Disinfection

Disinfection is broadly defined as the removal, deactivation, or killing of disease-causing (pathogenic) microorganisms. It may be accomplished by filtering out harmful microorganisms, the use of radiation (ultraviolet light), and adding disinfectant chemicals. Disinfection differs from sterilization in that disinfection is the process of eliminating or reducing harmful microorganisms to the extent that they cannot cause infection, whereas sterilization involves the killing of all microorganisms. Drinking water is disinfected but not sterilized. Recharge water is disinfected to:

- avoid potential health impacts from the consumption of recharge water
- meet regulatory requirements (pathogen concentration standards for recharge)
- prevent or minimize biological clogging.

The main disinfectants in use are:

- chlorine
- chloramines
- ozone
- ultraviolet light

The choice is disinfectant for a given application is based on effectiveness, cost, and the potential for disinfectant byproduct (DBP) formation. Some DBPs are known or suspected health hazards and drinking water standards or guidelines for their concentration in drinking water have been established.

12.8.1 *Chlorine*

Chlorine is historically the most commonly used disinfectant for water and has the advantages of being highly effective and relatively inexpensive. Chlorine used for water treatment is generally obtained as liquefied chlorine gas, sodium hypochlorite solution, or is generated on site by electrochlorination (passing of electric current through a salt solution). Chlorine gas reacts with water to produce hypochlorous acid (HOCl) and hydrochloric acid (HCl), which are dissociated to hydrogen and hypochlorite (OCl^-) ions:

$$Cl_2 + H_2O \rightarrow HOCI + HCI \tag{12.3}$$

Hypochlorous acid is a strong oxidizing agent and is the active ingredient of chlorine solutions.

In addition to being used as a primary disinfectant, chlorine is also added to potable water supply systems to provide a disinfectant residual in the distribution network. A main disadvantage of chlorination is that chlorine can react with naturally occurring organic matter (NORM) to form DBPs, particularly trihalomethanes (THMs) and haloacetic acids (HAAs). The USEPA maximum contaminant level (MCL) for total THMs is 0.080 mg/L or 80 ppb. In relatively organic-rich waters (e.g., reclaimed water and surface waters), a technical and regulatory challenge is optimizing the chlorine dose so that microbial water quality standards are met, while not forming DBPs at concentrations that exceed standards for THMs, and lesser commonly HAAs. Chemical disinfectants, such as chlorine, are poorly effective against *Cryptosporidium* and *Giardia* oocysts.

12.8.2 *Chloramines*

Chloramines are used as an alternative to free chlorine in municipal water disinfection because they are more stable, dissipate more slowly in the water distribution systems, and have a lesser tendency to form regulated DBPs. Chloramine, specifically monochloramine (NH_2Cl), is formed by the reaction of hypochlorous acid and ammonia (NH_3)

$$NH_3(aq) + HOCI \rightarrow NH_2Cl + H_2O \tag{12.4}$$

Chloramines are less effective as a disinfectant than chlorine but have the benefit of producing much less THMs and HAAs. Chloramination can produce the DBP N-nitrosodimethylamine (NDMA) but at concentrations well below those documented to cause health impacts.

12.8.3 Ozone

Ozone (O_3) is a strong and effective disinfectant and has the advantage that it also is effective in reducing the concentrations of organic chemical contaminants. Ozone is generated onsite because it is unstable and decomposes to elemental oxygen. Ozone is more effective than chlorine in destroying viruses and bacteria and, as it is generated onsite, reduces safety problems associated with shipping and handling (USEPA 1999a). The main disadvantages of ozone include relatively high costs and uneconomically high doses and costs for waters with high levels of suspended solids, biochemical oxygen demand (BOD), chemical oxygen demand (COD), or total organic carbon. Ozone also reacts with bromide in water to produce the DBP bromate BrO_3^-, which is suspected carcinogen. The USEPA MCL for bromate is 0.010 mg/L or 10 ppb.

12.8.4 Ultraviolet Radiation

UV irradiation is increasingly being used to disinfect waters because it avoids the need to handle and store dangerous chemicals, it minimizes the potential for DBP production, and it is effective in inactivating *Cryptosporidium*, *Giardia*, and other pathogenic protozoa (USEPA 1999b). It has the disadvantages of higher costs than chlorine and being less effective in waters with high turbidities, color, and suspended solids concentrations, and where pathogens are associated with particles. Higher levels of pretreatment may be required before UV disinfection. UV radiation also does not provide a disinfectant residual. When UV destroys living organic material, it can rupture the cell walls, thereby releasing free amino acids and other organic compounds that are utilized as food by some naturally-occurring groundwater bacteria. Enhanced growth rates of these naturally-occurring bacteria can lead to the formation of biofilms, contributing to clogging.

12.8.5 Disinfection Strategies

Disinfection is often a challenge for MAR systems that are used to store surface and reclaimed water because these waters often have relatively high concentrations of dissolved organic compounds (including disinfection-byproduct precursors), which may necessitate higher disinfectant doses and result in high levels of DBP formation. Disinfection procedures must be rigorous enough to meet target pathogen concentrations (e.g., applicable water quality standards), while at the same time also meeting water quality standards for DBPs. Bench-top testing is often necessary to evaluate various pretreatment and disinfectant options and determine optimum doses.

The National Research Council (2008) recommended that alternatives to chlorination be considered to meet primary disinfection requirements of MAR systems, such as UV, ozone, or membrane filtration, to minimize the formation of halogenated DBPs. However, the presence of residual chlorine may be necessary in at least some MAR systems to control biological clogging. Experimental studies by Fox et al. (1998) demonstrated the importance of maintaining a free chlorine residual in recharged water to control biofilm growth and minimize clogging of sand aquifers. A free chlorine residual of 2 mg/L was found to be able to effectively control biological activity near the injection point in the first 0.9 m (0.3 ft) of the experimental aquifer. A biologically active zone developed in the experimental aquifer after the free chlorine concentration decayed to levels below 0.6 mg/L. Chloramines apparently did not prevent biological clogging, but were capable of inhibiting biological growth, resulting in a greater distribution of biological activity.

The importance of maintaining a free chlorine residual in water recharged in wells is evident in the operational data from the Fountain Hills Sanitary District, Arizona ASR system. The ASR wells experienced a rapid decline in well performance (specific injectivity) of 53–64% after the disinfection system was changed from chlorination to ultraviolet disinfection (Small et al. 2007). The reintroduction of residual chlorine resulted in an immediate increase in well performance, reaching specific injectivity values similar to the values before removal of chlorine from the system.

Disinfection strategies are often influenced (dictated) by regulatory requirements, which may not adequately consider the natural attenuation of both pathogens (Sect. 7.2) and DBPs (Sect. 7.3) in the soil and groundwater environment. Chlorination may be the preferred disinfectant to maintain an effective residual in recharge wells where generated THMs are either naturally attenuated in the aquifer or their presence is accepted as there is no plausible scenario for consumption of the recharged water.

12.9 Chemical Pretreatments

Chemical pretreatments involve adjusting the chemistry of recharge water usually to avoid either clogging or fluid-rock interactions that could adversely impact water quality. Chemical pretreatments are most commonly needed for MAR systems that utilize wells because of their greater susceptibility to clogging and oftentimes large chemistry differences between recharge water and native groundwater in confined aquifers. In general, recharge waters tend to be fresh, have high DO concentrations, and are undersaturated with respect to most minerals than can cause clogging.

The main chemical pretreatments that have been applied to MAR systems are:

- pH reductions to prevent calcium carbonate precipitation in wells
- DO removal (oxidation-reduction potential reduction) to prevent the oxidative dissolution of chemically reduced minerals (e.g., iron sulfides) and associated arsenic and metals release
- aquifer treatment for clay dispersion management
- pH and dissolved DO adjustments to manage iron and manganese leaching.

12.9.1 pH Adjustments

Recharge water may be supersaturated with respect to calcium carbonate (calcite and aragonite) under well, infiltration basin, and aquifer physicochemical conditions, which can result in geochemical clogging. Saturation state with respect to carbonate minerals can be increased by decreases in the partial pressure of carbon dioxide (P_{CO_2}), increases in temperature (which decreases the solubility of CO_2 and other gases), and other processes that increase pH. Photosynthetic activity in infiltration basins can decrease P_{CO_2} to the point of calcium carbonate precipitation.

The saturation state of recharge water with respect to calcite (and other carbonate minerals), and thus scaling potential, can be lowered by decreasing the pH of the recharge water by an acid feed. Basic geochemical modeling can be used to determine the calcite saturation state but the modeling requires accurate measurement of pH. CO_2 can rapidly degas from water samples, and hence pH should be measured using a flow-through system in which the tested water is not in contact with the atmosphere.

The choice of acid to be used of pH adjustments is based on considerations of cost, safety, impacts to stored water quality, and regulatory acceptability. Strong acids, such as hydrochloric or sulfuric, are economical and effective, but have safety concerns over their storage and handling. Carbonic acid is an excellent option for modest pH adjustments. Carbonic acid is formed by the dissolution of CO_2 in water. Liquid CO_2 is widely available and CO_2 gas can be readily added to the recharge water flow using a bubbler apparatus. Carbonic acid is used to managed calcium carbonate scaling in some Florida ASR and production wells. A carbonic acid system tested at the Peace River Manasota Regional Water Supply Authority ASR system reduced the pH to between 6.9 and 7.6, compared to a typical finished-water pH of 8.2 (Eckmann et al. 2004). The system at the Marco Island Utilities Marco Lakes ASR system reduces the pH from about 7.5 to 7.0 (Poteet et al. 2017). Carbonic acid treatment has long been used to rehabilitate the production wells of the Island Water Association (Sanibel Island, Lee County, Florida) and more recently City of Apopka, Florida (Hahn et al. 2005), and Collier County, Florida.

12.9.2 Dissolved Oxygen Removal

Dissolved oxygen removal has been employed for some ASR systems as a means of controlling arsenic leaching, which appears to be due primarily to the oxidative dissolution of arsenic-bearing iron sulfide minerals (Sect. 6.5). The objective is to sufficiently lower the oxidation-reduction potential of recharged waters so that the water is in chemically equilibrium with the sulfide minerals. Pretreatment options for the removal of DO in Florida ASR systems were reviewed by ASR Systems (2006) and CH2M Hill (2007) in studies prepared for the Southwest Florida Water Management District. The DO removal options that have been identified as being potentially technically and economically viable are uncatalyzed chemical reduction and volatilization.

Uncatalyzed chemical reduction is most commonly performed using reduced sulfur compounds, such as a sulfide (S^{2-}), sulfite (SO_3^-) or thiosulfite ($S_2O_3^{2-}$). Uncatalyzed chemical reduction has the advantage of being a long-established, proven technology. It has the disadvantages of adding dissolved solids to the recharged waters, it involves the transport, storage and handling of reactive chemicals, and the reaction with DO may be incomplete at surface and aquifer temperatures (ASR Systems 2006). Several times more reactant may be needed for complete DO removal than indicated by reaction stoichiometry, which can significantly increase costs.

Volatilization uses either a carrier gas or negative pressure to strip oxygen out of solution. Its advantages are that it involves no chemical addition and the systems have a relatively small footprint. The use of a carrier gas can result in unwanted chemical changes in the injected water and potentially degassing of the carrier in the subsurface and associated gas binding (ASR Systems 2006). CH2M Hill (2007) evaluated both membrane contactors (Liqui-Cel®) and the GDT™ centrifugal vortex methods for oxygen stripping.

Membrane contactors utilize hollow fiber membranes in which a vacuum and/or sweep gas is applied to the center (lumen) of the fibers. Pressure is applied to the fluid to be treated, which surrounds the fibers, forcing dissolved gases through the membrane and into the lumen from which it is carried away. Membrane contactors have the advantages of relatively small footprints and installation costs, and a modular nature, which allows for ready expansion of capacity.

Bell et al. (2009) performed geochemical modeling to evaluate pyrite stability at various recharge water DO concentrations and mixing ratios for the City of Sanford, Florida, ASR system. To maintain reducing conditions at low mixing ratios, the DO of recharge water was found to have to be maintained below 0.06 mg/L, which is a 2-log removal from the recharge water value of 8 to 9 mg/L. The recommended pretreatment of sodium bisulfite for residual chlorine removal and membrane degasification (Membrana Liqui-Cel® contactors) for DO removal was subsequently installed.

Fouling of the membrane contactors occurred during cycle tests 3 and 4 (Poole et al. 2010; Camp Dresser and McKee 2012). Initially, the sodium bisulfite used for dechlorination was suspected as causing the fouling of the membrane contactors. After discontinuing the use of the sodium bisulfite dechlorination system, the membrane contactors still failed, which led to the conclusion that the source of the fouling was organic materials in the finished potable water used for initial testing. Membrana tested various cleaning agents and procedures in its internal laboratory, and established a cleaning procedure for the membrane contactors that was successfully used during cycle test 4.

The Seminole County, Florida, Markham ASR system also employed sodium bisulfite for dechlorination and Membrana Liqui-Cel® membrane contactors for DO removal (Camp Dresser and McKee 2011). Norton et al. (2012a, b) documented the results of a degasification test at the Bradenton (Florida) potable water ASR facility. Pretreatment consisted of dechloramination using sodium bisulfite and DO removal using hydrophobic membranes under vacuum. The testing results were a 99.93% removal of DO from the oxygen saturated source water and a 94% reduction of the total arsenic released at the Bradenton ASR site.

Pearce and Waldron (2011) documented the pilot testing of sodium hydrosulfide (NaHS, also called sodium bisulfide) for DO removal in the City of Deland, Florida, ASR system. The advantage of adding sulfide directly to the solution is that in addition to removing DO, it suppresses the dissolution of pyrite based on Le Chatelier's principle of equilibrium (Peace and Waldron 2011). Reported operating costs were estimated to range between US $20,000 and $30,000 per 100 million gallons (378,500 m^3).

12.9.3 Iron and Manganese Management

Changes in aquifer redox state caused by MAR can increase iron and manganese concentrations to values above drinking water standards. Leaching of iron and manganese may occur as the result of the oxidative dissolution or alteration of minerals containing the metals in a chemically-reduced state or by the reductive dissolution of minerals containing the metals in an oxidized state. Recharge of organic-rich water can result in the consumption of DO and the establishment of chemically reducing conditions.

ASR systems in the New Jersey Coastal Plain that use parts of the Potomac-Raritan-Magothy (PRM) Aquifer System as a storage zone have experienced leaching of iron and manganese into stored water as the result of the introduction of DO. The PRM aquifer is composed of sandy siliciclastic deposits in which ferrous (Fe^{2+}) iron is present in the minerals pyrite (FeS_2), pyrrhotite (Fe_7S_8) and siderite ($FeCO_3$; Lucas and McGill 1997). Recharged oxygenated water reacts with sulfide minerals resulting in a lowering of pH, increases in dissolved iron (Fe^{2+}) and sulfate concentrations, and decreases in alkalinity.

Rather than managing Fe and Mn leaching by removing DO, the strategy employed in some ASR systems has been to stabilizes the iron sulfides by forming a protective ferric hydroxide coating on the mineral surfaces (Lucas et al. 1994; Lucas and McGill 1997, 2006; Lucas 2007). Mineral surfaces can be stabilized by upward pH adjustments to about 8.5 from 7.2, increasing the DO concentration to 8.5, or by using permanganate (MnO_4^-) as an additive. The slightly negative charge of the coatings make them adsorptive of metals, including free iron ions migrating in the groundwater. The large sorptive capacity of the conditioned aquifer was reported to allow for recovery volumes 3–6 times that of the recharge volumes with low iron concentrations that meet the 0.3 mg/L New Jersey drinking water standard.

Elevated concentrations of manganese were present in water recovered during the initial stages of operation of the Chesapeake, Virginia, ASR system. At times, manganese concentrations exceeded 1 mg/L, which is well in excess of the secondary drinking water standard of 0.05 mg/L (Ibison et al. 1994). Manganese was found to occur in the storage zone in a manganese-bearing siderite (iron carbonate). Bench-top experiments indicate that the pH of the recharge water needs to be maintained above 8 to keep manganese concentrations low. Field testing in which the pH of the recharge water was increased to values in the 8.2–8.6 range by using a sodium carbonate feed resulted in a short-term decrease in manganese concentration. However, the pH of the storage water decreased to values below 8 during storage with a concomitant increase in manganese concentration. Ibison et al. (1994) suggested that the decrease in pH was due to prior injection of low (6.75–7.50) pH water. The City of Chesapeake installed a treatment system to lower the manganese concentration of the recovered water, which is then disinfected and blended with finished water from the Lake Gaston Water Treatment Plant.

Managing the leaching of metals by stabilizing the mineral surfaces was proposed for siliciclastic aquifers in the southern Netherlands and tested at the Herten pilot ASR system (Stuyfzand and Doomen 2004; Stuyfzand et al. 2006). The ASR storage zone is a deep, sandy, anoxic aquifer, and water quality changes were caused by redox reactions with organic matter, pyrite, and manganous siderite. During successive ASR cycles, reduced phases were "inactivated" by leaching and coating with iron (oxy)hydroxides. To prevent dissolution of pyrite and other reduced Mn and Fe minerals, O_2 and NaOH pretreatment was recommended (Stuyfzand et al. 2006). NaOH prevents the reduction in pH caused by oxidation processes. The pretreatment would act to

- coat reactive iron minerals with iron (oxy)hydroxides
- keep aquifers oxic for longer period of time
- buffer acidifying action of oxidation reactions.

In situ iron removal (ISIR) is a related process for managing the concentration of dissolved iron and other metals. The ISIR process involves cyclic injection of oxygenated water into an aquifer and the subsequent withdrawal of a greater volume of injected water and native groundwater in which the iron (and manganese) concentrations are less than those in the native groundwater (Appelo et al. 1999; Appelo and de Vet 2002). The operational success of in situ iron removal can probably be explained by the oxidation of exchangeable and sorbed Fe^{2+} to form iron (oxy)hydroxides (Appelo et al. 1999). During withdrawal, the exchange sites, including newly precipitated iron (oxy)hydroxides, sorb Fe^{2+} within groundwater as it flows past. Appelo et al. (1999) and Appelo and de Vet (2002) suggested that the efficiency of the process is limited by the amount of oxidant in the injected water, the exchange capacity of the aquifer, and the amount of exchangeable Fe^{2+} that is capable of consuming the oxidant during the injection stage.

12.9.4 Clay Dispersion Management

Clay dispersion can cause rapid, severe, and largely irreversible reductions in permeability when freshwater is injected into a "water sensitive" aquifer containing brackish or saline groundwater. Clay dispersion can be managed by exchanging monovalent cations in the electric double layers of clay particles with divalent or trivalent cations, which reduces the thickness of the double layer. Early experiments by Brown and Silvey (1973, 1977) demonstrated that clay dispersion can be managed by pretreating an aquifer with a preflush of calcium chloride (0.1 and 0.2 N). Cation exchange results in the uptake of calcium and the release of sodium. A limitation of aquifer treatment with divalent cation solutions, such as calcium chloride, is that the treatment is not permanent. It was noted that it was important not to recover stored water to a point where the formation water is brought back into the treated formation because the calcium for sodium exchange is reversible.

Trivalent cation treatment, such a polymeric hydroxyl aluminum solutions, is a more permanent treatment of water sensitive formations against clay swelling and dispersion because the ions adhere tenaciously to both the external and interlayer surfaces of clay minerals (Reed 1972). Polymeric hydroxyl aluminum solutions are formed by the reaction of an aluminum salt (typically $AlCl_3$) solution with sodium hydroxide (NaOH). The effectiveness of hydroxyl aluminum treatment has been demonstrated in both laboratory and field (oil field) tests (Reed 1972; Reed and Coppel 1972).

Experimental results by Reed (1972) on the highly water sensitive Berea Formation indicates that a 0.1 M solution of $AlCl_3$ treated with NaOH to a OH/Al ratio of approximately 2.0 was effective in preventing permeability loss due to water sensitivity. The most effective treatment method was reported to follow polymeric hydroxyl aluminum solution injection with an overflush of freshwater, and then have a two day to one week aging period. Reed (1972) and Reed and Coppel (1972) reported the hydroxyl aluminum treatment may also stabilize formations and reduce sand production. Care must be taken in the preparation of the treatment solution and subsequent aquifer treatment to avoid formation of aluminum hydroxide, which can clog formations.

The mechanisms and controls of the loss of permeability by clay dispersion was reviewed by Torkzaban et al. (2015), including the limitations of inorganic pretreatments. Chitosan, a natural biodegradable nontoxic biopolymer made by treating the chitin shells of shrimp and other crustaceans, was investigated as an alternative to traditional clay stabilizers (Torkzaban et al. 2015). Laboratory tests demonstrated its likely effectiveness in preventing well clogging when freshwater is injected into a brackish aquifer. A 10 ppm (or greater) solution injected at a low pH limited the reduction in permeability to less than 5% when the core was subsequent flushed with RO treated water. Chitosan treatment benefits were reported to last for months.

It is critical where significant clay dispersion is a possibility, that the potential for clay dispersion be evaluated early in a project. If there is a potential for significant clay dispersion, then strong consideration should be given to coring the storage or recharge zone and performing bench-top testing. Inasmuch as clay dispersion can cause permanent aquifer damage, pretreatment should be performed before the start of any injection.

12.10 Multiple-Element Pretreatment Systems

MAR systems can have multiple water quality requirements for recharged water, which often necessitate multiple-element pretreatment systems. Commonly initial steps focus on reducing turbidity and suspended solids concentration. Later steps may address organic matter concentration, disinfection, and chemistry adjustments. Follows are summaries of actual or proposed multiple-element pretreatment systems to illustrate some of the range of pretreatment options.

12.10.1 CERP Surface Water Treatment Systems

A proposed key element of the Comprehensive Everglades Restoration Plan (CERP) for South Florida is the large-scale implementation of ASR in which surface water is stored during the summer wet season and released during the winter and spring dry season. The CERP ASR project has been by far the most intensely scrutinized ASR project ever with all aspects of the project subject to review by multiple government agencies and independent scientific review panels.

The surface water to be stored in the CERP ASR system is generally of good quality and meets applicable primary drinking water and groundwater standards except for microbiological parameters. The main technical challenges are reducing suspended solids concentrations to minimize clogging and disinfection of the relatively organic-rich source water while not exceeding groundwater standards for disinfection byproducts. The initial screening of surface water treatment options considered five fatal flaws (CH2M Hill 2003):

- any process that is complex in nature compared to other processes and would require a high level of manned staffing and operator and maintenance attention
- any treatment process that produces levels of disinfection byproducts exceeding promulgated standards
- scalability, the treatment process must be expandable to the target system capacity
- any process that is high profile and has structures that would be considered public eyesores

- any process not suitable for high organic loads, algae, and high turbidities that are common characteristics of the surface waters of South Florida.

The use of free chlorine and chlorine dioxide were judged to have a fatal flaw because of disinfection byproducts formation. After the fatal flaw screening, the remaining treatment options were evaluated based on eight criteria (CH2M Hill 2003):

- the impacts of residuals on the environment and their management effort and costs
- disinfection byproduct formation
- pathogen removal and inactivation effectiveness
- operational considerations (operational, power, and monitoring requirements)
- potential for aquifer plugging
- potential for the removal of metals from recovered water
- process uncertainty
- environmental impacts.

Two pilot ASR systems have been constructed to date, the Kissimmee River ASR System and Hillsboro ASR System, which have been described in great detail (USACOE and SFWMD 2013). The surface water pretreatment system for the Kissimmee River ASR System consist of a 48-in. (122-cm) diameter T-shaped cylindrical passive-intake screen with a 1-mm screen mesh size and a design flow velocity of 0.25 ft/s (0.076 m/s). The screen is designed to prevent fish larvae and debris from entering the system. An air-burst system is used to periodically remove fouling material from the intake screen.

Filtration is provided by a dual-media pressure filter that consists of four cells in a 62-ft (18.9 m) long by 10-ft (3.05 m) diameter steel tank. Each cell can be backwashed separately and the system is designed to be operated at a surface loading rate of 6 gpm/ft^2 (4.07 l/s per m^2) at 5 MGD (18,900 m^3/d) with one cell out of service. Microbial inactivation, as required by the Florida Department of Environmental Protection for injection, is provided by a UV disinfection system.

The Hillsboro River ASR system also uses a passive screen with an air-burst cleaning system for its intake. Filtration is provided by an automatic self-cleaning disk-filter system that consists of eight screen filters (14-in., 36-cm) diameter. The skid was constructed to accommodate four additional units if needed in the future. Initially 30-μm screens were used, but the size of the screens was subsequently increased to 80-μm after early tests showed filter clogging. A UV disinfection system is also used for the Hillsboro River ASR system.

A combination of MIEX for organic carbon removal and disinfection using chloramination was proposed for the Moore Haven pilot ASR system, which has not been constructed to date.

12.10.2 Wastewater Treatment Prior to Recharge

Wastewater used for aquifer recharge via surface spreading in the United States and for aquifer recharge using wells commonly involves tertiary treatment. Secondary-treated wastewater is often further treated by granular-media filtration and chlorine disinfection (Asano and Cotruvo 2003). Reclaimed water stored in ASR systems is typically treated to unrestricted reuse standards, with the flow either sent to the reuse system, when needed, or to the ASR system during periods of excess supply. For example, the reclaimed water stored in the Destin Water User reclaimed water ASR system (northwest Florida) receives secondary treatment followed by dual-media filtration and chlorination (Maliva et al. 2013).

The water used during the testing of the intensely studied Bolivar (South Australia) ASR system was secondary-treated wastewater from the Bolivar Wastewater Treatment Plant, which was further treated by holding in a stabilization lagoon for a minimum of two weeks, followed by dissolved air flotation, dual-media filtration and chlorination. The extent and cost of treatment employed for the Bolivar system were dictated by the requirement to avoid clogging of the ASR well rather than the needs of end users of water (Dillon et al. 2006). The recharge water was suitable for unrestricted irrigation (non-potable) use but still had a substantial nutrient concentration and thus clogging potential.

12.10.3 Stormwater and Surface Water Pretreatment

Stormwater is referred to herein was surface runoff that has entered a stormwater conveyance during or following a precipitation event but has not yet entered a surface water body (e.g., river, creek, lake or dam). The main operational issue associated with stormwater MAR systems is that they tend to clog with fine sediments and debris resulting in reduced infiltration and recharge rates. Stormwater is often conveyed directly to infiltration basins without pretreatment. Alternatively, systems may be designed so that water first passes through a sedimentation basin or forebay to minimize deposition of fine suspended sediments in the infiltration basin. Greater pretreatment is needed where recharge is performed using wells, although there is still a preference for passive treatment technologies.

Systems that use dry or wet wells for recharge are particularly susceptible to clogging. Stormwater drainage wells in the United States are typically dispersed (constructed at or near the stormwater generation site) and utilize passive pretreatment elements. Stormwater drainage wells often pretreat water using some combination of screens and grates, sedimentation chambers or sumps, grit- and oil-and-grease separators, filter strips, and swales (Sect. 17.6; USEPA 1999c). The stormwater is not disinfected.

Surface-water ASR is practiced in India and other developed and newly industrialized countries by allowing stormwater to flow into existing wells during wet (monsoon) periods. The filtration provided is necessarily low technology and ideally passive. Holländer et al. (2009) documented pilot testing of ASR in a brackish aquifer in eastern India (state of Orissa). Four filter systems were evaluated (Holländer et al. 2009): single-gravel filter, two-gravel filters, single-gravel filter with compressed rice straw, and two-gravel filters with compressed rice straw in between. The filters were found to achieve 70–90% reductions in total suspended solids. However, the filtration was inadequate where the source water had very high suspended solids concentrations.

Sultana et al. (2014) investigated low-cost ASR in Bangladesh using monsoon water collected in ponds. A typical ASR system design used for pilot testing utilizes a double-chambered tank for pretreatment. One chamber contains a sand filter and the other holds filtered water. The sand filter has two layers (coarse sand overlain by local fine sand) and is covered with a "Geojute" canvas mesh. The well itself is completed as a second filter by being filled with gravel and a fine sand cap. Maintenance involves weekly washing of the Geojute canvas and less frequent replacement of the fine sand cap in the wells.

Macia and Lluria (2001) described the development of a pretreatment system for aquifer recharge using injection wells as part of the Salt River Project (SRP) in the Phoenix (Arizona) metropolitan area. The recharge water is surface water from the SRP canal system, which is of better quality than native groundwater. Nevertheless, pretreatment is necessary to remove biological and particulate components before recharge to manage clogging. Existing wells were equipped for recharge. This study is notable as four treatment systems were sequentially tested on site:

Phase 1: Commercially available rotating micro-screen drum filter followed by gas chlorination. The system was very effective in removing particles down to 10 μm but was very costly to run and had a low flow rate (100 gpm; 6.31 L/s).
Phase 2: One-stage filtration. Allowed for high flow rates (1000 gpm, 63.1 L/s) but frequently clogged from moss and algae in the canal water.
Phase 3: Three-stage filtration and hydrogen peroxide disinfection (self-cleaning rotational filter in the canal and two cylindrical filters). The system provided effective filtration but had a low flow rate (250 gpm, 15.8 L/s).
Phase 4: Two-stage filtration (self-cleaning rotational in the canal and 50 μm cylindrical) and hydrogen peroxide. Phase 4 has a capacity of over 1,000 gpm (63.1 L/s) and was the selected final option.

Hamadeh et al. (2014) reviewed the combination of constructed wetlands (CW) and MAR for wastewater treatment. Constructed wetlands serve to reduce the concentrations of suspended solids, pathogens, and ammonia through nitrification. The reduction in suspended solids concentrations in the CW reduces clogging potential. CW & MAR hybrid systems were proposed to provide a cost-effective, sustainable, and efficient treatment technology (Hamadeh et al. 2014).

The Andrews Farm ASR system in South Australia is a well-studied example of a stormwater ASR system that uses three interconnected stormwater wetland/detention basins for passive treatment. The water then passes through a 100 μm screen and geotextile filter prior to injection without disinfection (Pavelic et al. 2006). The recharged water during the initial testing had high MFI values (400–2,600 s/L^2), but clogging was not a significant issue due to the carbonate storage zone (Pavelic et al. 2006). The recovered water is used for the irrigation of parks and gardens. Comparison of the experiences of the Andrews Farm and the failed Urrbrae Wetland stormwater ASR systems illustrates that aquifer hydrogeology can be much more important than recharge water quality in controlling well clogging (Pavelic et al. 2008).

Surface water treated to potable water standards is stored in numerous ASR systems. The question arises as to whether this level of treatment is necessary, particularly where recovered water is not sent directly to the potable water system (i.e., the water is retreated). The Llobregat ASR system in Barcelona, Spain used surface water that after conventional treatment underwent further treatment split between two trains: (1) ozonation and granular activated carbon (GAC) filtration and (2) coagulation with $FeCl_3$, UF, UV irradiation and RO (Camprovin et al. 2017). The results of laboratory simulation of ASR recharge indicate that the conventional treatment of pre-ozonation, coagulation and flocculation, sedimentation, and sand filtration could be feasible without causing unacceptable clogging (Camprovin et al. 2017).

12.10.4 Full Advanced Treatment

Pretreatment for indirect potable reuse systems in the United States involves treating the water to essentially drinking water or higher standards. The state of the art is full advanced treatment (FAT), which includes MF/UF followed by RO plus an advanced oxidation process (AOP). AOP commonly includes a UV-based advanced oxidation process for removal of some organic compounds that passed through the RO membranes (e.g., NDMA) and inactivation of pathogens. Ozone treatment may also be used either before the MF/UF units to prevent membrane fouling or downstream of the RO units for additional organic removal or degradation. FAT and alternatives are further discussed in the context of indirect potable reuse (Sect. 22.6.1).

FAT produces water that is often of better quality than potable water produced by conventional treatment of surface water sources. Hence, it is a legitimate question as to whether FAT is employed primarily to address real health risks or to assuage public concerns and perceptions over indirect potable reuse.

12.11 Conclusions

Numerous pretreatment technologies are available for MAR, which vary greatly in their specific water quality benefits, effectiveness, technological sophistication, costs, and degree of active human intervention required. Hence, MAR project teams should include an engineer with expertise in water treatment technologies so as to identify the most cost-effective treatment process for a given system.

It is also important to recognize that local financial and technical resources vary greatly between regions. Pretreatment strategies employed in Orange County, California, for example, are clearly not economically and technically feasible in rural India, Bangladesh, or Africa. Hence Dillon et al. (2014) in "A Water Quality Guide to Managed Aquifer Recharge in India" presented interim guidelines with the goal of assisting "those recharging aquifers to take actions that will make water safer, but without a guarantee that recovered water will be safe for its intended uses, especially for drinking, without further treatment". The interim goal is to incrementally improve public health, recognizing that further improvements will hopefully follow.

Dillon et al. (2013) recognized that Australia risk management approach is too data hungry for ready application in India, particularly with regards to evaluating microbiological quality of source and recovered water. They suggested instead that a simplified approach needs to be developed consistent with World Health Organization (Davison et al. 2005) water safety plans. Separate plans should be developed for drinking water and non-drinking water systems.

High levels of pretreatment can also impact MAR viability in developed countries as MAR competes with other water supply, storage, and treatment options. The author has observed that a disconnect may occur between regulatory requirements and actual health and environmental risks. For example, underground injection control regulations in the United States require that water stored in ASR systems using brackish storage zones meet primary (and in some states secondary aesthetic-based) drinking water standards, even though the water cannot be directly consumed and would require RO treatment for potable use, which would remove the contaminants of concern. While such conservative standards could be argued as prudent from a public health perspective, they have a cost in that they can obstruct projects that provide needed water resource supply benefits. However, regulations and policies are evolving to allow for a zone of discharge in which water quality improvements by natural contaminant attenuation processes may occur.

References

Allhands, M. N. (2005). Removing solids with automated self-cleaning filters. In *15th Annual produced water seminar*, January 19–21, 2005. Houston, TX.

Appelo, C. A. J., & de Vet, W. W. J. M. (2002). Modeling in situ iron removal from groundwater with trace elements such as As. In C. A. J. Welch & K. G. Stollenwerk (Eds.), *Arsenic in ground water, geochemistry and occurrence* (pp. 381–401). New York: Springer.

Appelo, C. A. J., Drijver, B., Hekkenberg, R., & de Jonge, M. (1999). Modeling in situ iron removal from ground water. *Ground Water, 37*(6), 811–817.

Asano, T., & Cotruvo, J. (2003). Groundwater recharge with recycled municipal wastewater: Health and regulatory considerations. In R. Aertgeerts & A. Angelakis (Eds.), *Health risks in aquifer recharge using reclaimed water—State of the art report* (pp. 154–170). Geneva: World Health Organization.

ASR Systems LLC. (2006). *Evaluation of arsenic mobilization processes occurring during aquifer storage recovery activities. Task 2—Technical memorandum, literature review, arsenic mobi-*

lization processes during ASR operations. Report prepared for the Southwest Florida Water Management District. Gainesville, FL: ASR Systems LLC.
Beecher, P., Rupp, H., Clanton, D., Sheldon, H., & Vam-Skyhwak, H. (2013) Low cost disk filtration provides cheap water for public reuse in Florida. In *Proceedings of WEFTEC 2013*, Chicago, Illinois (pp. 6396–6407). Alexandria, VA: Water Environment Federation, October 5–9, 2013.
Bell, K. Y., Wiseman, L., & Turner, L. A. (2009). Designing pretreatment to control arsenic leaching in ASR facilities. *Journal American Water Works Association, 101*(6), 74–84.
Brown, D. L., & Silvey, W. D. (1973) Underground storage and retrieval of fresh water from a brackish-water aquifer. In J. Braunstein (Ed.), *Underground waste management and artificial recharge* (pp. 379–419), Proceedings of a Symposium held at New Orleans, September, 1973. Menasha, WI: G. Banta Company.
Brown, D. L., & Silvey, W. D. (1977). *Artificial recharge to a freshwater-sensitive brackish-water sand aquifer, Norfolk* (p. 939). Geological Survey Professional Paper: Virginia. U.S.
Camp Dresser and McKee. (2011). *Expanded executive summary, Seminole County, Markham regional water treatment plant, aquifer storage and recovery system, Special Publication SJ2012-SP8*. Palatka: St Johns River Water Management District.
Camp Dresser and McKee. (2012). *Expanded executive summary, City of Sanford Auxiliary (No. 2) Water Treatment Plant aquifer storage and recovery system*. Special Publication SJ2012-SP6. Palatka: St. Johns River Water Management District.
Camprovin, P., Hernández, M., Fernández, S., Martín-Alonso, J., Galofré, B., & Mesa, J. (2017). Evaluation of clogging during sand-filtered surface water injection for aquifer storage and recovery (ASR): Pilot experiment in the Llobregat Delta (Barcelona, Spain). *Water, 9*(4), 263.
CH2M Hill. (2003). *Aquifer storage and recovery pilot project surface facilities design for Lake Okeechobee and Hillsboro Canal sites*. Report prepared for the U.S. Army Corps of Engineers, Jacksonville District. CH2M Hill.
CH2M Hill. (2007). *Arsenic mobilization in two Suwanee Limestone aquifer storage recovery systems*. Final Technical Report submitted to the Southwest Florida Water Management District (August 2007). CH2M Hill.
City of Tucson. (n.d.). About sweetwater wetlands and access. https://www.tucsonaz.gov/water/sweetwater-wetlands. Accessed May 22, 2018.
City or Orlando. (n.d.). Orlando Wetlands Park. http://www.cityoforlando.net/wetlands. Accessed February 24, 2017.
Cleasby, J. L. (1990). Filtration. In F. W. Pontius (Ed.), *Water quality and treatment, a handbook of community water supplies* (4th ed., pp. 455–560). New York: McGraw Hill.
Crittenden, J. C., Trussell, R. R., Hand, D. W., Howe, K. J., & Tchobanoglous, G. (2012). *MWH's water treatment: Principles and design*. Hoboken: Wiley.
Davison, A., Howard, G., Stevens, M., Callan, P., Fewtrell, L., Deere, D., et al. (2005). *Water safety plans: Managing drinking-water quality from catchment to consumer*. Geneva: World Health Organization.
Dillon, P. J., Pavelic, P., Toze, S., Rinck-Pfeiffer, S., Martin, R., Knapton, A., et al. (2006). Role of aquifer storage in water reuse. *Desalination, 188*, 123–134.
Dillon, P., Page, D., Vanderzalm, J., & Chadha, D. (2013). Water quality considerations in managed aquifer recharge: From artificial recharge to managed aquifer recharge in India. *Journal of Groundwater Research, 2*(2), 8–15.
Dillon, P., Vanderzalm, J., Sidhu, J., Page, D., & Chadha, D. (2014). *A water quality guide to managed aquifer recharge in India*. Canberra: CSIRO Land and Water Flagship and UNESCO International Hydrological Program.
Eckmann, D. H., Hahn, W., Morris, K., & Weech, G. N. E. (2004). Improving the performance of ASR wells with CO_2 injection. *Florida Water Resources Journal, 56*(2), 38–40.
Edzwald, J. K. (Ed.). (2011). *Water quality and treatment: A handbook on drinking water* (6th ed.). New York: McGraw-Hill.

Fox, P., Wang, L., Johnson, P. C., Houston, S., Houston, W. N., & Brown, P. (1998). Chlorination for control of biological activity during recharge of tertiary effluent. *Water Science and Technology, 38*(6), 55–62.

Gearheart, R. A., & Finney, B. (1999). The use of free surface constructed wetlands as an alternative process treatment train to meet unrestricted water reclamation standards. *Water Science and Technology, 40*(4–5), 375–382.

Hahn, W., Burgess, K., & Godlewski, V. (2005). Well stimulation using CO_2 injection at the city of Apopka. *Florida Water Resources Journal, 57*(2), 30–31.

Hamadeh, A. F., Sharma, S. K., & Amy, G. (2014). Comparative assessment of managed aquifer recharge versus constructed wetlands in managing chemical and microbial risks during wastewater reuse: a review. *Journal of Water Reuse and Desalination, 4*(1), 1–8.

Hendricks, D., Barrett, J. M., Bryck, J., Collins, M. R., Janonis, B. A., & Logsdon, G. S. (1991). *Manual of design for slow sand filtration*. Denver: AWWA Research Foundation.

Holländer, H. M., Mull, R., & Panda, S. N. (2009). A concept for managed aquifer recharge using ASR-wells for sustainable use of groundwater resources in an alluvial coastal aquifer in Eastern India. *Physics and Chemistry of the Earth, Parts A/B/C, 34*(4), 270–278.

Howe, K. J., Hand, D. W., Crittenden, J. C., Trussell, R. R., & Tchobanoglous, G. (Eds.). (2012). *Principles of water treatment*. Hoboken: Wiley.

Huisman, L., & Wood, W. E. (1974). *Slow sand filtration*. Geneva: World Health Organization.

Ibison, H. A., Sanders, F. A., Jr., Glanzman, R. K., & Drinfield, D. G. (1994). Manganese in recovered water from an ASR well. In A. I. Johnson & R. D. G. Pyne (Eds.), *Artificial recharge of ground water, II: Proceedings of the Second International Symposium on Artificial Recharge of Ground Water* (pp. 539–547). Reston, VA: American Society of Civil Engineers.

ITRC (Interstate Technology & Regulatory Council Wetlands Team). (2003). *Technical and regulatory guidance document for constructed treatment wetlands*. Washington, DC: Interstate Technology & Regulatory Council.

Kadlec, R. H., & Wallace, S. D. (2009). *Treatment wetlands* (2nd ed.). Boca Raton: CRC Press.

Knapp, T., & Tucker, D. (2006). Side-by-side pilot testing of two disk filter manufactures at the City of Merced wastewater treatment plant. Proceedings of WEFTEC 2006, Dallas Texas (pp. 3378–3386). Alexandria, VA: Water Environment Federation, 21–14, October 2006.

Knight, R. L. (1997). Wildlife habitat and public use benefits of treatment wetlands. *Water Science and Technology, 35*(5), 35–43. http://www.cityoforlando.net/wetlands/history/.

Lin, E., Page, D., Pavelic, P., Dillon, P., McClure, S., & Hutson, J. (2006). *Evaluation of roughing filtration for pre-treatment of stormwater prior to aquifer storage and recovery (ASR), Report 03/06*. Canberra: CSIRO Land and Water Science.

Lucas, M. (2007). Aquifer storage and recovery: Technology update: New Jersey AWWA Water Supply and Re-Use Seminar, Newark, New Jersey, October 16, 2007.

Lucas, M., & McGill, K. (1997). Aquifer treatment methods for preventing iron concentrations in the recovered water from aquifer storage and recovery wells in clayey sand aquifers of the Atlantic Coastal Plain. In D. R. Kendall (Ed.), *Conjunctive use of water resources: Aquifer storage and recovery, Proceedings AWRA Symposium, Long Beach, California* (pp. 33–41). Middleburg, VA: American Water Resources Association.

Lucas, M., & McGill, K. (2006). Sustaining development while minimizing groundwater withdrawals by converting existing production wells to aquifer storage and recovery. *Geological Society of America Abstract with Programs, 38*(7), 193.

Lucas, M., McGill, K., & Glanzman, D. (1994). Controlling iron concentrations in the recovered water from aquifer storage and recovery (ASR) wells. In A. I. Johnson & R. D. G. Pyne (Eds.), *Artificial recharge of ground water, II: Proceedings of the Second International Symposium on Artificial Recharge of Ground Water* (pp. 574–587). Reston, VA: American Society of Civil Engineers.

Macia, N. F., & Lluria, M. R. (2001). Development of an extensive well recharge and recovery system for a large water utility. In *Proceedings Tenth Biennial Symposium. Artificial Recharge of Groundwater*, Tucson, AZ, June 2001, pp. 1–10.

Maliva, R. G., Griswold, R. F., & Autrey, M. M. (2013). Prototype for a reclaimed water aquifer storage recovery system benefits and operational experiences. *Florida Water Resources Journal, 65*(3), 54–59.

Missimer, T. M., Hogan, T. W., & Pankrantz, T. (2015). Passive screen intakes: Design, construction, operation, and environmental impacts. In T. M. Missimer, B. Jones, & R. G. Maliva (Eds.), *Intakes and outfalls for seawater reverse-osmosis desalination facilities innovations and environmental impacts* (pp. 79–104). Berlin: Springer.

National Research Council. (2008). *Prospects for managed underground storage of recoverable water*. Washington D.C: National Academy Press.

Noble, C., Horan, T., Brown, E., Shaffer, M., & Lopez, J. (2003). Microfiltration pilot studies for aquifer storage and recovery pretreatment. *Florida Water Resources Journal, 55*(11), 33–37.

Norton, S., Ellison, D., Kohn, S., Arthur, J., Fischler, C., Annable, M., & Harris, W. (2012a). Lessons learned from 10 + years of ASR research: Review of the Bradenton potable ASR degasification pilot project and comparative reactive transport modeling. Abstracts American Water Resources Association, 2012 Annual Water Resources Conference.

Norton, S., Ellison, D., & Kohn, S. (2012b). Minimizing arsenic mobilization during aquifer storage and recovery by source water degasification. abstracts 2012 NGWA Ground Water Summit, May 7–9, 2012, Orlando, Florida.

Page, D., Wakelin, S., Can Leeuwen, J., & Dillon, P. (2006). *Review of biofiltration processes relevant to water reclamation via aquifers, Report 47/06*. Canberra: CSIRO Land and Water Science.

Pavelic, P., Dillon, P. J., Barry, K. E., & Gerges, N. Z. (2006). Hydraulic evaluation of aquifer storage and recovery (ASR) with urban stormwater in a brackish limestone aquifer. *Hydrogeology Journal, 14,* 1544–1555.

Pavelic, P., Dillon, P., Barry, K., Armstrong, D., Hodson, A., Callaghan, J., et al. (2008). *Lessons drawn from attempts to unclog an ASR well in an unconsolidated sand aquifer. Australia*. Canberra: CSIRO Land and Water.

Pearce, M. S., & Waldron, M. (2011). Addressing the mobilization of trace metals in anaerobic aquifers. In *Proceedings 2011 Georgia Water Resources Conference*, April 11–13, 2011. Athens: University of Georgia.

Poole, B. R. Turner, L. A., & Wiseman, L. (2010). Solving the ASR arsenic problem: Pilot-tested approach shows promising results. Abstracts 2010 NGWA Ground Water Summit, April 11–15, 2010. Denver, Colorado.

Poteet, J., Weinstein, B., & Mayhut, J. (2017). Aquifer storage and recovery system enhancement through reduced operating and capital costs. *Florida Water Resources Journal, 68*(2), 30–31.

Randtke, S. J., & Horsley, M. B. (Eds.). (2012). *Water treatment plant design* (5th ed.). New York: McGraw Hill.

Reed, M. G. (1972). Stabilization of formation clays with hydroxyl-aluminum solutions. *Journal of Petroleum Technology, 24,* 860–864.

Reed, M. G., & Coppel, C. P. (1972). *Sand stabilization with hydroxyl aluminum solutions: Society of petroleum engineers of AIME, 43rd annual California regional meeting*. California: Bakersfield.

Seelaus, T. J., Hendricks, D. W., & Janonis, B. A. (1986). Design and operation of a slow sand filter. *Journal American Water Works Association, 78*(12), 35–41.

Slunjski, M., Cadee, K., & Tattersall, J. (2000). MIEX® resin water treatment process. In *Proceedings of Aquatech Amsterdam 2000* (pp. 26–29). The Netherlands.

Small, G. G., Huber, R. D., Gastelum, J., & Ault, C. (2007). Fountain Hills Sanitary District: ASR well rehab—it's a must. In P. Fox (Ed.), *Management of aquifer recharge for sustainability, Proceedings of the 6th International Symposium on Managed Aquifer Recharge of Groundwater* (pp. 412–422). Phoenix: Acacia Publishing.

Stuyfzand, P. J., & Doomen, A. (2004). *The Dutch experience with MARS (Managed Aquifer Recharge and Storage), a review of facilities, techniques and* tools, Water Research Publication 05.001. KIWA.

Stuyfzand, P. J., Wakker, J. C., & Putters, B. (2006). Water quality changes during aquifer storage and recovery (ASR): Results from the pilot Herten (Netherlands) and their implications for modeling. In *Recharge systems for protecting and enhancing groundwater resources, Proceedings of the 5th International Symposium on Management of Aquifer Recharge*, 11–16 June 2005 (pp. 164–173). Berlin, Germany, Paris: UNESCO.

Sultana, S., Ahmed, K. M., Mahtab-Ul-Alam, S. M., Hasan, M., Tuinhof, A., Ghosh, S. K., et al. (2014). Low-cost aquifer storage and recovery: Implications for improving drinking water access for rural communities in coastal Bangladesh. *Journal of Hydrologic Engineering, 20*(3), B5014007.

Torkzaban, S., Vanderzalm, J., Treumann, S., & Amirianshoja, T. (2015). *Understanding and quantifying clogging and its management during injection of CSG water permeates, brines and blends*. Canberra: CSIRO.

USACOE and SFWMD. (2013). *Final technical data report, comprehensive everglades restoration plan, aquifer storage and recovery pilot project, Kissimmee River ASR System and Hillsboro ASR System*. Jacksonville: U.S. Army Corps of Engineers, and West Palm Beach: South Florida Water Management District.

USACOE and SFWMD. (2014). *Central & South Florida Comprehensive Everglades Restoration Plan, draft regional aquifer storage and recovery technical data report*. Jacksonville: U.S. Army Corps of Engineers, and West Palm Beach: South Florida Water Management District.

USEPA. (1999a). *Wastewater technology fact sheet—ozone disinfection (EPA 832-F-99-063)*. Washington DC: United States Environmental Protection Agency, Office of Water.

USEPA. (1999b). *Wastewater technology fact sheet—ultraviolet disinfection (EPA 832-F-99-064)*. Washington DC: United States Environmental Protection Agency, Office of Water.

USEPA. (1999c). *The Class V underground injection control study. Volume 3. Stormwater drainage wells* (September 1999). Washington, DC: USEPA, Office of Ground Water and Drinking Water.

USEPA. (2010). *Constructed wetlands treatment of municipal wastewaters (manual)*. Cincinnati, OH: U.S. Environmental Protection Agency, Office of Research and Development, National Risk Management Research Laboratory.

Visscher, J. T. (1990). Slow sand filtration: Design, operation, and maintenance. *Journal American Water Works Association, 82*(6), 67–71.

Wegelin, W. (1996). *Surface water treatment by roughing filters*. St. Gallen, Switzerland: Swiss Centre for Development Cooperation in Technology and Management.

White, J. R., Ramesh Reddy, K., & Moustafa, M. Z. (2004). Influence of hydrologic regime and vegetation on phosphorus retention in Everglades stormwater treatment area wetlands. *Hydrological Processes, 18*(2), 343–355.

Zhang, X., Li, F., & Zhao, X. (2012). Application of a magnetic resin (MIEX®) in wastewater reclamation for managed aquifer recharge. *Water, Air, and Soil Pollution, 223*(8), 4687–4694.

Zhang, X., Zhao, X., Gao, Y., & Zhang, M. (2015). Combination of MIEX® resin and ozonation for reduction of dissolved organic matter during aquifer recharge using municipal effluent. *Environmental Earth Sciences, 73*(12), 7781–7787.

Chapter 13
ASR and Aquifer Recharge Using Wells

13.1 Introduction

The potential benefits of managed aquifer recharge (MAR) using wells have long been recognized. Artificial recharge experiments using wells were conducted as early as the 1890s by the East London Water Works Company in response to the depressurization of the Chalk and basal sands aquifer in England, although no records of these early experiments were apparently preserved (O'Shea 1994; O'Shea et al. 1995). The "Annotated Bibliography of Artificial Ground Water Recharge Through 1954" prepared by the U.S. Geological Survey (Todd 1959) indicates that artificial aquifer recharge was being widely investigated and implemented by the middle of the 20th century. The main areas for this early research on, and the implementation of, MAR using wells were in southern California, New York (Long Island), Kentucky (Louisville area), and New Jersey, with the goals of preventing, reducing, or reversing aquifer depletion.

MAR systems using wells can be roughly divided into three main classes (with considerable variability within each class and some overlap between classes):

- **Aquifer storage and recovery (ASR):** Injection of freshwater into an aquifer and its later recovery using either the same well or, less commonly, a nearby well
- **Aquifer recharge using wells:** Injection of water into an aquifer to increase aquifer water levels (or pressures) or to improve water quality
- **Vadose wells (also referred to as dry wells):** Water is injected into the vadose zone above the aquifer to be recharged.

There has been growing interest in ASR and groundwater banking over the past several decades as it has been recognized that the great storage capacity of aquifers can be cost-effectively used to manage variations in water supply. ASR does not provide new water but rather allows for the better management of existing water resources. The first successful test of an actual ASR system appears to have been performed at Camp Peary, near Williamsburg, Virginia, in April and May 1946. The test involved storing fresh surface water in an aquifer containing brackish water

R. G. Maliva, *Anthropogenic Aquifer Recharge*,
Springer Hydrogeology, https://doi.org/10.1007/978-3-030-11084-0_13

(Cederstrom 1947, 1957). The Cederstrom test is remarkable because it was successful and identified and addressed some of the basic issues that still impact the design and operation of ASR systems. Cederstrom (1957), for example, described the basis for his 1946 experiment as follows:

> The writer believed that, if fresh water was poured down a well reaching beds saturated with brackish water, a complete mixing of the fresh water with the brackish water would not necessarily result. In the first place, fresh water is less dense than brackish water and would have a tendency to "float" on the heavier water. Furthermore, since the movement of water through the interstices of sandy sediments is extremely slow and turbulence is lacking except in the immediate vicinity of the well screen, the recharge water, regardless of its specific gravity, might tend to push back the ground water, maintaining a rather narrow zone of diffusion between.

The degree of sophistication of ASR and MAR systems varies greatly between countries. In developing and rural areas of newly industrialized countries, aquifer recharge is performed in some areas by allowing excess water to flow into existing wells in an unregulated manner (Shah et al. 2003; Shah 2009; WaterAid in Nepal 2011). Conversely, in the United States and other developed countries, underground injection is strictly controlled through promulgated regulations and a permitting process, which includes water quality requirements for recharged water.

ASR and MAR using wells has been addressed in great detail in dedicated books by Pyne (1995, 2005) and Maliva and Missimer (2010). These volumes review the historical implementation of ASR, summarize experiences at numerous systems, and provide overviews or regulatory, design, and operational issues. A wealth of information on MAR using wells has accumulated (unfortunately not always in a readily accessible manner) since the groundbreaking early studies of Cederstrom and others. The most important factors impacting the performance of ASR and MAR systems using wells are:

- clogging, which impacts the rates of injection and recovery (Chap. 11)
- geochemical processes after recharge, which can impact the quality of stored water (Chap. 6)
- hydrogeology, which impacts the movement and mixing of recharged water, and in the case of ASR, the recoverability of water
- well and wellfield design and operation, which ultimately control the degree to which system performance approaches what is theoretically possible in a given hydrogeological setting.

This chapter addresses the key hydrogeological issues that impact the implementation of ASR and MAR using wells. MAR using wells for groundwater banking is in further addressed in Chap. 14. Treatment-type MAR systems are discussed in Chap. 18.

13.2 Definitions, System Types, and Useful Storage

The storage of water underground in aquifers, where hydrogeologically feasible, has the compelling advantages of enormous storage capacities, minimal surface footprints, and often much lesser costs and environmental impacts than other storage options, such as tanks and surface reservoirs. For the recharge of water using wells (or other means) to be of storage value, it must create an additional water resource that is recoverable in the future and would not otherwise be available. Maliva and Missimer (2008) referred to the recharge of water that creates an actual new or augmented source of recoverable freshwater as "useful storage." It is recognized that MAR can have benefits other than storage. For example, recharge using wells is performed to control saline-water intrusion and can mitigate land subsidence.

Aquifer storage and recovery was originally defined by Pyne (1995) as

> The storage of water in a suitable aquifer through a well during times when water is available, and the recovery of the water from the same well during times when it is needed.

Pyne's definition of ASR includes three main components: (1) water is stored underground, (2) the water is emplaced underground using wells, and (3) the water is recovered using the same well as was used for emplacement (Maliva and Missimer 2010, 2012). Most ASR systems use the same wells for injection and recovery because it is less expensive than constructing separate dedicated injection and recovery wells. Well pumps used for recovery can also be used for well rehabilitation (periodic backflushing). However, in some circumstances, use of dedicated separate injection and recovery wells may have overriding operational advantages. If the injected water is migrating laterally or vertically due to buoyancy, then a greater amount of the stored water may be recovered by having a dedicated recovery well located down gradient from the injection well and/or completed only in the top of the storage zone (Maliva and Missimer 2010).

To broaden the definition of ASR to include situations were injection and recovered is not performed using the same wells, Maliva and Missimer (2010) modified Pyne's definition as

> ASR is the storage of water in a suitable aquifer through a well during times when water is available, and the recovery of the same or similar quality water using a well during times when it is needed.

The essential feature of ASR, as the term is now commonly used, is that ASR involves the local storage of water.

Wells are also used in MAR systems with the goal of increasing aquifer water levels (heads) or arresting or decreasing the decline in water levels from over pumping. Recharge is performed at the site of excess available water or a hydrogeologically preferred location. The recharged water is recovered primarily elsewhere in the aquifer, at points of demand, rather than at the point of injection. The aquifer is used, in essence, to convey water from the point of recharge to the point of use. Strictly speaking, local recharge increases aquifer-wide pressures. Locations distant from the point of recharge benefit from higher aquifer pressures long before the

Table 13.1 ASR system types and useful storage

ASR system types	Useful storage
Physical storage	Injection of freshwater increases the volume of water physically stored in an aquifer
Chemically bounded	Injection of freshwater displaces poorer quality native groundwater
Interface management	Injection of freshwater controls the position of an interface (freshwater-saline boundary) allowing for additional freshwater withdrawals
Regulatory storage	Injection of freshwater confers that the right to later withdraw additional fresh groundwater

actual recharged water physically reaches the recovery site, especially in confined aquifers. Pressure pulses in aquifers travel much more rapidly than the transport of water molecules.

MAR systems using geographically separate injection and recovery wells do not meet either the Pyne (1995) or Maliva and Missimer (2010) definitions of ASR. There is not a widely-used term specific to aquifer recharge using wells. To avoid creating even more jargon, aquifer recharge using phreatic wells is simply referred to as "managed aquifer recharge using wells" or "aquifer recharge using wells."

The use of separate injection and recovery wells may be the preferred strategy were water quality improvement is the primary goal of a system. Rinck-Pfeiffer et al. (2006) defined "aquifer storage transfer and recovery" (ASTR) as the use of separate injection and recovery wells for the purpose of chemical and microbial contaminant attenuation. ASTR has the advantages of providing consistent and longer aquifer residence times, and better takes advantage of the filtration and biogeochemical removal of contaminants provided by aquifers. In the case of ASR systems, the last injected water is the first recovered, and thus can have a much shorter residence time than the first injected (and last recovered) water.

Maliva and Missimer (2008, 2010) noted that there are several distinct types of ASR systems that differ in how they achieve the useful storage of water (Table 13.1). Physical-storage ASR systems achieve the useful storage of water by causing an increase aquifer water levels (heads) that persists until the planned time of recovery. The net increase in storage is the product of the water level increases and aquifer storativity integrated over the aquifer area. Physical-storage ASR systems commonly involve the recharge of freshwater into freshwater aquifers.

Maliva and Missimer (2008) introduced the "the myth of residual pressure," which is the mistaken belief that local recharge into a geographically extensive confined aquifer results in a persistent local increase in pressure. Field experiences (Maliva and Missimer 2010) and modeling results (Maliva 2014) illustrate that once recharge stops, local pressure mounds quickly dissipates in the same manner as the local drawdown cone from pumping quickly (within hours to weeks) dissipates once pumping is terminated (Fig. 13.1). If local aquifer water levels are same after injection as before, then there has been no net physical storage of water. As discussed

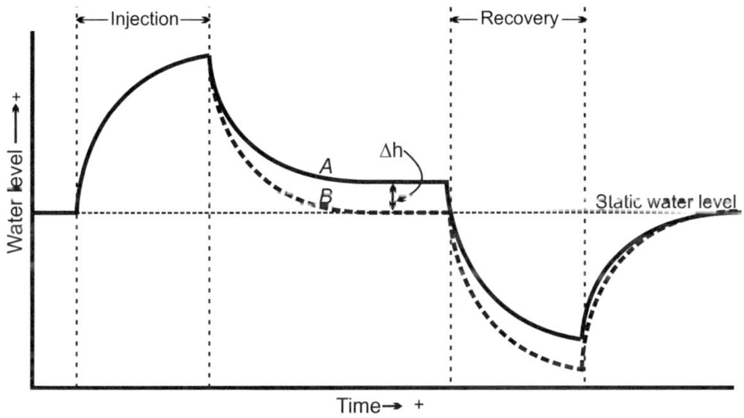

Fig. 13.1 Conceptual diagram illustrating hydraulic responses to injection and recovery. Large dynamic increases in water levels may occur near the site of injection, which dissipate once injection is terminated. An increase in local storage of water occurs only if there is an increase in head (Δh) that persists until the time of recovery (track A). Injection results in no local net storage if water levels recover to static levels after injection is terminated (track B)

by Maliva and Missimer (2010), some ASR systems have been constructed that inject freshwater into freshwater aquifer with the stated purpose of locally storing water, but have failed to meet that objective because there has been no persistent residual pressure increase after injection.

Useful storage is achieved in physical-storage ASR systems only where an aquifer has a limited geographic extent and the recharged volume materially increases aquifer-wide water levels (Fig. 13.2). Perhaps the best example of a successful physical-storage ASR system is the Las Vegas, Nevada system, which utilizes a largely closed intermontane basin as a storage zone (Maliva and Missimer 2010). Physical storage might also be achieved in some unconfined aquifers with a low hydraulic diffusivity (transmissivity/storativity), which results is lower dissipation rates of head changes. Physical-storage ASR system using closed basins and aquifer recharge using wells systems are hydrologically the same, but differ operationally depending upon whether the wells used for recharge are also used for recovery. In practice, this distinction is blurred. For example, the Las Vegas Nevada ASR system has both dual-use wells and dedicated recharge wells, and other aquifer uses benefit from the higher water levels.

A common type of chemically-bounded ASR system stores freshwater in brackish aquifers (Fig. 13.3). The injected freshwater flows radially outward from the injection well, displacing and partially mixing with the ambient native groundwater in the storage aquifer. A transition or mixing (buffer) zone develops between the injected and ambient waters. During recovery, the stored water, mixing zone, and ambient water are drawn back toward the ASR well. Useful storage is achieved if at least

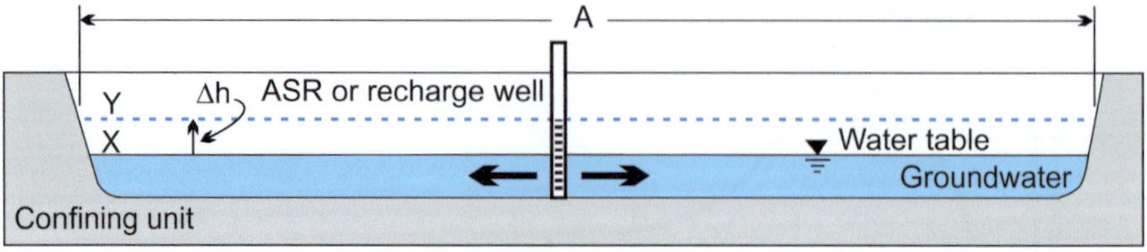

Fig. 13.2 Conceptual diagram of a physical storage-type ASR system. The amount of stored water (ΔS) is the product of the change in water level (Y−X, ΔS), the aquifer area (A), and aquifer storativity (S)

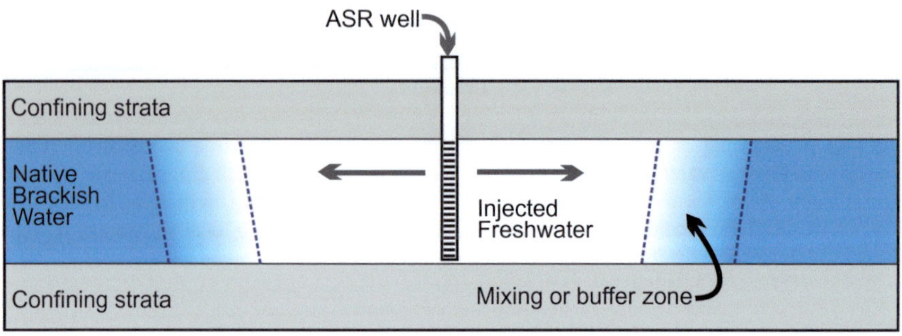

Fig. 13.3 Conceptual diagram of an ASR system using a brackish storage zone. Injection of freshwater laterally displaces the native brackish groundwater. Stored freshwater and native brackish water are separated by a buffer or mixing zone. During recovery (not shown) the mixing zone and native groundwater are drawn back toward the ASR well

some of the injected freshwater can be recovered at a quality suitable for its intended use.

Interface-management ASR systems are an uncommon class of systems in which local aquifer water quality is imperiled by saline-water intrusion or, theoretically, the movement of other water quality interfaces. At locations near the coastal saline-water interface, purely extractive groundwater pumping may cause saline-water intrusion and would thus not be sustainable. However, periodic injection of freshwater may allow for sustainable use of the aquifer by balancing extractions. The Wildwood, New Jersey, ASR system, which is the oldest operating ASR system in the United States, is an example of an interface-management ASR system (Maliva and Missimer 2010).

The last common type of ASR system is regulatory-storage systems in which injection of water confers the right to later pump additional groundwater, which would not otherwise be allowed. Depending upon the jurisdiction, the system owner may either obtain a 100% credit for injected water or a reduced credit to account for lost water or for the benefit of the aquifer. The rationale behind regulatory storage is that the injection and recovery of water does not adversely impact long-term aquifer water levels (i.e., contribute to over exploitation). However, the operation of regulatory storage ASR systems can still have adverse water resources and environmental impacts by locally increasing groundwater withdrawals during dry periods. Environmental impact assessments need to consider the geographic and temporal impacts of both injection and recovery. Regulatory-storage systems, which are also referred to as groundwater banking systems, are further discussed in Chap. 14.

13.3 Recovery Efficiency

Inasmuch as ASR is primarily a storage technology, the performance of ASR systems is typically evaluated in terms of the fraction of recharged water that can be recovered for use when needed at a quality suitable for its intended used. The most widely used measure of ASR system performance is recovery efficiency (RE), which was defined by Kimbler et al. (1973) as the quantity of water recovered divided by quantity of water injected:

$$\mathrm{RE}(\%) = 100\left(\mathrm{V}_{\mathrm{rec}}/\mathrm{V}_{\mathrm{inj}}\right) \tag{12.1}$$

where

V_{rec} volume of water recovered at a useable quality (m^3 or gal)
V_{inj} volume of water injected (m^3 or gal)

Recovery efficiency can be calculated over the entire operational history of a system (system recovery efficiency; SRE) or over an individual operational (recharge and recovery) cycle (operational recovery efficiency; ORE) (Sheng et al. 2007). ORE tends to improve over time (operational history) as the native groundwater is flushed from the vicinity of ASR wells by repeated operational cycles.

13.3.1 RE of Chemically Bounded (Brackish or Saline Aquifer) ASR Systems

Recovery efficiency is usually used to evaluate the performance of ASR systems that store freshwater in brackish or saline aquifers. For systems used for potable supply, the upper salinity limit for recovery in the United States is commonly the secondary drinking water standards of 250 mg/L for chloride or 500 mg/L for total dissolved solids (TDS). Recovery of water with higher salinities may be acceptable for non-potable uses or if the recovered water is blended with lower salinity waters. Salinity-based recovery efficiency values depend upon the amount of the recharged water that is actually recovered and the degree of blending of recharged water and native groundwater that is acceptable, which depends largely upon native groundwater salinity. For example, if 100 mg/L chloride water is recharged into a storage zone containing 400 mg/L chloride water, then a blend containing 50% native groundwater would meet the 250 mg/L secondary drinking water standard. However, if the same water were stored in a zone containing 4,000 mg/L of chloride, then only a 3.8% blend of native groundwater would meet the standard. Hence, low-salinity storage zones are very favorable for high recovery efficiencies. Recovery efficiency based on water quality is not meaningful where the native groundwater is of useable quality and, therefore, RE would necessarily always be 100%. RE could theoretically also be defined in terms of other, non-salinity-related parameters that impact the usability of recovered water.

Non-recoverability of a portion of stored water (i.e., a recovery efficiency of less than 100%) is construed by some of the general public (non-experts) as evidence that ASR does not work, is unreliable, or is wasteful. Unless there is only a small difference between the quality (e.g., salinity) of the injected water and native groundwater, recovery efficiencies will usually be less than 100% in brackish and saline aquifers (Maliva and Missimer 2010, 2012; Zuurbier 2015). A reasonable long-term ORE target for ASR systems using brackish aquifers as storage zones is 70–80%. OREs of 100% may be achieved where a storage zone is mildly brackish and blends of recharged and native groundwater meet water quality thresholds for use. OREs of 100% may also be achieved in some systems were scavenger wells are used to intercept the upward migration of more saline waters (Freshmaker systems; Zuurbier 2015). Setting unrealistic expectations, such as a 100% RE, can result in systems that are performing well (e.g., have a 70–80% RE), as allowed by local hydrogeological conditions, being improperly perceived as underperforming (Maliva and Missimer 2010).

RE efficiency depends upon both storage zone hydraulic and transport properties and water quality. Additionally, RE depends upon the operation of ASR systems. RE values depend upon net stored water volumes. Indeed, a strategy to increase RE is to first condition the storage zone by injecting a large volume of freshwater (target storage volume approach; Sect. 13.4). Long storage periods between recharge and recovery tend to result in lower RE values for a given stored water volume.

13.3.2 RE of Physical-Storage ASR Systems

The RE of physical-storage and interface-management systems cannot be defined in terms of water quality parameters, as freshwater is usually recharge into a freshwater aquifer. The storage benefits of physical-storage ASR systems necessarily must be defined in terms of the amount of net additional water in storage at the time of recovery. Maliva and Missimer (2010) proposed two manners for evaluating the hydrological performance of physical storage ASR systems. RE (fractional) can be defined as the ratio of the volume of water recovered until aquifer heads return to pre-injection (background) levels (V_{rec}) to the volume of injected water (V_{inj}):

$$RE = \frac{V_{rec}}{V_{inj}} \tag{12.2}$$

Alternatively, system performance can be quantified in terms of the net storage ratio (NSR), which is the ratio of the increase in aquifer storage (V_s) to the injected volume:

$$NSR = \frac{V_s}{V_{inj}} \tag{12.3}$$

where the increase in aquifer storage is estimated as follows:

$$V_s = A \cdot S \cdot \Delta h \tag{12.4}$$

and

A aquifer surface area (L^2)
S storativity (dimensionless)
Δh change in aquifer heads (L)

Changes is water storage can be estimated through groundwater modeling or from pre- and post-recharge potentiometric surface maps. If a persistent increase in aquifer heads does not occur, then the RE and NSR are necessarily zero. The term "persistent" incorporates a time element into the evaluation of system performance in that the head changes that are of concern are those that persist until the time of recovery. Groundwater modeling conducted in the feasibility stage of projects should be able to evaluate the magnitude, geographic extent, and persistence of water level (pressure) buildups resulting from MAR.

As further addressed in Chap. 14 with respect to groundwater banking, evaluating the storage benefits of ASR and aquifer recharge becomes more complicated when aquifer heads are already changing in response to factors other than the operation of the ASR or recharge system. There are definite system benefits if recharge offsets water level declines that would otherwise occur from overdraft.

13.3.3 RE of Regulatory Storage

The concept of recovery efficiency has little meaning for regulatory-storage ASR systems because the basic premise of the system type is that injection, or recharge of water by other means, confers the right to recover 100% of the water minus any local regulatory water "tax." (Maliva and Missimer 2010). From an owner/operator perspective, regulatory storage is a very attractive concept. ASR systems are typically not proposed as regulatory-storage ASR systems, but are rather described in physical-storage terms, e.g., recharge will increase aquifer water levels and thus avoid impacts from additional groundwater pumping during dry or high demand periods. As reported in a review of historical ASR implementation by Maliva and Missimer (2010), in the absence of local pressure or water level increases that persist until the time of recovery, some purported physical storage ASR systems are de facto regulatory storage systems.

13.4 Aquifer Conditioning and Target Storage Volume

Aquifer conditioning is the initial injection of a large volume of water at the start of operation of an MAR system to improve the RE of subsequent operational cycles. Aquifer conditioning appears to have been first proposed with respect to aquifer thermal energy systems (Schaetzle et al. 1980). Injection of either hot or cold water (depending upon whether the system was operated for heating or cooling) would increase or lower the temperature of the aquifer solids (rock or sediment) and allow subsequent operational cycles to operate at close to 100% recovery efficiency.

Pyne (2005) applied the concept of aquifer conditioning to ASR systems and introduced the term "target storage volume" (TSV), which is defined as the volume of water required to be emplaced in an ASR system so that projected water demands during recovery can be met while meeting flow, volume, and water quality goals with an acceptable level of reliability. Initial emplacement of the TSV would allow all subsequently injected water to be recovered at a high (approaching 100%) recovery efficiency (Pyne 2005). The TSV includes the stored freshwater volume needed for both recovery during an operational cycle and to establish the surrounding buffer zone, which separates the stored water from ambient groundwater. It is recognized that where the native groundwater in the storage zone has a relatively high salinity, and thus a significant density difference exists between stored and native groundwater, buoyancy-driven water movements will limit RE to values below 100% irrespective of the emplacement of a TSV. Nevertheless, the TSV strategy is attractive for ASR systems whose hydrogeology is favorable for high recovery efficiencies because it allows the system to quickly reach a long-term operational mode. The earlier practice of conducting numerous short-term injection and recovery (cycle tests) provided a lot of data, but delayed achieving a final high-recovery operational mode. The emplacement of a TSV should be considered part of the initial construction cost for an ASR system.

Determination of the TSV of an ASR system prior to any recharge is at best an educated guess, and the hydrogeological conditions of some systems may limit potential recovery efficiency to a value well below 100%. Experiences from other hydrogeologically similar sites can provide some guidance. Groundwater (solute-transport) modeling can also provide guidance on the size of the TSV, but such modeling tends to be poorly constrained in the absence of some operational data for model calibration.

The recommended approach is to perform an initial cycle test with a duration of perhaps 30 to 60 days, which would allow for the evaluation of geochemical changes during storage and clogging processes, testing of equipment, and provide some operational data for model calibration. Predictive simulations can then be performed to determine the approximate TSV and system recovery efficiency. If the initial TSV volume turns out to be too small, than it can be increased by additional recharge.

Injection of very large volumes of water cannot overcome adverse hydrogeological conditions that are unfavorable for acceptable recovery efficiencies. If hydrogeological conditions at a site are unfavorable for ASR, then the TSV volume may be largely lost. Modeling calibrated to an initial cycle test can provide valuable insights on eventual system operational performance.

13.5 Controls on RE in Brackish or Saline Aquifer ASR Systems

When freshwater is stored in brackish or saline aquifers, water become unrecoverable due to mixing with native groundwater and movement driven by density differences (buoyancy) or local hydraulic gradients. The ultimate controls over recovery efficiency can be divided into three categories:

- aquifer hydrogeological and hydraulic conditions
- aquifer chemical conditions (including salinity)
- design and operational issues.

Buoyancy stratification results in the upward and outward movement of stored freshwater and the migration of saline water into the bottom of a storage zone (Fig. 13.4). The latter can drastically impact the quality of water recovered in wells open to the entire thickness of the storage zone. Buoyancy stratification is controlled by:

- density difference (salinity of the storage zone)
- vertical hydraulic conductivity
- elapsed time since recharge.

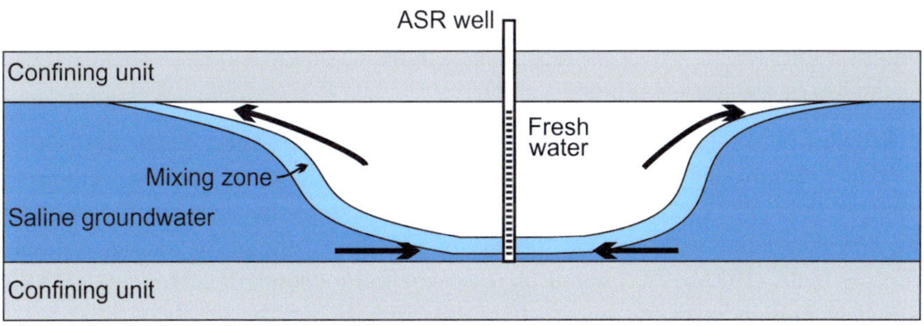

Fig. 13.4 Conceptual diagram of buoyancy stratification. Less dense freshwater will tend to rise and spread out beneath overlying confining strata. Denser native groundwater will flow toward the ASR well at the bottom of the storage zone

Numerous studies have been performed on the effects of various hydrogeological and operational variables on the recovery efficiency of ASR systems that store freshwater in brackish aquifers. The studies differ in the baseline hydrogeological conditions considered, particularly the salinity of the storage zone, and whether the simulations were density dependent. Follows are summaries of many of the important investigations of the controls over ASR system performance that can provide useful insights to guide site and aquifer selection and feasibility assessments of ASR projects. The studies are presented in chronological order to provide a historical perspective on the development of ASR.

13.5.1 Louisiana State University Studies

Louisiana State University conducted pioneering studies on the feasibility and controls of the storage of freshwater in saline aquifer ("salaquifers"). The studies included both computer modeling and physical modeling using laboratory mini-aquifers constructed of blasting sand fused with epoxy resin (Kimbler et al. 1975). Recovery efficiency was found to be effected by mixing (diffusion and dispersion) and gravity segregation (Esmail and Kimbler 1967). For saline storage zones, gravity segregation was found to be much more detrimental to freshwater recovery than mixing. Gravity segregation (tilting of the saline-freshwater interface) is a function of hydraulic conductivity and the density difference between waters, and decreases with the development of a mixing zone.

Kimbler et al. (1973, 1975) identified the key variables controlling recovery efficiency:

- dispersion coefficient
- difference in density
- rate and duration of injection, storage, and recovery
- aquifer dip
- direction and rate of groundwater flow under undisturbed conditions
- homogeneity of aquifer
- aquifer thickness
- duration of the storage period.

Kimbler et al. (1975) proposed that the movement of stored water due to pre-existing groundwater movement (i.e., ambient potentiometric gradient) could be managed through the creation of a zone of stagnant water by strategically pumping and injecting saline water (using "bounding wells") outside of the storage volume. Kimbler et al. (1973, 1975) noted the overwhelming economic advantage of ASR over steel storage tanks.

13.5.2 U.S. Geological Survey Miami-Dade ASR Investigation

Merritt (1985, 1986) performed modeling studies of the hydrogeological controls over recovery efficiency based on an ASR test site in Hialeah, Miami-Dade County, Florida. The modeling results provided basic insights into the hydrogeological controls over recovery efficiency in systems using brackish aquifers as storage zones. The modeling was performed using the SWIP (Subsurface Waste Injection Program) code (INTERCOMP Resource and Development 1976; INTERA Environmental Consultants 1979). The ASR test aquifer was 150 ft (45.8 m) thick and was divided into five layers, each assigned a percent of the total flow based on flowmeter log data. Merritt addressed the potential impacts of numerical dispersion on simulated recovery efficiencies. The main results of Merritt's investigation are:

- Buoyancy stratification has a variable impact on RE, which depends on both the density difference between the injected water and ambient groundwater, and permeability. As the density difference increases, RE decreases for a given permeability and low permeabilities are required to achieve an acceptable recovery efficiency.
- Longitudinal dispersivity has a very strong effect on RE in the 2–30 ft (0.6–9.0 m) range.
- Native groundwater salinity has a very strong effect (inverse relationship) on RE.
- Porosity has only a modest impact on RE. An approximately 7% change in simulated RE (64–71%) occurred over a 20–50% porosity range.
- Increasing the thickness of the most permeable zone from 12 to 48 ft (3.7–14.6 m) resulted in a modest (6%) decrease in RE.
- The importance of hydraulic gradient and storage period depends of the diameter of the freshwater mass relative to the downgradient movement. Increased storage time allows for more movement to occur.
- Large increases in RE occurred with increasing injected volume for small volumes. The simulated rate of increase in RE with volume was small at large volumes ($\geq$5 million ft^3, $\geq$141,600 m^3).
- Simulations of multiple cycles with recovery terminating at a chloride concentration of 250 mg/L showed major improvement in RE over the first several cycles. The improvement in RE continued at a reduced rate in later cycles.

Merritt (1985, 1986) modeled the effects of different well configurations and schedules of operation on the RE of multiple well ASR systems. The highest simulated RE was for sequential injection with a centered configuration (central well surrounded by peripheral wells) operated as follows:

- injection first in the central well
- starting injection at the peripheral wells when the injected water reaches the wells and injection is performed at the same rate as the central well
- equal withdrawal rated for all wells
- peripheral wells are shut in once their chloride concentration reaches 250 mg/L (drinking water standard).

13.5.3 USGS Cape Coral, Florida, ASR Modeling

Quiñones-Aponte and Wexler (1995) performed theoretical modeling of the potential effects of hydrogeology on a proposed ASR system in Cape Coral, Florida. The modeling was performed using a modification of the SUTRA code (Voss and Provost 2002). The baseline simulation was the injection of 49,055 m^3 (12.96 Mg) of water with a chloride concentration of 50 mg/L into an aquifer with a 500 mg/L concentration. Recovery efficiency was quantified using a threshold of 250 mg/L of chloride. The simulation results indicate that RE is sensitive to storage zone salinity, decreasing from 64% at a chloride concentration of 500 mg/L (baseline scenario) to 22% at a concentration of 2,000 mg/L. A counter-intuitive result was that RE decreased with injected volume. This result appears to be due to the system being able to accommodate considerable mixing of injected water and native groundwater with the blend still meeting the 250 mg/L of chloride threshold. A limitation of the Quiñones-Aponte and Wexler (1995) modeling investigation is that injected volumes were at least an order of magnitude less than volumes that would be injected in an actual ASR system.

13.5.4 CDM Missimer SEAWAT Modeling of Effects of Flow Zones

Missimer et al. (2002) simulated the effects of preferred flow zones on the movement and mixing of freshwater injected into brackish and saline aquifers using the U.S. Geological Survey (USGS) SEAWAT code (Guo and Langevin 2002). Flow zones (corridors) allow the freshwater plume to travel laterally much further than would occur in a uniform aquifer. The density-dependent simulations showed that injected freshwater that reaches a vertical flow corridor will move upward while ambient saline water above the corridor sinks downward. An unstable "fingering flow" occurs as a consequence of the unstable fluid stratigraphy. As the storage periods proceeds, saline water locally fills the flow corridor, both isolating bodies of non-recoverable freshwater and mixing with the freshwater. Isolation of freshwater and mixing both act to reduce RE. The modeling results indicate that if the TDS concentrations of a storage zone exceeds 20,000 mg/L, then it is extremely difficult to achieve a reasonable RE.

13.5.5 Clare Valley (South Australia) Fractured-Rock ASR Tracer Testing

Harrington et al. (2002) documented the results of tracer tests performed at two fractured-rock ASR systems (Wendouree Winery and Watervale Oval) in the Claire

Valley of South Australia. The tracer recovery after the recovery of 3–4 times the injected water volume was 46% (helium) and 48% (bromide) for the Wendouree Winery test, and 46% (fluorescein) and 78% (bromide) for the Watervale Oval test. The results of the tracer testing and experiences at other fractured-rock ASR systems in South Australia indicate that injected water will rapidly mix with native groundwater and be transported far from the ASR (injection) well due to the naturally high flow velocities in fractures. The RE in fractured and other dual-porosity ASR systems may, therefore, be much lower than that for single-porosity media (Harrington et al. 2002).

13.5.6 Maliva et al. (2005) Theoretical SEAWAT Modeling

A model of a hypothetical ASR system was developed using the SEAWAT code (Guo and Langevin 2002) that had 18 layers, of which 10 simulated a 100 ft (30.5 m) thick storage zone (Maliva et al. 2005). The model also included underlying and overlying confining units and aquifers. The simulations incorporated the target storage volume (TSV) approach in that the aquifer was conditioned by the initial recharge of 300 million gallons (MG; 1.14×10^6 m^3) of freshwater. The TSV emplacement was followed by a 360-day storage period and then a 200 Mg (757,000 m^3) operational cycles. The injected water had a TDS concentration of 200 mg/L and RE was calculated based on recovery of water to a TDS concentration of 500 mg/L. The baseline storage zone TDS concentration was 3,000 mg/L. The effects of varied hydraulic and transport parameters were as follows:

- Storage zone salinity had a strong effect on RE, which ranged from 87% for a TDS concentration of 1,500 mg/L to 11% for 12,000 mg/L (Table 13.1).
- Longitudinal dispersivity has a strong effect on RE at very high values. Recovery efficiency ranged from 78% at 10 ft (3.05 m) to 10% at 100 ft (30.5 m).
- Aquifer heterogeneity has a strong negative effect on RE. RE ranged from 46 to 65% (depending on hydraulic conductivity) for uniform conditions to 6–9% where 90% of flow was concentrated in one layer.
- Confining zone leakance had a moderate effect on RE over a great range of values. RE ranged from 35–40% for a leakance of 0.1 day^{-1} to 59–88% for a leakance of 0.0001 day^{-1}.
- Effective porosity had only a moderate effect on RE. Low porosity values ($\leq$10%) are unfavorable.
- Storage zone hydraulic conductivity had a minor effect on RE.

The modeled recovery efficiencies for different TDS and dispersivity values are provided in Table 13.2. Dispersivity is a measure of aquifer heterogeneity. Relatively homogeneous siliciclastic aquifers tend to have low dispersivity values, whereas highly heterogeneous aquifers (e.g., karstic or fracture dominated carbonate aquifers) have very high values. Table 13.2 indicates that high recovery efficiencies may be

Table 13.2 Effects of TDS and dispersivity of RE in ASR systems using brackish aquifers

Storage zone total dissolved solids				
Concentration (mg/L)	1,500	3,000*	6,000	12,000
Recovery efficiency (%)	87	46	23	11
Dispersivity				
Value (m (ft))	3 (10)	9 (30)*	30 (100)	90 (300)
Recovery efficiency (%)	78	46	10	0

*Baseline simulation values
Source Maliva et al. (2005)

expected in relatively homogenous aquifers with low storage-zone salinities. However, recovery efficiencies will be very poor as conditions move to the right (high salinities and high degrees of heterogeneity) on the table. Review of the historic performance of ASR systems supports the modeling results of high degrees of aquifer heterogeneity being a characteristic of some ASR systems that have had a very poor RE (Maliva and Missimer 2010).

13.5.7 Brown Doctoral Dissertation (University of Florida)

Brown (2005) performed theoretical modeling of the impacts of various hydrogeological and operational variables on RE using a model based on the Hillsboro ASR Pilot Project site in Palm Beach County, Florida. The modeling was performed using the MODFLOW (McDonald and Harbaugh 1988) and MT3DMS (Zheng and Wang 1999) codes and, in a limited number of simulations, the SEAWAT (Guo and Langevin 2002) code. Brown (2005) concluded that for ambient TDS concentrations of less than 4,000 mg/L and minimal storage duration, density independent predictions are practically the same as those including density. The baseline simulation was the recharge of water with a chloride concentration of 5 mg/L into an aquifer with an ambient chloride concentration of 1,000 mg/L. Recovery stopped once a chloride concentration of 250 mg/L was reached. The simulation results are summarized as follows:

- Dispersivity had a very strong effect on RE. Increasing dispersivity values (10 s of ft or m) results in greatly reduced RE.
- Storage zone thickness has a modest effect on RE. Thinner storage zones favor higher recovery efficiencies.
- Increasing recharge volume improved recovery efficiency but eventually a point of diminishing returns is reached (at about 10 Mg or 37,850 m^3 in these simulations).

- Increasing the local hydraulic gradient to about 0.0001 adversely impacted simulated RE.
- SEAWAT simulations indicate RE is reduced with longer storage times.
- Storage zone transmissivity had a minimal effect on RE.
- Degree of anisotropy had a minimal effect (density effects were not simulated).
- RE in insensitive to porosity in 25–40% range. A 10–12% reduction in RE occurs at low porosities (e.g., 5%).

Multiple-cycle simulations in which each cycle is terminated once a chloride concentration of 250 mg/L is reached showed a progressive increase in RE that asymptotically approaches a constant value that depends upon ambient salinity and dispersivity. Lateral heterogeneity (low and high transmissivity zones near the ASR well) resulted in a simulated reduction in RE in the early operational cycles.

Brown (2006) simulated the effects of operational cycles on ASR system performance. End member scenarios simulated were progressive emplacement of water over multiple recharge and recovery cycles ("pore volume approach") and emplacement of water in a massive single cycle ("TSV approach"). Modelling results (MODFLOW and MT3DMS) indicate that the difference in performance among operational schemes is minor after many years of multiple ASR cycles. Injecting a large volume of water at the start of operations has the advantage of more quickly achieving high operational recovery efficiencies although the cumulative recovery efficiency may be less. The effectiveness of the TSV was found to dissipate over time.

13.5.8 Theoretical Modeling of ASR in Aquifer Types of Wisconsin

Lowry and Anderson (2006) investigated the controls of RE by theoretical modeling based on three storage types found in Wisconsin: confined sandstone aquifer, glacial drift sand-and-gravel aquifer, and fractured-dolomite aquifer. The simulations were performed using MODFLOW, MODPATH (Pollock 1994) and MT3DMS. The simulations did not investigate density-dependent stratification. The mixing of waters was simulated by assigning the injected water a concentration of 0 and the ambient groundwater a concentration of 1000. The main results of this investigation are:

- Advection only simulations had a 30% higher RE than advection-dispersion simulations.
- Longitudinal dispersity has a large effect on RE, which decreases with increasing dispersivity values.
- Simulations were sensitive to hydraulic gradient for small injected volumes and low effective porosity.
- Aquifer heterogeneity had a modest effect on RE. Increasing the hydraulic conductivity of one layer initially resulted in a small increase in RE and then a decrease.
- Increasing injected volume increased RE, which asymptotically approached a maximum value.

13.5.9 USGS Review of ASR System Performance in South Florida

Reese (2002) reviewed the performance of ASR systems in South Florida that use brackish aquifers as storage zones, and proposed that the following hydrogeologic characteristics of storage zones are important for recoverability:

- ambient salinity (buoyancy and dispersive mixing)
- aquifer permeability
- aquifer thickness
- confinement
- ambient hydraulic gradient
- structural setting.

Recovery efficiency is maximized with relatively thin storage zones, moderate transmissivities (less than about 30,000 ft^2/day; 2,800 m^2/d) and moderate ambient salinities (chlorides less than 3,000 mg/L; Reese 2002). Reese (2002) also noted that structural setting may be important because of the potential for up-dip migration of stored water. Structural lows, areas of structural complexity, and areas with high dips should be avoided (Reese 2002).

Reese and Alvarez-Zarikian (2007) subsequently compared the performance of ASR systems in South Florida with respect to four hydrogeological and design factors: storage zone thickness, transmissivity, and ambient chloride concentration of the storage zone, and the thickness of the portion of the aquifer above the top of the storage zone. Threshold values of 150 ft (46 m), 30,000 ft^2/d (2,800 m^2/d), 2,500 mg/L, and 50 ft (15 m) respectively, were chosen for these factors to represent the approximate values above which RE could be adversely affected.

Increased permeability and transmissivity in carbonate aquifers, such as the Upper Floridan Aquifer of South Florida, typically translate to greater dispersive mixing with brackish and saline ambient ground water. High values for storage zone thickness could result in decreased RE because of the greater vertical extent of the transition zone along which mixing occurs and increased potential for dispersive mixing. An aquifer thickness above the top of the storage zone of more than 50 ft (15 m) could lower RE, depending on the vertical hydraulic conductivity of the aquifer and ambient salinity. The buoyancy of the injected freshwater in saline ambient groundwater could cause part of the freshwater "bubble" to migrate above the level of the top of the storage zone (base of casing) where it may be more difficult to recover (Reese and Alvarez-Zarikian 2007).

13.5.10 Dual-Domain Simulations

Dual-porosity systems contain an immobile and at least one mobile domain. The immobile domain is the relatively low hydraulic conductivity aquifer matrix that constitutes the bulk of the aquifer. The mobile domain is flow zones (e.g., fractures, dissolution conduits, and very permeable beds) that have relatively high hydraulic conductivities but constitute a small percentage of the total aquifer volume. Solute-transport is controlled by advection and dispersion in the mobile domain and mass transfer (mostly by diffusion) between the mobile and immobile domains. Dual-domain conditions can adversely impact the performance of ASR systems if poor quality (e.g., more saline) water in the immobile domain diffuses into freshwater stored in the mobile domain. Mass transfer between domains might be evidenced by salinity rebounds during storage (Culkin et al. 2008).

In an early solute-transport modeling study by Gale et al. (2002), matrix porosity was found to have a great impact on system performance as far as the recovery of freshwater. Large matrix porosities result in a large volume of ambient water that needs to be flushed out of the system.

Culkin et al. (2008) developed a calibrated solute-transport model of a short-term (11 day) ASR cycle test performed at Charleston, South Carolina, using the MODFLOW and MT3DMS codes. The chemical reaction package of MT3DMS (Zheng and Wang 1999) was used to simulate solute transfer between two porosity domains, which were conceptualized as a fracture network containing mobile water in which advective transport and dispersion occurs and an immobile domain (aquifer matrix) where advective transport does not occur. The rate of mass transfer between the domains was determined by flow velocity, permeability, diffusive length scale, and molecular diffusion (Culkin et al. 2008). Transfer between the domains was incorporated into the model through the "solute exchange rate coefficient" (β), which has the unit of time^{-1}. Fractures were represented in the model using the continuum approach by increased hydraulic conductivity (above the matrix value) and a low mobile phase porosity (f_m) rather than the aquifer total porosity (f_t).

Culkin et al. (2008) were able to obtain a better calibration, particularly for the storage period, using the DDM model than was possible using a single-domain model. Sensitivity analysis results indicate that simulated water quality is sensitive to the value of β and the f_m/f_{im} ratio, where f_{im} is the immobile phase porosity. A key factor affecting stored-water quality is whether the volume of immobile water present is large enough to produce a salinity rebound. Limitations of the DDM approach are that aquifers can have multiple domains and variable rates of exchange between domains (i.e., non-constant β values; Culkin et al. 2008). The MODFLOW/MT3DMS simulations also did not consider buoyancy stratification, which was believed to not be a significant process for storage zones with TDS concentrations below 5,000 mg/L (Culkin et al. 2008).

Lu et al. (2011) examined the effects of mass transfer in a dual-domain system. The key variables found to affect RE are the capacity ratio (ratio immobile to mobile domain porosity) and the mass transfer time scale T_{im}, which is defined a 1/α, where α is a first-order mass transfer rate coefficient. RE decreases with increasing capacity ratio (size of immobile domain). In aquifers with a large capacity ratio and slow mass transfer, the immobile phase may serve as a long-term contaminant source that has a negative impact on ASR RE. With increasing operational cycles, the immobile domain will eventually cease to be a significant contaminant source.

13.5.11 Aquifer Heterogeneity Simulations

Aquifer heterogeneity, specifically variations in transmissivity and porosity between layers, can impact the vertical distribution and horizontal extent of recharged water, and the amount of mixing between recharged water and native groundwater. Extreme aquifer heterogeneity, resulting in the concentration of most of the flow of recharged water into a thin zone, was identified as the primary cause of the very poor RE experienced in some South Florida ASR systems (Maliva and Missimer 2010).

Pavelic et al. (2006) investigated the effects of aquifer heterogeneity on ASR RE at the intensively studied Bolivar ASR facility in South Australia through the development of a calibrated solute-transport model using the FEFLOW code (Diersch 1998). This modeling study is noteworthy because of the large number of monitoring points (16) available for model calibration. The injection zone (T2 Aquifer) was subdivided into four main layers based on electromagnetic flowmeter, dissolved chloride breakthrough, and temperature data.

The results of the Pavelic et al. (2006) study demonstrated that the large dispersivity values estimated in calibrated solute-transport models are largely the product of macroscopic-scale aquifer heterogeneity (Maliva et al. 2006). Incorporation of the four layers into the solute transport model resulted in substantially lower dispersivity values being needed during model calibration than for the single-layer simulation.

Guo et al. (2014) simulated the effects of aquifer heterogeneity on ASR system performance using a SEAWAT model of hypothetical 30.5-m thick storage zone divided into 10 layers. Simulations were performed for a single operational cycle in which 200 mg/L TDS water was stored in an aquifer containing 3,000 mg/L TDS water (similar to some actual ASR systems in Florida). The storage zone was simulated as being either homogenous or with 50 or 90% of transmissivity (and thus flow) concentrated in one layer (Table 13.3).

The simulation results show that aquifer heterogeneity is unfavorable for recovery efficiency. Greater heterogeneity results in a greater contact area and mixing of recharged freshwater with native groundwater. The effects of dual-porosity conditions were simulated using the oilfield simulator Eclipse. The simulation conditions were 200 mg/L TDS water stored in a 5,000 mg/L TDS aquifer. The dual-porosity condition simulation partitioned 10% of porosity and 90% of the transmissivity into

Table 13.3 Simulated of aquifer heterogeneity on RE values using the SEAWAT code

Transmissivity and flow distribution	Recovery efficiency (to TDS = 500 mg/L)		
	Average K = 3.05 m/d	Average K = 30.5 m/d	Average K = 305 m/d
Uniform	46	61	65
50% in one layer	38	43	44
90% in one layer	6	9	10

Source Guo et al. (2014)

fractures. Recovery efficiency was 46% under single-porosity conditions and only 24% under the simulated dual-porosity conditions. Dual-porosity conditions resulted in a greater flow velocity and wider mixing zone. Runs using a higher background salinity (20,000 mg/L TDS) show a greater upward movement of freshwater due to the buoyancy effect in fractures.

13.5.12 Short-Circuiting and ASR RE

Short-circuiting occurs when confining strata, usually below the storage zone, contain zones of enhanced vertical hydraulic conductivity that allow for the rapid vertical migration of more saline (or otherwise poorer quality) water into ASR wells during recovery. It can be the result of natural hydrogeological conditions (e.g., fractured or karst zones) or anthropogenic conditions (e.g., improperly constructed or abandoned wells). A characteristic feature of short-circuiting is a very rapid increase in salinity, much quicker than would be expected assuming an intact confining zone, with the rate of salinization being independent of the injected volume (Zuurbier 2015; Zuurbier and Stuyfzand 2017).

Maliva and Missimer (2010) documented suspected short-circuiting at the abandoned Northwest Hillsborough County (Florida) Dechlorination Facility reclaimed water ASR system. The system had a very poor RE, which did not improve over successive operational cycles and with increased injected volume. The rate of salinity increase during recovery was approximately the same as occurred during a pumping test performed prior to any injection. Analysis of the recovered water chemistry data using a three-component mixing model indicates that the increase in salinity was due to the introduction of a small fraction of saline water with ion ratios similar to seawater.

Zuurbier (2015) and Zuurbier and Stuyfzand (2017) documented short-circuiting in the Westland ASR system in the coastal Netherlands. The system stored rainwater surplus from 270,000 m^2 of greenhouse roof into a local shallow (27–37 m bls) brackish (3,783–4,644 mg/L chloride) aquifer. Injection and recovery were performed using two multiple partially penetrating wells (MPPWs). A rapid, unexpected increase in salinity occurred in the first days of recovery. Field measurements, hydrochemical data, and modeling results using the SEAWAT code indicate that the rapid increase in salinity was due to upward migration of saline water through the borehole of an aquifer thermal energy storage (ATES) well located near an ASR well.

A key lesson is that old boreholes are unreliable and their presence should be avoided in siting new ASR well sites (Zuurbier 2015). With increasing distance between ASR wells and nearby conduits, the amount of mixing should decrease and the time of arrival increase (Zuurbier 2015; Zuurbier and Stuyfzand 2017).

13.5.13 Summary

Studies on the controls of the RE of ASR systems that store freshwater in brackish or saline aquifers, performed from the late 1960s onward, provide a coherent framework of the conditions favorable for high recovery efficiencies. The most important conditions favorable for a high RE are a relatively low salinity (chloride concentration < 1,000 mg/L) and low to moderate degree of aquifer heterogeneity (i.e., groundwater flow is not dominated by a thin flow zone within the storage zone). Higher salinities adversely impact RE through a greater sensitivity to the mixing of stored water and native groundwater and a greater tendency for buoyancy stratification. High degrees of aquifer heterogeneity results in greater dispersive mixing. Well-developed dual-porosity condition (e.g., fracture and karst-dominated flow conditions) are also very unfavorable for ASR. Other factors, such as confining and storage zone thickness, transmissivity, and porosity, are of secondary importance.

However, it must be stressed that ASR can still be successfully implemented using storage zones with suboptimal hydrogeological conditions. ASR can still economically supply needed water during high demand periods even at lower recovery efficiencies. The economic value of ASR depends upon the difference in the value of water at the time of storage versus its value at the time of recovery. If excess water with a little or no value is stored (e.g., reclaimed water and surface water during low demands period) and later recovered during high demand periods (e.g., droughts), then an ASR system with a low RE may still be economically viable. Additionally, design and operational options are available to overcome some suboptimal hydrogeological conditions (Sect. 13.8).

13.6 ASR Screening Tools

The hydrogeological factors that impact the RE of ASR systems are now well understood. Various quantitative multi-criteria numeric screening tools have been developed to evaluate ASR feasibility at a given site and as a site-screening tool for selection of the preferred site for an ASR system.

13.6.1 Weighted Scoring Systems

Weighted scoring systems are a type of multiple criteria decision analysis (MCDA) that are widely used for evaluation of water and wastewater infrastructure options. The simplest and most understandable method is the weighted sum method, in which a performance score for scenario “i” (Z_i) is calculated as

$$Z_i = \sum_{j=1}^{n} w_j z_{ij} \quad (12.5)$$

where

w_j weight factor for criteria "j"
z_{ij} performance value for criteria "j" in scenario "i"

Performance values are assigned for each scoring criteria (e.g., storage-zone salinity) with the optimal conditions receiving the highest scores. Decreasing scores are assigned with increasing departures from the optimal condition. The weight factor reflects the relative importance of each criteria. There is an inherent subjectivity in developing scoring criteria and weight factors and the systems are subject to abuse. Weighted scoring systems can be devised so that a given result is obtained. Nevertheless, when objectively applied, they can be a valuable tool for screening sites (geographic locations and aquifer combinations) for relative MAR feasibility.

CH2M Hill (1997a) developed an ASR feasibility screening tool for the St. Johns River Water Management District (SJRWMD) in Florida, which is based largely on the confined and semi-confined aquifers in Florida. The feasibility screening tool considers facility planning factors (which determine whether there is a storage need that could be met by ASR), technical factors (hydrogeological and environmental impact), costs, and regulatory issues. The technical feasibility evaluation considers the following seven criteria (CH2M Hill 1997a):

- storage zone confinement ("aquitard" leakance)
- storage zone transmissivity
- storage zone gradient and direction
- recharge water quality
- native water quality (salinity and contaminants)
- overall physical, geochemical, and design interactions (clogging potential)
- interfering uses and impacts.

The CH2M Hill technical scoring criteria are summarized in Table 13.4. CH2M Hill (1997a) cautioned that the screening tool was not designed to be utilized as an absolute yes-no decision tool but rather to enlighten the users on the factors that are issues for an ASR system. In the CH2M Hill (1997a) system, ASR at a site will never be scored as not feasible, because it was believed that a low-scored ASR system potentially might still provide a much needed resource to a utility. Maliva and Missimer (2010) rejected this approach and recognized that some site conditions may be so unfavorable as to render ASR practically unfeasible. Maliva and Missimer (2010) proposed instead that the first element of a site-screening process should be a fatal flaw analysis in which sites or ASR options are rejected if they have one or more fatal flaw.

Table 13.3 is provided as an example of a screening tool and is not an endorsement. Weighting factors and performance criteria and scoring will vary with system type, objectives, local hydrogeology, and geographic area.

Table 13.4 ASR feasibility numerical assessment criteria (after CH2M-Hill 1997)

Criterion 1: leakance (weight factor = 10)

Rank	Leakance (days)$^{-1}$
1	$>1.2 \times 10^{-6}$
2	$1.2 \times 10^{-6} < L < 5.8 \times 10^{-7}$
3 (default)	$5.8 \times 10^{-7} < L < 2.9 \times 10^{-7}$
4	$2.9 \times 10^{-7} < L < 1.2 \times 10^{-7}$
5	$<1.2 \times 10^{-7}$

Criterion 2: storage zone transmissivity (weight factor = 10)

Rank	Transmissivity (ft^2/d)			
	Potable water		Untreated surface water	
	(gpd/ft)	(m^2/d)	(gpd/ft)	(m^2/d)
1	<8,000	<99	<80,000	<993.5
2	8,000–15,000	99–186	80,000–250,000	993–3,105
3	15,001–40,000	186–497	250,001–400,000	3,105–4,967
4	40,001–50,000	497–621	400,001–500,000	4,967–6,209
5	50,001–80,000	621–994	500,001–1,000,000	6,209–12,420
4	80,001–120,000	994–1,490	1,000,001–1,150,000	12,420–14,280
3	120,001–200,000	1,490–2,484	1,150,001–1,400,000	14,280–17,390
2	200,001–400,000	2484–4,967	1,400,001–2,000,000	17,390–24,840
1	>400,000	>4,967.4	>2,000,000	>24,840

Criterion 3: storage zone aquifer gradient and direction ranking (weight factor = 1)

Rank	Aquifer gradient	Direction criterion
1	Many strong gradients exist	Extreme artificial gradient, reevaluate location of ASR system
2	Several strong influences	Exaggerated gradient, investigation needed.
3 (default)	Multiple minor influences	Affected gradient worth investigating
4	Single minor influence or abnormal natural gradient	Minor investigation of existing data search
5	No influence	No influence

Criterion 4: recharge water quality (weight factor = 2)

Rank	Chloride (mg/L)	TDS (mg/L)
1	>200	>450
2	200–171	450–351
3	170–101	350–201
4	100–50	200–100
5	<50	<100

(continued)

Table 13.4 (continued)

Criterion 5: storage zone native water quality (weight factor = 10)		
Rank	**Chloride (mg/L)**	**TDS (mg/L)**
1	>6,000	>10,000
2	6,000–3,001	10,000–5,001
3	3,000–801	5,000–1,301
4	800–400	1,300–700
5	<400	<700

Criterion 6: overall physical, geochemical, and design interactions (weight factor = 5)		
Subcategory	**Points**	**Criterion**
Total suspended solids	1	>2.0 mg/L
	2	2.0–0.05 mg/L (default)
	3	<0.05 mg/L
pH	1	7.8–8.6 (default)
	2	>8.6
	3	<7.8
Total phosphorous	1	>0.1 mg/L
	2	0.1–0.05 mg/L (default)
	3	<0.05 mg/L
Nitrate as N	1	>1 mg/L
	2	1–0.5 mg/L (default)
	3	<0.5 mg/L
Dissolved organic carbon	1	>5 mg/L
	2	5–2.5 mg/L (default)
	3	<2.5 mg/L
Total iron	1	>1 mg/L
	2	1–0.3 mg/L (default)
	3	<0.3 mg/L
Dissolved oxygen	1	>3 mg/L
	2	3–1.5 mg/L
	3	<1.5 mg/L
Rank	Total points	
1	7–10	High potential for plugging
2	11–12	
3	13–16	Moderate potential for plugging
4	17–18	
5	19–21	Low potential for plugging

(continued)

Table 13.4 (continued)

Criterion 7: interfering uses and impacts (weight factor = 5)		
Subcategory	**Points**	**Criterion**
Distance to domestic or public supply wells	1	0.10–0.25 miles (0.16–0.41 km)
	2	0.26–5 miles (0.42–8.05 km)
	3	>5 miles (> 8.05 km)
Distance to contamination source	1	0.10–0.25 miles (0.16–0.41 km)
	2	0.26–1 miles (0.42–1.61 km)
	3	>1 miles (> 1.61 km)
Overall interfering uses and impact rank	**Total points**	
1	2	High use/impact
2	3	
3	4	Moderate use/impact
4	5	
5	6	Low use/impact

ASR feasibility score		
Total score (sum of ranks * weight factor)	**Feasibility level**	**Type of study recommended**
180–225	High confidence	General—confirm assumptions
100–179	Moderate confidence	Focused—investigation specific factors
<99	Low confidence	Detailed—evaluate impacts of critical factors

13.6.2 Lumped-Parameter Methods

Lumped-parameter methods, as the name implies, combines two or more parameters into a new parameter that provides a more diagnostic measure of ASR feasibility. Pavelic et al. (2002) proposed a lumped-parameter method to estimate the recovery efficiency of the initial cycle of ASR systems. The two main factors affecting recovery efficiency considered were dispersion and regional movement of the injectate plume. The lumped parameters were scaled to the dimensions of the injected water plume in terms of the radial extent of an idealized cylindrically-shaped plume around the ASR well (r_m)

$$r_m = \sqrt{\frac{V_i}{\pi n_e b}} \tag{12.6}$$

V_i injected water volume (m^3)
n_e effective porosity (m^3/m^3)
b aquifer thickness (m)

Relative dispersivity α_{rd} (m) is defined as

$$\alpha_{rd} = \frac{\alpha}{r_m} \tag{12.7}$$

where α = longitudinal dispersivity (m). Relative drift (l_{rd}) is defined as

$$l_{rd} = \frac{Kit}{n_e r_m} \tag{12.8}$$

where

K aquifer hydraulic conductivity (m/d)
i regional hydraulic gradient (m/m)
t mean residence time of injected water in the aquifer (d)

Modeling (theoretical) and field trials results show that RE is strongly related to α_{rd}. High RE values (≥0.5; 50%) occur when α_{rd} is ≤0.1. RE is affected by l_{rd} at values in excess of 0.1 (a drift of at least 10% of the plume diameter). The principle disadvantage of this lumped-parameter approach is that "the appropriate value of α is highly uncertain, and in general, cannot be estimated to the degree of accuracy required without undertaking ASR" (Pavelic et al. 2002).

The recovery efficiency of ASR systems that store freshwater in brackish or saline aquifers is effected by advection, free convection (density/buoyancy-induced flow), and dispersivity. Ward et al. (2007, 2008, 2009) evaluated the conditions in which the above factors will have a significant adverse impact on ASR system performance. An "ASR Performance Ratio" (R_{ASR}) was proposed, which is the sum of four subsidiary ratios:

$$R_{ASR} = R_{TV} + R_{DISP} + M + R_{ST} \tag{12.9}$$

where

R_{TV} technical viability ratio
R_{DISP} dispersivity ratio
M mixed convection ratio
R_{ST} storage tilt ratio

A system is likely to fail (i.e., have negligible recoverable freshwater) is R_{ASR} exceeds 10, while a system with a value of less than 0.1 is likely to be successful. The individual terms of the ASR performance ratio are:

$$R_{TV} = \left| \frac{K_{x,ave} I t_{storage}}{\epsilon X_{i,upstream}} \right| \tag{12.10}$$

$$R_{DISP} = \frac{\beta_L}{X_{i,upstreatm}} \tag{12.11}$$

$$M = \frac{K_{z,ave}\alpha}{\left|\frac{Q}{2\pi b X_{i,upstream}}\right| - \left|\frac{K_{x,ave} I}{\varepsilon}\right|} \tag{12.12}$$

$$\propto = \frac{(\rho(C_s) - \rho_0)}{\rho_0} \tag{12.13}$$

$$R_{ST} = \frac{K_{z,ave} \propto b t_{storage}}{\in \left(X_{i,upstream}\right)^2} \tag{12.14}$$

where (using consistent units)

$K_{x,ave}$	average horizontal hydraulic conductivity (L/T)
$K_{z,ave}$	average horizontal hydraulic conductivity (L/T)
I	regional hydraulic gradient (dimensionless)
$t_{storage}$	storage duration (T)
$X_{i,upstream}$	location of the fresh-salt water interface upstream from the well at the end of injection (L)
ε	effective porosity (unitless)
β_L	longitudinal dispersivity (L)
b	aquifer thickness (L)
Q	pumping rate (L^3/T)
α	density difference ratio
ρ	density at concentration (C_s) (kg/m^3) (native groundwater)
ρ_0	reference density (1000 kg/m^3 for freshwater)

Bakker (2010) presented another feasibility screening tool, based on the radial Dupuit approximation, to predict the performance of ASR systems in saline aquifers. During injection and recovery, the sharp interface between the injected freshwater and saline native groundwater is displaced from an ASR well. The straight interface tilts over time as the less dense freshwater flows upwards and saline water flows toward the well at the bottom of the storage zone. Resistance to flow in the vertical direction and mixing are neglected. The Bakker (2010) method is based on modeling of changes in the position and shape (tilt) of the interface with recovery being terminated when the toe of saline water reaches the ASR well.

Recovery efficiency is estimated using a dimensionless parameter "*D*" (Bakker 2010) where

$$D = \frac{Q}{k v b^2} \tag{12.15}$$

$$v = \frac{\left(\rho_s - \rho_f\right)}{\rho_f} \tag{12.16}$$

Q discharge (L^3/T)

k horizontal hydraulic conductivity (L/T)
v dimensionless density difference
b aquifer thickness (L)
ρ_s, ρ_f densities of salt water (native groundwater) and freshwater (injected water) (m/L^3)

Zuurbier et al. (2013) assessed ASR performance predicted by the Ward et al. (2009) and Bakker (2010) methods on existing ASR systems using brackish aquifers in the Westland-Oostland areas of coastal Netherlands. The performance of the existing ASR systems were found to show a good agreement with predicted performance, with deviations attributed to simplifications of the conceptual model and uncertainties in hydrogeological and hydrochemical conditions. However, the hydrogeological conditions in the coastal Netherlands study area (e.g., siliciclastic aquifers with high effective porosities and low anisotropy ratios) are particularly favorable for ASR.

13.7 Modeling of ASR Systems

Applications of numerical modeling to the development of MAR systems was reviewed by Kloppmann et al. (2012). The main questions that are addressed by modeling are:

- recovery efficiency
- residence time
- quality of recovered water relative to target quality.

Theoretical modelling based on hypothetical aquifers has been used to examine the effects of hydrogeological variables, MAR wellfield configurations, and operating procedures (e.g., injected volume) on recovery efficiency (Sect. 13.5). Groundwater modeling of existing or proposed ASR systems is performed to either predict future performance or to develop a better understanding of the hydraulics, solute-transport, and/or geochemistry of existing ASR systems.

An important objective of groundwater modeling is to be able to predict ASR performance (e.g., RE) in advance of system construction to support feasibility assessments and to assist in system design. The limiting factor in predictive modeling performed prior to system construction is usually a paucity of data. Realistic estimates of RE require data from a recharge and recovery trial to calibrate a solute-transport model (Pavelic et al. 2006; Maliva and Missimer 2010; Kloppmann et al. 2012; Parsons et al. 2012). The primary data deficiency for solute-transport modeling is quantitative information on aquifer heterogeneity. Theoretical modeling results show that dispersivity, which is a measure of aquifer heterogeneity, is a primary control on RE. As noted by Maliva and Missimer (2010)

> Assumed dispersivity values are often used in the solute-transport models of ASR systems, but that introduces a large element of circularity into the system analysis. If an average dispersivity value is used based on aquifer type, then the modeling results may indicate

an average system performance, because modeled system performance depends to a large degree on the dispersivity values used. The uncertainty associated with dispersivity values can be reduced somewhat by incorporating aquifer heterogeneity into the solute-transport models and recalibrating models as operational data become available.

Solute-transport modeling is typically used to predict the salinity (or a related parameter) of recovered water and thus RE. Reactive solute-transport models combine a solute-transport model with a geochemical model to simulate water chemistry changes related to geochemical reactions. Reactive solute-transport models includes both equilibrium geochemistry and reaction kinetics, and may include some biogeochemical (e.g., bacterially mediated) reactions. For example, reactive transport-modeling might be used to simulate arsenic and metals leaching in ASR systems.

13.7.1 Solute-Transport Modeling of ASR Systems

The hydraulic response of aquifers to pumping or recharge can be simulated if data on basic aquifer hydraulic parameters (e.g., transmissivity, storativity, and confining strata leakance) are available. Hydraulic modeling is not sensitive to aquifer heterogeneity as bulk transmissivity values obtained from pumping tests integrate the transmissivities of both high-transmissivity flow zones and low-transmissivity semi-confining strata within the tested interval.

Solute-transport modeling results, on the contrary, are highly sensitive to aquifer heterogeneity. Lateral movement of injected water will be greatly impacted if most of the water enters a thin flow zone versus entering evenly throughout the entire aquifer thickness. Hence, it is important to incorporate as much of the heterogeneity of an aquifer as practicable into solute-transport models to improve their predictive ability. Pavelic et al. (2006) demonstrated how the division of a 50-m thick storage zone into four layers was sufficient to allow for a satisfactory model calibration and avoided the need to use unrealistically large dispersivity values for model calibration. This relatively coarse-scale division of a storage zone can be performed using flowmeter log data.

Solute-transport modeling can be performed using either density-dependent or density-independent codes. Density-dependent models codes simulate the effects of differences in fluid density on groundwater flow, which can be important where freshwater is injected into an aquifer containing water with a significantly greater salinity and thus density. Some density-dependent codes can also simulate the effects of variations in temperature on fluid density and, in turn, fluid flow.

It has long been understood that density-induced fluid-migration (i.e., free convection or buoyancy stratification), can adversely impact the RE of ASR systems, with the degree of impact depending upon the salinity (and thus density) difference between the recharge water and native groundwater, vertical and horizontal hydraulic conductivities, and storage duration (time between recharge and recovery).

Ward et al. (2007) performed a theoretical analysis of the effects of mixed convection on ASR systems. Mixed convection refers to the combination of forced convection (i.e., flow caused by natural or induced hydraulic gradients) and free convection (i.e., flow caused by density differences). Ward et al. (2007) concluded that free convection could impact all three phases of ASR systems (injection, storage, and recovery) and that there is no one density difference threshold that marks the point where free convection will have a material impact on ASR system performance. In addition to density differences, controlling parameters on RE include hydraulic conductivity, pumping rate, storage, and dispersivity.

Ward et al. (2007) concluded that the decision as to whether to consider fluid density in ASR solute-transport models should be based on a full mixed convection analysis. However, moving from a density-invariant solute-transport model to a density-dependent solute-transport model is not that great a step in terms of time and effort (Maliva and Missimer 2010). The more pertinent question is whether there is an overriding reason not to routinely perform at least some density-dependent simulations for any solute-transport modeling of an ASR system that uses an aquifer that is more than mildly brackish (TDS > 3,000 mg/L) as a storage zone.

13.7.2 Reactive Solute-Transport Modeling

The capability to quantitatively predict a range of potential geochemical changes in water recharged in ASR systems would be of obvious value. However, the accuracy of predictive solute-transport modeling is limited by the large number of variables (e.g., kinetic rate orders, constants, and coefficients) whose values are poorly constrained. Hence, Greskowiak et al. (2006) observed that

> Generally, the benefits of mechanistic multi-component reactive transport models is seen primarily in their capacity to constrain or reject hypotheses on interactions of physical, chemical and biological processes and to a lesser extent in their predictive capabilities.

There have been several studies in which reactive solute-transport models were developed and calibrated against operational data to gain insights into the geochemical processes that impacted stored water chemistry.

Petkewich et al. (2004) developed a reactive solute-transport model using the PHAST code (Parkhurst and Kipp 2002) for a test ASR system in Charleston, South Carolina. The model was used to evaluate the geochemical processes during cycle testing and potential RE. The main geochemical process identified were calcite dissolution and precipitation, and cation exchange. The modeling results suggest that calcite dissolution occurs near the injection well and calcite precipitation occurs further along the flow path at the mixing zone between the injected freshwater and native brackish water.

The most widely used reactive solute-transport code now appears to be PHT3D (Prommer et al. 2003), which is a general purpose reactive 3D multicomponent model that combines the widely used MODFLOW/MT3DMS (Zheng and Wang 1999) codes with a batch-type geochemical model PHREEQC-2 (Parkhurst and Appelo 1999).

Greskowiak et al. (2005, 2006) developed a calibrated geochemical model of the Bolivar, South Australia, reclaimed water ASR system using the PHT3D code. An important question addressed was how to simulate the microbial growth and decay processes associated with the injection of reclaimed water. Both a simple model, which assumed steady-state microbial concentrations, and a more complex model, in which microbial growth and decay processes were simulated, provided a mass-conservative representation of essentially all existing hydrochemical observations that were made during the trial period.

Prommer and Stuyfzand (2005a, b, 2006) developed a PHT3D model to simulate water quality changes during pilot testing of an ASTR site (i.e., separate injection and extraction wells were used) located near Someren, Southern Netherlands. The PHT3D model simulated a reaction network that included:

- complexation reactions
- dissolution/mass transfer of sediment-bound organic carbon
- kinetically controlled DOC mineralization
- ion exchange reactions
- kinetically controlled oxidation of pyrite

The model was calibrated to much of the data on spatial and temporal changes in major ion and redox chemistry. Pyrite oxidation was identified as the dominant process with respect to the removal of dissolved oxygen and nitrate. Pyrite oxidation rates were found to exhibit a strong temperature dependency.

Wallis et al. (2010) developed a PHT3D model of the Langerak ASTR trial site (The Netherlands) in which freshwater was injected into a deep anoxic (methanogenic) freshwater fluvial sand aquifer. A flow model was first developed using MODFLOW. The flow and non-reactive solute-transport model were then jointly calibrated using Cl tracer data as the primary constraint. The model was next extended to include reactive multicomponent transport. Reactive transport simulations provided a detailed description of the processes affecting the spatial and temporal hydrochemical changes that occurred in the deep well experiment.

PHT3D was subsequently used to model the Bradenton, Florida, potable water ASR system, which uses a brackish (1,200 mg/L TDS) anoxic aquifer (Suwannee Limestone) as a storage zone (Wallis et al. 2011). The simulations support the arsenic-leaching geochemical model of:

- arsenic is released by the oxidative dissolution of pyrite
- released As is adsorbed onto newly form hydrous ferric oxides (HFOs)
- reversal of flow causes a restoration of anoxic conditions
- As is released during recovery by (1) reductive dissolution of HFOs and desorption, and (2) a reduction of the sorptive capacity of As on HFOs due to the elevated TDS concentration of native groundwater.

The two models of arsenic release are not mutually exclusive.

13.7.3 Inverse Geochemical Modeling

Mass balance-based inverse modeling, using software such as PHREEQC and NETPATH (Plummer et al. 1991, 1994), has been performed to determine the geochemical processes that likely resulted in the measured chemistry of recovered waters. For a given recovered water chemistry, the contributions of different source waters (of known compositions) and geochemical reactions (e.g., dissolution, precipitation, cation exchange) can be calculated.

Inverse modeling using NETPATH was performed to analyze water quality data from the Myrtle Beach and Charleston, South Carolina, test ASR systems (Castro 1995; Castro and Gardner 1997; Campbell et al. 1997; Mirecki et al. 1998). Mirecki (2006) used PHREEQC to model geochemical changes in ASR systems in South Florida. Inverse-geochemical modeling requires that the source water chemistries be accurately known. It is also important to screen the results for implausible geochemical reactions. For example, NETPATH modeling by Campbell et al. (1997) indicated the dissolution of halite and gypsum, which are unlikely to be present in an aquifer containing only mildly brackish water.

13.8 Innovative ASR System Designs

The standard ASR system consists of one or more vertical wells that are used for both recharge and recovery. A variety of modifications of the standard design have been proposed and tested in attempts to improve system performance in response to site hydrogeological conditions. Although technically sound solutions for improved performance are available, their implementation faces non-technical barriers including (Zuurbier et al. 2016):

- lack of demonstration of long-term viability
- lack of analysis of the hydrological impacts on surroundings
- lack of knowledge of and ability to construct and operate systems
- inherent conservatism due to a lack of a local track record of successful implementation.

It has been the author's observation that the inherent conservatism in the utility industry often impedes implementation of innovative technologies. "Tried and true" solutions, even if not optimal, tend to be preferred over more innovative and perceived as riskier solutions because of the ramifications of an expensive under-performing or failed system (even at a low probability).

13.8.1 Multiple-Well Systems

The design of wellfields for chemically bounded ASR systems (e.g., systems that store freshwater in brackish or saline aquifers) needs to consider the pattern of displacement of native groundwater. Trapping of native groundwater between ASR wells could negatively impact RE. Merritt (1985, 1986) demonstrated how solute-transport modeling can (and should) be used to evaluate the effects of various well configurations and operational strategies on RE. Some general rules are (Bouwer et al. 2008):

- to control mixing, a closer well spacing should be used than that of a conventional wellfield, with spacing related to the lateral extent of the stored water around each well
- closer wells spacings result in greater drawdowns and pressure increases (well interference)
- too close well spacings results in large drawdowns, which can promote vertical movement and limit recovery rates
- native groundwater can become trapped between wells, which should be avoided through wellfield design and operational procedures
- where advective movement is significant, ASR wellfields should be oriented parallel to the regional groundwater flow direction, and injection should be preferentially performed upgradient and recovery downgradient.

13.8.2 Dedicated Recovery Wells

The basic design for ASR wells using brackish or saline aquifers consists of wells screened or completed with open holes through the entire thickness of the storage zone. Water is injected and recovered from essentially the same depth interval. Design and operational innovations to improve RE involve measures to provide finer-scale depth control over the injection and recovery of freshwater and to control the movement of the freshwater/saline-water interface. Where buoyancy stratification significantly impacts recovery efficiency, RE may be improved by selectively recovering water from the top of a storage zone and not producing from the bottom of the zone where the intrusion of more saline water may preferentially occur (Fig. 13.5). The optimal system design may be to recover using a dedicated well completed only in the upper part of the storage zone (Maliva et al. 2005; Maliva and Missimer 2005, 2010). The well should also be completed downgradient of wells used for recharge. The recovery well would essentially skim freshwater off the top of the storage zone. A horizontally drilled well completed in the top of the storage zone might be an even more effective solution.

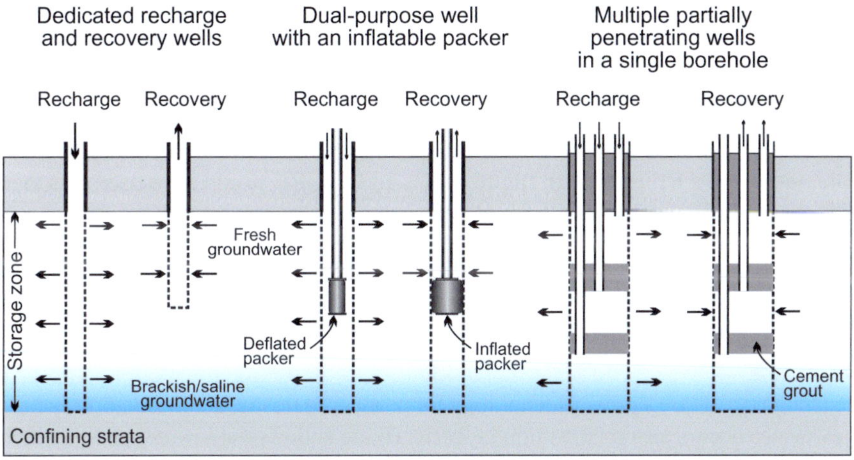

Fig. 13.5 Strategies for improving recovery efficiency in ASR systems in which buoyancy stratification occurs. Freshwater can be skimmed of the top of the storage zone using either dedicated shallower recovery wells, isolating the bottom of the storage using a packer (or one-way valve, not shown), or pumping only the shallower of multiple partially penetrating wells within a single borehole

13.8.3 Preferentially Recovery from the Top of ASR Wells

Preferential recovery from the top of ASR wells is a less expensive option than installation of dedicated wells. Restriction of recovery to the top of the storage zone can be accomplished by the use of a one-way valve or an inflatable packer (Maliva and Missimer 2010; Fig. 13.5). The packer would be deflated during recharge and inflated during recovery. The use of packers to isolate the upper part of a well during recovery has the great advantage that it is a non-permanent intervention. The packer can later be removed or reset at another depth.

13.8.4 Multiple Partially Penetrating Wells

Multiple partially penetrating wells (MPPWs) consist of multiple casings with individual screens or open holes installed in a single borehole (Zuurbier et al. 2014a, 2016; Fig. 13.5). MPPWs would allow for the optimization of freshwater recovery under conditions were buoyancy negatively impacts RE by allowing freshwater to be skimmed off top of the aquifer and offering operational flexibility. MPPWs have been successfully applied in a Dutch coastal greenhouse horticultural area in Nootdorp, the Netherlands (Zuurbier et al. 2014a, 2016). The storage zones are unconsolidated Pleistocene and Holocene fluvial deposits that are salinity stratified with relatively fresh water on top (118 mg/L Cl) and more brackish water at the base

(860–1,001 mg/L Cl). There are strict limits on the acceptable salinity of recovered water (Cl ≈ 18 mg/L). The tested ASR wells have four screens in a 350 mm diameter borehole. Density-dependent solute-transport (SEAWAT) modeling was performed comparing an MPPW to fully penetrating well (FPW) and single partially penetrating well (SPPW) designs. The SEAWAT model, calibrated to the first operational cycle data, indicated a RE of 60% in future cycles and that RE would have been less than 20% for a conventional FPW ASR system.

13.8.5 Scavenger Wells (Freshkeeper and Freshmaker)

Scavenger wells are a proven technology for producing freshwater in situations where freshwater occurs as a relative thin layer overlying saline groundwater and in which pumping would induce upconing of saline water. The concept is that pumping of saline water below the freshwater-saline water interface prevents the upward movement of the interface. Freshwater and saline groundwater are pumped through separate outlets, which can be either separate wells constructed at different depths above and below the freshwater-saline water interface (dual-bore systems) or single bores with pumps installed at different depths. By keeping the freshwater/saline water interface as a flow line, mixing is prevented in the aquifer and the geometry of the system is stable (Stoner and Bakiewicz 1993). The produced saline water is disposed to waste. Separate wells may be more efficient, but use of the same well may be a more economical solution, especially when an existing well is used (Zack 1988). The intake for the production well is placed as far above the saline water interface as possible.

The "Freshkeeper" system (Zuurbier et al. 2017) pumps brackish water from the lower part of an aquifer to manage the freshwater/brackish water interface. Produced water is desalinated by reverse-osmosis and the concentrate injected into a deeper zone. A Freshkeeper system was successfully applied at Noardburgum, province of Friesland, the Netherlands (Zuurbier et al. 2017). Field testing and modeling results indicate that the Freshkeeper concept could be used to reopen a wellfield that was abandoned due to saline water intrusion.

The "Freshmaker" system combines ASR and the Freshkeeper concept (Zuurbier 2015; Zuurbier et al. 2014a, b, 2015, 2016, 2017). The system involves a shallower ASR well(s) and a deeper abstraction/interceptor well constructed below the freshwater/saline-water interface, which is used to control buoyancy-induced flow. A test Freshmaker system was constructed in Ovezande, in the Zealand province of the Netherlands (Zuurbier 2015; Zuurbier et al. 2014b, 2015, 2016, 2017). The system uses two horizontal directionally drilled wells (HDDWs) completed in a fine to medium sand aquifer. The HDDWs were utility-type wells with a depth profile recorded in the field using a directional-drilling locating system (Digi Track) and GPS. SEAWAT modeling of a conventional ASR system using a HDDW predicted a limited freshwater recovery (50%), whereas the recovered freshwater volume from the Freshkeeper system would equal to the injected volume with the recovered water being a mixture of native fresh groundwater and injected water (Zuurbier 2015).

13.8.6 Direct Push Wells

Händel et al. (2014) proposed that the recharge of shallow unconfined aquifers could be performed using a battery of small diameter wells installed using direct-push (DP) technology as a cost-effective alternative to infiltration basins. Recharge would be performed under gravity and the wells were not proposed to be used for recovery. Hence, the system is technically aquifer recharge and opposed to ASR. A main advantage of the use of shallow wells is that infiltration rates will be more strongly dependent on horizontal hydraulic conductivity rather than the typically lower vertically hydraulic conductivity that controls recharge by surface spreading. Thin low hydraulic conductivity layers that can have a large impact on infiltration rates in basins would have a lesser impact on a system using wells.

Modeling using the Hydrus code was performed comparing the DP wells and an infiltration basin system. The baseline model indicates that the recharge rate for a 10 m by 6 m basin was equivalent to the rate from 1.5, 12-m long, 0.05-m diameter wells, which cost less than a quarter that of the basin (Händel et al. 2014).

A key issue is clogging. Händel et al. (2014) noted that clogging of surface basins is easier to treat and that the DP well system would be suitable for only high-quality water. The actual O&M requirements of the small diameter wells remains an unresolved issue. However, it was noted that due to the low costs of the DP wells, new replacement wells may be the most economical solution for addressing clogging issues (Händel et al. 2014).

Liu et al. (2016) documented field testing of a DP injection well system. The test site is an Early Pleistocene unconsolidated alluvial aquifer (Belleville Formation) in the Lower Republican River Basin of north-central Kansas. The depth to groundwater at the test site is 10.5 m or greater. An infiltration basin and single injection well recharge test were performed at the site. The surface recharge test was performed on a 6 by 10 m basin. The infiltration rate ranged from 420 to 73 m^3/d (110,000–19,300 gallons per day; gpd) at the end of the 30-h test. Infiltration was limited by a shallow clay layer that created perched conditions, which resulted in the movement of the infiltrated water becoming primarily horizontal.

The DP recharge test was performed using a 5.2 cm inner diameter (ID) PVC well with a total depth of 18.3 and 13.7 m of 0.05 cm-slot screen. Water was injected under gravity using a 3.5 cm ID injection tube. The recharge rate was controlled by the head difference between the water storage tank and injection well. No clogging was detected. The injection rate at the end of the 17.8-h test was 124 m^3/d (32,800 gpd). Water levels did not rise above the shallow clay layer.

Cost analyses using recharge rates at the end of the tests were reported to be US $5.8 per m^3/d for the injection well versus $34.2 per m^3/d for the surface infiltration basin. Hence, DP recharge wells may be a cost-effective alternative to infiltration basins (Liu et al. 2016). However, a major limitation of the study is that the potential impacts of clogging could not be addressed during the relatively short duration of the field tests (Liu et al. 2016).

Händel et al. (2016) evaluated MAR using a DP injection well at a test site (Pirna) located approximately 20 km southeast of Dresden in the state of Saxony, Germany. The tested aquifer consists mostly of fluvial fine to coarse sands and gravels. A medium term (14 day) injection test was performed on a 1-in. (2.5-cm) ID, approximately 12 m deep existing observation well using groundwater obtained from the tested aquifer. A mean water level rise of 0.31 m was observed at an average recharge rate of 0.75 L/s. No trend of rising water levels was observed that would be indicative of clogging. However, it was noted that biological and chemical clogging might occur in longer test and for water of a different quality.

13.8.7 Horizontal Directionally Drilled Wells

Horizontal directional drilling has revolutionized the oil-and-gas industry and thousands of horizontal wells have been completed for environmental remediation purposes (Fournier 2005). However, relatively few wells have been installed for potable water production and even fewer for MAR. For groundwater remediation applications, screen lengths have commonly reached approximately 1500 ft (450 m) long (Fournier 2005). Horizontal directionally drilled (HDD) wells have several advantages compared to vertical wells:

(1) a single horizontal well may have the capacity of multiple vertical wells
(2) multiple horizontal wells could be installed radiating outward from a central pumping station, reducing surface footprint
(3) horizontal wells may reduce local aquifer drawdowns with lesser drawdowns spread over a greater area
(4) longer screen lengths can result in low entrance velocities and thus slower clogging rates and associated maintenance requirements (Fournier 2005)
(5) horizontal wells could be constructed to produce from (or recharge into) a thin freshwater zone
(6) screens can be installed without disrupting overlying facilities and structures (Fournier 2005).

Horizontal wells can be attractive for use as production and ASR wells because of the potential for increased yields and the reduction of the surface footprint of systems, particularly if multiple wells are drilled radiating outward from a single node (well site). Horizontal wells have been proposed for use in ASR systems because they disperse the drawdown over a much wider area, and thus achieve significantly higher production rates (Pyne and Howard 2004; Pyne 2005, 2006). However, a very fundamental misconception concerning horizontal wells is that water production will occur evenly along the length of the screen or open hole (Maliva and Missimer 2010). A disproportionate amount of the water produced from horizontal wells may enter near the pumped end of the well with progressively lesser amounts entering towards the distal end of the well (Maliva and Missimer 2010). Hence, HDD wells may be suitable for systems storing freshwater in a freshwater aquifer or as dedicated recovery wells for skimming freshwater off the top of salinity (density) stratified systems but may not be effective as dual recharge and recovery wells for systems storing freshwater in brackish or saline aquifers.

The two main types of horizontal directionally drilled (HDD) wells are referred to herein as utility-type HDD wells and oilfield-type horizontal wells. Horizontal directional drilling technology has been used in the utility industry since the 1970s for long subsurface crossings of underground pipes, conduits, and cables. Boreholes are drilled at a shallow angle from an entry (launch) point to a distant exit point (Fig. 13.6). The usual drilling procedure is that a small diameter pilot hole is first drilled from the entry to exit point. The pilot hole is next reamed to achieve the target borehole diameter. Finally, the pipe, conduit or cable, is installed by pulling from the exit point to the entrance point. Shallow horizontal wells can be constructed using this method by installing a screen in the borehole. Wells are constructed a short distance below grade (usually <50 ft or 15 m). The main technical challenges with utility-type HDD production and recharge wells is emplacing an adequate filter pack around the screen and achieving proper development.

Oilfield-type horizontal wells are initially drilled using rotary drilling in same manner as vertical wells. The entire drill string is rotated during drilling. Once the depth at which directional drilling is to begin is reached, the directional drilling assembly is either tripped in or engaged. Directional drilling in performed using a mud motor with a bend of up to several degrees. The mud motor is powered by the pressure of drilling fluid pumped down the drill string. The mud motor rotates the drill bit while the drill string (pipe) is not rotated. The orientation of mud motor (both direction and bend) is known and controllable from land surface. Tools utilized in oilfield directional drilling include whipstocks (devices used to sidetrack out of cased wellbores), various bottom-hole assembly (BHA) configurations, three-dimensional measuring devices, mud motors, and specialized drill bits. Oilfield horizontal drilling technology has become very sophisticated and it is possible to geophysically log a borehole while drilling to precisely map and control the path of the well.

HDD drilling has some promise for MAR wells but may not be practical for many systems due to high costs (particularly the use of oilfield technology). Application of oilfield-type horizontal drilling technology may also be constrained by the unavailability of equipment-sized for the typically larger diameter wells used for water injection and production wells.

13.8.8 Low-Cost, Small-Scale ASR Systems

The benefits of domestic-scale ASR include: (1) reduction in demands on public potable water supplies, (2) more efficient water management with stormwater being managed at its source, and (3) slightly reduced urban runoff into the sea or receiv-

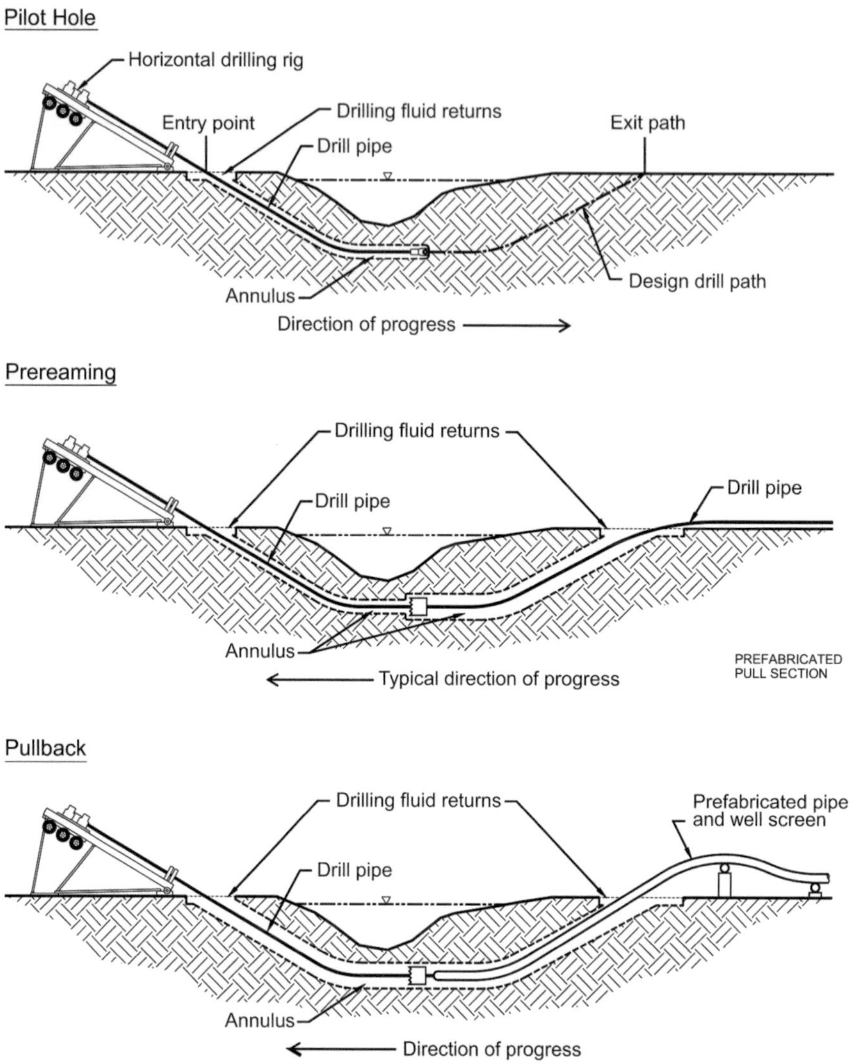

Fig. 13.6 Typical drilling sequence for utility-type horizontal directionally drilled (HDD) wells (Modified from U.S. Fish and Wildlife Service)

ing waters (Dillon and Molloy 2006). Potential problems that might be associated with large-scale implementation of domestic-scale ASR include that resulting higher groundwater levels might cause basement flooding, increase saline-water ingress into sewers, damage building foundations and roads, and cause submergence of underground utility infrastructure. Domestic-scale ASR could also increase the potential for pollution of shallow aquifers and for human contact with polluted groundwater. Dillon and Molloy (2006) therefore recommended that domestic-scale ASR not be undertaken in shallow aquifers where the water table is located or would rise to within about 5 m (16 ft) of land surface or where structures or utilities exist that could be impacted by rising water tables.

The viability of storing domestic roof runoff for use in garden irrigation was investigated at the Domestic Scale Rainwater ASR Demonstration Site in Kingswood, South Australia. Run-off from a single residential dwelling was stored in a shallow alluvial aquifer that contains brackish water (Barry and Dillon 2006; Barry et al. 2007). The initial results of the testing were unfavorable because the salinity of the recovered water quickly increased beyond values acceptable for irrigation use. The poor recovery efficiency may have been due to mixing losses, the effects of regional groundwater flow, and to density stratification (Barry and Dillon 2006). The injected volume over the three years of testing (452 m^3; 119,000 gallons) was likely insufficient to establish a buffer zone that would allow for high recovery efficiencies (Barry et al. 2007). Well efficiency was reduced by clogging attributed to particles and biofilms. Purging improved performance but there was still a gradual decline in injectivity over time (Barry et al. 2007). The operation and maintenance requirements were at a level beyond that which could reasonably be expected of normal householders (Barry and Dillon 2006).

Household ASR wells is currently not viable in the United States because the underground injection control regulations do not differentiate systems based on their size and capacity. A household system could face the same construction standards and monitoring requirements as a large-capacity utility system, which would render the system uneconomical.

Village-scale ASR may improve water management in developing areas by providing storage of seasonally available excess surface water. In the southwestern coastal districts of Bangladesh, the shallow aquifer that is used for drinking water supply is mostly brackish and may contain elevated concentrations of arsenic. Sultana et al. (2015) documented an investigation of small-scale ASR at 13 sites. The systems stored water that accumulated in ponds during the monsoon season. Each system consisted of a sand filter for pretreatment and four or more large-diameter (30.5 of 56 cm) screened infiltration wells located in a 3-m radius around a central 5-cm diameter extraction well equipped with a hand pump. Management of clogging is a key issue. The infiltration wells are filled with gravel topped with a 15-cm layer of fine sand to provide additional filtration. The wells are rehabilitated by manually removing and washing the upper 1.5 m of sand and gravel. After one year, there was a highly variable reduction in infiltration rates from initial values, ranging from 6 to 80%, with a majority of the sites having a reduction of greater than 40%. Manual washing of the gravel markedly improved infiltration rates but not back to initial values.

The recovered water had significantly reduced *E. coli* concentrations, but the bacteria were reported to still be present in about half of the samples, which may mean that disinfection is required. Some of water samples also had elevated arsenic concentrations (>100 μg/L), which indicates that As risks need to be carefully managed and require further investigation (Sultana et al. 2015). Sultana et al. (2015) noted several general limitations to the use of ponds for ASR source water:

- pond water is viewed as a scarce, valuable resource when available and, therefore, another source of freshwater is needed (rainwater)
- power outages are common, so systems should operate under gravity or perhaps solar power
- management of clogging is time consuming and labor intensive, and thus requires sustained community engagement
- costs of the systems should be commensurate with people's ability to pay for them
- institutional arrangements need to be in place, especially for water quality monitoring.

Cavity-tube wells are widely used for water supply in the Indian Subcontinent. Existing cavity-tube wells in India are being used as ASR wells by farmers (Taneja and Khepar 1996; Malik et al. 2002, 2006). The wells can be inexpensively retrofitted to allow for injection whenever excess rain or canal water is available. The main cost elements are construction of a connecting channel to convey canal water and a settling and filter tank (Goyal et al. 2008). The injection of stormwater can locally improve groundwater quality and is a simple solution for improving the management of groundwater resources. Goyal et al. (2008) reported that cavity wells were found to not clog even when injecting freshwater with large sediment loads. The cavities have large surface areas and therefore low entrance velocities, which would be expected to result in a low susceptibility to physical clogging. Economic analysis by Goyal et al. (2008) indicates that ASR in brackish-water aquifers using cavity wells is economically affordable.

13.9 Gravity Drainage Wells

Gravity drainage wells are used in some areas for the disposal of stormwater but can also have a primary or secondary aquifer recharge function. The wells can be either vadose (discharge into the unsaturated zone) or phreatic (discharge directly into an aquifer). The defining feature is that water is allowed to flow into the wells under gravity (with or without some pretreatment), as opposed to being pumped into the wells under pressure. As a generalization, vadose wells are used where the underlying aquifer contains freshwater, whereas phreatic wells tend to be used where the underlying aquifer contains saline water and is not considered a potential drinking water source. Vadose recharge wells are addressed in Sect. 17.6.

13.9.1 Florida Gravity Drainage Wells

The state of Florida has abundant drainage wells. Most of the wells are located in coastal Miami-Dade and Broward counties and discharge into saline groundwater, and thus serve a disposal rather than aquifer recharge function. Gravity drainage wells have been used in central Florida since 1904 to provide surface drainage and flood control with the ancillary benefit of providing recharge to the underlying Floridan Aquifer, a locally freshwater aquifer that is the primary water source in the region (Kimery and Fayard 1984; CH2M Hill 1997b, 1998). MAR gravity drainage wells are most abundant in Orange and Seminole counties, which have an estimated 400 wells (CH2M Hill 1997). The wells had previously been also used for the disposal of domestic sewage and other liquid wastes, but this practice has long been discontinued.

Large parts of central and northern Florida are closed-basin karst terranes with no surface water outflow (Kimery and Fayard 1984). Most of the drainage wells are used for urban drainage and lake control. Recharge occurs under gravity when water level rises above a weir at the intake structure. The wells discharge into the upper part of the Upper Floridan Aquifer (UFA). The main production zone of the UFA is located deeper than the injection zone of the drainage wells, although the shallower zone is also used for potable supply (Schiner and German 1983). The health concern is that the drainage wells might allow contaminants in the essentially untreated water to migrate into production wells.

It is clear that the recharge provided by the drainage wells makes an important contribution towards reducing the impacts of the intense groundwater pumping in this rapidly growing region. An estimated 40% of the recharge in central Florida is through recharge wells (Sheffield et al. 1995). CH2M Hill (1997b, 1998) estimated that the average annual recharge rate in Orange and Seminole counties is between 39 and 52 MGD (147,600 and 196,800 m^3/d), and that it is technically possible to at least double that recharge rate. However, with the implementation of federal and state Underground Injection Control (UIC) regulations in the 1970s and early 1980s, it is for all practical purposes not possible to permit new drainage wells. Existing wells may be rehabilitated or, in some instances, be replaced by a similar well ("like for like"). Existing wells were grandfathered (authorized by rule). Drainage wells are considered Class V injection wells and it is required under current regulatory interpretation of the rules that recharged water must meet primary and secondary drinking water standards at the wellhead. Surface waters invariably exceed water quality standards for total coliform bacteria (4 cfu/100 mL). UIC rules are focused on environmental protection, particularly the potential for aquifer contamination. The UIC permitting process does not consider the water resources management benefits of the recharge associated with drainage wells (CH2M Hill 1998).

Drainage wells pose some environmental risk as they provide a direct connection between surface waters and aquifers. Wells could potentially receive chemicals from accidental spills and intentional discharges. There have been rare example were contamination of a production well can be directly related to a specific discharge well. CH2M Hill (1997b, 1998) observed that the drainage wells have been recharging water containing total coliform bacteria for now over a century, but yet there is not widespread contamination of the UFA with coliform bacteria. Sheffield et al. (1995) provided water quality data for stormwater in Orange County. The main water quality concerns are nutrients (nitrogen and phosphorous), suspended sediment, and bacteria and viruses. Relatively few documented cases of severe aquifer pollution have been detected in public water supplies (Kimrey and Fayard 1984).

Schiner and German (1983) evaluated the quality of water in drainage wells in the Orlando area. With the exception of bacteria, water from drainage wells on the average, without treatment, meet the maximum contaminant standards for chemicals established for drinking water. However, the high total and fecal coliform and fecal streptococci bacteria in drainage wells indicate a potential for contamination of supply wells by drainage-well recharge if a supply well and a drainage well are hydraulically connected (Schiner and German 1983). Bacteria were less abundant in wells that receive lake water versus urban runoff. No serious health hazards were noted in water samples from supply wells. The median total nitrogen of drainage well water was reported to be 1.0 mg/L versus 0.29 mg/L for supply wells. Trace pesticides (2,4-D, 2,4-TP (silvex), diazinon, dieldrin, chlordane, 2,4,5-T) were detected in drainage wells, but at concentrations below drinking water standards. Some drainage well samples exceeded secondary (aesthetic-based) drinking water standards for color, hydrogen sulfide, iron, and manganese). The National Research Council (2008) reported elevated arsenic along flow paths, which may be due to the introduction of dissolved oxygen from air entrainment.

The water quality concerns over drainage wells can be addressed by pretreating the water. Sheffield et al. (1995) provided some pretreatment option. However, CH2M Hill (1998) noted that treating water to meet drinking water standards is difficult and expensive. Treatment systems used intermittently tend not to be cost effective in terms of costs per unit volume of water treated. However, the additional aquifer recharge that may be permitted by treating surface water to potable standards may be more economical than other alternative water supply options.

13.9.2 Qatar Drainage Wells

A pronounced surface feature of Qatar is the large number ($\approx$850) of shallow depressions, which are the surface expression of subsurface collapse caused by the dissolution of underlying calcium sulfate (anhydrite) and calcium carbonate (limestone) formations (Eccleston and Harhash 1982). After rainfall events that are sufficient to satisfy the surficial soil moisture deficit, runoff reaches the depressions by incipient drainage channels or overland sheet flow. The depressions are significant natural recharge sites, but the accumulation of fine sediments reduced recharge rates, which is evidenced by a slow decline in water levels that approaches the rate of open-water evaporation (Eccleston and Harhash 1982). The government of Qatar implemented a program of installing recharge wells in depressions in order to increase the rate of recharge and decrease evaporative losses (Al-Rashed and Sherif 2000; United Nations Environment Programme 2001). Hashim (2006) reported that 341 wells were drilled since 1986. A typical well is constructed with a perforated casing above land surface that is surrounded by a thick gravel lining to protect it from contamination (Fig. 13.7). The United Nations Environment Programme (2001) reported that the results of the initial testing indicated a 30% increase in groundwater recharge.

Fig. 13.7 Qatar drainage well. A perforated steel casing is surrounded at land surface by a filter consisting of gravel surrounded by a wire mesh

However, there is no available data on the long-term performance of the wells, particularly their susceptibility to clogging.

13.9.3 Agricultural Drainage Wells

Agricultural drainage wells (ADWs) can be defined as "constructed subsurface disposal systems used to accelerate the drainage of agricultural surface runoff and/or subsurface flow" (Ludwig et al. 1990). The primary purpose of most ADWs is to dispose of excess water, but ADWs may also have an intended purpose to recharge aquifers for irrigation water use or have combined objectives. Generally, ADWs consist of a buried collection basin or cistern, one or more tile drains, and a drilled or dug well (Fig. 13.8). Cisterns, depending on the setting, may receive irrigation return flows or field drainage from precipitation or floods (Ludwig et al. 1990). ADWs vary in the particulars of their designs. For example, systems in Texas may be constructed with the top of wells located approximately 2 ft higher than the base of the cistern, to allow for sediment to settle, and with screens to prevent coarse materials from entering the well (USEPA 1999). ADWs also includes improved sinkholes, which are considered injection wells under USEPA regulations (USEPA 1999b).

The USEPA (1999) estimated that 2842 ADWs existed in the USA at the time of the survey but also noted that thousands of more wells may exist. Landowners may not even be aware that a well exists on their property. The impacts of ADWs on groundwater quality were reviewed by the USEPA (USEPA 1999; Ludwig et al. 1990). Contaminants that can enter ADWs include suspended solids, pesticides, fertilizers, salts, metals, and microbes, including pathogens. Pesticides may enter wells either in solution or sorbed onto particles. The primary constituents in ADW injectate that are likely to exceed groundwater standards are nitrate, boron, sulfate, coliform bacteria. The concentrations of some pesticides may also exceed drinking

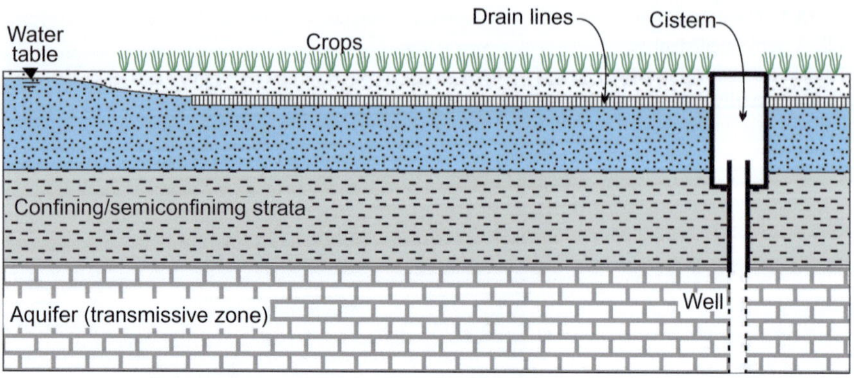

Fig. 13.8 Conceptual diagram of an agricultural drainage well system. The drain lines may be constructed of tile, plastic, perforated steel pipe, or other materials. Collected water enters a cistern and then flows under gravity to an underlying aquifer

water maximum contaminant levels or health advisory levels (USEPA 1999). Total dissolved solids often exceed the aesthetics-based secondary drinking water standard (USEPA 1999).

Surface runoff can be expected to have the greatest potential negative effect on groundwater quality because of its poorer quality, the absence of soil filtration, and greater volumes. Subsurface drainage is unlikely to have high suspended solids and bacteria due to soil filtration. Contamination potential is greater for wells completed in the saturated zone versus the vadose zone. ADWs are vulnerable to spills and illicit discharges, and land uses associated with high contaminant levels, such as large livestock confinement facilities with manure production and storage (USEPA 1999). Household septic systems were also found to have been connected to tile drains feeding ADWs.

Contamination of drinking water supplies from ADWs was reported in Idaho and Iowa (Ludwig et al. 1990). Elevated nitrate concentrations were documented in farm supply wells near drainage wells in Iowa and excessive coliform bacteria (above drinking water standards) was detected in domestic wells in an area of Idaho with high concentrations of ADWs. Drainage wells, to be effective in terms of capacity, need to discharge into high-transmissivity aquifers, such as karstic limestones and fractured basalts (USEPA 1999). These aquifers can accept large quantities of water with little susceptibility to clogging (Libra and Hallberg 1993), which is an important factor for MAR systems that do not involve significant active management. Secondary porosity features in these aquifers make them susceptibility to rapid contaminant transport (Ludwig et al. 1990).

The USEPA (1999) proposed a number of best management practices (BMPs) to reduce environmental risks associated with ADWs. One option is to abandon the wells and construct alternative drainage outlets. In some states of the United States, ADWs are not permitted and, if found, must be closed. However, the challenge is that there may be no record or knowledge of some wells constructed at the turn of the century (later 1800s—early 1900s). The BMPs relate mainly to reducing the actual or potential amount of contaminants entering the systems, such as improved fertilizer, pesticide, irrigation, nutrient, and livestock waste management.

ADWs were common in parts of north-Central Iowa (particularly Humboldt, Pocahontas, Floyd, and Wright Counties), which have poorly drained soils that overlie shallow limestones. Ponding in these soils severely limits farm production. The least expensive option identified to drain the soils was to drill wells through the surficial low-permeability glacial tills into the underlying, high-transmissivity fractured limestone, which created highly productive cropland areas (Glanville 1985; Baker et al. 1985). ADWs were constructed in the late 1800s and early 1900s to open areas for farming in places where there was no natural outlet for water. Closing ADWs has been in practice in Iowa for three decades. A permit was required to drill a new ADW starting in 1957. By 1983, exemptions for grandfathered wells was removed. Only 48 ADWs are known to still exist in Iowa, which occur in areas where installing a new drain, channel, or outlet can be very expensive, such as areas with shallow bedrock (Schwaller 2015).

Libra and Hallberg (1993) pointed out that natural features, such as sinkholes, have a similar impact on groundwater quality as ADWs and that in Iowa there are 30 times more sinkholes than ADWs. Closure of ADWs may simply divert water from one entry into the aquifer to another (i.e., sinkholes and losing streams), although the alternatives may be preferred in that they provide some natural attenuation of contaminants (Libra and Hallberg 1993). Where the overlying confining unit is thin, contaminants may enter the limestone aquifer through natural routes and it may not be possible to definitively link elevated contaminant concentrations to ADWs.

Baker et al. (1985) reported that 85% of the Iowa drainage water samples contained more than 10 mg/L NO_3-N (10 to 30 mg/L range) and that groundwater from farm water supply wells in the vicinity of drainage wells (within 2 km) have elevated nitrate concentrations. Pesticides were sometimes detected in injected water, but at levels less than 1 μg/L for subsurface drainage. Higher levels (up to 80 μg/L) were detected when surface runoff was being recharged. Recharge of surface water also introduced microorganisms into the aquifer. Current pesticide water quality criteria were not exceeded (Baker et al. 1985).

13.10 ASR and MAR Well Design Issues

Well construction and operational issues associated with ASR and MAR using wells were discussed in detailed by Pyne (2005) and Maliva and Missimer (2010). Follows are some lesson learned from global ASR and MAR experiences.

Clogging and well rehabilitation are normal parts of ASR and recharge well operations. Production wells require periodic rehabilitation to maintain specific capacity, although the frequency varies greatly between wells. ASR and other injection wells are much more prone to loss of well performance due to clogging because the flow of water is into the well and formation. The injected water may contain entrained suspended solids and is usually not in chemical equilibrium with aquifer rock and native water. The rehabilitation of ASR wells often involves periodic back-flushing and less frequent major rehabilitation. The need for well rehabilitation should be considered a normal part of well operation and maintenance, and should be incorporated into the initial system design and economic analysis. Wellheads should be designed to facilitate anticipated well rehabilitation activities.

Optimize ASR well design and construction. ASR and recharge wells are prone to the loss of performance due to clogging. It is far better to incur the modest additional expense to moderately over-design ASR wells to maximize well efficiency than risk an under-performing system. Ideally wells should have some excess capacity so that they can still be used to inject and recover water at their design rate after some clogging. The incremental, additional costs to construct a larger diameter well, use more corrosion resistant materials, and install a more efficient screen and filter pack, represent a small fraction of total system cost. Screen slot-size should be the maximum acceptable based on formation material, and the filter pack material should be carefully selected. Glass beads have been used in some ASR wells instead of natural sand because of their greater sphericity and more uniform size (e.g., Nutter and Gin 2016).

Segalen et al. (2005, 2006) reviewed the effects of well drilling and completion methods on the performance of ASR wells in unconsolidated aquifers. Performance data were compared for wells completed in the same formation and site using alternative well construction techniques. The results of the study include:

- the most critical issue affecting well performance is the removal of residual mud from the well
- wells drilled using the cable tool method perform better that wells drilled using a reverse circulation method because cable-tool drilling left a lesser thickness of mud in the wells
- removal of the sludge (mud) layer from the well by reaming the well with a slightly larger bit while circulating clean water can dramatically improve well performance
- larger screen apertures result in less well clogging
- wells drilled with polymer muds perform better than those drilled with bentonite mud.

Rigorous well development is critical, especially for screened wells. Well development is performed to remove residual drilling fluids and repair formation damage. Development is particularly critical for screened wells used for injection because residual drilling fluids and entrained fine solids are pushed into the formation during injection, contributing to well clogging. Well development should involve rigorous repeated alternating inward and outward flow through the screen and filter pack at velocities greater than those expected to occur during well operation. It is important that the entire screened interval be developed through interval development using packers or other means. Methods that involve surging the entire screen at once may develop just the more transmissive intervals of a well through which flow is concentrated. Jetting and swabbing are also effective development techniques. The performance (specific injectivity and capacity) of wells completed in carbonate aquifers may be increased by acidification. Additional more thorough and rigorous development can be the most cost-effective option for increasing the hydraulic performance of ASR and recharge wells.

Cascading should be avoiding. Air entrainment caused by the cascading of water down a well can cause clogging and introduce excess dissolved oxygen into the storage (recharge) zone. Wells, wellhead, and piping should be designed to maintain a positive pressure during injection. Where the depth to water is great, positive pressure conditions can be achieved using injection tubes and/or downhole flow control valves.

Stored water quality changes have been the principal geochemical challenge of many ASR systems. Potable and treated surface and reclaimed waters injected in ASR systems typically have moderate to high dissolved oxygen concentrations, whereas native groundwaters are often chemically reducing. This difference in oxidation-reduction potential can cause a variety of redox reactions during injection and storage. The leaching of arsenic and some metals appears to be caused by the introduction of DO and chlorine into aquifers containing reducing conditions and arsenic-bearing iron sulfide minerals. Arsenic leaching in stored water was a major surprise, despite the fact that it was well known in the aquatic geochemistry literature that the introduction of oxygenated water into aquifers containing reducing conditions can result in the mobilization and concentration of metals present in reduced minerals.

Managing redox reactions can be a major operational issue for ASR systems. Deoxygenation of injected water is an option, but it involves considerable additional capital and operational costs. Pre-treatment of aquifers to oxidize reduced minerals has been successfully performed in some systems to control iron and manganese concentrations. The amount of labile chemically reduced minerals that contain arsenic and metals of concern in some aquifers is small and will eventually be exhausted, and thus the amount of arsenic and metals leaching will gradually decrease over time. The potential for adverse fluid-rock interactions should be evaluated through a geochemical assessment, which should include a mineralogical evaluation and at least basic geochemical modeling.

Realistic expectations of recovery efficiency. ASR systems that store freshwater in brackish or saline aquifers will usually have a less than a 100% RE. Depending upon the system (particularly native groundwater salinity), a 70 to 80% RE is a more realistic goal. ASR systems should not be oversold by setting unrealistic RE expectations. Once expectations are set and agreed upon by all parties involved in a project, system performance should be objectively evaluated relative to those objectives. "Moving goalposts" should be avoided (i.e., changing objectives over time).

Installing a proper buffer zone. A considerable amount of water needs to be "invested" in ASR systems utilizing brackish or saline storage zones to flush native groundwater away from ASR wells. The term "invest" is used in that the water used to create a buffer zone should be viewed as a capital cost of the system. If a proper buffer zone is not installed, then recovery efficiency from cycle tests will be low. Pyne (2015) concluded that too many ASR projects have been stalled or terminated due to the implementation of cycle testing programs that failed to initially form and maintain a buffer zone. Low initial RE can lead to a loss of public support.

Excessive aquifer heterogeneity can profoundly impact ASR recovery efficiency. Review of global ASR system performance indicates that excessive aquifer heterogeneity, particularly the concentration of flow into a thin flow zone and well-developed secondary porosity (e.g., flow dominated by fractures or karst conduits), can result in very poor RE and associated system abandonment or repurposing (e.g., converting ASR wells into brackish-water production wells). Evaluation of aquifer heterogeneity is a critical part of feasibility assessments and aquifer characterization programs.

Groundwater modeling should be an integral and on-going process in ASR system development. Groundwater flow and solute-transport modeling can bring value to ASR and recharge projects by allowing for initial evaluations of potential system performance. Groundwater-flow modeling can be used to evaluate hydraulic impacts (i.e., magnitude, extent, and duration of head changes) during injection and recovery. Solute-transport modeling can be used to evaluate the moving and mixing of recharged waters. Density-dependent solute-transport modeling should be performed where the recharged water has a significantly different salinity (and thus density) than native groundwater. Solute-transport and more complex reactive solute-transport models have a limited predictive ability at the start of projects because the values of many key variables are poorly constrained. The process of calibrating a model against initial operational testing results provides insights into the hydraulic and geochemical processes that affect system performance.

Groundwater modeling should not be viewed as a one-time process. Instead groundwater models of ASR systems should be calibrated and recalibrated (as necessary) as operational data become available. Calibrated groundwater model development provides a means for developing an understanding of system performance, predicting future system performance, and evaluating the effects of changes in operating protocols and options for system expansion.

References

Al-Rashed, M. F., & Sherif, M. M. (2000). Water resources in the GCC countries: An overview. *Water Resources Management, 14,* 59–75.

Baker, J. L., Kanwar, R. S., & Austin, T. A. (1985). Impact of agricultural drainage wells on groundwater quality. *Journal of Soil and Water Conservation, 40*(6), 516–520.

Bakker, M. (2010). Radial Dupuit interface flow to assess the aquifer storage and recovery of saltwater aquifer. *Hydrogeology Journal, 18*(1), 107–115.

Barry, K., & Dillon, P. (2006). Domestic-scale ASR with rainwater at Kingswood, South Australia. In *Recharge systems for protecting and enhancing groundwater resources, Proceedings of the 5th International Symposium on Management of Aquifer Recharge*, Berlin, Germany, 11–16 June 2005 (pp. 656–662). Paris: UNESCO.

Barry, K., Dillon, P., & Pavelic, P. (2007). Mixing and clogging constraints on domestic-scale ASR at Kingswood, South Australia. In P. Fox (Ed.), *Management of aquifer recharge for sustainability.*

Proceedings of the 6th International Symposium on Managed Aquifer Recharge of Groundwater (pp. 61–73). Phoenix: Acacia Publishing.
Bouwer, H., Pyne, R. D. G., Brown, J., St Germain, D., Morris, T. M., Brown, C. J., et al. (2008). *Design, operation and maintenance for sustainable underground storage facilities*. Denver: AWWA Research Foundation.
Brown, C. J. (2005). Planning decision framework for brackish water aquifer, storage and recovery projects (unpublished Ph.D. dissertation). Gainesville, University of Florida.
Brown, C. J. (2006). Operational considerations and recommendations for the Everglades restoration ASR scheme. *Journal of Environmental Hydrology, 14,* 1–13.
Campbell, B. G., Conlon, K. J., Mirecki, J. E., & Petkewich, M. D. (1997). *Evaluation of aquifer storage recovery in the Santee Limestone/Black Mingo Aquifer, near Charleston, South Carolina, 1993–1995*. U.S. Geological Survey Water-Resources Investigations Report 96-4283.
Castro, J. E. (1995). *Aquifer storage and recovery, Myrtle Beach, South Carolina, Phase II: A hydrologic, geochemical and economic investigation,* Report 7. Columbia, SC: South Carolina Water Resources Division.
Castro, J. E., & Gardner, L. R. (1997). A geochemical model for aquifer storage and recovery project at Myrtle Beach, SC. In D. R. Kendall (Ed.), *Conjunctive use of water resources: Aquifer storage and recovery*, Proceedings AWRA Symposium, Long Beach, California (pp. 201–210). Middleburg, VA: American Water Resources Association.
Cederstrom, D. J. (1947). Artificial recharge of a brackish water well. *Commonwealth, Virginia Chamber of Commerce, 12*(12), 31, 71–73.
Cederstrom, D. J. (1957). *Geology and ground-water resources of the York, James Peninsula, Virginia*. U.S. Geological Survey Water-Supply Paper 1361.
CH2M Hill. (1997a). *Water supply needs and sources assessment, alternative water supply strategies investigation. A tool for assessing the feasibility of aquifer storage recovery*, Special Publication SJ97-SP4. Palatka: St. Johns River Water Management.
CH2M Hill. (1997b). *Water Supply Needs and Sources Assessment Alternative Water Supply Strategies Investigation: Artificial recharge of the Floridan Aquifer through drainage or injection wells in Orange and Seminole counties,* Special Publication SJ97-SP14. Palatka: St Johns River Water Management District.
CH2M Hill. (1998). *Central Florida artificial recharge demonstration program: Alternative water supply strategies in the St Johns River Water Management District*. SJWRWD Special Publication SJ98-SP11. Palatka: St. Johns River Water Management District.
Culkin, S. L., Singha, K., & Day-Lewis, F. D. (2008). Implications of rate-limited mass transfer for aquifer storage and recovery. *Ground Water, 46,* 591–605.
Diersch, H. J. (1998). *FEFLOW: Interactive, graphics-based finite-element simulation system for modeling groundwater flow, contaminant mass and heat transport processes*, User's Manual, Reference Manual, Version 4.6. Berlin, Germany: WASY, Institute for Water Resources Planning and System Research Ltd.
Dillon, P., & Molloy, R. (2006). *Developing aquifer storage and recovery (ASR) opportunities in Melbourne, Technical guidance for ASR* (Client Report prepared for the Smart Water Fund). Canberra: CSIRO Land Water.
Eccleston, B. L., & Harhash, I. E. (1982). *The hydro-geology of Qatar*, Water Resources and Agricultural Development Project Phase III. Qatar: Ministry of Industry and Agriculture.
Esmail, O. J., & Kimbler, O. K. (1967). Investigation of the technical feasibility of storing freshwater in saline aquifers. *Water Resources Research, 3,* 683–695.
Fournier, L. B. (2005). Horizontal wells in water supply applications. *Water Well Journal, 59*(6), 34–36.
Gale, I. N., Williams, A. T., Gaus, I., & Jones, N. K. (2002). *ASR—UK: Elucidating the hydrogeologic issues associated with aquifer storage and recovery in the UK.* UK Water Industry Research Report Ref. No. 02/WR/09/2. Nottingham: British Geological Survey.
Glanville, T. D. (1985). *Agricultural drainage wells in Iowa*. Ames: Iowa State University.

Goyal, V., Jhorar, B. S., Malik, R. S., & Streck, T. (2008). Performance evaluation of aquifer storage recovery for conjunctive water management as influenced by buffer storage volume and storage time. *Current Science, 94,* 465–472.

Greskowiak, J., Prommer, H., Vanderzalm, J., Pavelic, P., & Dillon, P. (2005). Modeling of carbon cycles and biogeochemical changes during injection and recovery of reclaimed water at Bolivar, South Australia. *Water Resources Research, 41,* W10418.

Greskowiak, J., Prommer, H., Vanderzalm, J., Pavelic, P., & Dillon, P. (2006). Quantifying biogeochemical changes during ASR of reclaimed water at Bolivar, South Australia. In *Recharge systems for protecting and enhancing groundwater resources*, Proceedings of the 5th International Symposium on Management of Aquifer Recharge, Berlin, Germany, June 11–16, 2005 (pp. 360–365). Paris: UNESCO.

Guo, W., & Langevin, C. D. (2002). *User's guide to SEAWAT: A computer program for simulation of three-dimensional variable-density ground-water flow*. U.S. Geological Survey Open-File Report 01-434.

Guo, W., Coulibaly, K., & Maliva, R. G. (2014). Simulated effect of aquifer heterogeneity on ASR system performance. *Environmental Earth Sciences, 31*(12), 7803–7809.

Händel, F., Liu, G., Dietrich, P., Liedl, R., & Butler, J. J. (2014). Numerical assessment of ASR recharge using small-diameter wells and surface basins. *Journal of Hydrology, 517,* 54–63.

Händel, F., Bimder, M., Dietze, M., Liedl, R., & Dietrich, P. (2016). Experimental recharge by small-diameter wells: The Pirna, Saxony, case study. *Environmental Earth Sciences, 75,* 930. https://doi.org/10.1007/s12665-016-5701-7.

Harrington, G. A., Love, A. J., & Sanford, W. E. (2002). Aquifer storage and recovery in a fractured rock aquifer of the Claire Valley, South Australia. In P. J. Dillon (Ed.), *Management of aquifer recharge for sustainability* (pp. 315–318). Lisse: A.A. Balkema.

Hashim, M. A. (2006). *Water, agriculture and environment in arid lands. Water and agriculture visions for Qatar by 2020*. Doha: Friends of Environment Centre.

INTERA Environmental Consultants. (1979). *Revision of the documentation for a model for calculating the effects of liquid waste disposal in deep saline aquifers.* U.S. Geological Survey Water-Resources Investigations 79–96.

INTERCOMP Resource Development and Engineering. (1976). *A model for calculating effects of liquid waste disposal in deep saline aquifers, Part 1—Development, Part 2—Documentation.* U.S. Geological Survey Water Resources Investigations 76–61.

Kimbler, O. K., Kazmann, R., & Whitehead, W. R. (1973). Saline aquifers—future storage reservoirs for fresh water. In J. Braunstein (Ed.), *Underground water management and artificial recharge,* Publication 110 (pp. 192–206). London: International Association of Hydrological Sciences.

Kimbler, O. K., Kazmann, R., & Whitehead, W. R. (1975). *Cycle storage of fresh water in saline aquifers.* Bulletin 10. Baton Rouge: Louisiana Water Research Institute.

Kimery, J. O., & Fayard, L. D. (1984). *Geohydrologic reconnaissance of drainage wells in Florida*. U.S. Geological Survey Water-Resources Investigations Report 84-4021.

Kloppmann, W., Aharons, A., Chikurel, H., Dillon, P., Gaus, I., Guttman, J., Kraitzer, T., Kremer, S., Masciopinto, C., Miotinski, K., Pavelic, P., Pettenati, M., & Picot-Colbeaux, G. (2012). In C. Kazner, T. Wintgens & P. Dillon (Eds.), *Water reclamation technologies for safe managed aquifer recharge* (pp. 311–250). London: IWA Publishing.

Libra, R. D., & Hallberg, G. R. (1993). *Agricultural drainage wells in Iowa: Hydrogeological settings and water-quality implications* (Technical Information Series 24). Iowa Department of Natural Resources.

Liu, G., Knobbe, S., Reboulet, E. C., Whittemore, D. O., Händel, F., & Butler, J. J. (2016). Field investigation of a new recharge approach for ASR projects in near-surface aquifers. *Groundwater, 54*(3), 425–433.

Lowry, C. S., & Anderson, M. P. (2006). An assessment of aquifer storage recovery using ground water flow models. *Ground Water, 44*(5), 661–667.

Lu, C., Du, P., Chen, Y., & Luo, J. (2011). Recovery efficiency of aquifer storage and recovery (ASR) with mass transfer limitation. *Water Resources Research, 47,* W08529.

Ludwig, R. D., Drake, R. L., & Strenitzke, D. A. (1990). *Agricultural drainage wells: Impact on ground water*. Ada, OK: U.S. Environmental Protection Agency Robert S. Kerr Environmental Research Laboratory.

Malik, R. S., Jhorar, B. S., Jhorar, R. K., Streck, T., & Richter, J. (2002). Long-term successful operation of existing brackish cavity wells for ASR to improve quality for irrigation by Indian farmers. In P. J. Dillon (Ed.), *Management of aquifer recharge for sustainability* (pp. 465–468). Lisse: A.A. Balkema.

Malik, R. S., Yadav, D. K., Jhorar, B. S., Jhorar, R. K., & Streck, T. (2006). Physical and chemical interactions in an ASR cavity well in a brackish aquifer. In *Recharge systems for protecting and enhancing groundwater resources*, Proceedings of the 5th International Symposium on Management of Aquifer Recharge, Berlin, Germany, June 11–16, 2005 (pp. 121–126). Paris: UNESCO.

Maliva, R. G. (2014). Groundwater banking: Opportunities and management challenges. *Water Policy, 16*(1), 144–156.

Maliva, R. G., & Missimer, T. M. (2008). ASR, useful storage, and the myth of residual pressure. *Ground Water, 46,* 171.

Maliva, R. G., & Missimer, T. M. (2010). *Aquifer storage and recovery and managed aquifer recharge using wells: Planning, hydrogeology, design, and operation*. Houston: Schlumberger Corporation.

Maliva, R. G., & Missimer, T. M. (2012). *Arid lands water evaluation and management*. Berlin: Springer.

Maliva, R. G., Guo, W., & Missimer, T. M. (2005). Hydrogeology and aquifer storage and recovery system performance. *Gulf Coast Associations of Geological Societies Transactions, 55,* 473–485.

Maliva, R. G., Guo, W., & Missimer, T. M. (2006). Aquifer storage and recovery: Recent hydrogeological advances and system performance. *Water Environment Research, 78,* 2428–2435.

McDonald, M. G., & Harbaugh, A. W. (1988). *A modular three-dimensional finite-difference ground water flow model*. U.S. Geological Survey Techniques of Water-Resources Investigation Report 06-A1.

Merritt, M. L. (1985). *Subsurface storage of freshwater in south Florida: A digital model analysis of recoverability*. U.S. Geological Survey Water-Supply Paper 2261.

Merritt, M. L. (1986). Recovering fresh water stored in saline limestone aquifers. *Ground Water, 24*(4), 516–529.

Mirecki, J. E. (2006). *Geochemical models of water-quality changes during aquifer storage recovery (ASR) cycle tests, Phase 1: Geochemical models using existing data,* Final Report ERDC/EL TR-06-8: Jacksonville: U.S. Army Corps of Engineers.

Mirecki, J. E., Campbell, B. G., Conlon, K. J., & Petkewich, M. D. (1998). Solute changes during aquifer storage recovery testing in a limestone/clastic aquifer. *Ground Water, 36,* 394–403.

Missimer, T. M., Guo, W., Walker, C. W., & Maliva, R. G. (2002). Hydraulic and density considerations in the design of aquifer storage and recovery systems. *Florida Water Resources Journal*, Feb 2002, 30–36.

Natural Research Council. (2008). *Prospects for managed underground storage of recoverable water*. Washington, DC: National Academies Press.

Nutter, N., & Gin, G. (2016). Aquifer storage and recovery well systems factory pumping injection testing: Is this necessary? International Symposium on Managed Aquifer Recharge (ISMAR9), Mexico City, June 20–24, 2016.

O'Shea, M. (1994). Borehole recharge—A drought management strategy in N. London, UK. In A. I. Johnson & R. D. G. (Eds.), *Artificial recharge of ground water, II: Proceedings of the Second International Symposium on Artificial Recharge of Ground Water* (pp. 210–219). Reston, VA: American Society of Civil Engineers.

O'Shea, M. J., Baxter, K. M., & Charalambous, A. N. (1995). The hydrogeology of the Enfield-Haringey artificial recharge scheme. *Quarterly Journal of Engineering Geology, 28,* S115–S129.

Parkhurst, D. L., & Appelo, C. A. J. (1999). PHREEQC (Version 2)—*A computer program for speciation, batch reaction, one-dimensional transport, and inverse geochemical calculations*. U.S. Geological Survey Water-Resources Investigations Report 99-42549.

Parkhurst, D. L., & Kipp, K. L. (2002). Parallel processing for PHAST—A three-dimensional reactive-transport simulator. In S. M. Hassanizadeh, R. J. Schotting, W. G. Gray & G. G. Pinder (Eds.), *Computational methods in water resources, Developments in Water Science* 47 (pp. 711–718). Amsterdam: Elsevier.

Parsons, S., Dillon, P., Irvine, E., Holland, G., & Kaufman, C. (2012). *Progress in managed aquifer recharge in Australia*; Waterlines Report Series No. 73. Canberra: National Water Commission.

Pavelic, P., Dillon, P. J., & Simmons, C. T. (2002). Lumped parameter estimation of initial recovery efficiency during aquifer storage and recovery. In P. J. Dillon (Ed.), *Management of aquifer recharge for Sustainability* (pp. 285–290). Lisse: A.A. Balkema.

Pavelic, P., Dillon, P. J., & Simmons, C. T. (2006). Multiscale characterization of a heterogeneous aquifer using an ASR operation. *Ground Water, 44,* 55–164.

Petkewich, M. D., Parkhurst, D. L., Conlon, K. J., & Mirecki, J. E. (2004). *Hydrologic and geochemical evaluation of aquifer storage recovery in the Santee Limestone/Black Mingo aquifer, Charleston, South Carolina, 1998–2002*. U.S. Geological Survey Scientific Investigations Report 2004-5046.

Plummer, L. N., Prestemon, E. C., & Parkhurst, D. L. (1991). *An interactive code (NETPATH) for modeling NET geochemical reactions along a flow PATH*. U.S. Geological Survey Water-Resources Investigations Report 91-4078.

Plummer, L. N., Prestemon, E. C., & Parkhurst, D. L. (1994). *An interactive code (NETPATH) for modeling NET geochemical reactions along a flow PATH–version 2.0.* U.S. Geological Survey Water-Resources Investigations Report 94-4169.

Pollock, D. W. (1994). *User's guide for MODPATH/MODPATH-PLOT version 2: A particle tracking post-processing package for MODFLOW, the U.S. Geological Survey finite-difference groundwater flow model*. U.S. Geological Survey Open-File Report 94-464.

Prommer, H., & Stuyfzand, P. J. (2005a). Identification of temperature-dependent water quality changes during deep well injection in a pyretic aquifer. *Environmental Science and Technology, 39,* 2200–2209.

Prommer, H., & Stuyfzand, P. J. (2005b). PHT3D modeling of water quality changes during deep well injection at Someren. In P. Dillon & S. Toze (Eds.), *Water quality improvements during aquifer storage and recovery. Volume 1: Water quality improvement processes*, Report 91056F, pp. 229–244. Denver: AWWA Research Foundation.

Prommer, H., & Stuyfzand, P. (2006). On the use of reactive multicomponent transport modeling for assessing water quality changes during managed aquifer recharge. In *Recharge systems for protecting and enhancing groundwater resources, Proceedings of the 5th International Symposium on Management of Aquifer Recharge*, Berlin, Germany, June 11–16, 2005 (pp. 415-420). Paris: UNESCO.

Prommer, H., Barry, D. A., & Zheng, C. (2003). MODFLOW/MT3DMS based reactive multicomponent transport modeling. *Ground Water, 41,* 247–257.

Pyne, R. D. G. (1995). *Groundwater recharge and wells*. Boca Raton: Lewis.

Pyne, R. D. G. (2005). *Aquifer storage recovery, a guide to groundwater recharge through wells*. Gainesville, FL: ASR Systems.

Pyne, R. D. G. (2006). Recent advances in ASR technology in the United States, In Recharge systems for protecting and enhancing groundwater resources. In *Recharge systems for protecting and enhancing groundwater resources*, Proceedings of the 5th International Symposium on Management of Aquifer Recharge, Berlin, Germany, June 11–16, 2005 (pp. 139–144). Paris: UNESCO.

Pyne, R. D. G. (2015). Aquifer storage recovery: An ASR solution to saltwater intrusion at Hilton Head Island, South Carolina, USA. *Environmental Earth Sciences, 73*(12), 7851–7859.

Pyne, R. D. G., & Howard, J. B. (2004). Desalination/aquifer storage recovery (DASR): A cost effective combination for Corpus Christi, Texas. *Desalination, 165,* 353–367.

Quiñones-Aponte, V., & Wexler, E. J. (1995). *Preliminary assessment of injection, storage, and recovery of freshwater in the Lower Hawthorn Aquifer, Cape Coral, Florida*. U.S. Geological Survey Investigations Report 94-4121.

Reese, R. S. (2002). *Inventory and review of aquifer storage and recovery in southern Florida*. U.S.G.S. Water-Resources Investigations Report 02-4036.

Reese, R. S., & Alvarez-Zarikian, C. A. (2007). *Hydrogeology and aquifer storage and recovery performance in the Upper Floridan Aquifer, Southern Florida*. U.S. Geological Survey Scientific Investigations Report 2006-5239.

Rinck-Pfeiffer, S., Pitman, C., & Dillon, P. (2006). Stormwater ASR in practice and ASTR under investigation in Salisbury, South Australia, In *Recharge systems for protecting and enhancing groundwater resources*, Proceedings of the 5th International Symposium on Management of Aquifer Recharge, Berlin, Germany, June 11–16, 2005 (pp. 151–159). Paris: UNESCO.

Schaetzle, W. J., Brett, C. E., Grubbs, D. M., & Seppanen, M. S. (1980). *Thermal energy storage in aquifers*. New York: Pergamon Press.

Schiner, G. R., & German, E. R. (1983). *Effects of recharge from drainage wells on quality of water in the Floridan Aquifer in the Orlando area, central Florida*. U.S. Geological Survey Water Resources Investigation Report 82-4094.

Schwaller, K. (2015, February 22). Ag drainage wells, a thing of the past. *The Messenger* (Fort Dodge, IA).

Segalen, A.-S., Pavelic, P., & Dillon, P. (2005). *Review of drilling, completion and remediation methods for ASR wells in unconsolidated aquifers*, Technical Report No. 04/05. Canberra: CSIRO Land and Water.

Segalen, A.-S., Pavelic, P., Dillon, P., & McCann, D. (2006). Review of the effects of drilling and completion methods of ASR wells in unconsolidated aquifers. In *Recharge systems for protecting and enhancing groundwater resources*, Proceedings of the 5th International Symposium on Management of Aquifer Recharge, Berlin, Germany, June 11–16, 2005 (pp. 160–163). Paris: UNESCO.

Shah, T. (2009). Climate change and groundwater: India's opportunities for mitigation and adaptation. *Environmental Research Letters, 4*(3), 035005.

Shah, T., Roy, A. D., Qureshi, A. S., & Wang, J. (2003). Sustaining Asia's groundwater boom: An overview of issues and evidence. *Natural Resources Forum, 27*(2), 130–141.

Sheffield, C.W., Rivero-deAguillar, C., & McCann, K. (1995, March). Drainage wells—A dual purpose. *Florida Water Resources Journal*, 51–55.

Sheng, Z., Arthur, J. A., Restrepo, J., & Allen-King, R. (2007). On recovery efficiency of managed underground storage of recoverable water systems. In P. Fox (Ed.), *Management of aquifer recharge for sustainability. Proceedings of the 6th International Symposium on Managed Aquifer Recharge of Groundwater* (pp. 84–95). Phoenix: Acacia Publishing.

Stoner, R. F., & Bakiewicz, W. (1993). Scavenger wells-1—Historic developments. In E. Custodio & A. Galofré (Eds.), *Study and modelling of saltwater intrusion into aquifers, Proceedings of the 12th Saltwater Intrusion Meeting*, Barcelona, Spain, 1–6 November 1992. Barcelona, Spain, pp. 545–556.

Sultana, S., Ahmed, K. M., Mahtab-Ul-Alam, S. M., Hasan, M., Tuinhof, A., Ghosh, S. K., et al. (2015). Low-cost aquifer storage and recovery: Implications for improving drinking water access for rural communities in coastal Bangladesh. *Journal of Hydrologic Engineering, 20,* B5014007.

Taneja, D. S., & Khepar, S. D. (1996). Effect of artificial ground-water recharge on aquifer parameters using cavity well. *Ground Water, 34,* 335–340.

Todd, D. K. (1959). *Annotated bibliography on artificial recharge of ground water through 1954*. U.S. Geological Survey Water-Supply Paper 1477.

United Nations Environment Programme. (2001). *Sourcebook of alternative technologies for freshwater augmentation in West Asia,* IETC Technical Publications Series Issue 8F. United Nations Environment Programme, Division of Technology, Industry and Economics.

USEPA. (1999). *The Class V underground injection control study. Volume 2. Agricultural drainage wells*. Washington DC: USEPA, Office of Ground Water and Drinking Water.

Voss, C. I., & Provost, A. M. (2002). *SUTRA, A model for saturated-unsaturated variable-density ground-water flow with solute or energy transport*. U.S. Geological Survey Water-Resources Investigations Report 02-4231.

Wallis, I., Prommer, H., Simmons, C. T., Post, V., & Stuyfzand, P. J. (2010). Evaluation of conceptual and numerical models for arsenic mobilization and attenuation during managed aquifer recharge. *Environmental Science and Technology, 44,* 5035–5041.

Wallis, I., Prommer, H., Pichler, T., Norton, S. B., Annable, M. D., & Simmons, C. T. (2011). Process-based reactive transport model to quantify arsenic mobility during aquifer storage and recovery of potable water. *Environmental Science and Technology, 45,* 6924–6931.

Ward, J. D., Simmons, C. T., & Dillon, P. J. (2007). A theoretical analysis of mixed convection in aquifer storage and recovery: How important are density effects? *Journal of Hydrology, 343*(3), 169–186.

Ward, J. D., Simmons, C. T., & Dillon, P. J. (2008). Variable-density modelling of multiple-cycle aquifer storage and recovery (ASR): Importance of anisotropy and layered heterogeneity in brackish aquifers. *Journal of Hydrology, 356*(1), 93–105.

Ward, J. D., Simmons, C. T., Dillon, P. J., & Pavelic, P. (2009). Integrated assessment of lateral flow, density effects and dispersion in aquifer storage and recovery. *Journal of Hydrology, 370*(1), 83–99.

WaterAid in Nepal. (2011). *Rainwater harvesting for recharging shallow groundwater*. Kathmandu, Nepal: WaterAid in Nepal.

Zack, A. L. (1988). *A well system to recover usable water from a freshwater-saltwater aquifer in Puerto Rico*. U.S. Geological Survey Water-Supply Paper 2328.

Zheng, C., & Wang, P. P. (1999). *MT3DMS: A modular three-dimensional multi-species model for simulation of advection, dispersion and chemical reactions of contaminants in ground water systems: Documentation and user's guide*, Report SERDP-99-1. Vicksburg, MS: U.S. Army Engineer Research and Development Center.

Zuurbier, K. G. (2015). Increasing freshwater recovery upon aquifer storage. A field and modelling study of dedicated aquifer storage and recovery configurations in brackish-saline aquifers. Ph.D. Dissertation, Technische Universiteit Delft.

Zuurbier, K. G., & Stuyfzand, P. J. (2017). Consequences and mitigation of salt water intrusion induced by short-circuiting during aquifer storage and recovery in a coastal subsurface. *Hydrology and Earth System Sciences, 21,* 1173–1181.

Zuurbier, K. G., Bakker, M., Zaadnoordijk, W. J., & Stuyfzand, P. J. (2013). Identification of potential sites for aquifer storage and recovery (ASR) in coastal areas using ASR performance estimation methods. *Hydrogeology Journal, 21*(6), 1371–1383.

Zuurbier, K. G., Zaadnoordijk, W. J., & Stuyfzand, P. J. (2014a). How multiple partially penetrating wells improve the freshwater recovery of coastal aquifer storage and recovery (ASR) systems: A field and modeling study. *Journal of Hydrology, 509,* 430–441.

Zuurbier, K., Zaadnoordijk, W. J., & Stuyfzand, P. (2014b). *The Freshmaker, An innovative tool for freshwater storage and abstraction in a salinized area using sophisticated horizontal drilled wells (HDDWs)*. Nieuwegein, Netherlands: KWR Water Research Institute.

Zuurbier, K. G., Kooiman, J. W., Groen, M. M. A., Maas, B., & Stuyfzand, P. (2015). Enabling successful aquifer storage and recovery of freshwater using horizontal direction drilled wells in a coastal aquifer. *Journal of Hydrologic Engineering, 20*(3), B4014003.

Zuurbier, K. G., Stuyfzand, P. J., & Hartog, N. (2016). Increasing freshwater recovery upon aquifer storage in brackish aquifers what can hydrological engineering bring? International Symposium on Managed Aquifer Recharge (ISMAR9), Mexico City, June 20–24, 2016.

Zuurbier, K. G., Raat, K. J., Paalman, M., Oosterhof, A. T., & Stuyfzand, P. J. (2017). How subsurface water technologies (SWT) can provide robust, effective, and cost-efficient solutions for freshwater Management in Coastal Zones. *Water Resources Management, 31*(2), 671–687.

Chapter 14
Groundwater Banking

14.1 Introduction

Groundwater banking broadly refers to the large-scale storage of water in aquifers for later use. The storage may be either seasonal or interannual. Interannual storage can provide a buffer against droughts and other interruptions in water supplies (e.g., disruptions of desalination facilities). Groundwater banking takes advantages of the general managed aquifer recharge (MAR) benefits of underground storage including:

- often lesser costs than surface storage options
- lesser environmental footprints than surface reservoirs and tanks
- avoidance of evaporative losses
- lesser susceptibility of stored water to contamination
- water may be recovered using wells at the point of use.

Groundwater banking is not synonymous with "water banking." MacDonnell et al. (1994) defined water banking as "an institutionalized mechanism specifically designed to facilitate the transfer of water use entitlements." O'Donnell and Colby (2010) similarly defined a water bank as "an institutional mechanism designed to facilitate transfers of water on a temporary, intermittent or permanent basis through voluntary exchange." Water banking serves to facilitate transfers of water from low-value to high-value uses by bringing buyers and sellers together (Frederick 1995; Clifford et al. 2004). Water banks may serve the roles of broker, clearinghouse, and/or market maker (Clifford et al. 2004; Maliva 2014). Brokers act to connect or solicit willing buyers and sellers to create sales. Clearinghouses primarily serve as a repository for bid and offer information. Market makers attempt to ensure that there are equal buyers and sellers in the market, and may increase liquidity by ensuring trades occur even when counter parties are not immediately available in the market.

Groundwater banking systems include both a physical component and a governance or institutional component. The physical component of groundwater banking involves direct recharge and in-lieu recharge. Direct recharge MAR is performed using either wells or surface-spreading methods. The in-lieu recharge concept gives

R. G. Maliva, *Anthropogenic Aquifer Recharge*,
Springer Hydrogeology, https://doi.org/10.1007/978-3-030-11084-0_14

credits for allowing groundwater that would normally have been pumped to remain in an aquifer by either not using water (e.g., curtailing irrigation) or by using an alternative water source (e.g., surface water). The in-lieu recharge concept recognizes that from an aquifer water-budget perspective, not withdrawing a volume of water that would otherwise have been pumped is equivalent to recharging the same volume of water. However, a key point is that there should be an actual reduction in groundwater extractions. The in-lieu recharge concept can be abused if credits are given for groundwater pumping that would not have otherwise occurred (Maliva 2014). For example, credits should not be given for not having irrigated with groundwater during a rainy period when there was no need for irrigation.

The institutional component of groundwater banking addresses regulatory issues, ownership of stored water, the amount, location and timing of withdrawals, and financial issues. Banked surface water is comingled with native groundwater. Reconciling groundwater rights is a critical institutional issue for groundwater banking systems. Preserving landowners' (and other existing users') right to access groundwater, while preserving stored surface-water for later extraction can be a complex undertaking (Pinhey 2003).

Groundwater banking ties into the conjunctive use of surface water and groundwater, which was defined by Morel-Seytoux (1985) "any scheme that capitalizes on the interaction of surface and ground waters to achieve a greater beneficial use than if the interaction were ignored." A common conjunctive use-groundwater banking scheme has surface water used for irrigation when available. Excess surface water is used to recharge a groundwater banking system. Groundwater is reserved for periods when surface water is either not available or is reserved for other uses (e.g., maintaining environmental flows).

Groundwater banking by direct recharge using wells could also be categorized as physical and regulatory storage ASR (Sect. 13.1). The goal of recharge is to increase the volume of water in storage, as manifested by an increase in aquifer water levels or pressures. Recharge of a given volume of water (i.e., a contribution to the groundwater bank) confers the right to later recover a volume equal to (or in some regulatory settings somewhat less than) the recharged volume.

Groundwater banking systems could be operated by and for the benefit of a single user, such as a governmental water management agency or utility. Alternatively, groundwater banking systems could have multiple participants in which credits are issued for water recharge, which are later cashed in for withdrawals. Systems could also be operated entirely or partially on a cash basis in which landowners with excess water would receive a cash payment for recharged water and users purchase a specified amount of stored water (Purkey et al. 1998). Some of the collected funds would be used to pay for the operation of the system.

Groundwater banking systems are diverse, but the systems constructed to date in the United States tend to share some characteristics:

- freshwater is recharged into aquifers containing freshwater or mildly brackish water that is still suitable for use
- water is stored in siliciclastic alluvial strata in basinal (bounded) aquifers (in the western USA)
- recharge is performed in overdrafted aquifers that have large available storage capacity
- recharge is performed by either surface spreading or injection
- recharge is usually performed using seasonally available excess surface water
- a single agency or user group owns and operates the system.

Groundwater banking systems are intended to operate in a general manner analogous to a financial bank. Just as a depositor to a financial bank expects to be able to recover his or her deposited money when needed, participants in a groundwater bank also expect to be able to recover recharged water when needed. Financial banks do not have cash on hand to redeem all deposits at once, and similarly, there are often physical, environmental, and regulatory limits on how much water can be pumped from an aquifer at any one time. To make one more analogy to a financial bank, the "solvency" of a financial or groundwater bank may not be evident until depositors try to make withdrawals (Maliva 2014). Aquifers have finite volumes of economically retrievable water, which may be exhausted if droughts are long and severe.

Depending upon the system, some or all of the recharged water may be imported from non-local sources and some of the recovered water may be used outside of the basin in which it is stored. The main operational issues facing groundwater banking systems are:

- less than 100% recovery of stored water
- drawdowns and associated hydrological impacts during recovery
- non-participants in systems extracting stored water.

Although well-executed groundwater banking makes sound water resources management sense, its implementation has not been without opposition. It has been claimed by opponents that groundwater banking is a plot to privative a public resource (Pitzer 2010). There are also legitimate concerns over the impacts of pumping during recovery. The issues of trust and control are central to implementing a successful groundwater banking program (Pinhey 2003). Target positive outcomes of groundwater banking systems are (Pinhey 2003):

- groundwater banks should store enough water to increase the reliability of the water supplies of participants
- groundwater bank operations should cause no harm to participants and non-participants
- groundwater bank operations should cause no harm to the aquifer or environment
- participant and nonparticipant water rights and rights of access should be preserved.

14.2 Aquifer Water Budget

Groundwater banking is intimately connected to the water budget of the storage aquifer. Fundamental technical issues are the immediate effect of recharge on the aquifer water budget and water levels (heads), and the amount of additional water available in the storage aquifer at the time of recovery. Aquifers are dynamic systems with multiple inputs and outflows, which need to be accounted for if a groundwater banking scheme is to be successful. Water budgets are the subject of Chap. 4. An aquifer water budget can be expressed using the following equation in which inflows and outflows are grouped in parentheses:

$$\Delta S = (R + AR + GH_{in} + GV_{in}) - (Q + D + GH_{out} + GV_{out}) \quad (14.1)$$

where (in units of volume/time)

ΔS	change in stored water volume
R	natural recharge
AR	anthropogenic recharge
GH_{in}	horizontal groundwater flow into the aquifer
GV_{in}	vertical groundwater flow into the aquifer
Q	groundwater pumping
D	discharges
GH_{out}	horizontal groundwater flow out of the aquifer
GV_{out}	vertical groundwater flow out of the aquifer

For an unconfined aquifer, inputs consist of local recharge (natural and anthropogenic) and any lateral or vertical flows into the aquifer. The main outflows are groundwater pumping, discharge (e.g., flows to springs, streams, and wetlands), and any flows out of the aquifer or basin.

A key point is that changes in the volume of water stored in an aquifer over any given time period depend on multiple inputs and outputs in additional to MAR. In some areas of California, for example, the lowering of the water table from historic pumping has resulted in a new equilibrium between pumping and recharge, as the lowering of the water table induced additional recharge from surface water bodies. Raising the water table through aquifer recharge may decrease induced recharge and result in a lesser net increase in aquifer storage (Purkey et al. 1998). It is important to have a firm understanding of how both MAR and subsequently recovery will impact an aquifer water budget. Some possible water budget scenarios include:

- Current aquifer water budget is balanced and recharged water would result in a corresponding increase in aquifer storage that persists until the time of recovery (best case scenario).
- Current aquifer water budget is balanced, but anthropogenic recharge impacts other water budget elements (e.g., increases discharge or leakage out of aquifer, or decreases natural recharge). The net increase in storage is less than the anthropogenic recharge volume.

- Aquifer water budget is out of balance (in overdraft) and anthropogenic recharge brings the water budget back into balance or decreases the rate of decline. At best, a net increase in storage occurs over time that is less than anthropogenic recharge volume. At worst, aquifer water levels continue to decline (albeit at a lesser rate) as aquifer storage continues to decline.
- Aquifer water budget is out of balance and anthropogenic recharge scheme encourages additional groundwater use.

The first option is usually the objective of groundwater banking systems; recharge is performed to increase the volume of water stored in an aquifer so that additional water is available at some time in the future when needed. Hence, monitoring data should demonstrate a corresponding increase in water levels over time. The increase in storage (ΔS) is estimated as the product of water level (head) increase (Δh) and storativity integrated over the area of the aquifer, and should roughly match the recharged volume (*AR*):

$$AR = \Delta S = \Delta h \cdot S \cdot A \tag{14.2}$$

Conceptual simulation results demonstrated the important concept that net recharge from a groundwater banking system can mask an aquifer overdraft from other aquifer users (Maliva 2014). Managed recharge can compensate for the overdraft and temporarily result in stable static water levels. However, once the extractions from the groundwater banking system start, greater drawdowns could occur from the combination of the existing aquifer overdraft and the additional extractions from the groundwater banking system. The conceptual simulation results show that large simulated dynamic head increases (mounding) can occur near a groundwater banking system wellfield during recharge, which will be quickly dissipated once recharge stops, revealing the previously masked effects of historic overdraft on aquifer static water levels (Maliva 2014).

The latter three unbalanced options are clearly suboptimal, but could still have value if they prolong the useful life of an aquifer or provide time to develop and implement sustainable water resources management plans. It is clearly important to have an accurate understanding of an aquifer water budget and realistic expectations as to the actual hydrological benefits of operation of a groundwater banking scheme.

14.3 Hydrological Impacts of Groundwater Banking Systems

Groundwater banking systems are not without potential third-party (non-participant) and environmental impacts. An important issue is avoiding injury to other groundwater users during recovery. Clear rules and limits on the recovery of water must be defined and the groundwater levels should not be allowed to drop below levels that would occur in the absence of a conjunctive use program (Pinhey 2003; Sandoval-

Solis et al. 2011). The effects of surface water withdrawals and groundwater-surface water interactions on the environment are another key issue. In California, for example, maintaining minimum environmental flows and temperatures required for the spawning of anadromous fish is a critical environmental issue. Increased drawdowns during recovery periods should not impact sensitive environments and stream flows upon which fish populations depend.

Morton (2015) performed an environmental and economic analysis of proposed groundwater banking schemes for the storing imported Colorado River Water by the Imperial Irrigation District (IID) of southern California. Unused annual allocation (referred to as "underruns") either goes to the beneficial use of junior water rights holders or remains in the main stem of the Colorado River. It was noted that the increased actual diversions of Colorado River water (although still within the IID allocation) would decrease river flows to the detriment of downstream ecosystems (e.g., Colorado River Delta). It was also observed that "soft path" solutions to future water scarcity, such as reduced irrigated area (fallowing) and increased irrigation efficiency, could have adverse impacts by reducing irrigation return flows, which are a water source for the Salton Sea.

Aquifers differ from water storage tanks in that where and when groundwater is recharged and recovered can result in hydrological impacts. Aquifers can be used, in essence, to convey water between hydrologically or logistically favored points of recharge and use. An important hydrological distinction is between dynamic aquifer responses to recharge and recovery and the aquifer-wide water budget responses (Fig. 13.1), which was illustrated by Maliva (2014) using a simple conceptual MODFLOW model. Local increases in aquifer water levels (heads) from injection and recovery are transient, and the occurrence of persistent residual local head increases is a "myth" (Maliva and Missimer 2008). Adverse impacts can occur locally even though the system is neutral in terms of the overall aquifer water budget (i.e., static water levels are not changed).

Perhaps the most critical hydrological issue for groundwater banking systems is that recovery will usually be constrained by dynamic aquifer responses during recovery (i.e., local drawdowns), rather than changes in static water levels related to changes in stored water volume. For example, consider a physical storage (freshwater to freshwater) ASR system once proposed in Florida, in which local groundwater pumping is constrained by dry season wetland impacts. The system concept was that recharge during the wet season, when excess water was available, would increase groundwater levels and allow for additional dry season withdrawals. However, in the absence of residual head increases (readily evident by basic groundwater modeling), operation of such a system would have actually increased dry season impacts by causing additional local drawdowns during recovery. A similar example would be where groundwater use is limited by seasonal spring flows.

Operational experiences have demonstrated that groundwater banking systems can have adverse hydrological impacts from large drawdowns during recovery. For example, impacts during recovery from the San Antonio Water System (SAWS) Twin Oaks ASR system in Texas were recognized and the Carrizo Aquifer Well Mitigation Program was implemented under an interlocal agreement between the

Evergreen Underground Water Conservation District and SAWS, even though mitigation was not required under Texas water law (Evergreen Underground Water Conservation District 2006). The well mitigation could involve lowering of pumps, drilling of replacement wells, or connection to an existing water purveyor. Similarly, some groundwater banking systems in California, such as systems in Kern County, experienced large drawdowns and associated impacts from recovery during a drought, which spurred multiple lawsuits (Barringer 2011). The 1995 "Memorandum of Understanding Regarding Operation and Monitoring of the Kern Water Bank Groundwater Banking Program (MOU)" includes mitigation measures for impacted overlying users, specifically lowering of pump bowls, provision of alternative supplies, and financial compensation (Pinhey 2003). The MOU holds that the project is subject to the "golden rule" whereby banking operations may not create conditions that are worse than would have prevailed absent the project. However, under drought conditions, the general question arises as to whether lower groundwater levels are due to "normal" groundwater extractions under drought conditions or recovery from the groundwater banking system (or a combination of both).

Contor (2009, 2010) proposed the use of surface water-aquifer response functions to equalize the hydrological value of recharge and abstraction with respect to time and location (Maliva 2014). Groundwater modeling may be employed to assess the impacts of proposed recharge and subsequent withdrawals on the water levels in surface water bodies, sensitive environments (e.g., wetlands), and/or spring flows. For example, "credits" (i.e., rights to later withdraw a given volume of water) granted for recharge would depend upon the degree to which the recharge at a specific location and time is demonstrated through modeling to result in an increase water levels or spring flows during a time period of concern (Maliva 2014). Similarly, the number of credits required to abstract a given volume of water would depend upon the modeled impacts of the extractions on water levels or spring flows during the period of concern. From an economic perspective, the application of aquifer-response functions turns water into a homogenous quantity as the withdrawal point and time no longer makes a difference with respect to credits issued or used (Contor 2010).

14.4 Water Accounting

As reviewed by Maliva (2014), adequately addressing institutional and management issues is critical for the success of groundwater banking systems. The system for assigning credits for recharged water and permitting withdrawals needs to be congruent with the actual supply of stored water and the amount of water that can safely be produced during a given time period. Using the water management parlance of the western United States, the amount of "paper water" (i.e., water that users have a right to withdraw) must be reconciled with the amount of "wet water" (i.e., water that can be safely extracted from the ground) in order for groundwater banking systems to be viable in the long term.

A key issue for a groundwater bank to be successful is that a bank's depositors must have assurance that water deposited in the bank will be available to them for withdrawal later when needed (Sandoval-Solis et al. 2011). Bank participants would clearly prefer to able to withdrawal any or all of their deposited water at their desired rate whenever they see fit. However, withdrawals may be limited by hydrological and environmental impact concerns. At a minimum, rules should be in place at the start of projects in which the rights of participants to withdraw water are clearly and firmly defined, so that an informed decision can be made as to whether to participate in the project. A water accounting system for groundwater banking systems serve to (Maliva 2014):

- protect the rights of participants to the banked water and ensure that they receive benefits commensurate with their deposits
- optimize the productivity of water by facilitating transfer of water credits
- ensure the sustainability of the system by preventing (or mitigating) adverse impacts associated with operation of systems.

Water banking systems assign credits in some form for recharged water and debit credits for water withdrawals by system participants. Contor (2009) proposed that the double-entry accounting method be used, in which every transaction is recorded as both a debit entry and a credit entry in separate ledger accounts. The recharge of a given volume of water by a system participant would be entered as a credit in the participants account and as a "liability" (i.e., water "owed") in the groundwater bank account. The basic requirement for a double-entry accounting, and the operation of a groundwater banks, is that both ledgers be balanced (Contor 2009).

The amount of credits that are "redeemable" should not exceed the "safe yield" of an aquifer, which was defined by Todd (1959) as

> the amount of water which can be withdrawn from it annually without producing an undesired result. Any draft in excess of safe yield is overdraft.

Safe yield is more meaningfully defined with respect to groundwater banking in terms of the amount of water than can be withdrawn during a given recovery period (rather than annually). Safe yield depends upon the ability to physically recover water and associated environmental and aquifer impacts (e.g., saline-water intrusion) that may limit groundwater withdrawals. Banked water that can be safely withdrawn also depends upon the withdrawals by other aquifer users who are not participants in a groundwater bank (Sect. 14.3). Hence, detailed monitoring of aquifer water levels (heads) are needed to quantify changes in storage over time and a well-calibrated groundwater model is required to evaluate the impacts of withdrawals and to quantity the safe yield.

In a properly functioning groundwater banking system, credits granted for future groundwater withdrawals are at least periodically reconciled to the volume of banked water that can be withdrawn from the system. Several solutions are available to address situations in which accumulated credits exceed an aquifer's safe yield (Maliva 2014). Where some recharged water is lost by leakage or other means, a discount could be applied to withdrawals (Contor 2009, 2010). A 4–10% loss (and/or contribution to the aquifer) is commonly assumed in California groundwater banking systems, which is subject to adjustment based on monitoring results (Thomas 2001; Pinhey 2003). Groundwater losses could be subtracted from the bank accounts proportionally to the amount of storage in each account (Sandoval-Solis et al. 2011). The discount could be applied either once at the time of recharge or there could be periodic (e.g., yearly) discounting of credits (Maliva 2014). When it is recognized that the accumulated credits exceed the safe yield of a system, all credits can be discounted (devalued) to bring the system back into balance.

Credits for earlier recharged water would undergo numerous devaluations over time, which creates a disincentive for long-term hoarding of water while maintaining incentives for new recharge intended for short and intermediate term use (Maliva 2014). Credits might alternatively be given a finite life and expire after a specified time if not used. Short credit lives (e.g., one year) minimize the risk of an imbalance between accumulated credits and system safe yield but reduce the value of a system to participants (Maliva 2014).

The total number of credits recovered in a given year can be restricted to an annual safe yield volume, which might be prorated amongst all credit holders based on the number of credits held (Maliva 2014). A drawback of this approach is that it reduces the value of accumulated credits and creates more uncertainty in a groundwater bank as a drought-proofing tool than participants may expect. Limiting annual abstractions would also not address the problem of a progressively increasing number of accumulated credits in a bank (Maliva 2014).

14.5 Institutional and Management Issues

Groundwater banking and other forms of MAR are relatively new concepts, and national, state, and local groundwater regulations may not adequately accommodate the concept. The testing, construction, and operation of elements of MAR systems, depending upon location, may come under the purview of a wide variety of governmental agencies and regulations. The types of regulations or regulatory issues that could impact MAR projects were reviewed by Maliva and Missimer (2010) and fall under the general categories of:

- water quality requirements for recharged water (underground injection controls rules)
- rights to source water used for recharge
- groundwater use regulations (as applied to recovery of stored water and pumping by other aquifer users)
- environmental protection regulations related to impacts of system operations (e.g., wetlands and endangered species)
- land development regulations
- building and construction codes.

For those not involved in land development projects, there can be surprising variety of regulations involved. For example, an unexpected issue that arose in obtaining approval for the construction of an ASR system in Collier County, Florida, was providing an adequate turnaround radius for fire-fighting vehicles on access roads to monitoring well sites.

Purkey et al. (1998) discussed the technical and regulatory issues associated with groundwater banking in the state of California. While many of the regulatory issues are state specific, the general concepts are broadly applicable. A basic requirement for any groundwater banking scheme is that some mechanism must be in place to prevent the stored water from being extracted by other aquifer users, particularly those who are not participating in the system. Use of the storage aquifer needs to be tightly regulated and not open to new allocations. There would be little value in a groundwater banking system if new users could freely withdraw additional water from the aquifer without contributing to the water stored in the system. Operation of the bank should also not impact the rights of existing native groundwater users.

Ward and Dillon (2009, 2012) addressed the coordination of MAR with natural resources management policies in Australia. Their Australia-centric ideas though have general applicability. In Australia, a robust separation of rights requires a three-tiered system of instruments to distribute and allocate volumes of water efficiently over time:

(1) **Entitlements**: define characteristics and number of unit shares of the defined consumptive pool and the distribution of shares among individual interests
(2) **Allocation**: process of periodically allocating the volume of water or aquifer storage space to each share
(3) **Use obligations**: establish the obligations of water use taking into account existing water users and third party effects.

One or all of the three instruments would be applied to each operational element of MAR systems as appropriate.

With respect to harvesting, which refers to the supply of recharge water, the system owner and operator needs legal access to the water. Where water is considered impaired (e.g., stormwater and reclaimed water), access to supplies has historically not been an issue as the waters were considered more of disposal problem rather than asset. In some areas, stormwater and reclaimed water are rapidly changing to an asset that has economic and commercial values (Ward and Dillon 2012). In Australia, jurisdictional rights to urban stormwater for MAR remains fragmented (Ward and Dillon 2012).

Recharge is controlled by environmental protection regulations that address both aquifer protection and access to underground storage space. Where MAR systems are sparse, access to underground storage space (provided that suitable aquifers are locally present) may not be an issue. However, as the density of systems increases, competition may arise over storage space. Ward and Dillon (2012) noted that in Australia, there is no fully specified and enforceable rights to storage space, which is the case in most other areas.

The recovery of recharged water is generally the most contentious issue. The right to extract water recharged in MAR systems in fully allocated and potentially overdrawn groundwater systems remains poorly or informally defined. There can be tension and conflict during times of water stress between MAR operators and native groundwater extractors (Ward and Dillon 2012). These periods of stress are precisely when water stored in MAR systems would be recovered to augment water supplies (Ward and Dillon 2009). The basic question is who has priority over limited water during times when water is scarce and there are thus both great demands on groundwater and limitations on how much groundwater can be produced. The question becomes more complex in the common situation where native groundwater extractors are senior to the MAR system operators (i.e., they were extracting groundwater long before the MAR system was constructed).

Ward and Dillon (2009, 2012) observed that to improve the security of water entitlements, MAR recovery entitlements are likely required to be institutionally differentiated from those governing entitlement to extract native groundwater. There is a little incentive to store water in an MAR system if the owner and operator cannot access the water during periods of scarcity when the water is most needed. Ward and Dillon (2012) proposed that entitlements to recharged water (recovery credits) should be endowed with a higher level of security than entitlements to native groundwater. While allowing groundwater banks to, in essence, jump to the head of the queue to withdraw water during periods of scarcity may be important for encouraging groundwater banking, this concept would be extraordinarily contentious where it has the net impact of reducing the ability of entrenched existing groundwater users to withdraw water when they need it.

Ward and Dillon (2012) also noted that the right to transfer credits for recovery is vital to the uptake of MAR as a groundwater management tool. Some restrictions on the amount of water than can recovered and the timing of recovery may be necessary. In systems in which freshwater is stored in brackish or saline water aquifer (i.e., chemically bounded systems; Maliva and Missimer 2008, 2010), the recovery efficiency of the systems limits the amount of water that can be recovered.

In systems in which freshwater is stored in an already utilized freshwater aquifer, the time in which recharged water may be recovered should reflect the hydraulic retention time, which can be considered the duration that recharged water will persist as additional storage in aquifers (Ward and Dillon 2012). Recovery of long-term water deposits are subject to a depreciation rate that reflects characteristics specific to the aquifer (Ward and Dillon 2009).

Groundwater use in the United States is regulated on the state level or intrastate regional level (e.g., by water management districts) and water use governance doctrines vary. Water use regulations tend not to fully address the needs and implications of MAR and, in some cases, even the realities of current water demands and uses. For example, the prior appropriation doctrine that is widely used to regulate surface water and, to a lesser degree, groundwater in the arid and semiarid western United States was adopted during the pioneer period when the population and water use were very low, and the primary objective was encouraging economic development rather than efficient water use.

14.6 Groundwater Banking in the Western USA

Groundwater banking schemes have been implemented in the semiarid to arid regions of the western United States in response to increasing water scarcity. The following summaries illustrate some of the policies and technical and regulatory issues facing groundwater banking systems. Despite the importance being placed on groundwater banking toward addressing water scarcity, there has been surprisingly little hard hydrological data published on the actual performance of systems, particularly on changes in stored water volume over the operational life of systems.

14.6.1 Arizona Groundwater Banking

Arizona is the most arid state in the United States with an average annual rainfall of 32.3 cm (12.7 in.) and even lower rainfalls in the more populated urban areas in the south-central part of the state. The central and southern parts of the state occur in the Basin and Range physiographic province in which groundwater occurs in thick siliciclastic strata in alluvial basins that are surrounded by essentially impervious bedrock mountains. Groundwater has historically been the primary water source in the region. Due to high irrigation use and low recharge rates, the aquifers are prone to overdraft with associated rapidly declining water levels (Anderson et al. 1992, 2007). Depletion of stored water is occurring but estimates of storage change based on water-balances and aquifer water levels have high uncertainties (Anderson et al. 2007).

The Arizona Legislature passed the Groundwater Management Act (GMA) in 1980 to address the serious groundwater overdraft that was occurring in some areas of the state, which were designated "Active Management Areas" (AMAs). Groundwater rights are quantified and regulated only in the AMAs. The primary goals of the AMAs for the most populous parts of the state were to achieve a safe yield by the year 2025, which was defined as achieving a long-term balance between annual groundwater withdrawals and natural and artificial recharge (Jacobs and Holway 2004; Eden et al. 2007; Pearce 2007; Megdal 2012). The GMA established the Arizona Department of Water Resources (ADWR) to implement and monitor compliance with the act. The GMA prohibits new agricultural irrigation using any water within AMAs.

The GMA mandated an Assured Water Supply (AWS) program in which new developments were allowed only if the developer could demonstrate an assured water supply for the next 100 years. The AWS program went into effect in 1995. To receive an AWS certificate, a proposed water supply must be physically, legally, and continuously available for the next 100 years and its use must be consistent with the goals of the AMAs, which include making substantial use of renewable supplies. The AWS program acts to prevent developers from mining groundwater and to provide protection to people purchasing or leasing subdivided land in AMAs by ensuring they will have an adequate supply and quality of water.

A 100-year supply of water can be met using groundwater, if groundwater use is offset by the recharge of renewable water, such as reclaimed water or surface water (Megdal 2012). A strong incentive for artificial recharge was created by giving credits for artificial recharge that can be used against groundwater withdrawals.

The Colorado River Compact of 1922 allocated Arizona 2.8 million acre-feet (MAF; 3,453.7 million m^3, MCM) annually of Colorado River water plus half of any surplus additional water in the lower Colorado River basin. At the time of the compact, Arizona had neither the demand for the water or a means to convey the water to the main irrigation areas in the central part of the state. California has the right to any water that Arizona leaves in the river.

The Central Arizona Project (CAP), fully completed in 1994, is a 541 km (336 mile) uphill canal that conveys water from the Colorado River to the Phoenix and Tucson metropolitan areas. The CAP began initial deliveries in 1985, but the water supply was substantially under-utilized into the early 1990s (Megdal et al. 2014). The construction of the CAP resulted in a strong impetus for Arizona to vigorously pursue groundwater banking to fully store any excess of its Colorado River allocation, which would otherwise be taken by California. The CAP is administered by the Central Arizona Water Conservation District (CAWCD).

The Arizona Water Banking Authority (AWBA) was created in 1996 to provide long-term underground storage of CAP water that Arizona was allotted but not yet using. The stored water may be used to protect municipal users from droughts or possible CAP disruptions, meet Native American water rights claims, assist in meeting local water management objectives, and facilitate interstate water banking (Jacobs and Holway 2004; August and Gammage 2007; Colby et al. 2007; Eden et al. 2007). The AWBA also stores water for the state of Nevada. In a time of shortage, Nevada would be able to use its credits for recharged water to withdraw additional surface water from the Colorado River, while Arizona would withdraw a corresponding lesser amount of surface water and use the banked water instead. As of 2013, aggregate of 3976 MCM (3.224 MAF) has been stored for intrastate purposes and 740 MCM (0.6 MAF) has been stored on behalf of the State of Nevada (Megdal et al. 2014).

The Central Arizona Groundwater Replenishment District (CAGRD) was established in 1993 to acquire excess Colorado River water and store it underground on behalf of developers without direct access to surface water (August and Gammage 2007; Peace 2007). The CAGRD also stores treated wastewater. Entities that cannot meet the requirements of the AWS program (100-year assured supply) have the option of paying a fee to the CAGRD for the groundwater a subdivision or water provider is using or will use in the future. The CAGRD takes responsibility for acquiring and replenishing water to offset the "mined" groundwater (Groundwater Awareness League n.d.). The CAGRD is unique in Arizona in that it is allowed to perform replenishment after the fact (i.e., after groundwater withdrawals; Megdal 2007). The CAGRD must find water and perform replenishment within three years of the excess groundwater use (i.e., use in excess of amounts allowable under the rules). A major challenge that the CAGRD faces is that it has virtually no access to firm supplies of water and is dependent on the availability of surplus water for recharge (Jacobs and Holway 2004; Megdal 2007).

Under Arizona statutes, there is no legal requirement that groundwater replenishment be hydrogeologically connected to pumping, although it must occur in the same AMA (Jacobs and Holway 2004; Megdal 2012). Depending on the depth to groundwater, the location of the pumping site relative to the recharge site, and aquifer hydraulics, local adverse impacts (e.g., unacceptably large drawdowns) may still occur. Impacts during recovery may be addressed through the recovery permitting process. A goal is to avoid recovery in areas where water levels are declining at more than a specified level (Megdal 2012).

Three types of permits are required for an MAR system, a facility permit, storage permit, and recovery permit. Facility permits are divided into underground storage facilities (USFs) and groundwater savings facilities (GSFs). USFs are constructed facilities (e.g., infiltration basins and injection wells) and facilities that use natural channels for MAR. GSFs are "in lieu" recharge facilities where surface or reclaimed water is used instead of groundwater, saving groundwater.

An underground storage facility (USF) permit allows the permit holder to operate a facility that stores water an aquifer. The criteria a USF must meet to be permitted include:

- the project must be hydrologically feasible
- the applicant must demonstrate financial and technical capability
- the applicant must agree in writing to obtain any required floodplain use permit from the county flood control district before beginning any construction activities
- the project may not cause unreasonable harm to other land or water users within the area of impact
- the project will continue to be monitored to ensure storage does not cause the migration of poor quality water.

A Water Storage (WS) Permit (Arizona Revised Statutes (A.R.S.) § 45-831.01) allows a permit holder to store water at a USF. Multiple entities may have permits to store water in a given USF. In order to store water, the applicant must provide to the Department evidence of its legal right to the source water proposed for recharge. A contract for CAP water must be submitted to the Department prior to storing CAP water obtained pursuant to that contract. The WS permit holder may choose to recover the water in the same calendar year or obtain long-term storage credits. A storer is entitled to a 95% credit for long-term storage of CAP water with the remaining 5% to remain as a cut to the aquifer (Eden et al. 2007; Megdal 2007). A 100% credit is granted if the water is recovered the same year that it is stored.

A Recovery Well (RW) Permit (A.R.S. § 45-834.01) allows a permit holder to recover long-term storage credits or to recover stored water annually. The impacts of recovering stored water in the proposed location must not damage other land and water users. An impact analysis is required under certain circumstances. Water does not have to be recovered at the same location at which it was recharged but it must occur within the same AMA. Recharge may occur in a part of a basin that is not experiencing the most significant declines in groundwater levels (due to land availability and proximity to surface water sources). The ADWR looks at the rate of decline at proposed recovery well sites versus rates established in respective AMA management plans (Megdal 2007).

Characteristics favoring groundwater banking for water security in Arizona include (Megdal et al. 2014):

- an awareness that augmentation of groundwater resources is necessary to address aquifer depletion and future imbalances between supply and demand
- availability of a water source that can used for intermittent or continuous recharge
- favorable hydrogeological conditions including suitable storage space and aquifer transmissivity
- a well-established regulatory framework that is adhered to by water users and ensures that system owners and participants obtain commensurate benefits from the water they recharged
- funding mechanisms that facilitate investments in water banking system planning, construction, and operation
- favorable institutional arrangements that links policy with investments.

14.6.2 Southern Nevada Groundwater Bank

The Southern Nevada Groundwater Bank (SNGB; also referred to as the Las Vegas Valley Water District artificial recharge or ASR system) is one of the largest groundwater banking systems in the world, and is perhaps the largest system that uses wells for recharge. The Las Vegas Valley Basin is a desert environment that receives an average annual rainfall of only 4 in. (10 cm) and has very minimal local natural recharge. The rapid development and associated increases in groundwater pumping resulted in a change from flowing artesian conditions, when the first well was constructed in the valley in 1907, to depths to water of more than 300 ft (100 m) in parts of the valley by 1980 (Wood 2000). The declining water levels necessitated a shift to surface water supplies. The Southern Nevada Water Authority (SNWA), a wholesale water provider for the Las Vegas/Southern Nevada metropolitan area, currently obtains about 90% of its water from the Colorado River.

The Las Vegas Valley Water District (LVVWD) and City of North Las Vegas began MAR in 1987. Treated Colorado River water from Lake Mead is recharged into the valley's primary alluvial aquifer in years when Nevada's Colorado River allocation exceeds demands. The LVVWD (2017) reported that it currently has 52 dedicated and dual-use recharge/recovery wells with a total injection capacity of about 100 million gallons per day (Mgd; 379,000 MCM/d).

The development of the SNGB was summarized by Pyne (1995, 2005), Katzer et al. (1998), Donovan et al. (2002) and Bloetscher et al. (2005). The primary goals of the artificial recharge program are (Katzer et al. 1998; Donovan et al. 2002; Bloetscher et al. 2005):

- optimization of the use of the LVVWD Colorado River allocation, particularly the capture of surface water rights that would otherwise be lost

- provision of a reliable long-term water supply
- restoration of aquifer water levels, which would reduce the costs of well pumping, deepening, and re-drilling
- minimization of the potential for land subsidence from pumping.

The SNGB has particularly favorable hydrogeological conditions for the successful operation of a groundwater bank. The City of Las Vegas and neighboring communities originally obtained their water supply from groundwater in the Las Vegas Valley Basin (Fig. 14.1). The geology and groundwater resources of the Las Vegas Valley Basin were discussed by Maxey and Jameson (1948), Malmberg (1965), Plume (1989), Zikmund (1996) and Johnson and Donovan (1998). The Las Vegas Valley Basin formed primarily by a middle Miocene extensional event and is filled with up to 5,000 ft (1,520 m) of mostly siliciclastic deposits that range in age from Miocene to Holocene. The upper 1,000–1,200 ft (300–360 m) of the valley-fill deposits consist predominantly of coarse-grained (gravel and sand), fine-grained (silt and clay), and mixed siliciclastic alluvial deposits that originated from erosion of the nearby mountains (Plume 1989).

Low-quality water is locally present near the top of the valley-fill deposits. The underlying confined aquifers can be subdivided into three general zones with depth intervals of approximately 0–200 ft (0–61 m), 200–700 ft (61–213 m), and 700–1,000 ft (213–305 m) below land surface (bls; Maxey and Jameson 1948; Malmberg 1965; Plume 1989). The deeper two intervals are referred to as the "principal aquifer system." The middle and lower zones are used as the storage zones for the SNGB artificial recharge system. The Las Vegas Valley basin is largely closed (underlain and surrounded by low permeability rock), so significant leakage of stored water does not occur. The use of confined storage zones provides protection of the stored water from surficial contamination.

The 1939 Nevada Underground Water Act granted the State Engineer total jurisdiction over all groundwater in the state. Nevada groundwater law follows the doctrine of prior appropriation, meaning that the first person to file on a water resource for beneficial use is typically considered first for a permanent right to the water, subject to the State Engineer's determination of available, unappropriated water. The Las Vegas Valley Basin is fully allocated and in 1992 the State Engineer ceased issuing even temporary water-right permits, except under certain circumstances (Katzer et al. 1998). Nevada state water law allows for the long-term banking of water, which made the SNGB feasible. Net recoverable aquifer recharge (AR) storage is defined by the Las Vegas Valley Water District (2008) as the sum of injected Colorado River water and recoverable in-lieu recharge minus recovered water. Recoverable in-lieu recharge is counted as 85% of the non-pumped groundwater allocation. As of early 2016, 336,787 AF (415.4 MCM) of water was reported to be stored in the SNGB (Entsminger 2016).

A cost-benefit analysis of the SNGB artificial recharge system demonstrated that the overall benefits of the artificial recharge system are greater than its costs. Operation of the system benefits all aquifer users by lowering energy costs, decreasing the need to deepen wells, lessening impacts from land subsidence, and providing

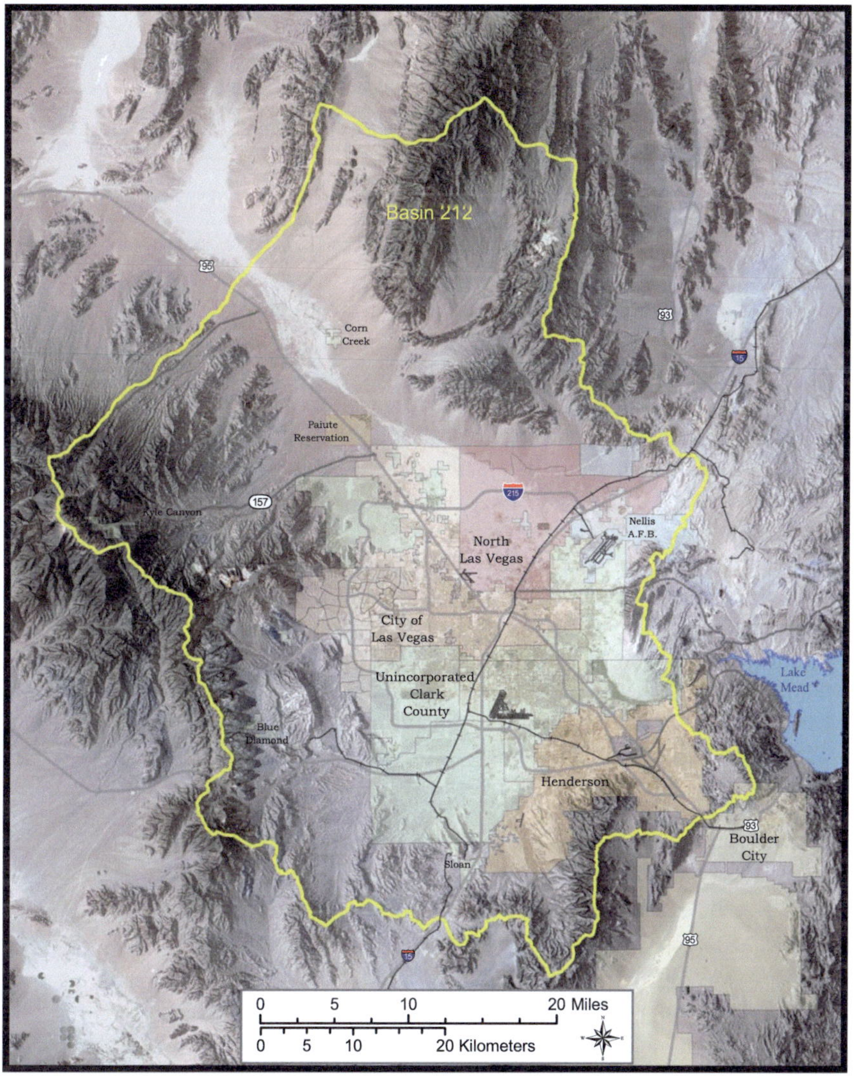

Fig. 14.1 Map showing the boundaries of the Las Vegas Valley Basin. Physical storage of water is possible because the basin in bounded by mountains composed mostly of low permeability rock (Courtesy of the Southern Nevada Water Authority)

additional water for the aquifer system (Katzer et al. 1998; Donovan et al. 2002). Non-municipal aquifer users were receiving free benefits from the system as they were not paying toward the system operation.

In 1997, the Nevada Legislature directed the SNWA to develop a program to protect and manage the Las Vegas Valley primary groundwater supply, which resulted in the creation of the Las Vegas Valley Groundwater Management Program. To fund the activities of the program, the SNWA bills an annual groundwater management fee to well owners and groundwater permit holders in the Las Vegas basin at a rate of $13 per year for domestic wells and $13 per acre-ft (AF) per year for all other types of wells. Approximately 60% of the fees are paid by municipal water purveyors and other government entities (the largest water rights holders in the valley).

14.6.3 California Groundwater Banking—Introduction

California faces severe water resources management challenges as the major population centers and agricultural areas of Southern California do not have adequate local freshwater resources to meet local demands. Many aquifers have long been in an overdraft condition, which has resulted in declining water levels and significant local land subsidence (Poland and Davis 1956; Poland 1961; Poland et al. 1975). Local water supplies are greatly augmented by fresh surface water conveyed from the wetter northern part of the state (Sierra Nevada mountains) and the Colorado River. Global climate change may have a serious impact on California's long-term water supply by reducing the thickness of the winter snow pack in the Sierra Nevada mountains, which supplies California's main rivers. California's overdrafted basin aquifers have ample storage capacity for groundwater banking.

California water law and regulatory policy with respect to MAR were reviewed by Parker (2007). The California Supreme Court adopted the doctrine of correlative rights with respect to percolating groundwater in 1903. Percolating groundwater is defined as groundwater that is not flowing as a subterranean stream (i.e., in a subsurface channel; California Department of Water Resources 2000). Under the correlative rights doctrine, the rights of all landowners over a groundwater basin to extract groundwater are coequal (usually based on land area owned) regardless of when first use was initiated, subject in California to the general constitutional requirement that the water use be reasonable and beneficial. If the ground water supply is inadequate to meet the needs of all users, then each user can be required to proportionally reduce use until the overdraft is ended.

Correlative shares of groundwater are not quantified in California unless the groundwater basin has been adjudicated. For an adjudicated groundwater basin, the court decides the groundwater rights of all landowners, how much groundwater well owners can extract, and who will be the Watermaster. The Watermaster ensures that a basin is managed according to the decrees of the court (California Department of Water Resources 2000, 2004). Groundwater management districts have been established that have the authority to limit or regulate groundwater extraction. Some cities and counties have also passed ordinances to regulate groundwater use. Cities may implement restrictions on new well installation to protect MAR systems.

Surface water is regulated by the state and a water rights permit is required for its use. For MAR using surface water, an underground storage supplement (USS) is required to be filed with the State Water Resources Control Board (SWRCB) that must document the point of diversion from the stream channel, the proposed diversions and conveyances, and the beneficial use of the water (Parker 2007).

14.6.4 California—Kern County

Kern County is a major agricultural area located at the southern end of the Central Valley (Fig. 14.2). The county is one of the major vegetable growing areas of the United States despite its arid climate. Bakersfield, the largest city in the county, receives on average only 16.5 cm (6.5 in.) of rainfall annually. The Kern County water sources include limited local surface water flows (Kern River), local groundwater, and imported water from the California State Water Project (SWP) and Central Valley Project (CVP). Groundwater banking has become so important to Kern County that almost every water district in the County participates in a banking program in some fashion (WAKC, n.d.). Kern County Groundwater Banking Programs include (WAKC, n.d.):

- Arvin-Edison Water Storage District Water Management Program
- Berrenda Mesa Property Joint Water Banking Project
- Buena Vista Water Storage District Water Management Program
- Buena Vista Water Storage District/West Kern Water District Water Supply Project
- Cawelo Water District/Dudley Ridge Water District Conjunctive Use Program
- Cawelo Water District's Modified Famoso Water Banking Project
- City of Bakersfield 2800 Acre Groundwater Recharge Facility
- Kern Delta Water District's Groundwater Banking Program
- Kern Water Bank
- North Kern Water District Groundwater Storage Project
- Thomas N. Clark Recharge and Banking Project
- Rosedale-Rio Bravo Water Storage District and Improvement District No. 4 Joint Use Groundwater Recovery Project
- Rosedale-Rio Bravo Water Storage District's Groundwater Banking Program
- Semitropic Groundwater Banking Project
- West Kern Water District's Groundwater Banking Program.

Christian-Smith (2011) provided an overview of groundwater banking in Kern County, which from 1962 to 2003 had a reported annual average net loss of groundwater of 1.4 M AF (1,727 MCM). Over the last four decades, the entire Central Valley was reported to have lost about 60 MAF (74,000 MCM) of groundwater. The associated drawdowns generated vast groundwater storage capacity in dewatered portions of the aquifer. MAR in Kern County started in the late 1970s and early 1980s using recharge ponds in alluvial fan deposits, which consist of high permeability sands well-suited for infiltration. As of 2011, there were ten groundwater banks in Kern County with a total maximum annual recharge capacity of 864,000 AF/year (1,066 MCM/year). The systems are recharged with water from local rivers and imported water from the SWP and CWP. In California, excess surface water is available for groundwater recharge in the winter and spring of some years from snowmelt.

The Kern Water Bank was reported to be able to store about 1.5 M AF (1850 MCM) of water, with a recharge capacity of about 500,000 AF/year (617 MCM/year) and recovery capacity of about half that (Patterson 2015). Jon Parker, General Manager of

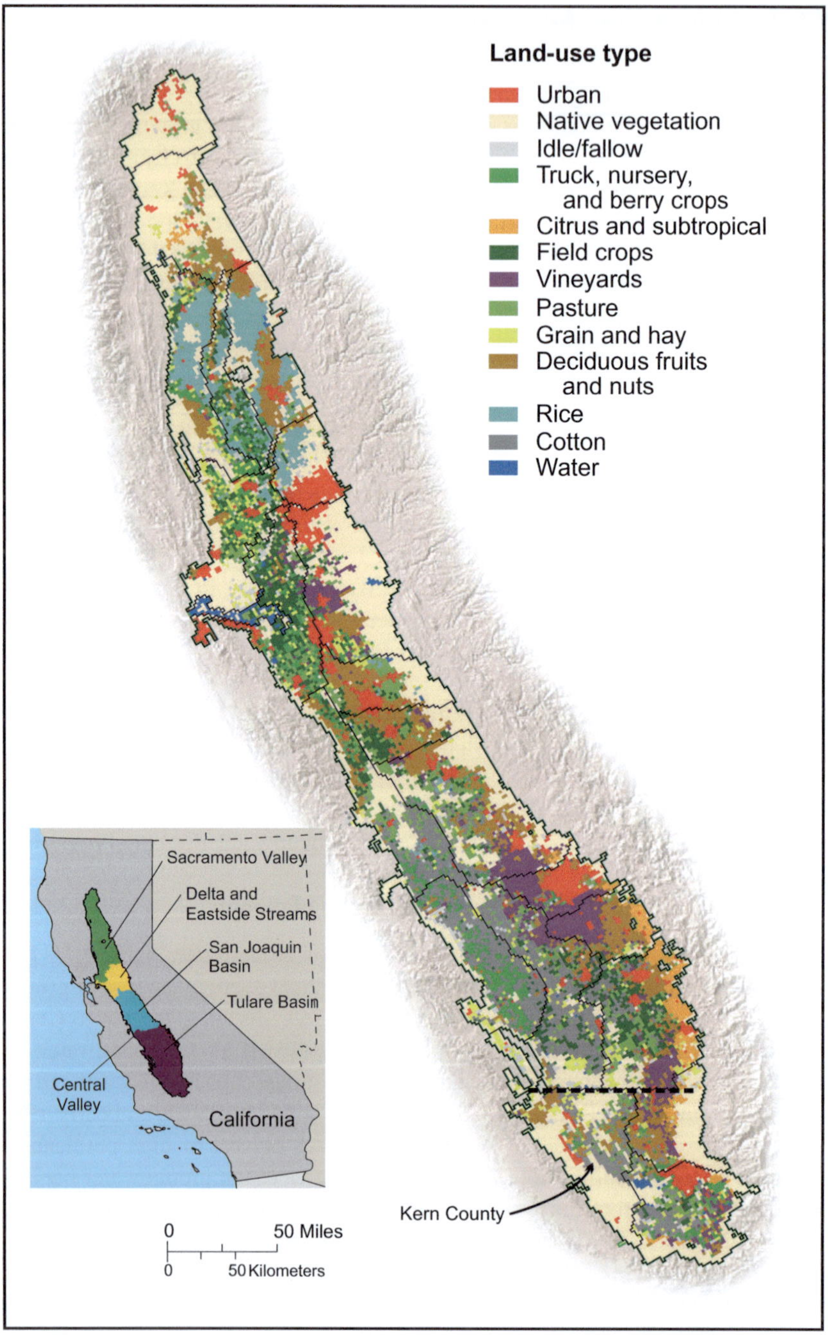

Fig. 14.2 Map showing the location of Kern County in the southern Central Valley of California and land uses (*Source* Faunt 2009)

the Kern Water Bank reported that the bank believes that 6% of the recharged water is lost and 94% is recoverable (Patterson 2015). Most of the Kern County water banks are storing water for wealthier out-of-basin interests, most notably the Metropolitan Water District, which serves Los Angeles and other southern California urban areas (Christian-Smith 2011). The Rosedale-Rio Bravo Water Storage District has a 2:1 bank requirement, which allows only 1 AF to be returned for every 2 AF of water banked in order to decrease local impacts and ensure that some water remains within the basin (Christian-Smith 2011).

A general challenge for groundwater banking in California is a lack of regulation on groundwater use in most of the state, which means that overlying owners may pump water from an underlying groundwater bank without permission or monitoring (Christian-Smith 2011). Christian-Smith (2011) observed that groundwater banking programs are best implemented as part of larger integrated planning efforts that include groundwater management.

The National Research Council (2008) summarized the Arvin-Edison Water Storage District groundwater banking system. The reported benefits were:

- stabilization of groundwater levels; pumping costs are less than they would otherwise have been with continued decline
- stabilization of groundwater levels allows users without access to surface water to continue to pump groundwater, avoiding the costs associated with extending a surface water system to them
- Groundwater is available in droughts when surface water is not available.

Arvin-Edison Water Storage District utilizes a hydrologically isolated aquifer in which there are no competing users in a position of reap benefits of the project as free riders.

Environmental impacts are a contentious issue concerning groundwater banks in Kern County. Recovery of water during a drought that began in 2007 was reported to have adversely impacted local water users leading to law suits (Barringer 2011). A Sacramento County Superior Court judge ruled in 2014 that the California Department of Water Resources never looked at the full ecological effects of running the Kern Water Bank when the state transferred the bank to private hands in 1997 (Burke 2014). Objections to the project include that it involves privatization of groundwater resources and that it would lead to unsustainable urban growth.

Pinhey (2003) examined the institutional arrangements that are essential for the successful conjunctive management of surface water and groundwater through groundwater banking in the Central Valley of California. The Kern Water Bank and Arvin-Edison Water Storage District systems were presented as examples of successful systems. Attributes of the Kern Water Bank that are favorable for successful groundwater banking include (Pinhey 2003):

- the Kern River Valley Basin is largely a closed basin that is surrounded on three side by essentially impermeable mountains and the other by low-permeability clayey strata
- the location of the system allows it to receive surface water from multiple sources

- the recharge area (Kern Fan Area) allows for high recharge rates (up to 15 cm/d, 6 in/d) via percolation ponds
- project participants and adjoining nonparticipants (stakeholders) are agriculture water, water storage, and irrigation districts that are governed by boards of local users that share common memberships in associations, which fosters communication and trust
- a shared belief that action needed to be taken to correct the overall water supply deficit in order to preserve the local economy (i.e., there are incentives to cooperate and collaborate)
- a shared desired to maintain local control of the groundwater supply and avoid adjudication, and to preserve rights to access the groundwater basin by overlying users.

Adequate trust and incentives are important for development of institutional arrangements for addressing uncertainty and thus gaining support for the system (Pinhey 2003). Pinhey (2003) used the initially proposed Madera Ranch Groundwater Bank Project (Madera County, Central Valley) as an example of an unsuccessful (abandoned) project. Opposition from local area farmers and user organizations stemmed from (Pinhey 2003):

- incomplete information
- lack of effective (early stakeholder) public involvement
- a perception that the project was top-down and politically driven
- hydraulic and water quality uncertainties
- a view that the project was a potential means for outside interests to gain access to the native groundwater and potentially surface-water entitlements
- a lack of trust in the project developers (U.S. Bureau of Reclamation and a private company).

The project was subsequently revived and a cooperative agreement between the Madera Irrigation District and U.S. Bureau of Reclamation was signed in 2011. It was observed by the local Congressman (Jim Costa) that the idea always had merit but it wasn't originally the right proposal (Western Farm Press 2011).

14.6.5 Las Posas Basin ASR Project

The Calleguas Municipal Water District (CMWD) was established in 1953 to provide southern Ventura County, California, with a reliable supply of high-quality supplemental water to meet the needs of a growing population and economy (Calleguas Municipal Water District 2008a). The CMWD has a semiarid climate and, like most of Southern California, now relies upon imported water to meet a large part of its water needs. The District obtains freshwater from the California SWP system of canals, reservoirs and pumping facilities, which convey water from the Sierra Nevada Mountains in Northern California to many regions throughout

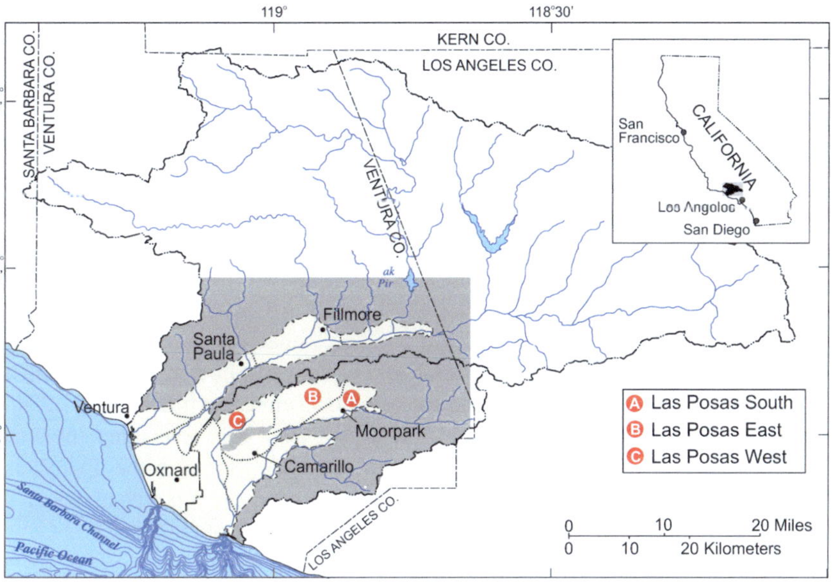

Fig. 14.3 Las Posas Basin in Ventura County, California (Modified from Hanson et al. 2003)

the state. To meet the long-term water demands of its service areas, the CWMD and Metropolitan Water District of Southern California (MWD) developed an ASR project in the Las Posas Groundwater Basin (Fig. 14.3) with the objective of storing up to 300,000 AF (370 MCM) of excess surface water obtained from the MWD. The stored water would be recovered if the state supply of water was reduced or disrupted. The water supply of the MWD is delivered through a single pipeline and, as such, is subject to a host of external forces, ranging from drought and earthquake damage to regulatory actions and water rights determinations (Calleguas Municipal Water District 2008b).

The Las Posas ASR system was summarized by Maliva and Missimer (2010). Water is injected using 18 dual-purpose injection and recovery wells. The injected water is treated to drinking water standards and the recovered water is treated again before it is distributed. The Las Posas Basin ASR system has an injection capacity of 40.7 Mgd (154,000 m^3/d) and a total extraction capacity of 58.2 Mgd (220,000 m^3/day). An average of about 1,500 AF/year (489 Mg/year; 1.85 MCM/year) was injected between 2002 and 2005 (Metropolitan Water District of Southern California 2007). The annual injected volume was thus a small fraction of the estimated total storage capacity (300,000 AF, 98,000 Mg, 370 MCM) of the groundwater basin.

The Las Posas Basin ASR system is a physical-storage type ASR system in that its expressed purpose is to increase the amount of water in storage in the Las Posas Groundwater Basin. Water use in the Las Posas Groundwater Basin is regulated by the Fox Canyon Groundwater Management Agency (FCGMA), which is an independent, special district that was established by the California Legislature in 1983 to oversee the Ventura County groundwater resources. The mission of the FCGMA is the preservation and management of the groundwater resources within the areas overlying the Fox Canyon Aquifer for the common benefit of the public and all agricultural, municipal and industrial users. The FCGMA has a water credit system in which credits are issued for water recharged using injection wells or for allocated water not used (in-lieu recharge). The credits are issued on a one-AF for one-AF basis and can be used in future years to offset overuse of the groundwater resources.

The historic accumulation of credits within the FCGMA has been steadily increasing, approaching 550,000 AF (179,200 Mg; 678 MCM) in 2006 (Fox Canyon Groundwater Management Agency 2007). The estimated total net credits balance in the East, West, and South Las Posas Basin at the end of calendar year 2006 was 116,002 AF (37,805 Mg; 143 MCM) compared to an annual extraction of 27,234 AF (8,875 Mg, 33.6 MCM). The accumulated credits were over four times the annual extraction rate. The volume of credits that accumulated through the operation of the ASR system and in-lieu recharge greatly exceeded the amount of water that could be extracted during a short-time period (e.g., major drought). The Fox Canyon Groundwater Management Agency (2007) warned that

> should there be an extended period with limited groundwater recharge by either natural or anthropogenic sources, a significant number of credits could be used in a short period of time, ultimately overstressing, and possible permanently damaging the resources. Thus, although the credit system represents a low-cost groundwater-use option that can assist individual operators during extended dry periods, it also represents a threat to the regional groundwater resource since, under the current Ordinance, it lacks limits that would mitigate cumulative regional overuse during these same periods.

It was noted that even a 5% use of the total amount of credits currently available would result in a net 24% increase in annual extraction, which could result in persistent depressions in groundwater elevations, land subsidence, and seawater intrusion (Fox Canyon Groundwater Management Agency 2007).

When a drought occurred in southern California starting in 2007, recovery from the system resulted in the anticipated large drawdowns and saline-water intrusion. It was subsequently determined that the storage capacity of the aquifer was on the order of only 50,000 AF (61.67 MCM). The system could not provide enough water to meet its commitment to the MWD. Therefore, CWMD reimbursed the MWD US $53 million for its investment in the construction of the project and the cost of non-recoverable water. The 150-million-dollar project was reported in the press to be a failure that was "marred by insufficient research, poor judgment and hollow assurances" (Blood and Spagat 2013).

It is recognized that the ASR system was oversold in that it could not meet its drought proofing objectives, but the system is still considered to provide value to the CWMD by providing protection from short-term interruptions in the water supply. A hydrogeological and numerical modeling investigation was contracted in 2016 to develop a better understanding of the basin hydrogeology and water budget, potential system performance, and impacts on basin water levels.

14.6.6 Pacific Northwest (U.S.A.)

ASR systems using basalt aquifers as storage zone in the Pacific Northwest of the United States were summarized by Maliva and Missimer (2010). The ASR systems use the Columbia River Basalt Group as a storage zone, which is locally in an overdraft condition with limitations placed on existing and future withdrawals. Maintaining environmental flows in streams and rivers is a major regulatory constraint on water use in the region. The basalt aquifers used as storage zones contain freshwater and the ASR systems are categorized as both physical and regulatory storage-type ASR systems and groundwater banking systems. The intent of the system is for recharge to increase the volume of water in storage to offset the hydrological impacts of recovery. The systems operate as regulatory storage ASR and groundwater banking systems in that recharge confers the right to later extract additional groundwater, which would otherwise not be permittable.

Aquifer recharge experiments were performed on the Columbia River Basalt Aquifer (basalt aquifer) in 1957 and 1958 at the City of Walla Walla, Washington (Price 1960). The objective of the artificial recharge projects was to determine if recharge of the basalt aquifer using seasonally available excess surface water from Mill Creek (the main water source for Walla Walla) could halt the decline of water levels within the basalt aquifer. The main operational issue was a 59% decline in specific capacity after the injection of 23 Mg (93,400 m^3) of water. Price (1960) concluded that the aquifer recharge program was worthwhile but cautioned that there was the risk of permanent damage to the production well used for injection. Several potential options for managing the aquifer clogging were suggested, such as periodic redevelopment and use of a foot valve to eliminate cascading during injection.

An ASR program was not implemented in Walla Walla until 1999. Faults or dikes compartmentalize the Columbia River Basalt Aquifer into distinct blocks, which are reported to have limited hydraulic connection with each other due to the development of low permeability fault gouge, secondary mineralization in the fault zone, and the offset of permeable interflow zones against the relatively low permeability basalt flow deposits (Banton and Klisch 2007). Groundwater modeling performed by Banton and Klisch (2007) indicates that the aquifer blocks are quite leaky. Only 0.42×10^6 m^3 (12.7%) of the modeled 3.32×10^6 m^3 of water recharged over a seven-month period goes into storage in the injected aquifer block. The remainder of the injected water leaks out into overlying unconsolidated aquifer strata and other parts of the basalt aquifer. The Walla Walla ASR system thus provided minimal local storage of water.

An important environmental issue in the Walla Walla Basin (eastern Oregon and Washington) is endangerment of fisheries resulting from decreased river flows during the summer peak agricultural demand period. MAR is being looked upon to seasonally replenish the aquifers to supply summer irrigation, allowing for increased summer flows in the Walla Walla River (Scherberg et al. 2014). Groundwater occurs primarily in two alluvial gravel aquifers and monitoring records show that aquifer water levels in the basin have declined at an average rate of 4.8 cm/year from 1950 to 2012 (Scherberg et al. 2014). MAR is performed by diverting winter and spring flows from the Walla Walla River into excavated basins.

Numerical modeling was performed to evaluate potential water management options including MAR. The key issue is the potential benefits and limitations of using MAR to augment seasonal groundwater levels to meet regional agricultural demands while withdrawals from the Walla Walla River are reduced during critical low flow periods. The modeling results indicate that the technical challenge is retaining water infiltrated in the basin in the spring and winter for summer use due to the high transmissivity of the gravel aquifers. The difference in mean water table elevation over the model area between the greatest amount of MAR and no MAR scenarios is about 1.5 m. The predicted increase in water elevations is most pronounced in the vicinity of the recharge basins and does not persist with distance away from the recharge source. Increased MAR results in increased discharge by seepage into springs and rivers.

The main conclusion of the modeling study is that increased MAR has the potential to stabilize groundwater levels and allow for increased use of groundwater, in lieu of surface water, during the summer. Without MAR, groundwater levels would continue to decline. A limitation is that some of the recharged water may flow out of the basin prior to the peak irrigation period in the later summer (Scherberg et al. 2014). Strategic siting of recharge basins may allow for increased effectiveness of MAR in the basin.

Foxworthy and Bryant (1967) documented artificial recharge and recovery tests performed on the Columbia River Group at The Dalles, Oregon. The Dalles area was identified as a Critical Ground Water Area because of a progressive decline in water levels in the basalt aquifer(s). The artificial recharge testing demonstrated that it was feasible to inject and recover large volumes of water in the basalt aquifers using existing production wells. A total of 81.4 Mg (308,100 m^3) of water was injected at an average rate of about 1,500 gallons per minute (gpm; 5,677 L/min). The injected treated water was about 5.5–13.9 °C cooler than the aquifer water temperature. The main operational issue was a decline in specific capacity that was attributed chiefly to the temperature-related increase in viscosity of the injected water and the release of air bubbles out of solution from the recharge water. Appreciable amounts of native groundwater remained in the aquifer near the well, which was suggested as being due to temperature-controlled density stratification and viscosity differences (Foxworthy and Bryant 1967). However, as freshwater was being stored in an aquifer that contains high-quality freshwater, recovery of the actual injected water was not a technical concern.

A key observation was that after injection was stopped, the groundwater mound near the recharge well quickly dissipated (within 56 min). There was thus no residual pressure build-up. The rationale presented for pursing aquifer recharge was that it would eliminate at least part of the annual overdraft that was responsible for the progressive decline of aquifer water levels in the project site vicinity. The testing results indicate that the aquifer is too transmissive for MAR to cause a persistent local increase in water levels.

The U.S. Geological Survey and Salem Heights Water District performed artificial recharge tests in the basalt aquifer of the Salem, Oregon area. The tests are amongst the earliest investigations of ASR in the United States. The artificial recharge tests were documented by Foxworthy (1970). Fluoridated and chlorinated surface water from the North Santiam River was injected into an existing production well. The injection tests were performed from March 20 to May 15, 1962. Three tests were performed with durations of 1, 5, and 15 days. The average injection rates ranged from 725 gpm (2,740 L/min; test 1) to 834 gpm (3160 L/min; test 3), and the total injected volume was 24.5 Mg (92,740 m^3). The buildup of heads in the main confined aquifer dissipated rapidly after the termination of injection. The injection tests were performed during a period of rising aquifer water levels so any persistence of the water level rises from managed recharge could not be clearly defined.

The U.S. Geological Survey and Salem Heights Water District artificial recharge tests demonstrated that ASR was feasible in the City of Salem vicinity, but no further work was apparently done for the next 30 years. Salem current operates a six well ASR system that stores treated water from the Geren Island Water Treatment Facility. The ASR system is decentralized and located in a park (Woodmansee Park) off the main transmission main. The recovered water is of a lesser quality than the water produced at the Geren Island WTF and initially had taste and odor problems. The exact cause of the taste and odor problems was never determined and the issues subsided to non-detect after three years. The operational complexity, lack of system-wide distribution, and the public relations problem that the ASR system created lead to the City to decide not to develop Phase 3 of the project (Mauldin 2004; Pulley 2008). The current system has a storage capacity of 450 Mg (1.70 MCM).

Other groundwater banking systems in the region operate under the regulatory storage concept. For example, the Beaverton, Oregon, ASR system stores drinking water (treated surface water) in the Columbia River Basalt Group Aquifer, which contains freshwater. Water injected into the ASR system is credited to a "storage account," which can later be debited by groundwater withdrawals (Eaton and Winship 2007). By injecting water during wet periods, the City of Beaverton obtains the right to withdraw additional groundwater during dry periods, which would otherwise not be allowed. The Beaverton ASR wells are also used for the extraction of native groundwater under an existing groundwater right. Under Oregon rules, water recovered from the ASR wells must first be debited against the ASR account.

14.7 Technical Lessons

Storing excess surface water in depleted aquifers makes eminent good sense and is typically much less expensive and more environmentally benign than surface storage options. Technical issues during recharge are usually related to water quality, shallow aquifer hydraulic responses (e.g., mounding), and achieving target recharge rates, which depends upon aquifer hydraulic properties and clogging rates. The technical issues that ultimately dictate the success of groundwater banking systems typically arise during recovery. As groundwater banking systems have storage objectives, their performance ultimately depends upon the ability to recover additional water when needed. Water recoverability often depends on drawdowns during extraction and associated impacts.

To objectively evaluate the performance of groundwater banking systems, and other types of MAR systems, it is critical that success criteria be clearly and quantitatively established (ideally memorialized in writing) at the start of a project (Maliva and Missimer 2010). If a system is intended to stabilize or increase aquifer water levels, then its performance should be evaluated by whether the expected changes in water levels (as evaluated using monitoring data) are actually achieved. Continuous monitoring of aquifer water level data and evaluation of storage changes are critical for successful long-term operation of groundwater banking schemes.

Groundwater banking systems may be developed to either allow for continued use of an aquifer or to allow for greater use during droughts or other periods of need. In the former case, extraction rates may not materially change (i.e., current extraction rates are maintained). In the latter case, aquifer-wide or local extraction rates may become significantly greater than historic rates. In the case of the Las Posas ASR system, it became clear once drought conditions occurred that the increased extractions, which were the objective of the system, could not be achieved without unacceptable impacts.

Hydrological impact analyses for groundwater banking systems need to focus on the impacts of systems on the overall aquifer water budget (i.e., static water levels) and local impacts during recovery. The "myth of residual pressure" (Maliva and Missimer 2008) is a critical issue for groundwater banking systems, as in the absence of persistent local head increases, recovery from groundwater banking systems will locally increase groundwater drawdowns beyond those that would occur in the absence of the system. Hydrological investigations of groundwater banking systems, therefore, should evaluate the magnitude, areal extent, and impacts of increased drawdowns during recovery. Both total drawdowns and change in drawdowns relative to a no-system baseline need to be quantified.

It is stressed that a no local impact criterion for a groundwater banking systems may not be reasonable. The benefits of groundwater banking systems can greatly exceed the costs of local impacts. Where local adverse impacts occur, then mitigation programs should be (and have been) implemented.

References

Anderson, T. W., Freethey, G. W. & Tucci, P. (1992). *Geohydrology and water resources in alluvial basins in south-central Arizona and parts of adjoining states*. U.S. Geological Survey Professional Paper 1406-B.

Anderson, M. T., Pool, D. R., & Leake, S. A. (2007). The water supply of Arizona: The geographic distribution of availability and patterns of use. In K. L. Jacobs & B. G. Colby (Eds.), *Arizona water policy: Management innovations in an urbanizing, arid region* (pp. 45–60). Washington, DC: Resources for the Future.

August, J. L., Jr., & Gammage, G., Jr. (2007). Shaped by water. An Arizona historical perspective. In K. L. Jacobs & B. G. Colby (Eds.), *Arizona water policy: Management innovations in an urbanizing, arid region* (pp. 10–25). Washington, DC: Resources for the Future.

Banton, D., & Klisch, M. (2007). Aquifer storage and recovery in a compartmentalized basalt aquifer system, Walla Walla, Washington. In P. Fox (Ed.), *Management of aquifer recharge for sustainability. Proceedings of the 6th International Symposium on Managed Aquifer Recharge of Groundwater* (pp. 307–320). Phoenix: Acacia Publishing, Phoenix.

Barringer, F. (2011, July 26). *Storing water for a dry day leads to suits*. New York Times.

Bloetscher, F., Muniz, A., & Witt, G. M. (2005). *Groundwater injection: Modeling, risks, and regulations*. New York: McGraw-Hill.

Blood, M. R., & Spagat, E. (2013, August 24). *Las Posas Basin aquifer failure illustrates risks of underground reservoirs*. Associated Press.

Burke, G. (2014, October 3). Judge: California must weigh water bank's impacts. *The San Diego Union-tribune (Associated Press)*.

California Department of Water Resources. (2000). *Groundwater management in California—six methods under current law.* Water Facts no. 8. Sacramento: California Department of Water Resources.

California Department of Water Resources. (2004). *Adjudicated groundwater basins in California*, Water Facts, no. 3. Sacramento: California Department of Water Resources.

Calleguas Municipal Water District. (2008a). District history and description. http://www.calleguas.com/about-calleguas-municipal-water-district/district-history.asp. Accessed 21 May 2018.

Calleguas Municipal Water District. (2008b). Las Posas Basin Aquifer Storage and Recovery Project. hhtp://www.calleguas.com/projects/lpbroc.pdf. Accessed 3 September 2008.

Christian-Smith, J. (2011). *Improving water management through groundwater banking: Kern County and Rosedale-Rio Bravo Water District*. Boulder: Pacific Institute.

Clifford, P., Landry C., & Larsen-Hayden, A. (2004). *Analysis of water banks in the Western States*, Publication No. 04-11-011. Olympia: Washington Department of Ecology.

Colby, B. G., Smith, D. R., & Pittenger, K. (2007). Water transactions. Enhancing water supply reliability during drought. In K. L. Jacobs & B. G. Colby (Eds.), *Arizona water policy: Management innovations in an urbanizing, arid region* (pp. 79–91). Washington, DC: Resources for the Future.

Contor, B. A. (2009). Groundwater banking in aquifers that interact with surface water: Aquifer response functions and double-entry accounting. *Journal American Water Resources Association, 45*(6), 1465–1474.

Contor, B. A. (2010). Status of ground water banking in Idaho. *Journal of Contemporary Water Research & Education, Issue, 144,* 29–36.

Donovan, D. J., Katzer, T., Brothers, K., Cole, E., & Johnson, M. (2002). Cost-benefit analysis of artificial recharge in Las Vegas Valley, Nevada. *Journal Water Resources Planning and Management, 128,* 356–365.

Eaton, L., & Winship, D. (2007). The City of Beaverton basalt-hosted ASR project: A successful case study. In P. Fox (Ed.), *Management of aquifer recharge for sustainability. Proceedings of the 6th International Symposium on Managed Aquifer Recharge of Groundwater* (pp. 494–506). Phoenix: Acacia Publishing, Phoenix.

Eden, S, Gelt, J., Megdal, S., Shipman, T., Smart, A., & Escobedo, M. (2007). Artificial recharge: A multi-purpose water management tool. *Arroyo* (Water Resources Research Center, College of Life Sciences, University of Arizona), Winter 2007.

Entsminger, J. J. (2016). *Southern Nevada Water Briefing. PowerPoint prepared for the Legislative Commission's Subcommittee to Study Water* (4-22-16). Southern Nevada Water Authority.

Evergreen Underground Water Conservation District. (2006). San Antonio Water System's Twin Oaks Aquifer Storage & Recovery Program status: powerPoint presentation. (January 16, 2006).

Faunt, C. C. (2009). *California's Central Valley Groundwater Study: A powerful new tool to assess water resources in California's Central Valley*. U.S. Geological Survey Fact Sheet 2009-3057.

Fox Canyon Groundwater Management Agency. (2007). *Annual Report for Calendar Year 2006*. Ventura, CA: Fox Canyon Groundwater Management Agency.

Foxworthy, B. L. (1970). *Hydrologic conditions and artificial recharge through a well in the Salem Heights area of Salem, Oregon*. U.S. Geological Survey Water-Supply Paper 1594-F.

Foxworthy, B. L., & Bryant, C. T. (1967). *Artificial recharge through a well tapping basalt aquifers at The Dalles, Oregon*. U.S. Geological Survey Water-Supply Paper 1594-E.

Frederick, K. D. (1995). Adapting to climate impacts on the supply and demand of water. *Climatic Change, 37*(1), 141–156.

Hanson, R. T., Martin, P., & Koczot, K. M. (2003). *Simulation of ground-water/surface-water flow in the Santa Clara–Calleguas ground-water basin, Ventura County, California*. U.S. Geological Survey Water-Resources Investigations Report 02-4136.

Jacobs, K. L., & Holway, J. M. (2004). Managing for sustainability in an arid climate: Lessons learned from 20 years of groundwater management in Arizona, USA. *Hydrogeology Journal, 12*, 52–65.

Johnson, M., & Donovan, D. (1998). *Hydrogeology and geology of the Las Vegas Valley Clark County*. Nevada: National Ground Water Association Field Trip Guide Book, Las Vegas, Nevada, December 16, 1998.

Katzer, T., Brothers, K., Cole, E., Donovan, D., & Johnson, M. (1998). *A cost-benefit analysis of artificial recharge in the Las Vegas Valley ground-water system, Clark County, Nevada*. Report prepared for the Southern Nevada Water Authority Ground Water Management Program.

Las Vegas Valley Water District. (2008). *Las Vegas Valley Water District 2007 artificial recharge annual report*. Las Vegas: Las Vegas Valley Water District.

LVVWD. (2017). Where your water comes from. Las Vegas Valley Water District. https://www.lvvwd.com/water-system/where-your-water-comes-from/index.html. Accessed September 19, 2017.

MacDonnell, L. J., Howe, C. W., Miller, K. A., Rice, T. A., & Bates, S. F. (1994). *Water banks in the west*. Boulder: Natural Resources Law Center, Univ. of Colorado School of Law.

Maliva, R. G. (2014). Groundwater banking: Opportunities and management challenges. *Water Policy, 16*(1), 144–156.

Maliva, R. G., & Missimer, T. M. (2008). ASR, useful storage, and the myth of residual pressure. *Ground Water, 46*, 171.

Maliva, R. G., & Missimer, T. M. (2010). *Aquifer storage and recovery and managed aquifer recharge using wells: Planning, hydrogeology, design, and operation*. Houston: Schlumberger Corporation.

Malmberg, G. T. (1965). *Available water supply of the Las Vegas ground-water basin Nevada*. U.S. Geological Survey Water-Supply Paper 1780.

Mauldin, F. (2004). *Sweet mountain water*. Corvallis OR: Oregon Oak Savanna Publishing.

Maxey, G. B., & Jameson, C. H. (1948). *Geology and water resources of Las Vegas, Pahrump, and Indian Spring Valleys, Clark and Nye Counties, Nevada*, Water Resources Bulletin 5. Nevada State Engineer.

Megdal, S. B. (2007). Arizona's recharge and recovery programs. In K. L. Jacobs & B. G. Colby (Eds.), *Arizona water policy: Management innovations in an urbanizing, arid region* (pp. 188–203). Washington, DC: Resources for the Future.

Megdal, S. B. (2012). Arizona groundwater management. *The Water Report*, Oct 15, 2012, (pp. 9–15).

Megdal, S. B., Dillon, P., & Seasholes, K. (2014). Water banks: Using managed aquifer recharge to meet water policy objectives. *Water, 6*, 1500–1514.

Metropolitan Water District of Southern California. (2007). *Groundwater assessment study, Chapter IV—Groundwater basin reports, Northwest Metropolitan Service Area—Ventura County Basins*. Los Angles: Metropolitan Water District of Southern California.

Morel-Seytoux, H. J. (1985). Conjunctive use of surface and ground water. In T. Asano (Ed.), *Artificial recharge* (pp. 35–67). Stoneham, MA: Butterworth.

Morton, S. (2015). Groundwater banking in Imperial Irrigation District: Planning for future water scarcity on the Colorado River. University of San Francisco, Master's Project.
National Research Council. (2008). *Prospects for managed underground storage of recoverable water*. Washington DC: National Academy Press.
O'Donnell, M., & Colby, B. (2010). *Water banks: A tool for enhancing water supply reliability*. Tucson: Department of Agricultural and Resources Economics. University of Arizona.
Parker, T. K. (2007). California's quandary: Managed aquifer recharge for increased water supply reliability under a very complex regulatory environment will it work? In P. Fox (Ed.), *Management of aquifer recharge for sustainability. Proceedings of the 6th International Symposium on Managed Aquifer Recharge of Groundwater* (pp. 109–125). Phoenix: Acacia Publishing, Phoenix.
Patterson, B. (2015). California dreaming: Developing an ATM for groundwater. *ClimateWire*. December 3, 2015.
Pearce, M. J. (2007). Balancing of competing interests: The history of state and federal water laws. In K. L. Jacobs & B. G. Colby (Eds.), *Arizona water policy: Management innovations in an urbanizing, arid region* (pp. 26–44). Washington, DC: Resources for the Future.
Pinhey, N. A. (2003). *Banking on the commons: An institutional analysis of groundwater banking programs in California's Central Valley*. Doctoral dissertation, University of Southern California, Los Angeles.
Pitzer, G. (2010). Saving it for later: Groundwater banking. *Western Water Magazine*, September 10, 2010.
Plume, R. W. (1989). *Ground-water conditions in Las Vegas Valley, Clark County, Nevada*. U.S. Geological Survey Water-Supply Paper 2320-A.
Poland, J. F. (1961). *The coefficient of storage in a region of major subsidence caused by compaction of an aquifer* (pp. 52–54). U. S. Geological Survey Professional Paper 424-B.
Poland, J. F., & Davis, G. H. (1956). Subsidence of the land surface in Tulare-Wasco (Delano) and Los Banos-Kettleman City area, San Joaquin Valley, California. *Transactions of the American Geophysical Union, 37,* 287–296.
Poland, J. F., et al. (1975). *Land subsidence in the San Joaquin Valley, California as of 1972*. U. S. Geological Professional Paper 437-H.
Price, C. E. (1960). A*rtificial recharge of a well tapping basalt aquifers, Walla Walla area, Washington*, Water Supply Bulletin No. 7. Olympia: Washington Division of Water Resources.
Pulley, J. (2008). ASR in Salem: An overview of Oregon's first ASR system. In *Aquifer storage and recovery and artificial recharge in Oregon: Overcoming technical, regulatory, and social challenges, symposium co-convened by OSU institute for water and watersheds and Oregon water resources department*, February 28, 2008. Corvallis, Oregon, PowerPoint presentation.
Purkey, D. R., Thomas, G. A., Fullerton, D. K., Moench, M., & Axelrad, L. (1998). *Feasibility study of maximal program of groundwater banking*. San Francisco: Natural Heritage Institute.
Pyne, R. D. G. (1995). *Groundwater recharge and wells*. Boca Raton: Lewis.
Pyne, R. D. G. (2005). *Aquifer storage recovery, a guide to groundwater recharge through wells*. Gainesville, FL: ASR Systems.
Sandoval-Solis, S., McKinney, D. C., Teasley, R. L., & Patino-Gomez, C. (2011). Groundwater banking in the Rio Grande Basin. *Journal of Water Resource Planning and Management, 137*(1), 62–71.
Scherberg, J., Baker, T., Selker, J. S., & Henry, R. (2014). Design of managed aquifer recharge for agricultural and ecological water supply assessed through numerical modeling. *Water Resources Management, 28*(14), 4971–4984.
Thomas, G. A. (2001). *Designing successful groundwater banking programs in the Central Valley: Lessons from experience* (p. 112). Berkeley (CA): The Natural Heritage Institute.
Todd, D. K. (1959). *Groundwater hydrology*. New York: Wiley.
WAKC (n.d.) Water banking. Water Association of Kern County. http://www.wakc.com/water-overview/sources-of-water/water-banking. Accessed 28 September 2017.

Ward, J. & Dillon, P. (2009). *Robust design of managed aquifer recharge policy in Australia. Facilitating recycling of stormwater and reclaimed water via aquifers in Australia*, Milestone Report 3.1. Canberra: CSIRO.

Ward, J., & Dillon, P. (2012). Principles to coordinate managed aquifer recharge with natural resources management policies in Australia. *Hydrogeology Journal, 20,* 943–956.

Western Farm Press. (2011, September 1). Madera Ranch water project gets big start. *Western Farm Press* (September 1, 2011). http://www.westernfarmpress.com/irrigation/madera-ranch-water-project-gets-big-start.

Wood, D. B. (2000). *Water use and associated effects on ground-water levels, Las Vegas Valley and vicinity, Clark County, Nevada, 1980–95, Water-Resources Information Report 35*. Carson City: State of Nevada Division of Water Resources.

Zikmund, K. (1996). *Extent and potential use of the shallow aquifer and wash flow in Las Vegas Valley, Nevada*. Las Vegas: Southern Nevada Water Authority.

Chapter 15
Surface Spreading System—Infiltration Basins

15.1 Introduction

Surface spreading is the simplest, oldest, and mostly widely used method of managed aquifer recharge (MAR; Todd 1980; Asano and Cotruvo 2004). Stormwater, river water, treated wastewater, and other waters are either applied to a land surface or locally impounded in infiltration basins, reservoirs, or modified stream channels. Typically, the water table is located below land surface, at least at the start of surface spreading, and the infiltrated water passes through the unsaturated zone. As infiltration progresses, the water table may rise to land surface at the spreading site. Surface spreading can be an efficient means of recharging shallow unconfined (water-table) aquifers where conditions are favorable. The most important requirement is that the strata between the land surface and the water table be sufficiently permeable to allow high infiltration and percolation rates. Clayey or other low permeability strata should not be present between the spreading surface and underlying water-table aquifer to avoided perched aquifer conditions. The water-table aquifer should also be sufficiently transmissive to avoiding mounding and associated waterlogging.

Surface spreading systems have the advantages of:

- usually lower costs than systems utilizing wells
- recharge systems (e.g., infiltration basins) may also provide some surface water storage
- maintenance tends to be easier to perform (relative to wells) because the clogging layer is usually present at land surface
- natural contaminant attenuation in the unsaturated zone may improve water quality
- well-designed surface spreading systems can result in environmental enhancement or restoration.

R. G. Maliva, *Anthropogenic Aquifer Recharge*,
Springer Hydrogeology, https://doi.org/10.1007/978-3-030-11084-0_15

The main disadvantages of MAR by surface spreading include:

- suitable available and affordable land may not be available (or affordable) in areas with hydrogeological and logistical conditions (e.g., water availability) favorable for recharge
- only unconfined aquifers can be recharged
- clogging is a major operational challenge where untreated or incompletely treated water is recharged
- recharge can mobilize contaminants in the soil zone
- water impoundment and recharge can create nuisances (e.g., water mounding at land surface, mosquito breeding).

Surface-spreading systems vary depending upon the type of land surface used, the degree and type of modification of the land surface, pretreatment provided, and the manner in which water is applied to the spreading surface. The most common types of anthropogenic aquifer recharge (AAR) involving surface spreading are summarized in Table 15.1. Irrigation return flows may be categorized as either managed or unmanaged recharge depending on whether it is planned and intentional. Infiltration basin systems are discussed in this chapter. Other types of surface-spreading systems are addressed in Chap. 16. Some of the basic hydraulic principles and design issues addressed herein for infiltration basin are applicable to other types of surface-spreading systems. Dune infiltration and sprinkler application methods are discussed in the context of aquifer recharge and recovery (ARR; Chap. 18).

Table 15.1 Surface spreading MAR types

System type	Description
Infiltration basins	Shallow constructed impoundments
Surface application	Water is applied to a largely unmodified land surface
Channel	Water is discharged to an ephemeral stream channel
Channel modification	A channel is modified to slow and retain flow and increase the wetted area
Reservoirs and lagoons	Impoundments constructed by damming channels
Dune infiltration	Water is applied to sand dunes for treatment and storage
Spray or sprinkler application	Water is applied to vegetated areas using sprinkler systems in excess of evapotranspiration requirements
Irrigation return flows	Application of water to cropped areas in excess of plant evapotranspiration requirements
Sand dams	Storage of surface water in artificial sand aquifers constructed behind low dams
Ditch and furrow systems	Diversion of water from a stream channel for recharge in off-channel ditches and furrows
Leaky wetland treatment system	Constructed wetlands with a pervious base

15.2 Infiltration Basins Introduction

Infiltration Basins are shallow, impounded areas designed to temporarily store, infiltrate, and treat water (Fig. 15.1). Infiltration basins have three main applications in water and wastewater management. The basins are most widely used as a stormwater best management practice (BMP). Stormwater infiltration basins are a passive technology in that they are operated without human intervention other than occasional maintenance activities. Rapid infiltration basins (RIBs) are used to treat wastewater and to recharge surficial aquifers. Soil-aquifer treatment (SAT) is a specific application of infiltration basins that is used primarily to treat wastewater (Chap. 19). Infiltration basins are used to recharge shallow aquifers with surface water, and less commonly treated wastewater, as part of groundwater banking and other aquifer recharge systems. Wastewater and surface water infiltration basins are typically divided into multiple cells and have actively controlled alternations between wetting and drying periods.

Infiltration basins share the same hydrogeological constraints as other surface-spreading systems. Hydrogeological characterization of potential system sites normally includes field testing of infiltration rates and evaluation of the hydraulic properties of the strata between the recharge system surface and the water table. The presence of intervening confining strata can result in a perched aquifer condition rather than recharge to the water table. Main issues that should be considered in determining the locations of infiltration basins and in basin design are (Bouwer 1978, 2002; Oaksford 1985):

- soils should be sufficiently permeable to yield acceptable infiltration rates
- thin, low permeable layers present at land surface (e.g., caliche layers) may be excavated and removed to increase infiltration rates
- surficial low-permeability material may be replaced by more permeable material, but this option may be cost prohibitive

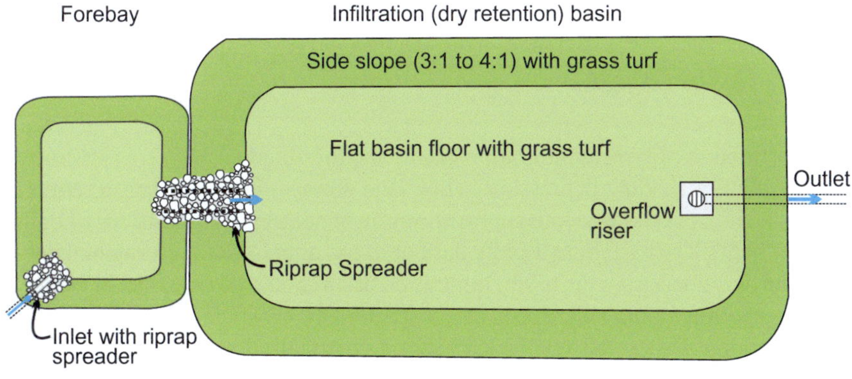

Fig. 15.1 Conceptual design diagram of a stormwater infiltration basin. Sedimentation forebay is optional

- the unsaturated zone should not contain low permeability layers that could result in perched groundwater conditions
- the regional water table should be sufficiently deep to keep the groundwater mound below the base of basins but not so deep so that large quantities of water are needed to wet the vadose zone before water reaches the water table
- the surficial aquifer should be unconfined and sufficiently transmissive to allow lateral movement of recharged water and prevent groundwater mounds from reaching the basin bottom and nearby land surfaces
- contaminants should not be present in the unsaturated zone that can be mobilized during infiltration
- if impaired (non-potable quality) water is recharged, then the direction, rate, and distance of transport of recharged water is a consideration, especially with respect to nearby potable water supply wells and sensitive environments.

The main operational challenge of infiltration basin systems is usually management of clogging. Low-permeability clogging layers tend to develop on the basin surface through some combination of the deposition of fine-grained sediments (silt and clay-sized materials), chemical (e.g., calcium carbonate) precipitation, and biological activities. The latter can reduce permeability through the formation of biofilm layers and the accumulation of biological debris.

The quality of runoff entering stormwater infiltration basins and other stormwater management systems (referred to as best management practices or BMPs in the United States) is also a design and operational concern. Infiltration basins and other BMPs are intended to provide some water treatment, but they can also facilitate the transfer of contaminants present in runoff (and other waters) into shallow aquifers. This chapter focuses on the hydraulic aspects of infiltration basins. Water quality issues associated with stormwater systems are addressed in Sect. 23.2 and with wastewater (soil-aquifer treatment) systems are addressed in Chap. 18.

15.3 Infiltration Basin Basics

15.3.1 Basin Design

The construction of typical infiltration basins is quite simple. Basin are constructed by excavating a shallow depression below land surface, berming (construction of embankments), or otherwise raising the elevation of adjacent land areas (or a combination of both). The basins are usually flat-bottom so as to have a uniform inundation depth and duration across the basin. Infiltration basin should be constructed with stable sides and maximum side slopes of 3:1 (preferably 4:1). Dense vegetation (e.g., grass) is commonly established on side slopes (especially for stormwater systems) to prevent erosion and sloughing. Infiltration systems may be constructed with either grass or sand bases. Where basins have a primary recharge function, vegetated bases may not be desirable because of additional evapotranspiration (ET) water losses. However, root growth and decay may promote higher infiltration rates.

Erosion can impact the integrity of a basin and allow for the suspension of fines from basin side slopes by wave action, which can contribute to clogging. Inflow points should be designed to provide protection from erosion (e.g., using flow spreaders, riprap, energy dissipators). Where clogging with suspended solids is a concern, a pretreatment stilling (sedimentation) basin or forebay can be used to capture sediment before it enters an infiltration basin.

The main design variables are:

- location of basins
- total basin area
- number of basin cells and their configuration
- depth of basin (below nearby land surface)
- operational maximum water depth
- pretreatment provided
- overflow elevation (if required).

Infiltration systems should be designed and operated to effectively manage clogging and there are opportunities for innovation. Infiltration basins are usually designed with flat bottoms in which clogging material accumulates across the bottom surface. An alternative furrow system design composed of closely-spaced sloped, flat-bottom ditches was described by Schiff (1957). Peyton (2001, 2002) described a ridge and furrow system for basin floors that was tested in California (Leaky Acres recharge basins in Fresno). The advantage of a furrowed, rather than flat, basin floor for maintaining infiltration rates is that fines tend to migrate off slopes into lower-lying areas when agitated by naturally occurring wave action. The fines-free slopes will better retain infiltration rates. Fines accumulate in furrows from which they are periodically removed. The main maintenance activity is inducing a mild wave action to resuspend and wash-off fine sediment deposited on the ridge slopes, which can be achieved by lowering the water level to the height of the ridges. Equipment, such as a Jet Ski, can be used to generate waves (Peyton 2003). The Leaky Acres system was reported to have furrows 6–8 in. (15–20 cm) deep, 10–15 ft (3.0–4.5 m) wide on 30 ft (9.0 m) centers, which were excavated using a motor grader. The 10–15 ft-wide furrows provide enough operating room to operate a paddle-wheel scraper.

Escalante (2013) documented the design and maintenance of MAR facilities at two sites in the Castille and Leon Regions of Spain. The MAR systems recharge the Arenales aquifer, which consists mostly of Quaternary sands. Furrowing of infiltration basins was found to be an effective means of increasing infiltration rates and managing clogging. The furrows increased the surface areas of infiltration ponds and allowed silt to settle in the furrow bottoms by gravity, while the furrow crests remained relatively clear. The infiltration rates in furrowed basins were reported to be double those in flat-bottomed basins.

Infiltration basins typically utilize the local native soils and sediments. Soil conditioning has been historically performed to improved infiltration and reduce clogging

(Huisman and Olsthoorn 1983). For example, calcium salts have been added to prevent deflocculation of clay particles, and organic matter has been added to increase microbial activity and pull particles together into aggregates (i.e., to improve soil structure; Huisman and Olsthoorn 1983). An active area of research is the use of various amendments to improve the removal of pollutants by increasing sorption and/or controlling the redox state of infiltrated water to increase nutrient removal. The primary interest has been to improve the pollutant removal efficiency of stormwater BMPs (Sect. 23.13).

15.3.2 Hydraulic Loading Rates and Basin Area

The design area and depth of infiltration basins depend upon temporary storage requirements, infiltration rates, and wet-dry cycling plans. Stormwater infiltration basins are designed to infiltrate the runoff of a catchment from a prescribed storm event within a specified period of time. Wastewater and surface water recharge infiltration basin systems are designed to infiltrate a specified maximum flow rate. Periodic drying of basins is required to manage clogging and avoidance of nuisance conditions. Alternation of drying and wetting is also important for the removal of some contaminants and nutrients, particularly nitrogen compounds (ammonium and nitrate). For stormwater infiltration systems, wetting and drying cycles naturally occur between precipitation events. Wastewater and surface water recharge infiltration basin systems are usually designed with multiple cells so that while one or more cells are receiving water, other cells are drying. Multiple-cell infiltration basins systems thus require greater total areas to infiltrate a given flow of water.

The USEPA (2006) "Process design manual for land treatment of municipal wastewater effluents" (which is based on Crites et al. 2000) provides a good overview of the basic design process for multiple-cell infiltration basin systems, as summarized below. Infiltration basin area (A) is determined by dividing the design flow rate (Q) by the hydraulic loading rate (L_w):

$$A = \frac{Q(0.0001)(365)}{L_w}(\text{metric}) \tag{15.1}$$

$$A = \frac{Q(3.07)(365)}{L_w}(\text{U.S customary}) \tag{15.2}$$

where:

A	area (ha [acres])
Q	average design flow (m^3/d [Mgd])
L_w	annual hydraulic loading rate (m/year [ft/year])
365	days per year
0.0001	metric conversion (ha-m to m^3/d)
3.07	U.S. customary conversion (acre-ft to Mgd).

Flow and hydraulic loading rates may vary seasonally. Infiltration system design equations can be easily modified for seasonal operation (e.g., Kallali et al. 2013). The systems need to be designed with a sufficient area to handle peak seasonal flows and/or minimum seasonal hydraulic loading rates.

Some short-term variation in flow above average design flows can be accommodated by basin storage. Infiltration rates will vary over time due to clogging, climatic variations, and maintenance activities. For example, infiltration rates are normally greatest after basin maintenances activities and then tail off as clogging progresses. Hence, for a constant inflow, infiltration basin systems need to be designed for worse-case lower rates that will occur toward the end of operational cycles (i.e., before the next basin maintenance or rehabilitation event occurs). The reliable infiltration rate (Ir) can be considered the infiltration rate threshold that would trigger rehabilitation activities. For large systems with multiple basins, average infiltration rates may suffice, if the system can be operated so that basins are at different stages in their operational/maintenance cycles.

For design purposes, hydraulic loading rate (L_w) can be estimated as the product of the reliable, long-term minimum (design) infiltration rate (Ir) and the fraction of time (f) in which a basin is receiving water:

$$L_w = Ir \cdot 365 \cdot f \tag{15.3}$$

Ir reliable infiltration rate (m/d or ft/d)
365 days per year
f wetting time fraction (days per year receiving water/365).

Hydraulic loading rate can alternatively be defined as the total infiltration per flooding event divided by the combined length of flooding and dying periods (Bouwer et al. 2008). Hydraulic loading rates account for down time, such as for normal drying cycles and other maintenance activities. Calculations of hydraulic loading rates may need to also consider the temperature dependence of viscosity and thus hydraulic conductivity. Hydraulic loading rates tend to be higher in the summer due to quicker drying and the temperature effect on viscosity. In areas with large seasonal temperature variations, viscosity effects alone can cause the infiltration rate in the winter to be close to half the summer rate (Bouwer and Rice 2001). On the contrary, greater biological activity in the summer may increase clogging rates and decrease infiltration rates.

Methods used to measure infiltration rates are discussed in Chap. 10. A fundamental challenge in designing surface-spreading systems is up-scaling measured infiltration rates from field testing values to average, long-term rates for full-scale operational systems. Infiltration rates obtained from small-area infiltration tests (e.g., double-ring infiltrometer tests) tend to overestimate large-area (basin-wide) infiltration rates due to the divergence of flow (Bouwer et al. 2008). Infiltration tests performed using progressively larger area test basins (initially 3 m by 3 m; Bouwer et al. 2008) can provide more representative rates, but cost and time are considerations.

Multiple scales of heterogeneity can impact infiltration test data. Where flow through macropores (e.g., fissures, borings, and burrows) is important in a vadose zone, the degree to which small-area testing represents basin-wide infiltration rates will depend on the degree to which the tests capture the effects of macropores. For example, high initial infiltration rates may occur where water infiltrates into the soil mainly through shrinkage cracks, which may subsequently close as the soil becomes saturated. Larger-scale heterogeneity in unsaturated zone properties can be caused by spatial variations in sediment composition. The relationship between aquifer heterogeneity and long-term infiltration rates can be complex. High-permeability features or zones may have greater initial infiltration rates, but the greater infiltration rates of the features can give them a greater load of clogging agents (e.g., suspended solids). As a result, high-permeability features may preferentially clog.

Operational infiltration rates will typically be considerably less than infiltration rates measured during initial site testing using clean water due to clogging. The decline of infiltration rates over time is system specific and cannot be accurately predicted in advance of system operation, or at least without pilot testing. Operational flexibility should thus be incorporated into system design. Hence, a safety factor or coefficient is applied to clean water infiltration rates to account for stormwater or wastewater-related reductions in permeability (Kallali et al. 2013). In state stormwater manuals, a safety factor of two and/or the use of the lowest measured infiltration test rate are commonly recommended for infiltration basin design.

The number of basins into which an infiltration area needs to be divided depends upon the final wet/dry ratio of the system (Crites et al. 2000; USEPA 2006). At a minimum, a system should have enough basins so that at least one basin can be flooded at all times. For example, if the application period is 3 days and the drying period is 4–5 days, then a minimum of 3 basins is required so that one basin can receive water at any time. The USEPA (2006, after Crites et al. 2000) tabulated the minimum number basins required for continuous wastewater applications for different wet/dry ratios (Table 15.2).

15.3.3 Water Depth and Infiltration Rates

Deeper basins have the advantage of allowing for greater water depths and thus providing a greater storage capacity for a given area. Bouwer (2002) addressed some basic principles of the effects of water depth and depth to the water table on infiltration rates. Infiltration rates after a soil has been flooded can be expressed by the equation (Bouwer 1978, 2002)

$$V_i = K_w \frac{\left(H_w + L_f - h_{we}\right)}{L_f} \tag{15.4}$$

Table 15.2 Minimum number of basins required for continuous wastewater application

Loading application periods period (days)	Cycle drying period (days)	Minimum number of infiltration basins
1	5–7	6–8
2	5–7	4–5
1	7–12	8–13
2	7–12	5–7
1	4–5	5–6
2	4–5	3–4
3	4–5	3
1	5–10	6–11
2	5–10	4–6
3	5–10	3–5
1	10–14	11–15
2	10–14	6–8
1	12–16	13–17
2	12–16	7–9
7	10–15	3–4
8	10–15	3
9	10–15	3
7	12–16	3–4
8	12–16	3
9	12–16	3

Source USEPA (2006)

where

V_i infiltration rate (m/d)
L_f depth to the wetting front (m)
K_w hydraulic conductivity at the wetted zone (m/d)
H_w ponding depth (m)
h_{we} capillary suction or negative pressure head at the wetting front (m).

The suction at the wetting front (h_{we}) is greatest (most negative) for unstructured fine-grained (clayey) soils and lowest for coarse sands. The value of h_{we} ranges from about -5 to -10 cm of water for medium to coarse sands, -35 cm for loams and structured clays, and -100 cm for dispersed clays (Bouwer 2002). From the Eq. 15.4, H_w has a lesser effect on V_i as L_f becomes greater. K_w is less than K_s (saturated hydraulic conductivity) due to air entrapment. As a rough estimate, $K_w = 0.5K_s$ for sandy soils and $0.25K_s$ for clays and loams (Bouwer 2002). A key observation from Eq. 15.4 is that as the wetting front moves downwards (and L_f increases), the ratio approaches unity and infiltration rates becomes equal to the hydraulic conductivity of the wetted zone (Bouwer 2002).

Infiltration rates are more often controlled by surficial clogging layers rather than by the properties of the underlying soil, except where the underlying soil is relatively fine-grained and infiltration rates are already low to begin with. When infiltration rates become less than the hydraulic conductivity of the soil below the clogging layer, the soil becomes unsaturated to a degree whereby the corresponding unsaturated hydraulic conductivity becomes numerically equal to the infiltration rate. The resulting unsaturated downward flow is then due to gravity with a hydraulic gradient of one (Bouwer 2002). Infiltration rates where a clogging layer is present is expressed by the equation (Bouwer 2002)

$$V_i = K_c \frac{H_w - h_{ae}}{L_c} \tag{15.5}$$

where (using consistent units)

K_c hydraulic conductivity of clogging later
H_w head (depth) of ponded water
h_{ae} negative pressure head below clogging layer (air entry value)
L_c thickness of clogging layer.

The properties of the clogging or restricting layer can be alternatively expresses in terms of the hydraulic resistance (R_c), which is defined as L_c/K_c with the dimension of time (Bouwer 2002):

$$V_i = \frac{(H_w - h_{ae})}{R_c} \tag{15.6}$$

Infiltration rates are controlled by the hydraulic resistance of the clogging layer and hydraulic head difference across the layer. Infiltration rates increase with increased ponding depths and greater (more negative) capillary suction below the clogging layer.

Three basic principles summarize the relationship between infiltration rates and water depth and groundwater level (Bouwer 2002):

- **Shallow depth to water**: Groundwater levels will rise to the water level in the basin and infiltration is controlled by lateral flow.
- **Great depth to water**: Flow from the infiltration basin is downward and controlled by gravity. The water content in the unsaturated zone establishes itself at a value whereby the corresponding unsaturated hydraulic conductivity is numerically equal to the infiltration rate. Inasmuch as the downward flow is due to gravity, the hydraulic gradient is about unity.
- **Clogged basin**: Infiltration rate almost linearly increases with depth of ponded water. Increases in water depth can compress the clogging layer, making it less permeable.

If the water table is more than 1 m below the bottom of a basin with a clogging layer, then infiltration rates are unaffected by changes in groundwater level (depth to the water table). Infiltration rates only decreases when the capillary fringe reaches the bottom of a basin (Bouwer 2002). Infiltration rates continue to decrease linearly with decreasing depth to groundwater until the water table has risen to the elevation of the surface water in the infiltration basin (Bouwer 2002).

The depth of ponded water can be increased to maintain infiltration rates. However, the benefits of the greater water depth may be short lived. Increases in the water depth in basins can result in compression of the clogging layer and an associated reduction in its permeability (Bouwer 1989; Bouwer and Rice 1989). The increases in infiltration rates associated with an increasing ponding depth may be much less than expected based on the water depth increase alone, and rates may actually decrease (Bouwer and Rice 1989). Bouwer and Rice (1989) also noted that decreases in infiltration rates caused by greater water depths can be especially severe where a decreased turnover rate of water (increased detention time) in infiltration basins causes increased growth of suspended algae. Algae can form a filter cake on the basin bottom and induce calcium carbonate precipitation by increasing the pH of the water through photosynthesis.

Shallow infiltration basins have the advantage of facilitating drying and other maintenance activities, such as disking and scraping. A maximum ponding depth of 1 m or less is commonly used for stormwater systems. Deep, permanently water-filled basins may experience large loses of infiltration capacity due to clogging caused by the settling and size segregation of suspended particles and compression of the clogging layer (Bouwer and Rice 2001).

15.3.4 Mounding and Basin Configuration

Mounding is the local rise of a regional or a perched water table toward and above land surface beneath and adjacent to an infiltration system (Fig. 15.2). Mounding is undesirable because it can reduce infiltration rates, result in saturated soil conditions (impacting land uses), cause the discharge of infiltrated water into low-lying areas (including building basements) and on slopes, and impact nearby surface-water bodies. The reduction in the thickness, or elimination, of the unsaturated zone can also decrease the natural attenuation of contaminants. Induced soil saturation could create habitat for wetlands fauna and flora, which may be desirable from an ecological perspective but could create liabilities to landowners.

Mounding reflects the balance between the downward flow of infiltrated water and lateral flow away from the infiltration site. The former depends upon the application rate and vertical hydraulic conductivity of the unsaturated zone strata. The latter depends upon the transmissivity of the water table aquifer. Perched aquifer mounding is the rise of water levels above confining strata within the unsaturated zone. Water levels in a perched aquifer will progressively rise until an equilibrium is reached

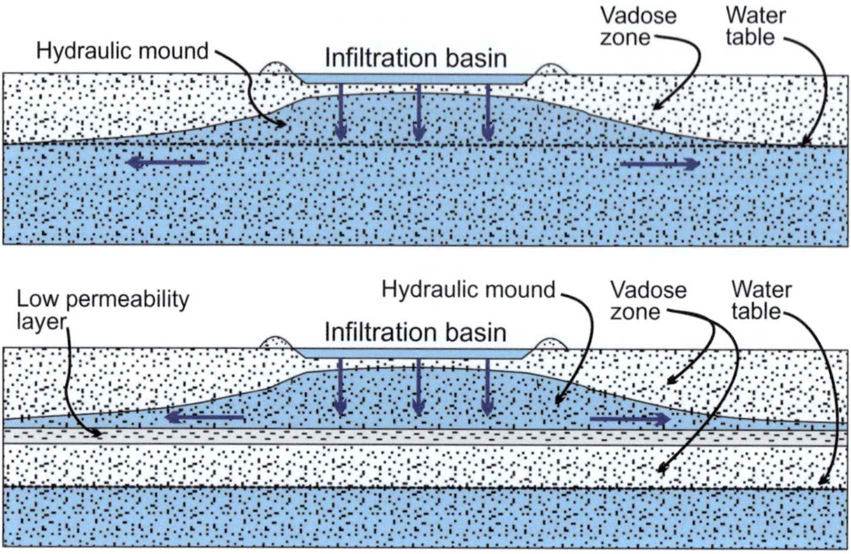

Fig. 15.2 Conceptual cross-sectional diagrams of an infiltration basin. Top: A groundwater mound develops atop of the water table whose size and shape depend on the rates of infiltration and lateral flow away from the basin. Recharge may cause the water table to locally rise to land surface (not shown). Bottom: The presence of an intervening low permeability layers can result in a perched-aquifer condition with little of the infiltrated water actually reaching the water table at the infiltration site

between water accumulating above the confining layer and water flow through or around the layer. Lateral spreading will also limit mound height.

The height, shape, and area of hydraulic mounds also depend upon the size, shape, and configuration of infiltration basins. Analytical equations are available for evaluating mounding beneath various geometries of surface-spreading systems, which were discussed by Bouwer (1978) and Huisman and Olsthoorn (1983). However, analytical approaches have become superseded by numerical computer modeling, which provides greater flexibility to incorporate aquifer heterogeneity, variations in basin shape, and multiple-cell system design and operation.

Carleton (2010), for example, utilized the MODFLOW code to evaluate the controls of groundwater mounding beneath hypothetical stormwater infiltration basins. Variables considered were soil permeability, aquifer thickness and specific yield, the magnitude of design storms, percentage of impervious area, infiltration-structure depth (maximum depth of standing water), and infiltration basin shape. The simulations did not include the delay and attenuation of flow in the unsaturated zone (Carleton 2010). Infiltrated water was simulated as recharging the underlying water table. MODFLOW modules are available to simulate unsaturated zone flow, such as the Unsaturated-Zone Flow (UZF1) package (Niswonger et al. 2006) and Variably Saturated Flow (VSF) package (Thoms et al 2006).

Numerical modeling requires data from a detailed aquifer characterization. The use of transmissivity values for an entire water-table aquifer obtained from pumping test can result in a serious underestimation of the rise of a mound because flow of recharged water is restricted to the upper active part of the aquifer (Bouwer 2002). Models also need to consider the increase in transmissivity of unconfined aquifer that occurs with an increase in saturated thickness.

Rastogi and Pandey (2002) simulated groundwater mound development below recharge basins of different shapes. For a given total area and recharge rate, groundwater mound height decreases as the basin perimeter increases. Circular basins had the greatest simulated mounding, elongate rectangular basins the least. Narrow rectangular basins are thus preferred where mounding needs to be controlled. More equant shapes are preferred when mound build up is desired, such as for salinity-barrier systems (Rastogi and Pandey 2002). The impact of the perimeter to volume ratio on mounding increases with water table elevation. As a mound height increases, and the depth to the water table decreases, groundwater flow becomes increasingly horizontal away from the mound rather than vertical. Once the underlying aquifer is fully saturated beneath the central part of a basin, the rate of vertical downward flow is equal to and controlled by the rate of outward horizontal flow.

15.3.5 Vadose Zone and Aquifer Heterogeneity

Where flow direction and travel time from an infiltration basin to a production well is an important issue (e.g., systems potentially involving indirect potable reuse), the effects of vadose zone flow and aquifer heterogeneity need to be considered. Residence time is an important factor controlling the attenuation of organic chemicals and pathogens.

O'Leary et al. (2012) investigated the movement of water recharged at a 40 ha detention basin located in the City of Stockton, California. Flowmeter log data from a nearby production well, tracer data, and groundwater modeling results indicate that most ($\approx$70%) of the flow to the well entered through screened intervals at 107–111 and 114–117 m below land surface. The study results demonstrated how aquifer heterogeneity (concentration of flow into thin flow zones) can result in high groundwater flow velocities and short-travel times to production wells, which has implications if subsurface residence is being relied upon for contaminant attenuation (O'Leary et al. 2012).

Improved ability to account for spatial heterogeneity by determining preferential flow paths and linking these to chemical interactions could be used to optimize water quality improvement (Parsekian et al. 2014). Parsekian et al. (2014) investigated the use of geophysical and hydrochemical data to improve aquifer characterization at an MAR site, the Aurora, Colorado, ARR project (Prairie Waters Project), in which aquifer recharge is performed using infiltration basins and water is recovered using extraction wells. Electrical resistivity tomography using Wenner arrays was used to identify likely flow units and conservative organic tracers were used to estimate travel times. Site-specific transforms based on core data enabled interpretation of sediment types within the aquifer from measured electrical properties. High resistivity values were associated with coarse particle textures. The method used could differentiate between sands, silts, and clays. The antiepileptic drugs carbamazepine and primidone were used as organic tracers because they are resistant to biodegradation and are present in the recharged water but not in the native groundwater.

The data indicate that certain areas of the infiltration basins have a greater hydraulic connection to extraction wells through preferential flow paths compared to other basin areas that are separated by fine-grained materials from their respective extraction wells. Conservative tracers provided information on the dilution of recharged water. Comparison of the concentrations of the conservative tracers with TOC and reactive tracers with varying degrees of biodegradability allowed for the evaluation of the attenuation rates of the latter (Parsekian et al. 2014). Flow path data can be used to optimize the design of an ARR system, such as to target the location of extraction wells to areas with more transmissive strata (Parsekian et al. 2014). An acknowledged limitation of the geophysical data is that it can detect only relatively large-scale sediment texture zones and thus flow paths (Parsekian et al. 2014). The combination of geophysical and hydrochemical data reduced the uncertainty regarding water quality changes during ARR (Parsekian et al. 2014).

15.3.6 Design and Operational Recommendations

The later Dr. Herman Bouwer had researched and written extensively on infiltration systems. Follows are a series of practical recommendations on the design and maintenance of infiltration systems from his publications (Bouwer 1985, 2002; Bouwer et al. 2008):

- Clogging on and in the soil should be minimized. Eventually clogging will restrict infiltration to the extent that all pressure head due to water depth in a basin is dissipated through the clogging layer. The underlying strata will remain unsaturated. Depending upon water quality, pretreatment may be necessary to control clogging.
- Erosion of the sides of basins should be avoided, especially when it may contribute fines to the clogging layer. Inner slide slopes may be covered with grass or less commonly with a geomembrane liner or riprap.
- Water (ponding) depths should be less than 1 ft (0.3 m) to allow for rapid draining and drying. Shallow ponding depths also minimize head loss across clogging layers and compression of the clogging layer.
- Basins should be properly graded so that there are no low spots where water can remain standing during drying periods.
- Suspended algae can settle and contribute to clogging. Higher turn-over rates reduce the time that suspended algae are exposed to sunlight and minimize their growth.

- Shallow low-permeability soils should be removed if deeper strata have a greater permeability.
- Flooding and drying cycles should be determined from operational experience.
- For water treatment (suspended sediment removal), a desilting basin with flocculant addition should be considered.
- The Golden Rule is to "start small, learn as you go, expand as needed."
- There is a difference of opinion on vegetation in basins. The negatives are that it could clog the soil, increase ET, make cleaning more difficult, and aggravate vector problems. Its benefits are that it may improve performance through shallow root channels (macropores) and maintenance of soil structure.
- Multiple-cell systems should be constructed in chains with the final basins receiving progressively cleaner water. Each basin should have its own input and output controls so that it can be operated independently. The elevation of basins should decrease in the downstream direction so that higher basins can drain under gravity into lower basins.
- Parallel-cell chains allow for continuous operation during maintenance activities.
- In the case of SAT (soil-aquifer treatment) systems for wastewater, the entire pretreatment, SAT, and post-treatment system should be designed together so that renovated water with the desired quality is produced at a minimum cost and minimum adverse environmental impact. Unfortunately, rapid-infiltration systems are often added as an afterthought.
- In arid regions with high evaporation rates, increases in salinity may occur depending upon the ratio of infiltration and evaporation rates.

15.4 Stormwater Infiltration Basins

15.4.1 Introduction

Basins (constructed depressions) are a key element of stormwater management because of the storage and water quality improvement they provide. Four main types of basins are utilized in stormwater management (Fig. 15.3):

- **Detention basins**: Low-lying areas that are designed to temporarily hold a set amount of water while slowly draining to another downstream location. They are used primarily for flood control by reducing peak discharges. Detention basins do not eliminate runoff. A basic design for a dry detention system is a dam with outlet pipe at its base with an invert level of the basin floor.
- **Retention (wet) ponds**: Depressions that are designed to hold a specific amount of water indefinitely. Water level in the pond fluctuates in response to precipitation and runoff.
- **Infiltration basins (dry retention basins)**: Depressions that are designed to store water until it can infiltrate into the ground. Unlike a retention pond, an infiltration basin does not have permanent standing water. Infiltration basins are designed

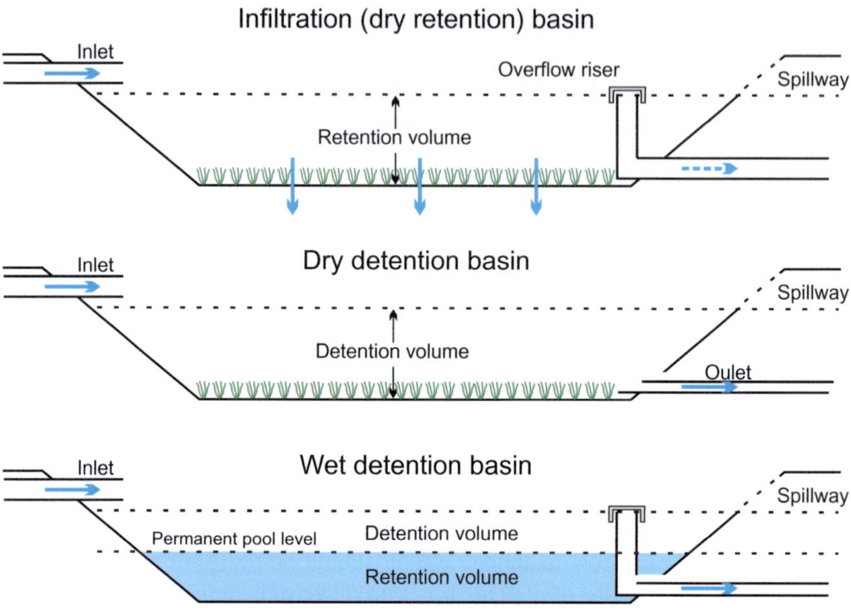

Fig. 15.3 Stormwater basin types

with overflows to divert flows beyond the storage capacity of the basin to other downstream elements of a stormwater management system (Fig. 15.4).

- **Enhanced detention basins (e.g., constructed wetlands)**: Systems with a primary goal of water quality improvement. Leaky constructed wetlands are also the site of local groundwater recharge.

Wet basins have a permanent pool (i.e., volume of water before a storm begins) that lengthens the residence time. Settling and biodegradation continue after a storm is over. Wet basins have both live and dead zones (Ferguson 1998). The live zone is the portion of the volume of a permanent pool on or near a direct line of flow between the inlet and outlet of the pond where replacement of old water with new water occurs. Within a pool of a given average depth, greater and lesser depths can perform different functions for water quality improvement and environmental enhancement. More sinuous flow paths result in higher treatment performance (Ferguson 1998).

Infiltration basins are closed basins in which the primary outflow is into the soil. Environmentally, they are most complete solution to the problem of urban stormwater (Ferguson 1998). Ponding time after a storm is the single most important aspect of the hydrological design of infiltration basins. Vegetation may actively maintain the porous structure of a soil. Aerated conditions in the soil supports deep-rooted vegetation and allows organic matter decomposers to form humus. Plants and animals build humus into the soils, aggregating it to form an open soil structure favorable for infiltration (Ferguson 1998).

Fig. 15.4 Infiltration basin overflow outlet, Celery Fields, Sarasota County, Florida. Overflow discharges to adjoining man-made wetlands of the Celery Fields Regional Stormwater Facility

All four types of stormwater basins can result in aquifer recharge depending upon the permeability of the base and sides of the basin. Infiltration basins, by definition, are primarily designed to infiltrate water and are thus of greatest importance for AAR.

15.4.2 Design Basics

Infiltration basins are a very commonly used element in stormwater management systems, particularly in suburban areas where suitable land is available. The basins are designed and constructed to accept runoff from impervious areas (roads, driveways, parking areas) and provide stormwater retention, groundwater augmentation, and water quality improvement. Stormwater infiltration basin systems are designed for passive operation (minimal human intervention). The basic components are an inlet, infiltration area, and an overflow.

For example, a stormwater infiltration basin within the author's community in Lee County, Florida, was constructed using berms and raising the elevation of an adjoining road and residential properties with fill (Fig. 15.5). The bottom of the berm is the original land surface (a sandy former farm field) and some native vegetation (slash pine trees) was preserved. Two basins are interconnected with a swale and receive runoff from the road. An overflow in one basin discharges to an adjoining wetland, which is connected through channels and lakes ultimately to tidal waters.

Fig. 15.5 Residential stormwater infiltration basin, Lee County, Florida

Stormwater infiltration basins are an element of "green infrastructure" and "low impact development," which seek to preserve or restore the predevelopment hydrology of sites (Chap. 23). The historical stormwater management practice has been to quickly convey water away from developed areas to prevent damage from flooding, whereas current emphasis is on reducing runoff from properties and locally infiltrating stormwater. Stormwater management system elements can be designed to also be a community amenity beyond their stormwater utility. Mowed grass-covered infiltration basin floors may be a recreational resource as playing fields (Fig. 15.6). Infiltration basins and other BMPs may serve an educational function and provide aesthetic richness, which is referred to as "artful rainwater design" (ARD, Echols 2007). ARD elements increase a landscape's attractiveness or value (Echols 2007; Echols and Pennypacker 2008, 2015). For example, basins can be landscaped to be more visually pleasing and interesting than a basic grass-covered basin.

The design, construction, and operation of stormwater infiltration basins are addressed in the United States in state and local (city) storm water manuals, construction standards, and guidance documents, most of which are now available on line. Stormwater manuals and construction standards from different jurisdictions share many similarities, which largely reflect common sources (e.g., U.S Environmental Protection Agency), as well as having some variations between states reflecting differences in local climatic and hydrogeological conditions. For example, the targeted separation of the base of infiltration basins from the seasonal high water table will necessarily be less in much of the state of Florida, where the water table during the wet season is close to land surface, than in more arid regions with greater depths to the water table. The storm water manuals also specify setback distances from buildings, potable water supply wells, and other facilities that might be impacted by the

Fig. 15.6 Residential infiltration basin in Sunrise, Arizona, which was made to be a community amenity (playfield)

systems. Recommendations and requirements from some stormwater manuals are provided below to illustrate some typical design considerations and processes.

Stormwater infiltration basins are sized based on the requirement to retain the runoff from design storms in their catchment areas and for the relatively rapid infiltration of the retained water. For most of the year, the basins are typically dry. The operation of stormwater infiltration basins is thus different from that of rapid infiltration basins (RIBs) used for the continuous (with operational drying periods) high-volume recharge of surface water and treated wastewater for aquifer recharge. The scale and technical sophistication of the hydrogeological investigation involved in the routine design of stormwater infiltration basin is typically much less than that involved with high capacity RIBs. The scope of work of hydrogeological investigation for stormwater infiltration basins may include evaluation of some or all of the following (Lowndes 2000):

- depth to high groundwater
- groundwater flow direction and rate of flow
- vertical and horizontal hydraulic conductivities
- presence and extent of perched aquifer conditions (confining strata above the water table)
- soil types
- field infiltration rates
- depth of bedrock and type of bedrock.

General infiltration basin design guidance include (Lowndes 2000):

- size and depth depends on design runoff storage volume; storage volume calculations are an exercise in geometry considering top and bottom dimensions and design depth (height of outflow above the basin floor)
- side slopes of 4:1 or gentler
- minimum separation of the basin floor from the seasonal high water table of 5 ft (1.5 m)
- infiltration times of no less than 6 h or more than 48–72 h
- the inflow should be designed to minimize erosion (e.g., using a rip-rap apron, level spreader, or grass apron)
- infiltration rates should not be so high as to minimize treatment
- soil hydraulic conductivities should be between 0.5 and 5.0 in./h (1.3–12.7 cm/h)
- some fines (clays) are desirable to increase the sorptive (treatment) capacity of soils.

Infiltration basins are designed to infiltrate a prescribed amount of water over a prescribed period of time. The amount of water from a prescribed (design) storm event is evaluated by surface-water flow modeling. Required basin floor area is calculated based on the design stored water volume, target drain time, and site infiltration rate, which is usually determined from a series of infiltration tests. Since basin-wide infiltration rates may vary from soil infiltration testing results and often decrease over time due to clogging, a safety factor (of at least 2) is usually applied to the lowest measured infiltration rate.

Massman (2003) proposed a rigorous step-by-step procedure for sizing infiltration basins:

(1) estimate the volume of stormwater that must be infiltrated over a prescribed period of time
(2) select a trial geometry and estimated water depth in the basin
(3) perform site characterization and data collection, including identification of small-scale layering that could affect vertical flow and cause mounding
(4) estimate saturated hydraulic conductivity (for example, from grain size analyses, infiltration tests, permeability measurements)
(5) estimate the vertical hydraulic gradient
(6) estimate the infiltration rate by multiplying the hydraulic gradient and hydraulic conductivity in accordance with Darcy's law
(7) apply correction factors (CFs) for biofouling, siltation, and pond geometry. The correction factors range from 0.9 to 0.2 depending on the susceptibility to biofouling and siltation and the degree of long-term maintenance (Massman et al. 2001)
(8) estimate flow rates
(9) design system
(10) conduct full-scale tests and adjust final design, if necessary.

It is important to sample and measure hydraulic conductivity of each layer and calculate an effective vertical hydraulic conductivity. Sampling of all layers between the basin bottom and water table is recommended, or, where the water table is very deep, down to twenty times the depth of ponding (Massman 2003).

The vertical hydraulic gradient can be calculated using a modification of the Green-Ampt equation (Massman and Butchart 2001):

$$Gradient(i) = \frac{(H_w + L_f - h_{we})}{L_f} \tag{15.7}$$

where (in units of length)

L_f depth to the wetting front
H_w ponding depth
h_{we} capillary suction negative pressure at the wetting front.

Massman (2003) proposed the following equations, based on modeling results, to estimate the steady-state hydraulic gradient beneath a medium-sized (0.6–6.0 acre) infiltration facility

$$Gradient(i) \approx \frac{H_{wt} + H_w}{138.62(K^{0.1})} CF_{size} \tag{15.8}$$

$$CF_{size} = 0.73(A_{pond})^{-0.76} \tag{15.9}$$

CF_{size} correction factor for size
H_{wt} depth from the base of the infiltration facility to the water table or first low-permeability layer (ft)
K saturated hydraulic conductivity (ft/d)
A_{pond} area of the pond (acres).

The equations are believed to be representative of facilities in areas where the depth to groundwater is from several feet to approximately 100 ft (30 m; Massman 2003). At greater depths, a gradient of 1 is recommended (Massman 2003). A layer is considered "low permeability" if its hydraulic conductivity is less than 10% of the value assigned to other layers and is less than the infiltration rate. More elongate basins tend to have greater infiltration rates. Massman and Butchart (2001) proposed the following computer simulation-derived correction factor for aspect ratio (CF_{aspect})

$$CF_{aspect} = 0.02A_{ratio} + 0.98 \tag{15.10}$$

where A_{ratio} is the aspect ratio of the pond (length/width). The *CF* should never exceed 1.4 (Massman 2003).

For layered strata beneath an infiltration facility, infiltration rates will be a function of the effective vertical hydraulic conductivity of the strata, which is the harmonic average of the saturated hydraulic conductivity (K_{sat}) of each layer. Saturated hydraulic conductivity can be estimated from grain size data, but it is important to recognize that hydraulic conductivity values depend on the degree of compaction of the sediment or soil. Strata that are well-compacted (e.g., from heavy equipment) or are heavily over-consolidated due to their geological history (e.g., glaciation) could have K_{sat} values an order of magnitude less than values for similar strata that have experienced minimal compaction (WSDOT 2014). Where the water table is deep, calculated effective hydraulic conductivity values could over estimate infiltration rates where deep layers have a small impact on infiltration rates or a low K_{sat} layer is present at a shallow depth below the infiltration basin or trench (WSDOT 2014).

15.4.3 Stormwater Infiltration Basin Performance and Maintenance

Stormwater infiltration basins typically have a passive operation and usually receive minimal maintenance. For example, the only maintenance that the infiltration basins in the author's community (Fig. 15.5) have received since their construction over 20 years ago is periodic mowing of the grass, trimming of trees, and removal of exotic vegetation for aesthetic purposes. Nevertheless, the basins have continued to operate as designed without incident and water has never risen to close to the outfall elevation even when hurricanes have hit the area.

From a hydraulic perspective, infiltration basin failure is either not initially achieving design infiltration rates or a subsequent reduction of infiltration rates to values significantly less than design rates due to clogging. Infiltration basins usually fail for one or more of the following reasons (Lowndes 2000):

- premature clogging (various physical and biological causes, and/or no or inadequate pretreatment)
- design infiltration rate used was greater than the actual rate (poor site characterization)
- sediment influx from the catchment
- compaction of the soil during construction
- upland soils or basin walls were not stabilized with vegetation and were a source of sediment delivered into the basin.

Clogging can be due to external sediment loading and internal sediment loading from poorly stabilized side slopes. Studies of infiltration basin performance suggest that limiting the flow that basins receive and avoiding overload conditions will improve long-term operation (Lowndes 2000).

Effective long-term operation of infiltration basins and other stormwater BMPs requires periodic inspections for clogging (e.g., sediment accumulation in inflow and pretreatment systems), erosion (gullying), health of vegetation, and infiltration times after storms. However, in many jurisdictions there is either not a formal requirement for regular inspections (and associated recording and reporting of the results) or the requirement is not enforced. Regular maintenance activities may include mowing of grass, trimming of vegetation, removal of unwanted (non-native) vegetation, and collection of garbage. Infiltration rates are rarely re-evaluated after construction unless clogging is manifestly evident by prolonged (>72 h) standing water after the end of large rainfall events.

Limited data are available on the performance of existing infiltration basins in terms of their current infiltration rates versus initial and design rates. Performance criteria vary, but a general requirement is that basins drain within a specified time. Bean and Dukes (2015) investigated whether stormwater basins in north-central Florida (Leon, Alachua, and Marion counties) were performing as designed. Retention basins (which include infiltration basins) in Florida are required by the Florida Administrative Code (Section 62-25.024 (4)) to provide the disposal capacity of the design volume of stormwater (i.e., water quality volume, WQV) within 72 h following the storm by percolation, evaporation, or evapotranspiration.

Two hundred fifty basins were screened from which 40 basins were selected that have either Department of Transportation (DOT) or residential uses. Design infiltration rates were obtained from permits and design documents, many of which included a safety factor of at least two. Current infiltration rates were measuring using a double-ring infiltrometer (DRI) based on ASTM Standard D3385-03. Standard ring diameters of 30 and 60 cm were used. Each basin had six test locations, except for two basins with nine locations and one basin with three locations. Soil core samples were also collected and analyzed for bulk density, organic matter, and texture.

DRI rates were significantly greater than the design rates for 14 basins (35%), significantly less for 15 basins (40%) and not significantly different for 10 basins (25%). However, 48 of the 250 basins considered in the initial screening were ponded for an extended period of time and, therefore, were clearly not functioning properly. It was estimated that only 48% of the basins considered were properly functioning, assuming the same performance ratios as the 40 tested basins (Bean and Dukes 2015). Basins with coarse soil textures (sands and sandy loams) were found to more likely have DRI rates greater than or equal to the design infiltration rates compared to basins with finer textured soils. A higher proportion of infiltration basins in the DOT watershed were determined to be functioning properly than in residential area (90% vs. 50%), which was suggested to be due to the DOT basins have a larger (deeper-rooted), more diverse vegetation and soil biota (Bean and Dukes 2015). Bean and Dukes (2015) recommended less frequent maintenance (mowing) to promote more diverse vegetation.

15.5 Rapid Infiltration Basins

15.5.1 Introduction

Rapid infiltration land treatment systems are used to treat wastewater and recharge surficial aquifers. Rapid infiltration systems include rapid infiltration basins (RIBs), in which water is applied to the land surface, and adsorption fields (also referred to as leach fields, tile fields, and disposal fields), in which water is applied to unsaturated soils using subsurface trench or gallery systems. The wastewater typically first receives at least secondary treatment (in the United States) or less commonly just primary treatment.

Rapid infiltration land systems have infiltration rates in the range of 6–90 m/year (20–300 ft/year) and provide treatment by physical and biological processes in the soil zone (USEPA 2003). The infiltrated water may either discharge to nearby surface water bodies or recharge the underlying aquifer. Slow-rate land treatment systems apply treated wastewater onto the land to support vegetative growth. Application rates are usually in the range of 0.6–6 m/year (2–20 ft/year; USEPA 2002). Applied water is both evapotranspired to the atmosphere and enters the groundwater. Treatment is provided by both vegetative (nutrient removal) and soil zone processes.

RIBSs can provide the benefits of recharging groundwater, providing further treatment to effluent, and reducing degradation of surface waters. Their main negative is that higher loading rates can alter groundwater flow patterns and potentially result in contamination of local groundwater. The most common contaminants associated with RIBs are nutrients (nitrogen and phosphorous), metals, pathogens, and organic compounds. A concern is that years of application of treated effluent with high concentrations of nutrients, pathogens, and organic compounds in RIBs will result in significant risks for environmental and public health problems (Türkmen et al. 2008). However, environmental and public health risks can be minimized through proper geographical and hydrogeological siting, design, and operation of systems.

Vadose zone processes play an important role in contaminant removal. A sufficiently thick vadose zone is needed for nitrogen and other contaminant removal processes active in the zone. State regulations in the United States vary in the minimum required vadose zone thickness for RIBs. The state of Delaware, for example, has a low thickness requirement of a 2 ft (0.6 m) separation of the base of the infiltration bed from the mounded water table (Türkmen et al. 2008). In the much more arid state of Nevada, the minimum design criteria include a depth to groundwater below the basin bottom of $\geq$10 ft (3 m) and a depth to an impermeable layer of $\geq$30 ft (9 m; Nevada Bureau of Water Pollution Control 1993). The State of Florida requires a demonstration that the groundwater mound will not intercept the ground surface during any portion of the loading cycle during any time of the year and that increases in groundwater elevations shall not interfere with reasonable uses of adjacent properties.

RIBs are sometimes considered synonymous with soil-aquifer treatment (SAT) systems in that both involve the infiltration of treated wastewater using basins (e.g., USEPA 2003). The design and operation of the RIBs and SAT basins can be essentially the same. However, the original concept of SAT is that it is a treatment technology in which the infiltrated water is recovered by the system operator from the immediate vicinity of the recharge site. The intent is to minimize impacts to groundwater by recovering all (or at least most) of the recharged water. Drainage (underdrains or wells) may be needed to maintain infiltration rates (reduce the decline in rates due to mounding) and to keep renovated water from mixing with native groundwater (if that is a system concern). RIBs refer herein to systems that have a primary aquifer recharge function. Water quality improvements obtainable using SAT are discussed in Sect. 19.3.

The general advantages and disadvantages RIBs were summarized by the USEPA (2003). The main advantages are:

- as a gravity distribution methods, the basins consume no energy and chemicals are typically not required
- they provide simple and economical treatment
- the process is not constrained by seasonal changes although the efficiency some microbially mediated processes may vary seasonally
- the process is very reliable with sufficient resting periods
- the process is suitable for small plants were operator expertise is limited.

The main disadvantages are:

- stringent regulatory requirements for the wastewater may result in the need for additional treatment with associated capital and operational costs
- the process may not meet stringent nitrogen concentration requirements for discharge to drinking water aquifers
- the systems require long-term commitment of a significant land area for the process
- management of clogging is required, which commonly involves occasional removal of accumulated deposits of organic material and the top few inches of the soil to expose clean material.

15.5.2 RIB Design

Key design and operational considerations for RIBs are similar to those for infiltration basins in general:

- hydraulic loading rates
- nitrogen and organics loading rates
- land area requirements
- flooding and drying cycle
- infiltration basin design
- groundwater mounding.

The USEPA design criteria for RIBs is summarized in Table 15.3. However, RIBs are regulated at the state level in the United States and requirements vary depending upon local hydrogeological conditions. Florida reuse rules (Florida Administrative Code Chapter 62-610) limit initial average hydraulic loading rates to 3 in./day (7.6 cm/d). A higher rate (up to 9 in./d or 22.9 cm/d) may be requested if technical justification is provided in an engineering report. RIBs are required in Florida to be designed with a minimum of three ft (0.9 m) of freeboard to protect the integrity of pond embankments and an emergency discharge device to prevent water levels from rising closer than one foot (0.3 m) from the top of the embankment or berm.

Table 15.3 USEPA design criteria for RIBs

Parameter	Recommended values
Basin infiltration area	0.3–5.5 ha/103 m^3/d (3–56 acres/Mgd)
Hydraulic loading rate	6–90 m/year (20–300 ft/year)
BOD loading	22–112 kg/ha/d (20–100 lb/acre/d)
Soil depth	at least 3–4.5 m (10–15 ft)
Soil permeability	at least 1.5 cm/h (0.6 in./h)
Wastewater application period	4 h–2 wks
Drying period	8 h–4 wks (Wet/dry ratio is always <1.0)
Soil texture	Coarse sands, sandy gravels (fine to medium sand is used in Florida)
Individual basin size (at least 2 basins in parallel)	0.4–4 ha (1–10 acres)
Height of dikes	0.15 m (0.5 ft) above maximum expected water level

Source USEPA (2003) from Crites et al. (2000)

Fig. 15.7 Sun Ray RIBs, near Frostproof, Florida

An example of a Central Florida RIB is provided in Fig. 15.7. Water enters from a central vertical pipe and a splash pad prevents erosion and spreads the water. The side walls are grass-covered, but other systems protect side walls from erosion using a plastic (high-density polyethylene) liner.

Crites et al. (2000) presented typical procedures for RIB design, which have been adopted (with some modifications) by the USEPA and many states:

(1) determine design infiltration rate from field testing
(2) determine hydraulic pathways of infiltrated water and discharge water quality requirements (i.e., where will the infiltrated water go and what are the applicable standards for discharge into the receiving aquifer or surface-water body)
(3) determine pretreatment requirements, which depend on operational consideration (e.g., clogging management), wastewater quality, expected pollutant attenuation, and regulatory discharge water quality requirements
(4) determine the hydraulic loading rate based on treatment needs (e.g., for nitrogen removal), infiltration rates, and wet/dry ratio
(5) calculate land requirements
(6) evaluate potential for unacceptable groundwater mounding
(7) select hydraulic loading cycles and number of basins
(8) calculate application rate and the final wet/dry ratio
(9) layout basins and ancillary infrastructure design
(10) determine and implement a monitoring program.

Selection of the hydraulic loading rate is the most critical step in the design process (Crites et al. 2000). Hydraulic loading rates are a function of infiltration and percolation rates, lateral flow (water-table aquifer transmissivity), depth to groundwater, quality of applied wastewater, and treatment requirements. On an annual basis, hydraulic loading rates are a small fraction of the measured clear water permeability of the most restrictive layer, allowing for drying periods and clogging (Crites et al. 2000). Crites et al. (2000) recommended that for RIBs, hydraulic loading rates should be no greater than 2–4% of the minimum measured rates from cylinder infiltrometer readings and no greater than 7–10% of measured rates from basin infiltration tests. The lower end of the range should be used when extended drying periods are required and the higher end of range should be used for high wet/dry ratios and mild climates.

Hydraulic loading rates are selected to maximize either the infiltration rate or nitrogen removal and may vary seasonally. Longer loading cycles and storage may be needed during the winter in cold climates due to lesser biological activity and freezing of the basins. The temperature dependence of hydraulic conductivity, and thus infiltration rates, also needs to be considered.

15.5.3 Water Conserv II and Reedy Creek Improvement District RIBs (Central Florida)

The state of Florida is a leader in the United States in the reuse of reclaimed water. RIBs are used to provide disposal capacity for excess flows and a beneficial use of the water by recharging underlying aquifers. Hydrogeological conditions in the central part of the state are particularly favorable for RIBs because of relatively high land-surface elevations and the presence of thick intervals of unsaturated sands above the main carbonate aquifer.

Water Conserv II is the largest wastewater reuse project of its kind in the world, combining agricultural irrigation with aquifer recharge using RIBs. The history of the Water Conserv II system was summarized by Estow (1996), Cook (2004), and Water Conserv II (2012). Water Conserv II was developed by Orange County and the City of Orlando in response to a court injunction to cease discharging treated wastewater into Shingle Creek, which flows into Lake Tohopekaliga, by March 1988. The selected option was a combination of citrus irrigation and RIBs. Construction commenced in 1983 and the project began operation on December 1986. Water Conserv II has 61 RIBs, each having 1–5 cells (Figs. 15.8 and 15.9). The RIBs are divided between seven sites, six in Orange County and one in Lake County, with the sites selected based

Fig. 15.8 Aerial photograph of part of Water Conserv II RIB Site 6A, western Orange County, Florida. Black border around the RIBs is the HDPE geomembrane liner on the side slopes (2011; *Source* USGS)

Fig. 15.9 Photographs of Water Conserv II RIBs showing HDPE geomembrane liner and maintenance (harrowing) of basins

upon percolation ability. Two additional sites in Lake County have been permitted but are not yet constructed. The total constructed RIB percolation (wetted bottom) area is 155.6 acres (63.0 ha) and corresponding permitted capacity is 35.0 Mgd (132,000 m^3/d). The expanded permitted system will have an infiltration area of 228.7 acres (92.6 ha) and a capacity of 43.3 Mgd (164,000 m^3/d). The recharged water receives at a minimum activated sludge treatment plus dual-media filtration and high-level disinfection with chlorine.

The Reedy Creek Irrigation District (RCID) serves the Disney World complex in Orlando. Reclaimed water is principally recycled for irrigation. Excess reclaimed water is used to recharge groundwater through a RIB system consisting of 85 one-acre (0.4-ha) basins with a total design capacity of 12.5 Mgd (548 L/s; McKim 2012). Operation of the RIBs began in September 1990.

The Water Conserv II and RCID RIBs are located in the Lake Wales Ridge physiographic region, which is characterized by thick sand deposits with a deep water table (locally greater than 100 ft or 30 m). The RIBs recharge the water-table aquifer, from which water is either lost to evapotranspiration, discharges to surface-water bodies, or, to the greatest extent, recharges the underlying Upper Floridan Aquifer, which is the primary potable water source in central Florida (O'Reilly 1998). A typical Water Conserv II RIB has the following features (O'Reilly 1998):

- construction by excavation in native sand with no soil profile modification
- each site (basin) has up to five adjacent cells that are connected with buried pipe
- the average RIB cell bottom area is 1.2 acres (0.5 ha)
- water discharges through a central pipe and is allowed to fill basins no more than 2–3 ft (0.6–0.9 m) deep
- a typically loading period is 1 week and followed by a 1–2 week resting period during which the groundwater mound dissipates.

O'Reilly (1998) simulated the impacts of the operation of the Water Conserv II and RCID RIBs using the MODFLOW and MODPATH codes for 1999 steady-state conditions. Significant water-table mounds occur under all RIB sites (except for one site that is rarely used). In the RIB site areas, reclaimed water application was found to dominate the hydrological system, exceeding the natural recharge rate by 10 times or more. Evaluation of the hydrological impacts of the RIBs was complicated by uncertainties over pre-application conditions. The simulated net natural leakage to the Floridan Aquifer System was 10.4 in./year (26.4 cm/year) averaged over the model area. The simulated recharge increased under 1995 conditions to about 11.6 in./year (29.5 cm/year), which represents about a 10% increase. The simulated minimum and average travel times from the RIBs to the top of the Floridan Aquifer System were approximately 1 and 10 years for Water Conserv II RIB reclaimed water and approximately 2 and 7 years for RCID RIB reclaimed water.

Sumner and Bradner (1996) investigated nutrient transport and transformation in the RCID RIB system. The basins were reported to be operated in 8 basin sets, each operated for one week, with 15–20 h application periods and 4–9 h rest periods. After each operational week, the basin sets have a four week rest period during which reclaimed water is routed to other sets of basins. The infiltration capacity of the basins ranges from 0.5 to 2.0 Mgd (1,890–7,570 m^3/d) and is sufficient so that ponded water does not cover the entire basin bottom. The basins are reported to be maintained by tilling in which the surface clogging layers are mixed into the underlying sediment.

The following observations were made on nutrient transport and transformations in the infiltrated reclaimed water (Sumner and Bradner 1996):

- total organic nitrogen decreased by roughly 50% in the upper 15 ft of the soil zone from concentrations in reclaimed water as a result of colloidal attachment, straining, and sedimentation
- significant nitrification occurred during the several week rest period as soils drained and became aerated
- ammonium concentration decreased by more the 50% in the upper 3 ft (0.9 m) of the soil profile
- short flooding periods maintained aerobic conditions and hindered denitrification (longer flooding periods were suggested)
- nitrification results in transient nitrate "spikes" in the shallow saturated zone, which were short-lived (a few hours) and dissipated by dispersion and dilution
- a 90% reduction in phosphorous occurred in the upper 15 ft (4.6 m) of the soil profile (0.25–0.02 mg/L) due to adsorption onto iron and aluminum oxyhydroxides
- some phosphorous (predominantly coarse, organic fraction) was carried to the water table where it accumulated under slacking pore water velocities. The phosphorous that reached the water table was immobilized by adsorption or precipitation reactions during basin rest periods.

15.5.4 Cape Coral, Massachusetts

Repert et al. (2006) investigated the long-term attenuation of carbon and nitrogen within a groundwater plume from discontinued wastewater RIBs on Cape Coral, Massachusetts. The RIBs at the Massachusetts Military Reservation site were used for the disposal of secondary-treated wastewater for over 60 years until discharges ceased in 1995. The resulting contaminant plume in the sand-and-gravel aquifer is more than 6 km long, 30 m thick and 1,000 m wide. The plume has a suboxic ammonium-containing core surrounded by an oxic to suboxic nitrate-containing outer zone. The contaminant plume core remained anoxic for the entire 10-year study period. In 2004, substantial amounts of total dissolved carbon (7 mol C m^{-2}) and fixed (dissolved plus sorbed) inorganic nitrogen (0.5 mol N m^{-2}) were still present in a 28-m vertical interval at the disposal site.

Relatively high groundwater flow rates resulted in the flushing of mobile constituents (e.g., boron). After 8 years, the mobile constituents advected downgradient beyond the 0.6 km study transect. On the contrary, the total dissolved organic carbon and nitrogen pools have remained stable despite the flushing. The data points to the long-term importance of sorption processes in contaminant removal during RIB operation and desorption processes after the cessation of recharge (Repert et al. 2006). Sorbed constituents have contributed substantially to the dissolved carbon and nitrogen pools and are responsible for the prolonged persistence of the contaminant plume (Repert et al. 2006). A key conclusion is that natural aquifer restoration in the discharge area will take at least several decades, even though groundwater flow rates and the potential for contaminant flushing are relatively high.

15.6 Surface Water Recharge Infiltration Basin Systems

15.6.1 Introduction

Infiltration basin systems used to recharge surface waters have similar constructions as infiltration basin used for stormwater and treated wastewater. An operational difference is that flooding and drying cycles are implemented primarily to manage clogging rather than to control redox state for nutrient removal. Infiltration basins are the primary means for recharging unconfined aquifers used for groundwater banking schemes in Arizona and California. Surface water infiltration basins have the basic requirement that the unsaturated zone strata should be sufficiently permeable to achieve high infiltration rates and for recharged water to reach the water table. Low-permeability layers within the vadose zone can result in perched aquifer conditions and reduce the amount of water that actually recharges the target unconfined aquifer. Mounding above perched layers can cause water logging and discharges at low-lying areas, which can adversely impact nearby property owners/occupants and be a loss of water.

Where recharged water is impaired (non-potable quality), such as surface waters impacted by urban and agricultural activities or blended with wastewater, the travel pathways, distance, and residence time from recharge sites to production wells are major technical and regulatory issues. Travel distance and time are a great concern where indirect potable reuse is a possibility or reality. For existing operational systems, travels times and flow pathways may be inferred from tracer and age data. Artificial tracers may be employed, but commonly recharged water has sufficient differences in composition from native groundwater so that its presence can be readily identified and often quantified.

Groundwater modeling is also a valuable tool for evaluating the transport of recharged water. However, modeling of solute transport requires detailed data on aquifer hydraulic and storage properties, which necessitates a detailed aquifer characterization. Surface geophysical methods have promise for evaluating the movement of infiltrated water in the vadose zone. For example, a time series of geophysical surveys (i.e., a 4-D survey), starting prior to the initiation of recharge, may be used to map changes in the water content of the vadose zone. Microgravity surveys can detect changes in the water content in the vadose zone from changes in mass (density) of saturated or partially saturated sediments. Resistivity-based methods can be used to map increases in water content from the associated decrease in resistivity.

The Orange County Water Department (OCWD) (California, USA) recharge system was simulated by Thompson et al. (1999). Age ($^3H/^3He$) data indicate that the age of water varies with depth. The Orange County Groundwater Basin consists of irregularly alternating sands, gravels, and conglomerates, and semi-permeable silts, siltstones and clays, which result in complicated flow paths. Groundwater flow modeling demonstrated the importance of aquifer heterogeneity and the need for high-resolution three-dimensional models to realistically represent flow pathways and travel times in this system. Geostatistical methods (transition probability/Markov indicator technique) were used to create a hydrogeological model that honored borehole information. The modeling and geochemical data indicated that groundwater entering production wells spanned a large range of ages (decades or more). Water in some production wells appears to be derived from multiple sources. The modeling results indicated that flow pathways are very sensitive to differences in hydraulic conductivity.

15.6.2 Arizona Infiltration Basin Systems

Infiltration basin systems are used in Arizona to recharge surface waters from in-state river systems, Colorado River water conveyed into the Phoenix and Tucson metropolitan areas by the Central Arizona Project (CAP) and, less commonly, treated wastewater. The ultimate goals of the recharge system are to fully utilize the state's Colorado River entitlement (which would otherwise be lost if not used) and to help the state reach an equilibrium between groundwater extractions and recharge (i.e., reach a "safe" or sustainable yield). Overdrafted basin aquifers in the state have huge storage capacities and undeveloped or agricultural land has been available outside of the urban areas for construction of infiltration basin systems. The largest infiltration systems are located in the Phoenix and Tucson Active Management Areas, which include the major population centers of the state, have experienced the greatest overdraft, and receive water from the CAP aqueduct (Table 15.4).

Basins were located, designed, and constructed to take advantage of coarse-grained (paleochannel and alluvial fan) deposits located near land surface. Proximity to water sources (particularly CAP aqueduct) was also a key consideration in locating systems. Arizona has a hot arid climate, which can be unfavorable for surface spreading-systems, but high infiltrations rates help keep evaporative losses to less than 5% (Megdal et al. 2014).

Granite Reef Underground Storage Project

The Granite Reef Underground Storage Project (GRUSP) is first major recharge facility in Arizona and one of the largest of its kind in the United States. The initial four basins were completed in 1994, and additional basins were completed in 1999, 2000, and 2005 (Fig. 15.10). The system has a total area of 197 acres (79.7 ha; Salt River Project n.d.). The recharge water is surface water from the Salt and Verde Rivers, CAP water, and reclaimed from the Mesa Northwest Water Reclamation Plant. The GRUSP is located in a channel of the Lower Salt River, downstream from the Granite Reef Dam. The system is underlain by a thick interval of high-permeability, coarse-grained unconsolidated sands and gravels. Recharge rates are 0.6–2 m/d (Lluria 2009).

Lluria (2008, 2009) provided a summary of the history of the GRUSP. More than 90% of project land is within the Salt River Pima-Maricopa Indian Reservation and leased from the tribal government. Every five years the land is reappraised and the rent adjusted to current value. The main land value determinant is the sand and gravel

Table 15.4 Major infiltration basin systems in Arizona

Project	Surface area [acres, (ha)]	Number of basins	Permitted annual storage capacity [acre feet, (MCM)]	Recharged water
Phoenix AMA				
Granite Reef USP[1]	197 (79.7)	7	93,000 (114.7)	River, CAP, reclaimed
New River—Agua Fria USP[1]	125 (50.6)	6	75,000 (92.5)	River, CAP, reclaimed
Agua Fria Project[2]	100 (40.5)	7	100,000 (123.3)	CAP
Tonopah Desert RP[2]	207 (83.8)	19	150,000 (185.0)	CAP
Hieroglyphic Mountain RP[2]	38 (15.4)	3	35,000 (43.2)	CAP
Tucson AMA				
Central Avra SRP[3]	317 (128.3)	11	80,000 (98.7)	CAP
Southern Avra SRP[3]	226 (91.5)	9	60,000 (73.8)	CAP
Pima Mine Road RP[2]	37 (15.0)	5	30,000 (36.9)	CAP
Lower Santa Cruz RP[2]	30	3	50,000	CAP

USP Underground Storage Project, *SRP* Storage and Recovery Project, *RP* Recharge Project
Sources [1]Salt River Project (n.d.), [2]CAP (n.d.), [3]City of Tucson (2017)

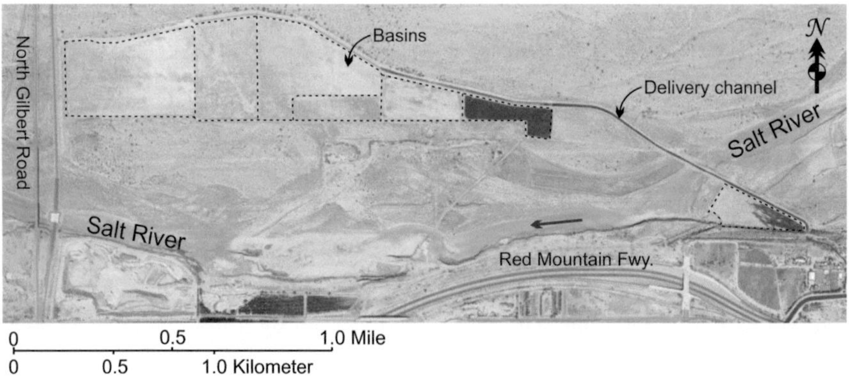

Fig. 15.10 Aerial photograph of the Granite Reef Underground Storage Project, Mesa, Arizona (September 2012; *Source* U.S. Geological Survey)

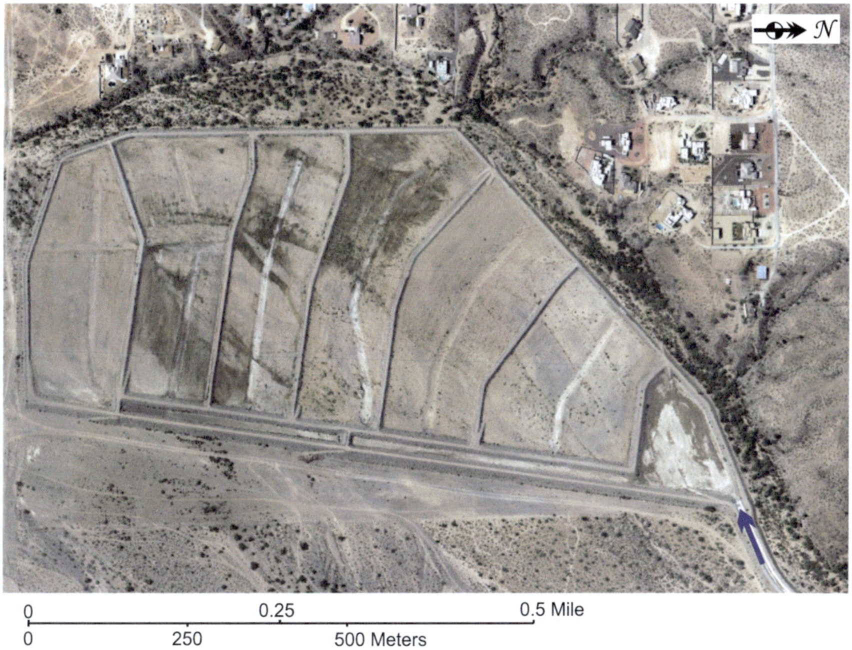

Fig. 15.11 Aerial photograph of the Agua Fria Recharge Project near Peoria, Arizona, in which the infiltration basin system is divided into cells (*Source* U.S. Geological Survey)

resources on the property, which are in high demand, resulting in steady substantial increases in rent and thus unit cost of recharge. The main operational issues have been:

- damage caused by storm water releases
- avoiding mounding at a downstream landfill
- suppressing seepage in nearby gravel mining operations.

Agua Fria Recharge Project

Agua Fria Recharge Project, located near the City of Peoria, consists of two operational components, a four mile section of the Aqua Fria River used for recharge and conveyance of water downstream, and a constructed infiltration basin system with an area of 100 acres (40.5 ha; Fig. 15.11; CAP n.d.) The total permitted recharge capacity is 100,000 AF (123.3 MCM) per year and construction was completed in May 2002. It is the only recharge project in Arizona to combine streambed recharge and infiltration basins at a single facility.

The constructed system consists of 7 spreading basins with the up-gradient basin (Basin A) serving as a sedimentation (desilting) basin (CAP n.d.). An earthen dam (headworks structure) captures surface flows from the river, from which they are conveyed to the spreading basins through a 4,000 ft-long, concrete-lined canal. A broad-crested weir is used to measure flow entering the recharge basins. Water from Basin A flows to the other 6 basins (B through G) via a distribution channel located on the west side of the facility. Basin A is 14 ft (4.3 m) deep and the remaining basins are 6 ft (1.8 m) deep.

Typical operations is wet cycles of 3–7 days and dry cycles of 5–10 days. Infiltration rates in the river section have ranged from 2.7 to 4.3 ft/d (0.8–1.3 m/d) with an average of 3.85 ft/d (1.17 m/d). Infiltration rates in the basins have ranged from 1.21 to 3.48 ft/d (0.37–1.06 m/d; CAP n.d.)

Tucson CAVSARP and SAVSARP

The City of Tucson Clearwater Renewable Resource Facility consists of the Clearwell Reservoir and two major aquifer recharge and recovery projects: the Central Avra Valley Storage and Recovery Project (CAVSARP) and the Southern Avra Valley Storage and Recovery Project (SAVSARP; City of Tucson 2017). CAVSARP includes 11 recharge basins totaling 317 acres (128.3 ha), 33 recovery wells, a reservoir, booster station, and pipelines (Fig. 15.12). The SAVSARP includes nine recharge basins with a total area of 226 acres (91.5 ha). Both projects store CAP water in the alluvial Avra Valley Aquifer and were constructed at the site of a paleochannel. The SAVSARP recharge basins were laid out to follow the curve of the natural paleochannel to maximize recharge potential (City of Tucson 2017). Both systems are designed to recharge the groundwater, allow for full utilization of the City of Tucson's CAP water entitlement, and provide a blend of CAP and groundwater to Tucson Water customers (PAG 2012).

Since 2007, the Pima Association of Governments (PAG), Tucson Water and the University of Arizona—Laboratory of Isotope Geochemistry jointly conducted stable isotope (hydrogen and oxygen) monitoring to track the movement of recharged CAP water in the aquifers below each of the recharge facilities (PAG 2012). CAP water and native groundwater have distinctly different isotope signatures. The stable isotope data indicate that recharged CAP water is reaching the aquifers beneath each recharge facility and is mixing with the native groundwater. However, the amount of CAP water in well samples near the facilities is quite variable, indicating a strong influence of subsurface hydrogeology on the flow of recharged CAP water (PAG 2012).

15.6.3 Orange County Water District (California)

The Orange County Water District (OCWD) aquifer recharge program was initiated in the 1950s and subsequently expanded to address overdraft within the Orange County Groundwater Basin. The aquifer recharge program started with the use of the natural channel of the Santa Ana River, and now includes extensive channel modifications, in-channel and off-channel recharge basins, and injection wells as

Fig. 15.12 Aerial photographs of the Central Avra Storage and Recovery Project located near Tucson, Arizona (May 2015; *Source* U.S. Geological Survey)

part of a salinity-barrier system. The history of the recharge system was reviewed by Mills (2002) and Hutchinson and Woodside (2002).

The OCWD aquifer recharge program is perhaps the most noteworthy system of its kind in the world for its scale, diversity of water sources and recharge methods used, and technically sophisticated and innovative approaches employed. Sources of water used for aquifer recharge are:

- Santa Ana River water, which includes natural flows, urban stormwater, and tertiary-treated wastewater discharged into the channel
- imported water
- recycled water that receives advanced water treatment (initially from Water Factory 21 and now the Groundwater Replenishment System).

The aquifer recharge program provides textbook examples of recharge by channel modifications (Sect. 16.3.1), a salinity barrier injection well system (Talbert Gap Salinity Barrier, Sect. 21.7.3), and off-channel infiltration basins.

The Santa Ana River has been divided into a flood channel (main river system) and a water conservation channel (off-river systems; Fig. 15.13). The main river system is an unlined channel in which infiltration is maximized by the construction of T-levees with bulldozers. The levees are temporary structures that are designed to be washed out during major floods. The levees act to minimize channelization, and increase the wetted surface area and transit time. The water conservation channel contains permanent structures that divide the channel into a series of shallow basins parallel to the main channel system. Three basins at the upstream end of the basin system are dedicated for desilting.

The off-river system consists of constructed recharge pits and a former sand and gravel mine. Clogging, as a result of nutrients and the suspended solids load of the river water, is addressed by occasional (approximately twice a year) draining, drying, and scraping of the clogging layer. The clogging layer is characterized as a hardened, calcium carbonate-cemented mixture of organic matter, silt, and clay. The clogging materials from the Anaheim, Kraemer, and Miller Basins was characterized as 92% mineral (montmorillonite, quartz, plagioclase, illite/mica, and calcite) and 8% organics (Hutchinson 2013).

Maintaining infiltration rates in large-scale surface-spreading systems is a critical issue for operators in general. Special machinery has been developed or adapted for the scarifying and scraping of infiltration basins, which are usually operated in dried basins (Huisman and Olsthoorn 1983). A more innovative approach was taken by the OCWD that involved the development of several generations of remotely operated subaqueous Basin Cleaning Vehicles (BCVs) that could remove clogging layers while basins remain in service (Hutchinson 2007). A self-contained, self-propelled beach cleaner was also found to be effective in selectively removing the clogging layer (Hutchinson 2007). Although the concept of in situ removal was found to be a valid approach, the BCVs, as designed, were not found to be cost effective. The OCWD is looking manage clogging by pretreatment instead (Hutchinson 2013).

A key operation issue is reducing system down times for drying and minimizing the costs associated with basin maintenance. The OCWD used a percolation rate decay model to determine the optimal basin cleaning program, which considered factors such as (Hutchinson 2007):

- lost recharge from down time when a basin is out of service
- costs associated with basin cleaning activities
- costs associated with draining deep basins.

Fig. 15.13 Santa Ana River, Orange County, California, in-channel and off-channel recharge facilities. A series of low berms in the main river channel (A) create a more tortuous and slower flow. Off-channel recharge basins parallel the main channel (B) and are present further to the north (*Source* U.S. Geological Survey)

Hutchinson et al. (2017) presented the results of an investigation of five potential pretreatment options to reduce suspended solids and thus clogging of the infiltration basins:

- riverbank filtration
- DAF (dissolved air flotation)
- cloth filers
- flocculation-sedimentation
- ballasted flocculation and sedimentation.

It was concluded that chemical methods can lead to clogging and would therefore not be used. Two riverbank filtration design options were investigated: (1) slotted PVC pipe packed with gravel and (2) stormwater "rain cells." The latter are plastic crates surrounded with a geotextile. Laterals of both designs were constructed across the river channel, about 1 m below the riverbed. RBF was found to achieve a 97%

removal of TSS, 50% removal of TOC and DOC, and 99% removal of total coliforms. The initial results indicated an increase in percolation rates. There were diminishing returns with decreasing lateral spacing. Too close laterals decreased the efficiency of the system.

15.6.4 Nassau County and Suffolk Counties, Long Island, New York

Recharge basins have been used in Long Island, New York, since 1935 for both drainage, flood control, and aquifer recharge (Seaburn and Aronson 1974). Over 3,000 recharge basins have been constructed in Nassau and Suffolk Counties (Ku and Simmons 1986). The basins are open, unlined pits, excavated in moderately to highly permeable glacial sands and gravels, and range in size from 0.1 to 30 acres (0.04–12 ha) with an average area of about 1.5 acres (0.6 ha). Most of the basins are about 10 ft (3 m) deep but some are as much as 40 ft (12 m) deep (Ku and Simmons 1986; Ku et al. 1992).

The basins vary depending upon whether they have an overflow structure to other basins or streams. Less than 10% of the basins were characterized as water-containing basins in that they held water for 5 days or longer after a 1-in. (2.5 cm) rainfall (Seaburn and Aronson 1974). Some basin design features are (Seaburn and Aronson 1974):

- settling basin are excavated in the basin floor; basins may have two or more levels with sediment and debris collected in the lowest level
- retention (sedimentation/settling) basins are connected to adjacent or nearby recharge basins
- diffusion wells are installed below some basin floors, which are constructed of porous concrete rings 8–10 ft (2.4–3.0 m) in diameter that penetrate through impermeable strata
- catch basins vary depending on whether they are sealed or have open bottoms and thus contribute to recharge.

Average infiltration rates at three studied sites ranged from 0.2 to 0.9 ft/h (0.06–0.27 m/h) with the slowest rate in the Deer Park basin attributed to a high percentage of silt, clay and organic debris washed in from the drainage area and a lack of plant growth on the basin floor. Plant roots are believed to keep the soil layer loose and permeable. At two basins, a close relationship was observed between infiltration and temperature (greatest rate in summer) due to viscosity effects.

Ku and Simmons (1986) examined the effects of recharge on water quality at five basins representative of different land uses:

- Centereach—commercial
- Huntington—shopping center and parking lot
- Laurel Hollow—low density residential

Table 15.5 Indicator bacteria concentrations in stormwater from five Long Island recharge basins

Parameter	Concentration (MPN/100 ml)	
	Range of median values	Maximum
Total coliforms	4,300–24,000	1,100,000
Fecal coliforms	930–9,300	240,000
Fecal streptococci	24,000–126,500	2,400,000

MPN most probable number

- Plainview—highway
- Syosset—medium-density residential.

The sampling was performed from 1980 to 1982. Concentrations of most measured constituents in stormwater were within Federal and State drinking water standards with few exceptions related to specific land uses and seasonal effects. Heavy metals were present in some stormwater samples with lead present in highway runoff up to 3,300 μ/L. Leaded gasoline was phased out in the United States in the 1970s. Chloride was found in parking lot runoff at up to 1,100 mg/L when salt is used for deicing. The following organic chemical were most commonly detected in stormwater (found in at least 50% of the samples):

- benzene
- bis(2-ethylhexyl) phthalate
- chloroform
- methylene chloride
- toluene
- 1,1,1 trichloroethane.

Only four priority pollutants constituents were detected above the then New York State guideline of 50 μg/L:

- P-chloro-m-cresol (75 μg/L, highway basin)
- 2,4-dimethylphenol (96 μg/L, highway basin)
- 4-nitrophenol (58 μg/L, parking lot basin)
- Methylene chloride (230 μg/L, highway basin).

The above chemicals were thought to be from point sources rather than general land use within the drainage areas.

Indicator bacteria were detected at high concentrations in the stormwater (Table 15.5). The indicator bacteria concentration in nearly all of the groundwater samples collected beneath recharge basins (sampled 1–2 days after a storm) were less than 3.0 MPN/100 mL.

Nassau County relies entirely on groundwater for its water supply. About 50% of the annual precipitation of 45 in./year (114 cm/year) recharges the groundwater system (Ku et al. 1992). Under natural conditions, almost all recharge occurs in the non-growing (dormant) season (mid-October to mid-May) when evapotranspiration is low (Ku et al. 1992). Nassau County can be divided into two regions based on stormwater management. In the northern and southern coastal areas, stormwater is routed mostly to storm sewers that discharge to streams and coastal waters. In the central part of the county, runoff is routed to recharge basins.

Urbanization and associated stormwater management resulted in spatial and temporal changes in recharge. In the recharge area, increased runoff that is conveyed to recharge basins increased recharge during the growing season. The average increase in recharge in the recharge area was estimated to be 12%. Increased runoff to tide in the "runoff areas" decreased recharge by an estimated 10%. Countywide, the increased recharge during the growing season is almost balanced by reduced recharge in the dormant season (Ku et al. 1992).

National Research Council (1994) concluded that studies have not shown any significant adverse impacts even after several decades of operation of the earlier basins and that the storage and flow accretions to the aquifer system appear to far outweigh any potential detrimental water quality impacts.

15.7 Infiltration Basin Clogging Management

Infiltration and percolation rates at infiltration basins depend upon the permeability of the underlying sediment, groundwater mounding, and clogging. Clogging reduces system infiltration rates and creates maintenance requirements and associated costs. Clogging of infiltration basins can occur as the result of the deposition of fine suspended particles, biological activity, chemical processes, and entrapped gases (Chap. 11).

Reductions in permeability may also occur due to clay dispersion (breaking apart of soil aggregates and releasing clay colloids that can clog pores) caused by the introduction of high-sodium and/or low-salinity water (Bouwer 1989; National Research Council 1994). Clogging of infiltration basins is usually a surficial process.

The physical clogging material includes the following organic and inorganic solids (Huisman and Olsthoorn 1983; Bouwer 2002; Hutchinson et al. 2013):

- suspended sediments present in the influent (recharge) water
- windborne dust
- material mobilized in spreading facilities by erosion
- particulate organic matter (e.g., plankton that settled out of suspension).

Clogging management consists of measures taken to reduce the rate of clogging and to remove or disrupt clogging layers after they form. Reduction of suspended solids concentration may be accomplished by having waters first enter a desilting (sedimentation) basin to allow fine sediments to settle out before they enter the main recharge basins.

Three main methods are used to maintain (periodically rehabilitate) infiltration basins (Huisman and Olsthoorn 1983):

- periodic drying of basins to crack or chip surface layers
- harrowing, plowing, disking, and other techniques to physical disrupt surficial layers and mix them within the underlying soil
- removal of accumulated fine sediments by scrapping.

All three of the above measures have their limitations. Periodic drying of basins is also effective in preventing or reducing the build-up of algae and other aquatic plant growth, but it does not result in the restoration of infiltration rates to their original values. Over time, the accumulation of clogging material in the basins increases, infiltration rates tend to progressively decrease, and basins tend to clog more rapidly (Schuh 1988), which necessitates more aggressive treatments to restore system capacity.

Harrowing, ripping, or plowing methods are effective in increasing short-term infiltration rates. Special machinery has been developed or adapted for the scarifying and scraping of dried infiltration basins (Huisman and Olsthoorn 1983). Harrowing and similar methods lead to a progressive increase in the concentration of fine sediment and organic matter in the mixed layer, which can lead to a long-term decline in infiltration rates and eventually necessitate removal of the mixed layer (Bouwer et al. 2008). A relatively thin clogging layer is mixed with underlying soil to form a thicker mixed layer that can be more expensive to remove and replace. A spring-tooth harrow was found to be the preferred technique in the Arizona CAP recharge basins because it tends to break up surficial soils in an upward motion as opposed to churning it deeper (Gorey and Dent 2007). Bouwer et al. (2008) recommended smoothing after disking and harrowing, which reduces the downward movement of fine particles through loose soil.

Bouwer et al. (2008) emphasized that is important to remove clogging material, which can be accomplished using some type of scraping device. The removed sediment may have to be replaced with clean fill or it can be reused after washing. The National Research Council (2008) concluding the disruption (disking or ripping) works best in shallow basins that are routinely rotated (dried) and silt loads are minimal. Draining and scraping is preferred for shallow basins receiving silt-laden water. In deep basins, sidewall infiltration may be much more significant than bottom infiltration. Drying of steep side walls by lowering water level can help restore capacity (National Research Council 2008).

Gette-Bouvarot et al. (2015) investigated ecological engineering alternatives to physical scrapping and other energy and resource-intensive methods for managing clogging of infiltration basins. Specifically, they explored regulation of algal biofilm growth by shading, an allelopathic macrophyte (*Vallisneria spiralis*), and an algae-grazing gastropod (*Viviparus viviparous*). An allelopathic plant releases chemicals that inhibit the growth of nearby plants. Test were performed using twelve, 30-cm diameter cylinders installed in an existing infiltration basin near Lyons, France. The grazing of the gastropod was found to be very efficient at reducing algal biomass and clogging. The macrophyte and shading did not significantly affect algal biomass and hydraulic conductivity. *Viviparus viviparous* is naturally present in the study area and thus does not pose the risks associated with the introduction of exotic species.

Experiences from other facilities can provide insights on the potential effectiveness of various maintenance options. Unfortunately there is little published, or otherwise readily available data, on actual clogging rates and the effectiveness of various basin treatment options. There have been very few comparative studies of treatment options. Mousavi and Rezai (1999) documented the effectiveness of five treatment options: (1) no treatment, (2) removal of deposited sediments, and (3) removal of deposited sediments plus 5 cm (2 in.), (4) 10 cm (4 in.), and (5) 15 cm (6 in.) of the underlying soil. Removal of deposited sediment and 15 cm (6 in.) of the underlying soil was the most effective method, restoring infiltration capacity to 68.3% of the original value. The Mousavi and Rezai (1999) study results show how infiltration of clay and silt-sized material into underlying soil can reduce infiltration rate. Scrapping or scratching the surficial deposited sediments was found to not be very efficient in restoring infiltration rates.

Determination of the optimal maintenance program necessarily involves site-specific adaptive management, which requires accurate data on basin infiltration rates both before and after maintenance activities. Infiltration rates can be calculated as the residual of the basin water budget, which requires data on flows into (and out of) basins, and precipitation and evaporation rates. Hence, for large systems, a site weather station (rain gauge and evaporation pan) is recommended. Reasonably accurate estimates of infiltration rates can be obtained from daily water level changes during periods with no rainfall and inflows, using evaporation rate data (pan-corrected) from a weather station in the general project site vicinity.

References

Asano, T., & Cotruvo, J. A. (2004). Groundwater recharge with reclaimed wastewater; health and regulatory considerations. *Water Research, 38*(8), 1941–1951.

Bean, E. Z., & Dukes, M. D. (2015). Evaluation of infiltration basin performance on coarse soils. *Journal of Hydrologic Engineering, 21*(1), 04015050.

Bouwer, H. (1978). *Groundwater hydrology*. New York: McGraw-Hill.

Bouwer, H. (1985). Renovation of wastewater with rapid-infiltration land treatment systems. In T. Asano (Ed.), *Artificial recharge of groundwater* (pp. 249–282). Boston: Butterworth Publishers.

Bouwer, H. (1989). Estimating and enhancing groundwater recharge. In M. L. Sharma (Ed.), *Groundwater recharge* (pp. 1–10). Rotterdam: A.A. Balkema.

Bouwer, H. (2002). Artificial recharge of groundwater: Hydrogeology and engineering. *Hydrogeology Journal, 10,* 121–142.

Bouwer, H., Pyne, R. D. G., Brown, J., Germain, D., Morris, T. M., Brown, C. J., et al. (2008). *Design, operation and maintenance for sustainable underground storage facilities*. Denver: AWWA Research Foundation.

Bouwer, H., & Rice, R. C. (1989). Effect of water depth in groundwater recharge basins on infiltration. *Journal of Irrigation and Drainage Engineering, 115*(4), 556–567.

Bouwer, H., & Rice, R. C. (2001). Capturing flood waters for artificial recharge of groundwater. In *Proceedings Tenth Biennial Symposium, Artificial Recharge of Groundwater* (pp. 1–10). Tucson, AZ, June 2001.

CAP. (n.d.). *Recharge program*. Retrieved October 11, 2017 from https://www.cap-az.com/departments/recharge-program.

Carleton, G. B. (2010). *Simulation of groundwater mounding beneath hypothetical stormwater infiltration basins.* U.S. Geological Survey Scientific Investigations Report 2010-5102.

City of Tucson. (2017). *Clearwater renewable resource facility*. Retrieved October 11, 2017 from https://www.tucsonaz.gov/water/clearwater.

Cook, J. (2004). *Innovative applications in water reuse: Ten case studies*. Alexandra, VA: WateReuse Association.

Crites, R. W., Reed, S. C., & Bastian, R. K. (2000). *Land treatment systems for municipal and industrial wastes*. New York: McGraw-Hill.

Echols, S. (2007). Artful rainwater design in the urban landscape. *Journal of Green Building, 2*(4), 101–122.

Echols, S., & Pennypacker, E. (2008). From stormwater management to artful rainwater design. *Landscape Journal, 27*(2), 268–290.

Echols, S., & Pennypacker, E. (2015). *The history of stormwater management and background for artful rainwater design: Creative ways to manage stormwater*. Washington, DC: Island Press.

Escalante, F. (2013). Practical criteria in the design and maintenance of MAR facilities in order to minimise clogging impacts obtained from two different operative sites in Spain. In R. Martin (Ed.), *Clogging issues associated with managed aquifer recharge methods* (pp. 119–154). Australia: IAH Commission on Managing Aquifer Recharge.

Estow, D. (1996). Water Conserv II: A model of cooperation and communication. *Florida Water Resources Journal, 59*(7), 29–30.

Ferguson, B. K. (1998). *Introduction to stormwater: Concept, purpose, design*. New York: Wiley.

Gette-Bouvarot, M., Volatier, L., Lassabatere, L., Lemoine, D., Simon, L., Delolme, C., et al. (2015). Ecological engineering approaches to improve hydraulic properties of infiltration basins designed for groundwater recharge. *Environmental Science and Technology, 49*(16), 9936–9944.

Gorey, T., & Dent, P. (2007). Maintenance techniques utilized by Central Arizona Project to optimize recharge. In P. Fox (Ed.), *Management of aquifer recharge for sustainability* (pp. 404–411). Phoenix: Acacia Publishing.

Huisman, L., & Olsthoorn, T. N. (1983). *Artificial groundwater recharge*. Boston: Pitman Advanced Publishing.

Hutchinson, A. (2013). Surface spreading recharge facility clogging—The Orange County Water District experience. In R. Martin (Ed.), *Clogging issues associated with managed aquifer recharge methods* (pp. 107–118). Australia: IAH Commission on Managing Aquifer Recharge.

Hutchinson, A., Milczarek, M., & Banerjee, M. (2013). Clogging phenomena related to surface water recharge facilities. In R. Martin (Ed.), *Clogging issues associated with managed aquifer recharge methods* (pp. 95–106). Australia: IAH Commission on Managing Aquifer Recharge.

Hutchinson, A. S. (2007). Challenges in optimizing a large-scale managed aquifer recharge operation in an urbanized area. In P. Fox (Ed.), *Management of aquifer recharge for sustainability* (pp. 346–354). Phoenix: Acacia Publishing.

Hutchinson, A. S., Rodriguez, G., Woodside, G. D., & Milczarek, M. (2017). Maximizing infiltration rates by removing suspended solids: Results of demonstration testing of riverbed filtration in Orange County, California. *Water, 9*, 119. https://doi.org/10.3990/w9020119.

Hutchinson, A. S., & Woodside, G. D. (2002). The Santa Ana River: The challenge of maximizing the use of an urban river. In P. J. Dillon (Ed.), *Management of aquifer recharge for sustainability* (pp. 509–514). Lisse: Swets & Zeitlinger.

Kallali, H., Yoshida, M., Tarhouni, J., & Jedidi, N. (2013). Generalization and formalization of the USEPA procedure for design of treated wastewater aquifer recharge basins: I. theoretical development. *Water Science and Technology, 67*(1), 131–138.

Ku, H. F. H., Hagelin, N. W., & Buxton, H. T. (1992). Effects of urban storm-runoff control on groundwater-water recharge in Nassau County, New York. *Ground Water, 30*(4), 507–514.

Ku, H. F. H., & Simmons, D. L. (1986). *Effect of urban stormwater runoff on groundwater beneath recharge basins on Long Island, New York*. U.S. Geological Survey Water-Resources Investigations Report 85-4088.

Lluria, M. R. (2008). Water spreading in the desert. *Southwest Hydrology*, 28–19.

Lluria, M. R. (2009). Successful application of managed aquifer recharge in the improvement of the water resources management of semi-arid regions: Examples from Arizona and the Southwestern USA. *Boletín Geológico y Minero, 120*(2), 111–120.

Lowndes, M. A. (2000). *Wisconsin storm water manual: Technical guidelines for storm water management practices*. Madison: Cooperative Extension of the University of Wisconsin-Extension.

Massman, J. W. (2003). *A design manual for sizing infiltration ponds*. Research report prepared for the Washington State Department of Transportation, Olympia, Washington.

Massman, J. W., & Butchart, C. D. (2001). *Infiltration characteristics, performance, and design of storm water facilities* (No. WA-RD 483.1). Olympia: Washington State Department of Transportation.

Massman, J. W., Butchart, C. D., & Stolar, S. (2001). *Infiltration characteristics, performance, and design of storm water facilities* (No. WA-RD 483.1). Olympia: Washington State Department of Transportation.

McKim, T. (2012). Reuse at Reedy Creek Improvement District. In *Guidelines for water reuse*, EPA/600/R-12/618 (pp. D-90–D-92). Washington, DC: U.S. Environmental Protection Agency.

Megdal, S. B., Dillon, P., & Seasholes, K. (2014). Water banks: Using managed aquifer recharge to meet water policy objectives. *Water, 6*, 1500–1514.

Mills, W. R. (2002). The quest for water through artificial recharge and wastewater recycling. In P. J. Dillon (Ed.), *Management of aquifer recharge for sustainability* (pp. 3–10). Lisse: A.A. Balkema.

Mousavi, S.-F., & Rezai, V. (1999). Evaluation of scraping treatments to restore infiltration capacity of three artificial recharge projects in central Iran. *Hydrogeology Journal, 7*, 490–500.

National Research Council. (1994). *Ground water recharge using waters of impaired quality*. Washington, DC: National Academy Press.

National Research Council. (2008). *Prospects for managed underground storage of recoverable water*. Washington, DC: National Academies Press.

Nevada Bureau of Water Pollution Control. (1993). *Guidance document for an application for rapid infiltration basins*. State of Nevada, Division of Environmental Protection, Bureau of Water Pollution Control.

Niswonger, R. G., Prudic, D. E., & Regan, R. S. (2006). *Documentation of the Unsaturated-Zone Flow (UZF1) package for modeling unsaturated flow between the land surface and the water table with MODFLOW-2005*. U.S. Geological Survey Techniques and Methods 6-A19.

Oaksford, E. T. (1985). Artificial recharge: Methods, hydraulics, and monitoring. In T. Asano (Ed.), *Artificial recharge of groundwater* (pp. 69–127). Stoneham, MA: Butterworth Publishers.

O'Leary, D. R., Izbicki, J. A., Moran, J. E., Meeth, T., Nakagawa, B., Metzger, L., et al. (2012). Movement of water infiltrated from a recharge basin to wells. *Ground Water, 50*(2), 242–255.

O'Reilly, A. M. (1998). *Hydrogeology and simulation of the effects of reclaimed-water application in west Orange and southeast Lake Counties, Florida*. U.S Geological Survey Water-Resources Investigations Report 97-4199.

PAG. (2012). *Clearwater Program stable isotope study 2012 monitoring report* (prepared for Tucson Water). Tucson: Pima Association of Governments.

Parsekian, A. D., Regnery, J., Wing, A. D., Knight, R., & Drewes, J. E. (2014). Geophysical and hydrochemical identification of flow paths with implications for water quality at an ARR site. *Groundwater Monitoring & Remediation, 34*(3), 105–116.

Peyton, D. E. (2001). Fines control system—Modified recharge basin floor maximizes infiltration efficiency. In *Proceedings Tenth Biennial Symposium, Artificial Recharge of Groundwater* (pp. 1–10). Tucson, AZ, June 7–9, 2001.

Peyton, D. E. (2002). Modified recharge basin floor to control sediment and maximize infiltration efficiency. In P. Dillon (Ed.), *Management of aquifer recharge for sustainability* (pp. 215–220). Lisse: Swets & Zeitlinger.

Peyton, D. E. (2003). Sediment solutions system—New maintenance operations and ridged recharge basin floors maximize infiltration efficiency. In *Proceedings 11th Biennial Symposium on Groundwater Recharge*, Phoenix, Arizona, June 5–7, 2003.

Rastogi, A. K., & Pandey, S. N. (2002). Influences of artificial recharge shapes on the mound underneath in phreatic aquifers. In P. Dillon (Ed.), *Management of aquifer recharge for sustainability* (pp. 295–297). Lisse: Swets & Zeitlinger.
Repert, D. A., Barber, L. B., Hess, K. M., Keefe, S. H., Kent, D. B., LeBlanc, D. R., et al. (2006). Long-term natural attenuation of carbon and nitrogen within a groundwater plume after removal of the treated wastewater source. *Environmental Science and Technology, 40*(4), 1154–1162.
Salt River Project. (n.d.). *SRP stores water for tomorrow*. Retrieved September 9, 2017 from https://www.srpnet.com/water/waterbankıng.aspx.
Schiff, L. (1957). The Darcy Law in the selection of water-spreading systems for ground-water recharge. In *Symposium Darcy, Dijon 20–26 Septembre 1956*, Publication No. 41 (pp. 99–110). International Association of Hydrological Sciences.
Schuh, W. M. (1988). In-situ method for monitoring layered hydraulic impedance development during artificial recharge with turbid water. *Journal of Hydrology, 101*(1–4), 173–189.
Seaburn, G. E., & Aronson, D. A. (1974). *Influence of recharge basins on the hydrology of Nassau and Suffolk Counties, Long Island, New York*. U.S. Geological Survey Water-Supply Paper 2031.
Sumner, D. M., & Bradner, L. A. (1996). *Hydraulic characteristics and nutrient transport and transformation beneath a rapid infiltration basin, Reedy Creek Improvement District, Orange County, Florida*. U.S. Geological Survey Water-Resources Investigations Report 95-4281.
Thompson, A. F. B., Carle, S. F., Rosenberg, N. D., & Maxwell, R. M. (1999). Analysis of groundwater migration from artificial recharge in a large urban aquifer: A simulation perspective. *Water Resources Research, 35,* 2981–2998.
Thoms, R. B., Johnson, R. L., & Healy, R. W. (2006). *User's guide to the Variably Saturated Flow (VSF) process for MODFLOW*. U.S. Geological Survey Techniques and Methods 6-A18.
Todd, D. K. (1980). *Groundwater hydrology* (2nd ed.). New York: Wiley.
Türkmen, M., Walther, E. F., Andres, A. S., Chirnside, A. A. E., & Ritter, W. F. (2008). *Evaluation of rapid infiltration basin systems (RIBs) for wastewater disposal: Phase I*. Newark, DE: Delaware Geological Survey.
USEPA. (2002). *Wastewater technology fact sheet. Slow rate land treatment*. Washington, DC: U.S. Environmental Protection Agency Municipal Technology Branch.
USEPA. (2003). *Wastewater technology fact sheet. Rapid infiltration land treatment*. Washington, DC: U.S. Environmental Protection Agency Municipal Technology Branch.
USEPA. (2006). *Process design manual for land treatment of municipal wastewater effluents*. Cincinnati: U.S. Environmental Protection Agency, Land Remediation and Pollution Control Division, National Risk Management Research Laboratory, Office of Research and Development.
Water Conserv II. (2012). *A world-renowned irrigation and rapid infiltration basin water reuse project*. http://waterconservii.com/downloads/Tour_Pack.pdf (revised 12-04-12).
WSDOT. (2014). *Highway runoff manual*, version M 31-16.04. Olympia, WA: Washington State Department of Transportation.

Chapter 16
Surface-Spreading AAR Systems (Non-basin)

16.1 Introduction

Infiltration basins are a widely used surface-spreading method for infiltrating stormwater, surface water, and reclaimed water (Chap. 15). A variety of other types of surface-spreading methods that do not employ basins are used for managed aquifer recharge (MAR; Table 15.1), which vary depending upon whether they occur in an unmodified channel, modified channel, or on off-channel land surfaces (e.g., adjacent flood plains, farm fields, and sand dunes). Irrigation return flows can be an important source recharge in spate irrigation systems, in which fields are intentionally flooded, and during irrigated agriculture in general (Sect. 24.9).

16.2 Ephemeral Stream Recharge

16.2.1 Ephemeral Stream Recharge Processes

Most (and in some settings, virtually all) recharge in arid and semiarid regions is indirect (i.e., focused) in that recharge occurs predominantly associated with storm events and at locations where runoff accumulates rather than at the site of precipitation. Ephemeral stream channels are referred to as wadis in the Middle East and arroyos in the western United States. Infrequent large storms are often the major source of recharge within arid and semiarid region basins, with wadi recharge playing an important role in regulating the total water resources of a watershed (Besbes et al. 1978).

Water tends to quickly infiltrate into alluvial sediments present in ephemeral stream channels during low to moderate intensity events. Transmission losses in channel alluvium are subject to several removal processes before the water can reach the water table and contribute to aquifer recharge, including evapotranspiration (ET) by near or in-channel vegetation, evaporation from wetted channel sediments, tem-

R. G. Maliva, *Anthropogenic Aquifer Recharge*,
Springer Hydrogeology, https://doi.org/10.1007/978-3-030-11084-0_16

porary storage in the underlying soil and river bank, and detention above impeding (semiconfining) layers within the vadose zone and subsequent downgradient discharge or loss to ET. Deep-rooted vegetation (e.g., phreatophytes) within and adjoining channels can be highly efficient in extracting infiltrated water long after flood events. Where the water table is deep and clay-rich beds present, little, if any, water infiltrated from floods may reach the water table. Hence, it should not be assumed that transmission loss volumes are equal or close to aquifer recharge volumes.

A simple anthropogenic aquifer recharge (AAR) method is the discharge of water to ephemeral stream channels. Ephemeral stream aquifer recharge can be either managed or unmanaged. Unmanaged recharge occur when water is discharged to ephemeral stream channels with a primary disposal purpose. For example, discharge of treated wastewater into Wadi Hanifa in Riyadh, Saudi Arabia, has transformed the ephemeral stream into a perennial stream with associated increased groundwater recharge. The Wadi Hanifa valley downstream of Riyadh has become the largest natural reserve in the kingdom and the created wetlands have become a stop for migratory birds and a popular area for recreation (Hassan 2017). Similarly, discharge of treated wastewater to an ephemeral wadi north of the Holy City of Al Madinah creates perennial flow supporting local agricultural and aquifer recharge.

Managed ephemeral stream recharge is the controlled release of water to an ephemeral channel and essentially involves the same processes as natural recharge. Transmission losses in ephemeral channels have been shown to be related to (Sorman and Abdulrazzak 1993; Stephens 1996; Cataldo and Pierce 2005; Blasch et al. 2004; Maliva and Missimer 2012):

- flow duration
- channel length and width
- channel geomorphology (distribution of riffle and pool sequences; Newman et al. 2006)
- antecedent moisture content (time since previous flood event)
- depth to water table
- peak discharge
- flow sequence
- thickness and characteristics of alluvium (e.g., hydraulic conductivity)
- presence of clogging layers and surface crusts (armoring)
- water temperature (through its effects on viscosity and hydraulic conductivity)
- suspended solids concentration of runoff
- watershed geometry
- location and intensity of the isohyets (rainfall contours) in the watershed.

The most important variables are flow volume and duration, peak flow rate, and wetted channel perimeter, as they control the duration and area afforded for infiltration (e.g., Walters 1990; Lane et al. 1971; Sorman and Abdulrazzak 1993; Wheater 2002; Cataldo and Pierce 2005). Methods used to estimate transmission losses and associated recharge in ephemeral channels were reviewed by the USDA (2007), Maliva and Missimer (2012), and Shanafield and Cook (2014).

Transmission losses can be most directly evaluated through a water budget analysis. Transmission losses to infiltration (*TL*) in ephemeral streams can be quantified as the difference in discharge between upstream (V_1) and downstream (V_2) gaging stations, plus any lateral (tributary) runoff (V_t) and baseflow (*BF*), and minus *ET*:

$$TL = V_1 - V_2 + V_t + BF - ET \text{ (units of either rate or volume)} \quad (16.1)$$

Direct precipitation into the channel reach is assumed to be negligible in Eq. 16.1 (but could added if locally significant).

Within ephemeral channels in arid or semiarid lands, baseflow will typically be negligible or non-existent. In the case of MAR into a dry channel (no upstream or lateral inflows), transmission losses within a channel reach will be equal to the volume or rate of anthropogenic discharge to the channel (V_d) minus downstream flow at the end of recharge (V_2) and *ET* losses

$$TL = V_d - V_2 - ET \quad (16.2)$$

If flow vanishes downstream, then

$$TL = V_d - ET \quad (16.3)$$

Transmission loss is commonly defined as water lost to infiltration. Evapotranspiration in the context of transmission loss is water that is lost before infiltration (e.g., surface water evaporation). ET after infiltration can significantly decrease the volume of water that actually reaches the water table. ET by riparian vegetation can play an important role in alluvial aquifer water budgets. Hence, evaluation of aquifer recharge resulting from managed releases requires the use of methods that evaluate the amount of water that actually reaches the water table (e.g., water-table fluctuation method, model calibration, chloride mass balance, and other tracer methods). Ideally, multiple methods should be employed.

16.2.2 Wadi Recharge of Floodwaters in the MENA Region

Rainfall in arid regions, such as much of the Middle East and North Africa (MENA) region, occurs mostly as intense short-duration events. Flash floods in wadis can be extremely dangerous, resulting in severe damage to physical structures and losses of life. Dams have been constructed in wadis for flood control, surface water storage, and aquifer recharge. The effects of wadi dam recharge can be dramatic. For example, Zeelie (2002) reported that the wadi dam and infiltration basin in the Omaruru Delta of Namibia increased the sustainable yield of the alluvial aquifer during the study period from 2.8 to 5.9 million m^3/d (MCM/d; 739–1,558 million gallons). However, it was noted that future sustainable yields could be reduced by a reduction in the dam storage capacity (e.g., due to siltation) and that Omaruru River can experience multiple year periods during which there is no runoff to be stored.

Recharge can be planned to occur both within the reservoir behind the dam (Sects. 16.4 and 16.5) and in the downstream wadi through controlled release of water. In the latter case, the reservoirs operate as large stilling (sedimentation) basins. Floodwaters typically have very high suspended solids concentrations and sediment deposition causes a loss of storage volume and reduction of infiltration rates by clogging of the sediment-water interface. Cleaner water discharged downstream of the dam more readily infiltrates into wadi alluvial sediments.

Three types of wadi dams are used in Oman: flood control dams, storage dams and recharge dams. Storage dams store water during times of excess flow for later irrigation and domestic use. Recharge dams serve a flood control function and store water to facilitate groundwater recharge by controlled releases to the downstream channel. Oman's dams are operated as siltation basins with the goal of limiting the areas effected by silting to the reservoirs (Ministry of Regional Municipalities and Environment and Water Resources 2005). Most of the recharge occurs by infiltration through the channel bed downstream of dams by controlled release of stored water (Abdalla and Al-Rawahi 2013). For example, the Al Khwad Dam is operated so that there is an approximately two day retention period to reduce the suspended sediment load before commencing spill to the wadi (Kiela 2010). A lamellar of clear water at the top of reservoirs is skimmed off for downstream wadi recharge.

Recharge resulting from the Wadi Al-Jizzi Dam, located 23 km south of Sohar, Oman, was investigated by Al-Kindi (2004). The Wadi Al-Jizzi Dam is an earthen dam, completed in 1989, with the objectives of capturing a significant part of floodwaters that would otherwise discharge to the sea, stabilizing groundwater conditions, and supplying additional water for new agricultural development. The hydrological effects of the dam were evaluated by the development of a MODFLOW model, which was calibrated for the 1985 through 1994 period and validated for the years 1995–2002. The model results indicated that pre-dam recharge was 1.825 MCM/year and post-dam recharge was 2.492 MCM/year, an increase of 36%. The recharge was evident by a rise of water levels in downstream wells after floods. However, the long-term decline in groundwater levels continued due to a tremendous increase in groundwater pumping that offset the positive impacts of the dam (Al-Kindi 2004).

The Alkhod Dam, located west of Muscat, was constructed between December 1983 and March 1985. Recharge was estimated from groundwater level data using the water table fluctuation (WTF) method with a specific yield of 0.2 calculated from pumping tests. The average annual run-off impounded by the dam was estimated to be 8.3 MCM/year and the estimated annual average recharge was 6.4 MCM/year, which is 77% of the impounded volume (Abdalla and Al-Rawahi 2013). Managed recharge and a reduction in wellfield pumping caused the saline-water interface to move back toward the coast and allowed groundwater levels to recover (Abdalla and Al-Rawahi 2013).

Wadi discharges can result in streambed erosion and fine sediment deposition as flow wanes. Sedimentation can impact infiltration rates through the formation of clogging layers and the accumulation of fine particles in macropores in underlying sediments. Prathapar and Bawain (2014) measured changes in infiltration rates over time at the same site at the Sahalanowt Dam, Oman, using a double-ring infiltrometer. A two orders of magnitude decrease in hydraulic conductivity occurred over 17 years. Sediments deposited after flood events were reported to be removed by the government and the decrease in hydraulic conductivity was suggested to be due to finer particles entering the soil profile (Prathapar and Bawain 2014). It was proposed that maintenance should include periodic removal of sediment and shallow ploughing or harrowing to improve infiltration rates.

Recharge results in an increase in down-gradient groundwater levels and an improvement in water quality. For example, analysis of water quality data using geostatistical methods in a GIS environment has demonstrated improvements in water quality (a decrease in total dissolved solids concentration) downstream of artificial recharge dams in Oman (Bajjali 2005). The recharged water acts to push saline-water seaward.

Ibn Ali et al. (2017) investigated MAR by controlled releases to an ephemeral riverbed in the Nadhour-Saouaf syncline of the Zaghouan District, Northern Tunisia. Infiltration of floodwaters through the beds of ephemeral streams is a major source of recharge. Infiltration rates from a hill dam were estimated for three gauged channel reaches using the water-budget method. Average infiltration rates in the reaches were 0.050, 0.127, and 0.043 m/d. The water table was observed to increase at a rate of 5 mm/d in the vicinity of the recharge site. However, water levels resumed their decline due to over exploitation. The benefits of MAR were concluded to be limited in time and space (Ibn Ali et al. 2017).

A geochemical evaluation of the effects of the Sidi Saad Dam in central Tunisia on the recharge of the local shallow (Zéround) aquifer was performed by Salem et al. (2012). The dam was constructed in 1982 and the geochemical data were from 2005 to 2007. MAR is performed by controlled releases to the downstream Zéround riverbed. Stable isotope, dissolved ion, and tritium data were used to differentiate between the different recharge water sources, which plot in different fields on a ^{3}H versus $\delta^{18}O$ plot. Recharged dam water was computed to constitute 13% and 152 MCM of the total aquifer volume in the sampled area. Since 1982 (through 2007), 390 MCM was reported to have been released from the dam. Salem et al. (2012) calculated that only 39% of the released water reached the aquifer, with the remaining amount evaporated and/or stored in the unsaturated zone. These calculations do not consider the amount of dam recharge water (and other sourced waters) that were extracted over the study period. The recharged percentage may be somewhat greater.

The fragmentary data on wadi channel releases indicate definite water resources benefits in terms of local groundwater recharge, although there is insufficient data to generalize as to what fraction of the released water actually reaches the water table. If the floodwater would otherwise be lost to tide, then even a low fraction of recharge could justify dam projects, especially if they also provide flood control benefits. It is clear that recharge dams alone, without controls on groundwater extractions, will usually be insufficient to address aquifer overdraft.

Optimizing of aquifer recharge in wadis is a research area that merits further investigation (Maliva and Missimer 2012). More data are needed on the both the infiltration rates and recharge rates for existing wadi dam systems. Additional research is particularly needed on the effectiveness of different options for increasing and maintaining recharge rates. Wadi dams need to be designed and constructed that have the capability to preferentially skim off the upper lamellar of low turbidity water for discharge downstream via gate, controlled spillway, or other types of systems. Because of the remote location of some wadi dams and highly sporadic nature of runoff events, automated or passive systems are preferred over options that require considerable on-site human intervention (Maliva and Missimer 2012).

16.2.3 Wastewater Recharge to Ephemeral Stream Channels

The Santa Cruz River in the Tucson, Arizona vicinity is a textbook example of a perennial stream that was changed to an ephemeral stream due to excessive groundwater pumping, and the later restoration of perennial flow through the discharge of treated wastewater. The Santa Cruz River was historically a perennial river in the Tucson area. The presence of the river, and associated lush vegetation and abundant wildlife, was the reason Native Americans lived in the area, which in turn was why the Spanish established a mission there that eventually grew into the city of Tucson (Regan 2001). Intense groundwater pumping lowered the water table, and by the 1940s perennial flow ceased with the river only flowing when torrential rains fell (Regan 2001).

Discharge of treated wastewater into the ephemeral river channel from the Tres Rios Water Reclamation Facility (WRF; formerly Ina Road WRF) and Roger Road WRF (replaced in January 2014 by the Agua Nueva WRF) resulted in a return of perennial flow and reestablishment of native riparian vegetation and habitats for birds and other wildlife. The Roger Road WRF first started operating in 1951 and the Ina Road/Tres Rios WRF began operations in 1977.

Thomure and Wilson (2003) provided an overview of the Santa Cruz River Managed Underground Storage Facility (SCRMUSF), which recharges treated wastewater from the Agua Nueva WRF (and formerly from the Roger Road WRF) into the Santa Cruz River channel (Fig. 16.1). The SCRMUSF includes approximately 5.1 miles (8.2 km) of unmodified stream channel downstream of the Roger Road WRF site to Ina Road. The system is passive and relies on natural scour during flood events to remove deposited organic matter and sediments. Recharge credits are determined from the net decrease in flow over the channel reach minus estimated ET losses. Credits are only accrued during time periods in which natural flow does not occur. Only 50% of recharged volumes is allowed to be accrued as recharge credits (Thomure and Wilson 2003). Under Arizona law, effluent users receive legal credits for 100% of what they recharge in constructed facilities, such as infiltration basins, and only 50% credit when recharge occurs in a river, which creates an incentive to stop recharge in rivers in favor of off-channel constructed facilities to the detriment of the riparian environment (Davis 2016a).

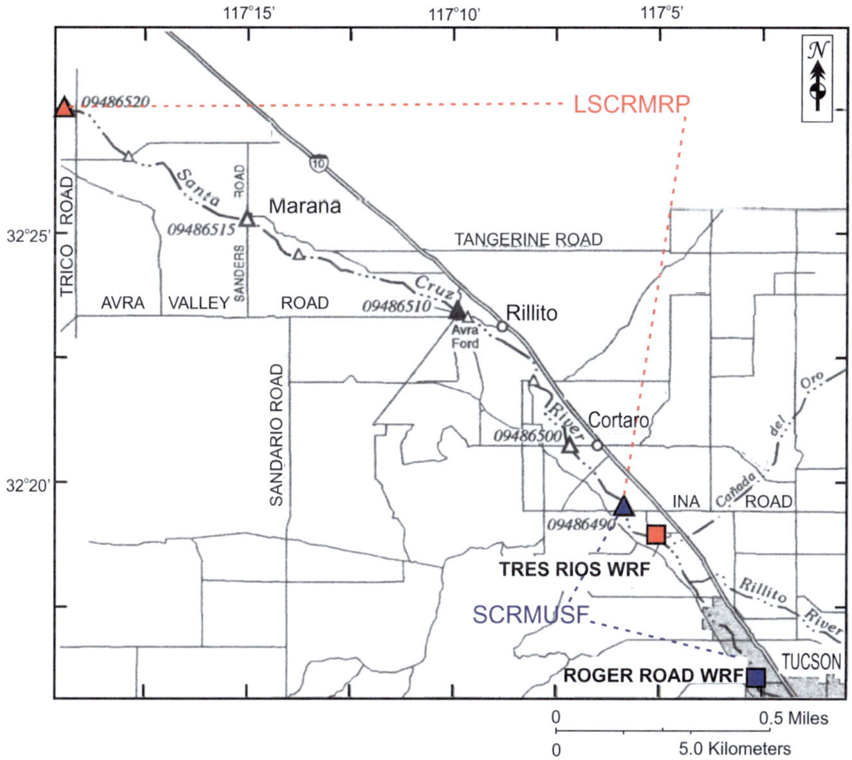

Fig. 16.1 Map showing the locations of the Santa Cruz Managed Underground Storage Facility (SCMUSF) and Lower Sana Cruz River Managed Recharge Project (LSCRMRP) in Pima County, Arizona. Flow direction is toward the northwest (Modified basemap from Galyean 1996)

Effluent conveyed beyond Ina Road is not recharged within the SCRMUSF. The Lower Santa Cruz River Managed Recharge Project (LSCRMRP) recharges treated wastewater from the Tres Rios WRF and Agua Nueva WRF into the Santa Cruz River channel downstream of the SCRMUSF. LSCRMRP is an approximately 17.9 miles (28.8 km) reach of the Santa Cruz River, from Ina Road to Trico Road (Prietto 2013). Effluent conveyed beyond Trico Road is not considered recharged within the LSCRMRP. The LSCRMRP recharged water is considered to be the combined flows from the Tres Rios WRF and Agua Nueva WRF, minus the recharge of the SCRMUSF, flows past Trico Road, and ET. Similar to the SCRMUSF, recharge is only counted in non-storm days and 50% of the calculated recharge is a cut to the aquifer as far as accruing recharge credits (Prietto 2013). The SCRMUSF is permitted to recharge 9,307 acre-feet per year (AF/year; 11.48 MCM/year) and recharged 6,100 AF (7.52 MCM) in 2012. The LSCRMRP is permitted to recharge 43,000 AF/year (53.0 MCM/year) and recharged 19,000 AF (23.4 MCM) in 2011 (Prietto 2013).

A U.S. Geological Survey study indicated that 88.4–90.2% of wastewater discharged from the Tres Rios WRF and Roger Road WRF over the March 23, 1990 to September 30, 1993 study period infiltrated into the Santa Cruz River channel in a 23 mile (37 km) reach downstream to Trico Road (Galyean 1996). A key observation is that infiltration significantly increased after storm flows due to alteration of the composition, structure, and geometry of the channel bed. Flood events appear to erode clogging deposits. Variations in ET rates had a minor impact on the channel water budget (Galyean 1996). Blasch et al. (2004) observed that in ephemeral channels erosion of the streambed can increase its hydraulic conductivity by two orders of magnitude or more in some instances.

The benefits of the discharge of reclaimed water to the Santa Cruz River have been so great that serious consideration is being given to pumping the water uphill so as to restore perennial flow of the river through downtown Tucson. The goal of the proposed "Agua Dulce" project is to restore the rich natural riparian habitat and create a "natural treasure" that could generate tourism and economic development (Davis 2016b).

16.2.4 Imported Surface Water Discharged to Channels

The U.S. Bureau of Reclamation (USBR 2009) Arroyos Ground-Water Recharge Project (AGWRP) involves the use of Central Arizona Project (CAP) surface water for in-channel recharge within the San Xavier District of the Tohono O'dham Nation (Native American reservation) in Pima County, Arizona (USBR 2009). Feasibility assessments were performed on six potential recharge sites. The investigations included both soil and infiltration testing.

A pilot testing program was performed with approximately 40 AF (0.05 MCM) of water recharged at an average rate of 400 gpm (25.2 L/s) for each drainage point. To manage erosion, riprap-filled excavations were used at discharge points and small check dams were installed to prevent down-cutting into the channel (Fig. 16.2). A groundwater rise of 68 ft (20.7 m) was observed in the central portion of the recharge area and 5 ft (1.5 m) of rise was reported 5,400 ft (1,646 m) down-gradient from the main arroyo discharge point. Surface infiltration rates averaged 8 ft/d (2.4 m/d).

The water quality impacts of recharge were evaluated as part of the environmental impact assessment. CAP water was reported to have an average TDS concentration of 680 mg/L versus a local groundwater concentration of about 300 mg/L. An initial spike in salinity occurred due to leaching of salts from the vadose zone. As recharge continued, aquifer salinity approached that of the CAP water.

Fig. 16.2 Riprap-filled excavation discharge structure (top) and check dam (bottom) used to manage erosion during in-channel recharge, Arroyos Ground-Water Recharge Project, Pima County, Arizona (*Source* USBR 2009)

16.3 Modified Channel Recharge Methods

Modified channel recharge methods consist of alterations of natural channels to increase local infiltration rates. Infiltration rates and volumes can be locally increased by:

- retaining water (e.g., within reservoirs behind dams of various sizes) so that there is a longer time for infiltration to occur.
- slowing the rate of flow or increasing the flow path so that there is more time for local infiltration.
- ponding water to increase the vertical hydraulic gradient.
- increasing the wetted area (area of infiltration).
- increasing the vertical hydraulic conductivity of riverbed sediments (or diverting flow into more conductive areas).

Modified channel recharge methods may involve the following:

- construction of permanent or temporary check dams to retain water and slow and/or widened the flow
- construction of dams (permanent or inflatable) or in-channel excavations to retain flood flows for later infiltration
- construction of levees within a channel to create a more tortuous flow path or spread flows
- construction of secondary recharge channels
- conditioning of the channel bed by scraping off fine surface sediments or crusts, or replacement of surficial sediments with more permeable, coarser-grained sediments
- leveling or widening of the channel.

Check dam and reservoir recharge are addressed in Sects. 16.4 and 16.5, respectively.

16.3.1 Temporary In-Channel Levees

The Orange County Water Department (OCWD; California, USA) in-channel recharge system in the main channel of Santa Ana River is an excellent example of the use of temporary sand levees to create a serpentine flow path (Fig. 16.3). The network of "T" and "L" levees increase the infiltration rate by lengthening the flow path, reducing the flow velocity, and increasing the wetted area. During major floods, the temporary levees are washed out and cease to be an impediment to the flow of floodwaters. Levees constructed using inflatable rubber dams are also used to divert water from the main channel into off-channel recharge basins (Markus et al. 1995).

16.3.2 Secondary Recharge Channels

Stream flow during low-flow periods tends to be restricted to the lowest point in a channel, referred to as the thalweg, which usually constitutes only a small part of the channel area. Diverting some of the flow to secondary channels would increase

Fig. 16.3 Temporary in-channel levees in the Santa Ana River, Anaheim, California (March 28, 2010, *source* U.S. Geological Survey)

the wetted area and infiltration rates (total infiltrated volume over time). The U.S. Bureau of Reclamation facilitated a collaborative effort with partners to construct and operate the "Enhanced Recharge Demonstration Project" (ERDP) to increase the recharge of treated effluent at the Lower Santa Cruz River Managed Recharge Project (LSCRMRP; Tosline et al. 2012). The LSCRMRP has historically recharged less than 50% of the permitted volume. The ERDP was developed to divert water from the Santa Cruz River channel into adjacent dry, abandoned secondary flow channels (thalwegs) to increase infiltration rates (Tosline et al. 2012). The channel modification was performed at the Powerline Grid Bar site and consisted of the excavation of two channels with average widths of about 10 ft (3 m) and depths of 0.5–3 ft (0.15–0.9 m). Transmission losses were determined using flumes with stilling wells located about 100 ft (30 m) downstream from the diversion point and near the ERDP outlet.

The construction of the channels and diversion inlet was reported to have been completed in 8 days (Tosline et al. 2012). Maintenance consisted of the removal of sediments that accumulated on the channel bottom, which was a residual from construction. Once the constructed-related sediment was removed, deposition of fine sediment was reported to be minimal. The project was washed out during summer monsoon floods, which included filling of the diversion inlet and burial of the flumes. However, after the washout, additional storm flow scoured the channels. High flow rates were reported to be beneficial in reducing biological activity, flushing fine particles, decreasing maintenance requirements, and improving hydraulic conductivity and infiltration rates (Tosline et al. 2012).

The infiltration rate (transmission loss) during the last 64 days of operation was 1.13 AF/day (1,394 m^3/d), which is equivalent to 2.7 ft/d (0.82 m/d) for a 10 ft (3 m) wide channel (Tosline et al. 2012). The ERDP involved a small construction effort, but the secondary channels are susceptible to damage from flood flow due to the dynamic nature of the Santa Cruz River (Tosline et al. 2012). The system can be restored after each flood.

16.3.3 Stream Bed Material Replacement

Replacement of channel sediments with more permeable material can increase infiltration rates. A limitation of this method is that bed sediments are subject to erosion and could experience rapid clogging from sediment-laden water. Ferreira and Leitão (2014) reported on a test riverbed infiltration system in the Rio Seco, Campina de Faro aquifer system, Southern Algarve Region (Portugal), which involved the construction of two 7-m deep, 20 m by 5 m (100 m^2) infiltration basins in the channel that were filled with washed gravel. A regional groundwater quality concern is diffuse pollution of aquifers caused by agricultural activities. Rainy month infiltration increased water levels and improved water quality, as indicated by reduced electrical conductivity, and nitrate and chloride concentrations. Electrical resistivity profiles show fresher waters with high resistivities below the channels from infiltration. The reported infiltration rate was 1.2 m/d.

Fig. 16.4 Dam with gated spillway. South Florida Water Management District, control structure S-190, Seminole Tribe of Florida Big Cypress Reservation

16.4 Check Dams and Weirs

16.4.1 Introduction

Dams and weirs can increase aquifer recharge and are thus considered an MAR technique where an intent of a project is to increase infiltration. Aquifer recharge is an incidental benefit for structures constructed for other purposes, such as flood control. The risks associated with dams and weirs increase with their size and the structures should be designed by a professional engineer with expertise in the field. Some inconsistency exists in the literature as to the definition of dams, weirs and check dams. Bligh (1915) defined a dam as "an impervious wall of masonry, concrete, earth, or loose rock which upholds a mass of water at its rear, while its face or lower side is free from the pressure of water to any appreciable extent." Dams are designed so that water flows through spillways rather than over their top. Small dam design and construction is addressed in books and manuals by the Bureau of Reclamation (1987), PFRA (1992), Stephens (2010), and USACOE (2004).

Small dams with gated spillways (water control structures; Fig. 16.4) are widely used in South Florida for flood control and to prevent over drainage of upstream areas. During dry periods, the structures maintain higher surface water levels in upstream areas, which contributes to increased groundwater recharge. During heavy rainfall periods, the gates are lowered to convey excess water downstream. Control structures are also used to prevent salt water from migrating up freshwater streams and canals.

Fig. 16.5 Weirs. Top: South Florida Water Management District West Weir, Seminole Tribe of Florida Big Cypress Reservation. Bottom: San Antonio River, San Antonio, Texas

Weirs (also referred to as overfall and overflow dams) differ from conventional dams in that water flows over the crest of the structure and tail water is formed below the dam (Bligh 1915). Weirs are essentially horizontal barriers constructed across streams that allow water to pond on the upstream side and regulate flow (Fig. 16.5). Weirs may be designed with a fixed or movable crest. The latter design allows the crest to be lowered during floods to allow more water to pass. Weirs can increase infiltration by increasing the wetted area of the channel and ponded water depth.

Fig. 16.6 Gabions used for ephemeral stream bank erosion control, northwestern Arizona

Check dams are small permanent or temporary barriers placed across perennial and ephemeral channels to reduce flow velocity and erosion, and to trap sediments. Check dams are commonly used as a construction and stormwater best management practice to reduce flow velocity in ditches and swales. Small, temporary reservoirs created by the check dams can increase infiltration and recharge. Temporary check dams are constructed of a variety of materials including rock, gravel bags, sand bags, wire gabions, logs, and lumber. Permanent check dams may be constructed of concrete or masonry. Gabions are typically metal cages (wire boxes) filled with rock (Fig. 16.6). Check dams usually have a notch or concave crest to control flow over the dam and avoid bank erosion. The dams can act as weirs during floods if they are overtopped. The leaky dam design (e.g., construction using gabions) retain some floodwater, with infiltration occurring in both the upstream ponded area and downstream by water that leaks through the structure. Log jams can be used as leaky dams that act to slow down water and provided addition area and time for infiltration to occur (Maxwell and Davidson 2016). Siltation upstream reduces suspended solids concentration downstream. Over time, structures will become less permeable as silt is deposited in gabions (GWP Consultants 2010).

A series (cascade) of check dams is often constructed in ephemeral channels, as opposed to a single large dam. The maximum slope and velocity reduction is achieved when the toe of the upstream dam is at the same elevation as the top of the downstream dam. Water that spills over the crest of the upper structure enters the pool behind the lower structure. The pool of the downstream dam should extend to the toe of the dam immediately upstream.

Check dams are a commonly used, standard method for the stabilization of gullies rather than solely for recharge. Gullies form by concentrated water flow on erosive soil. The adverse impacts of gullying include (Keller and Sherar 2003):

- taking land out of production
- lowering of the groundwater table
- being a major source of sediments.

Basic design issues for check dams are (Keller and Sherar 2003):

- dams should be constructed with a weir, notch or "U" shaped top on the structure to keep water flow concentrated in the middle of the channel.
- the structure should be keyed into the adjacent banks tightly to prevent erosion around the ends of the structure.
- scour protection should be provided at the outlet of the structure (e.g., protective layer of rock).

16.4.2 Check Dams in South Asia

Check dams are used in South Asia to hold water during the monsoon season. The increased contact time of the surface water with the river bed acts to increase infiltration. Published studies on check dams, mostly in India and Pakistan, demonstrated that check dams (Parimalarenganayaki and Elango 2013):

- increase local groundwater recharge, as evidenced by rising groundwater levels
- improve water quality by dilution of the salinity and contaminant concentrations of local groundwater
- have definite socioeconomic benefits by allowing for increased agricultural production and thus additional earnings.

A number of case studies document the water supply and quality benefits of check dams in India. Gale et al. (2006) researched three check dam sites in India (Karnampettai, Saflasana, Chikhalgaon). Water was available for capture during the monsoon season, with availability defined with respect to capturability without consideration of downstream impacts. The check dams increased recharge by 3–23% (managed vs. natural). The hydraulic effects of dams were found to be small relative to watershed water use. Infiltration depended on aquifer storage space and infiltration rate (permeability).

Permeability determines the shape and rate of flattening of hydraulic mounds (Gale et al. 2006). Low permeabilities favor higher mounds that persists longer near the recharge area to the benefit of local farmers. High permeabilities result in a flatter mounds that dissipate more quickly. Flatter mounds may benefit more farmers to a lesser degree. Benefits may still be present but they accrue to the region as whole rather than to specific wells. Volumes recharged by structures are often much larger than the volume of water that remains in the form of a distinct groundwater mound that could be tapped by specific beneficiaries (Gale et al. 2006).

Muralidharan et al. (2007) provided data on a small check dam system in a granitic area of India. The catchment has an area of approximately 350 m^2 and the dam reservoir has a storage of 150 m^3. Recharged water volume was estimated from the water level rise and previously determined specific yield. Recharge rates were estimated to be 27–40% of the total rainfall.

Water quality changes associated with a check dam constructed on the Arani River, northwest of Chennai, Tamil Nadu, India, were investigated by Parimalarenganayaki and Elango (2014, 2015). The Arani River check dam was documented to be 260 m long, with a 3.5 m crest from the riverbed, and a storage capacity 0.8 MCM. Recharge was estimated from daily changes in storage and potential ET data. Over the October 2010–May 2012 monsoon, 2 MCM of water was harvested and an estimated 1.3 MCM infiltrated (63%). The benefits of recharge (higher groundwater levels) extended 1.75 km from dam. Groundwater levels increased by 1.0–3.5 m. The study demonstrated that recharge from the check dam reduced the electric conductivity, total dissolved solids and ion concentrations of groundwater to levels suitable for irrigation and domestic use. The stored water had a high microbial load (>150 cfu/100 mL). Lesser concentrations were detected in the groundwater, and the water may not be suitable for direct domestic use. Parimalarenganayaki and Elango (2015) suggest locating a production well 400 m to the east of the river to achieve a 60-day minimum transit time deemed necessary for effective removal of the microbial load.

The benefits of check dams, in terms of increasing groundwater availability, may be indirectly assessed from changes in groundwater-dependent activities, such as irrigated agriculture. Anecdotal evidence has been reported on the benefits of check dams to local agricultural communities. Raghu and Reddy (2011) documented that after check dams were constructed at several sites in the Anantapur District, Andhra Pradesh, India, water levels in wells increased, abandoned wells were rejuvenated, new wells constructed, and irrigated areas increased. For example, after the construction of a check dam near Kotanka village, irrigated area increased from 8.4 to 13.8 ha and that dug wells that had seasonally dried up during the winter "Rabi" season held 2.5–3.5 m of water during the study period.

A check dam building program by the Sadguru Foundation in western India allowed for the conversion of drylands into productive farmlands and economically enhanced the local population (Agoramoorthy et al. 2008). The check dams result in local groundwater recharge and storage of water, which is pumped up to nearby farms (lift irrigation). The small dams were found to hold sufficient water during the dry season and neither displaced people or destroyed nature.

16.4.3 Check Dams in the MENA and Mediterranean Region

Recharge options for Wadi Madoneh, Jordan, were investigated Abu-Taleb (2003). Three design options were identified as being suitable for the area:

- 2.5-m dikes (check dams) with in-channel recharge downstream. Excess runoff would discharge through a concrete spillway and water would be released to the downstream wadi channel by under-dike pipes equipped with a valve.
- 5-m small dam with in-channel recharge downstream. Discharge to the downstream wadi channel would be performed using under-dam gated pipes or ideally a floating intake system to skim clearer water off the top of the reservoir. The latter has been reported to be difficult to maintain and is subject to vandalism.
- Water harvesting by hillside terracing.

The total average volume reported to be available for recharge for a pilot study at Wadi Madohen was 25,000 m^3 per rainfall event (Abu-Taleb 2003). Four small in-channel dams with heights of 3 to 6 m were subsequently constructed in Wadi Madoneh (de Laat and Nonner 2012). The dams were found to be effective in capturing all of the runoff from the catchment.

Alderwish (2010) evaluated the use of a series of check dams versus single large gravity dams to increase groundwater recharge in the Sana'a Basin of Yemen. The dams would be constructed in the upper reaches of wadis to avoid poor quality runoff that occurs lower in the wadi systems. Predictive modeling results suggest that average recharge rates for a series of check dams would be significantly greater (at least 36%) than for a gravity dam. Check dams have the benefits of (Alderwish 2010):

- reducing the velocity of flood flows
- forming small reservoirs that would increase infiltration time and area
- removing suspended solids through settlement
- lesser costs associated with lesser foundation requirements.

The hydrogeological efficiency of the incidental recharge of a series of check dams and gravel pits in the Almeria province of southeastern Spain was examined by Martín-Rosales et al. (2007). Check dams were designed and constructed for flood control with the goal of reducing or preventing sediment transport in the intensely deforested region. Although the structures are referred to as check dams, a figured dam is a masonry structure 13 m high. Gaging data were not available. Instead, runoff was estimated by modeling using the HEC-HMS code. Infiltration rates were estimated using theoretical flood hydrographs, reservoir height-volume and height-infiltration ratios using generic (as opposed to reservoir specific) hydraulic conductivity values obtained from pumping tests. Recharge at the check dams was estimated to be 10% of runoff.

16.4.4 Check Dams in the United States

Check dams are widely used in the United States for erosion control. Temporary check dams are commonly used to retain sediments at construction sites and in drainage systems. Check dams are much less often used for a primary groundwater recharge function.

Normal et al. (2016) investigated the effects of numerous small rock check dams on the hydrological response of streams to precipitation. The study site was El Coronado Ranch, located in the Chiricahua Mountains of southeastern Arizona, in which thousands of check dams were constructed by hand in small channels (Fig. 16.7). A comparison was made of the runoff between a subwatershed treated with check dams and an adjacent untreated sub-watershed. The treated watershed had a greater runoff normalized by drainage areas (0.61 cm) than the untreated sub-watershed (0.44 cm). It was posited that the check dams resulted in higher water tables and increased runoff rates (rejected recharge). The deposition of loose sandy soils with high infiltration capacities is believed to have increased local groundwater storage, which ultimately increases and extends baseflow. The treated watershed had a reduction in its average rate of flow and decreases in the size and duration of peak flows. Normal et al. (2016) concluded that check dams can be a valuable water management tool for maintaining baseflow in arid regions, which is critical for creating and maintaining functioning watersheds and for the survival and/or expansion of aquatic and riparian ecosystems.

Weirs (check dams) are typically constructed across channels but may also be constructed across lower lying areas of pasture land to increase infiltration and groundwater recharge. Fennessey et al. (2009) documented the installation and performance of two low-head weirs constructed in an urbanizing watershed (Fox Hollow Watershed) in central Pennsylvania. Urbanization results in both increased runoff and losses of natural recharge areas. The project site area was identified as a critical area for stormwater recharge. The weirs are earthen mounds covered with interlocking concrete pavers. During the 2003–2008 study period, which included two of the wettest years on record, less than 30% of the rainfall events resulted in significant outflow from the weirs and associated infiltrated areas. For smaller rainfall events, almost all (89–100%) of the runoff delivered to the weir facility infiltrated. The low-head weirs and infiltration zones are very effective in maintaining infiltration and recharge, and reducing potential stormwater runoff (Fennessey et al. 2009). Over time, the weirs have become almost visually indistinguishable from the surrounding pasture, which is left unmowed (Fennessey et al. n.d.).

16.4.5 Inflatable Dams

Inflatable (rubber) dams have the advantage that they can be inflated to retain water during low flow periods and deflated to allow flood waters to pass. Inflatable (rubber) dams are used in MAR to increased water levels and wetted areas, and to divert water from channels into off-channel infiltration facilities. The Orange Counter Water District (Southern California) uses two inflatable dams to divert river flows from the main channel into off-river recharge basins (Markus et al. 1995, 2000). Both inflatable dams were reported to be 2.1 m high and stretch 97.5 m across the Santa Ana River. Typical inflatable dam installation consists of a reinforced concrete foundation stretched across a riverbed with a rubber bladder anchored to the foundation. Each dam can be fully inflated or deflated to completely block or open the channel or can

Fig. 16.7 Check dams in the Chiricahua Mountains, Arizona (*Source* U.S. Geological Survey)

be set to operate at intermediate heights. Systems operation can be automated with remote monitoring and operational telemetry.

The Sonoma County Water Agency (California) operates an inflatable dam on the Russian River in the Mirabel area to increase water production capacity during peak demand months (Sonoma County Water Agency 2017). The dam is used to divert water to a series of infiltration ponds that are constructed adjacent to three horizontal collector wells. The dam also acts to increase the recharge area of the river behind the dam. Permanent fish ladders allow for fish passage around the rubber dam when it is inflated.

Rubber dams are widely used in arid parts of North China to increase groundwater recharge. Dong et al. (2014) reported the hydrological and water quality impacts of five dams on the Luohe River, the first of which was installed in 2000. The dams were reported to have almost completely alleviated the historic overdraft problem. The increased recharge initially resulted in an improvement in water quality, particularly a decrease in total dissolved solids and nitrate concentrations. As groundwater levels recovered, TDS, nitrate and ammonium concentrations increased, which was attributed to increased groundwater evaporation and a decreased travel distance and time to the water table. Some nitrate and ammonium that had been adsorbed in the vadose zone entered the aquifer.

16.5 Reservoir Recharge

Permanent dams on ephemeral streams (wadis) can have a variety of construction designs, similar to dams constructed on perennial rivers, including rock and earth-fill embankment dams, masonry dams, gravity dams, and arch-gravity dams (U.S. Department of the Interior 1987). Dams must be constructed in accordance with standard engineering design principles to protect downstream communities. Smaller dams may be of less robust earthen construction. The reservoirs behind dams serve three main functions with respect to water supply:

- storage of surface water for direct recharge
- groundwater recharge beneath the reservoir
- functioning as stilling or sedimentation basins for downstream recharge.

It has long-been recognized that siltation imperils the long-term storage capacity of reservoirs (e.g., Hudson et al. 1949). Deposition of thick layers of sediment in reservoirs (Fig. 16.8) also seals the bottom of reservoirs, reducing infiltration rates. Although dams have been constructed in which groundwater recharge is a proposed benefit (usually in addition to flood control), there are limited reliable data on actual recharge rates from reservoirs.

Infiltration rates in reservoirs are most commonly evaluated as the residual of the water-budget:

$$I = (Q_{in} - Q_{out}) + P - ET - \Delta S \quad (16.4)$$

Fig. 16.8 Mud-cracked recently deposited sediments in the Murwani Dam reservoir, western Saudi Arabia

where in units of rate (m/d) or volume over a period of time (m^3)

I	infiltration
$(Q_{in} - Q_{out})$	net surface water inflow (inflows – outflows)
P	precipitation
ET	evapotranspiration
ΔS	change in storage (negative for declining water levels).

In the case of ephemeral stream channels during times with no surface water inflows and outflows and precipitation, Eq. 16.4 simplifies to:

$$I = -ET - \Delta S \tag{16.5}$$

The changes in storage are estimated from changes in reservoir water levels and a reservoir-specific water level-lake area relationship, which is obtained from a detailed topographic survey of the area of a reservoir basin in which water level changes occur. Infiltration may be greater than actual groundwater recharge due to unsaturated zone flow and storage, and subsequent losses to discharge and evapotranspiration.

Wadi recharge dams have the same operations and maintenance issues as other surface-spreading recharge systems, particularly declines in recharge rates caused by the accumulation of clay, silt, and organic matter. Management options for recharge dams in Saudi Arabia were investigated by Al-Muttair et al. (1994) in a comparative study:

- release of water to downstream beds
- release to downstream artificial recharge basins

- silt removal
- scratching the reservoir beds using a plow with chisels.

Infiltration rates were measured from the decline in water levels corrected for evaporation, as measured using Class A evaporation pans at each of the studied dam sites. All four management methods provided benefits. Wadi channel recharge was determined to be a cost-effective option due to the naturally high infiltration rates of the coarse wadi sediments. Scratching was effective in the reservoir with a modest silt accumulation in which the chisels penetrated through the entire thickness of the silt layer. Discharge to recharge basins showed the least benefits, which was attributed to site-specific conditions at the study site. In this case, the reservoir already had a high infiltration rate and the infiltration rate of the basin happened to be relatively low.

A key issue in evaluating reservoir recharge rates is that infiltration rates will decline over time as silt accumulates in the reservoir. Hence, recharge rates estimated near the beginning of operation of recharge dams will likely be considerably greater than long-term rates. A large range of infiltration rates in reservoirs has been reported in the literature. It is not possible to independently verify results, and high ($\geq$80%) rates are suspect as far as being long-term, sustainable rates.

16.5.1 Percolation Tanks (India)

The primary recharge method in India is surface-spreading using percolation tanks (impoundments) and check dams constructed across streams and drainage channels to impound and retain runoff and to increase the opportunity time for recharge (Sakthivadivel 2007). Percolation tanks are very widely used in India to harvest surface water runoff during the monsoon period. Tanks are commonly simple earthen dams, several hundred meters (feet) long and several meters (feet) high, that are constructed across ephemeral stream channels (Sukhija 2008).

There was reported to be more than 1.5 million tanks, ponds, and earthen embankments in 660,000 villages in India (Pandey et al. 2003). The tanks act both as surface-water storage reservoirs and to recharge underlying aquifers. India receives on the average 1170 mm (46.1 in.) of rainfall each year. However, 80% of the rainfall is confined to the monsoon months of June to September and is restricted to 30–60 days during the season. The rainfall also occurs in intense short-duration events of an hour or less, which results in most of the water running off (Radhakrishna 2004).

Glendenning et al. (2012) emphasized that field measurements of actual recharge rates from rainwater harvesting systems is critical, but noted that there is a paucity of data due to the expense and difficulties of quantifying recharge. Calculated recharge rates tend to depend on local hydrogeological conditions, project scale, measurement location, and the methods used. Glendenning et al. (2012) reported that they could find no field studies that evaluated rainwater harvesting in India on a watershed scale.

The chloride mass balance (CMB) method was used to estimate infiltration rates in four percolation ponds (tanks) in the Andhra Pradesh and Gujarat states of India (Sukhija 2008). The method assumes that chloride behaves as a conservative tracer, the only sink for chloride is infiltration, and that increases in chloride concentration over time reflects evaporative concentration. Sukhija (2008) applied the method at the end of the monsoon period in which no further surface water inflows and outflows occurred. The raw data was a time series of measurements of chloride concentration versus reservoir volume measurements. The method assumes uniform chloride concentration in the stored water and uses the following equations (Sukhija 2008):

$$C_1V_1 = C_2V_2 + (1 - f)C_p(V_1 - V_2) \tag{16.6}$$

$$(1 - f) = (C_1V_1 - C_2V_2)/C_p(V_1 - V_2) \tag{16.7}$$

$$C_p = \sum C_iV_i / \sum V_i \tag{16.8}$$

V_1 pond volume at the end of the monsoon period (T_1)
V_2 pond volume at time T_2
f fraction of volume change ($V_1 - V_2$) due to water that evaporated
$(1 - f)$ fraction of volume change ($V_1 - V_2$) due to water that infiltrated
C_1, C_2 chloride concentrations at times T_1 and T_2
C_p average weighted chloride concentration of water during the T_2 to T_1 period.

Geological conditions have a greater impact on recharge efficiency than climate, as indicated by different recharge efficiencies between ponds in the same climate area. Sukhija (2008) observed a recharge efficiency of 60% in a pond on sandstone bedrock in the Saurashtra region of India, compared to an efficiency of only 20% in a pond constructed on basalt bedrock. The greatest rates occur where tanks are located on sediment or rock with a high hydraulic conductivity (Sukhija 2008).

Lamaye (2002) discussed the use of percolation tanks in the semi-arid hardrock areas of western India for aquifer recharge. In this region, the adoption of irrigation using wells by thousands of farmers has resulted in aquifer depletion. A variety of measures have been implemented to increase groundwater recharge. Percolation tanks are constructed of earthen bunds across stream channels. A typical bund was reported to be 6–10 m high, 100–150 m long, with submerged areas commonly of 5–10 ha. Percolation efficiency was calculated by deducting weekly evaporation losses from changes in the volume of stored water. Overall efficiencies range between 40 and 70% with the rate decreasing over time because of the accumulation of low permeability silt. The hydrogeologically preferred location of the bunds is areas where the bed of the stream has 2–3 m of sand and weathered rock above the bedrock, which are usually productive farm land (Lamaye 2002). Villagers instead prefer the systems to be constructed in poor-quality land that is not ideal for aquifer recharge (i.e., where percolation rates are very low).

The percolation rate in a typical small (7 ha) percolation tank near Sangapur village, Andhra Pradesh, was investigated by Massuel et al. (2014) over a two-year period. The local geology is weathered granite (saprolite), commonly 10–15 m thick, that is underlain by a 25–40 m thick layer of fissured granite. Percolation rates were calculated as the closing term of daily water budgets during the dry season and in days in which run off, overflow, and other water losses did not occur. Net inflow, a critical term in the water budget, was calculated as the closing term of the global water budget, as flash floods from short-term rainfall dominates the daily water budget (net inflow is close to the change in stored water volume during short-duration flow events).

Change in storage was calculated from reservoir head-to-area and head-to-volume curves. Reservoir evaporation was estimated using a Class A evaporation pan and a pan coefficient of 0.8. Evaporation rates were also confirmed using oxygen isotope fractionation. A site specific empirical relationship was established to determine the fraction of the water evaporated (f) from the oxygen isotope ratio ($\delta^{18}O$):

$$f = \frac{\delta^{18}O - 1.631}{23.381} \tag{16.9}$$

The estimated percolation efficiency was 57–63%, with the rest lost to evapotranspiration. The estimated percolation efficiency is greater than that reported from other sites, which was attributed to the calculations capturing percolation during rainfall events (Massuel et al. 2014).

Boisson et al. (2014) proposed a method for assessing of the impact of small recharge structures on groundwater based on two independently computed water balances: a surface-water balance and a groundwater balance. The study was performed on a percolation tank representative of a semi-arid crystalline setting of southern India (Tumulur percolation tank located 35 km south of Hyderabad). The surface water balance used to calculate the infiltration rate is

$$\Delta V = (A_{tank} \cdot P) + (R \cdot a) - (A_{tank} \cdot E) - (A_{tank} \cdot q_{swb}) - U \tag{16.10}$$

where

ΔV change in volume of water stored in tank (L^3/T)
A_{tank} tank flooded area (L^2)
q_{swb} infiltration rate (L/T)
P precipitation (L/T)
R runoff (L/T)
a "effective" drainage area of runoff (L^2)
E evaporation rate from tank (L/T)
U uptake by direction irrigation or livestock consumption (L^3/T).

The infiltration rate (q_{swb}) was determined during recession and R was determined during the filling period (estimated from precipitation rates using SCS-CN method).

The groundwater balance method is based on the water table fluctuation method. The change in stored groundwater (ΔS) is calculated as

$$\Delta S = S_y \cdot \Delta h \cdot A_{Total} \tag{16.11}$$

where

ΔS	change in storage (L^3/T)
S_y	specific yield
Δh	water table fluctuation (L)
A_{Total}	area considered for calculation (L^2)

$$\Delta S = \left(A^* \cdot D_p\right) + \Delta V_{gwb} - \left(A_{total} \cdot q_{farm}\right) + (Q_{in} - Q_{out}) - \left(A^* \cdot E_{gw}\right) \tag{16.12}$$

$A*$	study area outside of tank (L^2)
D_p	deep percolation (L/T)
ΔV_{gwb}	percolated water from the tank to the aquifer (L^3/T)
q_{farm}	farm abstraction rate (L/T)
$(Q_{in} - Q_{out})$	net flow of groundwater into the study area (L^3/T)
E_{gw}	evaporation from groundwater.

The results of the investigation are:

- Infiltration rate = 5.5 m/year
- Ratio of infiltration to total stored water volume = 63%
- 53–88% of theoretical infiltrated volume percolated to become recoverable groundwater.

The results demonstrated delayed recharge of the aquifer resulting from temporary storage and slow movement of water in the unsaturated zone. Only a limited number of farmers were found to benefit from the tank recharge despite a large number of pumped boreholes in the tank vicinity.

16.5.2 MENA Region Reservoir Recharge

Missimer et al. (2015) reported, based on a literature review, that up to 80% of the water in old wadi reservoirs is lost to free-surface evaporation before infiltration and recharge can occur. Recharge rates will vary based on local hydrogeological conditions and the age of dams.

The Hautat Bani Tamim (HBT) reservoir in central Saudi Arabia was constructed in 1982 (Al-Othman 2011). Six artificial recharge structures were constructed in the reservoir, which are wells with openings at different heights that are operated through the opening and closing of values. The surface area of the reservoir is known at 0.5 m intervals, which allows for calculation of changes in storage over the elevation intervals. Infiltration rates were calculated from the change in storage minus the evaporation rate.

The effectiveness of the recharge structures was evaluated by comparing infiltration over the same 2 m elevation interval both before and after the construction of the recharge structures. The recession from an April 4, 2009 rainfall event resulted in a calculated infiltration of 2.09 MCM versus an evaporative loss of 0.50 MCM. The recession from a May 6, 2010 event resulted in an estimated recharge of 2.33 MCM and evaporative loss of 0.26 MCM. The recharge percentage increased from 80.7 to 90.0%. Al-Othman (2011) cautioned that evaluation of the recharge wells occurred shortly after their construction and thus did not consider the potential impacts of clogging. The calculated infiltration rates are sensitive to the accuracy of the reservoir storage calculations.

Water budget estimates for two recharge reservoirs in central Saudi Arabia indicate that a high percentage (82 and 94.5%) of water in the reservoirs was taken into the soil (Al-Turbak and Al-Multair 1989). Al-Turbak (1991) and Al-Hassoun and Al-Turbak (1995) evaluated the amount of water that infiltrates through the reservoir bed at the Al-Amalih Dam, in central Saudi Arabia, that can be effectively used downstream. The dam was constructed in 1982. Reservoir infiltration (I_r) was measured as the change in storage minus evaporation. Storage was calculated from depth-volume and depth-area curves established from a detailed topographic survey of the reservoir. Evaporation rates were obtained from pan evaporation rates using a pan coefficient of 0.8 for the months of November to March and 0.70 for April to October. Subsurface groundwater flow into and out of the reservoir (Q) was calculated based on Darcy's law

$$Q = KAI \tag{16.13}$$

where K = hydraulic conductivity, A = cross section area of flow, and I = hydraulic gradient using consistent units. The change in stored water volume due to the rise and fall of the water table (ΔSwt) was calculated as

$$\Delta S_{wt} = I_r + Q_{in} - Q_{out} \tag{16.14}$$

The data from Al-Amalih Dam indicate both high rates of infiltration and that a large percentage of the infiltrated water appears in the shallow aquifer downstream (Table 16.1). The peak in downstream water levels occurred within two weeks of the peak in infiltration rates. Limitations of the analysis are that calculated subsurface flow rates are sensitive to the hydraulic conductivity value used (43.95 m/d), which was based on a single pumping test. The reservoir infiltration rates are also for early in the history of reservoir and will likely decrease over time due to clogging. The study results suggest that surface reservoirs are very effective in infiltrating water, but that a significant fraction of the water remains far away from where it is needed. A critical issue is that where there is a great depth to the water table, a large amount of water that infiltrates through the bottom of a reservoir may not locally recharge the underlying aquifer.

Table 16.1 Infiltration and recharge percentages Al-Amalih Dam, in central Saudi Arabia

Season	1895/1986	1986/1987	1987/1988
Reservoir infiltration (%)	93.3	94.5	95.8
Infiltrated water that appears as groundwater downstream (%)	66.6	75.9	94.4

Source Al-Hassoun and Al-Turbak (1995)

A MODFLOW modeling investigation of the dam on Wadi Tawiyaen, located in the Emirate of Fujairah, United Arab Emirates, compared the recharge with and without the dam (Sherif et al. 2011). Over a modeled 6-year, 3-month period, total recharge was estimated to be 4.22 MCM. The modeled recharge for a non-dam scenario was 1.23 MCM, which gives an estimated additional groundwater storage due to construction of the dam of 2.97 MCM or 70% of the total recharge. The total storage and groundwater recharge due to ponding during the validation period were estimated to be 13.6 and 2.97 MCM, giving an effective recharge rate of 22%. The low efficiency was attributed to high evaporation losses over a storage period that might reach 150 days.

Stable isotope studies of recharge at three dams in the UAE indicated recharge rates of 47, 22, and 31% for the Ham Dam, Tween Dam, and Bih Dam, respectively (MAF and ICBA 2012).

Kazemi et al. (2002) reported infiltration rates from two recharge projects in the Shahrood District of Northeast Iran. The Biarjomand Recharge Project is a 17 MCM earth dam that captures Chariano River runoff. Due to the high evaporation rates in the region and a high suspended solids content of the runoff, the percent infiltrated was 23% (77% evaporative loss) in the first year, decreasing to 6% in year 5. The additional function of the system is flood control. Kazemi et al. (2002) noted that even if the recharge projects increases local water resources by only a few percent, the systems might still be considered a success.

Ibrahim (2009) documented how recharge dams across intermittent streams in Gadarif City, Sudan, provided important benefits to the community. The first dam was constructed in 1995 for flood control purposes, but it was noted that water stored behind the dam had infiltrated into the soil and raised the water table by a more than 2 m. A second dam was constructed in 1998 on another stream in the city to further increase aquifer recharge. The recharge dams resulted in the renewed availability of water in some previously dry wells and allowed the villagers to drill new boreholes and wells to obtain water for drinking and farming. Recharge from the dams resulted in 2,200 m^3/d (580,800 gpd) of new groundwater resources, which is approximately 12% of the total demand.

16.5.3 United States

Groundwater recharge in reservoirs and playas in the Southern High Plains of Texas was investigated by Scanlon et al. (2003). The study included two reservoirs, SCS3 (operational on November 29, 1976) and SCS4 (operational on February 2, 1982) in Hale County, Texas. The study area is semiarid with a mean annual precipitation of 17.3 in. (44 cm). Surface water inflow into the reservoirs was estimated using the Soil and Water Assessment Tool (SWAT) and evaporation was estimated from pan evaporation data using a lake coefficient of 0.7. The average infiltration rate was calculated to be 0.5 in./day (1.3 cm/d). In SCS4, 155 AF (0.19 MCM) inflowed during a 6 month period, of which approximately 35% evaporated and 65% infiltrated. Some of the infiltrated water may have subsequently evaporated from the soil (Scanlon et al. 2003).

The U.S. Geological Survey (Heilweil and Marston 2011, 2013; Marston and Heilweil 2016) investigated infiltration in Sand Hollow, a 50 km^2 basin in Washington County, Utah, that is underlain by the Navajo Sandstone. The off-stream reservoir facility receives water from the Virgin River. Climate conditions dictate the amount of water that is available for recharge. Sand Hollow Reservoir is currently managed to maximize groundwater recharge. Recharge was calculated from water budget:

$$R = I_{sw} + I_{dr} - O_{sw} + P - \Delta S - E \tag{16.15}$$

where (in units of volume)

I_{sw} surface water inflow
I_{dr} drain and spring return flows
O_{sw} surface water outflow
P precipitation falling directly on reservoir
ΔS change in surface water storage
E evaporation.

PET was estimated from the McGuinness and Bordne (1972) version of the Jensen-Haise method, using air temperature and solar radiation:

$$PET = \{[((0.01T_a) - 0.37)Q_s\}0.000673\}2.54 \tag{16.16}$$

PET cm/day
T_a air temperature (°F)
Q_s solar radiation (cal per cm^2 per day).

During the 2002–2014 period, the net inflow pumped to the reservoir was 216,000 AF (266 MCM) and the estimated recharge was 127,000 AF (157 MCM; 58.8% of inflow). The estimated total reservoir evaporation was 70,000 AF (86 MCM). Major and minor dissolved inorganic ions, tritium, DOC, dissolved gases (FCFs, sulfur hexafluoride, noble gases) were used to evaluate the timing of recharge and the location of recharged water in the aquifer. The tracers often had different peak arrival times due to their different behavioral characteristics (adsorption, retardation, dispersion, and gas dissolution). Groundwater flows radially outward from reservoir. Recharge resulted in a local groundwater rise of as much as 40 m near the reservoir.

16.6 Sand Dams

Sand dams (also referred to as trap dams, sand-storage dams, and barrier dams) are a type of check dam that are constructed across ephemeral river valleys to intentionally trap sand and gravel and thus create an artificial aquifer (Baurne 1984; Nilsson 1988; Van Haveren 2004; Aerts et al. 2007; RAIN 2008; Stern and Stern 2011). Sand dams are different from subsurface (groundwater underground) dams, which are impervious barriers constructed within fluvial aquifers to arrest groundwater flow and increase the volume of water locally stored. Sand dams may serve a dual function of also intercepting local groundwater flow (Nilsson 1988). Subsurface dams can increase local water storage but are not considered herein as they do not cause recharge.

Sand dams are constructed above ground and are commonly up to 4–6 m (13.1–19.7 ft) high. The dams can be constructed of either reinforced concrete, stone masonry, compacted earth with a concrete spillway, stone gabions, or blocks either sealed with a clay layer or having a clay core (Nilsson 1988). Ideally, locally available materials are used to the degree practical. A key design issue is to fill the reservoir with coarse sediments and avoiding silt and clay deposition, which could impede vertical flow and thus the recharge of the reservoir. Coarse sand allows for high infiltration rates and easier extraction of stored water. Larger dams should be constructed in stages to ensure that only sand and gravel are retained behind the dam.

Upon completion, the created sand and gravel aquifer is recharged during flow periods with the bulk of the flow passing over the dam. The stored water is later captured by digging scoop holes, construction of vertical wells, or use of a horizontal outlet (delivery) pipe installed through the dam (Fig. 16.9). Scoop holes are the least expensive option but are vulnerable to contamination. Outlet pipes have the advantage that the water flows under hydrostatic pressure, but they also have the disadvantages that they may weaken the dam structure, maintenance is complicated, and it is a more expensive option (RAIN 2008).

Sand dams can be implemented on various scales. On the smaller scale, Sivils and Brock (1981) and Bleich and Weaver (1983) documented how sand dams have been used to provide water for livestock and wildlife. The two sand dams described by Sivils and Brock (1981) were about 1 m high and anchored in bedrock across ephemeral streams. The reservoirs included a buried network of 30.5 cm diameter aluminum pipe to provide increased water storage. The reservoirs were covered with coarse-gravel and small rocks and then sand to reduce evaporation.

Both sand dams and groundwater dams are increasingly being examined as tools for community-scale water supply in developing countries (Nilsson 1988). There are practical limits on the quantity of water that can be stored up-gradient of both types

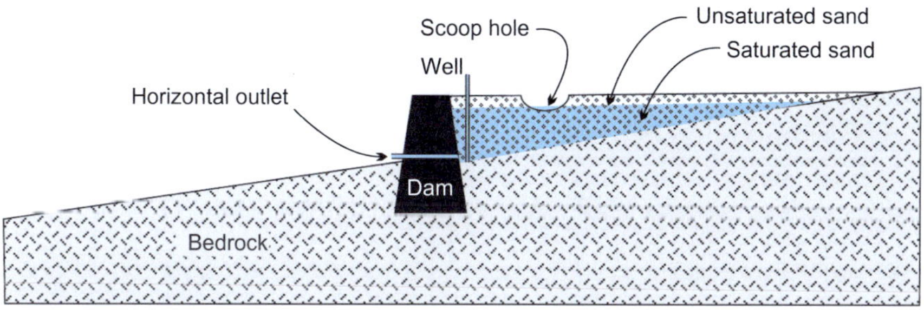

Fig. 16.9 Conceptual diagram of a sand dam

of dams, which limits their potential contribution to large urban areas. However, the volume of water that can be stored in sand and groundwater dam impoundments can significantly contribute to the domestic and irrigation water supply of villages. Sand dams in developing countries are a relatively inexpensive community-based project (Van Haveren 2004). The importance of community involvement throughout the site-selection, construction, operation, and maintenance of a project is strongly stressed (RAIN 2008; Lasage et al. 2008). As is the case for water storage projects in general, some arrangement is necessary for the equitable use of the stored water.

Sand dams have a very long history of creation of domestic, irrigation, and stock water supplies (Nilsson 1988; Agarwal and Narain 1997; Van Haveren 2004; Lasage et al. 2015). The main advantages of sand dams are that they:

- have greatly reduced evaporative losses compared to surface reservoirs, which can allow for more seasonal water storage in a given reservoir area.
- infiltration through the trapped sediments can improve water quality by filtering out of pathogenic microorganisms.
- can increase regional aquifer recharge and water levels, moderate stream flow (i.e., provide flood control benefits), and allow for sand harvesting and rehabilitation of gullies.
- tend to occupy low value land, are low maintenance, and are a low sophistication-level technology that is inexpensive enough to be implemented by local communities with available resources (Mutiso 2003).
- do not serve as a breeding ground for mosquitos (and other pests) as the water is stored underground.

Sand dams, if well constructed, should require little or no maintenance. A community organization should be capable and willing to perform whatever maintenance tasks are necessary, which might include repairing cracks and weak points in the dam and abutments, cleaning the well or outlet, and removing silt from the top of the aquifer (RAIN 2008).

The dams obstruct downstream groundwater flow through the riverbed. The volume of water that is available for extraction is considerably greater than the volume present in the artificial aquifer because of the additional water stored in riverbanks and baseflow (RAIN 2008). The amount of extractable water (V_e) depends upon the volume of sediment behind a dam, its specific yield (S_y), and any net inflows to and outflows from the aquifer from baseflow and leakage. Extractable water volume can be estimated as follows (after Aerts et al. 2007):

$$V_e = \left(X \cdot W \cdot D \cdot S_y\right) + B - L_v - L_{out} - L_{in} \quad (16.7)$$

where (in metric units),

X length of the river segment (m)
W average width of the river segment (m)
D average thickness of the riverbed sediments (m)
B baseflow into the river sediments from riverbanks (m^3)
S_y specific yield (unitless)
L_v vertical leakage out of the bottom of the aquifer (m^3)
L_{out} horizontal leakage out of the aquifer downstream of the dam (m^3)
L_{in} horizontal leakage into the aquifer from upstream of the dams (m^3).

The benefits of water storage in sand dam systems may also include higher groundwater levels in surrounding areas. Hut et al. (2008) examined the effects of sand storage dams on local groundwater levels. After a flash flood, reservoirs fill and water starts to infiltrate into the riverbed. It may take many flash floods or even many seasons before water levels in the surrounding soil come into equilibrium with water levels immediately upstream of the dam. Multiple dams close together may result in a "global" rise in groundwater levels in the dam area (Ertsen and Hut 2009). Due to the overlap of areas of influences, the closer dams are to each other, the lower the volume of water stored compared to the stored volume of the same number of dams with a greater separation (Ertsen and Hut 2009). The effectiveness of sand dam systems depends on retention and water use patterns (Hut et al. 2008). Where systems are used for household supply and withdrawals are modest, higher groundwater levels may occur in the dry season in both the reservoir and riverbanks. On the contrary, minimal infiltration effects occur when reservoirs are used for intensive irrigation.

The Rainwater Harvesting Implementation Network (RAIN 2008) prepared a useful guide to sand dam implementation. Dams must be designed by an experienced engineer so that they are anchored in the bedrock and stream banks to prevent failure during floods, underflow losses from the created aquifer, and erosion of the river bank (Nilsson 1988). The selection of locations of suitable catchments and riverbeds needs to consider the following factors (RAIN, 2008; Stern and Stern 2011):

- proximity to potential water users
- valuable property and land should not become submerged
- site should be accessible for construction, use, and maintenance
- river width should be no more than 25 m (82 ft)
- river slope gradient (0.3–4%, preferably 2–4%)
- availability of coarse sediments in the catchment
- maximum flood height

- capability of the riverbed to store water; dam should be built on bedrock or an impermeable layer
- riverbank height; both banks should be high enough so that the river will never overflow its banks during maximum flood events.

If the river slope gradient is too low, then coarse-grained material is normally not transported and reservoir will be filled with predominantly fine-grained materials. Where the slope is steep, the storage volume for a given dam height is less or a higher dam must be constructed to achieve storage goals. If the bedrock below a dam is relatively permeable (for example, from the presence of fractures), then efforts should be made to seal the fractures, such as by pressure grouting or pouring very thin mortar into the fracture network (Nilsson 1988). Stern and Stern (2011) also recommended a primary spillway that discharges normal flow into center of the river channel during the rainy season, and a secondary spillway (1 m above the primary) for heavy flows and storms. Wings should be constructed around the sides of dams to keep floodwaters from going around the sand dam and causing erosion and eventual undercutting of the dam walls.

Mutiso (2003), Aerts et al. (2007), Lasage et al. (2008), and Ertsen and Hut (2009) discussed sand dams in the Kitui District of eastern Kenya, which has the greatest density of sand dams in the world. Approximately 500 dams were constructed in 10 years, which were estimated to store 3.8% of the runoff during the April–October wet season and 1.8% during the November to March wet season (Aerts et al. 2007). The dams are typically 1–4 m high stone masonry structures constructed on ephemeral rivers (Ertsen and Hut 2009). The construction and maintenance of the dams is dependent on inputs and commitments of local communities in conjunction with technical and financial support from an NGO (Sahelian Solution Foundation; SASOL). Socioeconomic indicators and hydrological data show that sand dams are a successful local adaptation for dealing with droughts. By providing a local supply of water, less time is needed to fetch water, which has benefits such as significantly increasing school attendance and making more time available for income producing activities (RAIN 2008). Currently, the existing dams capture only a small fraction of the annual steam flow and, therefore, do not have a significant impact on downstream water resources. However, if global climate change increases the variability in rainfall, then during some dry years the water captured by the dams could cause downstream water shortages. More frequent water shortages would also occur if the number of dams were significantly increased.

Quilis et al. (2009) performed numerical modeling of a single sand storage dam and a cascade of three dams based on the Kitui District (Eastern Kenya) system. The main conclusions from the modeling and monitoring data are:

- groundwater levels respond rapidly to precipitation events whereas recession during dry periods is slower
- during the wet season, flow is from the riverbanks towards river
- during the dry season, flow is from riverbed towards banks
- a rise in groundwater levels occur both upstream and downstream of the dams; higher groundwater levels upstream induce downstream flow around dams

- model results are most sensitive to the thickness and hydraulic conductivity of the sediment layer on the riverbank
- cascades of dams close together results in lesser storage than would occur in three single dams at non-overlapping (area of influence) distances.

A concern for sand dam projects, in general, is that the capturing of surface water and obstruction of the groundwater flow in fluvial aquifers will decrease downstream water availability. Lasage et al. (2015) evaluated the potential downstream impacts of sand dams in the Dawa Catchment of southern Ethiopia. A concern is that decreased downstream water availability would offset increased upstream productivity. Downstream impacts were simulated using the STREAM (Spatial Tools for River basins and Environment and Analysis of Management Options) model. Adverse impacts are considered flows below the minimum environmental flow level, defined as one standard deviation above and below mean discharge. The simulation results indicate the under current climate conditions, a moderate implementation (613 dams) would not lead to a change in the number of months with flows below the environmental flow requirement. The high implementation strategy (2,190 dams) would result in a 4–9% increase in low flow months, which would occur before the start of the rainy seasons. The effects of climate change were simulated using three climate-change scenarios in downscaled general circulation models (GCMs). Under the most extreme climate scenario, the high implementation strategy would reduce downstream flows by only 4.5%. Combining the high implementation strategy and most extreme climate change scenario gives a simulated increase in low flow months of 4–50%. Lasage et al. (2015) concluded that sand dams are a feasible adaptation to present scarce water resources and for improving water security under climate change.

16.7 Spate Irrigation (Floodwater Harvesting)

16.7.1 Spate Irrigation Basics

Spate irrigation was defined by Mehari et al. (2007) as "a resource system, whereby floodwater is emitted through normally dry wadis and conveyed to irrigatable fields" (Van Steenbergen et al. 2010). The term spate irrigation is derived from the English word "spate," meaning a flood or inundation. Spate irrigation systems are, in general, characterized by a very large upstream catchment (200 ha–50 km^2) and "catchment area: cultivated area" ratio of 100:1 to 10,000:1 (Van Steenbergen et al. 2010). FAO Irrigation and Drainage Paper 65 (Van Steenbergen et al. 2010) provides an excellent detailed guideline addressing spate irrigation design and management from which much of this summary was derived.

Spate irrigation technology is very old, dating back to at least 3000 B.C., from which time remains of diversion dams are present in Iran and Pakistan (Van Steenbergen et al. 2010). It is estimated that at least 3.3 million ha are under spate irrigation globally, with schemes found in West Asia, Central Asia, the Near East, North Africa, the Horn of Africa, and Latin America (Van Steenbergen et al. 2010). From an MAR perspective, spate irrigation contributes to groundwater recharge. It can be either managed aquifer recharge, where groundwater augmentation is an intended purpose of the system, or unmanaged recharge, in which groundwater augmentation is an incidental benefit of the system.

There are two basic types of spate irrigation systems: floodwater harvesting and floodwater diversion systems (Van Steenbergen et al. 2010). In floodwater harvesting systems, turbulent channel flow is collected and spread throughout a wadi (ephemeral stream channel) in which the crops are planted. Cross-wadi dams are constructed with stones, earth, or both, and are often reinforced with gabions. In floodwater diversion systems, floodwaters are diverted into adjacent embanked (bunded) fields for direct application. A stone or concrete structure raises the water level within the wadi and diverts water to nearby cropping areas. Crops are planted typically after sufficient irrigation has occurred using residual moisture stored in deep alluvial soils.

Spate irrigation systems are characterized by (Van Steenbergen et al. 2010):

- an arid environment
- unpredictable and often short duration and intense flows
- very high sediment loads
- complex social organizations involving multiple farmers and substantial local wisdom and experience
- relatively low economic returns; it is mainly a subsistence activity
- sedimentation being a major factor with systems “growing” their owns soils and being susceptible to blocking of intakes and channels.

Systems vary in their scale and technical complexity from small schemes, under farmer management using traditional diversion practices, to large and technically complex systems involving external (i.e., governmental) management and technical and financial support (Van Steenbergen et al. 2010). Spate irrigation systems have declined, and disappeared in some areas, because of the high labor input required from farmers and a preference for more rewarding livelihoods. Ghebremariam and Van Steenbergen (2007) concluded that considerable social capital needs to be nourished and carefully considered in the development of spate irrigation systems. Transition to perennial cropping using groundwater or surface water provided by reservoirs allows for higher value crops to be grown. Where groundwater development has been possible, farmers have taken advantage of it as groundwater-based irrigation is more productive and predictable than traditional spate irrigation, with some estimates by a factor of six (Van Steenbergen et al. 2010). However, the abandonment of spate irrigation has resulted in the loss of groundwater recharge benefits. Spate irrigation (flooding of farm fields) is being rediscovered in some areas as a planned means of increasing aquifer recharge.

16.7.2 Hydrology and Sediment Transport

Spate irrigation systems do not exist in isolation. When spate irrigation diverts a substantial part of a wadi flow, potential impacts to groundwater recharge occurs downstream with potential negative implications for communities reliant of groundwater (Van Steenbergen et al. 2010). A river basin approach is, therefore, necessary to ensure that spate irrigation diversions result in an overall increase in crop production and avoids significant losses for downstream water users (Van Steenbergen et al. 2010). Key issues are (Van Steenbergen et al. 2010):

- a great variation in the size and frequency of floods, which directly impacts the availability of water for agriculture
- wadi floods are characterized by very high sediments load, which provides benefits by increasing the soil supply but can have negative impacts by clogging of intakes and channels
- systems can be damaged by large floods
- a scale factor occurs in data availability and resources for studies.

Although much data are needed to optimize the design of systems, it typically is not available for smaller systems. Small-scale systems are commonly designed and constructed by local farmers using historical knowledge with very little or no external technical and financial support. The following data should ideally be available for designers of spate irrigation systems (Van Steenbergen et al. 2010):

- annual volume of water available at diversion point(s)
- probability distribution of spate runoff events (peak flow and flood volumes) and the proportion of annual flows that occurs in different flow ranges
- distribution of flows during runoff events, particularly the shape of the recession limb of hydrographs
- wadi bed infiltration rates
- magnitude and return periods of extreme discharges for the design and protection of permanent works
- sediment concentrations and the size range of sediments transported in floods of various sizes
- sediment transporting capacity of existing canals.

It must be stressed that much of the above listed data are seldom available. Very commonly, stream gaging data are not available for ephemeral streams. Various surface water modeling techniques can be used to estimate flood flows from watershed characteristics and precipitation data, but even this approach may be applied for only large-scale systems. Furthermore, studies have demonstrated that there is actually a poor correlation between observed rainfall and runoff (Van Steenbergen et al. 2010). Often the only information available on flood frequency and water levels may be local farmers' knowledge, which should not be discounted (Van Steenbergen et al. 2010).

16.7.3 Spate-Irrigation System Design

It is of paramount importance to understand farmers' irrigation practices, priorities, and risk management strategies. Small systems should be simple enough for farmers to maintain with indigenous skills and locally available material. An important consideration is the balance between the costs of constructing more robust structures versus the time and costs to maintain and reconstruct structures damaged or washed out by floods. A low economic value of crops can warrant only low-cost improvements. However, poverty reduction and groundwater recharge and associated improved water supplies are also considerations. The basic design procedure is as follows (Van Steenbergen et al. 2010):

- calculate the mean annual runoff volume from the catchment.
- calculate the ephemeral stream flow duration curve (discharge vs. time).
- determine the design flood discharge; the proportion of the mean annual flood discharge to be diverted.
- estimate the sediment load.

Some design generalizations are (Van Steenbergen et al. 2010):

- sediment concentrations up to 5% by weight are common
- the fine fraction (silt and clay) typically constitutes most of the total annual sediment load
- the sediment fraction that will be diverted to spate canal in suspension is relatively fine (generally between 0.1 and 1.0 mm in diameter)
- coarse bed load represents only about 5% of the total annual sediment load but is this component tends to settle out and block canals and intakes.

Water diversion and control structures should be designed to split and guide flood flows, avoid excessive sediment loads in the spate system, minimize the potential for damage to canals and fields during large floods, be able to cope with large changes in wadi conditions, and respect established systems of water allocation. The main types of diversion structures are spur-type deflectors and diversion bunds.

Spur-type deflectors are a spur constructed within a channel to split the flow with the greater part continuing downstream (Fig. 16.10). They are usually made of wadi material reinforced with more durable material. Spur-type deflectors are usually constructed in the outside of channel bends, where the channel is deeper and lower flows are concentrated during flood recession. They are intended to take as much of low and medium flows as possible but only a small portion of large floods. Traditional systems are designed to be damaged or completely swept away by large floods, which reduces the potential for damage to the irrigation system.

Bund-type deflectors are large bunds (embankments) constructed across wadi beds to divert the available flow to canals on one or both banks. All of the wadi flow is diverted until the bund is overtopped and scoured out by a large flood or is deliberately cut by a farmer. A more robust weir may be constructed instead of a bund using more durable material (e.g., gabions, masonry and concrete).

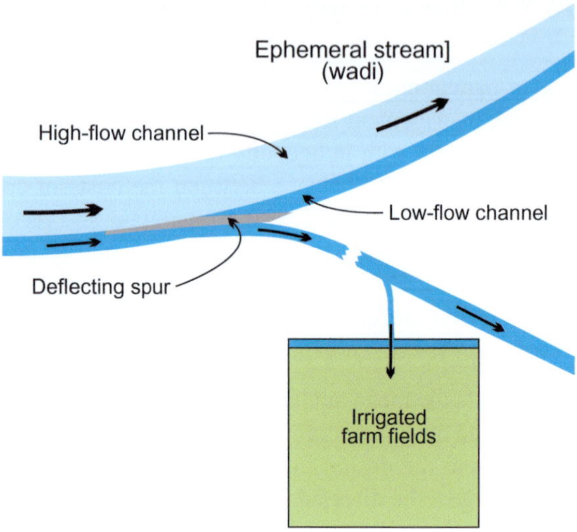

Fig. 16.10 Conceptual diagram of a spate irrigation system using a spur-type deflector

Spate irrigation is as much about sediment management as it is about water management (Van Steenbergen et al. 2010). Lawrence (2009) provides a very good review of sediment management practices for spate systems. The concentration of suspended sediments in wadi flash floods can be extremely high, reaching and exceeding 10%, with grain size ranging for suspended clays and silts to boulders and cobbles. Typically clay and silt content ranges between 50 and 90% of the annual sediment load (Lawrence 2009).

During very large floods in which very coarse sediments can be transported, the intakes are usually washed away reducing the volume of water and sediments diverted to canals. Coarse sediment tends to settle in wadi channels and canals, with finer sediment deposited in the fields where they are welcomed by farmers. Sediment deposition rates in spate irrigated fields was reported to range between 1 to more than 50 mm/year (Lawrence 2009). A goal is to design systems to exclude as much coarse sediment as possible from entering and settling in canals while still allowing the transport of fine sediment to the fields (Lawrence 2009). A rise of command levels in fields due to sedimentation may result in some parts of upstream irrigated areas no longer receiving water.

If the canal slope from the diversion point to fields is too low, then there will be limited sediment transporting capability and severe canal sedimentation problems may occur. A solution to sedimentation in canals is the use of sedimentation basins in which flow is slowed and coarser sediments are deposited. Sediments may be flushed back into the wadi or mechanically removed. Lawrence (2009) advises to accept the need for canal de-silting and plan for it. However, farmers tend to strongly object to what they perceive as a waste of water that could be diverted for irrigation and, therefore, may not want to operate sediment control measures if it reduces flows to the fields (Lawrence 2009). Farmers, therefore, need to be educated so that they understand that sediment control measures are beneficial to them.

16.7.4 Modernization of Spate Irrigation

Investment in spate irrigation has been neglected with respect to perennial irrigation largely because the latter is perceived as having (and often actually has) much greater economic value. However, over the past two decades, there has been increased recognition that spate irrigation can be a major asset for bettering the lives of poor communities and is worth investing in (Mehari et al. 2011). Renewed interest has focused mainly on interventions tailored at improving floodwater diversion efficiency. Mehari et al. (2011) proposed a series of measures that could improve field water management and soil moisture conservation. Excess irrigation (additional "turns") was found to not necessarily increase net soil water storage, but does deprive downstream farmers of needed water (Mehari et al. 2011). It was observed that modernization of systems could lead to a breakdown of long-established agreements on water distribution between upstream and downstream users. Recommendations to increase soil infiltration and holding capacity include pre- and post-irrigation tillage, mulching, and combined sowing and ploughing tillage practices.

16.8 Off-Channel Canal and Surface-Spreading Recharge

Spate irrigation systems are designed primarily to store floodwaters in the soil zone for growing crops. Off-channel recharge systems are similar in that water is diverted from perennial or ephemeral channels to recharge the adjoining aquifer. Oaksford (1985) briefly documented several types of off-channel recharge systems, which were noted to be the oldest type of MAR systems. In ditch and furrow systems, water is diverted into one more canals that follow the land contours, or into a network of lateral or dendritic canals extending outward from a channel. Infiltration occurs within the canals and, depending upon the system design, the non-infiltrated water may be returned to the main channel downstream. Alternatively, water diverted from the canal can be made to spread on the land surface.

Floodwater harvesting and spate irrigation are ancient techniques in Iran, where a system is called a "Bandsar" (Samani et al. 2014). A Bandsar is described as a basin surrounded by levees that is constructed along topographic contour lines and is used as a farm field. Water from ephemeral stream channels is diverted into the basin. Some transverse subsidiary walls with outlets are constructed for balancing the water spreading. During each flood event, the soil moisture increases to allow for crop cultivation and the basins receives fine-grained sediments that increase their fertility. A similar ancient floodwater harvesting system is the "Khooshab," which is described as a cross-dam constructed of soil and rock across a channel bed. The

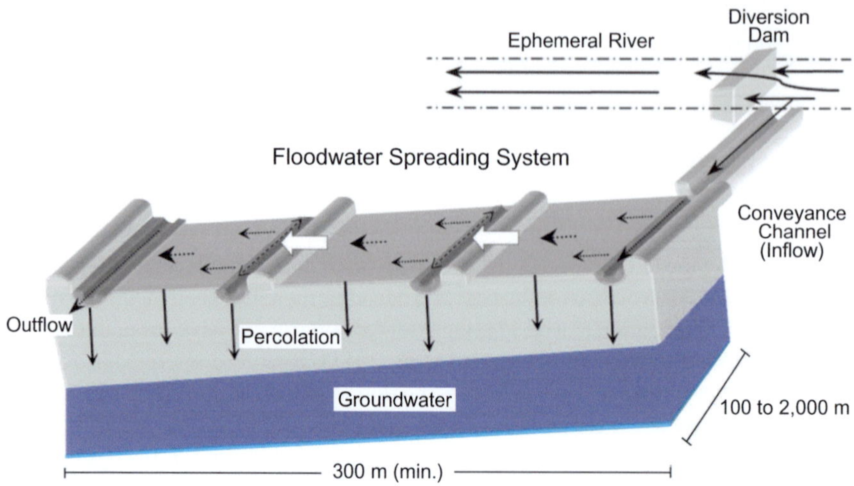

Fig. 16.11 Conceptual diagram of a three-basin flood-water spreading system (Modified from Hashemi et al. 2015)

dam acts to both collect sediment and harvest rainwater. Crops are cultivated in the moist soils behind the dams (Samani et al. 2014). Modern floodwater harvesting and artificial recharge by water spreading have been practiced at 36 multipurpose floodwater spreading (FWS) sites in Iran since 1983 (Hashemi et al. 2015).

Hashemi et al. (2013a, 2015) described a floodwater harvesting (FWH) system on the Gareh-Bygone Plain (GBP) of Southern Iran. Groundwater in the study area has been the main source of freshwater, and aquifer overexploitation was reported to have resulted in an about a 10 m decline in the water table over the previous 10 years. Floodwater is diverted to a conveyance channel, in which it flows under gravity to a series of leveled terraces of progressively decreasing elevation. Water overflows from the conveyance-spreader channel into the uppermost basin and is retained behind a downstream embankment (Fig. 16.11). Gateways in the embankment allow for the controlled flow of water into the next, down-gradient conveyance-spreader channel and basin. The process continues until almost clear water enters the last basins designed as infiltration ponds. Excess water from the last basin is returned to the river or a downstream FWS system (Hashemi et al. 2015).

Field experiments were performed to evaluate the improvement in agricultural yield using floodwater harvesting. The results showed a 2.5 fold increase in the yield of cultivated barley inside of the FWH compared to a similar cultivated plot outside of the area (Ghahari et al. 2014). Modeling results also demonstrated that the FWH system results in significantly increased groundwater recharge. Scarcity of data in many arid regions, especially in the Middle East, has necessitated the use of combined mathematical models and field observations to estimate recharge (Hashemi et al. 2013a, 2015).

The operation and the water balance of an FWS in the Gareh-Bygone Plain (GBP) of Iran were described by Hashemi et al. (2013a, 2015). Hashemi et al. (2013b) demonstrated the use of inverse-modeling using the MODFLOW and PEST codes to quantify the recharge from a floodwater spreading system and natural river channel recharge in the GBP. Steady-state calibrations were first performed for time periods in which measured hydraulic head differences were negligible. Transient calibrations were next performed to estimate specific yield values. The inverse modeling results showed that MAR from FWS is the main (80%) source of recharge in the study area.

Modeling results show a large variation between the inflow into the systems and recharge rates, and that there is not a linear relationship between recharge rates and the amount of diverted floodwater. Other factors are involved such as seasonal fluctuations of the soil cover and the duration of floods (Hashemi et al. 2015). Modeled recharge rates varied between several hundred thousand to 45×10^6 m^3/month for the rainy season (Hashemi et al. 2015). Despite the FWS system, groundwater levels in the GBP were still declining due to an increase in the number of irrigation wells (Hashemi et al. 2015). The Iranian FWS systems are a low cost, passive technology (Hashemi et al. 2015). However, continuous maintenance is required to repair damage from erosion, consolidation (settling), and clogging of the systems (Hashemi et al. 2015).

16.9 On-Farm Flood Capture and Recharge (California)

Application of water to farm fields during non-irrigation periods at rates exceeding dormant period ET rates is being investigated in California as an MAR method (RMC 2015). Bachand et al. (2012, 2014) documented field testing of on-farm flood capture and recharge (OFFCR) on 1,000 acres (405 ha) of farm land (part of the Terranova Ranch) in the San Joaquin Valley of California. The study site was divided into 11 areas ("checks") separated by berms. Four types of fields were investigated: fallow (before summer row crops planted), alfalfa, pistachio orchards, and wine grapes. Recharge rates were reported to average 4.2 in./d (10.7 cm/d) in the checks, with a range between checks from an average (for multiple infiltration events) of 2.6 in./d (6.8 cm/d) to 16 in./d (40 cm/d). The standard practice of deep ripping of a cemented layer near the soil surface was found to increase infiltration rates. Infiltration rates asymptotically decreased over time from an initial rate of about 5 in./d (12.7 cm/d) to 2–3 in./d (5.1–7.6 cm/d) after 2 days of inundation, with only a slight further decrease over longer inundation. In the wine grape and pistachio fields, about 50–75% of the applied water was calculated to go directly to recharge.

The field data demonstrated that prolonged flushing increased the salinity in the root zone. The conceptual model developed for the site has repeated flooding causing a salinity and nitrate front (pulse) to migrate down the soil zone. The nitrate and salinity concentrations in the groundwater may initially increase followed by a long-term decline in concentrations due the very good quality of the floodwater (Bachand et al. 2012, 2014).

An economic analysis suggests that OFFCR is cost competitive with other recharge options. The estimated cost of over a 25-year period was \$36 (USD) per AF (1 AF = 1233.5 m^3), compared to a range of \$5–\$97 (median \$51) per AF for engineered basin systems (Bachand et al. 2014). Additional regional benefits would also occur from reduced flood damage. The low costs would be paid for through savings from reduced groundwater pumping (Bachand et al. 2012).

RMC (2015) modeled the water resources benefits of flooding agricultural lands with excess winter river flows in the San Joaquin Valley, California, using the California Department of Water Resources C2Vsim integrated surface water and groundwater model. It was estimated that on average 79,200–130,000 AF/year (97.7–160.3 MCM/year) could be diverted using existing infrastructure. Of this diversion, the modeled increase in storage in the project area is 31,000–52,000 AF/year (38.2–64.1 MCM/year). The remaining recharge will act to increase stream baseflow (34,200–55,500 AF/year; 42.2–68.5 MCM/year) and increase groundwater storage elsewhere in the San Joaquin Valley. The additional modelled aquifer recharge is equivalent to 12–20% of estimated overdraft of 250,000 AF/year (308.4 MCM/year).

Key issues for agricultural flooding are (RMC 2015):

- land suitability for recharge (infiltration rates)
- crop recharge suitability—ability of crops to tolerate ponded conditions for an extended period
- recharge water availability (physical and water rights).

OFFCR technology is attractive because of its low costs as it uses existing agricultural lands and infrastructure to divert and convey water to the lands.

The University of California, Davis, performed experiments to determine whether flooding of almond orchards can be used to restore groundwater flows. One gallon (3.79 L) of water is needed to grow a single almond. The goal of the experiments was to see if flooding of groves in the winter can help pull California out of its chronic groundwater overdraft. Unused water would otherwise be lost to tide. University of California, Davis, research indicate that on-farm flooding can quickly raise groundwater levels without damage to crops. Diverting water to farm fields also reduces downstream flooding (Stormwater Report 2016). Key issues are whether tree damage and seepage of fertilizers into the groundwater would occur and ownership of floodwater (Quinton 2016; Stormwater Report 2016).

16.10 Overbank Floodplain Recharge

Overbank floodplain recharge (OFR) is an often overlooked component of the water budget in continental and global-scale models (Doble et al. 2014). As reviewed by Doble et al. (2011), OFR occurs when river stages exceeds bank height and water flows in large sheets across low lying areas. It is distinctly different from bank storage and transmission loss processes. The results of fully coupled, surface water-groundwater modeling using the HydroGeoSphere code indicate that OFR is limited by (Doble et al. 2011, 2014):

(1) infiltration through the soil surface, which is controlled by inundation depth and area, duration of floods, and, most critically, the soil clogging layer
(2) available storage in the unsaturated zone, which is limited primarily by the depth to groundwater
(3) ability of the aquifer to transport water away from the flooded areas, which is a function of the transmissivity of the aquifer
(4) area of inundation and stream depth (stage of flooding)
(5) presence of local depressions, which extend the time of inundation beyond the time when the flood wave retreats back to within the river banks.

Modeling of overbank flood recharge in Australia indicates that OFR in one catchment (Loddon Catchment) constitutes at least 4% of the total change in storage and 15% of the riparian recharge over the model period (Doble et al. 2014). These proportions are likely underestimated by the method used (Doble et al. 2014). Accurate estimation of OFR remains a challenge. Modeled OFR (using a fully couple surface water-groundwater numerical model) was found to under estimate OFR compared to point recharge measurements and estimates based on catchment-scale changes in groundwater storage.

The University of California Water Security and Sustainability Research Initiative (WASSRI) is investigating the benefits of moving levees back from river channels to increase the width of floodplains and thus area available for groundwater recharge (University News 2015). The main constraint is high costs; considerable time and money are involved in environmental permitting. Initial results suggest that moving levees can reduce flood risks to neighbors, allow row-crop agriculture to persist on portions of the flood plains, and increase recharge of the local aquifer.

References

Abdalla, O. A. E., & Al-Rawahi, A. S. (2013). Groundwater recharge dams in arid regions as tools for aquifer replenishment and mitigating seawater intrusion: Example of AlKhod, Oman. *Environmental Earth Sciences, 69,* 1951–1962.

Abu-Taleb, M. F. (2003). Recharge of groundwater through multi-stage reservoirs in a desert basin. *Environmental Geology, 44*(4), 379–390.

Aerts, J., Lasage, R., Beets, W., de Moel, H., Mutiso, G., Mutiso, S., & de Vries, A. (2007). Robustness of sand storage dams under climate change. *Vadose Zone Journal, 6*(3), 572–580.

Agarwal, A., & Narain, S. (1997) *Dying wisdom. Rise, fall and potential of India's traditional water harvesting systems*. New Delhi: India Centre for Science and Environment.

Agoramoorthy, G., Ghaudhary, S., & Hsu, K. J. (2008). The check-dam route to mitigate India's water shortages. *Natural Resources Journal, 48,* 565–583.

Alderwish, A. M. (2010). Induced recharge at new dam sites—Sana'a Basin, Yemen. *Arabian Journal of Geosciences, 3*(3), 283–293.

Al-Hassoun, S., & Al-Turbak, A. S. (1995). Recharge dam efficiency based on subsurface flow analysis. *Water International, 20,* 40–45.

Al-Kindi, R. S. (2004). *Assessment of groundwater recharge from the dam of Wadi Al-Jizzi, Sultanate of Oman (M.Sc. thesis)*. United Arab Emirates University, Al Ain, Abu Dhabi.

Al-Muttair, F. F., Sendil, U., & Al-Turbak, A. S. (1994). Management of recharge dams in Saudi Arabia. *Journal of Water Resources Planning and Management, 120,* 749–763.

Al-Othman, A. (2011). Enhancing groundwater recharge in arid regions—A case study from central Saudi Arabia. *Scientific Research and Essays, 6*(13), 2757–2762.

Al-Turbak, A. S. (1991). Effectiveness of recharge from surface reservoir to an underlying unconfined aquifer. In G. Schiller, R. Lemmela, & M. Spreafico (Eds.), *Hydrology of natural and man-made lakes. Proceedings of the Vienna Symposium, August, 1991*, Publication No. 206 (pp. 191–196). London: International Associations of Hydrological Sciences (IAHS).

Al-Turbak, A. S., & Al-Muttair, F. F. (1989). Evaluation of dams as a recharge method. *Water Resources Development, 5*(2), 119–124.

Bachand, P. A. M., Horwathm, W. R., Roy, S., Choperena, J., & Cameron, D. (2012). *Implications of using on-farm flow capture to recharge groundwater and mitigate flood risks along the Kings River, CA*. Report prepared for the U.S. Department of Agriculture, Bachand & Associates.

Bachand, P. A., Roy, S. B., Choperena, J., Cameron, D., & Horwath, W. R. (2014). Implications of using On-Farm Flood Flow Capture to recharge groundwater and mitigate flood risks along the Kings River, CA. *Environmental Science & Technology, 48*(23), 13601–13609.

Bajjali, W. (2005). Model the effect of four artificial recharge dams on the quality of groundwater using geostatistical methods in GIS environment, Oman. *Journal of Spatial Hydrology, 5*(2), 1–15.

Baurne, G. (1984). Trap dams: Artificial subsurface storage of water. *Water International, 9,* 2–9.

Besbes, M., Delhomme, J. P., & de Marsily, G. (1978). Estimating recharge from ephemeral streams in arid regions: A case study a Kairouan, Tunisia. *Water Resources Research, 14,* 281–290.

Blasch, K., Ferré, T., Hoffmann, J., Pool, D., Bailey, M., & Cordova, J. (2004). Processes controlling recharge beneath ephemeral streams in southern Arizona. In. J. F. Hogan, F. M. Phillips, & B. R. Scanlon (Eds.), *Groundwater recharge in a desert environment: The southwestern United States* (pp. 69–79). Washington, DC: American Geophysical Union.

Bleich, V. C., & Weaver, R. A. (1983). Improved sand dams for wildlife habitat management. *Journal of Range Management, 36*(1), 133.

Bligh, W. H. (1915). *Dams and weirs. An analytical and practical treatise on gravity dams and weirs; arch and buttress dams; submerged weirs; and barrages*. Chicago: American Technical Society.

Boisson, A., Baisset, M., Alazard, M., Perrin, J., Villesseche, D., Dewandel, B., et al. (2014). Comparison of surface and groundwater balance approaches in the evaluation of managed aquifer recharge structures: Case of a percolation tank in a crystalline aquifer in India. *Journal of Hydrology, 519,* 1620–1633.

Bureau of Reclamation. (1987). *Design of small dams* (3rd ed.). Washington, DC: U.S. Department of the Interior Bureau of Reclamation.

Cataldo, J., & Pierce, R. J. (2005). *An analysis of transmission losses in ephemeral streams: A case study in Walnut Gulch Experimental Watershed, Tombstone, AZ*. A Report to the National Center for Housing and the Environment.

Consultants, G. W. P. (2010). *Managed aquifer recharge (MAR): Practical techniques for the Caribbean*. Charlbury, UK: GWP Consultants LLP.

Davis, T. (2016a, February 27) River recharge efforts under a disadvantage in state law. *Arizona Daily Star*.

Davis, T. (2016b, November 11). Santa Cruz River could flow again through downtown Tucson. *Arizona Daily Star*.

de Laat, P. J., & Nonner, J. C. (2012). Artificial recharge with surface water; a pilot project in Wadi Madoneh, Jordan. *Environmental Earth Sciences, 65*(4), 1251–1263.

Doble, R., Crosbie, R., Peeters, L., Joehnk, K., & Ticehurst, C. (2014). Modelling overbank flood recharge at a continental scale. *Hydrology and Earth System Sciences, 18,* 1273–1288.

Doble, R., Crosbie, R., Smerdon, B., Peeters, L., Chan, F., Marinova, D., & Andersson, R. S. (2011). Examining the controls on overbank flood recharge for improved estimates of national water accounting. In *Proceedings of the 19th International Congress of Modelling and Simulation*, Perth, Australia, 12–16 December 2001 (pp. 3931–3937).

Dong, D., Liu, B., Liu, H., Wang, S., & Wang, L. (2014). Impacts of groundwater recharge from rubber dams on the hydrogeological environment on Luoyang Basin, China. *The Scientific World Journal, 2014*, Article 183457.

Ertsen, M., & Hut, R. (2009). Two waterfalls do not hear each other. Sand-storage dams, science and sustainable development in Kenya. *Physics and Chemistry of the Earth, 34,* 14–22.

Fennessey, L., Hamlett, J. M., & Cooper, R. (n.d.). *Installation and hydrologic response of the Penn State low head weirs (PowerPoint).* University Park, PA: The Pennsylvania State University. Retrieved January 2, 2017 from http://www.opp.psu.edu/services/eng-resources/weirs-presentation.pdf/view.

Fennessey, L., Hamlett, J. M., & Ralston, M. A. (2009). Low-head weirs for enhancing infiltration: Central Pennsylvania case study. *Journal of Hydrologic Engineering, 14*(8), 862–871.

Ferreira, J. P. L., & Leitão, T. (2014). Demonstrating managed aquifer recharge as a solution for climate change adaptation: Results from Gabardine project and ASEMWaterNet coordination action in the Algarve region (Portugal). *Acque Sotterranee—Italian Journal of Groundwater, AS10040*, 15–22.

Gale, I. N., Macdonald, D. M. J., Calow, R. C., Neumann, I., Moench, M., Kulkarni, H., et al. (2006). *Managed aquifer recharge: An assessment of its role and effectiveness in watershed management*. Final report for DFID KAR project R8169, Augmenting groundwater resources by artificial recharge. Keyworth, Nottingham: AGRAR, British Geological Survey.

Galyean, K. C. (1996). *Infiltration of Wastewater Effluent in the Santa Cruz River Channel, Pima County, Arizona*. U.S. Geological Survey Water-Resources Investigations Report 96-4021.

Ghahari, G., Hashemi, H., & Berndtsson, R. (2014). Spate irrigation of barley through floodwater harvesting in the Gareh-Bygone Plain, Iran. *Irrigation and Drainage, 63*(5), 599–611.

Ghebremariam, B. H., & Van Steenbergen, F. (2007). Agricultural water management in ephemeral rivers: Community management of spate irrigation in Eritrea. *African Water Journal, 1*(1), 48–65.

Glendenning, C. J., van Ogtrop, F. F., Mishra, A. K., & Vervoort, R. W. (2012). Balancing watershed and local scale impacts of rain water harvesting in India—A review. *Agricultural Water Management, 107,* 1–13.

Hashemi, H., Berndtsson, R., & Persson, M. (2013a). Floodwater harvesting for artificial recharge and spate irrigation in arid area. *VATTEN—Journal of Water Management and Research, 69*, 93–103.

Hashemi, H., Berndtsson, R., & Persson, M. (2015). Artificial recharge by floodwater spreading estimated by water balances and groundwater modelling in arid Iran. *Hydrological Sciences Journal, 60*(2), 336–350.

Hashemi, H., Berndtsson, R., Kompani-Zare, M., & Persson, M. (2013b). Natural vs. artificial groundwater recharge, quantification through inverse modeling. *Hydrology and Earth System Sciences, 17,* 637–650.

Hassan, R. (2017, June 28). Wadi Hanifa: A desert oasis running through Riyadh. *Arab News*.

Heilweil, V. M., & Marston, T. M. (2011). *Assessment of managed aquifer recharge from Sand Hollow Reservoir, Washington County, Utah, updated to conditions in 2010*. U.S. Geological Survey Scientific Investigations Report 2011-5142.

Heilweil, V. M., & Martsen, T. M. (2013). Evaluation of potential gas clogging associated with managed aquifer recharge from a spreading basin, southwestern Utah, USA. In R. Martin (Ed.), *Clogging issues associated with managed aquifer recharge methods* (pp. 84–94). Australia: IAH Commission on Managing Aquifer Recharge.

Hudson, H. E., Jr., Brown, C. B., Shaw, H. B., & Longwell, J. S. (1949). Problem of reservoir siltation. *Journal American Water Works Association, 41*(10), 913–915.

Hut, R., Ertsen, M., Joeman, N., Vergeer, N., Winsemius, H., & van de Giesen, N. (2008). Effects of sand storage dams on groundwater levels with examples from Kenya. *Physics and Chemistry of the Earth, 33,* 56–66.

Ibn Ali, Z. I., Triki, I., Lajili-Ghezal, L., & Zairi, M. (2017). A method to estimate aquifer artificial recharge from a hill dam in Tunisia. *Journal of Arid Land, 9*(2), 244–255.

Ibrahim, M. B. (2009). Rainwater harvesting for urban areas: A success story from Gadarif City in central Sudan. *Water Resources Management, 23,* 2727–2736.

Kazemi, G. A., Mehdizadeh, H., & Ashouri, R. (2002). Evaluation of groundwater recharge schemes in the Shahrood District, Northeast Iran. In P. J. Dillon (Ed.), *Management of aquifer recharge for sustainability* (pp. 225–229). Lisse: A.A. Balkema.

Keller, G., & Sherar, J. (2003). *Low-volume roads engineering. Best management practices field guide* (Produced for U.S. Agency for International Development). Washington, DC: U.S.D.A. Forest Service.

Kiela, H. C. (2010). *The application of vertical infiltration systems (VIS) in the Sultanate of Oman (Master's thesis).* Universiteit Utrecht.

Lamaye, S. D. (2002). Aquifer recharge management for sustainable water supply in semi-arid, hard rock region of western India. In P. J. Dillon (Ed.), *Management of aquifer recharge for sustainability* (pp. 309–313). Lisse: A.A. Balkema.

Lane, L. J., Diskin, M. H., & Renard, K. G. (1971). Input-output relationship for an ephemeral stream channel system. *Journal of Hydrology, 13,* 22–40.

Lasage, R., Aerts, J., Mutiso, G.-C. M., & de Vries, A. (2008). Potential for community based adaptation to drought: Sand dams in Kitui, Kenya. *Physics and Chemistry of the Earth, 33,* 67–73.

Lasage, R., Aerts, J. C., Verburg, P. H., & Sileshi, A. S. (2015). The role of small scale sand dams in securing water supply under climate change in Ethiopia. *Mitigation and Adaptation Strategies for Global Change, 20*(2), 317–339.

Lawrence, P. (2009). *Managing sedimentation in spate irrigation schemes.* Lecture note, short course on spate irrigation and water management under water scarcity, Delft, The Netherlands, UNESCO-IHE Institute for Water Education.

MAF, & ICBA. (2012). *Oman salinity strategy, Annex 1, Assessment of salinity problem.* Dubai: Ministry of Agriculture and Fisheries (MAF) and Sultanate of Oman International Center for Biosaline Agriculture (ICBA).

Maliva, R. G., & Missimer, T. M. (2012). *Arid lands water evaluation and management.* Berlin: Springer.

Markus, M. R., Thompson, C. A., & Ulukaya, M. (1995). *Enhanced artificial recharge utilizing inflatable rubber dams.* In A. I. Johnson & R. D. G. Pyne (Eds.), *Artificial recharge of ground water II. Proceedings of the Second International Symposium on Artificial Recharge of Ground Water* (pp. 129–128). Reston, VA: American Society of Civil Engineers.

Markus, M. R., Thompson, C. A., & Ulukaya, M. (2000). Aquifer recharge enhanced with rubber dam installation. *Roads & Bridges*, December 28, 2000. http://www.roadsbridges.com/aquifer-recharge-enhanced-ruber-dam-installations.

Marston, T. M., & Heilweil, V. M. (2016). *Assessment of managed aquifer recharge at Sand Hollow Reservoir, Washington County, Utah, updated to conditions through 2014.* U.S. Geological Survey Open-File Report 2016-1078.

Martín-Rosales, W., Gisbert, J., Pulido-Bosch, A., Vallejos, A., & Fernández-Cortés, A. (2007). Estimating groundwater recharge induced by engineering systems in a semiarid area (southeastern Spain). *Environmental Geology, 52*(5), 985–995.

Massuel, S., Perrin, J., Mascre, C., Mohmamed, W., Boisson, A., & Ahmed, S. (2014). Managed aquifer recharge in South India: What to expect from small percolation tanks in hard rock? *Journal of Hydrology, 512,* 157–167.

Maxwell, C., & Davidson, R. (2016). Supporting-ecological resiliency through managed aquifer recharge practices in the Southwestern United States. In *International Symposium on Managed Aquifer Recharge (ISMAR9)*, Mexico City, June 20–24, 2016.

McGuinness, J. L., & Bordne, E. F. (1972). *A Comparison of lysimeter-derived potential evapotranspiration with computed values*. Technical Bulletin 1452. Washington, D.C.: Agricultural Research Service, US Department of Agriculture.

Mehari, A., van Steenbergen, F., & Schultz, B. (2007). Water rights and rules and management in spate irrigation systems in Eritrea, Yemen and Pakistan. In B. Van Koppen, M. Giordano, & J. Butterworth (Eds.), *Community-based water law and water resource management reform in developing countries. Vol. 5: Comprehensive Assessment of Water Management in Agriculture Series* (pp. 114–129). Wallingford, UK, CAB International.

Mehari, A., Van Steenbergen, F., & Schultz, B. (2011). Modernization of spate irrigated agriculture: A new approach. *Irrigation and Drainage, 60*(2), 163–173.

Ministry of Regional Municipalities, & Environment and Water Resources. (2005). *Water resources of Oman*. Ministry of Regional Municipalities, Environment and Water Resources, Sultanate of Oman.

Missimer, T. M., Guo, W., Maliva, R. G., Rosas, J., & Jadoon, K. Z. (2015). Enhancement of wadi recharge using dams coupled with aquifer storage and recovery wells. *Environmental Earth Sciences, 73*(12), 7723–7731.

Muralidharan, D., Andrade, R., & Rangarajan, R. (2007). Evaluation of check-dam recharge through water-table response in ponding area. *Current Science, 92*(10), 1350–1352.

Mutiso, S. (2003). The significance of subsurface water storage in Kenya. In A. Tuinhof, & J. P. Heederick (Eds.), *Management of aquifer recharge and subsurface storage*, Publication No. 4 (pp. 25–31). Netherlands National Committee—International Association of Hydrogeologists (NNC-IAH).

Newman, B. D., Vivoni, E. R., & Groffman, A. R. (2006). Surface water—Groundwater interactions in semiarid drainages of the American southwest. *Hydrological Processes, 20,* 3371–3394.

Nilsson, A. (1988). *Groundwater dams for small-scale water supply*. Bourton-on-Dunsmore, UK: ITDG Publishing.

Norman, L. M., Brinkerhoff, F., Gwilliam, E., Guertin, D. P., Callegary, J., Goodrich, D. C., et al. (2016). Hydrologic response of streams restored with check dams in the Chiricahua Mountains, Arizona. *River Research and Applications, 32*(4), 519–527.

Oaksford, E. T. (1985). Artificial recharge: Methods, hydraulics, and monitoring. In T. Asano (Ed.), *Artificial recharge of groundwater* (pp. 69–127). Stoneham, MA: Butterworth Publishers.

Pandey, D. N., Guota, A. K., & Anderson, D. M. (2003). Rainwater harvesting as an adaptation to climate change. *Current Science, 85*(1), 46–59.

Parimalarenganayaki, S., & Elango, L. (2013). A review of managed aquifer recharge by check dams: A case study near Chennai, India. *International Journal of Research in Engineering and Technology (IJRET), 2,* 416–423.

Parimalarenganayaki, S., & Elango, L. (2014). Impact of recharge from a check dam on groundwater quality and assessment of suitability for drinking and irrigation purposes. *Arabian Journal of Geosciences, 7,* 3119–3129.

Parimalarenganayaki, S., & Elango, L. (2015). Assessment of effect of recharge from a check dam as a method of managed aquifer recharge by hydrogeological investigations. *Environmental Earth Sciences, 73,* 5349–5361.

PFRA. (1992). *Small dam design and construction manual*. Agriculture Canada, Prairie Farm Rehabilitation Administration, Prairie Resources Service.

Prathapar, S. A., & Bawain, A. A. (2014). Impact of sedimentation on groundwater recharge at Sahalanowt Dam, Salalah, Oman. *Water International, 39*(3), 381–393.

Prietto, J. (2013). *Stakeholder incentives for effluent utilization in the Tucson Metropolitan Region and recharge in the Santa Cruz River*. Central Arizona Project. https://www.cap-az.com/documents/education/2013-Prietto.pdf.

Quilis, R. O., Hoogmoed, M., Ertsen, M., Foppen, J. W., Hut, R., & de Vries, A. (2009). Measuring and modeling hydrological processes of sand-storage dams on different spatial scales. *Physics and Chemistry of the Earth, 34,* 289–298.

Quinton, A. (2016). *Scientists flood almond orchards to restore groundwater in California*. Capital Public Radio, January 20, 2016. Retrieved July 1, 2016 from http://www.capradio.org/articles/2016/01/20/scientists-flood-almond-orchards-to-restore-groundwater-in-california.

Radhakrishna, B. P. (2004). Man-made drought and the looming water crisis. *Current Science, 87*(1), 20–22.

Raghu, V., & Reddy, M. K. (2011). Hydrogeomorphological mapping at village level using high resolution satellite data and impact analysis of check dams in part of Akuledu Vanka Watershed, Anantapur District, Andhra Pradesh. *Journal Indian Geophysical Union, 15*(1), 1–8.

RAIN. (2008). *A practical guide to sand dam implementation*. Amsterdam: Rainwater Harvesting Implementation Network.

Regan, M. (2001, May 3). A river ran through it. *Tucson Weekly*.

RMC. (2015). *Creating an opportunity: Groundwater recharge through winter flooding of agricultural land in the San Joaquin Valley*. Summary Report (October 2015). RMC Water and Environment.

Sakthivadivel, R. (2007). The groundwater recharge movement in India. In M. Giordano & K. G. Villholth (Eds.), *The agricultural groundwater revolution: Opportunities and threats to development* (pp. 195–210). Wallingford, UK: CAB International.

Salem, S. B. H., Chkir, N., Zouari, K., Cognard-Plancq, A. L., Valles, V., & Marc, V. (2012). Natural and artificial recharge investigation in the Zéroud Basin, central Tunisia: Impact of Sidi Saad Dam storage. *Environmental Earth Sciences, 66,* 1099–1100.

Samani, A., Khalighi, S., Arabkhedri, M., & Farzadmehr, J. (2014). Indigenous knowledge and techniques of runoff harvesting (Bandsar and Khooshab) in Arid and Semi Arid Regions of Iran. *Journal of Water Resource and Protection, 6*, 784–792.

Scanlon, B., Angle, E. S., Christian, B., Pi, J., Martinez, K., Reedy, R., et al. (2003). *Evaluation of enhanced recharge potential to the Ogallala Aquifer in the Brazos River Basin, Hale County, Texas*. Report 355. Austin: Texas Water Development Board.

Shanafield, M., & Cook, P. G. (2014). Transmission losses, infiltration and groundwater recharge through ephemeral and intermittent streambeds: A review of applied methods. *Journal of Hydrology, 511,* 518–529.

Sherif, M. M., Al Mulla, M. M., & Ebraheem, A. A. (2011). Assessment of the effectiveness of dams on the recharge of the shallow aquifer in the northern area of UAE. In R. E. Beighley & M. W. Killgore (Eds.), *Proceedings World Environmental and Water Resources Congress 2011: Bearing Knowledge for Sustainability, May 22–26, Palm Springs, CA* (pp. 813–822). Reston, VA: American Society of Civil Engineers.

Sivils, B. E., & Brock, J. H. (1981). Sand dams as a feasible water development for arid regions. *Journal of Range Management, 34*(3), 238–239.

Sonoma County Water Agency. (2017). *Water supply*. Retrieved October 18, 2017 from http://www.scwa.ca.gov/water-supply/.

Sorman, A. U., & Abdulrazzak, M. J. (1993). Infiltration-recharge through wadi beds in arid regions. *Hydrological Sciences—Journal des Sciences Hydrologiques, 38*(3), 173–186.

Stephens, D. B. (1996). *Vadose zone hydrology*. Boca Raton: CRC Press.

Stephens, T. (2010). *Manual on small earth dams. A guide to siting, design and construction* (FAO Irrigation and Drainage Paper 64). Rome: Food and Agriculture Organization of the United Nations.

Stern, J. H., & Stern, A. (2011). *Water harvesting through sand dams*. North Fort Myers: ECHO.

Stormwater Report. (2016). *Farmers flood fields to replenish groundwater*. Water Environment Federation Stormwater Report. Retrieved July 1, 2016 from http://stormwater.wef.org/2016/01/farmers-flood-fields-replenish-groundwater/.

Sukhija, B. S. (2008). Adaptation to climate change: Strategies for sustaining groundwater resources during droughts. In W. Dragoni & B. S. Sukhija (Eds.), *Climate change and groundwater*. Special Publication 288 (pp. 169–181). London: Geological Society.

Thomure, T., & Wilson, W. R. (2003). Managed (in-channel) recharge of effluent in the Tucson Active Management Area, Arizona. In *11th Biennial Symposium on Groundwater Recharge*, Tempe, AZ, June 5–7, 2003.

Tosline, D., Prudhom, B., Falcon, D., Holler, F. E., & Romero-Krzystek, D. (2012). *Enhanced recharge demonstration project increasing effluent recharge rates in the Santa Cruz River, Tucson, Arizona*. Phoenix: U.S. Bureau of Reclamation.

University News. (2015). *Floodplain management can increase groundwater supply*. University News, University of California—Merced. Retrieved July 1, 2016 from http://www.ucmerced.edu/news/2015/floodplain-management-can-increase-groundwater-supply.

USACOE. (2004). *General design and construction considerations for earth and rock-fill dams*. Washington, D.C.: U.S. Army Corps of Engineers.

USBR. (2009). *Arroyos Ground-Water Recharge Project, San Xavier District, final environmental assessment*. Phoenix: U.S. Bureau of Reclamation.

USDA. (2007). Chapter 19 Transmission losses, Part 630. In *National engineering handbook*. United States Department of Agriculture, Natural Resources Conservation Service. https://directives.sc.egov.usda.gov/OpenNonWebContent.aspx?content=17756.wba.

Van Haveren, B. P. (2004). Dependable water supplies from valley alluvium in arid regions. *Environmental Monitoring and Assessment, 99,* 259–266.

Van Steenbergen, F., Lawrence, P., Haile, A. M., Salman, M., & Faurè, J. M. (2010). *Guidelines on spate irrigation*. FAO Drainage Paper 65. Rome: Food and Agricultural Organization of the United Nations.

Walters, M. O. (1990). Transmission losses in arid regions. *Journal of Hydraulic Engineering, 116,* 129–138.

Wheater, H. S. (2002). Hydrological processes in arid and semi arid area. In H. Wheater & R. A. Al-Weshah (Eds.), *Hydrology of wadi systems*, IHP-V, Technical documents in hydrology, no. 55 (pp. 5–22). Paris, UNESCO.

Zeelie, S. (2002). OMDEL dam and recharge ponds to enhance recharge in the Namib Desert. In P. J. Dillon (Ed.), *Management of aquifer recharge for sustainability* (pp. 387–392). Lisse: A.A. Balkema.

Chapter 17
Vadose Zone Infiltration Systems

17.1 System Types, Advantages, and Disadvantages

Surface infiltration systems are generally preferred for managed aquifer recharge (MAR) because they offer the best opportunity for clogging control and water quality improvements through contaminant attenuation processes in the vadose zone (Bouwer 2002). Recharge into the saturated zone using wells can be an attractive option where a storage zone is present with a sufficiently high transmissivity to accept target flow rates. Equivalent horizontal hydraulic conductivity values of strata are often orders of magnitude greater than equivalent vertical hydraulic conductivity values, hence greater aquifer recharge rates can, in some instances, be obtained by injecting or infiltrating directly into an aquifer or vadose zone rather than by surface spreading.

Vadose zone infiltration systems, by definition, release water below land surface and above the water table and include:

- infiltration trenches (including French drains)
- infiltration galleries
- infiltration shafts and pits (including soakaways)
- dry (vadose) wells.

Vadose zone infiltration systems have the advantages of:

- bypassing low permeability material present at or close to land surface
- having largely subsurface constructions, which allow for the use of overlying land areas
- providing some temporary water storage
- providing an opportunity for contaminant attenuation processes in the unsaturated zone to occur
- lower costs than phreatic injection well systems
- often being more readily permittable than phreatic injection well systems
- a lesser potential for unsupervised or uncontrolled contact with the recharged water
- no potential for mosquito breeding and other nuisances.

R. G. Maliva, *Anthropogenic Aquifer Recharge*,
Springer Hydrogeology, https://doi.org/10.1007/978-3-030-11084-0_17

The main disadvantage of vadose zone infiltration systems is that they are prone to clogging and clogging remediation is more difficult to perform than for surface-spreading and phreatic zone injection wells. Clogging layers of surface-spreading systems typically occur at land surface and are thus relatively accessible. Phreatic injection wells can be rehabilitated by pumping and surging (along with other techniques), which are often not possible in vadose zone infiltration systems.

Vadose zone infiltration systems are most commonly used for stormwater management, and much information on their design, operation, and maintenance is found in governmental stormwater management design and guidance documents. The systems can involve either unmanaged or managed aquifer recharge depending on whether they have a primary disposal function or are intended to augment local groundwater resources.

17.2 Infiltration (Recharge) Trenches

17.2.1 Infiltration Trench Basics

A trench is generally defined as a long, narrow excavation in the ground. The United States Occupational Safety and Health Administration (OSHA) more specifically defines a trench as "a narrow underground excavation that is deeper than it is wide, and is no wider than 15 ft (4.5 m)" (OSHA, n.d.). The main advantages of trenches are a small surface footprint, the ability to bypass shallow impermeable layers, and that suspended sediments tend to settle on bottom with vertical walls remaining relatively free of sediment. Trenches also provide temporary storage while water infiltrates into soils.

Infiltration trench design and regulatory requirements in the United States are addressed with respect to stormwater systems by the USEPA (1999a) and city, county and state stormwater management design manuals. Stormwater trenches are designed to infiltrate episodic runoff from rain fall events rather than continuous flows. They are also intended to provide water quality improvement benefits through contaminant attenuation process that occur during infiltration into the trench and exfiltration into the surrounding and underlying soils. The limited storage capacity of trenches allows them to provide runoff quantity control for only small events.

Infiltration trenches fail if they receive large sediment loads and, therefore, erosion control or pretreatment are critical. Schueler et al. (1992) reported that approximately fifty percent of stormwater infiltration trenches have partially or completely failed within five years. In an early survey of the performance of infiltration trenches in Maryland, Lindsey et al. (1991) reported that 53% of trenches were operating as designed, 36% were partially or totally clogged, and 22% showed slow infiltration. The USEPA (1999a) noted that trench rehabilitation may be required every 5–15 years.

The main types of infiltration trenches are (Hannon 1980, and others):

- unsupported open cuts with side slopes
- vertical open-sided excavations where side support is not necessary (e.g., trenches excavated in rock) and the trench is covered with a slab
- excavations backfilled with aggregate (gravel)
- excavations in which plastic-crate systems are used instead of (or in addition to) aggregate backfill.

A conceptual diagram of a typical gravel-filled infiltration trench is provided in Fig. 17.1. Water is applied through screened (slotted or perforated) pipe. The gravel (aggregate) fill serves mainly to support the trench (i.e., prevent collapse of the trench wells) and to provide water storage. The discharge pipe and gravel fill should be designed to equalize water levels within the trench. Head losses within the pipe and gravel should not result in mounding at one end of the trench. Geotextile fabric liners can be used to separate the fill and native aquifer material. Where the surficial sediments are unconsolidated, temporary sheet pilings may be required to prevent collapse of walls during excavation and construction.

Several infiltration trench design modifications have been proposed. Bouwer (2002) proposed a seepage trench design variation in which a trench is "T" shaped to provide a greater surface area for infiltration into the trench. Finer material is placed in the upper, wider "T" layer to obtain better removal of suspended solids. A modification of aggregate-filled trench systems is the emplacement of large-diameter perforated steel or concrete pipe within a trench to increase its storage capacity to handle large-volume rainfall events (Hannon 1980). The large diameter pipe provides an economic benefit by reducing the aggregate requirement.

Open trenches must be self-supporting under a given load and can be covered with slabs of concrete, steel, or aluminum without the need to backfill. Where possible, open trenches have the advantages of a greater storage capacity for a given trench volume and more ready access for rehabilitation. Rehabilitation of clogged aggregate-filled trenches usually involves reconstruction (excavation and replacement or cleaning of the aggregate). Since trenches are relatively inexpensive to construct, they can be replaced when their useful life comes to an end.

17.2.2 Stormwater Infiltration Trench Design

The USEPA (1999a) summarized basic design issues for infiltration trenches used for stormwater management. Infiltration trenches are used to capture and treat small amounts of runoff (first flush), but do not have the capacity to control peak hydraulic flows. Instead, infiltration trenches have been used in conjunction with detention ponds to provide both water quality and peak flow control. The principle operational challenge impacting the operation of infiltration trenches is management of clogging and the potential for groundwater contamination. Runoff that may contain high concentrations of sediments and hydrocarbons (oil and grease) that could clog a trench is

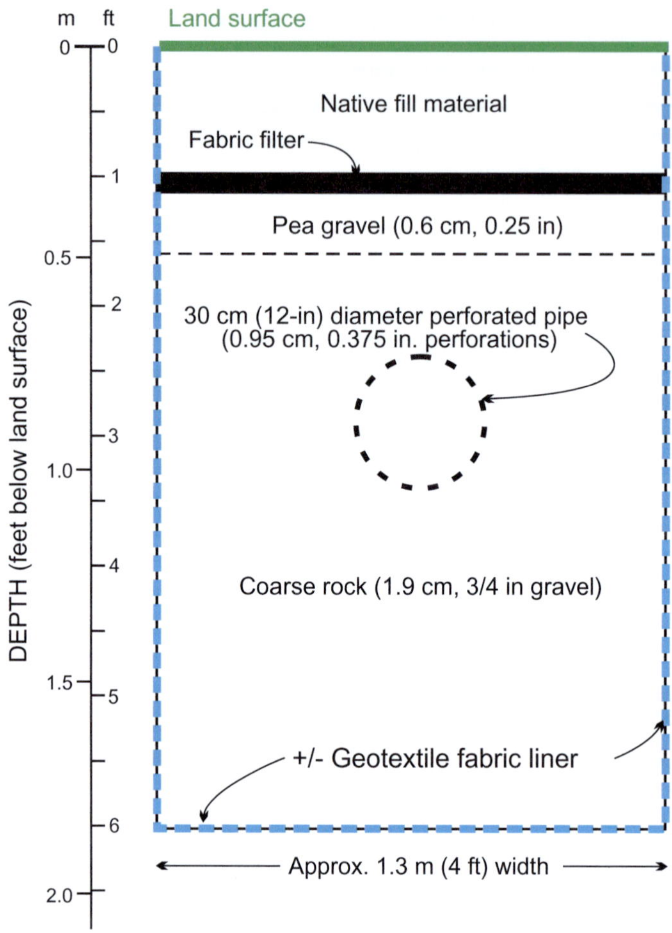

Fig. 17.1 Example of a basic infiltration trench design

often pretreated using grit chambers, sediment traps, swales, or vegetated filter strips. Reported typical pollutant removal efficiencies for stormwater infiltration trenches are provided in Table 17.1

A basic stormwater infiltration trench design is an excavation 3–12 ft (0.9–3.7 m) deep, backfilled with a stone aggregate (gravel) storage media, and lined with a non-woven geotextile filter fabric (USEPA 1999a). The filter fabric acts to prevent sediments from the sides of a trench from clogging the aggregate. A replaceable filter fabric may be placed 6–12 in. (15–30 cm) below ground surface, above the aggregate, to prevent (or minimize) suspended solids from clogging the storage media from above.

Trenches may be designed for either diffuse input (i.e., by infiltration through the top of the trench) or a concentrated input. Slotted pipe in the aggregate/gravel can both

Table 17.1 Typical pollutant removal efficiencies of stormwater trenches

Pollutant	Typical removal efficiency (%)
Sediment	90
Total phosphorous	60
Total nitrogen	60
Metals	90
Bacteria	90
Organics	90
Biochemical oxygen demand	70–80

Source Schueler et al. (1992), USEPA (1999a)

distribute water throughout the trench and provided additional storage. For diffuse input, runoff can be captured by depressing the trench surface or by constructing a berm downgradient of the trench (USEPA 1999a). Trenches can be topped with top soil and sod. A pea-gravel layer placed above the stone aggregate can improve both filtration and pollutant removal (Fig. 17.1). The pea gravel is separated from the underlying aggregate (coarse gravel) using a geotextile layer. Clogging can be remediated by removing and replacing the pea gravel layer and geotextile (USEPA 1999a; Lowndes 2000; Atlanta Regional Commission 2001). A layer of organic matter placed above trenches appears to enhance the removal of metals and nutrients (USEPA 1999a).

Follows are some basic design recommendations for stormwater infiltration trenches (USEPA 1999a; Lowndes 2000; Atlanta Regional Commission 2001; Metropolitan Council 2001; VDOT 2013):

- The bottom of the trench should be located sufficiently above the water table (minimum of 1 m or 3 ft) to allow for filtration in the soil zone. Areas with shallow water tables are not suitable.
- Soil infiltration rates should be 0.5 in./h (1.3 cm/h) or greater. Areas with relatively impermeable soils, sediment, and rock should be avoided.
- Gravel/aggregate should be clean (washed) and not contain soil. The recommended aggregate size is 1.5–3.5 in. (4–9 cm).
- A ±6-in. (15-cm) sand layer at the bottom of a trench can improve drainage and minimize compaction of the underlying soil when gravel is added.
- Plastic sheets may be placed against clay zones.
- Low hydraulic conductivity layers below a trench may limit infiltration if they are not excavated.
- Areas with shallow bedrock may not be suitable due to greater excavation costs.
- At least one observation well should be installed within the trench and one outside of the trench.
- Locations with fine-grained soil types that may be mobilized and transported into trenches should be avoided. Areas with steep slopes also tend to be unsuitable for infiltration trenches.

- Trenches are not appropriate for commercial and industrial sites where large contaminant releases are possible.
- In cold climates, part of the trench should be constructed below the frost line.
- A minimum drainage time of 6 h should be provided to ensure satisfactory pollutant removal.
- Trenches should drain prior to the next storm event (maximum drainage time should be 72 h).
- Construction techniques should avoid smearing of the trench wall and over-compaction of the aggregate (storage media) and surrounding soil. The sides of the trench should be scarified before emplacement of the gravel.
- Trenches should be located away from shallow wells used for potable water supply and from building foundations.
- A bypass system is needed to convey high flows (in excess of the storage capacity) around the trench.
- Vegetative buffers around trenches can minimize clogging.

Stormwater trench design is based on the design volume of water to be initially stored and the percolation rate through the bottom of the trench. Commonly used area calculations are conservative in that they do not consider infiltration through the sides of the trench. Infiltration rates decreases over each operational event as water level in the trench and wetted perimeter area progressively decrease. Over-designing trenches (by not considering lateral flow) provides a safety factor for declines in infiltration rates due to clogging.

The design equation, using metric units

$$A = \frac{100V}{Knt} \tag{17.1}$$

where

A area of the bottom of the trench (m^2)
V design runoff volume to be filtered (m^3)
K percolation rate (cm/h)
n porosity of the storage media (fractional)
t retention time (hours, maximum 72 h).

For U.S. customary units

$$A = \frac{12V}{Knt} \tag{17.2}$$

A area of the bottom of the trench (ft^2)
V design runoff volume to be filtered (ft^3)
K percolation rate (in/h)
n porosity of the storage media (fractional)
t retention time (hours, maximum 72 h).

Variations of Eqs. 17.1 and 17.2 that consider storage are (Atlanta Regional Commission 2001):

Metric units:

$$A = \frac{V}{nd + Kt_f/100} \tag{17.3}$$

U.S. customary units:

$$A = \frac{V}{nd + Kt_f/12} \tag{17.4}$$

d trench depth (m or ft)
t_f time for trench to fill with water (hours).

A porosity value of 0.31 and fill time of 2 h are recommended default values.

Hannon (1980) present a methodology used in Miami-Dade County, Florida, in which infiltration rates used for trench design are based on lateral flow through the wall of an auger boring:

- a 9-in. (33-cm) diameter hole is drilled to the anticipated trench bottom or to at least 2 ft (0.6 m) below the low-water elevation expected at the site.
- an 8-in. (20-cm) diameter screened casing is installed in the borehole.
- water is introduced into the casing until the water surface elevation is equal to the design water elevation of the trench.
- the time is recorded for 6-in. (15-cm) drops in water level.

Infiltration rate per linear foot of trench wall is calculated as twice the value for the auger hole circumference to account for infiltration occurring through two sides of the trench.

There has been very little detailed investigation of the long-term operation of individual stormwater infiltration trenches. Bergman et al. (2011) investigated clogging in two infiltration trenches that had been in operation in central Copenhagen for over 15 years (since 1993). Both trenches receive roof runoff from a housing area. The trenches have dimensions of 16 m × 0.8 m × 0.8 m and the porosity of the filling material is estimated to be 0.38. The trenches overflow to the sewer system. Even through the trenches have identical designs and are located only 7 m apart, the field saturated hydraulic conductivity (K_{fs}) in the southern trench was estimated to be a tenth that of the northern trench.

The K_{fs} of the bottom and sides of the southern trench were estimated from recession data. Water levels in the northern trench remained so low during the monitoring period that hydraulic conductivity values could not be calculated. The evolution of K_{fs} (μm/s) is expressed by the equations

$$K_{fs,sides} = \frac{1}{0.73 + 1.9t} \tag{17.5}$$

$$K_{fs,bottom} = \frac{1}{2.8 + 0.71t} \tag{17.6}$$

where t is years since 1994.

The greatest decrease in K_{fs} occurred in the first years of operation. To address the decline in performance over time, either maintenance is required, which is difficult as trenches are underground and not easily accessible, or the decline should be accounted for in design plans (Bergman et al. 2011).

17.2.3 Aquifer Recharge Trenches

Heilweil and Watt (2011) documented an infiltration trench test that was performed at the site of a previous infiltration pond experiment. The study site, Sand Hollow Reservoir, Utah, is underlain by the Navajo Sandstone, a regional fractured eolian sandstone. The Navajo Sandstone has an upper weathered zone in which the fractures are filled with caliche (calcium carbonate). The weathered zone is overlain by a caliche layer and then a soil zone. The hydraulic conductivities of the weathered and unweathered sandstone are greater than that of the soil and caliche layer.

An infiltration pond test performed on a 30 m-radius pond gave a hydraulic conductivity of 0.05 m/d. The infiltration trench had a 1-m width and 3-m depth, and penetrated into the fractured Navajo Sandstone. The trench was filled with gravel with a perforated pipe installed 1 m below the top. The total trench bottom area was approximately 90 m^2. The infiltration rates during a 48-day test in which 2000 m^3 of groundwater was recharged were as follows:

- Phase 1 (0–5 days; saturation of sandstone near trench): 1.8–7.6 m/d
- Phase 2 (5–40 days; wetting front progressed toward the water table): 0.51 ± 0.015 m/day
- Phase 3 (40–48 days; trench hydraulically connected to the water table): 0.39 ± 0.009 m/day.

The infiltration rate for the trench system was about an order of magnitude greater than that for the infiltration pond experiment. About 30% of the difference may be due to the temperature effect on viscosity and hydraulic conductivity (water recharged in the trench was about 5 °C warmer). The greater infiltration rate of the trench was attributed to a greater contact with the more transmissive Navajo Sandstone, particularly the part of the formation with open-fractures. Trenches also avoid the evaporative losses of ponds if water levels are keep below the top of the trench.

Heilweil et al. (2015) examined the controls over infiltration rates in trenches by variable-saturated numerical modeling using the VS2DI code (Hsieh et al. 2000). The model was based on and calibrated to test data from the Sand Hollow Reservoir. A general observation is that infiltration rates are highest at the start of infiltration and the rate decreases upon connection to the regional water table. Greater rates occur were water table is deeper. The main modeling results and conclusions are (Heilweil et al. 2015):

- Infiltration rates are primarily governed by the saturated hydraulic conductivity of the vadose zone strata and the initial depth to the water table. Infiltration rates will be much lower when the initial water table is shallow and a hydraulic connection quickly occurs between trench and aquifer.
- Trench width and depth have a lesser impact on infiltration rates.
- Infiltration rates increase with larger spacings between parallel trenches.
- Infiltration through trench walls is more important than through the trench bottom (hence the small effect of increasing trench width).

Increasing the trench depth from 3 to 6 m increased modeled infiltration rates by 9–13%. The benefits of increased infiltration may be offset by the increased costs of additional excavation, materials, and long-term maintenance (Heilweil et al. 2015).

An Empirical equation was proposed for use as a planning tool (Heilweil et al. 2015):

$$I_f = K_{sat}(2.79\ln(d) - 5.47) \tag{17.7}$$

where I_f = final volumetric infiltration rate (m^3/d per m^2) and d is the depth to water (m). A limitations of the modeling is that it does not account for clogging. Infiltration rates will be lower if clogging occurs. However, sediment will tend to settle on the trench bottom rather than walls.

Infiltration trenches in the Ruhr region of Germany were described by Hantke and Schlegel (1995). The trenches have a basic design of a 1 m (3.3 ft) width and depths of up to 6 m (19.7 ft). The trenches are filled with coarse sand and covered with either wood, corrugated steel, or concrete slabs. Clogging was not a significant problem when high quality water (e.g., river Rhine water purified to drinking water standards) was recharged.

17.2.4 Trench Safety

Trenches are vulnerable to collapse and, as a result, pose a significant risk to workers who enter them and to people who accidentally fall into them during construction. Trench accidents result in numerous death in the United States in each year. The United States OSHA has strict trench safety regulations, which include that trenches 5 ft (1.5 m) deep or greater are required to have a protective system unless the excavation is made entirely in stable rock. If a trench is less than 5 ft deep, then a competent person may determine that a protective system is not required (OSHA n.d.). A competent person is defined as an individual who is capable of identifying existing and predictable hazards or working conditions that are hazardous, unsanitary, or dangerous to workers, soil types and protective systems required, and who is authorized to take prompt corrective measures to eliminate these hazards and conditions. Protective measures to prevent cave ins include benching, sloping, shoring, and shielding.

Failure to follow OSHA or other applicable local or national regulations could result in injury or death to workers, and fines and criminal charges against project owners and managers. Hence, it is critical that project teams are fully aware of, understand, and follow all regulations related to trench safety.

17.3 Infiltration Galleries

Galleries differ from trenches in that they are wider. There is no generally accepted maximum width or length-to-width ratio that distinguishes a trench from a gallery. Galleries constructed below the water table are used to withdraw and filter water from overlying surface water bodies. Infiltration galleries refers herein to systems constructed above the water table to recharge the underlying aquifer. The aggregate-filled gallery design is similar to the trench design, with the main differences being that galleries are wider and that instead of a single perforated or screened pipe, a series of parallel pipes are often used to more evenly distribute the water. Both the pipe system and aggregate should have a sufficiently high hydraulic conductivity so the infiltration rate is close to being equal across the bottom area of the gallery and localized mounding does not occur.

Design options other than a gravel fill are available and have been implemented. For example, the Atlantis Infiltration Tank System (developed by Atlantic Water Management) uses modular polypropylene crates (Matrix tank modules) as a substitute for gravel. The Atlantis system has the benefits of rapid construction, light weight, high strength, and high storage capacity (90% void space). The Rainstore system (developed by Invisible Structures, Inc.) consists of stackable thin-walled cylindrical columns constructed of either high-impact polypropylene (HIPP) or high-density polyethylene (HDPE) plastic. The systems can be used for water storage (detention) through the use of an impermeable liner or for recharge (retention) through the use of a permeable geotextile liner. Plastic crate recharge systems have the additional advantage that if the systems clogs, the crates could be excavated, cleaned, and reused.

Another design option is open bottom chambers, such as the concrete Terre Arch and corrugated plastic ChamberMaxx systems, manufactured by Contech Engineered Solutions, the plastic StormTech system, manufactured by StormTech (a division of ADS, Inc.), and the StormChamber system manufactured by Hydrologic Solutions. Bower (2011) presented the initial results of pilot testing of infiltration gallery designs for the Walla Walla Basin MAR Project (Oregon, USA). Four types of subsurface infiltration galleries were tested:

(IG-1) 4″ (101 mm) perforated, corrugated, flexible drainage pipe
(IG-2) 4″ (101 mm) perforated PVC pipe

(**IG-3**) Stormtech open-bottom infiltration chambers
(**IG-4**) Atlantis raintanks

Average instantaneous infiltration rates ranged from 0.4 cfs (11.3 L/s) to 1.67 cfs (47.3 L/s). Design option IG-2 had the highest infiltration rates.

Field testing was performed in Western Australia of secondary-treated wastewater recharge using infiltration galleries constructed with either a gravel-fill or the Atlantis system design (Bekele et al. 2009, 2013). The gallery constructed using the Atlantis system was found to be less prone to clogging. The cause of clogging of the gravel-filled trench was uncertain but may have been related to the presence of plant roots. Although the Atlantis system was more expensive to construct, its superior performance in terms of less clogging and reduced maintenance costs far outweighed its greater construction cost (Bekele et al. 2009, 2013). Testing of infiltration galleries using secondary-treated wastewater in Australia indicated a 3-log reduction in microorganisms (Bekele et al. 2009).

Bekele et al. (2011) investigated water quality improvement during vadose zone transport at a pilot infiltration system constructed at the CSIRO Centre for Environment and Life Sciences in Floreat, Western Australia. Secondary-treated wastewater was recharged using two infiltration galleries. One was a conventional gravel-filled system and the other was constructed using the Atlantis modular tank system. The water infiltrated through 9 m of calcareous sand, and groundwater was recovered from a well (BH1) located 2.3 m from the west gallery. The travel time was reported to be 3.7 days through the vadose zone and 0.5 day through the aquifer to BH1, based on tracer tests results.

Transport through the vadose zone did not result in a significant change in total nitrogen concentrations, although the concentrations of TKN and ammonia decreased and the nitrate concentration increased due to nitrification. Phosphorous and fluoride concentrations decreased, but as the decrease was likely to adsorption, the process may not be sustainable (Bekele et al. 2011). Iron decreased by 60% and total organic carbon decreased by 51%. The concentrations of the pharmaceuticals oxazepam and temazepam decreased, but the amount of reduction was difficult to quantify due to variations in their concentrations. Carbamazepine was persistent during vadose zone transport.

The concentrations of thermotolerant coliforms and enterococci were reduced compared to concentrations in wastewater, but they were still detected in both BH1 and an upgradient well. It was hypothesized that excreta from grazing sheep were the source of these bacteria (Bekele et al. 2011). The concentrations of the other pathogens were also reduced compared to the wastewater. F+ bacteriophage, which is commonly used as surrogate enteric virus, was detected in 96% of the wastewater samples, but in only 4% of the samples from well BH1. It was also detected in 6% of ambient (background groundwater) samples. Vadose zone transport was thus demonstrated to provide significant improvements in water quality, but the recovered water may still require additional treatment depending upon its intended use (Bekele et al. 2011). Greater improvements in water quality may also be obtained with a greater groundwater residence time (Bekele et al. 2011).

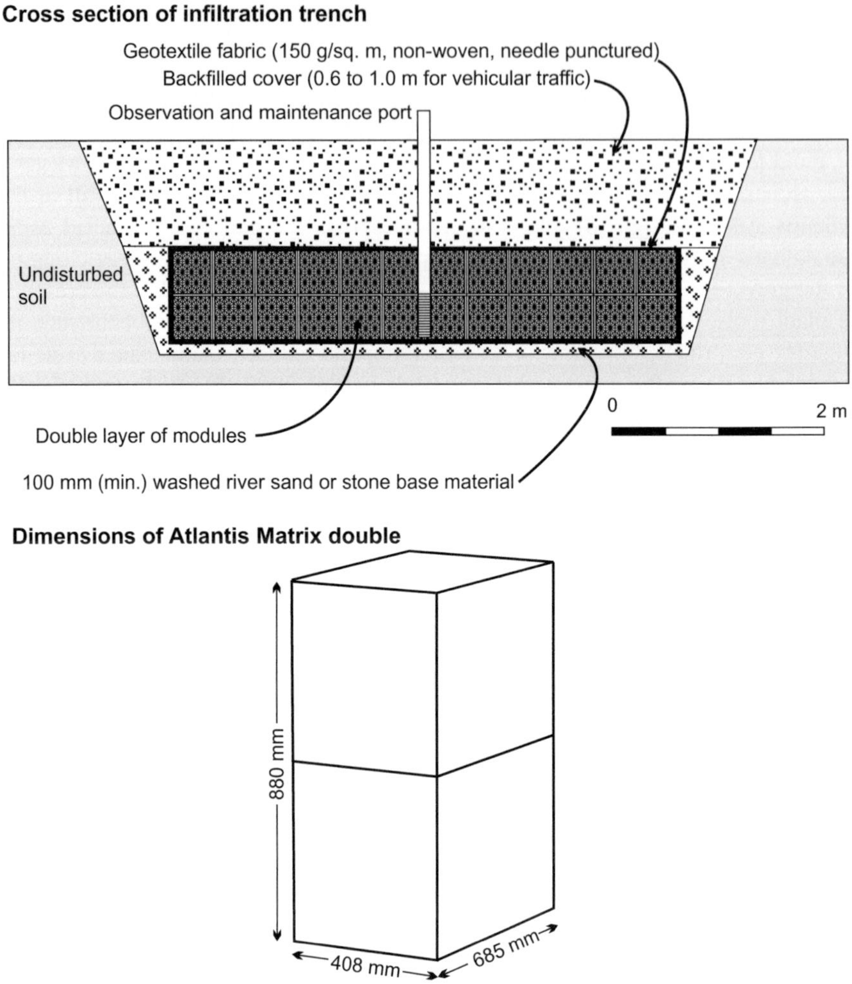

Fig. 17.2 Infiltration gallery design using a modular plastic crate (Atlantis) system (*Source* Schlumberger Water Services 2013)

Schlumberger Water Services (2013) investigated infiltration gallery design concepts as an option for the recharge of treated coal seam gas produced water into the Central Condamine Alluvium (CCA), Queensland, Australia. The proposed test gallery design utilizes a modular polypropylene crate system with a 40 m length, 5 m width, and 2 m depth (Fig. 17.2). The recommended elongate design was chosen so that the galleries could be installed within and oriented parallel to existing road right-of-ways to minimize impacts to adjoining agricultural operations. The indicative design of the final, full-scale infiltration system is modular and consists of individual infiltration galleries (10 m × 50 m each) separated by 100 m end-to-end spacing. The final configuration of the trenches (length, width and depth) would be determined based on site hydrogeology and land availability constraints. A more elongate gallery or trench (5 m × 100 m) might be required to fit the system into a narrower strip of land.

17.4 Infiltration (Recharge) Shafts and Pits

Infiltration (recharge) shafts and pits are large-diameter holes that are dug or drilled to bypass lower permeability material present near land surface. The large diameter provides water storage and a greater surface area for lateral infiltration. Shafts are filled with coarse gravel or cobbles. As is the case for vadose zone infiltration techniques in general, shafts completed above the static water table can be difficult to rehabilitate because they often cannot be pumped unless persistent mounding occurs (e.g., under perched aquifer conditions above a semiconfining unit in the unsaturated zone). A design option to better manage clogging is to place a layer of finer gravel and sand at the top of the coarse gravel or cobbles to create an inverted filter. The upper sand layer will trap fine materials and is periodically removed and replaced.

McCormick (1975) reported on an early study of recharge shafts constructed at the Leaky Acres Recharge Facility site in Fresno, California. Semi-perched conditions were reported to occur at the site. Two shafts 1.2 m (4 ft) in diameter were excavated with a bucket auger to the top of a confining layer located at about 17.7 m (58 ft) below land surface. Both shafts were equipped with a centered 25.4 cm (10 in.) conductor pipe with a 1.2 m (4 ft) perforated interval at its base. One shaft (no. 1) was completed with a pea gravel filter to land surface. The other shaft (no. 2) had a dual-filter design in which the conductor casing was surrounded by a screen 0.9 m (3 ft) in diameter with an intervening pea gravel filter pack. The annulus between the screen and unlined shaft was filled with sand. Both shafts experienced clogging of the filter pack, which was remediated by backflushing using a submersible pump in shaft no. 1 and a centrifugal pump in shaft no. 2. The pumps were turned on and off to create a surging action. After the initial clogging event, shafts nos. 1 and 2 were restored to 109 and 78% of their original infiltration rate.

Bianchi et al. (1978) documented a horizontal collector drain system piloted tested at the Leaky Acres Recharge Facility site. During surface spreading in infiltration basins, water elevation in the shallow strata rises toward the basin floor. Horizontal drains installed above the uppermost low-permeability unit collected the filtered water, which then flowed under gravity to a recharge well. The drains were constructed of perforated plastic drain lines (20.3 cm, 8 in. in diameter) surrounded by a thick sand envelope.

A main design concern for the recharge well is managing clogging and maintaining injectivity. Colloidal clay dispersed by the low electrolyte recharge water was believed to be a primary cause of clogging. It was recognized from experience elsewhere that "hydraulic mining" of the sand during construction and well rehabilitation greatly improved recharge well performance (Bianchi et al. 1978). The solution employed was to emplace a coarse gravel pack in the well that would both support the formation, transmit aquifer sands, and move downwards into the borehole to fill voids. In general, the main design issues for gravity recharge wells is achieving target initial capacities and maintaining those capacities. The integrated drain and recharge well system provides pretreatment by filtration and has a redevelopment ability (Bianchi et al. 1978).

Chadha (2003) documented recharge shafts (dry wells) in India, which are vertical shafts 2–3 m (3.3–6.6 ft) in diameter and up to 6 m (19.7 ft) deep that are excavated into relative high-permeability granular strata. Shafts are filled with sand and gravel packs that act as an inverted filter so as to provide silt-free water for recharge. The Central Ground Water Board (2000) reported numerous examples of recharge shafts and pits used in India. Recharge shafts have been constructed with injection wells drilled through their bottom to pierce through layers of impermeable material and reach into the underlying aquifer.

Recharge shafts and dry wells can be installed inside infiltration basins to create hybrid systems (Oaksford 1985; Bouwer 2002). The basin provides storage and some water quality improvement (settling of fines), while the shafts or wells allow for greater infiltration rates by bypassing less conductive strata located near land surface.

17.5 Dug Well Recharge

The Central Ground Water Board (2000) reported that there are thousands of dug wells in India that have either gone dry or the water levels have declined considerably, and can be repurposed for aquifer recharge. Storm or surface water can be

Fig. 17.3 Large-diameter, hand-dug water well, Wadi Qidayd, Saudi Arabia

diverted into these structures to recharge the local aquifer. The quality of source water, including the silt content, should be such that water quality in the aquifer is not adversely impacted and rapid clogging does not occur (Central Ground Water Board 2000). Similar dry dug wells are common in the Western Coastal Plain of Saudi Arabia (Fig. 17.3) that could also be used for aquifer recharge. Missimer et al. (2015) proposed that wadi reservoirs be used as sedimentation basins from which clean water is conveyed under gravity flow to downstream existing wells for aquifer recharge. The major attraction of dug well recharge is that it has relatively low costs as existing, no longer usable wells are utilized.

17.6 Vadose (Dry) Wells

Dry or vadose wells are widely in used in some areas for stormwater management and, much less commonly, for the managed aquifer recharge of treated wastewater or surface water. The advantages of vadose wells are that they can bypass low-permeability surficial layers that impede the percolation of water from surface-spreading facilities, have a small surface footprint, provide contaminant attenuation before infiltrated water reaches the water table, and may be less expensive to construct than phreatic dry wells. The main disadvantages of vadose wells are that they a prone to clogging, can be difficult to rehabilitate, and favorable hydrogeological conditions for injection may not be locally present in the vadose zone. The typically much thicker phreatic zone offers greater flexibility in the selection of injection zones. Vadose zone wells could be economical (i.e., offer considerable cost savings over wet/phreatic injection wells) even if they have to be replaced every 5 years (National Research Council 2008; USEPA 2012).

The basic design options for vadose wells are

- conventional water well design with either a screened or open hole completion
- gravel-filled borehole with a screened pipe extending the entire length of the hole
- gravel-filled borehole that is only partially screened
- precast concrete cylinder with ports (interior empty or filled with gravel).

Vadose wells are most commonly used for stormwater management where the practice is allowed and hydrogeological conditions are favorable. The USEPA (1999b) estimated that in 1999, there was over 247,000 drainage wells in the United States and that this figure may be an underestimation. Approximately 81% of the wells occurred in seven western states (Arizona, California, Washington, Oregon, Idaho, Montana, and Utah). Fiver other states (Ohio, Florida, Michigan, Maryland, and Hawaii) have about 15% of the total. Stormwater drainage wells are most prevalent in areas with poor surface drainage, intermittent high intensity rainfall, the absence of adequate storm sewer systems, and rapid urban development that has out-paced infrastructure development (USEPA 1999b). Stormwater drainage wells inject under gravity (and are thus also referred to as gravity drainage wells) and a large majority of them are vadose wells. Phreatic stormwater drainage wells, were allowed, usually

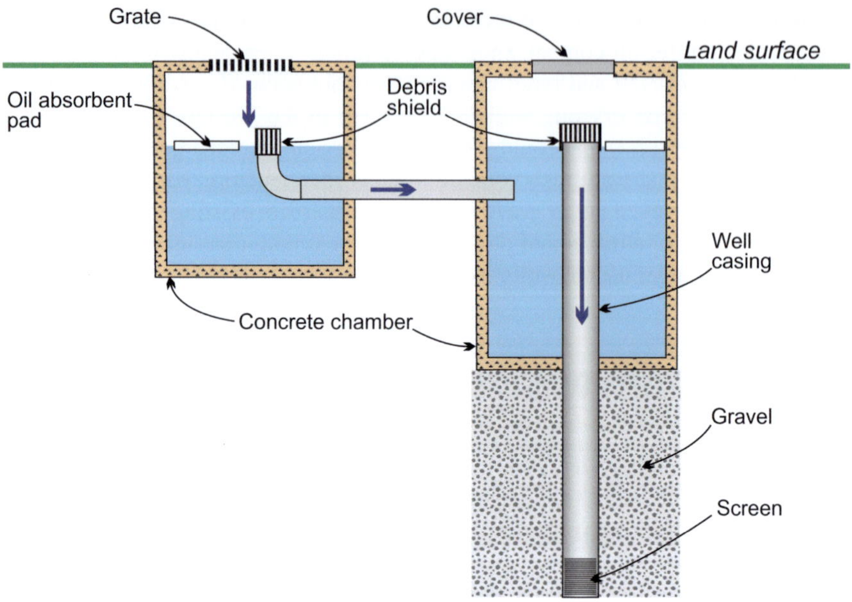

Fig. 17.4 Diagram of a catch basin with a settling system. Oil and grease can be removed using absorbent pads that require periodic replacement (*Source* USEPA 1999b)

either recharge into saline waters or are older systems that have been "grandfathered." Existing wells are allowed to continue to operate but could not be permitted under current rules (e.g., central Florida lake drainage wells; Sect. 13.8).

Stormwater vadose wells usually have some type of passive pretreatment, which may involve a sedimentation chamber, oil-water separator, and perhaps absorbent pads (Fig. 17.4). Systems can be designed as a single unit with a screen or grate at its top, or with a separate intake/pretreatment unit. Systems are commercially available in which the sedimentation chamber and screen are located in a single borehole.

Follows is a summary of the experiences with, and regulatory policies for, vadose wells in the United States. Because water is infiltrated deeper into the vadose zone, an important issue has been whether the wells pose a substantial risk to local groundwater quality.

17.6.1 City of Scottsdale (Arizona) Water Campus

A notable example of the use of vadose wells primarily for aquifer recharge is the City of Scottsdale (Arizona) Water Campus, where a networks of wells are used for the recharge of highly treated wastewater. The Water Campus (Fig. 17.5) is

Fig. 17.5 Aerial photograph of the Scottsdale Water Campus (Arizona) showing locations of vadose wells

an advanced water purification facility located in Maricopa County, Arizona, in the Phoenix metropolitan area. The treatment history of the plant was reviewed by Alexander et al. (2014). The original treatment process consisted of (Fig. 17.6):

- conventional nitrification and denitrification activated sludge process
- tertiary filtration with cloth media filters
- disinfection by chloramination
- MF and RO followed by lime stabilization.

During the latest expansion (2009–2012), the disinfection system was changed to an ozone system, the RO system was upgraded to large-diameter (16-in.) membranes, increasing the system capacity to 20 million gallons per day (Mgd; 90,920 m^3/d), and a UV treatment system was added for additional oxidation of organic compounds (Alexander et al. 2014).

Dry wells were initially chosen over deep injection wells because they were much less expensive to construct and could be replaced more economically should it become necessary due to clogging or diminished infiltration rates (Marsh et al. 1995). Recharge began in 1997. The initial system consisted of 27 wells with total depths of approximately 55 m (180 ft). The system was subsequently expanded to 27 recharge standard (RS) and 28 recharge emergency (RE) wells. The RS wells are operated on a daily basis and the RE wells are designed to temporarily accommodate wet-weather flows (Gastélum et al. 2009). The RS wells have a screened completion and injection is performed using either an injection tube with a fixed orifice plate or a downhole flow control valve. The Water Campus recharge system currently has 63 recharge wells (City of Scottsdale n.d.).

The original 27 injection wells experienced an overall decrease in specific injectivity between 1999 and 2007 ranging from 12 to 89% with a system average decrease of 48%. The RS wellfield capacity decreased from 123,500 m^3/d (32.63 Mgd) in 1999 to 65,300 m^3/d (17.26 Mgd) in 2007 (Gastélum et al. 2009). The reduction in specific injectivity is attributed to physical clogging from suspended solids and chemical clogging (calcium carbonate scale) (Lluria 2009; Gastélum et al. 2009).

17.6.2 Arizona Stormwater Vadose Wells

Vadose wells are used in areas of Maricopa County (Phoenix metropolitan area) and Pima County (Tucson metropolitan area) without stormwater systems. Disposal is the primary objective, but the wells also serve to recharge the uppermost aquifer. Dry well use in Maricopa County was reviewed by Graf (2010). A typical well design is a prefabricated unit installed inside a 4 ft (1.2 m) diameter auger hole. Recharge is performed through a gravel packed screen. The upper part of the well contains a settling chamber with a concrete or geotextile base. A riser pipe extends upwards through the settling chamber and is capped with an inlet screen and debris shield. A conceptual diagram of typical, commercially available dry well that has been widely installed is provided in Fig. 17.6. The bottom of wells is required to be 10 ft (3 m) above the water table. Dry wells have been constructed in Maricopa County as deep as 180 ft (55 m) although the majority are less than 100 ft (30 m) deep. Well design (depth) depends on the depths of transmissive strata.

Wells in industrial areas are constructed with a pretreatment interceptor to better manage flow and trap "first flush" constituents (Graf 2010). Dry wells are commonly combined with a retention basin (Fig. 17.7) to meet the Maricopa County requirement of no ponded water after 36 h (Graf 2010; Lacy 2016). The design dry well disposal capacity is assumed to be 0.1 cfs (2.83 L/s) unless a higher rate is supported by a percolation test or testing performed on a completed well (Lacy 2016).

Past contamination incidents were related to disposal of contaminants into wells through ignorance, accident, or perhaps indifference (Graf 2010). Modern technology and operation and maintenance procedures allow wells to operate safely and even improve groundwater quality (Graf 2010). The Arizona Department of Environmen-

Fig. 17.6 Conceptual diagram of a basic Arizona stormwater dry well design

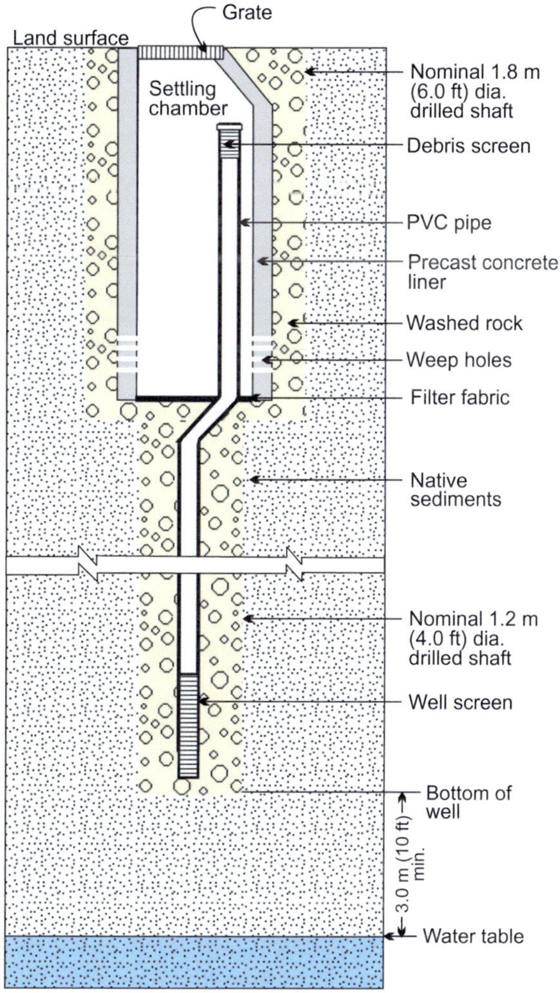

tal Quality recommends flow control and pretreatment technologies for dry wells. Pretreatment includes a settling chamber, an absorbent (hydrophobic petrochemical) sponge to provide removal of pavement oils and grease, a raised inlet (several inches), and a debris shield/screen (Lacy 2016).

Bandeen (1984, 1987) performed a groundwater modeling investigation of dry wells using a model based on the hydrogeology of the University of Arizona Water Resources research site in the Tucson area of Arizona. The site is underlain by relatively high hydraulic conductivity sands and gravels to about 30 ft (9 m) below land surface, and then less permeable clay-rich strata. The modeling results support the earlier observation of Wilson (1983) that low permeablity strata may result in greater lateral spreading of infiltrated water, and thus a greater degree and duration of

Fig. 17.7 Dry retention area with a vadose well, Sunrise Community Park, Arizona

exposure of drainage water to vadose zone soils. The greater extent and duration of exposure to vadose zone soils is thought to result in greater contaminant attenuation. Bandeen (1987) recommended that:

- areas be avoided that are underlain by uniform, high permeability soil materials between the base of the dry wells and the water table.
- dry wells should be located in areas where subsurface conditions are characterized by multi-layer soil materials, some of which are predominantly clay in composition.

The above two recommendations allow for maximum attenuation of waterborne pollutants in the vadose zone. Clay-rich layers should still be permeable enough to allow for some vertical drainage and have sufficiently low hydraulic conductivity to promote subsurface lateral flow.

Wilson et al. (1990) presented the results of a University of Arizona investigation from 1986 to 1988 of the groundwater recharge and contamination potential of dry wells in Pima County, Arizona. Three dry well sites were chosen for intensive investigation, which were located in asphalt-paved driveways at industrial, commercial, and residential sites. The sampling program included:

- run-off that flowed into the dry wells
- sediments within the dry well collection chamber
- vadose zone water samples collected within 3 ft (0.9 m) of the dry well
- groundwater from the water table also within 3 ft (0.9 m) of the dry well.

The run-off entering the dry wells and sediment samples contained variety of base/neutral semivolatile compounds (polynuclear aromatic hydrocarbons) and heavy metals (As, Cd, Cr, Cu, Hg, Ni, Pd, and Zn). However, there was no evidence for the accumulation of volatile and semivolatile organic compounds or pesticides in the vadose zone. Ethyl benzene and toluene (constituents of gasoline) were detected at the water table at the commercial site, but below USEPA drinking water limits. The concentrations of heavy metals in the run off exceeded drinking water standards, but with the exception of manganese, do not appear to be mobile in the vadose zone. Wilson et al. (1990) concluded that dry wells do not appear to be a major source of groundwater pollution in Pima County, but recommended the installation and monitoring of additional monitoring wells at other selected dry well sites.

Arizona makes a regulatory distinction between dry wells that receive only stormwater and wells draining areas where hazardous substances are used, stored, loaded, or treated. Dry wells are to be used for the sole purpose of disposal of stormwater with the exception of the following activities listed in the Arizona Revised Statutes (ARS) §49-245 (B)(23):

a. fire-fighting system testing and maintenance
b. potable water sources, including waterline flushings
c. irrigation drainage and lawn watering
d. routine external building wash down without detergents
e. pavement wash water where no spills or leaks of toxic or hazardous material have occurred unless all spilled material has first been removed and no detergents have been used
f. air conditioning, compressor and steam equipment condensate that has not contacted a hazardous or toxic material
g. foundation or footing drains in which flows are not contaminated with process materials
h. occupational safety and health administration or mining safety and health administration safety equipment.

If other fluids have been directed to a drywell, then it is subject to the Aquifer Protection Program (APP) and/or closure requirements, and may be considered an underground injection well that requires both Arizona Department of Environmental Quality (ADEQ) and USEPA permitting. Spills to the drywell may also trigger permitting, clean closure, or enforcement actions.

17.6.3 Washington State Stormwater Vadose Wells

Dry wells are used in Washington State for stormwater disposal. The typical Washington State Department of Transportation (WSDOT) design utilizes precast concrete barrels with seepage ports (Fig. 17.8). Either a single or two stacked barrels (double-barrel design) are installed inside of a downward-tapered, gravel-filled excavation (Massmann 2004).

Massmann (2004) used the VS2DH 3.0 code to estimate infiltration rates for single- and double-barrel dry wells in various hydrogeological conditions, specifically different depths to groundwater and vadose zone hydraulic conductivities. Unsaturated hydraulic characteristics were represented by the Van Genuchten equation. The modeling results indicate that infiltration rates are linearly proportional to hydraulic conductivity if the depth to the water table is held constant. Steady-state infiltration rates increase with depth to the water table with the effect being most pronounced if the depth to the water table is less than 30 ft below the bottom of the dry well. Depth to the water table was found to have little effect on steady-state infiltration rates if the depth to the water table is greater than 30 ft.

Massmann (2004) evaluated three analytical equations used to estimate flow from open vertical boreholes, the USBR and Hvorslev deep and shallow flow field equations:

Fig. 17.8 Precast concrete dry well sections (*Source* www.peerlessconcrete.com/product-catalog.html)

USBR equation

$$Q = \frac{2\pi K H^2}{\ln\left[\frac{H}{r} - \sqrt{1 + \left(\frac{H}{r}\right)^2}\right] - \frac{\sqrt{1+\left(\frac{H}{r}\right)^2}}{\frac{H}{r}} + \frac{1}{H/r}} \tag{17.8}$$

Hvorslev deep flow field equation

$$Q = \frac{2\pi K L H}{\ln\left[\frac{2L}{r} + \sqrt{1 + \left(\frac{2L}{r}\right)^2}\right]} \tag{17.9}$$

Hvorslev shallow flow field equation

$$Q = \frac{2\pi K L H}{\ln\left[\frac{4L}{r} + \sqrt{1 + \left(\frac{4L}{r}\right)^2}\right]} \tag{17.10}$$

where (using consistent units):

Q discharge rate (L^3/t)
K saturated hydraulic conductivity (L/t)
H height of water in borehole (L)
r radius of borehole (L)
L length of screen portion of well (L).

The USBR and Hvorslev deep equations were recommended for a deep (>35 ft, 10.7 m) water table and the Hvorslev shallow equation for a shallow water table (Massmann 2004). Massmann (2004) developed two regression equations relating state-state flow rates to saturated hydraulic conductivity and depth to the water table for WSDOT type dry wells:

$$\text{Double-barrel wells: } Q = K[3.55\ln(D_{wt}) + 12.32] \tag{17.11}$$

$$\text{Single-barrel wells: } Q = K[1.34\ln(D_{wt}) + 8.81] \tag{17.12}$$

where

Q infiltration rate (cfs)
K saturated hydraulic conductivity (ft/min)
D_{wt} depth to the water table from the bottom of the dry well (ft).

Infiltration rates may be significantly greater if less than the design volume of water is infiltrated and steady-state conditions are not reached.

The recommended design procedure includes the following steps (Massmann 2004):

- perform a site evaluation to identify soil layers and collect samples and determine the depth to the water table
- estimate saturated hydraulic conductivity of each layer from grain size data or other means, and calculate an effective hydraulic conductivity value
- estimate uncorrected steady-state infiltration rates using above equations
- estimate stormwater volumes and inflow rates that must be infiltrated
- apply a correction factor to infiltration rates for siltation (0.5 or less)
- monitor performance after construction.

An additional safety factor is prudent due to uncertainties in the data, particularly hydraulic conductivity values.

The regulatory requirements of stormwater dry wells in Washington State are addressed in the Washington State Department of Ecology document "Guidance for UIC Wells that Manage Stormwater" (WSDOE 2006). In Washington State, Underground Injection Control program (UIC) wells may be used to manage stormwater when pollutant concentrations that reach groundwater are not expected to exceed Washington State groundwater quality standards. UIC stormwater wells are dry wells that take advantage of vadose zone processes for contaminant attenuation. The wells must not discharge stormwater directly into groundwater at any time, even where the water table has risen as a result of UIC discharges. UIC stormwater wells are prohibited to receive stormwater from a variety of specified types of areas where pollutants may be present (e.g., vehicle maintenance, repair, and service areas). Registrants for wells are required to show that the non-endangerment standard is met. The non-endangerment standard can consider contaminant attenuation in the vadose zone.

Some key design and construction requirements for stormwater UIC wells are:

- wells must be located at least 100 ft (30 m) from drinking water wells
- wells must be capable of handling the "water quality design runoff treatment volume," which is the amount of runoff predicted from the 6-month, 24-h storm
- complete drainage of ponded water should occur within 48–72 h after flow to the well has stopped
- a minimum vertical separation of 5 ft (1.5 m) should be present from the base of the well and the seasonal high water table, bedrock, hardpan, or other low-permeability layer.

Greater separation distances may be required based on vadose zone material type. A lesser separation (down to 3 ft, 0.9 m) may be allowed if a system analysis indicates overtopping will not occur and site suitability criteria are met.

Pollution control may be met through source controls and pretreatment. Source control is means to reduce stormwater exposure to pollutants and includes spill containment, management and storage of products to avoid spills, spill response planning, and limitations on the use of potential pollutants. Pretreatment includes filtration, biofiltration, dual-well designs (catch basin or other pre-settling or spill control structures), catch-basin inserts (e.g., absorbents), and other facilities that can provide treatment of expected pollutants.

Pretreatment and minimum separation from the seasonal high water table depend upon the vadose zone treatment capacity and expected pollutant loading based on land uses and activities. In Washington State vadose zone geological materials are categorized as having either high, medium, low, or no treatment capacity. High treatment capacity materials are fine grained with high capacities to filter discharged water and remove pollutants by chemical means (e.g., cation exchange and adsorption). Low treatment capacity materials, such as sands and gravels, must have a minimum 25 ft (7.6 m) separation from the bottom of the well to the seasonal high water table. No treatment capacity materials include clean gravels, boulders and cobbles, and fractured rock.

17.6.4 Oregon Stormwater Vadose Wells

A best management practice manual for stormwater injection wells in Oregon focuses on municipal injection wells that inject only stormwater runoff from residential, commercial or industrial facilities and roadways (URS 2003). Stormwater injection wells include vertical and horizontal dry (vadose) wells. A basic requirement is that systems should be sited to provide adequate horizontal and vertical separation between the stormwater injection systems and the underlying aquifer and drinking water wells, such that the systems do not become conduits for the migration of contaminants into the groundwater or drinking water supplies.

Systems are authorized by rule (may be constructed and operated without an individual permit) if a series of conditions are met including (URS 2003):

- the wells receive only stormwater
- no domestic wells are located within 500 ft (152 m)
- no public water-supply wells are located within 500 ft or the 2-year travel time contour, whichever is more protective
- discharge does not occur directly into groundwater or below the seasonal high water table
- hazardous or toxic materials shall not be used, handled, or stored in areas draining stormwater to municipal stormwater injection wells
- preparation and implementation of a stormwater management plan
- soil or groundwater contamination should not be present that would be impacted by the operation of the system
- confinement should exists between the bottom of the injection well and seasonal high water table, or natural or engineered filtration medium should be placed between the bottom of the injection well and seasonal high water table if pollution control or treatment BMPs are not used
- the system should be designed to drain within 48–72 h of the design storm.

A minimum separation of between 4 and 10 ft (1.2 and 3.0 m) between the bottom of the injection system and seasonal high groundwater level is recommended with a greater separation recommended when the system is underlain by coarse-grained soils.

The most common dry well and sump design is perforated concrete cylinders or barrels with the annulus backfilled with gravel, or simple holes filled with gravel. Wells are sized based on infiltration rate and wetted surface areas when the dry wells are filled to their maximum design elevation. It is recommended that gravel fill be completely enclosed with geotextile to prevent mixing of fine solids with the high porosity aggregate.

Oregon (DEQ) owners of stormwater UIC wells (UICs), which include drywells, soakage trenches, drill holes and infiltration galleries, are required to show that a UIC is protective of groundwater if the UIC is located within 500 ft (152 m) of a water well or within the two-year time-of-travel of a public water supply well (Oregon DEQ 2015). Demonstration of protectiveness must show that the vertical separation distance (distance from the bottom of the UIC to the seasonal high groundwater) or horizontal separation distances are large enough so that pollutants in the stormwater will not endanger underground sources of drinking water. The protectiveness demonstration can be made through site-specific modeling or using the results of existing groundwater protectiveness demonstrations from a site in the same jurisdiction and with similar geology as the UIC site. A table is provided with the protective vertical and horizontal separation distances for a number of locations and geologies (Oregon DEQ 2015).

The City of Portland currently has approximately 9000 UICs used to recharge the shallow aquifer with stormwater from public right-of-ways (City of Portland 2008). The typically stormwater dry well system design consists of a sedimentation manhole and the UIC. The stormwater manhole receives water from catch basins and consists of a concrete cylinder 3–4 ft (0.9–1.2 m) in diameter and 10 ft (3 m) deep. Stormwater manholes provide pretreatment by allowing sediment to settle and prevent floatables (e.g., debris, and oil and grease) from entering the UIC (City of Portland 2008). The UICs are constructed of precast concrete and are generally 4 ft (1.2 m) in diameter and 2–40 ft (0.6–12.2 m) deep (mostly approximately 30 ft or 9 m deep). The UICs have a solid bottom and generally a 2 ft (0.6 m) deep sediment sump to collect coarse particulate matter (City of Portland 2008).

The City of Portland (2008) developed the Groundwater Protectiveness Demonstrations (GWPD) tool to evaluate urban right-of-way dry wells, which is a spreadsheet that estimates the reduction of stormwater pollutant concentrations during flow through the unsaturated zone. The pollutant selection process for fate and transport analysis was based on the frequency of detection in stormwater, mobility, persistence in the environment, toxicity to humans, and representativeness of broad chemical categories. The pollutants selected were:

- volatile organic compounds: Toluene
- semivolatile organic compounds: PCP, DEHP
- polynuclear aromatic hydrocarbons: Benzo(a)pyrene, naphthalene
- pesticides/herbicides: 2,4-D, methoxychlor
- metals: Copper, lead.

The GWPD tool is based on a one-dimensional, constant-source advection and dispersion equation that incorporates sorption, degradation (biotic and abiotic), and dispersion. Analyses were based on average conditions representing stormwater characteristics, soil characteristics, and degradation rates for the City of Portland. The model was not intended to be applied to spills of hazardous substances, large-inflows of petroleum products, and runoff from heavily industrialized properties. Tool input parameters are:

- pore water velocity
- porosity
- soil moisture content
- fraction of organic carbon in soils
- organic carbon portioning coefficients
- degradation rates (from literature review).

The GWDP tool was applied to develop a range of generic stormwater pollutant concentrations and environmental conditions protective of groundwater for City-owned UICs with separation distances of ≥5 ft (1.5 m), which is presented as a look-up table. If a UIC or group of UICs meets the conditions (e.g., geology, ≥5 ft separation, and stormwater concentrations within the range of input concentrations identified in the table), then the UIC is considered to be protective of groundwater (City of Portland 2008).

17.6.5 New Jersey Dry Wells

Dry wells are to be used in New Jersey to collect and temporarily store clean runoff from roof tops; treatment of runoff from other surfaces is prohibited. The NJDEP (2016) dry wells stormwater best management practices note that the basic design used is an open-bottomed chamber with a filter fabric on top and around the sides. Infiltration rate is based on flow through the open bottom and at least one inspection port is required. The design criteria are:

- **design volume**: water-quality design storm
- **maximum drainage time**: 72 h using the slowest design permeability
- **minimum design infiltration rate**: 0.5 in./h (1.3 cm/h)
- **maximum design infiltration rate**: 10 in./h (25.4 cm/h)
- **minimum distance between dry well bottom and seasonal high water table**: 2 ft (0.3 m)
- **soil permeability**: testing is required and the design rate is the lowest measured rate with a factor of 2.
- **mounding analysis**: required to demonstrate that systems will not cause surface ponding, flooding of basements, and interference with subsurface sewage disposal (septic) systems.

Talebi and Pitt (2014) examined the performance of dry wells for stormwater disposal in Millburn Township, New Jersey. A 1999 Township Ordinance required increased runoff from new impervious areas to be directed into seepage pits (dry wells) in order to minimize increases in surface water flows and reduce associated local erosion and drainage problems. The most common construction is precast concrete cylinders or barrels with seepage ports and open bottoms resting on 0.6 m (2 ft) of crushed stones and with 0.6 m (2 ft) of crushed stone surrounding the dry well. Water enters the wells through open (grated) covers and/or subsurface pipes for roof runoff.

Existing wells were evaluated by short-term falling head tests or continuous water level monitoring. Water levels were recorded at 10 min intervals and the data analyzed using the Horton (1940) and Green and Ampt (1911) methods. It was observed that some dry wells experience periodic or continuous long-term standing water, which indicates permanent or seasonal high water table conditions or clogging. Dispersion of clay was thought to be a cause of clogging in some wells due to a high sodium concentration in runoff from salt used for de-icing.

Water samples were collected both within and below some dry wells. The dry wells were found to not significantly change the water quality for most parameters. A statistically significant increase in total coliforms (attributed to regrowth) and significant decreases in *E. coli* and COD were observed. The samples frequently exceeded groundwater quality standards for total coliforms, *E. coli*, and lead. It was concluded that if influent quality is good, dry wells can be a safe disposal method for stormwater, but additional treatment may be required if an aquifer is critical (Talebi and Pitt 2014).

17.6.6 Modesto, California Dry Wells

Dry wells have been used extensively throughout the Modesto, California, vicinity since the 1950s as means to rapidly route and distribute storm water to the subsurface in agricultural and urban areas. Hannon (1980) reported that there were then over 6,500 individual dry wells in the area and that performance experience had varied. Jurgens et al. (2008) subsequently reported that over 11,000 dry wells have been constructed in the Modesto incorporated area and that they could affect the quality of shallow ground water in the area.

In the Modesto urban area, dry wells are drilled using large-diameter bits (about 1 m) to depths of 15–25 m. An outer perforated casing is installed that is filled to the surface with rock aggregate. An inner perforated casing, approximately 6 m long and 15 cm wide, is placed in the center of the borehole for the first 6 m below land surface (bls) to promote infiltration (Jurgens et al. 2008). Catch basins that act as sedimentation chambers are constructed adjacent to the dry wells to capture surface runoff and deliver water to the dry well when the basins overflow. The base of the wells is a minimum of 10 ft (3 m) above water table (Hannon 1980).

17.6.7 Hawaii Dry Wells

Dry wells are widely used in Hawaii to dispose of stormwater from roads. Most of the Department of Public Works dry wells are excavations about 5 ft (1.5 m) in diameter and have an average depth of 22 ft (6.7 m) with 90% of the wells between 10 and 30 ft (3 and 9 m) deep (Izuka 2011). Izuka (2011) performed a modeling study of the potential impacts of the dry wells on water quality. Dry wells result in a concentration of stormwater flow, which can result in a pulse of contaminants delivered to the aquifer. Concentrations are reduced by dilution, which depends upon the thickness of the unsaturated zone, the hydraulic properties of the aquifer (transmissivity), and the rate of regional groundwater flow. The modeling results indicate that in all simulations, the maximum concentration is less than 1% of the infiltrated water value within 700 ft (213 m) downgradient of the well and less than 0.1% of the value within 0.5 miles (0.8 km) downgradient (Izuka 2011).

17.6.8 Soakaways (United Kingdom)

Soakaway is a British term used for stormwater disposal systems that consists of pits or trenches that are either filled with rubble, lined with dry-jointed brickwork, or constructed of pre-cast perforated concrete ring units surrounded by a suitable granular backfill (BRE 2003). Systems are designed to handle the runoff from a design storm through a combination of storage and infiltration during the runoff event. The systems must drain sufficiently quickly to provide necessary capacity to receive the runoff from the subsequent storm (BRE 2003). It is recommended that systems discharge from full to half volume within 24 h to ensure readiness to cope with inflows from subsequent flow events (BRE 2003; Chen et al. 2008).

BRE (2003) Digest 365 summarizes the design of soakaway systems. The required storage, which is provided by both the chamber and/or porosity of the rubble or granular fill, is calculated as (BRE 2003):

$$S = I - O \tag{17.13}$$

where

S required storage in the soakaway (L^3)
I inflow volume from impermeable surface(s) (L^3)
O outflow into the soil during the rainfall event (L^3).

The system storage capacity, in the case of the common perforated precast concrete ring design, includes both the chamber volume and the porosity with the granular fill material.

The inflow volume is the product of the impermeable area and the total rainfall in the design storm (10-year storm is recommended). Outflow during a rainfall event is calculated as:

$$O = as50 \cdot f \cdot D \tag{17.14}$$

where (using consistent units)

$as50$ internal surface area of the soakaway to the 50% effective depth excluding the bottom area. Effective depth is the difference in elevation between the bottom of the structure and the invert of the drain discharging into the soakaway,
f soil infiltration rate (ideally determined from a test pit),
D storm duration.

The soil infiltration rate can be determined by filling a trial pit and recording the time for the water level to fall from 75 to 25% effective depth:

$$f = Vp(75{-}25)/(ap50 \cdot tp(75{-}25)) \tag{17.15}$$

$Vp(75–25)$ effective storage volume of water in the trial pit between the 75 and 25% effective depth
$ap50$ internal surface of the trial pit to the 50% effective depth including the base area
$tp(75–25)$ time for the water level to fall from the 75 to 25% effective depth.

The above design equations include conservative elements that provide a safety factor to accommodate reductions in performance from clogging including (BRE 2003):

- 100% of the rainfall is assumed to enter the soakaway
- no allowance is made for the time for run-off to reach the soakaway
- outflow from the soakaway is overestimated because higher infiltration rates occur at greater depths of storage.

The BRE (2003) recommends use of a geotextile to separate the granular fill from the surrounding soil. The concern is that migration of soil into the fill could lead to local subsidence. Similarly, the top of the fill should also be covered with a geotextile. It is also recommended that runoff from paved surfaces should be passed through a suitable oil interception device prior to discharge to soakaways. Design considerations include the slope of land, the potential for down-gradient water logging, and the instability of soil and fill materials. Site investigation and system design should involve a hydrogeologist and geotechnologist.

Chen et al. (2008) proposed a method for evaluating the performance of existing soakaways in which the soil infiltration rate is calculated from the time required for water levels to fall from 75 to 25% of the maximum filling depth. The geometry and storage volume of existing systems in which the construction details are not available is estimated from the volume of water required to fill the system, assuming that the amount of outflow into the surrounding ground is relatively small during the filling period. Chen et al. (2008) concluded, based on their testing results, that most existing systems were in good condition (i.e., not excessively clogged) and could remain on site for future use.

Chen et al. (2008) sampled soils near soakaways and water inside chambers for total petroleum hydrocarbons (TRPH) and heavy metals. It was observed that soils near soakaways and water inside systems that have been serving industrial or commercial areas may be contaminated with petroleum hydrocarbons and metals. Tested soakaways near residential areas appeared to have no obvious contamination with petroleum hydrocarbons and metals. It was recommended that oil interceptors be used in soakaways serving industrial areas (Chen et al. 2008).

17.6.9 Dry Well Contamination Issues

Edwards et al. (2016) performed a literature review of dry well performance for stormwater management and groundwater quality control. The potential types and concentrations of contaminants present in stormwater depend up local land uses and activities. Industrial sites would be expected to have a greater contamination potential than residential sites and vacant land. The USEPA (1999b) reported that

> Three distinct types of contamination incidents associated with storm water drainage wells are described in the literature. The first type occurs when residents or commercial businesses intentionally misuse the storm water wells. The second type of contamination incident occurs when industries unintentionally misuse storm water drains and the wells become contaminated. The final type involves the contamination of storm water wells located at or near industrial sites; these wells are contaminated because of the nature of the runoff.

Reported contaminants associated with stormwater in dry wells include (USEPA 1999b; Edwards et al. 2016):

- metals (As, Cd, Cr, Cu, Fe, Mg, Hg, Ni, Th, U, Zn)
- pesticides and herbicides (e.g., 2,4-D, methoxychlor, atrazine, simazine)
- nutrients (nitrate, phosphate)
- organics (volatile and semivolatile organic compounds, polynuclear aromatic hydrocarbons, phenols)
- oil and grease and petroleum hydrocarbons
- salts (sodium and chloride)
- trace organic compounds
- pathogens.

Hydrogeological factors that affect the performance and pollution potential of dry wells identified from previous studies include (Edwards et al. 2016):

- Greater vadose thicknesses are favorable because they allow for greater vertical travel distances to the water table (and thus opportunity for vadose-zone contaminant attenuation process to occur). Greater vadose zone thicknesses also allow for greater screen or open hole lengths, which favor greater well capacities.
- Homogeneous strata with high hydraulic conductivity values allow for rapid infiltration and percolation of water and contained contaminants, with potentially insufficient contaminant attenuation.
- High hydraulic conductivity flow zones can result in more rapid flows and greater lateral extents of recharged water, and thus risk to nearby drinking-water supply wells (if present).
- Low hydraulic conductivity (e.g., clayey) strata are favorable for contaminant attenuation, but infiltration rates may be too low to handle required stormwater volumes.
- Layered geology (sand, loam, and clay) could provide both a sufficiently high effective hydraulic conductivity and attenuation of contaminants.

The USEPA (1999b) observed with respect to stormwater wells in general (vadose and phreatic) that contamination related to storm water drainage wells has been reported to various degrees in Ohio, Kansas, Wisconsin, California, Washington, Arizona, Oklahoma, Tennessee, New York, Indiana, Florida, Kentucky, and Maryland. Several studies, however, did not clearly distinguish between contamination from storm water drainage wells versus more general, nonpoint source pollution.

Edwards et al. (2016) concluded that

> Although contaminants are detected in stormwater samples in all of the examined dry-well studies, the conclusion of the majority of studies is that when conducted properly and allowing for a sufficient separation distance and subsurface pollutant attenuation, dry well infiltration of stormwater does not pose a threat to groundwater and drinking water sources.

and

> Although some contaminants were detected in stormwater samples above regulatory levels, these contaminants were rarely detected in groundwater at similar levels.

Clearly an important part of stormwater drainage well programs is minimizing the risk of pollutants entering the wells. Pretreatment may also reduce the contamination risk and improve long-term system performance by reducing clogging rates. The USEPA (1999b) noted with respect to the potential impacts on Underground Sources of Drinking Waters (USDWs) that

> A variety of best management practices (BMPs) can be implemented to minimize the potential for contamination of USDWs resulting from storm water drainage wells. The BMPs can be organized into the following five general categories: (1) siting, (2) design, (3) operation, (4) education and outreach to prevent misuse, and (5) proper closure, plugging and abandonment. The proper design and siting of the storm water drainage well minimizes the likelihood of both accidental and routine contamination resulting from either poor operational practices or misuse.

References

Alexander, K., Nunez, A, Talabi, B., Faber, D, & Filteau, G. (2014). Large-diameter reverse osmosis facility defines reuse. *Opflow*, February 2014, 16–20.

Atlanta Regional Commission. (2001). *Georgia stormwater management manual* (Vol. 2). Atlanta: Atlanta Regional Commission.

Bandeen, R. F. (1984). *Case study simulations if dry well drainage in the Tucson basin*. Tucson: University of Arizona Water Resources Research Center.

Bandeen, R. F. (1987). *Additional case study simulations of dry well drainage in the Tucson basin*. Tucson: University of Arizona Water Resources Research Center.

Bekele, E., Toze, S., Patterson, B., Devine, B., Higginson, S., Fegg, W., & Vanderzalm, J. (2009). Chapter 1. In *Design and operation of infiltration galleries and water quality changes*. Canberra: CSIRO, Water for a Healthy Country National Research Flagship.

Bekele, E., Toze, S., Patterson, B., Fegg, W., Shackleton, M., & Higginson, S. (2013). Evaluating two infiltration gallery designs for managed aquifer recharge using secondary treated wastewater. *Journal of Environmental Management, 117*, 115–120.

Bekele, E., Toze, S., Patterson, B., & Higginson, S. (2011). Managed aquifer recharge of treated wastewater: Water quality changes resulting from infiltration through the vadose zone. *Water Research, 45*, 5764–5772.

Bergman, M., Hedegaard, M. R., Petersen, M. F., Binning, P., Mark, O., & Mikkelsen, P. S. (2011). Evaluation of two stormwater infiltration trenches in central Copenhagen after 15 years of operation. *Water Science & Technology, 63*(10), 2279–2286.

Bianchi, W. C., Nightingale, H. I., & McCormick, R. L. (1978). Fresno, Calif., subsurface drain collector-deep well recharge system. *Journal American Water Works Association, 70*(8), 427–435.

Bouwer, H. (2002). Artificial recharge of groundwater: Hydrogeology and engineering. *Hydrogeology Journal, 10*(1), 121–142.

Bower, R. (2011). Infiltration gallery design for integration into dual purpose irrigation systems: Managed aquifer recharge in Walla Walla Basin. In: *NWRI Managed Aquifer Recharge Symposium*, January 25–26, 2011, Irvine, California.

BRE. (2003). *Soakaway design (Digest* 365). Walford, UK: Building Research Establishment.

Central Ground Water Board. (2000). *Guide on artificial recharge to groundwater*. New Delhi: Central Ground Water Board, Ministry of Water Resources.

Chadha, D. K. (2003). State of art of artificial recharge applied on village level schemes in India. In A. Tuinhof & J. P. Heederick (Eds.), *Management of aquifer recharge and subsurface storage*, Publication No. 4 (pp. 19–24). Utrecht: Netherlands National Committee—International Association of Hydrogeologists (NNC-IAH).

Chen, H. P., Stevenson, M. W., & Li, C. Q. (2008). Assessment of existing soakaways for reuse. *Proceedings of the Institution of Civil Engineers-Water Management, 161*(3), 141–149.

City of Portland. (2008). *Decision making framework for groundwater protectiveness demonstrations*. City of Portland Bureau of Environmental Services.

City of Scottsdale. (n.d.). Recycled water. *City of Scottsdale*. Retrieved May 25, 2018 from http://www.scottsdaleaz.gov/water/recycled-water.

Edwards, E. C., Harter, T., Fogg, G. E., Washburn, B., & Hamad, H. (2016). Assessing the effectiveness of drywells as tools for stormwater management and aquifer recharge and their groundwater contamination potential. *Journal of Hydrology, 539,* 539–553.

Gastélum, J., Lluria, M., & Small, G. G. (2009). Vadose zone recharge wells: Ten years later at the City of Scottsdale's Water Campus Facility. In *Arizona Hydrological Society 2009 Annual Water Symposium Proceedings*.

Graf, C. (2010). Drywells: One county's novel approach to stormwater management and disposal. *Southwest Hydrology, 9*(1), 22–34.

Green, W. H., & Ampt, G. A. (1911). Studies on soil physics: I. Flow of air and water through soils. *Journal of Agricultural Science, 4,* 1–24.

Hannon, J. B. (1980). *Underground disposal of storm water runoff* (FHWA-TS-80-218). Washington, DC: U.S. Department of Transportation Federal Highway Administration.
Hantke, H. H., & Schlegel, W. (1995). The use of seepage trenches for artificial ground water recharge. In A. I. Johnson & R. D. G. Pyne (Eds.), *Artificial recharge of ground water II. Proceedings of the Second International Symposium on Artificial Recharge in Ground Water* (pp. 177–187). New York: American Society of Civil Engineers.
Heilweil, V. M., Benoit, J., & Healy, R. W. (2015). Variably saturated groundwater modelling for optimizing managed aquifer recharge using trench infiltration. *Hydrological Processes, 29*(13), 3010–3019.
Heilweil, V. M., & Watt, D. E. (2011). Trench infiltrations for managed aquifer recharge to permeable bedrock. *Hydrological Processes, 25*(1), 141–151.
Horton, R. E. (1940). An approach toward a physical interpretation of infiltration capacity. *Soil Science Society of America Proceedings, 5,* 399–417.
Hsieh, P. A., Wingle, W., & Healy, R. W. (2000). *VS2DI–A graphical software package for simulating fluid flow and solute or energy transport in variably saturated porous media*. U.S. Geological Survey Water-Resources Investigations Report 99-4130.
Izuka, S. K. (2011). *Potential effects of roadside dry wells on groundwater quality on the island of Hawai'i—Assessment using numerical groundwater models*. U.S. Geological Survey Scientific Investigation Report 2011-5072.
Jurgens, B. C., Burow, K. R., Dalgish, B. A., & Shelton, J. L. (2008). *Hydrogeology, water chemistry, and factors affecting the transport of contaminants in the zone of contribution of a public-supply well in Modesto, eastern San Joaquin Valley, California*. U.S. Geological Survey Scientific Investigations Report 2008-5156.
Lacy, M. (2016). *Optimization model for the design of bioretention basins with dry wells* (Master of science thesis). Arizona State University, Tempe, Arizona.
Lindsey, G., Roberts, L., & Page, W. (1991). *Storm water management infiltration*. Baltimore: Maryland Department of the Environment, Sediment and Storm Water Administration.
Lluria, M. R. (2009). Successful application of managed aquifer recharge in the improvement of water resources management of semi-arid regions: Examples from Arizona and the southwestern U.S.A. *Boletin Geológico y Minero, 120,* 111–120.
Lowndes, M. A. (2000). *Wisconsin storm water manual: Technical guidelines for storm water management practices*. Madison: Cooperative Extension of the University of Wisconsin-Extension.
Marsh, F. L., Dueker, L. L., & Small, G. G. (1995). Recharge well technology at the Water Campus Project, Scottsdale, Arizona. In A. I. Johnson & R. D. G. Pyne (Eds.), *Artificial recharge of ground water II. Proceedings of the Second International Symposium on Artificial Recharge in Ground Water* (pp. 220–230). New York: American Society of Civil Engineers.
Massmann, J. (2004). *An approach for estimating infiltration rates for stormwater dry wells*. Final Research Project prepared for Washington State Department of Transportation, Olympia, Washington.
McCormick, R. L. (1975). Filter-pack installation and redevelopment techniques for shallow recharge trenches. *Ground Water, 13*(5), 400–405.
Metropolitan Council. (2001). *Minnesota urban small sites BMP manual: Stormwater best management practices for cold climates* (pp. 3-169–3-180). St. Paul, MN: Metropolitan Council of the Twin Cities Area.
Missimer, T. M., Guo, W., Maliva, R. G., Rosas, J., & Jadoon, K. Z. (2015). Enhancement of wadi recharge using dams coupled with aquifer storage and recovery wells. *Environmental Earth Sciences, 73*(12), 7723–7731.
National Research Council. (2008). *Prospects for managed underground storage of recoverable water*. Washington, D.C.: National Academies Press.
NJDEP. (2016). Dry wells. In *New Jersey stormwater best management practice manual* (Section 9.3). Trenton: New Jersey Department of Environmental Protection.
Oaksford, E. T. (1985). Artificial recharge: Methods, hydraulics, and monitoring. In T. Asano (Ed.), *Artificial recharge of groundwater* (pp. 69–127). Stoneham, MA: Butterworth Publishers.

Oregon, D. E. Q. (2015). *Groundwater protectiveness demonstrations for underground injection control permits*. Portland: Oregon Department of Environmental Quality Underground Injection Control Program.

OSHA. (n.d.). OSHA Quick card. *Trenching and excavation safety*. Retrieved October 20, 2017 from https://www.osha.gov/Publications/trench_excavation_fs.html.

Schlumberger Water Services. (2013). *Field program design for injection trials in the Central Condamine Alluvium*. Final report (January 2013). Report prepared for the State of Queensland Department of Natural Resources and Mines. https://www.dnrm.qld.gov.au/__data/assets/pdf_file/0006/106089/cca-injection-trial-field-program.pdf.

Schueler, T., Kumble, P., & Heraty, M. (1992). *A current assessment of urban best management practices: Techniques for reducing non-point source pollution in the coastal zone*. Technical guidance to implement Section 6217(g) of the Coastal Zone Management Act, Report prepared for US EPA. Washington, DC: Metropolitan Washington Council of Governments.

Talebi, L., & Pitt, R. E. (2014). Evaluation and demonstration of stormwater dry wells and cisterns in Millburn Township, New Jersey. *Journal of Water Management Modeling, C376m*. https://doi.org/10.14796/jwmm.c376.

URS. (2003). *Underground injection wells for stormwater—Best management practices manual*. Portland: Oregon Association of Clean Water Agencies.

USEPA. (1999a). *Storm water technology fact sheet. Infiltration trench* (EPA 832-F-99-019). Washington, DC: United States Environmental Protection Agency, Office of Water.

USEPA. (1999b). *The Class V Injection well control study. Vol. 3: Storm water drainage wells*. Washington, DC: U.S. Environmental Protection Agency Office of Ground Water and Drinking Water.

USEPA. (2012). *Guidelines for water reuse* (EPA/600/R-12/618). Washington, D.C.: U.S. Environmental Protection Agency.

VDOT. (2013). Infiltration trench (Chapter 8). In *VDOT BMP design manual of practice*. Richmond: Virginia Department of Transportation. http://www.virginiadot.org/business/resources/LocDes/BMP_Design-Manual/Chapter_8_Infiltration_Trench.pdf.

Wilson, L. G. (1983). *A case study of dry well recharge*. Tucson: University of Arizona Water Resources Research Center.

Wilson, L. G., Osborn, M. D., Olson, K. L., Maida, S. M., & Katz, L. T. (1990). The ground water recharge and pollution potential of dry wells in Pima County, Arizona. *Ground Water Monitoring & Remediation, 10*(3), 114–121.

WSDOE. (2006). *Guidance for UIC wells that management stormwater*, Publication no. 05-10-067. Olympia: Washington State Department of Ecology.

Chapter 18
Recharge and Recovery Treatment Systems

18.1 Introduction

Managed aquifer recharge (MAR) is used to either store or treat water (or a both). MAR systems that have a primary treatment function include:

- **Aquifer storage transfer and recovery**: Water is injected into an aquifer and recovered using different, nearby wells to naturally treat water through filtration, sorption, and biodegradation processes.
- **Soil-aquifer treatment**: Wastewater is infiltrated into shallow basins to improve its quality by vadose and phreatic zone processes. The flow of water is controlled and recharged water is restricted to a limit area of an aquifer.
- **Aquifer recharge and recovery**: Recharge of surface water by land application and local recovery, usually using wells, to improve water quality.
- **Dune filtration**: Infiltration of water into sand dunes and recovery with the goal of improving its quality.
- **Riverbank filtration (bank filtration)**: Pumping groundwater near surface water bodies to induce additional recharge and improve the quality of water as it passes across the sediment-water interface and flows through an aquifer.

Soil-aquifer treatment and riverbank filtration are addressed in Chaps. 19 and 20, respectively. Indirect potable reuse often involves recharge and recovery of water. The technical, regulatory, and health risks associated with indirect potable reuse are addressed in Chap. 22.

Aquifer recharge and recovery systems, by definition, involves MAR and recovery of recharged water, typically using production wells. Water quality improvement occurs as the water infiltrates into the vadose zone (including through biologically active surficial clogging layers), and flows through the vadose and phreatic zones. The efficacy of physical processes (e.g., straining and filtration) and biogeochemical processes in removing pathogens and chemical contaminants depends upon both the properties of the vadose zone and aquifer materials and residence times. Longer residence times provide more time for natural contaminant removal processes to

R. G. Maliva, *Anthropogenic Aquifer Recharge*,
Springer Hydrogeology, https://doi.org/10.1007/978-3-030-11084-0_18

occur. For example, pathogen inactivation rates are commonly expressed as $\log_{10}$ removal rates (i.e., number of days for a 90% reduction in concentration). The longer recharged water remains in an aquifer, the more $\log_{10}$ removals that will occur. Hence, a key technical issue is aquifer heterogeneity. If flow occurs mainly through a thin high-transmissivity zone, then the rate of flow between recharge wells or areas and production wells will be greater and the residence time less than would occur in a more homogeneous aquifer.

Pathogen and chemical attenuation rates also depend upon the geochemical conditions encountered between recharge and recovery, particularly the oxidation reduction potential (ORP, eH, or redox state). Individual pathogens and chemicals may have more rapid removal rates in either oxic or anoxic conditions. Overall contaminant removal may be optimized where the recharged water passes through both oxic and anoxic geochemical environments.

In recharge and recovery MAR systems, flow is unidirectional and recharged water often sequentially passes through different geochemical environments. A reversal of flow directions, such as occurs between recharge and recovery in aquifer storage and recovery (ASR) systems, does not occur. Contaminants sequestered near a well during recharge are not susceptible to remobilization as flow is reversed during recovery. Hence, an important advantage of recharge and recovery systems is that unidirectional flow results in a greater predictability of contaminant removal processes.

18.2 Aquifer Storage Transfer and Recovery

Aquifer storage transfer and recovery (ASTR) refers to MAR systems that use of separate injection and recovery wells for chemical and microbial contaminant attenuation (Fig. 18.1; Dillon 2005; Rinck-Pfeiffer et al. 2006). Water quality improvement occurs due to physical and biogeochemical processes that occur along the flow path from the injection well to the recovery well. Aquifer residence provides time for biodegradation processes to occur. There is a somewhat blurred line between aquifer recharge systems using reclaimed water, ASTR, and indirect potable reuse systems because some recharged water may eventually enter potable water-supply wells. For example, some water injected into the Talbert Gap salinity barrier in Orange County, California, migrates landward and will eventually enter wells used for potable water supply. The characteristic feature of ASTR is that both injection and extraction wells are part of an integrated system. ASTR systems are operated so that the injected water is recovered and the intended purpose is improved water quality.

The primary advantage of ASTR over aquifer storage and recovery (ASR), which can also provide water quality improvements, is that it provides more uniform residence times and travel distances in an aquifer, which allow for more predictable chemical and microbial attenuation of contaminants (Pavelic et al. 2006; Rinck-Pfeiffer et al. 2006). In ASR systems using a single dual injection and recovery well, the last injected water is the first recovered, and thus has a shorter total travel distance and residence time than the first injected water. The key design and operational

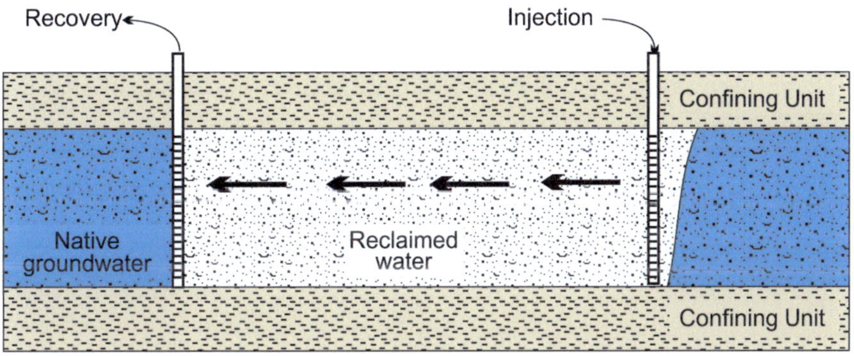

Fig. 18.1 Conceptual diagram of an ASTR system in a confined aquifer. Water quality is improved through natural contaminant attenuation processes as the water flows from the injection wells to the recovery wells

issues for ASTR system are (1) the fraction of recharged water that is present in the recovered water and (2) the residence time (also referred to as retention or transport time) in the aquifer. The former is important where the native groundwater is of poor quality (e.g., brackish). As is the case for ASR, recovered water must be of a quality suitable for its intended use.

18.2.1 Hueco Bolson Recharge Project

The El Paso (Texas, USA) and Ciudad Juarez (Mexico) area is an arid environment with a very limited water supply. The primary local water source is the Hueco Bolson Aquifer, which has experienced overdraft since the initial development of the aquifer in the early 1900s. Heavy water use has resulted in drawdowns in the El Paso area of more than 200 ft (60 m) and increases in the salinity of the aquifer water (Heywood and Yager 2003). MAR using reclaimed water is recognized to be one component of the long-term management of the Hueco Bolson Aquifer. The recharge would not stop the overdraft, but would contribute towards extending the lifetime of the aquifer.

The El Paso Water Utilities (EPWU) Hueco Bolson Recharge Project is an excellent example of a long operational ASTR and indirect potable reuse project. Highly-treated wastewater from the Fred Hervey Water Reclamation Plant is injected into the upper Hueco Bolson Aquifer and is later recovered for potable use (National Research Council 1994; Sheng 2005). Knorr and Cliett (1985) documented the initial development of the Hueco Bolson Recharge Project. Key issues in the project design were:

- The primary (65%) water source of the City of El Paso is the Hueco Bolson aquifer, which is being mined and was projected to be 97% exhausted by 2040.

- Recycled wastewater was identified as the least costly large-volume supply available to the city over the long term.
- During the planning stage of the project, it was decided that the recharged treated wastewater should meet all primary drinking water standards.
- Wells would be spaced and flow controlled so that the residence time would be at least two years prior to production. Two years was a conservative value based on known virus inactivation rates.
- A downgradient spacing of 1210 ft (368.9 m) to the nearest production well was determined via numerical computer modeling to provide a two-year residence time.
- The hydraulic gradient is towards the production wells, which increases the probability that all the recharged wastewater would be recovered.
- Injection was planned into the upper part of the Hueco Bolson aquifer, which contains freshwater (<1,000 mg/L TDS).
- The initial system would have 10 injection wells each with a capacity of 700 gpm (2,646 L/min). The design considered some loss of capacity due to clogging. Injection was to be performed through 4-in. (10 cm) diameter injection tubes to minimize air entrainment. Wells are to be equipped with a 1,000 gpm (3,785 L/min) vertical turbine pump to allow for backflushing.
- Energy recovery was investigated in which vertical turbine pumps are run in reverse rotation as a power turbine during recharge.

The treatment process consists of:

- screening, degritting, primary settling and equalization
- two-stage PACT (powdered activated carbon added to activated sludge reactors)
- lime treatment (for heavy metals, phosphorous and virus removal)
- sand filtration
- ozone disinfection
- GAC filtration (polishing step for removal of residual organic compounds).

The reclamation plant has a design capacity of 10 Mgd (37,850 m^3/d) and produces water of a quality that is comparable to that of the water currently present in the aquifer and meets all USEPA and Texas Department of Health (TDH) primary and secondary drinking water standards (National Research Council 1994). Ten injection (recharge) wells were completed in 1984, recharge commenced in May 1985 (Fig. 18.2).

The basic objective of the EPWU Hueco Bolson Recharge Project is to increase potable water supplies with the lowest practicable risk (National Research Council 1994). The recharge and recovery wells have a minimum spacing of 2,566 ft (782 m) to provide an adequate aquifer residence time to assure complete inactivation of viruses in the recovered water. Purification within the aquifer (i.e., natural aquifer treatment) was a design goal of the project (National Research Council 1994) and the well spacing was chosen to provide a minimum two-year travel time. The injection zone consists of fine and medium-grained sand with interbedded lenses of clay, silt, gravel, and caliche (Sheng 2005), which is favorable for a predominance of matrix

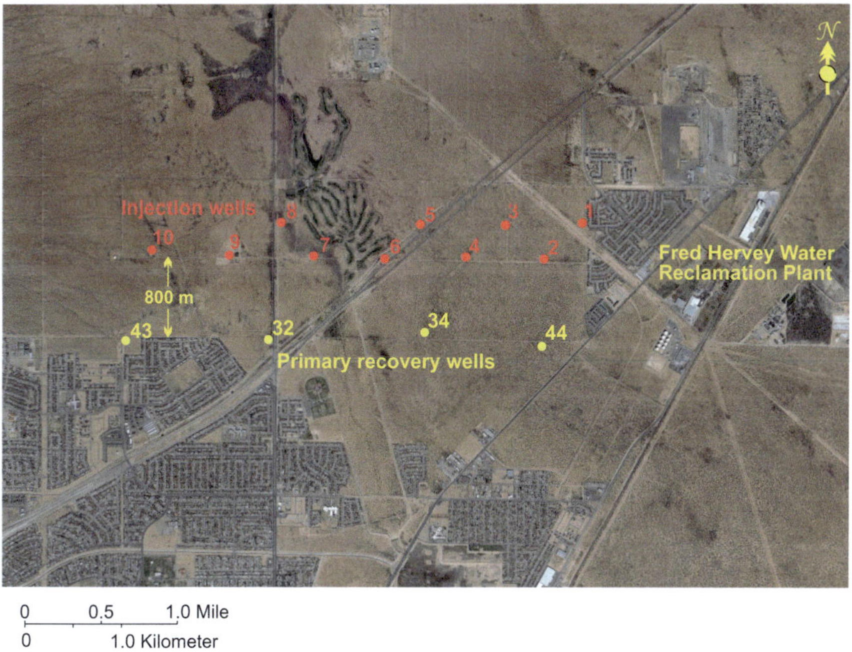

Fig. 18.2 El Paso Water Utilities Hueco Bolson Recharge Project injection and production wells locations

flow, a critical factor for the filtration of injected water during transport through the aquifer.

The wells have a screened completion and injection was performed initially through injection tubes to reduce the potential for significant air entrainment. Failure of some wells has occurred due to electrochemical corrosion. The National Research Council (1994) reported that monitoring of breakthrough of reclaimed water at observation and recovery wells has proved to be difficult because of the similarity between recharge and formation water qualities. Monitoring data suggest that the minimum residence time is closer to 7 years (National Research Council 1994). Buszka et al. (1994) reported that the tracer breakthrough velocity ranged up to 1.3 ft/d (0.4 m/d) and that the tracer data indicate that the residence time of injected water in the aquifer may be less than 6 years. Treated wastewater is now also being recharged using infiltration basins, which were found to be less expensive to operate than injection wells.

18.2.2 Salisbury, South Australia Studies

Pavelic et al. (2004, 2006) demonstrated the use of groundwater modeling, using FEFLOW and a semianalytical model, to evaluate different wellfield configurations and operational strategies for a proposed ASTR trial at Salisbury, South Australia. Treated fresh surface water is stored in a Tertiary-age, marine siliceous calcarenite aquifer ("T2" aquifer). The basic design constraints are:

- a minimal mixing fraction of recharged water of 0.9 (less than 10% native groundwater).
- a minimum effective residence time of 300 days to allow for several $\log_{10}$ removals of the most persistent pathogens.

Variables considered in the modeling included (Pavelic et al. 2004):

- the volume of water available for recharge from the treatment wetland
- the timing of the demand for recovered water
- an injection sequence to maximize flushing of the "transfer zone" in the first stage of operation
- uncertainties in aquifer properties (notably porosity and dispersivity)
- the effect of operational scheduling on travel time
- the orientation of the well field with respect to the regional groundwater flow
- the effect of the ambient hydraulic gradient
- the ratio of recovered to injected volumes.

The preferred configuration was a total of six wells with two recovery wells in the center surrounded by four peripheral injection wells, and inter-well separation distances of 75 and 100 m. Effective porosity and dispersivity strongly influence ASTR viability but these values are poorly constrained. The impacts of uncertainty in parameter values were evaluated through a sensitivity analysis. Effective porosity is an important variable in that it impacts travel time. The potential for significant preferential flow in the aquifer represents the single greatest risk to ASTR viability since it challenges both design constraints by reducing travel times and mixing fractions (Pavelic et al. 2006). Simulations of measured solute-breakthrough data at the nearby Parafield ASR site suggest that the target aquifer is heterogeneous and that effective porosity may be less than the valued used in the base-case model scenario for the ASTR wellfield, which would imply shorter travel times between recharge and recovery wells (Pavelic et al. 2004).

Miotliński et al. (2014) subsequently calibrated a dual-domain FEFLOW model to the first three years of operation (1,420 days, September 2008–June 2012) of the Salisbury ASTR system, which was constructed with two recovery wells surrounded by four injection wells in a rhombic pattern with a 50-m well spacing. A high-transmissivity zone at the bottom of the aquifer was avoided as it would result in excessive dispersive mixing, reduce residence times to less than the minimum acceptable duration, and significantly reduce recovery efficiency. The calibrated numerical model predicted the mixing fraction of the flushing period, but by the third recovery cycle it underestimated the mixing fraction by 0.05–0.10 (Miotliński et al. 2014). The discrepancy in modeled mixing fractions was believed to be due more to modeling limitations than physical processes, such as matrix diffusion or aquifer heterogeneity combined with ambient groundwater flow (Miotliński et al. 2014).

Miotliński et al. (2014) concluded that

> The results of this investigation demonstrate how a combination of detailed aquifer characterization and solute transport modeling can be used to optimize the design of ASTR systems using brackish aquifers in terms of maintaining acceptable salinities in recovered water and achieving adequate residence times and minimum travel distances to obtain target natural treatment of recharged water. ASTR using brackish aquifers can be used to reduce treatment costs and take advantage of aquifers that might locally not be otherwise beneficially used.

18.2.3 Proposed Santee Basin (California) ASTR Project

There has been increasing interest in de facto ASTR, although the systems are often not described as such. The U.S. Bureau of Reclamation investigated aquifer recharge options in the Santee Basin aquifer in San Diego, California (Blatchford 2011). Advanced-treated recycled water would be injected into an alluvial aquifer incised into the bedrock. The stated goal of the project is to augment the groundwater supply. However, the system would in essence be ASTR in that the recycled water would be injected and recovered using dedicated wells.

Key technical and regulatory issues are the capacity of the injection zone and injection wells, and the travel time to the nearest potable water supply well (Blatchford 2011). The California Code of Regulation Title 22 specified minimum residence times (9–24 months) based on the method used to estimate the times (Blatchford 2011). Since the Blatchford (2011) report was published, Title 22 of the California Code of Regulations was updated. Groundwater Replenishment Recharge Projects (GRRP) are required to achieve at least a 12-log enteric virus reduction, 10-log *Giardia* cyst reduction, and 10-log *Cryptosporidium* oocyst reduction. Virus log reduction credits can be achieved per month of groundwater residence time, with the credits received depending on how residence time is calculated (Table 18.1). More credits are obtained for methods considered most accurate, with tracer tests using introduced tracers considered to provide the most accurate data. Additional groundwater testing was recommended to better determine hydraulic parameters and the on-site depth to alluvium.

Table 18.1 California methods used to estimate residence time and virus removal credits

Method used to estimate residence time to the nearest downgradient drinking water well	Virus log reduction credit per month
Tracer study using an added tracer	1.0
Trace study utilizing intrinsic tracer	0.67
Calibrated numerical model	0.5
Analytical modeling using academically accepted equation (e.g., Darcy's law)	0.25

California Code of Regulations, Title 22, Division 4, Environmental Health, Article 5.2. Indirect Potable Reuse: Groundwater Replenishment—Subsurface Application, Table 60320.108 (July 16, 2015)

18.2.4 ASTR System Design

Compared to ASR, there is much less operational ASTR project experience to draw upon for guidance for future implementation. ASTR recharge wells are essentially the same as other phreatic injection wells and, therefore, management of clogging will be an important issue. As is the case for injection wells in general, well and wellhead designs should accommodate well rehabilitation activities. Consideration should be given to either installing a permanent pump or air-line system for periodic backflushing.

The key technical issue for ASTR system is achieving target residence times and fractions of recharged water in the recovered water. Evaluation of residence times between recharge and recovery is particularly critical where underground storage is being relied upon for pathogen removal. Residence time and mixing with native groundwater will depend on the extent and type of aquifer heterogeneity. Hence, a detailed aquifer characterization is a critical element of ASTR projects. The aquifer characterization program should evaluate the degree to which flow is concentrated into high-transmissivity flow zones and whether dual-porosity conditions will dominate flow, which can result in a lower effective porosity and greater flow velocities.

Solute-transport modeling, based on a detail aquifer characterization, is needed to quantify transport time to recovery wells and mixing of recharged and recovered water. However, a basic limitation of solute-transport modeling is that aquifer characterization programs seldom provide sufficient information on solute-transport parameters (e.g., effective porosity and dispersivities) for accurate predictions. Field data on actual solute transport, such as from tracer testing, are needed to calibrate solute-transport models. In practice, the effects of uncertainty in solute-transport parameters on potential residence times can be evaluated through a sensitivity analysis or perhaps a stochastic approach. Ideally pilot testing should be performed to provide some operational data (breakthrough curves) that can be used for model calibration. The calibrated model can then be used for the design of a full-scale system and evaluation of operational strategies.

18.3 Dune MAR

Eolian sand dunes have several characteristics that are favorable for MAR: primary-porosity dominated flow systems, low degrees of heterogeneity, relatively uniform grain sizes, and in coastal dunes, usually a higher degree of compositional maturity (i.e., predominantly quartz compositions). Three main types of MAR systems utilize sand dunes:

- aquifer recharge and recovery (ARR) systems, in which water is recharge in dune fields and recovered using nearby production wells (Fig. 18.3).
- filtration systems in which passage through dune sediments is used to treat stormwater before recharge into surface water bodies.
- storage systems using internal (non-coastal) dunes in arid regions for water storage.

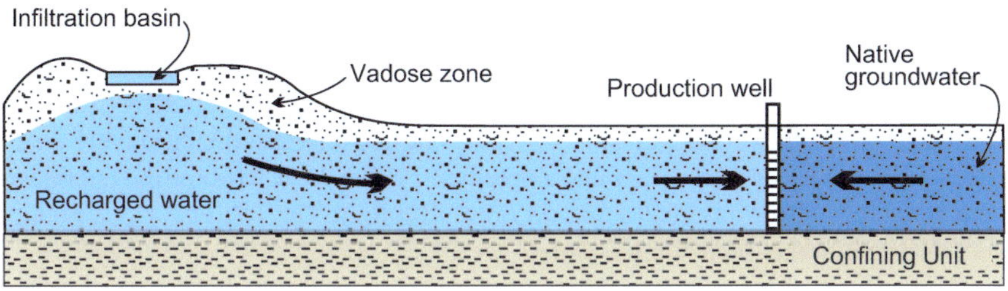

Fig. 18.3 Conceptual diagram of a dune filtration system

In the latter case, flow barriers (e.g., slurry walls) could be installed around an area of sand dunes to, in essence, create a container to store desalinated water (Maliva and Missimer 2012; Lopez et al. 2014).

18.3.1 Dune ARR in The Netherlands

Dune ARR is used in the Netherlands as a treatment step for producing potable water from river waters. A key feature of the systems is that the water receives considerable treatment both before recharge and after recovery.

Piet and Zoeteman (1980, 1985), Olsthoorn and Mosch (2002), and Tielemans (2007) summarized the history of dune recharge in North Holland and water quality changes during dune filtration. Dune groundwater has been used along the Dutch coast for the drinking water supply for Amsterdam and The Hague since the mid-nineteenth century. Amsterdam Water Supply utilized the Amsterdam dune area since 1853. In the late 1930s, increased water consumption led to the premature exhaustion of wells and saline-water intrusion toward water supply wells.

Dune recharge began in the 1950s to restore the equilibrium between fresh and saline waters in the dune aquifers. Water from the Rhine River is piped to the dune area and recharge is performed using ponds and channels. Due to deterioration in Rhine water quality, clogging of the 40 recharge ponds of the Amsterdam system became a problem, requiring scraping of the ponds every 2–3 years (Olsthoorn and Mosch 2002). The ponds were reported to be about 40 m wide, 80 cm deep, and have a total combined area of about 86 ha. Pretreatment systems were upgraded from sand filtration at the intake to include coagulation with $FeCl_3$ and settling before rapid-sand filtration. The improved pretreatment was reported to have reduced the cleaning frequency to about every 20 years.

Recovery is performed now mostly using wells, as open systems are being phased out due to recontamination from the feces of birds and animals living in the dunes (Tielemans 2007). The minimal travel time is about 30–60 days, which was originally based on *E. coli* removal (Tielemans 2007). Water quality improvements include attenuation of pathogenic microorganism and adsorption and degradation of some micropollutants. Treatment of the recovered water is system specific and includes some of the following elements: softening, membrane filtration, powdered activated carbon, rapid-sand filtration, slow-sand filtration, and disinfection by ozonation (Tielemans 2007). Over 500 substance were identified and semiquantitatively determined in Rhine River water (Piet and Zoeteman 1980, 1985). Most of the chemicals were removed during pretreatment and dune filtration. The following compounds were detected in the dune-filtered water:

- o-dichlorobenzene (1.1 ± 0.1 μg/L)
- p-dichlorobenzene (0.2 ± 0.1 μg/L)
- bis(2-chloroisopropyl)ether (2.9 ± 0.1 μg/L).

The dune area is a national reserve and environmental issues have become of upmost political importance (Olsthoorn and Mosch 2002). The water supply system is operated to maintain more natural water level fluctuations and to restore the hydrological system to a more natural state. The water supply system has become more environmentally sustainable, but it has become harder to change, extend, and renew infrastructure in the dunes. Water utilities face the charge of trying to meet multiple societal and political objectives (Olsthoorn and Mosch 2002).

Anaerobic conditions are needed for the removal of nitrate and some pesticides. One unanticipated risk is that the oxidizing nature of recharged water will progressively reduce the reductive capacity of aquifers (De Jonge et al. 2002). Nitrate and dissolved oxygen (DO) in recharged water consumes organic matter and iron sulfides. As a result of the oxidization of aquifers, nitrate and relevant pesticides may spread further from recharge basins (De Jonge et al. 2002). Redox front displacement velocities were calculated to be 0.4–2.3 m/year (De Jonge et al. 2002).

Stuyfzand (2015) investigated trace element patterns in recharged Rhine River water between recharge basins and a recovery canal system in a dune aquifer system. Data are presented from a long slow-flow (0.01–0.1 m/d) transect and a shorter fast-flow (0.3–2 m/d) transect. A vertical redox zonation is evident from suboxic (O_2 and NO_3 reducing), to anoxic (Fe_{3+} reducing) and deep anoxic (SO_4 reducing) conditions. Downgradient, trace element concentrations decrease due to sorption and precipitation in the suboxic zone. The available data suggest that ferrihydrite (present as coatings on grains) can be an effective barrier to some trace elements. Clay minerals also retard the transport of positively charged trace elements because of their negative surface charge.

At the suboxic-anoxic boundary, DO and nitrate have been reduced and the reductive dissolution of the abundant ferrihydrite coatings on sand grains begins (Stuyfzand 2015). Reductive dissolution of ferrihydrite was identified as a potential arsenic source. Peak arsenic concentration (50 μg/L) occurred in the anoxic zone (Stuyfzand 2015).

Migration of As and Mo is clearly halted when sulfate reduction takes place, suggesting that they are coprecipitated with pyrite (Stuyfzand 2015) Uranium is also practically immobile in deep anoxic zone (immobilized as UO_2) with the onset of sulfate reduction. As, Mo, U show redox-dependent behavior. Other trace elements

(B, Ba, Co, Cs, Cu, Li, Mo, Rb, Sb, Sr, U, W) show a clear sorptive behavior (Stuyfzand 2015).

Schijven et al. (1999) investigated the removal of viruses in the Castricum dune artificial recharge system in The Netherlands using the bacteriophages MS2 and PRD1 as surrogates. High concentration of the bacteriophages were added to the recharge water and sampled for in down-gradient monitoring wells. The experiment results indicate that an at least 8 $\log_{10}$ reduction is achieved with 30 m passage, which corresponds to a travel time of about 25 days. An approximately 3 $\log_{10}$ removal occurred within the first 2.4 m and the remaining 5 long_{10} removal occurred in a linear fashion in the following 27 m. The rates of reduction during transport were much greater than the field measured inactivation rates of free bacteriophages of 0.12 and 0.030 day^{-1} for PRD-1 and MS2, respectively, which correspond to $\log_{10}$ removal times of 8.3 and 33.3 days. The main removal process was thus found to be attachment.

18.3.2 Belgium Dune Aquifer Recharge and Recovery (St-André System)

Freshwater resources in the St Andrè and Westhoek catchments of the Flemish Coast, Belgium, are only available in small dune ridges along the coast, which have a limited capacity (Van Houtte et al. 2012). Vandenbohede et al. (2008a, b) described the dune ARR and indirect potable reuse system at St-André. Groundwater exploitation in the dune area began in 1947. Extraction rates in the dunes are limited by saline-water intrusion and the environmental sensitivity and ecological value of the dune environment. Artificial recharge began in 2002 using reclaimed water that additionally receives MF and RO treatment and UV disinfection. MAR replenishes the drinking water aquifer, levels out seasonal variations in water availability, and prevents saline-water intrusion (Van Houtte et al. 2012).

Intermunicipal Water Company of the Veurne Region (IWVA) is responsible for production and distribution of drinking-water within six communities in the western part of the Flemish coast near the French border. It is also responsible for the collection of wastewater in this area (Van Houtte and Verbauwhede 2008). A multi-barrier approach is applied to protecting public health including (Van Houtte et al. 2012):

- conventional activated sludge wastewater treatment at the Wulpen WWTP
- advanced wastewater treatment at the Torreele Water Reclamation Plant (at the Wulpen WWTP)
- groundwater infiltration
- conventional potable water treatment (aeration and rapid-sand filtration).

The wastewater treatment system was described by Van Houtte and Verbauwhede (2008), Van Houtte (2012), Van Houtte et al. (2012), and Onyango et al. (2014). Wastewater from the Wulpen wastewater treatment is additionally treated at the Torreele facility by pH and antiscalant adjustment (NaOCl and NH_4Cl), ultrafiltration

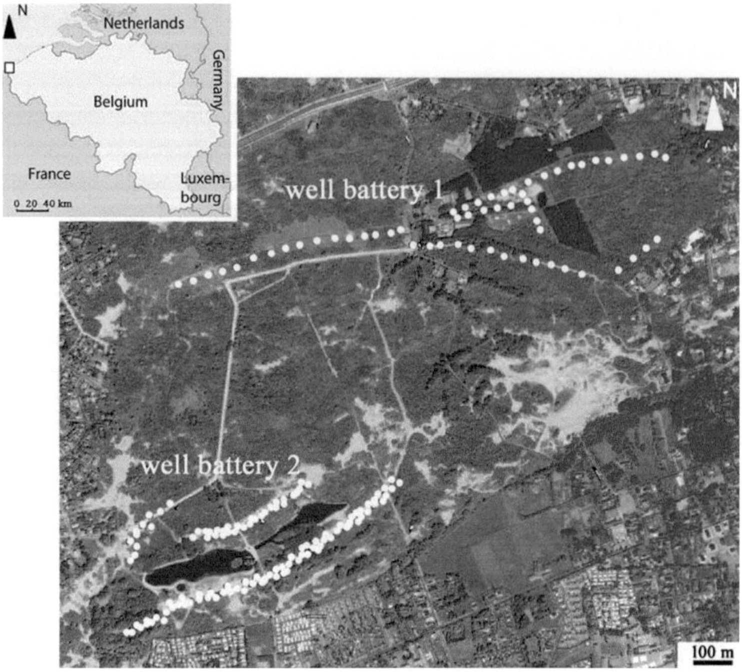

Fig. 18.4 St-André, Belgium, dune aquifer recharge system layout showing the locations of the two recharge ponds between wells at well battery 2 (*Source* Vandenbohede et al. 2008b)

pretreatment, reverse-osmosis (DOW n.d.), and ultraviolet disinfection. Two RO trains have capacities of 185 m^3/h each.

The water is discharged into two ponds and is recovered through two batteries of previously constructed and operated wells (Fig. 18.4). By 2002, the two batteries contained a total of 182 active wells (Vandenbohede et al. 2008a, b). The recovery wells are from 8 to 12 m deep with a spacings of 33–153 m (average 59 m) from the infiltration ponds (Van Houtte et al. 2012). Groundwater modeling results indicate that the extracted water is a mix of waters with variable residence times ranging from slightly less than 30 days to almost 5 years depending upon the flow path (Vandenbohede et al. 2008a, b; Van Houtte et al. 2012). The recovered water is treated by aeration and filtration to remove iron and manganese, and is then suitable for potable use. Up to 2.5 million m^3/year (MCM/year) of highly treated wastewater can be recharged and recovered, in addition to the 1.7 MCM/year of natural dune water that can be extracted by the wells (Vandenbohede et al. 2008a).

During passage through the aquifer the water experiences the following geochemical changes (Vandenbohede et al. 2009):

- mixing of recharged water and native groundwater
- increases in calcium and to a lesser extent magnesium, potassium, and fluoride concentrations from mineral dissolution

- change from oxic to anoxic conditions with the consumption of DO and nitrate
- increases in iron, manganese, and sulfate concentrations from the oxidation of iron sulfide and manganese minerals.

The extracted water consists of about 80% artificial recharge water and 20% native dune groundwater, with the mixing ratio being highly dependent on recharge and extraction rates (Vandenbohede et al. 2009). Operation of the system has reduced natural groundwater extractions, resulting in rising groundwater levels and restoration of natural flows in the dune field (Vandenbohede et al. 2008a).

The recovered water is treated by aeration, slow-sand filtration, chlorination, and UV treatment before being sent to the distribution system (DOW n.d.; Van Houtte and Verbauwhede 2008; Van Houtte 2012 and Onyango et al. 2014). Aquifer recharge with the very highly treated reclaimed water has increased the quality of the final drinking water over the years (Dow n.d.).

18.3.3 Proposed Dune ARR in Western Saudi Arabia

Lopez et al. (2014) proposed that water stored in reservoirs in western Saudi Arabia could be conveyed under gravity to dunefields, where it would be stored as an emergency water source for downstream communities. The specific example described would convey water from the Al Murwani dam to a dunefield ASR system that could serve as an emergency water source for local communities and/or the city of Jeddah. An important advantage of dune storage is avoidance of the huge evaporative storage losses of reservoirs. A key design issue is containing the stored water. Boundaries to the sand reservoir would be provided by the underlying low-permeability desert pavement, the natural land gradient, and a downstream slurry wall. Water would be recovered using wells.

18.3.4 Dune Filtration

Stormwater often has high concentrations of pathogens (e.g., fecal coliform bacteria) and its discharge to surface water bodies can create a human health risk. Open (pipe) discharges along the shoreline can impact beach water quality, which could have economic consequences if beaches have to be closed to swimming. Dune infiltration systems (DISs) using infiltration galleries were field tested in Kure Beach, North Carolina, as a means of reducing health risks at the beach associated with exposure to polluted stormwater (Bright et al. 2011; Burchell et al. 2013; Price et al. 2013). The goal of the DISs is to recapture natural, pre-development processes of stormwater infiltration into sand and natural filtration. The basic concept is to use sand dunes as giant natural sand filters.

Three systems were installed at Kure Beach, the first two in 2006 and the third in 2009 (Burchell et al. 2013). The Kure Beach DISs consist of open-bottomed chambers (StormChamber systems) installed beneath sand dunes (Bright et al. 2011; Burchell et al. 2013; Price et al. 2013). The systems were constructed of inverted halfpipes composed of corrugated high-density polyethylene (HDPE). The chambers are 3.5 ft (1.1 m) high, 5.0 ft (1.5 m) wide, 8.2 ft (2.5 m) long and installed in series parallel to the shore. The systems are installed atop a 6–12-in. (15–30-cm) deep layer of gravel to provide a stable base and enhance infiltration. The chambers are covered on their top and sides with a geotextile fabric to reduce sand intrusion. Upon completion of construction, the trenches were backfilled, and the site was restored to natural grade and planted with American beach grass (*Ammophila breviligulata*) and sea oats (*Uniola paniculata*).

The DISs were designed to capture peak stormwater runoff associated with a 13 mm (0.5 in.) one-hour storm (Bright et al. 2011). The systems were reported to have captured from 80 to 100% of stormwater flows and indicator bacteria (enterococci) were reduced by 98% (Burchell et al. 2013). The DIS also achieved a 3-log removal of fecal coliform bacteria (Burchell et al. 2013).

Basic design issues are that the dunes should be high enough (>4 m) so that there is a sufficient depth of sand (>3 m) above the water table and there should be sufficient separation (>60 m) of the dunes from the beach (Bright et al. 2011). DISs appear to be appropriate for installation in small watersheds (<4 ha, 10 ac), though further research was recommended to identify bacterial removal processes and residence times, and to quantify the lateral extents of the water table mounds (Price et al. 2013).

The Kure Beach DISs are not ARR in that the infiltrated water was not recovered and used. However, similar subsurface DIS systems could be used in more inland areas to augment shallow freshwater supplies and serve as the recharge element of ARR systems.

18.4 Aquifer Recharge and Recovery

Aquifer recharge and recovery (ARR) includes a wide variety of system types with the essential feature that they involve both aquifer recharge and recovery for subsequent use (National Research Council 2008). The definition of ARR is narrowed herein to describe MAR systems involving recharge by land application and some aquifer flow between the recharge point and recovery wells.

18.4.1 ARR Systems in Finland

Jokela and Kallio (2014) reported that there are 26 MAR plants in Finland. A typical Finnish MAR operates to remove natural organic matter from surface waters. Lake water, or less commonly river water, is infiltrated into an esker or other glaciofluvial sand and gravel formation, and recovered using wells located a few hundred meters downgradient (Helmisaari et al. 2006; Jokela and Kallio 2014; Jokela et al. 2017). Infiltration is performed using either basins, wells, or surface sprinkling. Basin infiltration has been used since the 1970s and sprinkling irrigation has also been used since the end of the 1990s (Helmisaari et al. 2006).

Helmisaari et al. (2006) documented total organic carbon (TOC) removal in the ARR systems. Removal was documented to occur in the aquifer as opposed to the unsaturated zone. The infiltrated water was transported through the 5–30 m thick unsaturated zones within a few hours. The key variable controlling TOC removal appears to be residence time. TOC removal was about 70% for residence times of 20–60 days. Large molecular organic carbon was preferentially removed relative to smaller molecules. Jokela et al. (2017) reported on six MAR plants located in southern and central Finland. The capacities of the plants range from 10,000 to 105,000 m^3/d and the residence times range from 0.5 to 3 months. The raw water TOC concentrations ranged from 6.5 to 11 mg/L and the concentration in the recovered water was approximately 2 mg/L, which corresponds to a reduction of 70–85%. Jokela and Kallio (2014) documented testing of a planned 20,000 m^3/d potable-water production system located near the city of Tamere in which infiltration will be performed using both wells and sprinkling.

Most Finish MAR systems do not pretreat the recharged water. Chemical pretreatment is performed where the raw water turbidity or NOM concentration is high or if there are rapid fluctuations in raw-water quality (Jokela et al. 2017). The chemical pretreatment at the Kuivala and Virttaankangas MAR plants consists of the addition of a coagulant (ferric sulfate or polyaluminum chloride), dissolved-air flotation, and rapid filtration using either sand or a dual media (Jokela et al. 2017).

High NOM concentrations could create more chemically reducing conditions favorable for mobilization for iron and manganese in the soil, but the mobilization was not reported in the systems documented by Jokela et al. (2017). Manganese and iron mobilization could be avoided by chemical pretreatment to reduce NOM concentrations (Jokela et al. 2017).

18.4.2 Prairie Waters Project (Aurora, Colorado)

Regnery et al. (2016) introduced the term "sequential managed aquifer recharge technology" or "SMART" for MAR treatment trains that combined multiple MAR systems to achieve sequential exposure to different geochemical environments for enhanced contaminant removal. Regnery et al. (2016) demonstrated that improved organic chemical contaminant removal can be achieved by a sequential configuration of carbon-rich and predominantly anoxic conditions followed by carbon-depleted and predominantly oxic conditions. The proposed treatment train is short riverbank filtration passage for depletion of biodegradable dissolved organic carbon (BDOC) and nutrients and a transition to reducing conditions, followed by reaeration during surface spreading. Reaeration would allow for removal of less-degradable compounds under low-BDOC and oxic conditions.

The Prairie Waters Project (Aurora, Colorado) is given as type example of a SMART system (Regnery et al. 2016). Surface water is obtained from the South Platte River by RBF and then sent to an ARR system consisting of an infiltration basin complex that is surrounded by slurry walls. The removal of 19 trace organic compounds was examined. The compounds were grouped into four bins based on their attenuation behavior:

(1) Compounds that were immediately attenuated during short RBF.

- atenolol
- diphenhydramine
- triclocarban
- trimethoprim

(2) Moderately biodegradable compounds that were not completely removed by RBF but were immediately attenuated during infiltration at the ARR site.

- caffeine
- diclofenac
- gemfibrozil
- naproxen

(3) Moderately degradable to relatively recalcitrant compounds that are 50–75% removed by RBF and additionally removed by 25% of more during ARR treatment.

- DEET
- dilantin
- meprobamate
- sulfamethoxazole
- TCEP
- TCPP
- TDCP

(4) Recalcitrant compounds.

- primidone
- carbamazepine

No difference in removal efficiency of moderately biodegradable compounds occurred between travel times of 2 weeks and more than 2 weeks, which suggests that most of the trace organic chemical removal occurs after about 1 m of infiltration. Total dissolved organic carbon (DOC) was reduced from 6.4 ± 0.3 mg/L in the South Platte River, to 3.1 ± 0.5 after RBF to 2.1 ± 0.1 after ARR. The native groundwater concentration was 2.2 ± 0.1 mg/L. The combined treatment reduced UV254 and specific ultraviolet absorbance (SUVA) from 13.1 ± 1.1 m^{-1} to 2.0 ± 0.2 L/mg·m, respectively, in the surface water, to 3.2 ± 0.1 m^{-1} and 1.5 ± 0.1 L/mg·m, respectively, after RBF, which are slightly below groundwater values.

18.4.3 Japanese ARR System

Hida (2009) documented ARR systems in Japan in which surface water from irrigation canals is recharged using infiltration basins in the upper and middle parts of alluvial fans. In the reported systems, water is recovered downgradient using a horizontal collector for industrial use and vertical wells for domestic use.

18.4.4 Proposed Reclaimed-Water Wadi ARR in the Middle East

De facto treated-wastewater ARR occurs in wadis downstream of some Middle Eastern cities (Riyadh, Al Madinah) where treated wastewater is discharged to wadi channels, where it infiltrates and is recovered by downstream agricultural users using wells. Missimer et al. (2012) proposed that wadi aquifers can be restored by a more engineered approach to wadi ARR. Systems could be designed and operated to prove target additional treatment as well as storage of reclaimed water. Reclaimed water would be introduced into a wadi aquifer and allowed to flow down-gradient within the channel. The systems would be designed to manage the flow and capture the reclaimed water (e.g., using slurry walls within the wadi channel) before it spreads out to a degree that it cannot be easily captured by wells or it mixes with water of an undesirable quality, such as highly saline water or water containing very high nitrate concentrations (Missimer et al. 2012).

References

Blatchford, D. B. (2011). *Santee Basin aquifer recharge study*. Temecula, CA: U.S. Bureau of Reclamation. Southern California Area Office.

Bright, T. M., Burchell, M. R., Hunt, W. F., & Price, W. (2011). Feasibility of a dune infiltration system to protect North Carolina beaches from fecal bacteria contaminated storm water. *Journal of Environmental Engineering, 137*(10), 968–979.

Burchell, M. R., Hunt, W.F., & Price, W. D. (2013). *Dune infiltration systems for reducing stormwater discharge to coastal recreational beaches*, Publication AG-781. Raleigh, NC: North Carolina Cooperative Extension.

Buszka, P. M., Brock, R. D., & Hopper, R. P. (1994). *Hydrogeology and selected water-quality aspects of the Hueco Bolson Aquifer at the Hueco Bolson Recharge Project area, El Paso, Texas*. U.S. Geological Survey Water-Resources Investigations Report 94-4092.

De Jonge, H. G., Kortleve, M. W., & Frapporti, G. (2002). Sustainable use of aquifers for artificial recharge in South-Holland. In P. J. Dillon (Ed.), *Management of aquifer recharge for sustainability* (pp. 41–48). Lisse: A.A. Balkema.

Dillon, P. (2005). Future management of aquifer recharge. *Hydrogeology Journal, 13*(1), 313–316.

DOW. (n.d.). *Indirect potable reuse via groundwater recharge in Belgium*. Retrieved May 25, 2018 from http://msdssearch.dow.com/PublishedLiteratureDOWCOM/dh_053c/0901b8038053c343.pdf?filepath=liquidseps/pdfs/noreg/609-02150.pdf&fromPage=GetDoc.

Helmisaari, H.-S., Derome, J., Hatva, T., Illmer, K., Kitunen, V., Lindroos, A.-J., et al. (2006). Artificial recharge in Finland through basin and sprinkling infiltration: soil processes, retention time and water quality. In *Recharge systems for protecting and enhancing groundwater resources. Proceedings of the 5th International Symposium on Management of Aquifer Recharge*, Berlin, Germany, 11–16 June 2005 (pp. 280–285). Paris: UNESCO.

Heywood, C. E., & Yager, R. M. (2003). *Simulated ground-water flow in the Hueco Bolson, an alluvial aquifer-basin aquifer system near El Paso, Texas*. U.S. Geological Survey Water-Resources Investigations report 02-4108.

Hida, N. (2009). Managed aquifer recharge by using spreading basin methods on alluvial fans: A general overview of the situation in Japan. *Boletín Geológico y Minero, 120*(2), 311–320.

Jokela, P., Eskola, T., Heinonen, T., Tanttu, U., Tyrväinen, J., & Artimo, A. (2017). Raw water quality and pretreatment in managed aquifer recharge for drinking water production in Finland. *Water, 9*(2), 138.

Jokela, P., & Kallio, E. (2014). Sprinkling and well infiltration in managed aquifer recharge for drinking water quality improvement in Finland. *Journal of Hydrologic Engineering, 20*(3), B4014002.

Knorr, D. B., & Cliett, T. (1985). Proposed groundwater recharge at El Paso, Texas. In T. Asano (Ed.), *Artificial recharge of groundwater* (pp. 425–480). Stoneham, MA: Butterworth.

Lopez, O., Stenchikov, G., & Missimer, T. M. (2014). Water management during climate change using aquifer storage and recovery of stormwater in a dunefield in western Saudi Arabia. *Environmental Research Letters, 9*(7), 075008.

Maliva, R. G., & Missimer, T. M. (2012). *Arid lands water evaluation and management*. Berlin: Springer.

Miotliński, K., Dillon, P. J., Pavelic, P., Barry, K., & Kremer, S. (2014). Recovery of injected freshwater from a brackish aquifer with a multiwell system. *Groundwater, 52*(4), 495–502.

Missimer, T. M., Drewes, J. E., Amy, G., Maliva, R. G., & Keller, S. (2012). Restoration of wadi aquifers by artificial recharge with treated waste water. *Groundwater, 50*(4), 514–527.

National Research Council. (1994). *Ground water recharge using waters of impaired quality*. Washington, D.C.: National Academy Press.

National Research Council. (2008). *Prospects for managed underground storage of recoverable water*. Washington, D.C.: National Academy Press.

Olsthoorn, T. N., & Mosch, M. J. M. (2002). Fifty years artificial recharge in the Amsterdam dune area. In P. J. Dillon (Ed.), *Management of aquifer recharge for sustainability* (pp. 29–33). Lisse: A.A. Balkema.

Onyango, L., Leslie, G., & Wood, J. G. (2014). *Global potable reuse case study 5. Torreele/St Andre, Koksijde Belgium*. Brisbane: Australian Water Recycling Center for Excellence.

Pavelic, P., Dillon, P. J., & Robinson, N. (2004). *Groundwater modelling to assist well-field design and operation for the ASTR Trial at Salisbury, South Australia*. Canberra: CSIRO Land and Water.

Pavelic, P., Dillon, P., & Robinson, N. (2006). Modelling of well-field design and operation for an aquifer storage transfer and recovery (ASTR) trial. In *Recharge systems for protecting and enhancing groundwater resources. Proceedings of the 5th International Symposium on Management of Aquifer Recharge*, Berlin, Germany, 11–16 June 2005 (pp. 133–138). Paris: UNESCO.

Piet, G. J., & Zoeteman, B. C. J. (1980). Organic water quality changes during sand bank and dune filtration of surface waters in The Netherlands. *Journal American Water Works Association, 72*(7), 400–404.

Piet, G. J., & Zoeteman, B. C. J. (1985). Bank and dune infiltration of surface water in The Netherlands. In T. Asano (Ed.), *Artificial recharge of groundwater* (pp. 529–540). Stoneham, MA: Butterworth.

Price, W. D., Burchell, M. R., Hunt, W. F., & Chescheir, G. M. (2013). Long-term study of dune infiltration systems to treat coastal stormwater runoff for fecal bacteria. *Ecological Engineering, 52,* 1–11.

Regnery, J., Wing, A. D., Kautz, J., & Drewes, J. E. (2016). Introducing sequential managed aquifer recharge technology (SMART)—From laboratory to full-scale application. *Chemosphere, 154,* 8–16.

Rinck-Pfeiffer, S., Pitman, C., & Dillon, P. (2006). Stormwater ASR is practice and ASTR (aquifer storage transfer and recovery) under investigation in Salisbury, South Australia. In *Recharge systems for protecting and enhancing groundwater resources. Proceedings of the 5th International Symposium on Management of Aquifer Recharge*, Berlin, Germany, 11–16 June 2005 (pp. 151–159). Paris: UNESCO.

Schijven, J. F., Hoogenboezem, W., Hassanizadeh, S. M., & Peters, J. H. (1999). Modeling removal of bacteriophages MS2 and PRD1 by dune recharge at Castricum, Netherlands. *Water Resources Research, 35,* 1011–1111.

Sheng, Z. (2005). An aquifer storage and recovery system with reclaimed water to preserve native groundwater resources in El Paso, Texas. *Journal of Environmental Management, 75,* 367–377.

Stuyfzand, P. J. (2015). Trace element patterns in Dutch coastal dunes after 50 years of artificial recharge with Rhine water. *Environmental Earth Sciences, 73,* 7833–7849.

Tielemans, M. W. M. (2007). Artificial recharge of groundwater in The Netherlands. *Water Practice & Technology, 2*(3). https://doi.org/10.2166/wpt.2007064.

Vandenbohede, A., Lebbe, L., & Van Houtte, E. (2008a). Artificial recharge of fresh water in the Belgian coastal dunes. In *20th Salt Water Intrusion Meeting (SWIM-20)* (pp. 282–285).

Vandenbohede, A., Van Houtte, E., & Lebbe, L. (2008b). Groundwater flow in the vicinity of two artificial recharge ponds in the Belgian coastal dunes. *Hydrogeology journal, 16*(8), 1669–1681.

Vandenbohede, A., Van Houtte, E., & Lebbe, L. (2009). Water quality changes in the dunes of the western Belgian coastal plain due to artificial recharge of tertiary treated wastewater. *Applied Geochemistry, 24*(3), 370–382.

Van Houtte, E. (2012). Water reclamation for aquifer recharge in the Flemish dunes. In *Guidelines for water reuse*, EPA/600/R-12/618 (pp. E-21–E-23). Washington, D. C.: U.S. Environmental Protection Agency.

Van Houtte, E., Cauwenberghs, J., Weemaes, M., & Thoeye, C. (2012). Indirect potable reuse via managed aquifer recharge in the Torreele/St Andrè project. In C. Kazner, T. Wintgens, & P. Dillon (Eds.), *Water reclamation technologies for safe managed aquifer recharge* (pp. 33–45). London: IWA Publishing.

Van Houtte, E., & Verbauwhede, J. (2008). Operational experience with indirect potable reuse at the Flemish Coast. *Desalination, 218*(1), 198–207.

Chapter 19
Soil-Aquifer Treatment

19.1 Introduction

The term soil-aquifer treatment (SAT) is used to refer to both high-rate domestic wastewater aquifer recharge and recovery (ARR) systems and, more broadly, the process of natural contaminant attenuation that occurs as applied water passes through the vadose zone. The former, original definition is used herein. SAT involves the application of partially-treated domestic wastewater to the soil surface within engineered infiltration basins. The vadose (unsaturated) zone is used as a natural filter to remove suspended solids, biodegradable organic matter, and pathogenic microorganisms. Significant reductions in nutrients, trace organic compounds (TrOCs), and heavy metals often also occur by sorption and a variety of biologically mediated reactions (Bouwer 1989, 1991; Pescod 1992; Fox et al. 2001a, b). Additional filtration and biogeochemical removal of contaminants occurs as the recharged water travels through the underlying aquifer. The same natural treatment processes active in SAT systems occur in other types of surface-spreading systems, and hence experiences at SAT systems are transferable.

A key features of SAT, as originally defined, is that recharged treated wastewater is recovered and its extent in an aquifer is controlled (Bouwer 1974, 1985, 1989, 1991). Where the underlying receiving aquifer contains freshwater, an integral part of the design and operation of an SAT system is controlling the flow of recharged water in the aquifer so that it is does not migrate away from the system and eventually enter wells used for potable water supply (Bouwer 1989, 1991).

Properly designed and operated SAT systems provide a clear, essentially odorless renovated water that can be used for unrestricted irrigation, primary contact recreation, and other purposes (Bouwer 1985). Post-treatment requirements for recovered water depend upon its quality and its intended use. Disinfection (chlorination) may only be required for some non-potable uses (e.g. public access irrigation). Some organic compounds and pathogens (at low concentrations) pass through SAT systems and additional treatment (e.g., granulated activated carbon and reverse osmosis) is required if the water is to be used for potable supply.

R. G. Maliva, *Anthropogenic Aquifer Recharge*,
Springer Hydrogeology, https://doi.org/10.1007/978-3-030-11084-0_19

The main advantages of SAT systems (Bouwer et al. 1983; Bouwer 1991) are:

- they are robust and usually fail safe
- lesser operator technical expertise is required for their operation
- they provide some underground storage to absorb seasonal or other differences in supply and demand
- they enhance the public acceptance of reuse by breaking the pipe-to-pipe connection between the treatment plant and reuse activity (i.e., the water loses its identity as sewage water).

SAT takes advantage of both renewable (sustainable) and nonrenewable processes to improve water quality (Bouwer 1991). The removal of biochemical oxygen demand (BOD), nitrogen, and microorganisms are renewable processes. Sorption processes are nonrenewable, but appear to occur at slow rates relative to system capacities so that their exhaustion does not impact the long-term operation of systems (Bouwer 1991).

Disadvantages of SAT systems include:

- they require large land areas, which can be a major constraint in already developed (e.g., suburban and urban) areas
- their performance depends upon local hydrogeological conditions, which may not be favorable
- clogging of basins can be a significant operational challenge and cost
- SAT does not removal all organic contaminants
- SAT does not remove salts, and a modest salinity increase can occur due to evaporation, leaching of salts in the soil, and the atmospheric deposition of salt as dust and aerosols.

The recommended pretreatment for municipal wastewater prior to SAT includes primary and secondary treatment or a stabilization pond. Tertiary treatment and disinfection are also commonly performed in the United States (Asano and Cotruvo 2004), where SAT serves more of a polishing function. Alternatively, the SAT is used in Spain and France as a tertiary treatment process to prepare water for unrestricted irrigation (Asano and Cotruvo 2004). Bouwer (1991) suggested that secondary treatment may not be needed. High dissolved organic carbon (DOC) concentrations may result in enhanced nitrogen (N) removal by denitrification (Lance et al. 1980; Bouwer 1991). Greater biological activity was suggested to enhance removal of some synthetic organic compounds through co-metabolism and secondary usage.

The use of primary effluent in SAT is an attractive option for developing countries because it does not require a sizeable investment in engineered treatment facilities and the extensive use of energy and chemicals. Abel (2014) noted that a main limitation of the use of primary effluent in SAT is rapid clogging of the infiltration basins due to the relatively high suspended solids concentration of the effluent. The results of laboratory-scale column experiments indicated that pretreatment of the effluent by coagulation (using aluminum sulfate and iron chloride) and settling improves both water quality and the operation of spreading basins. The use of coagulants increased the removal of DOC, phosphorus, and pathogens (Abel 2014).

19.2 SAT Design Basics

SAT infiltration basin systems are designed with multiple cells to allow for continuous operation while individual basins are alternated between flooding and drying. Flooding and drying cycling functions to manage clogging and control the redox state of infiltrated waters, which is important for nitrogen removal. Bouwer (1985) observed that the most critical factor for the successful operation of an SAT system is unquestionably having an adequate basin area to handle the design flow and to allow for a system-specific optimal wetting and drying schedule. System design must, therefore, consider inevitable reductions in infiltration and hydraulic loading rates due to clogging.

Bouwer (1974, 1989, 1991) presented several conceptual design types of SAT systems. Infiltration creates local hydraulic mounds and recovery creates cones of depression, which are used to control the movement of infiltrated water. A key design consideration is configuring the infiltration and recovery locations and rates so that the movement of the effluent plume is controlled. Groundwater modeling is thus an essential tool for system design.

Two main practical design options are to recover water using a central line of production wells paralleled on both sides by recharge basins or have recharge basins in the center of the system surrounded by production wells (Fig. 19.1). In the former case, the production wells create a cone of depressions that draws the infiltrated water inward. In the latter design, the production wells capture both in the infiltrated water and some groundwater from the opposite direction. Basic design considerations for SAT systems are (Bouwer 1974, 1989, 1991; Pescod 1992):

- A sufficiently high infiltration (hydraulic loading) rate should be achieved so that target effluent flows can be accommodated in an economically-sized infiltration basin system.
- Soil and aquifer sediments should allow for sufficient filtration and residence time for natural attenuation processes to achieve water quality goals. Very high loading rates may result in a lesser degree of treatment.
- Infiltration basins should be located and designed so that the basin floors are at least one meter (3 ft) above the water table at all times (Pescod 1992). Sites with shallow water tables are unsuitable for SAT because they have an inadequately thick vadose zone.
- If the applied sewage effluent is of significantly lesser quality than the native groundwater, then the system will have to be designed and operated to prevent migration of the effluent outside of the part of the aquifer used for the SAT system.
- System-specific optimal flooding and drying schedules must be determined to achieve treatment goals, such as meeting nutrient concentration targets and effective management of clogging.
- Pretreatment of wastewater (TSS removal) should be sufficient to allow for an acceptable (manageable) clogging rate.
- Biological activity in the infiltration basins should be controlled. Excessive algae growth could cause accelerated clogging through the development of nearly imper-

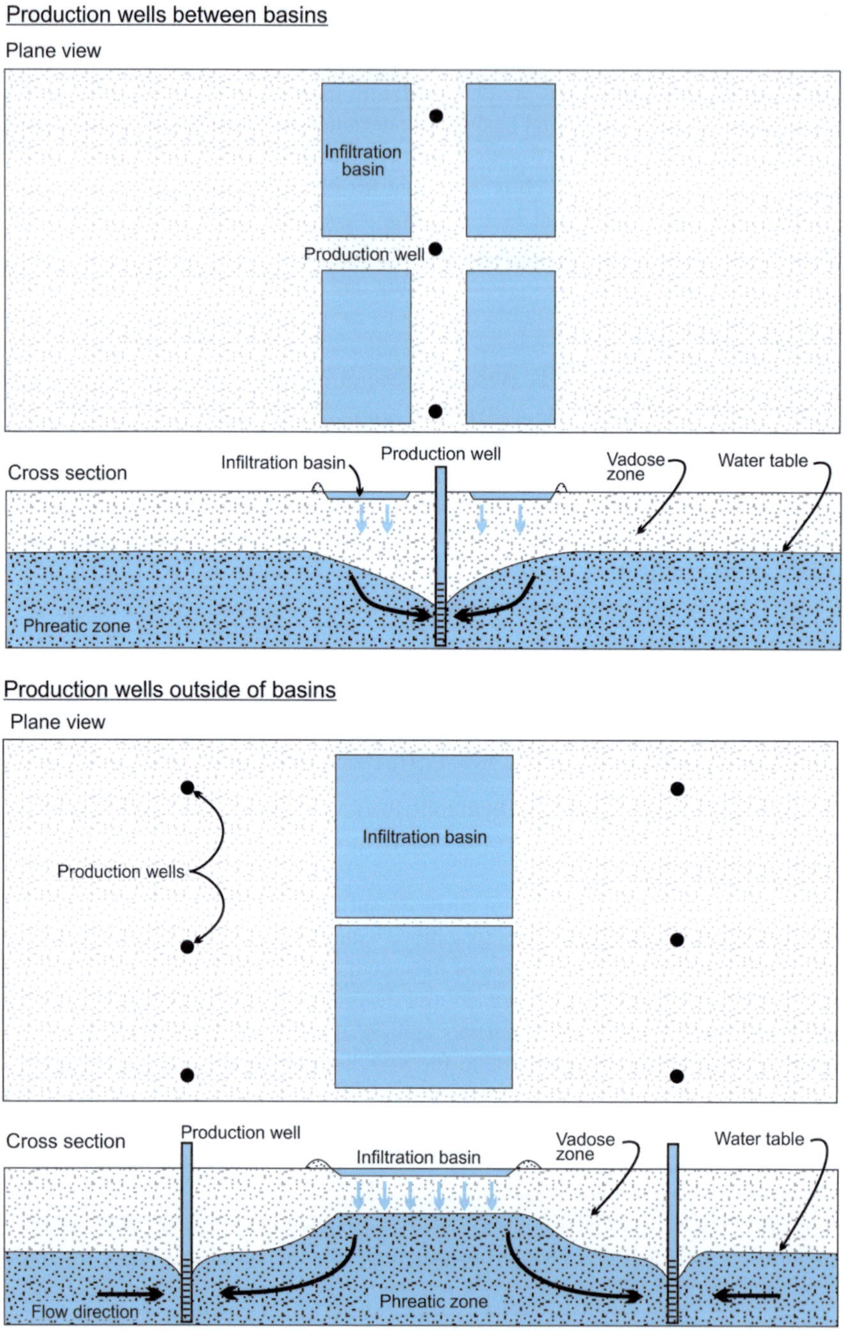

Fig. 19.1 Conceptual diagram of SAT system designs with production wells located between basins (top) and outside of basins (bottom)

vious biofilm layers, accumulation of organic matter, and induced calcium carbonate precipitation by the removal of carbon dioxide (Bouwer 1991; Asano and Cotruvo 2004).

Site selection considerations for an SAT system in Alice Springs, Australia, included (Dillon et al. 2006):

- proximity to the source of water and locations of potential demand
- preference to operate on government land
- avoidance of sacred sites
- avoidance of densely populated areas
- avoidance of areas that are flood prone or have high reliefs
- presence of a sufficiently thick vadose zone (storage space)
- avoidance of salinization and the creation of groundwater dependent ecosystems
- presence of high vertical hydraulic conductivities to avoid perched conditions
- avoidance of areas were aquifers are locally used for potable supply (the selected site had a salinity above drinking water guidelines).

Design options include the:

- size, depth, and configuration of basins (depths may be increased to access high permeability soils)
- number and location of recovery wells (distance from basins affects transit/residence times)
- depths and screened intervals of the recovery wells.

Greater travel times and distances to recovery wells, in general, result in a better quality of the recovered water, at least up to a certain limit (Bouwer 1974). Most of the improvement in water quality occurs in the upper 1–2 m (3.3–6.6 ft) of the infiltrated soil. There are thus diminishing returns concerning water quality improvement with increasing residence time and travel distance. A trade off-occurs between infiltration rates and treatment. High infiltration and percolation rates can result in rapid flow through the unsaturated zone and, as a result, lesser time for biogeochemical contaminant reduction processes to occur.

SAT systems should be constructed in granular soils that have an adequate permeability to allow for sufficiently high infiltration rates but yet be fine enough to provide good filtration. The best soils are in the fine sand, loamy sand, and sandy loam range (Bouwer 1985, 1989, 1991; Pescod 1992). Ideally, fine sands should be underlain by coarser materials. The vadose zone and underlying shallow aquifer should not contain beds of very low permeability material, such as clays, that would impede the vertical flow of water. Undesirable soil properties for SAT include (National Research Council 1994):

- high degree of spatial variability of properties
- different residence times, which can locally exhaust cation exchange capacity (CEC) and sorptive capacity
- low-permeability clogging layers that may limit vertical flow irrespective of underlying soil properties

- high-permeability zones that may receive greater flow and preferentially clog
- preferential flow through structured soil (e.g., macropores, variations in packing), which can result in rapid clogging of high permeability pathways.

Pretreatment is additionally important in controlling the redox state of SAT systems. High organic carbon concentrations result in high total oxygen demands and, in turn, the removal of dissolved oxygen (DO) and creation of anaerobic conditions in the saturated zone (Fox et al. 2001a, b). Low total oxygen demands are necessary in order to maintain aerobic conditions. Anoxic conditions can have the adverse impact of mobilizing metals (e.g., Fe and Mn). However, periods of anoxic conditions are desirable for denitrification.

19.3 Water Quality Improvement Processes During SAT

Operational data and laboratory experiment results demonstrate that SAT is an effective method for improving the quality of treated wastewater. SAT can reduce the concentrations of pathogens, nutrients, metals, dissolved organic matter, and organic pollutants. General pollutant removal processes are discussed in this section. Operational data from specific systems are summarized in Sect. 19.4.

19.3.1 Pathogen Removal

Pathogen attenuation occurs during infiltration, percolation, and phreatic zone flow as a result of a variety of physical and biological processes, including physical retention (e.g., filtration, straining, sedimentation, adhesion), inactivation (dying off and predation), and dilution (Sect. 7.2.1). SAT has been documented to be highly effective in removing pathogens (e.g., Wilson et al. 1995). Pathogen attenuation has not received as much attention as chemical attenuation in the intensely studied SAT systems because the reclaimed water used was already of high quality (commonly tertiary treated with disinfection) and had very low pathogen concentrations.

Fox et al. (2001a, b) reported that field data from the Montebello Forebay (California, USA) SAT site suggest that 30 m (100 ft) of subsurface travel is sufficient for a 7-log removal of a bacteriophage that was used as a virus tracer. Testing of infiltration galleries using secondary-treated wastewater in Australia indicate a 3-log reduction in microorganisms (Bekele et al. 2009).

In general, passage through the vadose zone is insufficient for complete pathogen removal. Pathogens are frequently detected in shallow monitoring wells installed near recharge basins. Pathogen concentrations decrease with increasing distance from the basins. Reduction in pathogen concentrations to non-detectable levels occurs during groundwater flow by filtration, sorption, and inactivation.

19.3.2 Nitrogen Removal

SAT systems are typically operated to remove dissolved organic nitrogen from wastewater by a combination of nitrification and denitrification. Nitrification is the biologically mediated process by which ammonia (NH_3) and ammonium (NH_4^+) are converted to nitrite (NO_2^-) and then nitrate (NO_3^-) under oxic conditions. Ammonium is the ionized form of ammonia with the ratio of species dependent on pH. The basic equations for the nitrification processes are (Stumm and Morgan 1996):

$$NH_3 + 1.5O_2 \rightarrow NO_2^- + H_2O + H^+ \tag{19.1}$$

$$NH_4^+ + 1.5O_2 \rightarrow NO_2^- + H_2O + 2H^+ \tag{19.2}$$

$$NO_2^- + 0.5O_2 \rightarrow NO_3^- \tag{19.3}$$

Denitrification is the bacterially mediated conversion of nitrate to nitrogen gas, which occurs under anoxic conditions, as follows

$$NO_3^- + 1.25\{CH_2O\} + H^+ \rightarrow 0.5N_2(g) + 1.75H_2O + 1.25CO_2 \tag{19.4}$$

Nitrification is an autotrophic processes, whereas denitrification is a heterotrophic process in that it requires a source of organic carbon, which in SAT systems is normally provided by the wastewater. Nitrogen removal is a function of residence time, biochemical oxygen demand (BOD) to N ratio, and the development of anoxic conditions (Crites et al. 2014). Nitrogen is also absorbed onto the soil to varying degrees during recharge, which provides time for later biological conversion.

Fox (2002) observed that at the Sweetwater Recharge Facilities SAT site (Tucson, Arizona) insufficient organic carbon for heterotrophic denitrification was present and that the anaerobic ammonium oxidation (ANAMMOX) process was likely active. The ANAMMOX process uses adsorbed ammonium as an electron donor instead of carbon (Van de Graaf et al. 1995):

$$5NH_4^+ + 3NO_3^- \rightarrow 4N_2 + 9H_2O + 2H^+ \tag{19.5}$$

Gable and Fox (2003) performed bench-top experiments using SAT system soil (from the Sweetwater Recharge Facilities) to investigate nitrogen removal mechanisms. The results demonstrated that the presence of ammonia enhances nitrogen production. Adsorbed ammonia can act as a sustainable supply of electron donors for the observed nitrate removal. From an operational perspective, wetting and drying cycles should be adjusted to maintain adsorbed ammonia in the soil (Gable and Fox 2003).

Column experiments by Fox and Shah (2006) demonstrated that when a mixture of nitrite and ammonia was added, the nitrogen removal rate was an order of magnitude greater compared to a mixture of nitrate and ammonia. The operative reaction is

$$NH_4^+ + NO_2^- \rightarrow N_2 + H_2O \tag{19.6}$$

which is mediated by bacteria belonging to the bacterial phylum Planctomycetes. Nitrite is typically present at much lower concentrations than nitrate in treated wastewater. Fox and Shah (2006) noted that their experimental results suggest that the conversion of nitrate to nitrite is a potential rate limiting step for nitrogen removal in SAT systems. They also noted the exact mechanism for the reduction of nitrate to nitrite had not yet been determined.

The form and concentration of nitrogen compounds in water passing through SAT systems depend upon the hydraulic loading rate and the flooding and drying schedule for the infiltration basins (Bouwer 1974, 1985, 1991; Pescod 1992). System-specific flooding and drying schedules need to consider such factors as (Bouwer 1991):

- ammonium and carbon concentrations in the sewage effluent
- infiltration rate
- cation exchange capacity of the soil
- exchangeable ammonium percentage in the soil
- depth of oxygen penetration into the soil during drying periods
- soil and effluent temperature.

Frequent, short-duration (several days long) flooding and drying periods promotes aerobic processes, which favor the nitrification of ammonium. Longer (one month or greater) flooding and drying periods lead to anaerobic conditions. Rapid infiltration systems that are continuously flooded maintain anaerobic conditions, which are not conducive for nitrogen removal (Benhan-Blair and Associates and Engineering Enterprises 1979). Intermediate (one to two week) flooding and drying periods can result in a succession of aerobic and anaerobic conditions in the upper part of the soil profile, which stimulates nitrification and then denitrification (Pescod 1992). Optimization of the operation of SAT systems is an exercise in adaptive management. The optimal site-specific schedule of flooding and drying, and the cleaning of basins, must be evaluated based on site-specific operational experiences (Pescod 1992).

19.3.3 Phosphorous Removal

SAT is effective in reducing the concentration of phosphorous through chemical precipitation (calcium phosphate) and adsorption. Adsorption is the more rapid process, but soils have a finite adsorptive capacity that may eventually become exhausted. However, non-sustainable processes (adsorption and chemical precipitation) can persist for a very long time (tens or hundreds of years; Idelovitch et al. 2003). The Dan Region Project (Shafdan) SAT system in Israel still had excellent and stable removal of trace elements and phosphorous after 25 years of operation (Idelovitch et al. 2003). Phosphorous removal rates by adsorption will depend upon soil properties and travel distance. Crites et al. (2014) noted that reported phosphorus removal from SAT systems ranged from 27 to 99% with most reported values greater than 50%. If phosphorous removal is a critical objective, then adsorption tests should be performed using site-specific soils (Crites et al. 2014). Actual long-tern phosphorous retention in SAT systems was reported to be 2–5 times greater than the values measured in standard 5-day jar adsorption tests (USEPA 1981; Crites et al. 2014).

19.3.4 Organic Carbon Removal

It was recognized from the earliest SAT research (e.g., Bouwer et al. 1974; Bouwer 1980) that SAT is effective in reducing the concentration of dissolved organic carbon (DOC) in wastewater. Wastewater effluent organic matter (EfOM) includes both natural organic matter (NOM) that was already present in the drinking water and soluble microbial products (SMPs) derived from the biological wastewater treatment process (Amy et al. 1987; Drewes and Fox 1999). NOM is dominated by humic and fluvic compounds. Data from operational SAT sites located in Mesa (Northwest Water Reclamation Plant) and Tucson, Arizona (Sweetwater Underground Storage and Recovery Facility) indicate a 50–70% removal of DOC after accounting for dilution (Fox et al. 2001a; Drewes et al. 2003; Amy and Drewes 2007). Organic carbon removal in MAR systems, in general (including SAT system), is summarized in Sect. 7.5.

Organic carbon removal in SAT systems occurs primarily by biodegradation processes, rather than by sorption, and is a sustainable process (Quanrud et al. 2003; Fox et al. 2005). Field and laboratory studies of DOC removal and changes in SAT systems indicate the following (Quanrud et al. 1996; Drewes and Fox 1999; Fox et al. 2001a; Drewes et al. 2003; Sattler et al. 2006; Amy and Drewes 2007; Drewes 2009):

- preferential early removal of ultra-hydrophilic compounds, such as amino acids, proteins, and polysaccharides, and dissolved organic nitrogen (DON)
- sustained biodegradation of more poorly degradable compounds occurs during longer-term SAT
- THM formation potential is greatly reduced
- Specific UV absorbance (SUVA), a measure of aromaticity, may initially increase reflecting the preferential biodegradation of non-humic EfOM
- Long-term SAT results in a loss of aromatic carboxylic character (removal of humic substances) and a relative increase in aliphatic compounds (decrease in SUVA).

19.3.5 Trace Organic Compounds

The effectiveness of SAT in the removal of TrOCs has been investigated through both laboratory experiments and field evaluations, which are summarized in Sect. 7.4.4. SAT is effective in reducing the concentrations of many TrOCs that survive conventional secondary and tertiary treatment processes. Some compounds are not completely removed during passage through the vadose zone, but are later attenuated in the groundwater environment. A small number of refractory compounds show little or no removal. Drewes et al. (2002) evaluated TrOC removal in treated effluent and monitoring well samples from five SAT systems in the United States. The main observations are:

- Anti-inflammatory drugs (e.g., diclofenac, ibuprofen, fenoprofen, ketoprofen, naproxen) are present in secondary and tertiary treated effluents. Their concentrations are significantly lower in facilities employing nitrification and denitrification.
- TrOCs were not detected in RO permeate from the Scottsdale Water Campus (Arizona).
- Antiepileptic drugs carbamazepine and primidone were detected in all downgradient monitoring wells. Significant removal of these compounds was not evident.
- SAT has a high potential for removing acidic drugs (analgesics/anti-inflammatory drugs and lipid regulators).

Field data from SAT systems in Mesa and Tucson, Arizona, documented reductions of 17β-estradiol, estriol, and testosterone concentrations to below detection limits (Amy and Drewes 2007). The only pharmaceutically active compounds (PHACs) detected after SAT were the antiepileptic drugs carbamazepine and primidone. The chlorinated flame retardant TCIPP (tris-(chloroisopropyl)-phosphate) was also detected after SAT in the most down-gradient monitoring well, but at a significantly reduced concentration.

The performance of SAT systems, in terms of removal of more refractory TrOCs, may improve over time as the biological activity of the systems increases with ripening. Column testing results showed that the removal of gemfibrozil, diclofenac, and bezafibrate increased from less than 20% in reactors ripened for five days to over 90% when the reactors were ripened for 240 days (Abel 2014). Phenacetin, paracetamol, ibuprofen and caffeine were easily removed under various operating conditions. Reduced removal of some TrOCs might occur after drying or scraping of SAT basins.

The potential health impacts of the low levels of TrOCs, including pharmaceutically active compounds (PHACs) remaining in reclaimed water after SAT must be put into perspective. Amy and Drewes (2007, p. 25) noted that

> The health effects of such low (ng/L) levels are still being debated but it is noteworthy that I70 values (lifetime intake via drinking water based on 70 years and 2L/day) are far less than daily therapeutic doses for almost all PHACs.

19.3.6 *Metals*

Metals occur in water in the dissolved form (free or complexed ions) and in the particulate form. Where metals are present as finely divided suspended solids (or sorbed onto such particles), their removal occurs mainly through straining and filtration. Dissolved metals are removed by ion exchange reactions, precipitation, and surface adsorption.

Short-term laboratory experiments by Lin et al. (2004) indicate that surface adsorption and precipitation on Fe oxides and/or carbonates may be the major mechanisms of the retention of Cu, Ni, and Zn in SAT soils. The establishment of reducing conditions can result in the reductive dissolution and mobilization of oxidized metals in SAT systems, particularly manganese (Oren et al. 2007) and iron.

19.4 Demonstration and Operational SAT Systems

Pioneering studies of SAT were performed in Phoenix, Arizona, as a cooperative project between the U.S. Water Conservation Laboratory, the Salt River Project, and the City of Phoenix. The first project was the Flushing Meadows pilot project, which was succeeded by a larger-scale SAT demonstration project, the 23rd Avenue Project. The Phoenix program is particularly noteworthy for the general insights it provided on the improvements in water quality possible using SAT and on the operational controls on the quality of renovated water.

19.4.1 *Flushing Meadows Project*

Bouwer (1974) described a pilot SAT system, the Flushing Meadows Project, that was constructed in 1967 in the bed of the Salt River (an ephemeral river) in the City of Phoenix. Secondary treated wastewater was pumped to six parallel basins, each 6.1 m by 214 m (20 by 400 ft), located 6.1 m (20 ft) apart (Fig. 19.2). The soil in the basins consisted of about 1 m (3 ft) of fine, loamy sand, followed at depth by coarse sand and gravel layers to the top of a clay confining unit at about 75 m (240 ft) below land surface (Bouwer 1980). The project site was damaged by a severe flood in March 1978 and was not restored.

The first five years of the project focused on maximizing the hydraulic loading rate and associated water quality improvements. The main thrust of the second five years of the project was research on maximizing nitrogen removal (Bouwer 1980). The system achieved an annual loading rate of 92–122 m (300–400 ft) at a water depth of 0.3 m (1 ft). The maximum hydraulic loading rates were obtaining with a flooding period of about 20 days followed by drying periods of 10 days in the summer and

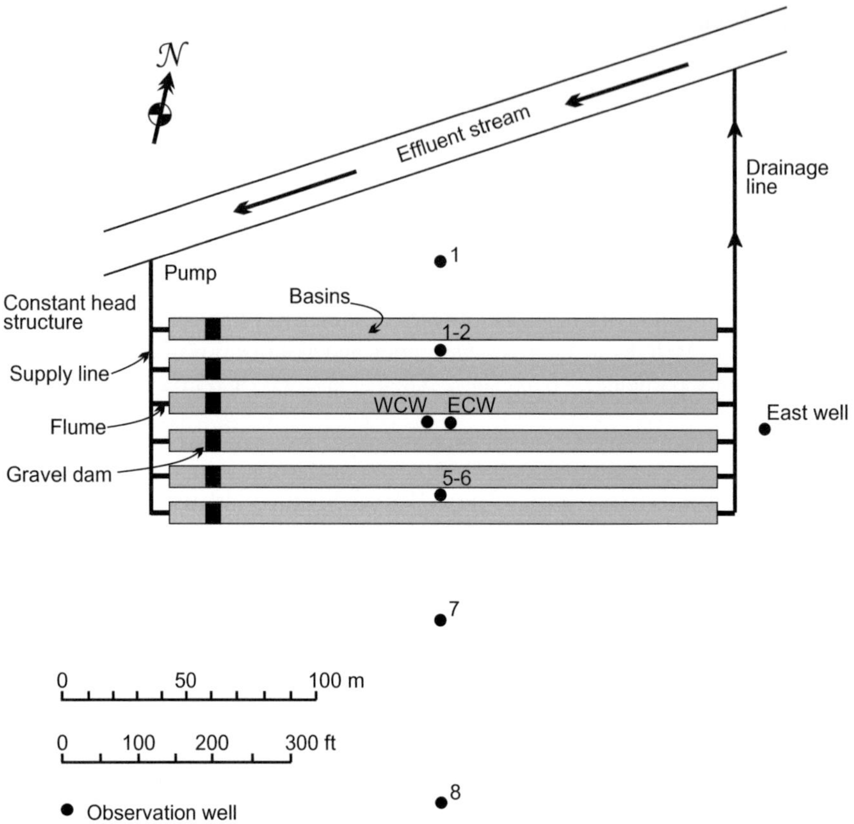

Fig. 19.2 Diagram of the Flushing Meadows Project, Phoenix, Arizona (modified from Bouwer et al. 1974; Bouwer 1980)

20 days in the winter. Renovated water was recovered from wells located between the basins (WCW and ECW).

The pilot SAT system achieved a substantial improvement in water quality. Overall total nitrogen removal was about 30% at the maximum hydraulic loading rate. Bouwer (1974) summarized the basic process of nitrogen removal from the wastewater, which contained about 35 mg/L of nitrogen (almost entirely as ammonium):

- During flooding, nitrification of ammonium initially occurs using DO in the recharged water. Ammonium is adsorbed onto clay and organic matter.
- During drying, the entry of oxygen allows for the nitrification of adsorbed ammonium and the capillary water becomes enriched in nitrate.
- During subsequent flooding, nitrate is pushed downward resulting in an initial nitrate peak.
- Denitrification occurs when water containing nitrate and organic carbon enters the anaerobic zone.

A key result of the second phase of research at the Flushing Meadows site is that the rate of nitrogen removal depended upon the wetting and drying cycle lengths and hydraulic loading rate. Nitrogen removal depended on the balance between the amount of NH_4–N that enters the soil during flooding and the amount of adsorbed NH_4–N that is nitrified during drying (Bouwer 1980). A change to a cycle of 9 days flooding and 12 days drying, and reduction of the hydraulic loading rate to 70 m/year increased the total nitrogen removal rate to 65% (Bouwer 1980; Bouwer et al. 1983). Too long flooding periods (1 month or more) resulted in the cation exchange capacity of the soils becoming saturated with respect to ammonium and increased ammonium concentrations in the recovered renovated water, reducing the nitrogen removal percentage.

Other water quality changes were (Gilbert et al. 1976; Bouwer 1980, 1985; Bouwer et al. 1983):

- **Total phosphate**: 50–90% removal, with the primary cause suggested to be calcium phosphate precipitation. Phosphorous removal increased with decreasing loading rates and in the groundwater with distance from the basins.
- **Salinity**: 2% increase due to evaporation.
- **BOD_5**: reduction from 10 to 20 mg/L to less than 0.5 mg/L in the renovated water.
- **COD**: reduction from 30 to 60 mg/L in the effluent to 10–20 mg/L in the renovated water.
- **TOC**: reduction from 5 to 2 mg/L, indicating that some refractory organic carbon (humic and fluvic) remained in the renovated water.
- **Fecal coliform bacteria**: reduced from 10^5 to 10^6/100 mL to usually 0–100/100 mL. No fecal coliform bacteria were detected beyond 91 m.
- **Viruses**: not detected in the renovated water.
- **Trace metal removal**: the decrease in renovated water varied between metals (in μg/L)
  - zinc: 193 → 35
  - copper: 123 → 6
  - cadmium: 7.7 → 7.2
  - lead: 82 → 66
- **Boron**: not removed.

The Flushing Meadows Project demonstrated that bacterial and viral pathogens are largely removed as the effluent percolates through the soil (Gilbert et al. 1976). Viruses, enteric bacterial pathogens, and pollutant indicator organisms were either eliminated or greatly reduced in the renovated water (Gilbert et al. 1976). The highest fecal coliform concentrations in the renovated water occurred at the beginning of flooding due to a clogging layer not yet being developed and lesser biological activity in the soil (Bouwer et al. 1983). Column experiments using secondary-treated sewage effluent with viruses added and soils from the Flushing Meadows Project demonstrated that most of the virus removal occurs as the effluent passed through the first few centimeters of the soil (Lance et al. 1976). Virus removal was found to be due primarily to adsorption. Virus desorption occurred when deionized water was applied within two hours of the removal of effluent-virus mixture. Most of the desorbed viruses appeared to be re-adsorbed lower in the column (Lance et al. 1976).

19.4.2 23rd Avenue Demonstration Project

The 23rd Avenue Demonstration Project was constructed in 1975 as a large-scale successor to the Flushing Meadows ASR pilot project. The demonstration project was also located in the Salt River bed in the Phoenix area. The project consisted of a 16 ha basin that was divided into 4 ha four cells. Water was recovered from a well located in the center of the basins. The depth of the water table was about 17 m and the hydraulic loading rate was about 100 m/year (Bouwer and Rice 1984; Bouwer 1991). Some of the changes in raw water quality between the recharged secondary effluent and renovated water are summarized in Table 19.1.

Volatilization was found to be important for the removal of low-molecular weight volatile compounds, which had removal rates of between 30 and 70% as water moved through the basins (Bouwer et al. 1984). Non-halogenated aliphatic and aromatic hydrocarbons present in the effluent showed high degrees of removal during percolation with reductions in concentration in the renovated water of 50–99% and measured concentrations near or below detection limits. Halogenated organic compounds tends to be more refractory with lesser removal rates (Bouwer et al. 1983, 1984). The concentrations of trichloroethylene, tetrachloroethylene, and pentachloroanisole

Table 19.1 Water quality changes at the 23rd Avenue (Phoenix) SAT demonstration project

Parameter	Secondary treated effluent (chlorinated)	Renovated water
TDS (mg/L)	750	790
TSS (mg/L)	11	1
Ammonium N (mg/L)	16	0.01
Nitrate N (mg/L)	0.5	5.3
Organic N (mg/L)	1.5	0.1
Phosphate	5.5	0.4
BOD (mg/L)	12	0
TOC (mg/L)	12	1.9
Fecal coliforms (no/100 mL)	3,500	0.3
Viruses (PFU/100 L)	1.3	0

Bouwer et al. (1983), Bouwer and Rice (1984), Bouwer (1991)

appeared to be significantly higher in the renovated water than in the basin water, which might have been due to variations in the concentrations in the effluent (Bouwer et al. 1984). The renovated water contained very low concentrations (typically μg/L level) of a wide variety of synthetic organic compounds and additional treatment (e.g., GAC) would be needed for potable use (Bouwer 1980, 1991). RO treatment of some of the water would also be required to reduce TDS concentrations to drinking water standards.

19.4.3 Dan Region Water Reclamation Project (Shafdan)

Dan Regional Reclamation Project (known by its Hebrew acronym Shafdan) is the largest wastewater treatment plant in Israel, and larger than any other plant in Europe and the Middle East, with an annual treated volume of more than 140 million m^3/year (MCM/year) from the Greater Tel Aviv area (Aharoni et al. 2011; Cikurel et al. 2012). Secondary-treated wastewater (using the activated sludge process) is treated by SAT to almost of drinking water quality and is transported south to the Negev region for unrestricted irrigation.

Banin et al. (2002) evaluated changes in the soil after 23 years of operation (since 1977) of the SAT system. The hydrogeology of the site consists of a shallow aquifer (Coastal Plain Aquifer), composed mainly of calcareous sandstones, that is overlain by coastal dune sands. Nine soil profiles were taken, each of which was sampled at 10 depths to a total depth of 4.0 m. The profiles were compared to a nearby pristine dune. The main changes in soil composition were:

- Total organic carbon accumulation was 5.82 kg per m^2 to a 4.0 m depth, which represents only ~3% of the total organic carbon input. The majority of the applied organic carbon was thus decomposed.
- Carbonate leaching occurring with up to 50% of the original carbonate removed in the upper horizons of the basin soils.
- Leaching of some metals occurred. About 50% of the manganese was removed by the reductive dissolution of manganese oxides. A lesser amount (0–10%) of the iron was removed by the reductive dissolution of various iron oxides.
- Accumulation of trace metals (Cu, Cr, Ni, Zn) occurred by sorption and ion exchange processes.
- Redox cycling. Anoxic/reduced conditions occurred during recharge (Eh ~ −100 to −250 mV, pe −2 to −4) and oxic/oxidized conditions occurred following draining (Eh ~ 500–600 mV, pe 8–10).

The Dan Region Project SAT system still had excellent and stable removal of trace elements and phosphorous after 25 years of operation (Idelovitch et al. 2003). Operational problems after 30 years (Cikurel et al. 2012) included:

- deterioration of infiltration rates due to clogging
- anoxic conditions in the aquifer cause releases of Mn and Fe to the reclaimed water that later form oxides that clog irrigation systems

- no more land is available for system expansion
- biofouling of effluent pipes.

Aharoni et el. (2011) discussed several hybrid SAT options potentially applicable at Shafdan and other facilities, which include additional pretreatment or post-treatment, short-term SAT (15–20 m of travel and 30-day residence time) and/or the use of dug wells for recharge. Specific options include:

- pretreatment by UF, rapid infiltration in a dug well, and short-term SAT
- short-term SAT as pretreatment for NF to polish for indirect potable reuse quality water
- pre-ozonation prior to dug well infiltration.

Potential advantages of pretreatment include:

- reduced clogging
- reduced buildup of organic matter and heavy metals accumulation in soils
- prevention of manganese dissolution
- more effective removal of micropollutants.

Short-term SAT could be used to remove more readily biodegradable compounds as a pretreatment to conventional UF-RO processes.

The declining infiltration rates and lack of room for new infiltration ponds prompted an investigation of hybrid SAT strategies (Cikurel et al. 2012). Short-term SAT consisting of pretreating secondary-treated effluent by UF and recharge through a series of dug wells located around the infiltration basins and closer to the extraction wells was pilot tested. The short-term SAT had a residence/travel time of up to two months versus 12 months in the existing operational SAT system. The pilot system dug wells had depths of 2.5 m, with 12–13 m of remaining underlying unsaturated zone, and recharge rates of 100–120 m^3/d.

The results of the short-term SAT testing were

- introducing organic-free and DO-rich water improved redox conditions
- effective removal of microorganisms
- DOC was reduced from about 10 to 2–3 mg/L, which is a slightly lesser removal than the 1–2 mg/L concentration achieved by conventional SAT
- similar nitrogen removal as conventional SAT
- relatively low phosphorous removal
- suspended solids were reduced at the 0.01 μm level by UF, which would help maintain infiltration rates.

Elkayam et al. (2015) noted that the experiences at the Shafdan SAT system demonstrate that SAT provides effective pathogen removal and that the recovered water is suitable for unrestricted irrigation uses (including crops to be eaten raw) without either pre-disinfection or post-disinfection. Since 1995, there was not a single positive test for fecal coliforms, and since 2001, enteroviruses had not been detected in observation wells located between the infiltration area and reclamation wells. Early detections were suggested to have been false positives. A key feature of the Shafdan system is a long retention time (average of 960 days), which provides ample time for pathogen attenuation to occur. Although disinfection of the recovered water is not necessary, it was noted that maintaining residual disinfection in the distribution system might be needed to prevent regrowth.

19.4.4 Northwest Water Reclamation Plant (Mesa, Arizona) SAT System

The City of Mesa Northwest Water Reclamation Plant (NWWRP) SAT system (Fig. 19.3) is particularly noteworthy because it has been the subject of a number of field investigations on water quality changes during SAT. The history of the NWWRP SAT system was reviewed by Fox et al. (2001a, b). Operation of the facility began in 1990 and the system currently consists of four recharge basins with a total area of about 28.9 acres (11.7 ha). The individual basin areas are 7.8, 7.4, 5.3, and 8.4 acres (3.2, 3.0, 2.1, and 3.4 ha). The water sent to the infiltration ponds is treated by primary sedimentation, biological nitrification/denitrification, chlorination, and secondary filtration. The system has an operational capacity of 4 Mgd (15,000 m^3/d). The flooding-drying cycle varies between basins depending upon infiltration rates

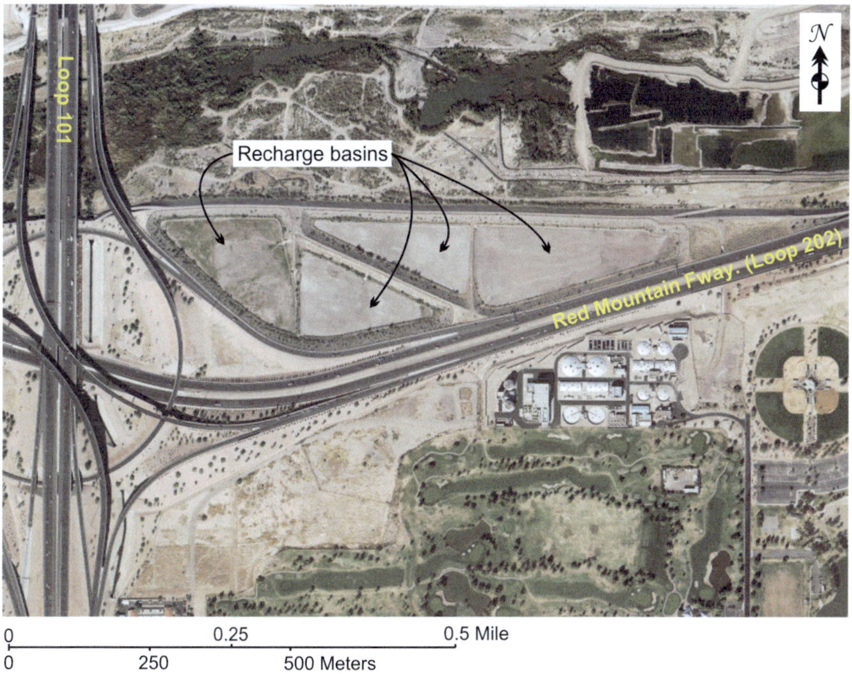

Fig. 19.3 Aerial photograph of the Mesa Northwest Water Reclamation Plant SAT system, Arizona (Photograph source: U.S. Geological Survey)

and soil properties. Fox (2011) reported that the system is operated with approximately 7 days of wetting and 21 days of drying. The SAT system was designed for a plant capacity of 8 Mgd (30,000 m^3/d), but fine layers of clay limited the plant's capacity to its current 4 Mgd and result in horizontal flows (Fox 2011). Recovery of water is minimal and the reclaimed water was reported to be moving to the southwest, where it is monitored by over 20 monitoring wells.

Investigations of the NWWRP include:

- dissolved organic carbon removal and compositional changes (Fox et al. 2001a, c, 2005)
- steroid and TrOC removal (Amy and Drewes 2007)
- fate of alkylphenol polyethoxylates (APEOs) metabolites (Montgomery-Brown et al. 2003).

19.4.5 Sweetwater Recharge Facilities SAT System (Tucson, Arizona)

The history of the Tucson Water (a department of the City of Tucson) Sweetwater Recharge Facilities (Fig. 19.4) was described by Kmiec and Thomure (2005) from which this summary was derived. The Demonstration Phase (1984–1989) consisted of the construction and testing of four small (≈0.75 acres, 0.3 ha each) recharge basins to determine infiltration rates, evaluate monitoring and measuring equipment, and evaluate water level and quality changes during recharge. The development phase (1989–1997) consisted of the construction of the initial four recharge basins (RB-001 through RB-004), which have a total area of 13 acres (5.3 ha). The basins were constructed to a depth of 10–15 ft (3.0–4.5 m) so as to penetrate more permeable strata present at these depths and thus increase infiltration rates. The system was permitted to recharge and recover approximately 3,200 acre-ft (3.95 MCM) per year of secondary-treated wastewater from the Pima County Roger Road Wastewater Treatment Plant. Infiltration rates were greatly reduced due to clogging caused by algal flocculation. Clogging was successfully addressed by decreasing the wetting phase to less than a week and increasing the length of the drying cycle to desiccate, shrink, and crack the algal clogging layer.

The full-scale phase extends from 1997 to present and includes the construction of Sweetwater Wetlands treatment system and four additional recharge basins (RB-005 through PB-008) on the opposite side of the Santa Cruz River with the approximately same area as the first four basins. Water is recovered using six extraction wells. Infiltration rates were reported to average approximately 2.3 ft/d (0.7 m/d) under full-scale operation. The permitted storage volume is 6,500 acre ft (8.02 MCM) per year and most of the recharged water is recovered each year. Infiltration capacity is maintained by wetting and drying cycling that disrupts the algal layers and creates desiccation cracks. Periodic more extensive maintenance involves ripping the basin floor to a depth of 1–3 ft (0.3–0.9 m) and construction of furrows to increase the surface area of the basins.

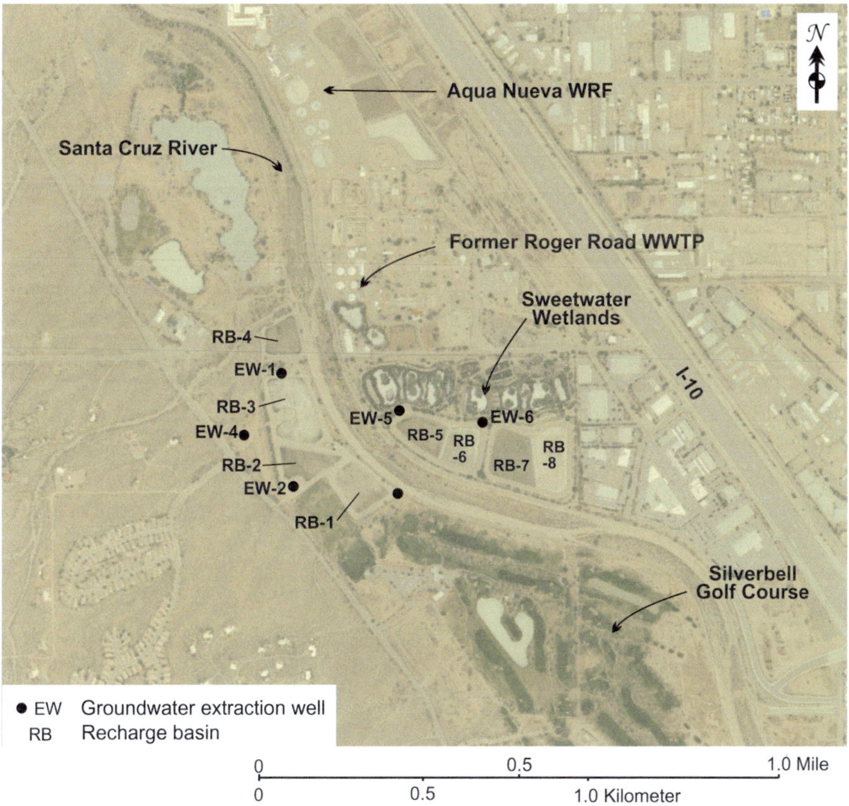

Fig. 19.4 Aerial photograph of the Mesa Sweetwater Recharge Facilities SAT system, Tucson, Arizona (Photograph source: U.S. Geological Survey)

Water quality improvements during the developmental phase of the SAT project were investigated by Wilson et al. (1995). Vadose zone and phreatic zone water samples from RB-001 were collected from the time period when it was recharged with tertiary treated water. The recharged water already had very low concentrations of pathogens. The SAT system is very effective in reducing the concentration of DOC, which was decreased by approximately 92% during percolation through the entire vadose zone to values within the natural background range in Tucson. The bulk of the removal occurs in the upper 3.1 m of the vadose zone below the basin. Total organic halides concentrations decreased by approximately 85%. Wilson et al. (1995) reported a reduction in total nitrogen concentration of 57%, and indicated that denitrification likely occurs within a perched layer that experiences alternating wet and dry conditions. Kmiec and Thomure (2005) reported a system-wide total nitrogen removal of 29%. Several other studies were performed at the site including nitrogen removal processes (Fox 2002; Gable and Fox 2003), organic carbon removal (Quanrud et al. 2003), and steroid removal (Amy and Drewes 2007).

19.4.6 Mandurah, Western Australia

Toze et al. (2004) documented water quality improvements resulting from infiltration and groundwater flow at a wastewater treatment facility in Mandurah, Western Australia. Secondary-treated wastewater is discharged into 2500 m^2 infiltration ponds, where it is allowed to infiltrate into a shallow fractured-limestone aquifer. Water is recovered from two bores installed approximately 80 and 100 m from the ponds. The recovered water is intended for irrigation of public open spaces and footpaths (i.e., public access reuse). The recovered water samples were analyzed for enteroviruses, coliphage, *E. coli*, TKN (total Kjeldahl nitrogen), TDS, ammonia, nitrate, phosphorous, and TOC. Survival experiments were also performed using the selected pathogens poliovirus, coxsackievirus, adenovirus, *Salmonella typhimurium*, *E. coli*, and coliphage MS7. Pathogens were not detected in the recovered water despite their presence in the treated wastewater. Based on TDS concentration data, it was determined that 77–81% of the recovered water was treated effluent. The results of a tracer test indicate a travel time of about 32 days from the basins to the recovery wells. In addition to the pathogen removal, there were decreases in TKN, TOC, and phosphate concentrations and a modest increase in nitrate concentration. However, nitrate concentrations were always below 6 mg/L.

19.5 Conclusions

SAT is a well-established wastewater treatment technology. Operational data have demonstrated that it is effective in substantially reducing the concentrations of pathogens, nutrients (nitrogen and phosphorus), DOC and TrOCs. More refractory TrOCs (e.g., carbamazepine) may pass through SAT systems, but there is no evidence that they pose a health or environmental risk at the concentrations in which they are found in recovered water. Water produced from SAT systems is suitable for most non-potable uses, with perhaps disinfection required for uses involving public contact. SAT treated water is not considered directly suitable for potable use, although SAT could be used as a pretreatment element for indirect potable reuse systems. The main limitations of SAT are its land requirements and that its performance is dependent upon site hydrogeological conditions, which may not be locally favorable. Favorable conditions include a deep water table, absence of confining strata in the vadose and shallow phreatic zone, and soils in the fine sand, loamy sand, and sandy loam range. Sands should not be so fine as to result in unacceptably low infiltration rates, nor be so coarse and permeable so as to result in too rapid percolation and inadequate residence time for contaminant attenuation processes to occur. The greatest value of SAT may now lie in developing countries, where it can provide a relative low-technology solution for reducing the health-risks associated with wastewater reuse.

References

Abel, C. D. T. (2014). *Soil aquifer treatment. Assessment and applicability of primary effluent reuse in developing countries* (Ph.D. dissertation). Delft University of Technology, The Netherlands.

Aharoni, A., Guttman, J., Cikurel, H., & Sharma, S. (2011). *Guidelines for design, operation and maintenance of SAT (and hybrid SAT) systems*. Merkot and UNESCO-IHE.

Amy, G. L., Bryant, C. W., & Belyani, M. (1987). Molecular weight distributions of soluble organic matter in various secondary and tertiary effluents. *Water Science and Technology, 19*(3–4), 529–538.

Amy, G., & Drewes, J. (2007). Soil aquifer treatment (SAT) as a natural and sustainable wastewater reclamation/reuse technology: Fate of wastewater effluent organic matter (EfOM) and trace organic compounds. *Environmental Monitoring and Assessment, 129,* 19–26.

Asano, T., & Cotruvo, J. A. (2004). Groundwater recharge with reclaimed wastewater; health and regulatory considerations. *Water Research, 38,* 1941–1951.

Banin, A., Lin, C., Eshel, G., Roehl, K. E., Negev, I., Greenwald, D., et al. (2002). Geochemical processes in recharge basin soils used for municipal effluents reclamation by the soil-aquifer treatment (SAT) system. In P. J. Dillon (Ed.), *Management of aquifer recharge for sustainability* (pp. 327–332). Lisse: A.A. Balkema.

Bekele, E., Toze, S., Patterson, B., Devine, B., Higginson, S., Fegg, W., et al. (2009). *Chapter 1. Design and operation of infiltration galleries and water quality changes*. Canberra: CSIRO: Water for a Healthy Country National Research Flagship.

Benhan-Blair, & Affiliates and Engineering Enterprises. (1979). *Long-term effects of land application of domestic wastewater: Milton, Wisconsin, rapid infiltration site*. Report EPA-600/2-79-14. Ada, OK: Robert S. Kerr Environmental Research Laboratory, U.S. Environment Protection Agency.

Bouwer, H. (1974). Design and operation of land treatment systems for minimum contamination of ground water. *Ground Water, 12*(3), 140–147.

Bouwer, H., Rice, R. C., & Escarcega, E. D. (1974). High-rate land treatment. I. Infiltration and hydraulic aspects of the Flushing Meadows project. *Journal Water Pollution Control Federation, 46*, 835–843.

Bouwer, H. (1980). Rapid infiltration research at Flushing Meadows Project. *Journal Water Pollution Control Federation, 52*, 2457–2470.

Bouwer, E. J., McCarty, P. L., Bouwer, H., & Rice, R. C. (1984). Organic contaminant behavior during rapid infiltration of secondary wastewater at the Phoenix 23rd Avenue Project. *Water Research, 18*(4), 463–472.

Bouwer, H. (1985). Renovation of wastewater with rapid-infiltration land treatment systems. In T. Asano (Ed.), *Artificial recharge of groundwater* (pp. 249–282). Stoneham, MA: Butterworth Publishers.

Bouwer, H. (1989). Ground water recharge with sewage effluent. In A. I. Johnson & D. J. Finlayson (Eds.), *Artificial recharge of ground water*, *Proceedings of the International Symposium*, Anaheim, California, August 23–27 (pp. 170–185). New York: American Society of Civil Engineers.

Bouwer, H. (1991). Groundwater recharge with sewage effluent. *Water Science and Technology, 23,* 2099–2108.

Bouwer, H., & Rice, R. C. (1984). Renovation of wastewater at the 23rd Avenue rapid infiltration project. *Journal Water Pollution Control Federation, 56*, 76–83.

Bouwer, H., Rice, R. C., & Bouwer, E. J. (1983). The vadose zone as a waste water treatment system. In *Proceedings of the NWWA/US EPA Conference on Characterization and Monitoring of the Vadose Unsaturated Zone* (pp 248–262). Dublin OH: NWWA.

Cikurel, H., Guttman, J., & Aharoni, A. (2012). Managed aquifer recharge for agricultural reuse in Shafdan, Israel. In: C. Kazner, T. Wintgens, & P. Dillon (Eds.), *Water reclamation technologies for safe managed aquifer recharge* (pp 83–120). Brussels: EU Commission & London: IWA Publishing.

Crites, R. W., Middlebrooks, E. J., Bastian, R. K., & Reed, S. C. (2014). *Natural wastewater treatment systems* (2nd ed.). Boca Raton: CRC Press.

Dillon, P., Pavelic, P., Toze, S., Rinck-Pfeiffer, S., Martin, R., Knapton, A., et al. (2006). Role of aquifer storage in water reuse. *Desalination, 188,* 123–134.

Drewes, J. E. (2009). Ground water replenishment with recycled water-water quality improvements during managed aquifer recharge. *Ground Water, 47*(4), 502–505.

Drewes, J. E., & Fox, P. (1999). Fate of natural organic matter (NOM) during groundwater recharge using reclaimed water. *Water Science and Technology, 40*(9), 241–248.

Drewes, J. E., Heberer, T., Rauch, T., & Reddersen, K. (2003). Fate of pharmaceuticals during groundwater recharge. *Ground Water Monitoring and Remediation, 23,* 64–72.

Drewes, J. E., Heberer, T., & Reddersen, K. (2002). Fate of pharmaceuticals during indirect potable reuse. *Water Science and Technology, 46,* 73–80.

Elkayam, R., Michail, M., Mienis, O., Kraitzer, T., Tal, N., & Lev, O. (2015). Soil aquifer treatment as disinfection unit. *Journal of Environmental Engineering, 141*(12), 05015001.

Fox, P. (2002). Soil aquifer treatment: As assessment of sustainability. In P. J. Dillon (Ed.), *Management of aquifer recharge for sustainability* (pp. 21–26). Lisse: A.A. Balkema.

Fox, P. (2011). A comparison of two different case histories in Arizona: The Tucson sweetwater underground storage and recovery facility and the Mesa Northwest Water Reclamation Plant. In *2011 Idaho Water reuse Conference*, May 24 and 25, Boise, Idaho.

Fox, P., Aboshanp, W., & Alsamadi, B. (2005). Analysis of soils to demonstrate sustained organic carbon removal during soil aquifer treatment. *Journal of Environmental Quality, 34*(1), 156–163.

Fox, P., Houston, S., & Westerhoff, P. (2001a). *Soil aquifer treatment for sustainable water supply*. Denver: American Water Works Association Research Foundation.

Fox, P., Houston, S., Westerhoff, P., Drewes, J. E., Nellor, M., Yanko, W., et al. (2001b). *Investigations of soil-aquifer treatment for sustainable water reuse*. Tempe, AZ: National Center for Sustainable Water Supply.

Fox, P., Narayanaswamy, K., Genz, A., & Drewes, J. E. (2001c). Water quality transformations during soil aquifer treatment at the Mesa Northwest Water Reclamation Plant, USA. *Water Science and Technology, 43*(10), 343–350.

Fox, P., & Shah, S. (2006). Anaerobic ammonia oxidation during sub-surface transport. In *Recharge systems for protecting and enhancing groundwater resources, Proceedings of the 5th International Symposium on Management of Aquifer Recharge*, Berlin, Germany, 11–16 June 2005 (pp. 269–273). Paris: UNESCO.

Gable, J., & Fox, P. (2003). Nitrogen removal by anaerobic ammonium oxidation in a soil aquifer treatment soil. In *Proceedings 11th Biennial Symposium on Groundwater Recharge*, June 5–7, 2003, Phoenix, AZ.

Gilbert, R. G., Gerba, C. P., Rice, R. C., Bouwer, H., Wallis, C., & Melnick, J. L. (1976). Virus and bacteria removal from wastewater by land treatment. *Applied and Environmental Microbiology, 32*(3), 333–338.

Idelovitch, E., Icekson-Tal, N., Avraham, O., & Michail, M. (2003). The long-term performance of soil aquifer treatment (SAT) for effluent reuse. *Water Science & Technology: Water Supply, 3*(4), 239–246.

Kmiec, J. P., & Thomure, T. M. (2005). Sweetwater recharge facilities: Serving Tucson for 20 years. *WaterReuse*. Retrieved November 1, 2017 from https://watereuse.org/wp-content/uploads/2015/09/Abstract-Arizona-Sweetwater-Recharge-Facilities.pdf.

Lance, J. C., Gerba, C. P., & Melnick, J. L. (1976). Virus movement in soil columns flooded with secondary sewage effluent. *Applied Environmental Microbiology, 32*(4), 520–526.

Lance, J. C., Rice, R. C., & Gilbert, R. G. (1980). Renovation of wastewater by soil columns flooded with primary effluent. *Journal Water Pollution Control Federation, 52*(2), 381–388.

Lin, C., Shacahr, Y., & Banin, A. (2004). Heavy metal retention and partitioning in a large-scale soil-aquifer treatment (SAT) system used for wastewater reclamation. *Chemosphere, 57*(9), 1047–1058.

Montgomery-Brown, J., Drewes, J. E., Fox, P., & Reinhard, M. (2003). Behavior of alkylphenol polyethoxylate metabolites during soil aquifer treatment. *Water Research, 37*(15), 3672–3681.

National Research Council. (1994). *Ground water recharge using waters of impaired quality*. Washington, DC: National Academy Press.

Oren, O., Gavrieli, I., Burg, A., Guttman, J., & Lazar, B. (2007). Manganese mobilization and enrichment during soil aquifer treatment (SAT) of effluents, the Dan Region Sewage Reclamation Project (Shafdan), Israel. *Environmental Science & Technology, 41*(3), 766–772.

Pescod, M. E. (1992). *Wastewater treatment and use in agriculture*. FAO Irrigation and Drainage Paper 47. Rome: Food and Agriculture Organization of the United Nations.

Quanrud, D. M., Arnold, R. G., Wilson, L. H., & Conklin, M. H. (1996). Effects of soil type on water quality improvement during soil aquifer treatment. *Water Science and Technology, 33*(10–11), 419–431.

Quanrud, D. M., Hafer, J., Karpiscak, M. M., Zhang, J., Lansey, K. E., & Arnold, R. G. (2003). Fate of organics during soil-aquifer treatment: Sustainability of removals in the field. *Water Research, 37*(14), 3401–3411.

Sattler, A., Amy, G., & Drewes, J. (2006). Fate of wastewater effluent organic matter (efOM) through soil aquifer treatment (SAT). In *Recharge systems for protecting and enhancing groundwater resources. Proceedings of the 5th International Symposium on Management of Aquifer Recharge*, Berlin, Germany, 11–16 June 2005 (pp. 573–580). Paris: UNESCO.

Stumm, W., & Morgan, J. J. (1996). *Aquatic chemistry* (3rd ed.). New York: Wiley-Interscience.

Toze, S., Hanna, J., Smith, T., Edmonds, L., & McCrow, A. (2004). Determination of water quality improvements due to the artificial recharge of treated effluent. In *IAHS Publications-Series of Proceedings and Reports* (Vol. 285, pp. 53–60).

USEPA. (1981). *Process design manual for land treatment of municipal wastewater*. Cincinnati: U.S. Environmental Protection Agency Center for Environmental Research Information.

Van de Graaf, A. A., Mulder, A., de Bruijn, P., Jetten, M. S. M., Robertson, L. A., & Kuenen, J. G. (1995). Anaerobic oxidation of ammonium is a biologically mediated process. *Applied and Environmental Microbiology, 63,* 1246–1251.

Wilson, L. G., Amy, G. L., Gerba, C. P., Gordon, H., Johnson, B., & Miller, J. (1995). Water quality changes during soil aquifer treatment of tertiary effluent. *Water Environment Research, 67*(3), 371–376.

Chapter 20
Riverbank Filtration

20.1 Introduction

Riverbank filtration (RBF), which is also referred to as bank filtration, induced infiltration, and induced recharge, is a very old technology for treating surface water. As opposed to pumping water directly from a river or other surface-water body, surface water is indirectly drawn using wells or galleries located on adjacent land (Fig. 20.1). RBF systems take advantage of the natural filtration that occurs as water passes through both riverbed sediments and underlying aquifer sediment or rock. Groundwater pumping lowers the pressure (head) in the aquifer, which creates a downward hydraulic gradient (or increases the natural downward gradient) from the surface water body into the aquifer. RBF technologies and their water quality benefits were reviewed in several dedicated books (Ray 2002a, b; Ray et al. 2002a; Hubbs 2006a), review papers and chapters (e.g., Huisman and Olsthoorn 1983; Kuehn and Mueller 2000; Tufenkji et al. 2002; Hiscock and Grischek 2002; Gollnitz et al. 2004), and numerous papers that document various aspects of the performance of existing systems. A plethora of information is available on all aspects of RBF that can be drawn upon to guide future projects.

RBF systems provide both an intake and initial treatment of surface waters. RBF systems can reduce the concentrations of particulates (suspended solids, turbidity), pathogenic and non-pathogenic microorganisms, dissolved organic carbon (DOC), and many (but not all) organic and inorganic compounds. Underground passage of infiltrated water may compensate for peaks in contaminant concentrations and shock loads (Kuehn and Mueller 2000). Contaminant concentration peaks, such as from a major pollutant discharge (e.g., spill), are smoothed out by the variable travel times from a surface-water body to RBF wells or galleries. RBF also provides a time buffer between a surface water contamination event and the contaminated water reaching a water treatment plant, usually at a lower concentration. RBF systems can provide substantial cost savings over conventional surface water treatment systems that include open intakes and various filtration, coagulation, and settlement steps.

R. G. Maliva, *Anthropogenic Aquifer Recharge*,
Springer Hydrogeology, https://doi.org/10.1007/978-3-030-11084-0_20

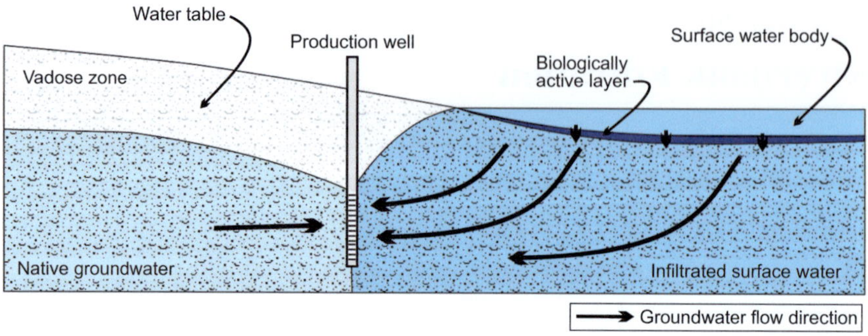

Fig. 20.1 Conceptual diagram of a riverbank filtration system using vertical wells. Groundwater pumping both induces infiltration from the nearby surface water bodies and draws in native groundwater

RBF has the important advantage in developing areas of being a simple technology that does not require a high degree of local technical sophistication and is often substantially less expensive than conventional surface water treatment options. It has thus been described as the "poor man's filtration system" in that it is no match for the efficiency of modern filters, but the systems excel in their sheer size and travel times that are not obtainable in a filtration plant (Deininger et al. 2002). In developed regions, RBF can be an effective barrier in a multiple barrier approach to defending against waterborne pathogens and chemical contaminants in drinking water (Emelko et al. 2010).

In developing area where resources are not locally available for modern water treatment plants, implementation of RBF can result in significant improvement in water quality compared to existing practices. Extracting water from a shallow well near a river, rather than the river itself, can result in a substantial incremental improvement in water quality and associated reduction in waterborne diseases, although the extracted water without further treatment may not be considered fully suitable for consumption [i.e., it does not meet World Health Organization (2017) guidelines]. For example, parts of India experience a high burden of diarrheal diseases related to unsafe potable water, which may be mitigated by riverbank filtration (Bartak et al. 2015). Sandhu and Grischek (2012) and Sandhu et al. (2016) reviewed riverbank filtration system design, capacity and pathogen removal efficiency in India. The main designs used are vertical wells, radial-collector wells, large (~10 m) diameter caissons, and "Koops," which are constructed with a central vertical pipe and four radial perforated pipes that are installed below or beside the stream bed. The RBF systems produce water with a significantly better quality than the surface water sources. In a site in northern India, total and fecal coliform concentrations were reduced by 1/3 to >5.2 log and 2.3 to >4.2 log, respectively. The reduction in total and fecal coliform concentrations in Koop systems were reported to be 1.0–3.4 log and 0.3–2.8 log, respectively (Sandhu and Grischek 2012). Increased and improved implementation of RBF is perhaps the most cost-effective MAR technique for improving drinking water quality, with associated health benefits, in developing and newly industrialized countries, particularly in rural areas.

20.2 History of RBF

The earliest known RBF system is reported to be the Glasgow Water Works system in Scotland, constructed 1810, which was a perforated pipe installed parallel to the River Clyde (Huisman and Olsthoorn 1983; Ray et al. 2002b). RBF has been used along the lower Rhine River in Germany for over a century (Sontheimer 1980). The Düsseldorf Waterworks system (Germany) dates to 1870 (Ray et al. 2002b). RBF technology likely goes back for centuries but has not been recognized as such (Kuehn and Mueller 2000). Production wells located close to rivers or lakes may benefit from induced recharge but not be recognized as RBF. There are no universally accepted criteria for when a water well or other groundwater extraction facility located near a surface water body is considered to be an RBF system in terms of the fraction of produced water derived from induced recharge.

In the United States, groundwater from a well is considered by the U.S. Environmental Protection Agency to be under the direct influence of surface water (UDI or GWUDI) if it contains remains of large microorganisms, algae, and other surrogate indicators of surface water, as determined by a microscopic particle analyses (USEPA 1992). However, the goal of RBF is to provide sufficient filtration so as to not be UDI.

RBF systems were initially developed to provide water that was suitable for direct human consumption without further treatment. The Düsseldorf Waterworks RBF systems was reported to have provided safe drinking water from the Rhine River with only disinfection for the first 80 years of its operation. Increasing pollution of the lower Rhine River in the 1950s resulted in a deterioration in the quality of water produced by RBF systems. The water quality deterioration included increases in iron and manganese concentrations and the breakthrough of odor and taste-causing substances and chemical pollutants (Sontheimer 1980). Today nearly all water utilities along the lower Rhine River treat produced water using granulated activated carbon (GAC), often combined with ozonation and filtration (Sontheimer 1980). The water utilities thus now use RBF as a pretreatment step prior to more advanced treatment (Ray et al. 2002b).

RBF can be used to extract water from other surface water bodies. For example, bank filtration is increasingly being used as an alternative intake for seawater desalination facilities (Missimer 2009; Missimer et al. 2015). Bank filtration-based alternative intakes can produce better quality raw water that reduces pretreatment requirements (and thus operational costs) and avoids some of the environmental impacts associated with conventional intakes, such as the impingement and entrainment of marine organisms.

20.3 RBF Basics

Lake-bed and streambed sediments are defined herein as recently deposited unconsolidated sediments that underlie lake and streams (rivers). In flowing waters, bed sediments are periodically remobilized and redeposited. Bed sediments can range in size from silt-sized material in lakes to cobble and boulder-sized material in very high-energy streams. The hyporheic zone was defined by White (1993) as "the saturated interstitial areas beneath the stream bed, and into the stream banks, that contain some proportion of channel water, or that have been altered by channel water infiltration". The hyporheic zone is of major importance for RBF as it forms the interface between surface water and groundwater (Massmann et al. 2008). The hyporheic zone includes both the bed sediments and some underlying aquifer strata.

The induced infiltration rate through river and lake bed sediments is a function of the vertical hydraulic conductivity (K_z) and thickness of the bed sediments and the hydraulic gradient. For bedded sediments, the effective or equivalent K_z value is the harmonic mean, weighted for bed thickness, of the K_z values of each bed. The K_z of a given material differs depending on whether the strata are saturated or unsaturated (i.e., whether the base of the water-filled part of the channel intersects the water table). If a channel is perched above the underlying water table, then the vertical flow rate will be controlled by the unsaturated K_z value, which is much lower than the saturated K_z value (Hoehn 2002). Pumping-induced drawdown will generally not cause significantly additional infiltration where the water table is located below the channel and downward flow is driven by gravity and moisture potential.

A key issue in bank filtration is that induced flow of water through bed sediments results in the formation of a clogging layer consisting of fine-grained sediments and organic matter. A biologically active layer, analogous to the "schmutzedecke" layer of slow-sand filters, can form at the riverbed surface. The surficial layer is also referred to as a "colmation layer" (Emelko et al. 2010). Colmation is defined as the clogging of surficial sediments by the deposition, accumulation, and storage of fine-grained particles. The fine-grained, organic-rich layer is an important element of the filtration process as indicated by substantial reductions in the concentrations of pathogens and organic compounds within the first few centimeters of the flow path (Ray et al. 2002b). Grischek et al. (2002a) reported that about 50% of the total DOC and chemical oxygen demand (COD) removal in RBF systems occurs in the first few decimeters of the riverbed. Permeability, and thus the hydraulic connection between the surface water and aquifer, may be enhanced by processes that disrupt the clogging layer (e.g., bioturbation and erosion) or be reduced by biofilm development and mineral precipitation.

Where a well-developed clogging layer is present on a streambed, the rate of infiltration will depend upon the thickness and K_z of the layer, and the hydraulic gradient across the layer, which will be approximately equal to the water depth (Bouwer 2002). Greater infiltration rates will tend to occur during floods. However, greater water depths (e.g., in the deep parts of channels) can cause compression of the clogging layer and a reduction in its hydraulic conductivity (Bouwer 2002).

Erosion (scouring) of the riverbed can occur during floods and also from turbulence caused by river traffic (Hubbs 2006b). Disruption of the clogging layer has a regenerative effect in restoring infiltration rates. If the biologically active layer becomes too thick, then it could adversely impact system performance by reducing the infiltration rate. Periodic scouring of the surface layer may be necessary to maintain system performance. Such scouring naturally occurs in many rivers. In locations with stagnant water (e.g., lakes), it may be necessary to periodically manually remove or disrupt the surface layer to maintain system yields.

Concerns have been raised that streambed scour during periods of high flow velocity could cause the loss of the fine-grained layer and jeopardize the filtration process (Berger 2002; Gollnitz 2002; Gollnitz et al. 2004). In slow-sand filters, a "ripening" period of usually around 2 days is required until the schmutzedecke layer is reestablished and peak filtering performance occurs (Cleasby 1990; Ray et al. 2002b).

With respect to *Cryptosporidium*, Berger (2002) noted that oocysts may be absent, or at least below regulatory limits, during periods of normal flow and may occur at concentrations of public health concern only during periods of high water stage when the fine-grained protective layer is removed. A challenge is that monitoring programs may not adequately detect short-period events. Studies of an RBF system near Cincinnati, Ohio (USA) indicate that streambed scour did not occur to a significant degree during high-stage events. The biologically active layer was reported to take only 2–3 days to be reestablished after river flow stabilizes at a system on the Elbe River at Torgau, Germany (Baveye et al. 2002). Gollnitz et al. (2004) noted that even if the surface layers were removed, the underlying aquifer should be still capable of removing particulate material and most of the larger active bacteria and oocysts.

The proportion of the water extracted in an RBF production well or gallery that originates from a surface water body depends upon the nature of groundwater-surface water interactions near the well. Important issues are whether a stream is gaining or losing and the hydraulic connection to the pumped aquifer. Pumped water may include water derived from the river, groundwater on the landward side of the RBF system, and groundwater drawn from the opposite side of the river.

Flow paths between rivers and abstraction wells may vary during RBF because of both the complexity of river and groundwater environments and the progressive development and erosion of a clogging layer at the sediment-water interface. Variations in flow path lengths and residence times often result in a balancing of seasonal and shorter-term fluctuations in source water quality. Water that enters a production well may contain waters that infiltrated into a riverbed at widely differing times (Schubert 2002a).

The primary objectives of RBF system design are to:

- obtain the target water volume
- obtain the target water quality improvements
- maximize the fraction of induced recharge water in the produced water
- balance seasonal fluctuations in source water quality

- mitigate shock loads of contaminants
- minimize adverse environmental impacts.

20.4 RBF System Types

A variety of water extraction technologies are available and have been used in RBF systems including (Hunt et al. 2002):

- vertical drilled wells
- pit or dug wells
- siphon well (well point) systems
- horizontally or obliquely drilled wells
- horizontal wells (subsurface collector pipes) installed in trenches
- horizontal-collector wells
- galleries constructed along shores
- water tunnels.

Vertical wells used for RBF systems have essentially the same construction as typical groundwater production wells. Pit or dug wells are large diameter wells with perforated walls. Siphon well systems consists of a series of small diameter wells connected via a discharge manifold to one or more suction pumps. Horizontal or obliquely drilled (slant) wells are used to install screens below a river (or other surface-water body) to reduce the flow path and obtain a greater fraction of the produced water from induced infiltration. Horizontal wells, trenches, and galleries have been constructed directly beneath river channels by the excavation and installation of gravel-packed screens and plastic tank systems (e.g., Camacho 2003; Hutchinson et al. 2017).

Horizontal-collector wells, which are also referred to as Ranney collector wells or collectors (after their inventor Leo Ranney), consist of a vertical concrete caisson and multiple screened or perforated laterals (Fig. 20.2). Horizontal-collector wells have been used in the United States for RBF since the middle 1930s (Hunt 2002). The construction of horizontal-collector wells was summarized by Hunt (2002). Caissons are made of reinforced concrete with segments (lifts) constructed at land surface. Lifts are emplaced by excavating the soil within the caisson and allowing the casisson to gradually settle into place. The first installed lower lift is fitted with openings through which the laterals will be installed.

Laterals may either radiate outward in all directions from the caisson or may located just toward the river. In the United States, laterals are often located below the river, whereas in Europe collector and vertical wells are constructed some distance from the river to increase travel times (Ray 2002a, b). Laterals are installed using either the perforated pipe, projection pipe, or gravel packing method. A robust perforated pipe (screen) can be jacked directly into the formation. The projection pipe and gravel packing methods involve first advancing (pushing) a heavy-duty projection

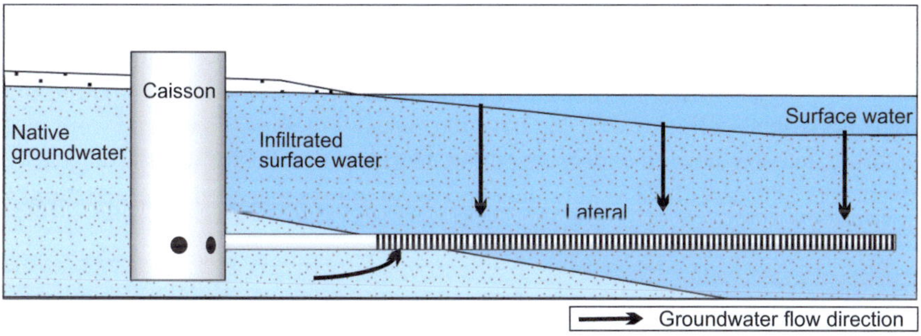

Fig. 20.2 Conceptual diagram of riverbank filtration system using a horizontal-collector well. Laterals beneath the river (shown) will normally produce mostly river water. Laterals oriented parallel and away from the river (not shown) may produce a larger fraction of native groundwater

pipe into the formation. The well screen and gravel pack (if used) are installed within the projection pipe, which is then removed.

Water tunnels are large diameter tunnels into which a series of wells or laterals drain. Hubbs et al. (2002), Ball (2012) and Hunt (2012) described the design of a "hard rock water tunnel" that was the selected option for the Louisville Water Company (Louisville, Kentucky) RBF system. The Riverbank Filtration Tunnel and Pump Station consists of four horizontal collector systems that drain under gravity into a 7,800 ft (2,377 m) long tunnel with a 10 ft (3.0 m) finished diameter. The tunnel is located approximately 150 ft (46 m) below ground. The horizontal collectors consist of a 13 ft (4.0 m) diameter caisson, approximately 100 ft (30 m) deep, each with eight 12-in. (30-cm) diameter laterals that extend as far as 260 ft (79 m) into the sand-and-gravel aquifer. The system has a single common pump station that is independent of the collectors. The system has a design capacity of 60 million gallons per day (MGD; 273,000 m^3/d) and went online in December 2010 (Ball 2012). As discussed by Ball (2012), a key factor in the decision to construct the water tunnel was to address public concerns associated with visual impacts from the construction of additional horizontal-collector systems.

The most commonly used RBF designs to date are vertical wells and horizontal-collector wells. For any new projects, the feasibility, advantages, disadvantages, and costs of the various options need to be carefully considered, which will vary depending upon site-specific hydrogeological and surface-water hydrological conditions. There is no one universal best design. The choice of design should consider system performance, construction (capital) costs, and operational and maintenance costs, including the need for backup capacity. Horizontal systems (horizontal-collector wells, infiltration galleries, and horizontally drilled wells) may be the preferred option where the production aquifer or aquifer zone is relatively thin, and as a result, vertical wells could have only limited exposure to (short screened intervals in) the production zone (Hunt 2002; Hunt et al. 2002).

Vertical wells may be less expensive to construct, are easier to rehabilitate than horizontal wells and galleries, and there are usually local well drillers available with the equipment and expertise to construct the wells. For large-capacity systems, horizontal-collector wells or galleries may be preferred over numerous vertical wells because of the large number of vertical wells that would be required. A single horizontal-collector well can have the capacity of multiple vertical wells and, therefore, may be the least expensive option. Horizontal-collector wells also have the advantages of a lesser land requirement and fewer pumps that need to be maintained.

20.5 RBF System Design

20.5.1 Design Basics

The key design issue for RBF systems is obtaining the optimal balance between system yield, water quality improvement, system reliability, and cost. RBF systems behave essentially as slow-sand filters. If the grain size is too coarse, then the rate of flow will be high but the degree of filtration may be too low. In coarse sediments, fines can penetrate deep and cause permanent clogging. The rates of induced infiltration, effectiveness of filtration and contaminant attenuation, and clogging management depend upon the following variables (Huisman and Olsthoorn 1983; Kuehn and Mueller 2000; Gollnitz 2002; Hoehn 2002; Hunt et al. 2002; Grischek et al. 2002a, b; Schubert 2002a; Gollnitz et al. 2004; Caldwell 2006; Hubbs 2010):

- river stage elevation
- water viscosity (i.e., temperature dependence of hydraulic conductivity)
- riverbed thickness
- riverbed hydraulic conductivity
- wetted streambed area
- stream flow velocity
- flooding frequency
- local soil and water chemistry
- aquifer hydraulic conductivity and thickness (transmissivity)
- stream and aquifer physical properties (e.g., grain and pore throat sizes)
- connectivity of the riverbed and aquifer (vertical leakance)
- vertical and horizontal separation of collection points from surface-water bodies
- residence (travel) time.

Addition general siting recommendations include (Schubert 2002a):

- avoid locating systems in areas where fine sediments are being deposited
- preferred locations are areas of "movable ground," such as the inner, point bars of channels
- avoid locating systems close to "fixed ground," such as riverbed pavements that form near the outer sections of river bends. Riverbed pavements may become fully clogged with suspended sediments.

Most RBF systems use alluvial sand-and-gravel aquifers with hydraulic conductivities greater than 1×10^{-4} m/s (8.64 m/d; 28.35 ft/day) (Grischek et al. 2002b, c). In Europe, distances between production wells and the bank of the river and lake are generally more than 50 m (167 ft). European RBF systems usually have travel times of >50 days, whereas travel times of <50 days have been reported in the United States (Grischek et al. 2002b). In European systems, the proportion of bank filtrate water varies between 10 and 100%, with >50% at most sites (Grischek et al. 2002c).

The optimal distance between wells and the surface water body depends on the (Huisman and Olsthoorn 1983; Grischek et al. 2002b):

- expected vertical and horizontal leakage rate of the river into the aquifer
- preferred flow path length
- preferred residence time
- acceptable groundwater drawdowns.

Considering all the variables that can affect the performance of RBF systems, it is clear that performance will vary greatly between systems. Prospective RBF systems must be independently evaluated on a case-by-case basis. Huisman and Olsthoorn (1983) provided analytical equations for the hydraulics of basic RBF systems. However, the current state of the art is the use of numerical (computer) flow and solute-transport models, which allow for the consideration of more complex system designs and hydrogeology (Sect. 20.5.2). However, the accuracy of numerical models depends upon the quality of the data used in their construction and population. A critical variable is the hydraulic conductivity of the riverbed material, which is difficult to measure and changes over time due to clogging and erosion. Tracer data are invaluable for quantifying the fractions of induced-recharged water and native groundwater in produced water, which information is needed for solute-transport model calibration.

The zone of contribution (or capture) is the areas of an aquifer that contributes water to a well (Gollnitz 2002). With continued pumping, the zone of contribution expands until the amount of recharge (natural and induced) equals the pumping rate. If there is a good hydraulic connection with a river, then the resulting high induced infiltration rate with limit the expansion of the zone of contribution and the fraction of produced native groundwater.

Wells installed near a surface water body will produce some local native groundwater. Some of the improvement in water quality in RBF systems is caused by dilution with native groundwater. Groundwater production in RBF systems could contribute to local environmental impacts, such as the dehydration of wetlands and land subsidence from the compaction of clays and peats (Stuyfzand et al. 2006).

If the hydraulic conductivity of the riverbed sediments is too low, then the induced infiltration rate will be low and a greater proportion of the produced water will be native groundwater. In some locations, native groundwater may be of poorer quality than river water. For example, native groundwater near rivers may have high natural dissolved iron concentrations or anthropogenic contaminants associated with industrial and agricultural activities along the rivers (Ray and Prommer 2006; Wett 2006). Krüger et al. (2006) reported that the RBF systems on the Elbe River near

Torgau, Germany, produced between 45 and 65% bank filtrate in the pumped water. An important factor in water quality management is the selection of well sites according to their landside catchment zone. Catchments differ in their DOC, Fe, and Mn concentrations. Significant improvements in raw water quality may be achieved by extracting from wells with landsides of the best quality (Krüger et al. 2006).

In operational systems, the ratio of riverbank filtered (RF) to native groundwater (GW) can often be accurately determined from field data if there is a consistent difference in the concentration of a conservative chemical parameter between the river water and native groundwater. The ratio may greatly vary between sampling points. For example, in the Louisville RBF system, the laterals of a horizontal-collector well closest to the Ohio River had a 10% dilution with groundwater after about one year, whereas the landward lateral had a 70% dilution (Wang 2002). The ratio may also change over time as the result of clogging and declogging of the riverbed. Solute-transport modeling can be used to estimate RF/GW ratios during the design phase, but the accuracy of the modeling is often constrained by the absence of suitable data for model calibration (Sect. 20.5.2).

20.5.2 Modelling of RBF Systems

Numerical groundwater flow modeling of RBF systems is used to perform the following functions:

- optimization of system performance; evaluation of various design and operational options for meeting target system capacity
- evaluation of the source of produced water (i.e., the balance between bank filtered water and native groundwater)
- evaluation of travel times from surface water bodies to production wells
- assessment of the potential environmental impacts from wellfield operation
- delineation of wellfield protection areas.

Clogging layers can be represented using the conductance term in the MODFLOW river package (McDonald and Harbaugh 1988) and through the leakance term if the clogging layer is simulated as a separate layer. Solute-transport and reactive solute-transport modeling have the potential for simulating water quality changes for different design and operational options (e.g., pumping rates and schedule).

If solute transport is an important concern, then aquifer heterogeneity needs to be adequately incorporated into a model. Aquifer heterogeneity can impact the performance of RBF systems in the following ways:

- High hydraulic conductivity flow zones can result in more rapid flow to production wells, reducing residence time and bypassing filtration.
- Low hydraulic conductivity confining strata can separate the production zone from the surface water body and greatly reduce the amount of water obtained by filtration of surface waters.

- Wells completed in relatively high hydraulic conductivity flow zones can have great well yields. The flow zones could act in a similar manner as a horizontal gallery installed below the surface-water body.

Temperature effects on water viscosity and thus, hydraulic conductivity, need to be assessed in areas that experience large seasonal temperature fluctuations. The viscosity of water increases with decreasing temperature, which can result in significant reductions in well yields and specific capacity during the winter. Temperature variation has had significant documented effects on the yield from some RBF systems (Caldwell 2006).

Non-calibrated models can provide useful initial insights into the behavior of RBF systems. However, a fundamental technical issue is the connection between a surface-water body and underlying aquifer, for which there is typically considerable uncertainty. The hydraulic properties of riverbed sediments can vary greatly depending upon the degree of development of the clogging layer. Some field data are needed for calibration of groundwater flow and solute-transport modeling, which in the absence of actual operation data can be provided by aquifer pumping testing data. Schafer (2006) used inverse modeling with MODFLOW to interpret the data from a 70-day pumping test of a horizontal-collector well installed by the Louisville Water Company on the Ohio River (USA). Laterals were represented as drain cells with specified head and conductance values.

Field tests should be of sufficient duration to detect flow from the surface water body. A model calibrated against pumping test (transient calibration) results can be used as a predictive tool to evaluate various design options (Schafer 2006). However, it must be stressed that data from pumping tests and initial system operation may not be indicative of long-term system performance because of temporal variations in clogging at the sediment-water interface (Schubert 2006a; Hubbs 2006a, b).

Groundwater flow modeling can be used to estimate the RF/GW ratio. For example, Shankar et al. (2009) developed a three-dimensional, finite-element groundwater flow model of the Grind RBF system, located near Düsseldorf, Rhine Valley, Germany. The RF/GW ratio was determined by the flow across control lines located along the Rhine River (RF flux) and across a peninsula upon which the wellfield is located (GW flux). The model results indicated an average RF/GW ratio of 75/25, and that the ratio varies depending on hydrological factors such as river stage. Grischek et al. (2002b) performed theoretical MODFLOW modeling that confirmed the expected relationship that the least mixing occurs when the RBF wells are located either on an island or the inside of a meander.

The RF/GW ratio can also be determined by solute-transport modeling using either actual water quality data or by arbitrarily assigning the two waters different concentrations. Non-reactive solute transport codes can be used to simulate the mixing of waters from different sources, if there is a distinct chemical difference between the waters. For example, nitrate was found to be a suitable tracer for modeling of a system in Austria in which the local groundwater concentration was much greater than that in the River Enns (Wett 2006).

Alternatively, mixing can be evaluated using the MT3DMS code (Zheng and Wang 1999) by assigning the surface water a concentration of 100 and the native groundwater a concentration of zero. The modeled concentration in the recovered water would be the percentage of bank-filtered water. Modeling of RBF systems can be taken a step further through reactive, multi-component solute-transport models, which allows for the simulation of redox reactions and contaminant attenuation processes (e.g., Ray and Prommer 2006). Again, it must be emphasized that the fundamental constraint in modeling solute-transport in RBF systems is not the modeling process itself but rather the availability and quality of hydrogeological data to populate the models.

A key issue for the reliability of modelling approaches is accurate determination of the necessary input parameters, particularly the properties of the clogging layers, and for simulation of the fate and transport of contaminants, parameters relevant to the dispersion, diffusion, biodegradation, sorption, dissolution, and precipitation processes (Hiscock and Grischek 2002). Hence, a sensitivity analysis is an important part of modeling of RBF systems, in which the effects of uncertainties in model parameter values on model results are systematically evaluated.

20.5.3 Geochemical (Redox) Processes

The chemistry of bank filtrate will change due to mixing with native groundwater, interactions with aquifer rocks, and biogeochemical processes that occur along the flow path from a river to production wells or galleries. Geochemical processes documented to operate in RBF systems in the Netherlands include (Stuyfzand et al. 2006):

- adsorption and degradation of dissolved organic compounds and metals
- redox reactions that result in a decrease in dissolved oxygen (DO) and increase in total inorganic carbon and sulfate
- carbonate dissolution, which results in an increase in calcium and bicarbonate concentrations.

Surface water typically contains DO, which is consumed after infiltration mainly by the bacterial oxidation of organic matter. Infiltrated water will thus tend to become more chemically reducing over its flow path. The rate of change in redox state will depend on the initial DO concentration and concentration of biodegradable organic carbon. The redox zonation pattern will also depend upon the infiltration rate and groundwater flow velocity.

Changes in redox state can be beneficial as some desirable reactions, such as denitrification and biodegradation of some trace organic compounds, occur under reducing conditions. The establishment of reducing conditions can also have adverse impacts on water quality. After the consumption of the DO and nitrate, the next utilized electron acceptors are Mn^{+3} and Fe^{+3}. The reduction of Mn and Fe, present as manganese oxides and iron (oxy)hydroxides, can release these dissolved metals into stored water. Elevated Mn and Fe concentrations may also be the result of their high concentrations in native groundwater produced from the landward side of RBF systems.

The development of anoxic conditions allows for the reduction of nitrate concentrations through denitrification. Long flow paths and travel times are beneficial for the removal of NO_3^- and organic compounds. Grischek et al. (1998) investigated denitrification in an SAT system on Elbe River near Torgau, Germany, using nitrogen stable isotope data. Denitrification preferentially removes nitrate containing ^{14}N leaving a residual with increased ^{15}N. Denitrification was observed in the upper layer of the aquifer as evidenced by an enrichment of the $\delta^{15}N$ value of the residual NO^{3-} of the groundwater. The source of oxidizable organic carbon to support denitrification is both dissolved organic carbon (DOC) in the infiltrating water and solid organic carbon (SOC) present in riverbed and aquifer materials. Grischek et al. (1998) suggested that the available SOC reservoir in the biologically active infiltration zone may become a limiting factor in sustaining denitrification.

Grischek and Paufler (2017) observed at RBF sites in Germany (e.g., Torgau system on the Elbe River) that a "wash out" effect occurs, resulting in a slow, but steady, decrease in iron concentration between the riverbank and extraction wells. The supply of iron-rich pore waters and/or leachable iron is gradually exhausted. Grischek and Paufler (2017) proposed that iron concentrations might be controlled by placing wells closer to the riverbank, thereby resulting in the production of a greater portion of bank filtrate water and decreasing the time until the wash out effects occurs. Higher pumping rates, and thus infiltration rates, can reduce the extent of Mn and Fe release, but may be detrimental for pathogen removal (Bourg et al. 2002).

Bourg and Berlin (1993, 1994) discussed redox processes active at an RBF system on the Lot River in southwestern France. Chloride was used a tracer for the mixing of relatively high chloride native groundwater with fresher river water. The chloride concentration data indicate that 80% to nearly 100% of the abstracted water originates from the River Lot. The concentration of DO and DOC were found to abruptly decrease due to microbial activity as the water passes through the first 15–20 m of the alluvial sediments. When DO consumption exceeds supply, nitrate-reducing conditions become established. Sulfate-reducing conditions were reportedly not reached. Manganese mobilization occurred in the zone of maximum DO depletion.

Further along the flow path from the reduced zone, DO was added, presumably from recharge through the overlying unsaturated zone. Mn was removed from solution by adsorption and oxidation reactions when the groundwater is reoxygenated. A seasonal variation in Mn concentrations was observed in wells located nearest the river, with concentrations starting to increase at the beginning of the summer and usually highest at the beginning of autumn. A threshold temperature of 10 °C appears to be necessary in order to trigger the biological activity that creates the reducing conditions necessary for Mn mobilization (Bourg and Berlin 1994).

Kedziorek and Bourg (2009) proposed that the susceptibility of potential RBF sites to iron and manganese reductive dissolution could be screened using the electron trapping capacity (*ETC*), which is defined as

$$ETC = 4(O_2) + 5(NO_3^-) \tag{20.1}$$

where (O_2) and (NO_3^-) are the molar concentrations of DO and nitrate. ETC is the sum of the concentration of each electron acceptor and the number of electrons involved in each redox reaction per molecule of electron acceptor. The greater the ETC, the lower the probability of the dissolution of manganese or iron oxyhydroxides. The threshold for dissolution was proposed to be about 0.2 mM, with systems unaffected at higher values.

The redox state of bank filtrated water and associated water quality changes also depends on mixing with native groundwater, which may be either oxic or anoxic. Shamrukh and Abdel-Wahab (2011) illustrated the impacts of mixing on redox state in prospective RBF systems along the Nile River in Egypt. There are no planned RBF systems along the Nile River, but some production wells are located close to the river and thus act as de facto RBF systems. Shamrukh and Abdel-Wahab (2011) documented water quality improvement in three production wells located within 50–70 m of the Nile River at the Naga Hammadi barrage, located 500 km south of Cairo. RBF was effective in reducing chemical oxygen demand (COD), biochemical oxygen demand (BOD_5), and total and fecal coliforms to concentrations below detection limits. The ambient groundwater has elevated concentrations of many chemical species. Elevated Mn and Fe concentrations due to a chemically reducing conditions is a common problem in the Nile Aquifer. A design challenge for RBF along the Nile River is that if wells are placed close to the river, then microbial removal will be reduced. However, if wells are placed further from the river, then elevated concentrations of Mn and Fe and other compounds may occur from groundwater mixing and the development of reducing conditions (Shamrukh and Abdel-Wahab 2011).

River and infiltration water chemistry often changes seasonally and in response to floods and droughts. Eckert et al. (2006) examined the effects of changes in river water quality and hydraulics on the RBF system at Düsseldorf, Germany. The study included an extreme low water event in the summer of 2003 and subsequent flood events. Yearly changes in river water temperature have a direct influence on the DO concentration of the Rhine River, which varies from between 11 and 13 mg/L in the winter to about 7 mg/L in the summer. Increased microbial activity during the summer leads to anaerobic conditions in the aquifer, which were associated with greater removal of micropollutants. Because of incomplete denitrification, dissolved iron and manganese were not detected in the raw bank filtrate.

Total organic carbon (TOC) in the river ranged between 2 and 4 mg/L and was stable at about 1 mg/L in the bank filtrate (Eckert et al. 2006). A notable exception was a January 2004 flood, which saw a spike in the TOC in the river water and bank filtrate, and detection of bacterial colonies in the raw water. It was suggested that the increased hydraulic gradient during flood events resulted in a mass flux of TOC that temporarily exceeded the microbial degradation rate. Greater engineered treatment of the raw water was implemented following the flood events. A key lesson of the study is that the impacts of temporal changes in river water quality and hydraulics on the natural purification processes of RBF systems need to be well understood and considered in the design and operation of engineered water treatment systems so that adequate multiple barriers are utilized (Eckert et al. 2006).

20.5.4 Clogging (Colmation) Layer Development

The downward flow of water across a riverbed (sediment-water interface) can result in clogging by the deposition of suspended fine solids, chemical precipitation, and biological activities (biofilm formation). Clogging of riverbed sediments decreases the amount of water produced by induced infiltration and, as result, the percentage of native groundwater in the produced water increases. Clogging has the beneficial effect of improving filtration and the removal of pathogens. Clogging processes are an impediment to accurate prediction of RBF system behavior because clogging is the result of complex and poorly predictable dynamic interactions between a river and an aquifer (Schubert 2006b). Unlike the case for surface-spreading MAR systems, it is generally not practicable to routinely remediate the clogging of a riverbed. Therefore, the potential for, and effects of, clogging need to be considered during the site selection and design phase of RBF projects.

The hydraulic effectiveness of the clogging layer depends upon its thickness (b) and vertical hydraulic conductivity (K_z), which are combined in the leakance (leakage coefficient) value (units of 1/time):

$$L = b/K_z \tag{20.2}$$

The term clogging coefficient (w) has also been used in RBF studies, which the reciprocal of leakance and has the units of time (Fischer et al. 2006). During floods, w values decrease, whereas long, low-flow periods result in increasing w values (Fischer et al. 2006).

Hubbs (2006a, b) proposed that riverbed clogging could be caused or increased by the development of local unsaturated conditions just below the sediment-water interface. Unsaturated conditions would result in an increase in effective stress (as the hydrostatic pressure is reduced), which would be expected to increase the rate of compaction of sediments in the clogging layer and thus reduce its porosity and permeability. Hubbs (2006b) provided pressure monitoring data and field observations that unsaturated conditions developed below a centimeter-thick clogging layer at the Louisville RBF site. Dewatering below the river reduces hydraulic conductivity to partially saturated values versus fully saturated values (Schafer 2006).

Clogging forces water to enter the aquifer at greater distances from production wells, exacerbating clogging at greater and greater distances (Schafer 2006). The primary recharge areas thus migrate outward. Hubbs (2006b) reported that the inner boundary of the primary zone of recharge may be evident in the field as the point of transition from hardpan to areas where relatively clean sand is consistently observed in the riverbed. Reduction in the K_z of a riverbed can cause an intensification of the cone of depression (i.e., increase drawdowns; Schubert 2006b). Deepening of the cone of depression can cause a spreading outward of the unsaturated zone, interrupting the direct contact between the river and aquifer. Deepening and expansion of the cone of depression will result in the development of additional infiltration area (Schubert 2006b).

The final stage of clogging is the total interruption of the connection between the riverbed and aquifer (Schubert 2006b). However, it appears that clogging rates asymptotically reached a steady-state, long-term value. For example, at a horizontal collector RBF system at Louisville, Kentucky, on the Ohio River, the clogging rate was greatest in the first year of system operation, and significantly declined over the next two years of operation (Schafer 2006).

The main factors that affect clogging rates (i.e., rate of reduction in the vertical hydraulic conductivity of the riverbed strata) are (Huisman and Olsthoorn 1983; Schubert 2002a, 2006a, b; Hubbs 2006a):

- suspended solids concentration of the river water (average, and temporal and spatial variations)
- infiltration rate, which depends upon the hydraulic gradient from the river to the aquifer (which is a function of the river stage and aquifer drawdown), pumping rates, and the location of wells and galleries
- effectiveness of self-cleaning mechanisms, such as bed load transport, scour, and biological activities.

Schubert (2006a, b) emphasized that clogging is unavoidable and needs to be considered during system design. Management of clogging is viewed by Schubert (2006a) as a "perpetual search for a balance correlated to the fluctuating river-aquifer interactions." Clogging rates may be reduced for a given system capacity by decreasing the induced infiltration rate, which requires either spreading out the wellfield (increasing area), or moving the wells further away from the river. The latter option may increase the contribution of native groundwater to produced water.

Conditions in the source water body play a critical role in clogging. Current velocity needs to be great enough to exert a sufficiently high shear force to result in bedload transport (i.e., rolling, skipping, and sliding of sand grains) and to keep very fine sediments in suspension (Schubert 2002a, b). Areas experiencing either erosion or the deposition of very fine sediments are prone to clogging. Areas of high flow velocity, in which the riverbed is paved with coarse (gravel and cobble-sized) material, may have a high susceptibility to clogging because the bed material is immobile. Clogging can occur by deep filtration whereby suspended material accumulates in pores below the sediment-water interface. The accumulated clogging materials are not readily removed by erosion because the sediment-water interface is armored by the coarse materials. Hence, sandy streambeds may allow for greater long-term induced infiltration rates than initially more permeable gravel beds because the clogging layer at the sediment-water interface can be more readily removed by scouring (Stuyfzand et al. 2006). Biological activity appears to be an important cleaning mechanism for restoring permeability in lake-bank filtration systems (Schubert 2006a, b; Dash et al. 2008).

Hubbs (2006a) discussed methods for estimating shear stress at the sediment-water interface. However, an important starting point for a RBF investigation should be a field sedimentological study of proposed sites. Basic field observations can provide important insights as to scour and sediment deposition and transport patterns.

Fischer et al. (2006) reported that the RBF systems at Dresden on the Elbe River experienced severe clogging of the riverbed in the 1980s due to high loads of organics from upstream pulp and paper mills. After improvements in river water quality in the 1990s, problems with clogging of the riverbed and bad taste have not been encountered. This demonstrates that prior poor water quality did not irrevocably impact future performance of the system.

Clogging rates should be monitored over the operational life of RBF systems. Clogging may be evident by increasing drawdowns (decreasing specific capacity) in wells and galleries, and a decrease in the RF/GW ratio. Infiltration rates through a riverbed can be estimated using temperature as a tracer, if there is a temperature difference between the river water and aquifer (Hubbs 2006b). Calculated flow velocities will be an underestimation if there is a horizontal element to the flow (Hubbs 2006b). Constantz et al. (2006) documented the use of diurnal and seasonal groundwater temperature patterns combined with well water levels to estimate spatial and temporal variations in streambed hydraulic conductivities at the Russian River RBF facility near Sonoma, California. Hydraulic conductivities were estimated by 2-D inverse-modeling of groundwater flow and heat transport using the VS2DHI software package (Hsieh et al. 2000). Theoretical temperature versus time curves for monitoring locations were fitted to field data.

20.6 Pathogen Removal

20.6.1 Introduction

Water quality improvement in RBF systems was reviewed in a series of papers edited by Ray (2002a, b), Ray et al. (2002a), and Ray and Shamrukh (2011). Pathogens are removed during infiltration and aquifer transport in RBF, and other MAR systems, by varying combinations of (Schijven et al. 2002):

- inactivation (die off)
- straining in pores
- sedimentation in pores
- colloidal filtration
- sorption.

Horizontal and vertical setbacks and groundwater travel times are key variables that affect pathogen removal (Schijven et al. 2002). Longer travel times provide more time for natural attenuation processes to operate. Pathogen removal during RBF is most effective in unconsolidated granular media. Consolidated or semiconsolidated media may not be suitable due to the presence of fractures that allow for rapid direct transport (Berger 2002). Most pathogen removal occurs during the first several days and meters of transport in the riverbed sediments with the clogging layer being a critical barrier. The efficiency of straining depends on the ratio of the diameter of the granular media (and thus pore and pore throat size) and the diameter of suspended particles. Straining is insignificant when the ratio is greater than 20 and no particle penetration occurs when the ratio is less than 10 (Berger 2002; Schijven et al. 2002). Straining is, therefore, significant for relative large pathogens (e.g., *Cryptosporidium* oocysts; diameters 4–6 μm) in fine-grained sediments.

Direct quantification of *Cryptosporidium* and *Giardia* removal rates is hampered by their usually low concentrations in natural waters relative to method detection limits. RBF systems are not amenable to microbial challenge (pathogen spiking) studies because of the difficulties and great costs of seeding large quantities of pathogens or surrogates into surface water bodies, especially flowing bodies (Gollnitz et al. 2004). Natural surrogates in river water, such as coliform bacteria, enterococci, total aerobic endospores, algae, diatoms, turbidity, and particle counts, have been used instead. The natural surrogates are present in greater concentrations than *Cryptosporidium* and *Giardia* oocysts and, therefore, are more conducive to quantitative analysis of their removal rates.

The pathogen removal performance of RBF systems has been evaluated based on either concentrations in the produced water or the efficacy of pathogen removal (e.g., percent or number of $\log_{10}$ reductions in concentration). The former directly relates to health risks but has the limitation that pathogen concentrations measured in bank filtrate depend upon the concentration in the source water, which may vary considerably over time.

Under current US Environmental Protection Agency (USEPA) rules, surface water treatment facilities must meet specified treatment goals, such as a 3-log (99.9%) removal of *Cryptosporidium* oocysts. In order for RBF systems to receive treatment credits, they must demonstrate a quantitative treatment capability with respect to microorganisms, such as *Cryptosporidium* and *Giardia*, rather than just their absence (Gollnitz et al. 2005). The USEPA (2010) Long Term 2 Enhanced Surface Water Treatment Rule (LT2ESWTR), promulgated in 2010, requires public water treatment plants without engineered filtration systems to achieve a 2 or 3 log removal of *Cryptosporidium* oocysts, depending on their concentration in the source water. Bank filtration systems are eligible for 0.5 or 1.0 *Cryptosporidium* removal credits if the following conditions are met (USEPA 2010):

- Wells must draw water from granular aquifers composed of clay, silt, sand, pebbles, or larger particles. Partially cemented aquifers may be acceptable it they can be shown not to be fractured. Fractured and karstic aquifers are not eligible for removal credits.
- Only horizontal and vertical wells are eligible (infiltration galleries are not eligible) for credits.

- Wells must be located at least 25 ft (7.6 m) from the surface water for 0.5 log removal credits and at least 50 ft (15.2 m) from the source water for a 1.0 log removal credit.
- An aquifer characterization must be performed that includes collection of a relatively undisturbed core to the bottom of the well screen depth. Each sampled interval should be at least 2 ft (0.6 m) in length and each sample should contain at least 10% grains with diameters less than 1.0 mm.
- Turbidity monitoring is required for the produced water with samples collected at least every 4 h. Turbidity should be less than 1 NTU.

The USEPA (2010) provided site selection and system design recommendations. The potential for scouring should be considered in the site selection process. Greater separation distances may be advisable if frequent scouring is a possibility.

A greater than 1 log *Cryptosporidium* removal credit may be granted by states based on a demonstration of performance (DOP). A DOP involves the collection of data on the removal of *Cryptosporidium* oocysts or a surrogate, and on related hydrogeological and water quality parameters over the full range of operating conditions. A DOP must include sampling of production wells and monitoring wells that are screened and located along the shortest flow path between the surface water source(s) and the production well(s). The study should include an evaluation of temporal variation in *Cryptosporidium* concentrations in the source water, travel time between the source water and production wells, and the effects of dilution on *Cryptosporidium* concentrations in production wells.

Log removals should be calculated using either *Cryptosporidium* oocysts or more abundant surrogate organisms similar to *Cryptosporidium* oocysts. Favorable surrogates should be (USEPA 2010):

- equivalent in size and shape to *Cryptosporidium* oocysts (i.e., 4–6 μm and oblate)
- sufficiently numerous in the surface water to allow for calculation of removal rates
- sufficiently long-lived in the subsurface so that inactivation during subsurface passage does not significantly affect removal rates calculations.

Potential surrogates are aerobic bacterial spores, anaerobic bacteria spores, coliform bacteria, and diatoms. DOPs typically also involve tracer testing and groundwater modeling. Follows are summaries of reported pathogen removal in operational RBF systems.

20.6.2 Charles M. Bolton Groundwater System (Greater Cincinnati Water Works)

The Charles M. Bolton (CMB) RBF system (Cincinnati, Ohio) extracts water from the Great Miami River using vertical wells. The pathogen removal performance of the CMB RBF system was investigated by Gollnitz et al. (2003, 2004) whose data and interpretations are presented below. The vertical production wells are completed

Table 20.1 Summary of MPA algae data from CMB flow path study

Sample site	Number of samples	Samples with algae	Lowest concentration (per 100 L)	Highest concentration (per 100 L)	Average concentration (per 100 L)
River	44	44	2.3×10^8	1.5×10^{10}	2.5×10^8
FP-1I	12	12	4.2×10^4	3.2×10^6	4.0×10^4
CMB-1	19	11	1	7.4×10^4	6.9×10^3
FP-8I	10	10	10	2.3×10^5	6.6×10^4
CMB-8	14	4	1	2.5×10^3	6.6×10^2

Source Gollnitz et al. (2003)

in the Great Miami River Buried Valley Aquifer, which consists of glacial outwash sand and gravels. The effectiveness of RBF as a microbial treatment process was evaluated in two phases. The initial study involved sampling of the Great Miami River and each of the ten production wells for *Giardia lamblia, Cryptosporidium parvum*, and microscopic particle analysis (MPA). The second phase of the investigation was a flow path study in which water quality was monitored at various points along two flow paths between the river and production wells. The flow paths were towards the well (CMB-1) which was judged to have the highest potential risk based on a relatively short flow path and previously performed MPA results, and a low risk well (CMB-8). Water samples were collected and analyzed for:

- *Giardia*
- *Cryptosporidium*
- MPA (including algae)
- aerobic spores
- particle counts
- total coliform bacteria.

Of the 36 river water samples, 12 tested positive for *Giardia* and 4 tested positive for *Cryptosporidium*. Maximum concentrations of cysts were 430/100 L for *Giardia* and 60/100 L for *Cryptosporidium*. The MPA algal data for water samples collected from the river, slant wells screened 5–8 ft below the river bed (FP-1I, and FP-8I), and production wells CMB-1 and CMB-8 are provided in Table 20.1.

Average total algae removal at CMB-1 was 3 logs in the riverbed and an additional 1.8 logs through the aquifer (4.8 logs total). Average total algae removal at CMB-8 was 3.7 logs in the river bed and an additional 2.5 logs through the aquifer (6.2 logs total). Of the other surrogates, there was a 5-log removal of aerobic spores and 3.7 log decrease in particle counts. Particle counts is conservative in that it includes silt and clay particles, some which may come from the aquifer matrix. The CMB RBF system investigations results indicate that RBF provides better particle removal than a conventional surface water treatment plant and is a highly effective treatment process for reducing the risks of *Giardia* and *Cryptosporidium* contamination from surface water (Gollnitz et al. 2003).

Table 20.2 Louisville RBF demonstration plant microbial surrogate removals ($\log_{10}$)

Surrogate	Removal ($\log_{10}$)	
	Range	Average
Total coliforms	0.9 to >6.0	3.8
Heterotrophic plate count (HPC)	0.4 to >4.0	2.2 (2.0 corrected for dilution)
Total aerobic spores	2.7–6.3	>3.9
Algae	4.6 to >8.3	>7.1
Diatoms	>5.7 to >7.4	>6.7
Spores	2.7–6.3	>3.9

Source Wang (2002)

20.6.3 Louisville Water Company

Wang (2002) documented water quality improvement during the first two years of operation of a 0.88 m^3/s (20 Mgd) RBF demonstration project on the Ohio River at Louisville, Kentucky. A horizontal collector system was installed with seven laterals. The more riverward laterals are located approximately 15 m below the river bed. The travel time to the collector was estimated to be about 4 weeks and the monitored lateral located closest to the river was found to collect mostly filtrate.

Turbidity decreased from a mean and median of 45 and 12.6 NTU to less than <0.4 NTU in the RBF filtrate, with the greatest reduction occurring in the first 0.6 m of transport. Microbial contaminant removal was evaluated using a series of surrogates (Table 20.2). The Louisville RBF system is highly effective in removing total coliform bacteria (3.8 log removal), HPC bacteria (2.0 log removal) and total aerobic spores (3.0 log removals). Significant (≈50%) removal of natural organic matter and disinfection byproduct precursors also occurs. An important observation is that much of the water quality improvement, particularly for particulate matter (turbidity and microorganisms), occurs during the first meter (3.3 ft) of travel (Wang 2002).

20.6.4 Midwestern United States RBF Systems

Weiss et al. (2002, 2005) evaluated microbial removal in RBF systems at Jeffersonville, Indiana (Ohio River), Terre Haute, Indiana (Wabash River), and Parkville, Missouri (Missouri River). Concentrations in the filtrate and $\log_{10}$ removals are provided in Table 20.3. Quantifying microbial removals was complicated by non-detections. *Clostridium* had a >2.9 to 3.4 log removal. Bacteriophage removal was >1.6 to >3.0 logs. Total trihalomethane formation potential (THMFP) decreased by 35–82% and total haloacetic acid formation potential (HAAFP) decreased by 47–80%. During the January 2002–July 2003 monitoring period, *Cryptosporidium* and *Giardia* were occasionally detected in river water, but never detected in well water (Weiss et al.

Table 20.3 Midwestern U.S. RBF systems field monitoring summary for pathogen indicators or surrogates

Site		Bacillus (CFU/L)	Clostridium (CFU/L)	Male-specific bacteriophage (PFU/L)	Somatic bacteriophage (PFU/L)
		(log_{10} removals are given in parentheses)			
Indiana-American Water Company (Jeffersonville, IN)	Ohio River	8.1×10^4	7.6×10^2	4.1×10^1	1.7×10^3
	Well #2 Well #9	1.7×10^2, 6.7×10^2 (2.7, 2.1)	$<1.1 \times 10^{-2}$, $<1.7 \times 10^2$ (>4.8, >4.7)	<0.11, <0.17 (>2.6, >2.4)	1.1, <0.17 (3.2, >4.0)
Indiana-American Water Company (Terra Haute, IN)	Wabash River	2.6×10^5	3.1×10^3	3.6×10^1	2.4×10^3
	Collector Well #3	1.8×10^3, $<2.0 \times 10^2$ (2.2, >3.1)	1.1×10^3, 1.6×10^1 (0.4, 2.3)	<0.11, <0.20 (>2.5, >2.3)	<0.11, <0.20 (>4.3, >4.1)
Ohio River Wells #2 and #9	Missouri River	3.9×10^5	8.9×10^5	3.4×10^1	2.6×10^3
	Well #4 Well #5	6.2×10^4, 1.0×10^3 (0.8, 2.6)	$<1.1 \times 10^{-2}$, $<1.5 \times 10^{-2}$ (>4.9, >4.6)	<0.11, <0.25 (>2.5, >2.1)	<0.11, <0.25 (>4.4, >4.0)

Source Weiss et al. (2002, 2005)

2005). The average turbidity of the well water was 0.1–0.5 NTU (max. 3.8 NTU) and total coliforms were detected in only 14 of 213 samples.

20.6.5 India RBF Systems

Dash et al. (2008) documented the performance of an RBF system at Nainital Lake in the state of Uttarakhand, northern India. Water is obtained from tube wells installed adjacent to the lake, whose water is non-potable water because of high concentrations of organic matter and coliform bacteria. The proportion of lake water in the produced water was estimated using oxygen isotope data to be about 80% in the monsoon season and 25–40% during the non-monsoon period. The water recovered from the tube wells did not contain coliform bacteria and was found to be of better quality than that produced by rapid-sand filtration at the water treatment plant. The superior performance of the RBF system lead to the closure of the water treatment plant and the installation of additional tube wells.

Dash et al. (2010) subsequently documented the performance of an RBF system located on banks of the River Ganga in western Uttarakhand state, northern India. Production well data indicated 2.5 and 3.5 $\log_{10}$ removals of total coliform and fecal coliform bacteria, respectively, during non-monsoonal periods, and 4.4 and 4.7 $\log_{10}$ removals of total coliform and fecal coliform bacteria, respectively, during monsoonal periods when the concentration of coliform bacteria in the river are greater. Dash et al. (2010) found that RBF followed by final disinfection with chlorine provides clean, safe drinking water. A subsequent survey by Essl et al. (2014) reported that about 50% of the respondents reported that the RBF water they received was safe and reliable, whereas the other 50% reported that the water could be of better quality. Sand and taste (due to salinity) were the reported complaints.

Bartak et al. (2015) reported on the application of the risk assessment framework of the Australian MAR Guidelines to a RBF system in Haridwar, India. Water is obtained from the Ganga river. As of 2013, the RBF system consists of 22 large-diameter wells that are typically 7–10 m deep, have diameters of approximately 10 m, and productivities of 789–7,526 m^3/d. The extracted water is treated only by disinfection with sodium hypochlorite. Travel time is reduced during monsoonal floods and thermotolerant coliform (TTC) bacteria concentration increases (Bartak et al. 2015).

A quantitative microbial risk assessment (QMRA) was performed. The WHO (2017) upper tolerance level of risk is 10^{-6} DALYs per person per year. An alternative baseline scenario of 0.00533 DALYs per person per year was adopted by the WHO for Southeast Asia as a less stringent and economically and technically viable health outcome. The RBF system achieved a >3.5 $\log_{10}$ removal of TTC, with a mean concentration of 18 TTC/100 mL during the monsoon and 1 TTC/100 mL during non-monsoon periods, compared to 10^4–10^5 TTC/100 mL in the Ganga. The average risk associated with *E. coli* was reduced to 0.00165 DALYs per person per year by

RBF and disinfection. Data were not available to quantify the risks associated with *Cryptosporidium* and rotavirus.

Risk management plan elements identified include:

- monitoring and measurements of disinfection throughout the distribution system.
- optimization of well operation, such as the preferential use of wells with longer travel times and higher portions of bank filtrate.
- regular well maintenance, wellhead sealing, and rehabilitation.
- wellhead sanitation and protection; prevention of unsanitary activities in the vicinity of RBF wells.

Risks for inorganic chemicals, salinity, nutrients, and turbidity were acceptable. RBF was concluded to be robust against monsoonal water quality effects compared to surface water intakes.

20.7 Chemical Contaminant Removal

RBF systems are effective in reducing dissolved organic carbon (DOC) concentrations. The reported reduction of DOC in RBF systems on the Rhine River is about 50% (Kuehn and Mueller 2000). Similar reductions in total organic carbon (TOC) concentrations were also documented at three RBF sites in the United States (Hoppe-Jones et al. 2010). The majority of the TOC reduction occurs very rapidly during the initial phase of infiltration. Non-aromatic TOC is preferentially removed. The amount of reduction in organic carbon (TOC and DOC) is related to the DO concentration of the infiltrating water. Schubert (2002c) reported that DOC, ultraviolet absorbance, and adsorbable organic halogens true removals (i.e., decreases in concentration corrected for dilution) at 0.6 m along the flow path were 38, 43 and 52%, respectively, at an RBF site at Düsseldorf, Germany.

During summer months, the Rhine River has lower DO concentrations, which limits the biological degradation of organic carbon and results in anaerobic conditions (Eckert et al. 2006). The anaerobic period may be beneficial in allowing for additional attenuation of some micropollutants (Eckert et al. 2006). RBF systems are also effective in reducing the concentrations of phosphate, iron, and some heavy metals. Chromium and arsenic concentrations were reduced in the Duisberg, Germany, RBF system by over 90%, whereas other metals (silver, selenium, beryllium, tin) had removal rates of 20% or less (Sontheimer 1980).

Kivimäki (2001) documented organic removal at the Nokia waterworks (Site 1) and Kangasala waterworks (Site 2), Finland, RBF systems. The estimated transit time from the main bank filtration site to the production wells was 3–4 months at site 1 and 1–1½ months at site 2. TOC was reduced in the production wells by 61–71% from lake water values, excluding the effects of mixing and dilution with native groundwater, with 27–41% of the reduction occurring at the very beginning of underground passage. Molecular weight (MW) fractions were measured by high-pressure size-exclusion chromatography. The highest MW fractions were removed at the beginning of filtration and the decrease in high and medium MW fractions continued along underground passage. The TOC that was left in the pumped water was composed mainly of the lowest MW fractions.

Total organic carbon (TOC) removal between the river and nearest horizontal collector system lateral (approximately 15 m vertical travel distance) at the Louisville, Kentucky RBF demonstration project averaged around 40%, fluctuating between 25 and 60% (Wang 2002). Biodegradable DOC decreased from about 0.56 mg/L to below detection limits over the same distance. THMFP, HAAFP, and total organic halides formation potential (TOXFP) reductions were 25, 45, and 35% after the first 2.75 m of travel, and 40, 60 and 45% after 15 m of travel.

20.8 Trace Organic Compounds

The removal of trace organic compounds (TrOCs) in RBF systems has received considerable study. TrOCs are present in treated effluent that is discharged to surface waters, and their concentrations are reduced in rivers by dilution and removal by physical and biological processes. TrOC concentrations in river water also depend upon the amount of wastewater in the river flow, which is a function of water use patterns. Hoppe-Jones et al. (2010) documented that TrOC compounds found in European rivers were not detected in surface water at the studied Ohio River and Cedar River RBF sites in the United States. It was suggested that their non-detection may be due to a greater per capita water use in the United States and thus greater dilution.

Sacher and Brauch (2002) investigated the removal of TrOCs using a "testfilter" system, which simulated aerobic biodegradation, and water quality data from Rhine River RBF systems. The presence of chemicals in raw RBF water depends upon their concentrations in the river water, biodegradation rates, and dilution with fresh groundwater. Recalcitrant compounds consistently present in river and RBF raw water include:

- EDTA
- 1,5-napthalenedisulfonate (and other two- and threefold sulfonated naphthalene compounds)
- carbamazepine
- MTBE (methyl-tertiary-butyl ether).

Compounds that were removed by RBF include:

- nitrilotriacetic acid
- 2-napthalenesulfonate
- diclofenac
- bezafibrate
- bisphenol A.

Other compounds are persistent in the "testfilter" but are uncommonly present in river water and, therefore, their removal behavior could not be evaluated.

Heberer et al. (2011) investigated the presence of antibiotics and other TrOCs in water produced by RBF facilities on the Platte River of Nebraska. A concern is that antibiotics are widely used in the region as growth-promoting and prophylactic agents for livestock production and could enter the water supply. The following compounds were detected in a well water sample:

- phenol
- bromacil
- cholesterol
- β-sitosterol
- sulfamethoxazole
- prometon
- 3-β-coprostanol
- bisphenol A
- triphenyl phosphate.

Metolachlor, caffeine, prometon, and sulfamethoxazole were detected in a drinking water sample.

Storck et al. (2012) examined the factors controlling TrOC removal at several RBF sites in Europe and North America. The investigation also included laboratory experiments. The key conclusion was that the majority of the compounds investigated were removed efficiently regardless of site-specific factors. Some compounds showed enhanced removal in either anaerobic/anoxic or aerobic settings. The overall effect of temperature and river discharge rate on removal was rather low, and only a minority of compounds was affected by these factors.

The greatest TrOC removal occurs near the infiltration zone. A prolonged residence time was found to improve the removal of some persistent compounds that have a long lag time before biodegradation and to contribute to the removal of remaining residues and transformation products. Sorption retards the transport of TrOCs and affects their removal by allowing for more time for biodegradation to occur and by reducing the bioavailability of the compounds (Storck et al. 2012).

Compounds that were found to be persistent (<50% removal) include

- ethylenediaminetetraacetic acid (EDTA)
- sulfamethoxazole
- carbamazepine
- amidotrizoic acid
- iopamidol
- atrazine
- naphthalene-1,5 disulfonate
- naphthalene-1,3,5 trisulfonate.

The removal of CECs at the Lake Tegel, Berlin, Germany, RBF systems was evaluated by Heberer et al. (2004) and Mechlinski and Heberer (2006). The following compounds were found in surface waters and were present at low (ng/L) concentrations within water supply wells:

- AMDOPH
- carbamazepine
- clofibric acid
- diclofenac
- primidone
- propyphenazone
- TCEP
- TCIPP.

Bezafibrate and indomethacin (indomethacin) found to be efficiently removed by RBF.

The effects of redox state on the removal of TrOCs at the Lake Wannsee, Berlin, RBF system were investigated Grünheid and Jekel (2006). Removal was evaluated during groundwater transport to monitoring wells. The studied compounds were absorbable organic halides (iodide—AOI, and bromide AOBr), iopromide (X-ray contrast medium), sulfamethoxazole (antibiotic), and naphthalene disulfonates. The results were as follows:

- **AOI**: 30–40% decrease under aerobic conditions, 64% decrease low redox conditions.
- **AOBr**: 35–58% decrease under oxic conditions, 81% decrease in the monitoring well with the lowest redox potential.
- **iopromide**: 99% removal under oxic conditions, 65% removal in the anoxic/anaerobic monitoring well.
- **sulfamethoxazole**: 46–64% removal under oxic conditions, 97% removal under anoxic/anaerobic conditions.

The removal of napthathalene disulfonates was variable. 1,5-napthalenesulfonic acid (1,5-NSA) was not efficiently degraded under either redox condition. 1,7 NSA and 2,7 NSA, on the contrary, were more efficiently removed during oxic infiltration.

The following TrOCs were detected in a water supply well at Lake Wannsee (Heberer et al. 2011):

- carbamazepine
- clofibric acid
- primidone
- propyphenazone
- o,p′-DDA, p,p′DDA
- TCEP
- TCIPP.

Compounds removed by RBF (i.e., present in surface water but not in the water supply well) are:

- bezafibrate

- diclofenac
- bentazone
- mecoprop (MCPP).

Fanck and Heberer (2006) examined the removal of antibiotics during bank filtration at Lake Tegel and Lake Wannsee. Clarithromycin, roxithromycin, trimethoprim, and acetyl-sulfamethoxazole were efficiently removed by bank filtration. Only sulfamethoxazole was detected in water supplies, but at trace concentrations that are way too low to cause any toxic effects in humans. RBF was thus determined to be an efficient antibiotic removal mechanism.

The behavior of the following synthetic and natural estrogens during bank filtration in Berlin was investigated by Zuehlke et al. (2004):

- estrone (E1)
- 17β-estradiol (E2)
- 17α-ethinylestradiol (EE2).

Only a few samples contained E1 above the limits of quantification (LOQ) of 0.1 ng/L and these came from a shallow monitoring well located very close to the bank of the source water pond. Sewage treatment by a municipal wastewater treatment removed from 76% (EE2) to 94% (E2) of the estrogens. E2 and EE2 were not present in Berlin surface water above the LOQ of 0.2 ng/L. Steroidal hormones have not been identified as being relevant for drinking production via RBF in Berlin despite the surface water containing a considerable treated wastewater component (Zuehlke et al. 2004).

Regnery et al. (2015) documented the start-up performance of the RBF system of the Prairie Water Project (Aurora, Colorado), which obtains water from the South Platte River using vertical wells. The full-scale system was commissioned in 2009 and has a daily extraction capacity of approximately 0.5 m^3/s. Water chemistry (anions and cations), conductivity, and temperature data were used to estimate the fractions of water obtained by induced infiltration from the river and from the landside. The removal of 11 TrOCs were evaluated, which were assigned to three bins based on their biodegradability:

Good removal (>90%)

- atenolol
- caffeine
- naproxen
- trimethoprim

Moderate removal (25–90%)

- gemfibrozil
- *N,N*-Diethyl-m-toluamide (DEET)
- diclofenac
- sulfamethoxazole

Poor removal (<25%)

- atrazine
- carbamazepine
- primidone.

Improved TrOC removal occurred in the summer (except for atenolol), which was attributed to increased microbial activity under higher temperatures. Wells with a greater landside contribution (RBF 50%) exhibited a greater percent removal, which was attributed to longer travel times and less reducing redox conditions. DOC removal in the RBF 90% wells ranged from 45 to 49% and UV254 was reduced by 22–39%.

The removal of pharmaceuticals by bank filtration and artificial recharge and recovery was reviewed by Maeng et al. (2013). The main controls over micropollutant removal were identified as:

- redox conditions—some micropollutants are more degradable under oxic than anoxic conditions and vice versa
- travel time—removal mechanisms are time dependent; longer travel times may result in water passing through different redox zones and regions with different mineralogies and geochemical conditions
- source water bulk organic matter content
- type of pharmaceutical compounds and their characteristics including molecular size, polarity, surface charge, solubility in water (hydrophobicity and hydrophilicity), and presence of functional group and structural features that facilitate or prevent enzymatic attack
- presence of competing micropollutants.

The reviewed data from multiple sites indicate that both RBF and ARR can be effective in removing or significantly reducing the concentrations of TrOCs. However, some recalcitrant compounds consistently show low (<25%) removal including (Maeng et al. 2013):

- sulfamethoxazole (antibiotic)
- roxithromycin (antibiotic)
- carbamazepine (anticonvulsant/antiepileptic)
- primidone (anticonvulsant/antiepileptic)
- clofibric acid (lipid regulator)
- iopamidol (X-ray contrast media).

20.9 RBF and Climate Change

Global climate changes are expected to impact hydrological systems through changes in annual precipitation and the frequency of extreme events. The direction and severity of impacts will vary geographically with some regions experiencing wetter conditions and other regions experiencing overall drier conditions and more frequent droughts. Climate change can impact water supplies through:

- changes in temperature
- changes in annual precipitation
- changes in seasonal precipitation pattern
- increases in the frequency and intensity of floods
- increases in the frequency, intensity, and duration of droughts.

The impacts of climate change on RBF system performance were reviewed by Schoenheinz and Grischek (2011). During low flow periods, the following changes might occur:

- a greater portion of wastewater in surface water due to less dilution
- higher temperatures can result in increased biomass production and greater clogging
- lower DO concentrations due to DO removal by microbial degradation of organic matter and a lower solubility of oxygen
- lower hydraulic gradients.

During high flow periods, the following may occur:

- increased loading of contaminants washed from land surfaces
- increased dilution of wastewater from point sources
- partial removal of clogging layer and associated shorter residence times and more rapid breakthrough of contaminants.

Bench-scale experiments were performed on DOC removal at different temperatures (5, 15, and 25 °C) using flow-through and circulating flow regimes (Schoenheinz and Grischek 2011). The main results were:

- greater DOC removal occurred at higher temperatures
- increased influent DOC concentrations resulted in higher effluent concentrations but no change in removal percentage
- greater removal rates occurred with longer residence times (slow migration velocity)
- in circulating system, the rate of DOC reduction greatly decreased after 7–10 days as easily degradable DOC is consumed.

Adaptations of RBF to climate changes include ensuring sufficient long flow paths to handle changes in surface water contaminant concentrations and the loss of the integrity of the clogging layer (Schoenheinz and Grischek 2011).

20.10 Limitations and Opportunities of RBF

RBF systems can efficiently remove microbial contaminants, but their efficiency can be diminished by short flow paths, high aquifer heterogeneity, high hydraulic gradients, and accompanying high groundwater flow velocities (Schijven et al. 2002). An important benefit of RBF processes is that they are always working to minimize contaminant concentrations. The liability is that a temporary RBF failure to completely remove pathogenic microorganisms could result in a difficult to recognize short period of moderate contaminant concentrations (Schijven et al. 2002). The performance of RBF systems is dependent on local hydraulic and hydrogeological conditions, which may not be favorable for achieving target system performance. Utilities in the United States and Canada are often hesitant about investing in intensive and costly site-specific testing needed to demonstrate RBF performance (Emelko et al. 2010).

The potential public health risk is greatest when the RBF water receives no other treatment other than disinfection and is lowest where RBF is one element in a multiple-barrier water treatment system. RBF systems are effective in removing or reducing the concentrations of pathogens and chemical contaminants, but some microorganisms and chemicals are resistant to removal. Even if the water produced by an RBF system does not meet potable standards and will receive additional treatment, RBF may still be a less expensive option than a conventional surface water intake and filtration system.

In developing countries, RBF may be a viable water treatment option where the technical and financial resources for advanced water treatment facilities are not available (Ray 2011). RBF may produce water that is of substantially better quality and safer than existing water supplies and can thus provide immediate health benefits. RBF would thus be an interim step towards producing water that fully meets drinking water standards and guidelines.

Opportunities also exist to use RBF as an element of wastewater reuse. Where wastewater is discharged to lakes and ephemeral stream channels, recovery of the water for reuse using nearby wells may result in a significant improvement in water quality and reduction in health risks. RBF may also be used to extract and initially treat water for storage or treatment in another MAR system. For example, the Prairie Waters Project (Aurora, Colorado) uses RBF to extract and treat water from the South Platte River, which is then sent to an ARR system consisting of an infiltration basin complex that is surrounded by slurry walls (Regnery et al. 2016). The Orange County Water District, California, uses RBF to pretreat water from the Santa Ana River for recharge using infiltration basins (Hutchinson et al. 2017).

References

Ball, K. (2012). Louisville launches innovative RBF project. *Journal American Water Works Association, 104*(3), 60–67.

Bartak, R., Page, D., Sandhu, C., Grischek, T., Saini, B., Mehrotra, I., et al. (2015). Application of risk-based assessment and management to riverbank filtration sites in India. *Journal of Water and Health, 13*(1), 174–189.

Baveye, P., Berger, P., Schijven, J., & Grischek, T. (2002). Research needs to improve the understanding of riverbank filtration for pathogenic microorganism removal. In C. Ray, G. Melin, & R. B. Linsky (Eds.), *Riverbank filtration. Improving source-water quality* (pp. 311–319). Dordrecht: Kluwer Academic Press.

Berger, P. (2002). Removal of cryptosporidium during bank filtration. In C. Ray (Ed.), *Riverbank filtration: Understanding contaminant biogeochemistry and pathogen removal, NATO Series VI: Earth and Environmental Sciences* (pp. 85–121). Dordrecht: Kluwer.

Bourg, A. C. M., & Berlin, C. (1993). Biogeochemical processes during infiltration of river water into an alluvial aquifer. *Environmental Science and Technology, 27,* 661–666.

Bourg, A. C. M., & Berlin, C. (1994). Seasonal and spatial trends in manganese solubility in an alluvial aquifer. *Environmental Science and Technology, 28,* 868–876.

Bourg, A. C. M., Kedziorek, M. A. M., & Darmendrail, D. (2002). Organic matter as the driving force in the solubilization of Fe and Mn during riverbank filtration. In C. Ray (Ed.), *Riverbank filtration: Understanding contaminant biogeochemistry and pathogen removal, NATO Series VI: Earth and Environmental Sciences* (pp. 43–54). Dordrecht: Kluwer.

Bouwer, H. (2002). Artificial recharge of groundwater: Hydrogeology and engineering. *Hydrogeology Journal, 10,* 121–142.

Caldwell, T. G. (2006). Presentation of data for factors significant to yield from several riverbank filtration systems in the U.S. and Europe. In S. A. Hubbs (Ed.), *Riverbank filtration hydrology* (pp. 299–344). Dordrecht: Springer.

Camacho, A. (2003). Evaluation of the existing performance of infiltration galleries in alluvial deposits of the Parapeti River. In G. Melim (Eds.), *Riverbank filtration: The future is now! Proceedings of the Second International Riverbank Filtration Conference*, Cincinnati, OH, September 16, 2003, National Water Research Institute.

Cleasby, J. L. (1990). Filtration. In F. W. Pontius (Ed.), *Water quality and treatment* (4th ed., pp. 455–560). New York: McGraw Hill.

Constantz, J., Su, G. W., & Hatch, C. (2006). Heat as a ground-water tracer at the Russian River RBF facility, Sonoma County, California. In S. A. Hubbs (Ed.), *Riverbank filtration hydrology* (pp. 243–258). Dordrecht: Springer.

Dash, R. R., Bhanu Prakash, E. V. P., Kumar, P., Mehrotra, I., Sandhu, C., & Grischek, T. (2010). Riverbank filtration in Haridwar, India: Removal or turbidity, organics, and bacteria. *Hydrogeology Journal, 18*(4), 973–983.

Dash, R. R., Mehrotra, I., Kumar, P., & Grischek, T. (2008). Lake bank filtration at Nainital, India: Water-quality evaluation. *Hydrogeology Journal, 16*(6), 1089–1099.

Deininger, R. A., Lee, J., Ancheta, A., & Somana, C. (2002). Public health aspects of riverbank filtration. In C. Ray (Ed.), *Riverbank filtration: Understanding contaminant biogeochemistry and pathogen removal, NATO Series VI: Earth and Environmental Sciences* (pp. 153–176). Dordrecht: Kluwer.

Eckert, P., Rohns, H. P., & Irmscher, R. (2006). Dynamic processes during bank filtration and their impact of raw water quality. In *Recharge systems for protecting and enhancing groundwater resources. Proceedings of the 5th International Symposium in Management of Aquifer Recharge, ISMAR5*, Berlin, Germany, 11–16 June 2005 (pp. 17–22). Paris: UNESCO.

Emelko, M. B., Tufenkji, N., Stone, M., Rudolph, D. L., & Marsalek, J. (2010). Mitigation of urban stormwater and polluted river water impacts on water quality with riverbank filtration. In B.-F. Chang (Ed.), *Effects of urbanization on groundwater* (pp. 164–198). Reston: ASCE.

Essl, L., Starkl, M., Kimothi, P. C., Sandhu, C., & Grischek, T. (2014). Riverbank filtration and managed aquifer recharge as alternative water supply technologies for India: Strengths–weaknesses–opportunities–threats analysis. *Water Science and Technology: Water Supply, 14*(4), 690–697.

Fanck, B., & Heberer, T. (2006). Transport and attenuation of antibiotic residues during river bank filtration in Berlin, Germany. In *Recharge systems for protecting and enhancing groundwater resources. Proceedings of the 5th International Symposium in Management of Aquifer Recharge, ISMAR5*, Berlin, Germany, 11–16 June 2005 (pp. 525–528). Paris: UNESCO.

Fischer, T., Bay, K., & Grischek, T. (2006). Sustainability of riverbank filtration in Dresden, Germany. In *Recharge systems for protecting and enhancing groundwater resources. Proceedings of the 5th International Symposium in Management of Aquifer Recharge, ISMAR5*, Berlin, Germany, 11–16 June 2005 (pp. 23–28). Paris: UNESCO.

Gollnitz, W. D. (2002). Infiltration rate variability and research needs. In C. Ray, G. Melin, & R. B. Linsky (Eds.), *Riverbank filtration. Improving source-water quality* (pp. 281–290). Dordrecht: Kluwer Academic Press.

Gollnitz, W. D., Clancy, J. L., McEwen, J. B., & Garber, S. C. (2005). Riverbank filtration for IESWTR compliance. *Journal American Water Works Association, 97*(12), 64–76.

Gollnitz, W. D., Clancy, J. L., Whitteberry, B. L., & Vogt, J. A. (2003). RBF as a microbial treatment process. *Journal American Water Works Association, 95*(12), 56–66.

Gollnitz, W. D., Whitteberry, B. L., & Vogt, J. A. (2004). Riverbank filtration: Induced filtration and groundwater quality. *Journal American Water Works Association, 96*(12), 98–110.

Grischek, T., Hiscock, K. M., Metschies, T., Dennis, P. F., & Nestler, W. (1998). Factors affecting denitrification during infiltration of river water into a sand and gravel aquifer in Saxony, Germany. *Water Research, 32*(2), 450–460.

Grischek, T., Macheleidt, W., & Nestler, W. (2002a). River bed specifics and their effect on bank filtration efficiency. In P. J. Dillon (Ed.), *Management of aquifer recharge for sustainability* (pp. 59–64). Lisse: A.A. Balkema.

Grischek, T., & Paufler, S. (2017). Prediction of iron release during riverbank filtration. *Water, 9*(5), 317. https://doi.org/10.3390/w9050317.

Grischek, T., Schoenheinz, D., & Ray, C. (2002b). Siting and design issues for riverbank filtration schemes. In C. Ray, G. Melin, & R. B. Linsky (Eds.), *Riverbank filtration. Improving source-water quality* (pp. 291–302). Dordrecht: Kluwer Academic Press.

Grischek, T., Schoenheinz, D., Worchm, E., & Hiscock, K. (2002c). Bank filtration in Europe—An overview of aquifer conditions and hydraulic controls. In P. J. Dillon (Ed.), *Management of aquifer recharge for sustainability* (pp. 485–488). Lisse: A.A. Balkema.

Grünheid, S., & Jekel, M. (2006). Fate of trace organic pollutants during bank filtration and groundwater recharge. In *Recharge systems for protecting and enhancing groundwater resources. Proceedings of the 5th International Symposium in Management of Aquifer Recharge, ISMAR5*, Berlin, Germany, 11–16 June 2005 (pp. 555–561). Paris: UNESCO.

Heberer, T., Mechinski, A., Fanck, B., Knappe, A., Massmann, G., Pekedger, A., et al. (2004). Field studies on the fate and transport of pharmaceutical residues in bank filtration. *Ground Water Monitoring and Remediation, 24*(2), 70–77.

Heberer, T., Verstraeten, I. M., Meyer, M. T., Mechlinski, A., & Reddersen, K. (2011). Occurrence and fate of pharmaceuticals during bank filtration—Preliminary results from investigations in Germany and the United States. *Journal of Contemporary Water Research and Education, 120*(1), Article 2.

Hiscock, K. M., & Grischek, T. (2002). Attenuation of groundwater pollution by bank filtration. *Journal of Hydrology, 266*, 139–144.

Hoehn, E. (2002). Hydrogeological issues of riverbank filtration—A review. In C. Ray (Ed.), *Riverbank filtration: Understanding contaminant biogeochemistry and pathogen removal, NATO Series VI: Earth and Environmental Sciences* (pp. 17–41). Dordrecht: Kluwer.

Hoppe-Jones, C., Oldham, G., & Drewes, J. E. (2010). Attenuation of total organic carbon and unregulated trace organic chemicals in U.S. riverbank filtration systems. *Water Research, 44*, 4643–4659.

Hsieh, P. A., Wingle, W., & Healy, R. W. (2000). *VS2DI—A graphical software package for simulating fluid flow and solute or energy transport in variably saturated porous media*. U.S. Geological Survey Water-Resources Investigations Report 99-4130.

Hubbs, S. A. (Ed.). (2006a). *Riverbank filtration hydrology*. Dordrecht: Springer.

Hubbs, S. A. (2006b). Evaluating streambed forces impact the capacity of riverbed filtration systems. In S. A. Hubbs (Ed.), *Riverbank filtration hydrology* (pp. 21–42). Dordrecht: Springer.

Hubbs, S. A. (2010). Laboratory-simulated RBF particle removal processes. *Journal American Water Works Association, 102*(12), 57–66.

Hubbs, S., Ball, K., Haas, D. L., & Robison, K. J. (2002). Riverbank filtration construction options considered at Louisville, Kentucky. In C. Ray, G. Melin, & R. B. Linsky (Eds.), *Riverbank filtration. Improving source-water quality* (pp. 49–59). Dordrecht: Kluwer Academic Press.

Huisman, L., & Olsthoorn, T. N. (1983). *Artificial groundwater recharge*. Boston: Pitman Advanced Publishing.
Hunt, H. (2002). American experience in installing horizontal collector wells. In C. Ray, G. Melin, & R. B. Linsky (Eds.), *Riverbank filtration. Improving source-water quality* (pp. 29–34). Dordrecht: Kluwer Academic Press.
Hunt, H. (2012). Unique riverbank filtration system provides 80 MGD at Louisville. In *Minnesota Ground Water Association Fall 2012 Conference and the 57th Midwest Ground Water Conference*, October 1–3, 2012, Minneapolis, MN.
Hunt, H., Schubert, J., & Ray, C. (2002). Conceptual design of riverbank filtration systems. In C. Ray, G. Melin, & R. B. Linsky (Eds.), *Riverbank filtration. Improving source-water quality* (pp. 19–27). Dordrecht: Kluwer Academic Press.
Hutchinson, A. S., Rodriguez, C., Woodside, G., & Milczarek, M. (2017). Maximizing infiltration rates by removing suspended solids: Results of demonstration testing of riverbed filtration in Orange County, California. *Water, 9*, 119. https://doi.org/10.3990/w9020119.
Kedziorek, M. A., & Bourg, A. C. (2009). Electron trapping capacity of dissolved oxygen and nitrate to evaluate Mn and Fe reductive dissolution in alluvial aquifers during riverbank filtration. *Journal of Hydrology, 365*(1), 74–78.
Kivimäki, A.-L. (2001). Removal of organic matter during bank filtration—Field studies. In *Artificial recharge of groundwater*, European Commission Project ENV4-CT95-0071 Final Report (pp. 142–146). Luxembourg: European Communities—European Commission Directorate-General for Research.
Krüger, M., Ende, K., & Grischek, T. (2006). Management of river bank filtration in the Elbe River basin near Torgau, Germany. In *Recharge systems for protecting and enhancing groundwater resources. Proceedings of the 5th International Symposium in Management of Aquifer Recharge, ISMAR5*, Berlin, Germany, 11–16 June 2005 (pp. 49–54). Paris: UNESCO.
Kuehn, W., & Mueller, U. (2000). Riverbank filtration: An overview. *Journal American Water Works Association, 92*(12), 60–69.
Maeng, S. K., Salinas Rodriguez, C. N. A., & Sharma, S. K. (2013). Chapter 13—Removal of pharmaceuticals by bank filtration and artificial recharge and recovery. In D. B. Mira Petrovic & P. Sandra (Eds.), *Comprehensive analytical chemistry* (Vol. 62, pp. 435–451).
Massmann, G., Nogeitzig, A., Taute, T., & Pekdeger, A. (2008). Seasonal and spatial distribution of redox zones during lake bank filtration in Berlin, Germany. *Environmental Geology, 54*(1), 53–65.
McDonald, M. G., & Harbaugh, A. W. (1988). *A modular three-dimensional finite-difference ground water flow model*. U.S. Geological Survey Techniques of Water-Resources Investigation Report 06-A1.
Mechlinski, A., & Heberer, T. (2006). Fate and transport of pharmaceutical residues during bank filtration. In *Recharge systems for protecting and enhancing groundwater resources. Proceedings of the 5th International Symposium in Management of Aquifer Recharge, ISMAR5*, Berlin, Germany, 11–16 June 2005 (pp. 543–547). Paris: UNESCO.
Missimer, T. M. (2009). *Water supply development, aquifer storage, and concentrate disposal for membrane water treatment facilities. Methods in Water Resources Evaluation Series No. 1*. Houston: Schlumberger Water Services.
Missimer, T. M., Jones, B., & Maliva, R. G. (Eds.). (2015). *Intakes and outfalls for seawater reverse-osmosis desalination facilities*. Berlin: Springer.
Ray, C. (2002a). Effect of biogeochemical, hydrogeological, and well construction factors on riverbank filtrate quality. In C. Ray (Ed.), *Riverbank filtration: Understanding contaminant biogeochemistry and pathogen removal* (pp. 1–16). Dordrecht: Kluwer Academic Publishers.
Ray, C. (Ed.). (2002b). *Riverbank filtration: Understanding contaminant biogeochemistry and pathogen removal, NATO Series VI: Earth and Environmental Sciences*. Dordrecht: Kluwer.
Ray, C. (2011). Riverbank filtration concepts and applicability to desert environments. In C. Ray & M. Shamrukh (Eds.), *Riverbank filtration for water security in desert countries* (pp. 1–4). Dordrecht: Springer.

Ray, C., Melin, G., & Linsky, R. B. (2002a). *Riverbank filtration. Improving source-water quality*. Dordrecht: Kluwer Academic Press.

Ray, C., & Prommer, H. (2006). Clogging-induced flow and chemical transport simulation in riverbank filtration systems. In S. A. Hubbs (Ed.), *Riverbank filtration hydrology* (pp. 155–177). Dordrecht: Springer.

Ray, C., Schubert, J., Linsky, R. B., & Melin, G. (2002b). Introduction. In C. Ray, G. Melin, & R. B. Linsky (Eds.), *Riverbank filtration. Improving source-water quality* (pp. 1–18). Dordrecht: Kluwer Academic Press.

Ray, C., & Shamrukh, M. (2011). *Riverbank filtration for water security in desert countries*. Dordrecht: Springer.

Regnery, J., Barringer, J., Wing, A. D., Hoppe-Jones, C., Teerlink, J., & Drewes, J. E. (2015). Start-up performance of a full-scale riverbank filtration site regarding removal of DOC, nutrients, and trace organic chemicals. *Chemosphere, 127,* 136–142.

Regnery, J., Wing, A. D., Kautz, J., & Drewes, J. E. (2016). Introducing sequential managed aquifer recharge technology (SMART)—From laboratory to full-scale application. *Chemosphere, 154,* 8–16.

Sacher, F., & Brauch, H.-J. (2002). Experiences on the fate of organic micropollutants during riverbank filtration. In C. Ray (Ed.), *Riverbank filtration: Understanding contaminant biogeochemistry and pathogen removal* (pp. 135–151). Dordrecht: Kluwer Academic Publishers.

Sandhu, C., & Grischek, T. (2012). Riverbank filtration in India—Using ecosystem services to safeguard human health. *Water Science and Technology: Water Supply, 12*(6), 783–790.

Sandhu, C., Grischek, T., Ronghang, M., Mehrotra, I., Kumar, P., Ghosh, N. C., et al. (2016). Overview of bank filtration in India and the need for flood-proof RBF systems. In T. Wintgens, A. Nättorp, L. Elango, & S. R. Asolekar (Eds.), *Natural water treatment systems for safe and sustainable water supply in the Indian context* (pp. 17–37). London: IWA Publications.

Schafer, D. C. (2006). Use of aquifer testing and groundwater modeling to evaluate aquifer/river hydraulics at Louisville Water Company, Louisville, Kentucky, USA. In S. A. Hubbs (Ed.), *Riverbank filtration hydrology* (pp. 179–198). Dordrecht: Springer.

Schijven, J., Berger, P., & Miettinen, I. (2002). Removal of pathogens, surrogates, indicators, and toxins using riverbed filtration. In C. Ray, G. Melin, & R. B. Linsky (Eds.), *Riverbank filtration. Improving source-water quality* (pp. 73–116). Dordrecht: Kluwer Academic Press.

Schoenheinz, D., & Grischek, T. (2011). Behavior of dissolved organic carbon during bank filtration under extreme climate condition. In C. Ray & M. Shamrukh (Eds.), *Riverbank filtration for water security in desert countries* (pp. 51–67). Dordrecht: Springer.

Schubert, J. (2002a). German experience with riverbank filtration systems. In C. Ray, G. Melin, & R. B. Linsky (Eds.), *Riverbank filtration. Improving source-water quality* (pp. 35–48). Dordrecht: Kluwer Academic Press.

Schubert, J. (2002b). Hydraulic aspects of riverbank filtration—Field studies. *Journal of Hydrology, 266,* 145–161.

Schubert, J. (2002c). Water-quality improvements with riverbank filtration at Düsseldorf Waterworks in Germany. In C. Ray, G. Melin, & R. B. Linsky (Eds.), *Riverbank filtration. Improving source-water quality* (pp. 267–277). Dordrecht: Kluwer Academic Press.

Schubert, J. (2006a). Significance of hydrologic aspects on RBF performance. In S. A. Hubbs (Ed.), *Riverbank filtration hydrology* (pp. 1–20). Dordrecht: Springer.

Schubert, J. (2006b). Experience with riverbed clogging along the Rhine River. In S. A. Hubbs (Ed.), *Riverbank filtration hydrology* (pp. 221–242). Dordrecht: Springer.

Shamrukh, M., & Abdel-Wahab, A. (2011). Water pollution and riverbank filtration for water supply along River Nile, Egypt. In C. Ray & M. Shamrukh (Eds.), *Riverbank filtration for water security in desert countries* (pp. 5–18). Dordrecht: Springer.

Shankar, V., Eckert, P., Ojha, C., & König, C. M. (2009). Transient three-dimensional modeling of riverbank filtration at Grind well field, Germany. *Hydrogeology Journal, 17,* 321–326.

Sontheimer, H. (1980). Experience with riverbank filtration along the Rhine. *Journal American Water Works Association, 72*(7), 386–390.

Storck, F. R., Schmidt, C. K., Lange, F. T., Henson, J. W., & Hahn, K. (2012). Factors controlling micropollutant removal during riverbank filtration (PDF). *Journal American Water Works Association, 104*(12), E643–E652.

Stuyfzand, P. J., Juhàsz-Holterman, M. H. A., & de Lange, W. J. (2006). Riverbank filtration in the Netherlands: Well fields, clogging, and geochemical reactions. In S. A. Hubbs (Ed.), *Riverbank filtration hydrology* (pp. 119–153). Dordrecht: Springer.

Tufenkji, N., Ryan, J. N., & Elimelech, M. (2002). The promise of bank filtration. *Environmental Science and Technology, 36*(21), 422A–428A.

USEPA. (1992). *Consensus method for determining groundwater under the direct influence of surface water using microscopic particulate analysis (MPA)*, EPA 910/9-92-029U.S. Washington, DC: Environmental Protection Agency.

USEPA. (2010). *Long term 2 enhanced surface water treatment rule toolbox guidance manual*. Washington, DC: U.S. Environmental Protection Agency Office of Water.

Wang, J. (2002). Bank filtration case study at Louisville, Kentucky. In C. Ray, G. Melin, & R. B. Linsky (Eds.), *Riverbank filtration. Improving source-water quality* (pp. 117–145). Dordrecht: Kluwer Academic Press.

Weiss, W. J., Bouwer, E. J., Aboytes, R., LeChevallier, M. W., O'Melia, C. R., Le, B. T., et al. (2005). Riverbank filtration for control of microorganisms: Results from field monitoring. *Water Research, 39*(10), 1990–2001.

Weiss, W. J., Bouwer. E. J., Ball, W. P., O'Melia, C. R., Aurora, H., & Speth, T. F. (2002). Reduction in disinfection byproduct precursors and pathogens during riverbank filtration at three midwestern United States drinking-water utilities. In C. Ray, G. Melin, & R. B. Linsky (Eds.), *Riverbank filtration. Improving source-water quality* (pp. 147–174). Dordrecht: Kluwer Academic Press.

Wett, B. (2006). Monitoring clogging of a RBF system at the River Enns, Austria. In S. A. Hubbs (Ed.), *Riverbank filtration hydrology* (pp. 259–280). Dordrecht: Springer.

White, D. S. (1993). Perspectives on defining and delineating hyporheic zones. *Journal of the North American Benthological Society, 12*(1), 61–69.

World Health Organization. (2017). *Guidelines for drinking-water quality* (4th ed.). Geneva: World Health Organization.

Zheng, C., & Wang, P. P. (1999). *MT3DMS: A modular three-dimensional multi-species model for simulation of advection, dispersion and chemical reactions of contaminants in ground water systems: Documentation and user's guide*. Report SERDP-99-1. Vicksburg, MS: U.S. Army Engineer Research and Development Center.

Zuehlke, S., Duennbier, U., Heberer, T., & Fritz, B. (2004). Analysis of endocrine disrupting steroids: Investigations of their release into the environment and their behavior during bank filtration. *Ground Water Monitoring and Remediation, 24*(2), 78–85.

Chapter 21
Saline-Water Intrusion Management

21.1 Introduction

Saline-water intrusion is the induced flow of saline or brackish water into freshwater aquifers. In coastal areas, horizontal saline-water intrusion involves the landward migration of the roughly wedge-shaped interface between saline and fresh groundwater (Fig. 21.1a). Vertical saline-water intrusion (referred to as up-coning) is the upward migration of more saline water into freshwater aquifers or aquifer zones (Fig. 21.1b). Horizontal and vertical saline-water intrusion are commonly caused by excessive groundwater extractions lowering the head (pressure) in a freshwater-containing area of an aquifer so that it becomes less than that in adjoining more saline areas, inducing the flow of the saline water toward the pumped area. Increases in groundwater salinity (i.e., salinization) in coastal areas may have causes other than vertical and horizontal saline-water intrusion. Therefore, it is important to first ascertain the actual cause(s) of observed salinity increases (Sect. 21.2).

Saline-water intrusion is an ever-increasing problem in many coastal areas because of a combination of:

- coastal urban areas continuing to experience rapid population growth and concomitant increases in water demands
- a paucity of alternative economical water sources, especially for agriculture
- land development activities that provide conduits for intrusion (e.g., canal construction) and reductions in recharge (e.g., increases in impervious area).

Arid and semiarid lands areas are particularly vulnerable to saline-water intrusion because of their low natural recharge rates. Even modest groundwater withdrawals can result in the development of a landward hydraulic gradient that can induce saline-water intrusion.

Saline-water intrusion in coastal settings has been intensely studied and a voluminous literature has been produced on the subject over the past several decades. Numerous conferences and conference sessions dedicated to saline-water intrusion and related coastal groundwater issues have been held. The Salt Water Intrusion

R. G. Maliva, *Anthropogenic Aquifer Recharge*,
Springer Hydrogeology, https://doi.org/10.1007/978-3-030-11084-0_21

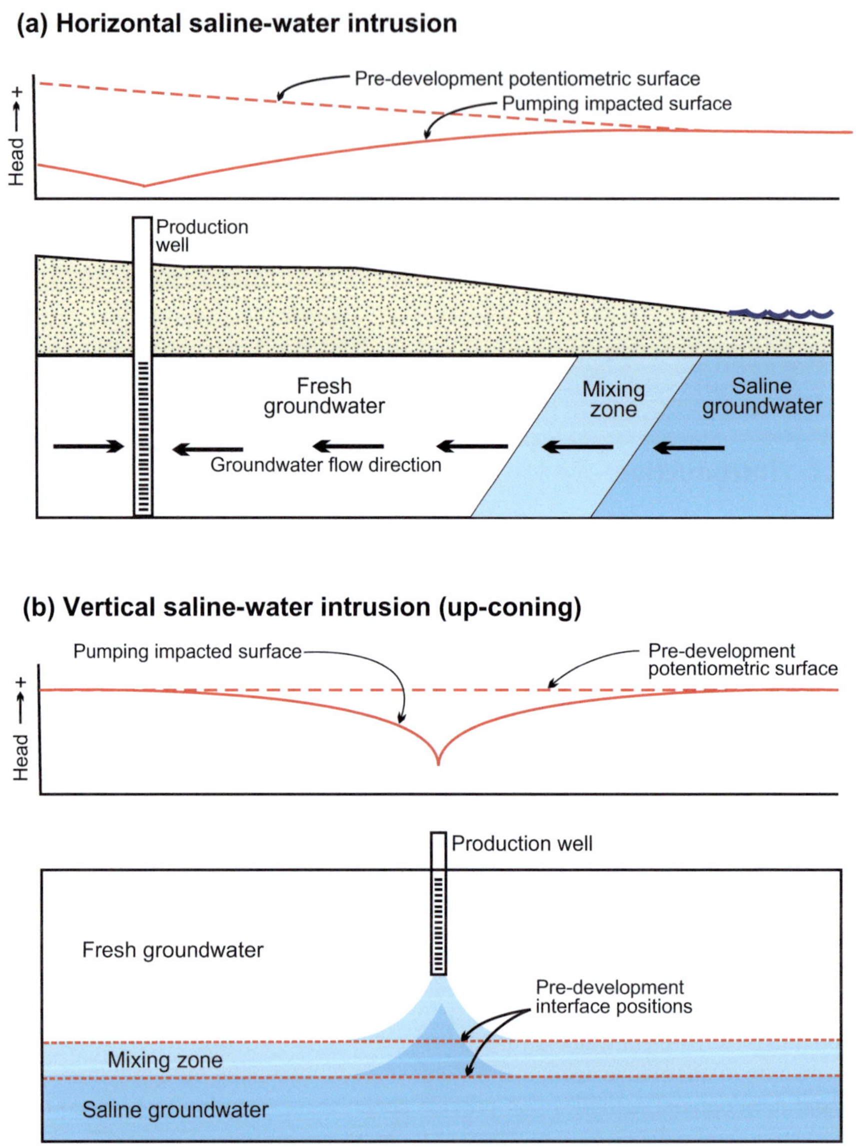

Fig. 21.1 Conceptual diagrams of **a** horizontal and **b** vertical saline-water intrusion (upconing). The lowering of heads at a pumped well results in a hydraulic gradient that draws saline groundwater toward the well

Meeting (SWIM) conference series has been held biennially since 1968. Numerous case study papers have been published and there are few (if any) major areas that are experiencing significant saline-water intrusion that have not already been investigated to some degree.
Management of saline-water intrusion involves the following elements:

- mapping and characterization the fresh water/saline-water interface, which is typically a transition zone rather than a sharp interface
- monitoring changes in the interface position over time using wells and geophysical methods
- development of analytical and numerical models to simulate and predict the behavior and movement of the interface
- identification and implementation of measures to prevent, slow, or reverse intrusion.

Saline-water intrusion can be managed using three basic strategies:

- reduce or eliminate the causes of intrusion (i.e., restore the local aquifer water balance)
- create a hydraulic barrier at or near the freshwater/saline-water interface to arrest the movement of the interface
- construction of a physical barrier between saline water and inland fresh groundwater.

The chapter provide an overview of saline-water intrusion causes, and investigative and mitigation measures, focusing on MAR options.

21.2 Causes of Groundwater Salinity Increases

Management of the salinization of aquifers requires knowledge of the specific cause(s) of an increase in salinity. Salinization of coastal aquifers may have multiple causes. For example, four possible causes of salinization were identified in Bahrain (Zubari 1999):

- horizontal seawater invasion
- upconing of brackish water from underlying aquifers
- migration of sabkha water
- infiltration of irrigation return flows (or salinization of soils).

Sea spray, saline irrigation return flows, evaporation of irrigation water from soils, and dissolution of evaporite minerals can also be significant salinity sources (Custodio 2004). Incorrect identification of salinity sources can lead to ineffective management (Zubari 1999; Custodio 2004; Milnes and Renard 2004; Trabelsi et al. 2007).

The source of salinity can be determined from the three-dimensional distribution of elevated salinity with respect to potential sources and environmental tracer data. Salt sources may have significant differences in major and minor element concentrations, stable isotope ratios, and anthropogenic chemical concentrations (e.g., fertilizers and pesticides). A seawater source would have ion ratios similar to those in seawater. Non-seawater sources may be indicated by ion ratios (e.g., sodium to chloride) significantly different from that of seawater. Tracer data can be evaluated and interpreted using linear mixing equations, more advanced multivariate statistical methods (e.g., Trabelsi et al. 2007), mixing curves (e.g., Buszka et al. 1994), and inverse modeling techniques (e.g., Campbell et al. 1997a, b).

Irrigation return flows can progressively increase the salinity of shallow (unconfined) aquifers through solute recycling. Part of the irrigation water leaves the soil-groundwater system by evapotranspiration (ET), whereas the solutes in the water are left behind and eventually enter the groundwater. Commercial fertilizers also commonly contain some salt, which adds to the overall salinity of irrigation return flows. The rate of salinity increase is related to the aquifer turnover time, which is a function of the aquifer extraction rate and water volume (Milnes and Renard 2004). Groundwater salinity increases can have multiple contributing causes. For example, saline irrigation return flows may be superimposed on the horizontal seawater intrusion process and could be recognized by a correlation of salinity parameters (e.g., chloride) with agriculture-related parameters (Miles and Renard 2004).

Horizontal (coastal) saline-water intrusion occurs if the hydraulic gradient at the freshwater/sea water interface is landward. Unlike the case for freshwater aquifers, the magnitude and direction of hydraulic gradients in areas with large differences in salinity cannot be determined directly from water level data from wells. Instead, water level data need to be converted to a common salinity reference value. Commonly, water level data are converted into equivalent freshwater heads (h_f), as follows (Fig. 21.2; Guo and Langevin 2002):

$$h_f = \frac{\rho}{\rho_f} h - \frac{\rho - \rho_f}{\rho_f} Z \tag{21.1}$$

where,

h head (water altitude) measured in a well above the base of aquifer datum (ft, m)
ρ density of water in well (g/cm^3)
ρ_f density of freshwater in well (g/cm^3, ≈1.0 at 4 °C)
Z elevation of the base of a well above the base of aquifer datum (ft, m)

Water levels measured in two wells could be identical, which, if the water salinities were equal, would indicate the absence of a hydraulic gradient. However, if one well is open to an aquifer zone containing saline water, its equivalent freshwater head would be greater than that of the well completed in a freshwater zone, and a hydraulic gradient would exist from the saline well toward the freshwater well.

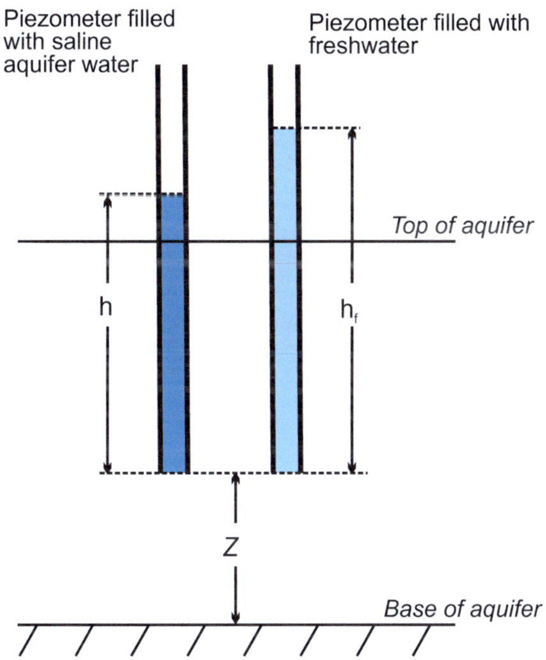

Fig. 21.2 Diagram showing relationship between measured head (h) in a saline aquifer and equivalent freshwater head (h_f)

21.3 Climate Change and Saline-Water Intrusion

Low-lying coastal regions are especially vulnerable to climate change through sea level rise and changes in precipitation. The important local issue is the change in relative sea level (i.e., change in sea level in relation to local land surfaces) rather than global sea level change. Relative sea level change includes global sea level change, natural changes in land surface elevation (e.g., isostatic rebound, tectonic uplift, or subsidence), and anthropogenic changes in land surface elevation (e.g., subsidence due to groundwater pumping or hydrocarbon production). Vulnerabilities of coastal communities to sea level rise include (Bloetscher et al. 2010):

- higher water-table elevations may result in a lesser capacity to store rainwater and increased runoff
- runoff rates exceeding stormwater drainage system capacity
- increased risk of groundwater contamination from seawater inundation and tropical storm surge
- rising groundwater levels in (or inundation of) land on which septic tanks provide on-site sewage treatment and disposal
- increased saline-water infiltration into wastewater systems, which may impair treatment and reuse systems

- increased saline-water intrusion and associated contamination of freshwater resources.

Rising sea level will result in both the inundation of low-lying areas and an increase in saline-water heads. The change in the position of the saline-interface will depend upon the magnitude of the sea level rise and local hydrogeological conditions. A key issue is the response of inland freshwater levels in coastal aquifers to sea level rise. Werner and Simmons (2009) made the distinction between flux-controlled and head-controlled coastal aquifer systems. In a flux-controlled system, the groundwater discharge to the sea is not changed; groundwater levels rise and the hydraulic gradient is maintained. In a head-controlled system, inland freshwater heads remain the same despite the sea level rise because of various surface and near-surface controls, such as drains and rivers.

Analytical modeling results indicate that minimal inland movement of the freshwater/saline water interface will occur in flux-controlled systems, whereas large changes may occur in head-controlled systems (Werner and Simmons 2009). The migration of the toe of the saline-water wedge will be greater with decreasing recharge rates, increasing hydraulic conductivity, and increasing depth of the aquifer below sea level (Werner and Simmons 2009). The potential impacts of sea level rise can be evaluated by numerical modeling.

21.4 Location, Characterization and Monitoring of Saline-Water Interface

Effective management of saline-water intrusion requires knowledge of the position and shape of the local freshwater/saline-water interface and it is changes over time. In coastal aquifers, fresh and saline groundwaters are separated by a transition (i.e., mixing) zone of varying thickness. The nature of the interface is of importance for understanding and simulating saline-water intrusion. The depth of the interface can be calculated using the well-known Ghyben-Herzberg relation:

$$z = \frac{\rho_f}{\left(\rho_s - \rho_f\right)} h \tag{21.2}$$

where,

z depth of the presumed sharp saline-water interface below sea level (m or ft)
h height of the freshwater zone above sea level (m or ft)
ρ_s density of seawater (g/cm^3; lbs/ft^3)
ρ_f density of freshwater (g/cm^3; lbs/ft^3)

The Ghyben-Herzberg relationship is frequently inaccurate because the underlying assumptions of a sharp interface and the absence of freshwater flow to the sea (i.e., hydrostatic conditions) are not met. The shape, position, and rate of movement of the saline-water interface depend upon multiple factors including geological and hydrogeological structure (particularly aquifer heterogeneity), and present and past hydrodynamic conditions (Custodio 2004).

The position and shape of the saline-water wedge and mixing zone are controlled by the variance in hydraulic conductivity, which in turn controls longitudinal and transverse macro-dispersivities (Kerrou and Renard 2010). Theoretical modeling results show that the penetration length of the toe of the saline-water wedge increases with increasing heterogeneity (Kerrou and Renard 2010). The presence of a flow zone (i.e., strata with a relatively high hydraulic conductivity) near the base of an aquifer would allow saline-water to migrate inland more rapidly and to a greater extent than would occur under more homogeneous conditions. Theoretical methods can provide some rough guidance on the location and shape of the freshwater/saline-water interface. However, accurate mapping of the interface requires a field investigation, which normally involves well and/or surface geophysical methods.

A multiple-element approach is preferred to determine the position and shape of freshwater/saline water interfaces and to detect changes in their positions over time. There is no adequate substitute for a monitoring well network from which direct data on groundwater salinity can be obtained. Monitoring well data may be augmented by borehole and surface geophysical data. Borehole water quality and geophysical data allow for the calibration and ground truthing of the surface geophysical data. Surface geophysical data can provide a much greater areal coverage than is practicably possible using wells alone.

21.4.1 Monitoring Well Methods

Aquifer characterization methods used to evaluate groundwater salinity and salinity changes over time were reviewed by Maliva (2016). The freshwater/saline-water interface can be located and monitored through sampling of a series of wells located perpendicular to the coast. Most production and many monitoring wells are open to relatively thick intervals. The salinity of water produced from wells open to large aquifer thicknesses is an average value that is weighted by the transmissivity of each penetrated bed. Salinity-versus-depth profiles are needed instead, which requires sampling of discrete, small-thickness depth intervals. One-time salinity-versus-depth profiles can be obtained during drilling or afterwards by performing off-bottom (single) packer or straddle-packer tests to collect depth-specific water samples. Some drilling methods (dual-tube and cable tool) allow for the collection of representative water samples while drilling as the strata above the drill bit are cased off during drilling.

In wells with long screened intervals or open holes, salinity versus-depth profiles can be obtained by allowing the well to stabilize (i.e., salinity inside the well equilibrates with salinity in the adjoining formation) and slowing lowering a conductivity probe or collecting water samples with a thief sampler. A preferred monitoring system design is to have the wells screened or open to small depth intervals (2–5 m;

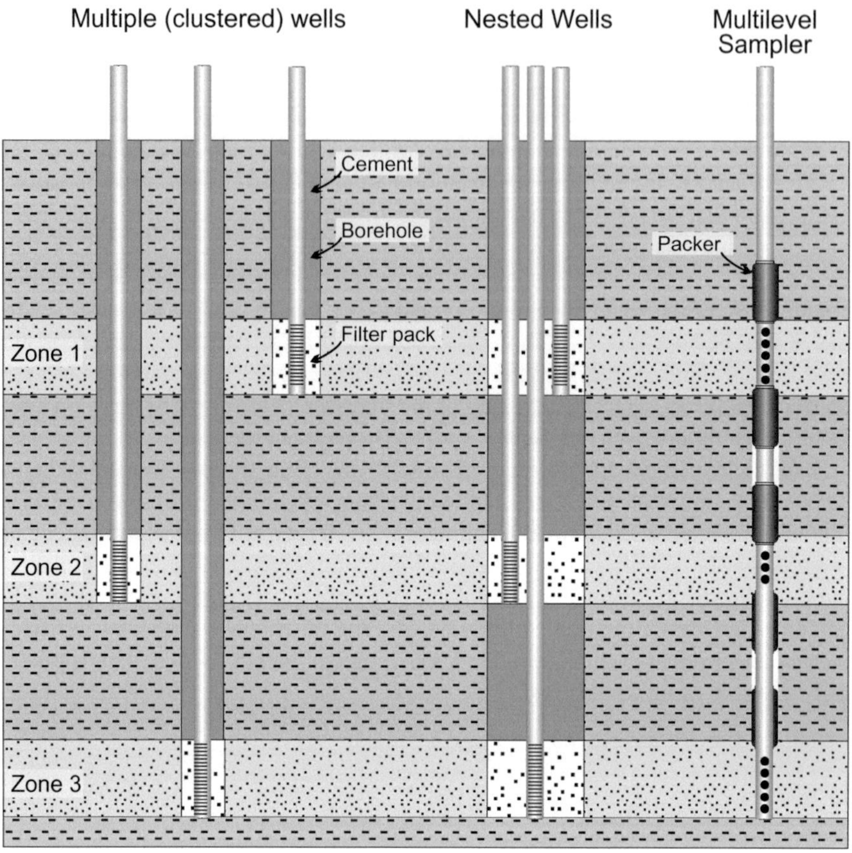

Fig. 21.3 Conceptual diagram of options for multiple zone monitoring

6.6–16.4 ft) rather than the entire aquifer thickness. An often used option is to construct a cluster of wells at a given location with each well open to a different depth. The wells are periodically sampled for salinity-related parameters and/or equipped with an electrical conductivity probe and datalogging system. Landward movement of the wedge-shaped interface would be manifested by increasing salinity in progressively shallower wells over time.

Installation of multiple separate monitoring wells can be quite expensive, especially for a deep aquifer. Dual- or tri-zone monitoring wells, in which a single well is open to multiple depth intervals can be less expensive than separate wells. Multiple-zone wells may have either a nested design (Fig. 21.3) or coaxial casing construction (Fig. 21.4). Multiple-zone monitoring wells are more complicated to properly construct so that the monitored zones are hydraulically separated. Maintaining the mechanical integrity of the wells is also a concern.

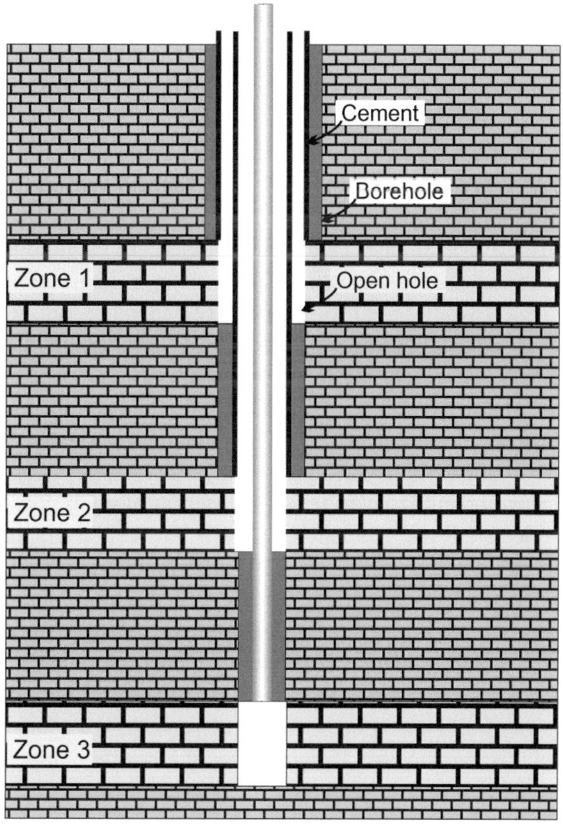

Fig. 21.4 Conceptual diagram of triple-zone monitoring well completed with open holes in limestone

Multilevel sampling (MLS) systems are an alternative to multiple separate or multiple-zone monitoring wells. MLS systems are constructed with multiple sampling points within a single casing. MLS systems were reviewed by Einarson (2006), Koch and Pearson (2007) and Cherry et al. (2015). Packer-based MLS systems include the Westbay system (NovaMetrix) and the Solinist Waterloo systems. The Westbay system consists of multiple packers installed in a single borehole that isolate selected zones of an aquifer. The Westbay system allows for the collection of both pressure data and water samples from each discrete monitored zone. The advantage of MLS systems is that they allow for a large number of monitoring points within a single well. MLS systems tend to be most economical for deep aquifers (where well drilling costs are high) and where a large number of monitoring points is required.

21.4.2 Borehole Geophysical Logging

Borehole geophysical logging is commonly used to obtain continuous salinity-versus-depth profiles. A combination of a deep resistivity (or induction) log and porosity (e.g., sonic) log are routinely processed to obtain profiles of formation water resistivity versus depth. Formation water resistivity data can be further processed to obtain a salinity-versus-depth profile. Most borehole geophysical logs are designed to be run on an open (uncased) borehole. In wells completed with open holes, a time series of logs can be run to detect changes in groundwater salinity over time.

Resistivity-based logs generally cannot be run on steel-cased boreholes. However, electromagnetic induction (EM) logs can be run on PVC-cased wells to detect changes in resistivity over time, which can be processed to obtain data on changes in total dissolved solids and chloride concentrations. For example, sequential EM logs were run on PVC-cased monitoring wells in the San Joaquin Groundwater Subbasin of Northern California to monitor changes in chloride concentration over time (Metzger and Izbicki 2013). Aquifer lithology is constant during sequential or time-series analyses, so changes in bulk EM resistivity can only be caused by changes in groundwater chemistry. Chloride concentration was estimated using an empirical relationship between electromagnetic resistivity and chloride concentration, which was obtained from log runs on screened intervals from which chloride concentration data were obtained by conventional water sampling and from water samples extracted from core samples. EM signals are attenuated through the PVC casing and cement grout. Calculated pore-fluid concentrations were judged to be best interpreted in the relative sense rather than in terms of absolute numbers (Metzger and Izbicki 2013).

21.4.3 Surface and Airborne Geophysical Methods

Surface resistivity geophysical methods are commonly used for mapping the vertical and geographic position of the freshwater/saline-water interface because the interface is typically marked by a sharp, readily detectable change in resistivity. Surface and airborne geophysical methods are typically less expensive and can be performed quicker than methods that require the drilling of multiple boreholes, which allows for a larger number of measurements and thus greater spatial coverage. However, the greater spatial coverage comes at the expense of lesser vertical resolution. Both DC-resistivity and time-domain electromagnetic (TDEM) surveys using land-based systems and airborne platforms have been successfully used in numerous studies to map the vertical and horizontal location of freshwater/saline water interfaces. EM methods are particularly well suited for mapping the interface because of the large contrast in resistivity (conductivity) between freshwater and saline water. A time-series of surface resistivity surveys can be performed to map changes in the position of the freshwater/saline-water interface over time. For example, Abdalla et al. (2010) documented an approximately 600 m (about 2,000 ft) seaward migration of a freshwater/saline-water interface in Oman using TDEM surveys conducted in 2002 and 2007. A limitation of resistivity and EM methods is that they are vulnerable to interference from anthropogenic features, particularly nearby or buried conductors (metallic objects), which can constrain their use in urban settings.

21.5 Simulation of Saline-Water Intrusion

Numerical groundwater modeling is now a critical element of saline-water intrusion management, serving four important purposes:

(1) the inverse modeling processes used to calibrate models can provide insights on local hydrogeology
(2) prediction of changes in salinity over time under current or future water use and climate scenarios
(3) a screening tool to evaluate various saline-water intrusion management options
(4) as a design tool to evaluate specific design and operational options.

Inverse modeling is essentially the determination of cause from effects. Data from measurements of observable parameters (e.g., aquifer heads and salinity) are used to infer the values of aquifer hydraulic and transport parameters. Model calibration is an exercise in inverse modeling whereby model parameter values are adjusted to achieve a better fit between simulated and observed values. Inverse modeling is particularly valuable for aquifer characterization because it has a very large volume of investigation, which consists, on a broad scale, of the entire model domain and, on a finer scale, the area around observation points (e.g., monitoring wells or piezometers). Inverse modeling can be used to evaluate both conceptual models and potential parameter (hydraulic conductivity, effective porosity, dispersivity) values.

Once a model has been constructed and acceptably calibrated, it can use used for predictive simulations, which might include a continuation of current practices (status quo) and various mitigation options. Various design and operational options may be evaluated. For example, for a salinity barrier system, the performance of various well location and injection rate options may be evaluated. Groundwater models inherently have considerable (and poorly quantified) inaccuracy due to uncertainties in the underlying conceptual model and parameter values. Hence, models should be developed and used as dynamic tools that are modified (recalibrated) as additional hydrogeological and operational (observation) data become available.

Density-dependent solute-transport codes are required to simulate groundwater flow and salinity changes in coastal aquifers because heads (pressure) depend upon both aquifer water levels (as measured in wells) and water density. Where significant density differences occur (due to differences in salinity and/or temperature), measured water level data need to be corrected for density, such as by converting all values to equivalent freshwater heads. A number of codes have been developed that can simulate density-dependent solute-transport, such as FEFLOW (Diersch 1998) and the U.S. Geological Survey SEAWAT code (Guo and Langevin 2002; Langevin et al. 2007), which is based upon on the widely used MODFLOW and MT3DMS codes.

Solute-transport modeling requires much more data than is needed for the simulation of bulk groundwater flow. Heterogeneity with respect to hydraulic conductivity and porosity needs to be evaluated and incorporated into models. Data are also needed on aquifer dispersivity values and the three-dimensional distribution of salinity. Detailed aquifer characterization, therefore, is critical for obtaining the data needed for model development. Aquifer characterization programs typically require aquifer performance testing (using pumping and monitoring wells) to determine aquifer bulk hydraulic properties (e.g., transmissivity, storativity, and leakance) and some combination of packer testing and borehole geophysical logging to quantify aquifer heterogeneity. Current salinity distribution can be evaluated from monitoring well, borehole geophysical, and surface geophysical data.

Most hydrogeological data available for model development is point data from wells. A variety of geostatistical methods are available for prediction of parameter values between wells. The basic method is to first identify and quantify the spatial structure of the variables of concern and then to interpolate or estimate the values of variables from neighboring values taking into account their spatial structure. Geostatistical methods have been used to obtain realizations of sedimentary facies distributions. The facies data typically requires upscaling to a coarser groundwater model grid and assignment of hydraulic parameters values for the various facies types based on field data. A promising approach is hybrid methodologies that combine facies models (including sequence stratigraphic interpretations) and other soft geological information with geostatistical methods.

The accuracy of models is typically limited by the availability of data to populate the model rather than limitations of the modeling codes. Current modeling capability allows for the evaluation of whether the current transition zone is equilibrium with present-day sea level and the overall movement of the interface over time. However, it is not possible to accurately forecast salinity changes at the level of a given well (Sanford and Pope 2010). Finer spatial and temporal discretization increases computation time. However, even if infinite computer capability were available, that would not guarantee that a simulation could forecast salinity increases at any particular well because there would be insufficient data to estimate hydraulic parameters at the required spatial detail (Sanford and Pope 2010). As is the case for groundwater modeling in general, it is critical to understand the limitations of models and to not abuse models by relying upon them beyond their capabilities.

21.6 Mitigation of Saline-Water Intrusion

Strategies to control horizontal and vertical saline-water intrusion were reviewed by Banks and Richter (1953), Bruington (1969), Todd (1974, 1980), Kashef (1977), Atkinson et al. (1986), Oude Essink (2001), and Maliva and Missimer (2012) among many, and include:

- reducing fresh groundwater pumping to restore the water balance of an aquifer toward its predevelopment conditions, which may entail developing alternative water sources
- relocating pumping inland, away from the freshwater/saline-water interface
- creating a positive salinity barrier (hydraulic mound) by managed aquifer recharge (MAR) on the freshwater side of the saline water interface (between the interface and production wells)
- MAR inland to increase the flow of freshwater toward the coast
- extraction barriers, which involves pumping on the saline-water side of the interface to pull the interface seaward
- subsurface physical barriers
- scavenger wells, which involves managing vertical saline-water intrusion (upconing) by pumping brackish or saline groundwater below freshwater production zones.

21.7 Reduction and Relocation of Pumping

Reduction and relocation of pumping are strategies aimed at restoring the hydraulic equilibrium at a freshwater/saline-water interface. From a technical perspective, reduction and relocation of pumping are obvious mitigation options. A strong regulatory framework needs to be in place that is capable of controlling local groundwater use, which is often not the case in developing countries. Ignoring water use practices that are unsustainable may be politically expedient, as opposed to the immediate discord that could arise from forcing local farmers (often with no other livelihoods) to stop irrigating. Reduction of groundwater pumping can be very difficult to implement where alternative water sources are either unavailable or are unaffordable. Any area of the world that has access to the coast, theoretically, has an essentially unlimited supply of water through seawater desalination. However, desalted seawater is typically too expensive for other than potable and high-end industrial uses. Reclaimed water is increasingly being used for non-potable uses and there is growing interest in both indirect and direct potable reuse.

The most common historical reaction to saline-water intrusion has been to relocate wells further inland. For example, in many Florida coastal communities, the initial production wells were located close to the coast where the earliest populations were concentrated. As water demands increased and increases in the salinity of production wells were detected, new wells were constructed progressively further inland and the use of more seaward wells was discontinued and/or entirely new inland wellfields were constructed. Moving water production inland may at least temporarily address a saline-water intrusion problem, but new challenges arose, particularly reductions in water levels in wetlands and lakes. A key issue is whether moving production landward allows for a new dynamic equilibrium to be established at the coast, stabilizing the position of the interface, or does it only provide some additional separation from an interface that is continuing to move inland. If the relocation of wells inland is coupled with an increase in production, then saline-water intrusion could be potentially exacerbated. Relocation of wellfields is usually an expensive option, particularly in densely developed areas where available land for new wellfield construction is limited.

21.8 Positive Hydraulic Salinity Barrier

21.8.1 Introduction

Inland migration of the freshwater/saline-water interface is driven by a landward hydraulic gradient. Positive hydraulic salinity barriers arrest or reverse horizontal saline-water intrusion by the restoration of a seaward hydraulic gradient at the freshwater side of the freshwater/saline-water interface by MAR. MAR is used to create a hydraulic mound inland of the interface (Fig. 21.5a). The recharge can be performed using wells or, for unconfined aquifers, infiltration basins or other surface-spreading methods. Recharge is typically performed using relatively low-value water, such as treated surface water or wastewater. Potable water is not used because of its high costs and the limited availability of excess water. However, in some systems, such as the Orange County Water District (OCWD) Talbert Gap salinity barrier, the wastewater is treated to such a high degree that it is of superior quality to many potable water supplies.

Positive hydraulic barrier systems should be designed to efficiently create an effective barrier. Saline water should not flow between or around the injection wells. The hydraulic mound at the barrier also results in an inland flow of recharged water. Hence, consideration needs to be given to the potential impacts of the recharged water on inland freshwater supplies. For example, some of the water injected in the Orange County Water District (California; OCWD) Talbert Barrier will eventually migrate inland and enter potable water supplies. The very high level of treatment that the wastewater receives ensures that any indirect potable reuse will not cause health risks. Where water that receives lesser treatment is recharged, the travel time to any supply wells should be sufficiently long to ensure removal of pathogens.

As is generally the case for MAR project in which solute transport is of concern, aquifer heterogeneity needs to be adequately characterized as high hydraulic conductivity flow zones can dominate the flow of recharged water. The focusing of groundwater flow into a thin high-transmissivity zone could result in greater landward migration of the recharged water and a less effective barrier if little of the recharged water enters the base of the aquifer to prevent movement of the toe of the saline-water interface.

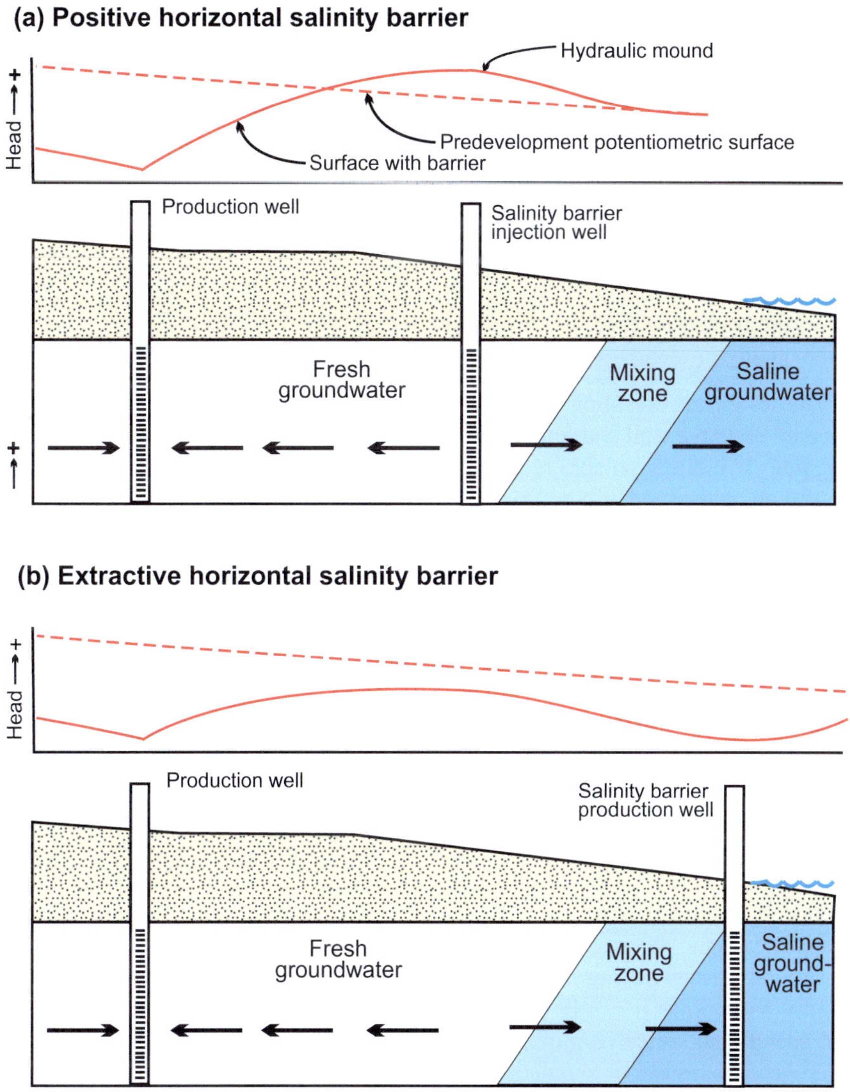

Fig. 21.5 Conceptual diagram of a **a** positive horizontal salinity barrier and an **b** extractive horizontal salinity barrier. Both types of systems can create a seaward hydraulic gradient at the interface between fresh and saline groundwater

There are numerous design and operational parameters that need to be considered in the design of positive salinity barrier systems including:

- the location of barriers with respect to the freshwater/saline water interface and inland aquifer users
- the type of infiltration systems (wells versus surface spreading)
- well depths
- well open hole or screened depths
- well spacing and the length of well wellfield
- recharge rate (total and per well or infiltration basin)
- the number, size, depth, and distribution of infiltration basins
- the quality of recharged water and pretreatment requirements.

The design and operation of positive salinity barriers systems also need to address standard recharge well and infiltration basin issues, particularly the management of clogging. If indirect potable reuse is a possibility, then recharged water needs to be of high quality and a monitoring program established to ensure that the potable water supply is not adversely impacted. Monitoring is especially important if natural attenuation processes in an aquifer are being relied upon to meet water quality standards. The potential health and environmental risks are far less when groundwater near a salinity barrier system has only non-potable uses. If the reclaimed water is of suitable quality for unrestricted irrigation, then its injection and subsequent recovery for irrigation should pose no additional risks.

Travel (aquifer residence) times from the salinity barrier to the nearest potable water supply wells (and any intervening non-potable wells) are critical issues for the evaluation of the potential risks of a salinity barrier system when the recharged water is not of potable quality. A detailed, accurate aquifer characterization is the critical foundation for evaluating the fate and transport of contaminants present in the recharged water. To increase travel times from the salinity barrier to production wells, it may be cost-effective to move, abandon, or replace some production wells located nearest to the coast and the proposed salinity barrier.

Groundwater modeling is a critical tool for the design and operation of positive salinity barriers, as it can be used to evaluate the effectiveness of various design and operational scenarios. Solute-transport modeling needs to incorporate aquifer heterogeneity, instead of simulating the aquifer in question as a homogeneous unit. Groundwater monitoring is also critical. A network of monitoring wells, including transects perpendicular to the saline-water interface and barrier is needed to monitor changes in the position of the interface and the movement of recharged water.

Typically, inadequate data are available at the start of a project (before actual recharge begins) to adequately calibrate a solute-transport model. Hence, for major salinity barrier systems, post-audits should be performed after the start of operation, which involves history matching (comparing model predictions to actual results), and recalibration of models against operational data, if necessary. Systems should be designed to allow for operation flexibility, such as changes in recharge rates between wells (or basins) based on monitoring results.

21.8.2 Regulatory Issues

The regulation of salinity barrier systems varies between countries and sometimes between states and districts within countries. In the United States, injection well systems are required to meet federal and state Underground Injection Control (UIC) regulations. Salt water intrusion barrier wells regulatory issues and environmental impacts were reviewed by the USEPA (1999). The basic UIC requirement (40 CFR 144.12(a)) is that owners are prohibited from engaging in any injection activity that allows for movement of fluids containing any contaminants into underground sources of drinking water (USDWs) if the presence of that contaminant may cause a violation of any primary drinking water regulation or may otherwise adversely affect the health of persons (USEPA 1999). A USDW is defined as non-exempt aquifer that contains less than 10,000 mg/L of total dissolved solids (TDS). Hence water injected into a USDW must be treated to meet primary (health-based) drinking water standards.

An alternative is to inject into a non-USDW part of an aquifer (i.e., zone that contains >10,000 mg/L of TDS), particularly the toe of the saline-water wedge. However, endangerment of a USDW could still occur if the injected water migrates out of the saline zone into a USDW (i.e., into a fresher zone). Injection into saline water near the freshwater/saline-water interface may cause some saline-water to migrate landward.

Salinity barriers in the United States may face additional regulations from state and local environmental programs, especially if wastewater is used and indirect potable reuse is a possibility or reality. That fact that a water meets federal and state numerical drinking water standards (maximum contaminant levels) does not necessarily indicate that it is safe to drink because a wide variety of pathogens and chemical contaminants might still be present. Hence, much more stringent water quality standards are applied to projects in which indirect potable reuse may occur.

Key regulatory issues for a proposed salinity barrier project in the City of Hallandale, Florida, (MWH 2013) and an earlier proposed project in the nearby City of Hollywood, Florida, involving injection of treated wastewater effluent were hydraulic impacts (mounding) and the fate and transport of potential contaminants into both adjacent aquifers and surface water bodies. The City of Hallandale project involves using 18 new stormwater drainage wells proposed to be constructed parallel to the coastline to also recharge treated wastewater during dry periods when stormwater is not available. Stormwater drainage wells in coastal southeastern Florida are designed to recharge into saline (non-USDW or G-IV) aquifers and reasonable assurance is required that the recharged water will not migrate upward and adversely impact USDW aquifers. The City of Hollywood project, proposed in the early 1990s, would have also injected into a non-USDW part of the Biscayne Aquifer (the toe of the saline water wedge), but at that time regulatory approval could not be obtained because of the specter of indirect potable reuse.

21.8.3 Orange County Water District (California) Talbert Barrier

The history and recent operation of the Talbert Barrier was summarized by Cook (2004), Herndon and Markus (2014), and Burris (2014). The Orange County Water District started pilot studies in 1965 on the feasibility of injecting effluent from an advanced wastewater treatment facility (AWT) into the aquifers at the Talbert Gap. Construction of the AWT facility, known as Water Factory 21, began in 1972 and injection of treated wastewater commenced in 1976 using 23 multi-aquifer (nested) injection wells with a total of 81 injection points (Cook 2004). The initial start-of-the-art Water Factory 21 treatment process involved GAC (granular activated carbon) with RO (reverse osmosis) introduced in 1977 (50:50 split between GAC and RO). Finally, the treatment process was upgraded to RO with UV disinfection and an advanced oxidation process (Cook 2004).

The Talbert Barrier (also referred to as the Talbert Gap Salinity Barrier, Talbert Seawater Intrusion Barrier and similar names) is perhaps the best known and a textbook example of an operational large-scale salinity barrier system. The Newport-Inglewood fault zone in coastal Orange County forms the southwestern boundary of the Orange County Groundwater Basin. Saline water is able to migrate inland into the Orange County Groundwater Basin through gaps in the fault zone where shallow aquifers connect the basin to the Pacific Ocean. The Talbert Gap is approximately 4 km (2.5 miles) wide and is located between the cities of Newport Beach and Huntington Beach. Geologically, the Talbert Gap is located between the Huntington Beach Mesa and the Newport Mesa (Fig. 21.6).

The salinity-barrier system recharges water into four injection zones (shallow aquifers), which are referred to, in descending order, as the Talbert aquifer, and alpha, beta, and lambda aquifers (Fig. 21.7). Each multi-zone injection well is open to 2–4 of the aquifers. The 23 "legacy" injection wells are constructed at approximately 180 m (600 ft) intervals along a city street (Ellis Avenue). Thirteen additional injection wells were constructed between 1998 and 2006 with a total of 28 injection zones. Eight of the injection zones are in the deeper main aquifer and are used for the purpose of aquifer recharge rather than saline-water intrusion control (Herndon and Markus 2014).

Since January 2008, the injected water consists primarily of highly treated reclaimed water produced by the state-of-the-art Groundwater Replenishment System (GWR) Advanced Water Purification Facility (AWPF), which was described by Markus (2009) and on the GWR website (http://www.ocwd.com/gwrs/). The GWR treatment process consists of microfiltration (MF) pretreatment followed by reverse osmosis (RO) and ultraviolet light and hydrogen peroxide treatment to break down remaining organic compounds through an accelerated decomposition (oxidation) process. The treatment process produces ultra-pure water. Such a high level of treatment is required because some of the recharged water will eventually enter the potable supply.

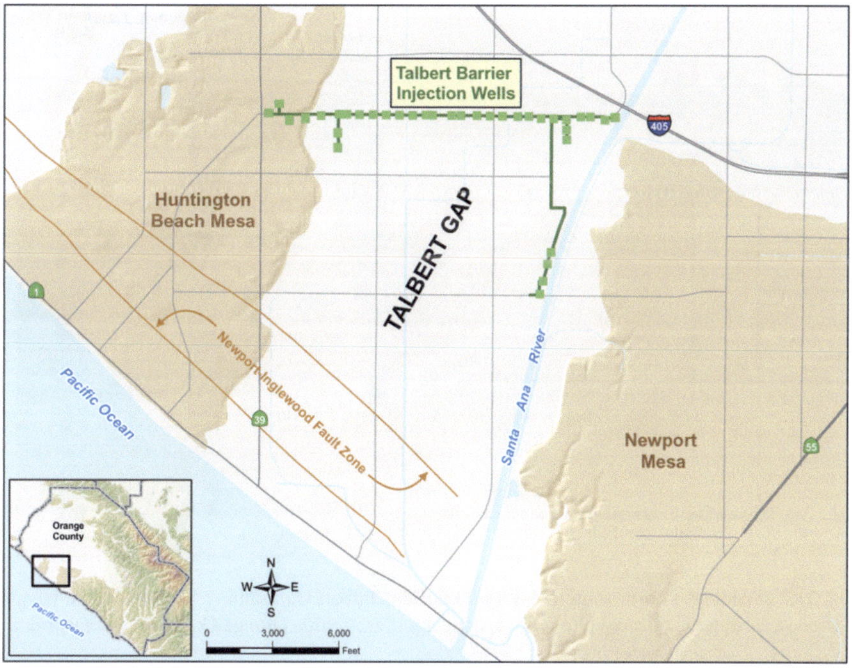

Fig. 21.6 Map showing the location of the Talbert Gap salinity barrier system, which consists of an alignment of injection wells between the Huntington Beach Mesa and the Newport Mesa. *Source* Orange County Water District

Injected volumes are dictated by aquifer water levels, particularly whether water levels in the gap areas are at or above protective levels needed to prevent saline-water intrusion into the Orange County Groundwater Basin. The required injected volumes were originally determined through groundwater modeling, and were subsequently found to be quite accurate (Herndon and Markus 2014). In 2014, the total injected volume was 10,737 million gallons (MG; 40.64 million m^3; MCM), which corresponds to an average daily flow rate of 29.4 MGD (0.111 MCM/d; Burris 2014). Nearly all of the water was purified recycled water from the GWR AWPF with a negligible volume of potable water injected. The distribution of injection by zones in 2014 was as follows:

45% shallow zone (Talbert and Alpha aquifers)
42% intermediate zone (Beta, Lambda, Omicron, and Upper Rho aquifers)
13% deep zone (Lower Rho and Main aquifers).

The individual wells are constructed with a 12-in. (30-cm) diameter 316L stainless casing and wire-wrapped screens. Pneumatic down-hole flow control valves are used to maintain full columns of water (Herndon and Markus 2014). Clogging is an operational issue despite the high quality of the recharged water. Analysis of the

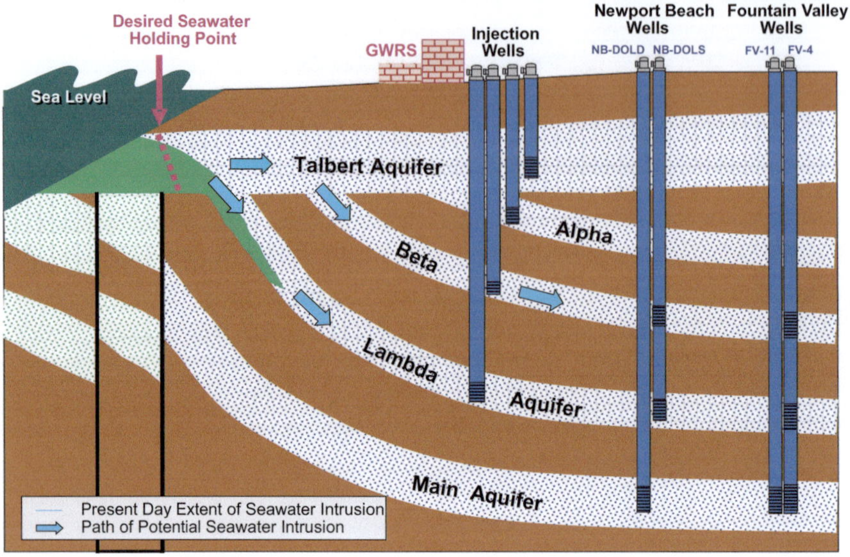

Fig. 21.7 Schematic north-south cross section of the Talbert Gap salinity barrier section. Highly treated wastewater is injected into four separate aquifers. *Source* Orange County Water District

clogging material indicates that it is composed of fine-particles of calcium carbonate, iron oxide, and aluminum silicate (Herndon and Markus 2014).

An extensive well rehabilitation program was initiated in 2009 and continued into the mid-2011 to restore the injection capacity of all of the legacy wells (Burris 2014). A redevelopment cycle of three years was reported to be sufficient to maintain injection flow rates in the legacy wells without significant reductions in well efficiency (Burris 2014). Wells are maintained through periodic major well rehabilitation and periodic backflushing. Three of the new wells are equipped with dedicated submersible pumps and the rest of the new wells are equipped with dedicated air lines (10 cm diameter) that join the blank well casing above the screen (Herndon and Markus 2014; Burris 2014). Air development of the legacy wells requires removal of the wellheads. Burris reported that wells are backflushed by airlifting after one to three months of operation [15–40 MG (0.057–0.015 MCM)] of injection). Alternating backflushing and injection (multiple reversal method) was found to be particularly effective.

Backflushing using submersible pumps is performed more frequently (after 9–10 MG; 0.034–0.038 MCM of injection) but for shorter-periods of time, and can be easily and remotely performed (Burris 2014). Burris (2014) reported that frequent short duration backflushing with a submersible pump tends to increase and maintain the efficiency of the injection well system better than the less frequent longer-duration airlift backflushing method, but noted that additional investigations are being performed by OCWD staff to determine the most cost-effective means of maintaining well performance.

21.8.4 Los Angeles County, California Salinity Barriers

Coastal saline-water intrusion from aquifer overdraft was first detected in the Central and West Coast Basins of Los Angeles County, California, in 1912 at Redondo Beach (Callison and Roth 1967; Johnson 2007). Initial testing of the injection of potable water to create a salinity barrier was performed in 1951 by the Los Angeles County Flood Control District (Lipshie and Larson 1995). Based on the successful results of the initial testing program, three major salinity barriers were constructed in Los Angeles County: West Coast Basin Barrier Project (WCBBP), Dominguez Gap Barrier Project (DGBP; operational in 1971), and Alamitos Barrier Project (ABP; operational in 1966). The later was developed in conjunction with the Orange County Water District. The construction and operation of these systems were documented by Callison and Roth (1967), Lipshie and Larson (1995), Johnson and Whitaker (2004), Land et al. (2004), Johnson (2007), Foreman (2014), and Los Angeles Department of Public Works (n.d.).

The WCBBP contains 153 injection wells over a distance of about 9 miles, and 296 observation wells. The DGBP has 41 injection wells, extending 12 miles, and 107 observation wells. The ABP has 43 injection wells, 4 extraction wells, and 226 observation wells (Los Angeles Department of Public Works n.d.). The WCBBP injects into three aquifers (200-Foot Sand, Silverado, and Lower San Pedro Aquifers) and uses both single and dual-zone injection wells in which water is injected through a conductor pipe (injection tube). In dual zone/aquifer wells, a packer filled with nitrogen gas isolates the two aquifers. Bray and Yeh (2008) developed a calibrated model of the ABP and predictive simulations were performed to evaluate the optimal scheduling of injection and well locations.

Imported water was used until 1996 when recycled water was introduced (Foreman 2014). Imported water supplies are being completely replaced by highly treated wastewater as demand for imported surface water has increased and less is available for recharge (Foreman 2014). The reclaimed water is treated by MF, RO, and advanced oxidation (UV and H_2O_2) in some cases. However, the rising costs of the barrier systems and aging infrastructure requires alternatives to the barriers to be developed. Reduction in groundwater pumping along the coast may increase aquifer water levels and decrease the amount of water injection required to maintain a protective barrier (Johnson 2007).

The main operation challenge for the WCBBP and DGBP is clogging (both physical and biological). Well performance has declined over time (Foreman 2014; Los Angeles Department of Public Works n.d.). About $4,000,000 per year was reported to be spent on well rehabilitation (Foreman 2014). New wells are equipped with air lines and dedicated back-flush collection and reuse systems. Airlifted water brought up during development is processed through a mobile treatment system before being discharged to the storm sewer. The treatment system removes suspended solids, adjusts pH, and lowers turbidity.

21.8.5 Salalah, Oman Salinity Barrier

A salinity barrier system was constructed at the city of Salalah, coastal Oman (Shammas and Jack 2007; Shammas 2008). The main source of bacteriological contamination of the groundwater in the city appears to be disposal of household sewage using septic tanks, soakaways, and cesspit systems. The Salalah Central Sewage Treatment Plant (STP) was constructed to provide centralized sewage treatment service to the community. The wastewater receives tertiary treatment and disinfection. The goals of the salinity barrier system were to (Shammas 2008):

- act as a barrier to saline-water intrusion
- increase the potentiometric surface of the coastal aquifer
- provide a source of water for crop irrigation.

The first operational phase of the Salalah salinity barrier system consisted of 40 injection wells and 40 observation wells (Shammas 2008). The wells are located 1.5–2.0 km (0.9–1.3 miles) from the shoreline and are located approximately 300 m (1,000 ft) part with a total depth of 40 m (120 ft). The salinity-barrier system is located near an area where the aquifer is used for agricultural water supply. The potable water wellfield is located further inland and is, thus, further separated from the salinity barrier by the agricultural wells. There are no readily available reports on the subsequent operational history of the system.

21.8.6 Llobregat Delta Aquifer Salinity Barrier

Llobregat delta aquifer salinity barrier was described by Ortuño et al. (2010, 2012). The Llobregat aquifer (Barcelona, Spain) is sand-and-gravel aquifer, which has experienced saline-water intrusion since the 1960s. The positive hydraulic barrier recharged highly treated wastewater that received tertiary treatment followed by UF, RO, and UV disinfection. Pilot testing started with four injection wells on March 26, 2007, and as of 2010, the system contained 15 wells with a total system capacity of 15,000 m^3/d. The wells are arranged in a 6 km alignment parallel to the coast and have depths of 70 m. Clogging was managed by backflushing every two weeks using a submersible pump (12 wells) and air lifting (3 wells). Operation of the system was reported to have been temporarily stopped in July 2011 (Ortuño et al. 2012).

21.9 Extractive Salinity Barriers

Saline-water intrusion may be controlled by pumping brackish or saline-water seaward of a freshwater/saline water interface (Fig. 21.5b). Theoretical discussions and the results of modeling studies of extractive salinity barriers are provided by

Coe (1972), van Dam (1999), Sherif and Hamza (2001), Todd (1980), Kacimov et al. (2009), and Pool and Carrera (2010). The extractive barrier concept has been considered for many years. An experiment was conducted in Oxnard County, California, in the late 1960s to access the feasibility of a strictly extractive barrier (California Department of Water Resources 1970). Key technical issues are the number and location of the production wells and their completion zone. The optimal design may involve extraction of water from near the base of the aquifer, at or near the toe of the saline-water interface (van Dan 1999).

Important operational issues are disposal of the produced water and minimization of the extraction of freshwater that will be drawn toward the production (extraction) wells. An obvious use of produced brackish or saline is as feedwater for desalination. A hybrid extractive barrier and desalination system has the benefits of both protecting fresh groundwater resources and providing additional potable water. Abd-Elhamid and Javadi (2011) evaluated what they referred to as ADR (abstraction, desalination, and recharge). Brackish water from the front of a saline water wedge is desalted and the produced water is recharged further landwards. Simulation-optimization modeling was performed in which the design parameters considered were the depth of extraction and recharge wells, locations of wells with respect to the freshwater/saline-water interface, and extraction and recharge rates. The model was based on the Biscayne Aquifer in southeastern Florida. The objectives were to minimize salt concentration in the aquifer, and construction and operation costs. Multiple sets of design parameters were evaluated in terms of fitness in competition with other solutions. The conclusion was that ADR is more cost-effective than abstraction or recharge alone.

Avital et al. (2010) described the Ashdod Nir-Am Eastern Interceptor under development in the South Coastal Aquifer of Israel. The final planned system would consist of a row of 34 production wells and 2 brackish water desalination plants. The well-field will be located along the eastern margin of the aquifer and will prevent saline water from migrating into the freshwater in the aquifer to the west. Modeling results indicate that the system will eliminate 15,000–17,000 metric tons (16,500–18,700 tons) of salt per year from flowing to the west. The interceptor system will serve a dual purpose of protecting existing freshwater resources and being a source of raw water for desalination. It was recognized that the interceptor system is not the sole solution to the rehabilitation and restoration of the Southern Coastal Aquifer. Water use will also have to be reduced to achieve a balance with the recharge rate.

As a strategy to minimize the loss of freshwater, Pool and Carrera (2010) theoretically investigated a double-pumping barrier system design. The negative barrier would consist of two wells (or rows of wells) configured with seawater wells located at the saline-water interface and freshwater wells located further inland. The freshwater wells would capture freshwater that would otherwise be captured by the seawater wells. The modeling results indicate that the double pumping barrier system is feasible. However, achieving long-term performance goals would require a well-balanced design that would be determined through groundwater modeling.

Hybrid extractive salinity barrier and desalination systems is a concept that warrants further investigation and application, as alternative seawater intakes are increasingly being used for raw water supply for desalination systems. Alternative intakes have the advantages of avoiding much of the environmental impacts associated with conventional open intakes (particularly entrainment and impingement of marine life) and reducing pretreatment requirements, as reviewed by Missimer et al. (2015) and Missimer and Maliva (2017). Hence, locating an alternative intake where it can also provide salinity barrier benefits makes good sense. Desalination plants are designed to treat water with a given envelope of water quality (salinity and some specific ion concentrations), so the chemical stability of the produced water is an important design consideration. An additional consideration is minimization of the extraction of freshwater. For example, a wellfield seaward of the OCWD Talbert Barrier was evaluated to supply seawater for the proposed Huntington Beach desalination system (ISTAP 2014). It was concluded that vertical and slant wells had the fatal flaws of interfering with the management of the salinity barrier and the interior freshwater basin. In particular, pumping seaward of the barrier would draw fresh groundwater through the barrier, causing an unacceptable loss of freshwater (ISTAP 2014).

21.10 Combined Positive Hydraulic Barrier and Extractive Barrier Systems

Both a positive hydraulic barrier and extractive barrier could be combined to create a system that is theoretically more effective in controlling saline-water intrusion than either type of barrier alone. The commonly proposed design for such systems have extraction wells located seaward of the freshwater/saline-water interface and the freshwater injection wells located landward of the interface (Todd 1974; Tsanis and Song 2001; Pool and Carrera 2010). An alternative design proposed by Sheahan (1977) has extraction wells located landward of the injection wells. Both are located inland of the saline-water interface. A proposed salinity barrier system (for Santa Clara, California) would use reclaimed water for injection. The rationale for the more inland freshwater extraction wells is that they would prevent any possible degradation of fresh groundwater supplies. The extraction wells would capture injected wastewater and prevent its inland migration. While combined barrier systems may offer technical benefits, the outstanding question is whether any additional saline-water intrusion control benefits would justify the additional costs. Extractive barriers, in general, can be a very cost-effective option if they are used to provide otherwise needed raw water for desalination facilities.

21.11 Scavenger Wells

Scavenger wells are, in essence, an extractive barrier against vertical saline-water intrusion. The concept is that pumping of saline water below the freshwater/saline-water interface will prevent the upward movement of the interface. Scavenger/production well couplets are used to produce freshwater in situations where freshwater occurs as a relative thin layer overlying saline groundwater and in which pumping would induce upconing of saline water.

The basic concept of scavenger wells is that freshwater and saline groundwater are pumped through separate outlets, which can be either separate wells constructed at different depths above and below the freshwater-saline water interface (dual-bore systems) or bores with pumps installed at different depths within a single well (above and below the interface). Separate scavenger and production pumps and drop pipes installed in a single well are ambiguously referred to as "wells" is some reports. By keeping the freshwater/saline water interface as a flow line, mixing is prevented in the aquifer and the geometry of the system is stable (Stoner and Bakiewicz 1993). The produced saline water is usually disposed to waste.

Separate wells may be more efficient, but use of the same well may be a more economical solution, especially when an existing well is used (Zack 1988). The intake for the production well is placed as far above the saline water interface as possible. Scavenger and production well pairs are not particularly cost-effective because of the additional construction and operational costs of an extra well (Zack 1988). When using a single well, the most effective operation of the couplet has usually been obtained by trial and error adjustments of the pump placements and pumping rates. Numerical modeling has been used to evaluate the scavenger well design concept and the optimization of system design (e.g., Saravanan et al. 2014).

Gregg (1971) described a vertical extractive salinity barrier system constructed in Glynn County, coastal Georgia. High rates of pumping of the principal artesian aquifer (Floridan Aquifer System) had resulted in upconing of brackish water. Upward migration appears to have occurred due to breaches in the confining strata between the fresh and brackish aquifer zones. In 1968, a well was installed in the brackish zone to lower the potentiometric surface and restore the equilibrium between the brackish-water zone and the overlying principal artesian aquifer. The monitoring data for the initial operational period indicated that the "protective pumping" system was successful in reversing the increase in chloride concentration in a nearby production well. However, a full-scale extractive salinity control was not constructed in the area. Chloride concentrations in two major industrial wellfields in the area have continued to rise despite modification of production wells to eliminate production from deep saline zones and decreases in pumping at both facilities (Cherry et al. 2011).

Beeson et al. (1994) documented pilot testing of scavenger wells at four sites in the Lower Indus Basin of southern Pakistan. The sites were selected with the aid of surface geophysical methods (VES and TDEM). The testing results demonstrated that scavenger wells are highly successful at recovering significant quantities of fresh water from lenses located above saline water within the permeable sandy alluvial aquifer that underlies parts of southern Pakistan (Beeson et al. 1994).

Zack (1988) documented testing of an existing abandoned well at Barrio La Trocha, near Vega Baja, Puerto Rico. There are numerous similar wells in Puerto Rico that could potentially also be brought back into production as scavenger/production well couplets. Based on testing results, a family of curves was developed for the Barrio La Trocha well that describe the relationship between scavenger and production well pumping and produced water chloride concentration.

Freshwater on the island of Cozumel, Mexico, accumulates from rainwater percolating through the soil zone and into cenotes. The fresh groundwater overlies saline groundwater and is produced from numerous shallow wells that are subject to increases in salinity from the upconing of saline water (Zack and Lara 2003). Vulnerability to saline-water intrusion was reported to limited production to about 1 L/s. A preliminary evaluation revealed that water quality and production could be markedly improved in 29 of 173 wells by installing scavenger wells (i.e. separate pumps and drop pipes). Field testing demonstrated the scavenger well extractions could allow fresh groundwater production to be increased to as much as 4.5 L/s while controlling the upward advance of saline water. The total cost of scavenger well technology was observed to be much less than the cost of desalination (Zack and Lara 2003).

Scavenger well systems in the Netherlands have been introduced as the "Freshkeeper" system in which brackish water is pumped from the lower part of aquifers to manage the freshwater-brackish water interface (Zuurbier et al. 2017). Produced water is desalted by reverse-osmosis and the concentrate is injected into a deeper zone. The Freshkeeper system was successfully applied at Noardburgum, province of Friesland, the Netherlands. Field testing and modeling results indicate that the Freshkeeper concept could be used to reopen a wellfield that was abandoned due to saline-water intrusion (Zuurbier et al. 2017).

21.12 Physical Barriers

Physical barriers are impervious or semi-impervious subsurface barriers constructed landward of the saline-water interface. Their main advantages are that they can be highly effective barriers (if properly designed and constructed) and have minimal operational and maintenance costs. The main disadvantages of physical barriers are:

- relatively high costs, which depend upon their depth, length, and construction materials and method
- low operational flexibility
- saline waters can become trapped on the upgradient sides of the barrier.

Construction methods for physical barriers include (Todd 1974; Atkinson et al. 1986; Oude Essink 2001; Luyun et al. 2011; Nurnawaty et al. 2016):

- **Slurry walls**: a slurry of water and bentonite clay is injected into a trench while excavation is progressing.
- **Grout cut offs or curtains**: a slurry of bentonite or cement is injected under pressure in the soil or rock through closely spaced boreholes.
- **Sheet pilings**: corrugated steel sheets driven into the ground.

- **Diaphragm walls**: in situ walls constructed by excavating a series of narrow trenches, which are supported by an engineered fluid (typically a bentonite mud) and then filled with concrete from the base upward.

Other physical barrier options have been proposed. Barcelona et al. (2006) proposed that a salinity barrier can be created in fractured carbonate rock by inducing gypsum precipitation. Gypsum precipitation would be induced by injected a solution super-saturated with respect to $CaSO_4$ and containing organic crystallization inhibitors.

The low operational flexibility of physical barriers requires that they be correctly located with respect to the freshwater/saline-water interface. Physical barrier system should be designed so that saline-water cannot migrate inland around the barrier into areas of concern.

A physical barrier was installed on New Providence Island, Bahamas, to limit the impacts of the construction of a marina on the position of the saline-water interface and to prevent the drainage of the local freshwater lens (Missimer Groundwater Science 2005). Solute-transport modeling showed that the barrier had to wrap around most of the landward perimeter of the marina basin and that a collection pipe system had to be located on the upgradient side of the barrier to collect the ponded freshwater for recovery and conveyance to an inland storage basin.

21.13 Optimization of Saline-Water Intrusion Management

The overriding objective for managing saline-water intrusion and other water management problems is to identify the optimal management strategy in terms of both performance and costs. Key issues to be decided are the type of salinity management system most appropriate for a given location, its design and construction, and operational procedures. Density-dependent solute-transport modeling is an important tool for evaluating the performance of different management strategies. There have been a number of published studies in which modeling was used to evaluate salinity barrier systems and optimize designs (e.g., Sherif and Hamza 2001; Tsanis and Song 2001; Mahesha 1996a, b, c; Ru et al. 2001; Rastogi et al. 2004; Bloetscher et al. 2005; Bray and Yeh 2008; Abd-Elhamid and Javadi 2011; Allow 2012). The general conclusion is that all salinity-control methods (e.g., reduce pumping, recharge, and extraction barriers) are potentially effective and that the challenge lies in the optimization of the design (Tsanis and Song 2001).

Typically, a series of management options and variations of options (e.g., different wellfield configuration and pumping or injection rates) are simulated to find a "best" solution. However, there are infinite combinations of plausible groundwater management options and, therefore, the optimal management strategy cannot be identified by sequentially evaluating all the different management schemes (Sreenkanth and Datta 2015). As reviewed by Sreekanth and Datta (2011, 2015) and applied by Park et al. (2008) and Kourakos and Mantoglou (2011), management models for sustainable utilization of groundwater in coastal and other settings can be developed by using numerical simulations within an optimization algorithm. The simulation-optimization approach includes two elements: a simulation model and an optimization algorithm. Simulation models are used to evaluate the effects of different management alternatives on the aquifer system. Optimization algorithms are used to perform an organized search for new and improved management strategies that are evaluated by the simulation models.

The simulation-optimization by external linking methodology is perhaps the conceptually simplest technique, whereby the simulation model is coupled as an external subroutine or subprogram to an optimization algorithm (Sreekanth and Datta 2015). Simulation-optimization using three-dimensional density-dependent solute-transport models of regional aquifers with numerous wells is challenged by a great computational burden and the scarcity of required data (Sreekanth and Datta 2011, 2015). Preliminary assessments might be performed using simplified 2-D and sharp interface models, but more precise simulations, for example, of concentration contours around a well may require a very fine-scale density-dependent model (Sreekanth and Datta 2015).

Surrogate or "meta-models," such as artificial neural networks (ANN), have therefore been used to replace numerical simulations within optimization algorithms (Sreekanth and Datta 2011, 2015). An AAN may be trained as an approximator for a 3-D density-dependent solute-transport model. Sreekanth and Datta (2011) used an ANN as a meta-model in an optimization algorithm for managing coastal saline-water intrusion. The goal was to develop an optimal strategy for operating a hypothetical coastal production wellfield and extractive salinity barrier. The modeling involved the following basic steps:

- a finite-element numerical solute-transport model was developed
- a neural network meta-model was trained to the numerical model
- candidate solutions for operation of the production and extraction wells were randomly generated
- solutions were sent to the neural network, which evaluated aquifer responses (salinity levels) at monitoring locations
- the optimization model evaluated whether salinity levels would be within acceptable levels
- candidates that meet salinity constraints were evaluated in terms of maximum pumping from production wells and minimum pumping from extraction wells.

The main limitation to surrogate approaches include the loss of accuracy incurred when replacing a numerical model with a surrogate and the additional uncertainty from the use of the surrogate (Sreekanth and Datta 2015). Stochastic approaches have been used for optimization of decision making under uncertainty for groundwater management applications. However, the application of stochastic modeling for the management of coastal aquifers is challenged mainly be the computation burden involved in the simulation of coupled flow and solute transport (Sreekanth and Datta 2015).

Simulation-optimization modeling is receiving considerable academic interest, but the question still remains as to its utility and cost-effectiveness in the applied realm. The question become whether it is providing new insights or confirming what one already knows or could more readily find out. A firm understanding of the dynamics of coastal aquifer systems and local aquifer hydrogeology can provide efficient guidance to constrain the envelope of conditions to be evaluated by numerical simulations. Simulation-optimization and stochastic modeling techniques are never a valid alternative to a thorough aquifer characterization program.

Optimization of salinity-intrusion management systems also requires an adaptive management approach, which allows for improvements in system performance based upon operational experiences. Adaptive management of salinity-barrier systems requires that they be designed and constructed to allow for operational flexibility. Pilot testing of a small-sale system is critical for the development of large-scale systems. The results of pilot testing results can provide a strong indication of whether the planned salinity-barrier system will likely perform as anticipated and provide information on potential operational problems, such as excessive clogging and adverse fluid-rock interactions and associated water quality impacts.

References

Abd-Elhamid, H. F., & Javadi, A. A. (2011). A cost-effective method to control seawater intrusion in coastal aquifers. *Water Resources Management, 25,* 2755–2780.

Abdalla, O. A. E., Ali, M., Al-Higgi, K., Al-Zidi, H., El-Hussain, I., & Al-Hinai, S. (2010). Rate of seawater intrusion estimated by geophysical methods in an arid region: Al Khabourah, Oman. *Hydrogeology Journal, 18,* 1437–1445.

Allow, K. A. (2012). The use of injection wells and a subsurface barrier in the prevention of seawater intrusion: A modelling approach. *Arabian Journal of Geosciences, 5*(5), 1151–1161.

Atkinson, S. F., Miller, G. D., Curry, D. S., & Lee, S. B. (1986). *Salt water intrusion: Status and potential in the contiguous United States*. Chelsea, MI: Lewis Publishers.

Avital, A., Lumlesky, S., Moss, D., Gev, I., Negev, I., & Oron, H. (2010). Desalination as a key component for the southern Israeli coastal aquifer rehabilitation and restoration plan. In *Proceedings, EuroMed 2010, Desalination for Clean Water and Energy*, Tel Aviv, October 3–7, 2010.

Banks, H. O., & Richter, R. C. (1953). Sea-water intrusion into ground-water basins bordering the California coast and inland bays. *Transactions. American Geophysical Union, 34*(4), 575–582.

Barcelona, M. J., Kim, M., Masciopinto, C., & La Mantia, R. (2006). A gypsum-barrier design to stop seawater intrusion in a fractured aquifer at Salento (Southern Italy). In *Proceedings 1st SWIM-SWICA Joint Saltwater Intrusion Conference*, Cagliari-Chia Laguna, Italy-September (pp. 24–29).

Beeson, S., Carruthers, R., & Wyness, A. J. (1994). Scavenger wells-2—field investigations and monitoring. In E. Custodio & A. Galofré (Eds.), *Study and modelling of saltwater intrusion into aquifers*, Proceedings of the 12th Saltwater Intrusion Meeting, Barcelona, Spain, November 1–6, 1992. Barcelona, Spain (pp. 557–571).

Bloetscher, F., Muniz, A., & Witt, G. M. (2005). *Groundwater injection: Modeling, risks and regulations*. New York: McGraw-Hill.

Bloetscher, F., Meerfoff, D. E., Heimlich, B. W., Brown, A. R., Batler, D., & Loucraft, M. (2010). Improving resilience against the effects of climate change. *Journal American Water Works Association, 102*(11), 36–46.

Bray, B. S., & Yeh, W. W.-G. (2008). Improving seawater barrier operation with simulation optimization in southern California. *Journal of Water Resources Planning and Management, 134,* 171–180.

Bruington, A. E. (1969). Control of sea-water intrusion in a ground-water aquifer. *Ground Water, 7*(3), 9–14.

Burris, D. L. (2014). *Groundwater replenishment system 2014 annual report*. Fountain Valley CA: Orange County Water District.

Buszka, P. M., Brock, R. D., & Hooper, R. P. (1994). *Hydrogeology and selected water-quality aspects of the Hueco Bolson Aquifer at the Hueco Bolson Recharge Project area, El Paso, Texas*. U.S. Geological Survey Water-Resources Investigations Report 91-4092.

California Department of Water Resources. (1970). *Oxnard Basin experimental extraction-type barrier, Bulletin 147-6*. Sacramento, CA: California Department of Water Resources.

Callison, J. C., & Roth, J. N. (1967). Construction geology of the west coast basin barrier project. *Engineering Geology, 4*(2), 42–54.

Campbell, B. G., Mirecki, J. E., Conlon, K. J., & Petkewich, M. D. (1997a). Evaluation of aquifer storage recovery in the Santee Limestone/Black Mingo Aquifer near Charleston, South Carolina, 1993–1995. In K. R. Kendall (Ed.), *Conjunctive use of water resources: Aquifer storage and recovery*, Proceedings AWRA Symposium, Long Beach, California (pp. 231–239). Middleburg, VA: American Water Resources Association.

Campbell, B. G., Conlon, K. J., Mirecki, J. E., & Petkewich, M. D. (1997b). *Evaluation of aquifer storage recovery in the Santee Limestone/Black Mingo Aquifer, near Charleston, South Carolina, 1993–1995*. U.S. Geological Survey Water-Resources Investigations Report 96-4283.

Cherry, G. S., Peck, M. F., Painter, J. A., & Stayton, W. L. (2011). *Groundwater conditions in the Brunswick–Glynn County area, Georgia, 2009*. U.S. Geological Survey Scientific Investigations Report 2011-5087.

Cherry, J. A., Parker, B., Einarson, M., Chapman, S., & Meyer, J., (2015). Overview of depth-discrete mutilevel groundwater monitoring technologies: Focus on groundwater monitoring in areas of oil and gas well stimulation in California. In B. K. Esser et al. (Eds.), *Recommendations on model criteria for groundwater sampling, testing, and monitoring of oil and gas development in California*, Final Report (LLNL-TR-669645). Sacramento: California State Water Resources Control Board.

Coe, J. J. (1972). Seawater extraction barrier. *Journal of the Irrigation and Drainage Division, 98*(3), 387–403.

Cook, J. (2004). *Innovative applications in water reuse: Ten case studies*. Alexandra, VA: WateReuse Association.

Custodio, E. (2004). Myths about seawater intrusion in coastal aquifers. In L. Araguás, E. Custodio, & M. Manzano (Eds.), *Groundwater and saline intrusion. Selected papers from the 18th Salt Water Intrusion Meeting*, 18 SWIM, Cartagena, Spain (pp. 599–608).

Diersch, H. J. (1998). FEFLOW: Interactive, graphics-based finite-element simulation system for modeling groundwater flow, contaminant mass and heat transport processes. In *Getting started; user's manual; reference manual, Version 4.6*. Berlin, Germany; WASY, Institute for Water Resources Planning and System Research Ltd.

Einarson, M. D. (2006). Multilevel ground-water monitoring. In D. M. Nielsen (Ed.), *Practical handbook of environmental site characterization and ground-water monitoring* (2nd ed., pp. 808–848). Boca Raton: CRC Press.

Foreman, T. L. (2014). Managed aquifer recharge (MAR) and design and construction of hydraulic barriers against seawater intrusion: The California case. *Boletín Geológico y Minero, 125*(2), 133–142.

Gregg, D. O. (1971). Protective pumping to reduce aquifer pollution, Glynn County. *Georgia. Ground Water, 9*(5), 21–29.

Guo, W., & Langevin, C. D. (2002). *User's guide to SEAWAT: A computer program for simulation of three-dimensional variable-density ground-water flow*. U.S. Geological Survey Open-File Report 01-434.
Herndon, R., & Markus, M. (2014). Large-scale aquifer replenishment and seawater intrusion control using recycled water in Southern California. *Boletin Geológico Y Mínero, 125*(4), 143–155.
ISTAP. (2014). Final Report: Technical feasibility of subsurface intake designs for the proposed Poseidon Water desalination facility at Huntington Beach, California. Report by the Independent Scientific Technical Advisory Panel (ISTAP) Convened and Facilitated by CONCUR, Inc.
Johnson, T. (2007). *Battling seawater intrusion in the Central & West Coast Basins*, Technical Bulletin 13. Lakewood, CA: Water Replenishment District of Southern California.
Johnson, T. A., & Whitaker, R. (2004). Saltwater intrusion in the coastal aquifers of Los Angeles County, California. In A. H. Cheng & D. Ouazar (Eds.), *Coastal aquifer management* (pp. 29–48). Boca Raton: Lewis Publishers.
Kacimov, A. R., Sherif, M. N., Perret, J. S., & Al-Mushikhi, A. (2009). Control of sea-water intrusion by salt-water pumping: Coast of Oman. *Hydrogeology Journal, 17,* 541–558.
Kashef, A. A. I. (1977). Management and control of salt-water intrusion in coastal aquifers. *Critical Reviews in Environmental Science and Technology, 7*(3), 217–275.
Kerrou, J., & Renard, P. (2010). A numerical analysis of dimensionality and heterogeneity effects on advective dispersive seawater intrusion processes. *Hydrogeology Journal, 18,* 55–72.
Koch, R. J., & Pearson, S. G. (2007). *Evaluation of sampling systems for multiple completion regional aquifer wells at Los Alamos National Laboratory,* LA-UR-07-4034, August 2007, EP2007-0486). Las Alamos, NM: Las Alamos National Laboratory.
Kourakos, G., & Mantoglou, A. (2011). Simulation and multi-objective management of coastal aquifers in semi—Arid regions. *Water Resources Management, 25*(4), 1063–1074.
Land, M., Reichard, E. G., Crawford, S. M., Everett, R. R., Newhouse, M. W., & Williams, C. F. (2004). *Ground-water quality of coastal aquifer system in the West Coast Basin, Los Angeles County, California, 1999–2002*. U.S. Geological Survey Scientific Investigations Report 2004-5067.
Langevin, C. D., Thorne, D. T., Jr., Dausman, A. M., Sukop, M. C., & Guo, W. (2007). *SEAWAT Version 4: A computer program for simulation of multi-species solute and heat transport*. U.S. Geological Survey Techniques and Methods Book 6, Chapter A22.
Lipshie, S. R., & Larson, R. A. (1995). The West Coast Basin, Dominguez Gap, and Alamitos seawater-intrusion barrier system, Los Angeles and Orange Counties. *California. AEG News, 38*(4), 25–29.
Los Angeles Department of Public Works. (n.d.). *Seawater barrier*. Retrieved May 29, 2018 from https://dpw.lacounty.gov/wrd/barriers/index.cfm.
Luyon, R., Jr., Momii, K., & Nakagawa, K. (2011). Effects of recharge wells and flow barriers on seawater intrusion. *Ground Water, 49*(2), 239–249.
Mahesha, A. (1996a). Transient effect of battery of injection wells on seawater intrusion. *Journal Hydraulic Engineering, 122,* 266–271.
Mahesha, A. (1996b). Steady-state effect of freshwater injection on seawater intrusion. *Journal Irrigation Drainage Engineering, 122,* 149–154.
Mahesha, A. (1996c). Control of seawater intrusion through injection-extraction well system. *Journal Irrigation Drainage Engineering, 122,* 314–317.
Maliva, R. G. (2016). *Aquifer characterization techniques*. Berlin: Springer.
Maliva, R. G., & Missimer, T. M. (2012). *Arid lands water evaluation and management*. Berlin: Springer.
Markus, M. R. (2009). Groundwater replenishment & water reuse: *The Water Report*, Issue 59 (January 15, 2009), 1–9.
Metzger, L. F., & Izbicki, J. A. (2013). Electromagnetic induction logging to monitor changing chloride concentrations. *Groundwater, 51,* 108–121.

Milnes, E., & Renard, P. (2004). The problem of salt recycling and seawater intrusion in coastal irrigated plains: An example from the Kiti aquifer (southern Cypress). *Journal of Hydrology, 288,* 327–343.

Missimer, T. M., Jones, B., & Maliva, R. G. (Eds.). (2015). *Intakes and outfalls for seawater reverse-osmosis desalination facilities*. Berlin: Springer.

Missimer, T. M. & Maliva, R. G. (2017). Environmental issues in seawater reverse osmosis desalination: Intakes and outfalls. *Desalination*. https://doi.org/10.1016/j.desal.2017.07.012.

Missimer Groundwater Science. (2005). Hydrogeology of the Albany House Development site and surrounding lands, New Providence Island, Bahamas, Unpublished report.

MWH. (2013). Preliminary assessment of a hydraulic salinity barrier for wellfield protection. Report prepared for the City of Hallandale, Florida.

Nurnawaty, M., Thaha, M. A., & Maricar, F. (2016). Seepage saltwater reduction by physical barrier at coastal aquifer. *International Journal of Applied Engineering Research, 11*(23), 11358–11362.

Ortuño, F., Molinero, J., Custodio, E., Juárez, I., Garrido, T., & Fraile, J., (2010). Seawater intrusion barrier in the deltaic Llobregat aquifer (Barcelona, Spain). In *Performance and pilot phase results. SWIM 21—21st Salt Water Intrusion Meeting*, Azores, Portugal, June 21–26, 2010 (pp. 135–138).

Ortuño, F., Molinero, J., Garrido, T., & Custodio, E. (2012). Seawater injection barrier recharge with advanced reclaimed water at Llobregat delta aquifer (Spain). *Water Science and Technology, 66*(10), 2083–2089.

Oude Essink, G. H. P. (2001). Improving fresh groundwater supply—Problems and solutions. *Ocean and Coastal Management, 44,* 429–449.

Park, N., Kim, S., Shi, L., & Song, S. (2008). Field validation of simulation-optimization model for protecting excessive pumping wells. In *Proceedings of the 20th SWIM*, Naples, Florida.

Pool, M., & Carrera, J. (2010). Dynamics of negative hydraulic barriers to prevent seawater intrusion. *Hydrogeology Journal, 18*(1), 95–105.

Rastogi, A., Choi, G. W., & Ukarande, S. K. (2004). Diffuse interface model to prevent ingress of seawater in multi-layer coastal aquifers. *Journal of Spatial Hydrology, 4,* 1–31.

Ru, Y., Jinno, K., Hosokawa, T., & Nakagawa, K. (2001). Study on effect of subsurface dam in coastal seawater intrusion. In *First international conference on saltwater intrusion and coastal aquifers—monitoring, modeling, and management*. Essaouira, Morocco.

Sanford, W. E., & Pope, J. P. (2010). Current challenges using models to forecast seawater intrusion: Lessons from the Eastern Shore of Virginia, USA. *Hydrogeology Journal, 18,* 73–93.

Saravannan, K., Kashyap, D., & Sharma, A. (2014). Model assisted design of scavenger well system. *Journal of Hydrology, 510,* 313–324.

Shammas, M. I., & Jacks, G. (2007). Seawater intrusion in the Salalah plain aquifer, Oman. *Environmental Geology, 53,* 575–587.

Sheahan, N. T. (1977). Injection/extraction well system—A unique seawater intrusion barrier. *Ground Water, 15*(1), 32–50.

Shammas, M. I. (2008). The effectiveness of artificial recharge in combating seawater intrusion in Salalah coastal aquifer, Oman. *Environmental Geology, 55,* 191–204.

Sherif, M. M., & Hamza, K. I. (2001). Mitigation of seawater intrusion by pumping brackish water. *Transport in Porous Media, 43,* 29–44.

Sreekanth, J., & Datta, B. (2011). Optimal combined operation of production and barrier wells for the control of saltwater intrusion in coastal groundwater well fields. *Desalination and Water Treatment, 32*(1–3), 72–78.

Sreekanth, J., & Datta, B. (2015). Review: Simulation-optimization models for the management and monitoring of coastal aquifers. *Hydrogeology Journal, 23,* 1155–1166.

Stoner, R. F., & Bakiewicz, W. (1993). Scavenger wells-1—Historic developments. In E. Custodio & A. Galofré (Eds.), *Study and modelling of saltwater intrusion into aquifers*, Proceedings of the 12th Saltwater Intrusion Meeting, Barcelona, Spain, November 1–6, 1992. Barcelona, Spain (pp. 545–556).

Todd, D. K. (1974). Salt-water intrusion and it control. *Journal American Water Works Association, 63*(3), 180–187.

Todd, D. K. (1980). *Groundwater hydrology* (2nd ed.). New York: Wiley.
Trabelsi, R., Zairi, M., & Ben Dhia, H. (2007). Groundwater salinization of the Sfax superficial aquifer, Tunisia. *Hydrogeology Journal, 15,* 1341–1355.
Tsanis, I. K., & Song, L.-F. (2001). Remediation of sea water intrusion: A case study. *Ground Water Monitoring and Remediation, 21*(3), 152–161.
USEPA. (1999). *The Class V underground injection control study. Volume 20. Salt water intrusion barrier wells*. Washington, D.C.: USEPA Office of Ground Water and Drinking Water.
Van Dam, J. C. (1999). Exploitation, restoration, and management. In J. Bear, A. Cheng, S. Sorek, D. Ouazar, & A. Herrera (Eds.), *Seawater intrusion in coastal aquifers: Concepts, methods, and practices* (pp. 73–125). Netherlands: Kluwer.
Werner, A. D., & Simmons, C. T. (2009). Impact of sea-level rise on sea water intrusion in coastal aquifers. *Ground Water, 47*(2), 197–204.
Zack, A. L. (1988). *A well system to recover usable water from a freshwater-saltwater aquifer in Puerto Rico*. U.S. Geological Survey Water-Supply Paper 2328.
Zack, A., & Lara, F. (2003). Optimizing fresh groundwater withdrawals in Cozumel, Quintana Roo, Mexico. In C. Voss & L. Konikow (Eds.), *Second International Conference on Saltwater Intrusion and Coastal Aquifers—Monitoring, Modeling, and Management*, Merida, Mexico, March 30–April 2, 2003 (pp. 1–10).
Zubari, W. K. (1999). The Dammam aquifer in Bahrain—Hydrochemical characterization and alternatives for management of groundwater quality. *Hydrogeology Journal, 7*(2), 197–208.
Zuurbier, K. G., Raat, K. J., Paalman, M., Oosterhof, A. T., & Stuyfzand, P. J. (2017). How subsurface water technologies (SWT) can provide robust, effective, and cost-efficient solutions for freshwater Management in coastal zones. *Water Resources Management, 31*(2), 671–687.

Chapter 22
Wastewater MAR and Indirect Potable Reuse

22.1 Introduction

Irrigation with wastewater has been practiced for millennia, with the earliest indications going back in time at least 5,000 years to the Minoan civilization of ancient Greece (Asano and Levine 1996; Vignesnaran and Sundaravadivel 2004). Irrigation with wastewater is practiced in areas facing water scarcity because it does not require water of the highest quality and wastewater contains nutrients needed for crops. Wastewater is also recycled for industrial uses and there is increasing interest in its use to augment potable water supplies. Indeed, some areas facing water scarcity are now investigating indirect and direct potable reuse of wastewater, which would have been unthinkable a couple of decades ago. Where current freshwater reserves are near or at their sustainable limits, recycled wastewater is often the only significant low-cost alternative water source for agricultural, industrial, and urban nonpotable purposes (Lazarova et al. 2000; Miller 2006). In many parts of the world, using water once simply is no longer an option (Levine and Asano 2004). An extensive literature has developed on all aspects of wastewater reuse.

Wastewater reuse has five main benefits:

(1) it can provide additional needed water in water scarce regions where other options are either unavailable or too expensive
(2) it can provide needed plant nutrients and thus reduce fertilization costs
(3) reuse of wastewater may prevent or reduce the adverse impacts associated with its disposal
(4) it is a reliable source of "new" water as its supply from urban areas is year round
(5) it is the only source of additional water that increases as population increases.

An important characteristic of reclaimed water is that it is a very reliable source of water. Both the production and quality of reclaimed water are relatively constant throughout the year and are almost constant between years (Dillon 2000; Friedler 2001). Reliability of supply is important for water users because it allows for greater confidence in agricultural investments (Friedler 2001). However, the quality

R. G. Maliva, *Anthropogenic Aquifer Recharge*,
Springer Hydrogeology, https://doi.org/10.1007/978-3-030-11084-0_22

of treated wastewater may not be ideal for agricultural irrigation and other uses. Treated wastewater in arid and semiarid lands and coastal areas often has elevated salinities caused by a poor quality of the freshwater supply, evaporative concentration, and seepage of saline waters into collection systems.

In addition to human health considerations, the suitability of treated wastewater for irrigation depends upon its salinity, sodium concentration (sodium adsorption ratio), trace elements, macronutrients and micronutrients concentrations, plant sensitivity and tolerance, soil characteristics, and irrigation management practices (Ayers and Wescott 1985). As addressed in the FAO (Food and Agriculture Organization of the United Nations) Irrigation and Drainage Paper 29, Water Quality for Agriculture (Ayers and Wescott 1985), elevated concentrations of some elements may be toxic to plants (i.e., phytotoxicity will occur) with sensitivity varying between plant types. For example, boron is essential for plant growth but is toxic at high concentrations. The FAO recommended limit for boron is 0.75 mg/L.

Wastewater reuse may pose a public health risk based on the presence of pathogenic microorganisms and, to a lesser degree, chemical contaminants of variable toxicity. The USEPA (2012) noted that the "key objective is to achieve a quality of reclaimed water that is appropriate for the intended use and is protective of human health and the environment." Large-scale use of sewage effluent for irrigation often requires that it be treated for unrestricted irrigation so that farmers can grow what they want and the water can be used for other purposes that may involve public contact, such as landscape irrigation (Bouwer 1991). However, the reality is that in many periurban areas of developing countries, wastewater is being used for agricultural irrigation with little or no treatment.

Proven wastewater treatment and purification processes currently exist to produce water of virtually any quality desired (Asano and Levine 1996; Mujeriego and Asano 1999; USEPA 2012). Wastewater can be treated to such a degree that it poses essentially no public health risk even if the water were to be directly consumed. However, such a high level of treatment is not a feasible template for the developing world because of its high construction and operational costs and required technical resources.

A politically and technically challenging question is the specific degree of treatment necessary for different types of wastewater recycling. In developed countries, over-treatment (i.e., treatment beyond that necessary to prevent material health risks) is sometimes performed based on the precautionary principle ("better safe the sorry") and to obtain political and public support for projects. It is important to recognize that over-treatment can unnecessarily increase project costs so that a water recycling project becomes economically unviable and its water resources benefits are lost. The USEPA (2012) observed that "more restrictive regulations, such as those in California and Italy, while amply protective, are potentially prohibitively expensive in some economic contexts without necessarily improving the public health outcome." Over-treatment also results in an inefficient use of what are often limited financial resources and is not viable in developing countries where financial and technical resources are limited.

Managed aquifer recharge (MAR) affords opportunities to improve the implementation of wastewater reuse and water resources management in general. Specific wastewater MAR opportunities include:

- augmentation of overall groundwater supplies (groundwater banking)
- additional treatment ("polishing") of wastewater through natural aquifer treatment processes (Chap. 7) prior to reuse
- seasonal storage of reclaimed water in ASR systems for later use in reuse systems
- a water source for positive hydraulic salinity-barrier systems.

Wastewater treatment MAR systems include soil-aquifer treatment (SAT) systems (Chap. 19) and aquifer storage transfer and recovery (ASTR) systems (Sect. 18.2). A key distinction is whether recharged water will enter the potable water supply (i.e., whether indirect potable reuse will occur).

The overriding technical, regulatory, and public perception concern with respect to wastewater MAR is potential public health impacts. The greatest risk from wastewater is pathogens because a one-time exposure to some pathogens (e.g., *Cryptosporidium* or *Giardia* oocycsts) can be sufficient to cause illness, whereas chemical contaminants are typically present in treated wastewater in such low concentrations that very long-term chronic exposure is required for an increased risk of adverse health impacts.

Potential or hypothetical health risk considerations have limited expansion of the use of reclaimed municipal wastewater for MAR. Some wastewater MAR projects have not received regulatory approval or were subject to onerous regulatory requirements that rendered them economically unviable. It is important to recognize that accidental or unplanned groundwater recharge with wastewater already widely occurs through irrigation and land applications, collection and conveyance system leaks, and disposal of municipal and industrial wastewater via percolation ponds and infiltration (Asano and Cotruvo 2004). Asano and Cotruvo (2004) noted that a properly planned and managed water reuse project can produce higher quality finished water than unplanned reuse as is currently common practiced. They also noted that that:

> The irony is that water derived from 'natural' but obviously imperfect sources, often receives only basic treatment (filtration and disinfection). The final product might not be as high quality as the reclaimed wastewater that has been subject to much more rigorous treatment, water quality control, and management.

22.2 Wastewater Terminology

A variety of terms are used in the technical literature with respect to wastewater (Table 22.1). Wastewater terminology has evolved over time. What were originally referred to as "sewage treatment" plants later became more politely referred to as "wastewater treatment" plants or facilities, and more recently as "water reclamation" and "water purification" facilities. The term "treated sewage effluent" (TSE) refers to the treated water produced by water reclamation facilities. The terms "reclaimed

Table 22.1 Wastewater terminology

Term	Meaning
Wastewater	Liquid wastes discharged from domestic and commercial premises to individual or municipal disposal or treatment systems. The term often refers to untreated (raw) liquid wastes
Municipal wastewater	Wastewater that is produced mainly by households and non-industrial commercial activities
Sewage	Mixtures of human excreta, water used to flush the excreta, and water used for domestic purposes
Graywater (greywater)	Household wastes from kitchens, baths, or laundries that generally do not contain excreta (toilet wastes)
Effluent	Liquid that flows out of a process or facility, such as a wastewater treatment plant
Primary treatment	Typically temporary holding of wastewater in a tank to allow heavier solids to settle out and lighter materials to float to the surface for removal
Secondary treatment	Primary treatment plus a biological process to remove dissolved and suspended organic compounds
Tertiary treatment	Additional steps beyond secondary treatment to improve effluent quality, such as filtration and/or biological nutrient removal
Excreta	Human waste (feces and urine)
Treated sewage effluent (TSE)	Treated water produced by a wastewater treatment facility
Reclaimed water Recycled water	Treated water produced by a sewage treatment facility that meets specific water criteria required for its being reused
Water reuse	Use of treated wastewater
Potable reuse	Planned augmentation of drinking water supplies with reclaimed water
De facto reuse	A situation where potable reuse of treated wastewater is, in fact, practiced but not officially recognized or described as such
Indirect potable reuse	Augmentation of drinking water source with reclaimed water in which an environmental buffer precedes drinking water treatment
Direct potable reuse	Introduction of reclaimed water (with or without retention in an engineered buffer) directly into a drinking water treatment plant or water distribution system without an environmental buffer
Sludge	Mixtures of solids and liquids that are byproducts of the wastewater treatment process
Water recycling	Alternative term for water reclamation and reuse, which is hoped to result in a better public reception as it does not include the word "waste"
Biosolids	Treated sludge

Sources Levine and Asano (2004), WHO (2006), USEPA (2012) and Maliva and Missimer (2012)

water" and "reuse water" refer to sewage effluent that has been treated to a quality suitable for reuse. However, treated sewage effluent and reclaimed water are commonly used interchangeably.

22.3 Wastewater Treatment Technologies

22.3.1 Introduction

Wastewater treatment processes are categorized into preliminary, primary, secondary, and tertiary or advanced treatment processes (Table 22.2). Concise summaries of wastewater treatment processes were provided by Prescod (1992), National Research Council (1994) and Mujeriego and Asano (1999). More detailed descriptions can be found in wastewater engineering text and reference books (e.g., Asano et al. 2007; Water Environment Foundation 2012; Drinan and Spellman 2013; Kariar and Christian 2013; Riffat 2013; Tchobanoglous et al. 2014). Natural and low-technology treatment systems suitable for developing countries were reviewed by Kivaisi (2001), Mara (2003), Jiménez et al. (2010), Sharma et al. (2012) and Crites et al. (2014).

The levels and methods of sewage treatment are dictated by uses of the reclaimed water, human health concerns, regulatory requirements, operational issues, and environmental priorities. Wastewater treatment for MAR systems may be dictated by regulatory requirements and operational concerns, particularly minimization of clogging. A major factor in determining the level of treatment required for wastewater is whether indirect potable reuse (Sect. 22.4) is expected or may potentially occur. Where indirect potable reuse is a possibility, then recharged wastewater may be required to be treated to essentially potable quality (i.e., meet drinking water standards).

Reclaimed-water MAR systems that store water for non-potable uses (e.g., irrigation) in aquifers in which indirect potable reuse will not occur require lesser degrees of treatment. Secondary treatment plus filtration and disinfection should be sufficient. Filtration is used for further removal of suspended solids and large microorganisms (e.g., *Giardia* cysts and *Cryptosporidium* oocysts). Meeting of disinfection-byproduct (DBP) standards can be an important regulatory issue for MAR systems recharging chlorinated reclaimed water because the relatively high dissolved carbon concentration in reclaimed water is favorable for DBP formation both before and after recharge.

The level of treatment wastewater is required to receive prior to MAR also depends on the degree to which natural aquifer treatment (NAT) processes will occur and whether "credit" is given for these processes. For example, in many states in the United States, water quality standards for injection well systems have to be met at the wellhead (with no credit given for NAT), as opposed to allowing for a zone of discharge (ZOD) or an attenuation zone in which NAT processes are allowed to operate. Where ZODs are allowed, groundwater quality standards are required to be met at the boundary of the ZOD rather than at the wellhead.

Table 22.2 Basic wastewater treatment processes

Treatment level	Objectives	Processes
Preliminary and primary treatment	Coarse screening and grit removal	• Grit chambers or channels • Sedimentation/clarification • Skimming of floating materials
Secondary treatment	Aerobic microbiological removal of biodegradable organics	• Activated sludge process • Trickling filters • Rotating biological contactors • Stabilization ponds • Membrane bioreactors • Surface aerated basins • Secondary sedimentation
Tertiary Treatment—Filtration	Additional fine (suspended) solids removal	• Depth filtration (sand, dual-media) • Surface filtration • Membrane filtration • Dissolved air floatation
Tertiary—advanced wastewater treatment	Additional nutrient and chemicals removal	• Coagulation and sedimentation • Nitrification and denitrification • Phosphorous removal • Granulated activated carbon • Reverse osmosis
Disinfection	Inactivation of pathogens	• Chlorine • Chloramines • Ozone • Ultraviolet light (UV)
Natural treatment processes	Natural alternatives (or additions) to conventional wastewater treatment facilities	• Waste stabilization ponds • Wastewater storage and treatment reservoirs • Constructed wetlands • Soil-aquifer treatment

Sources National Research Council (1994), WEF and AWWA (1998), USEPA (2012), Maliva and Missimer (2012)

22.3.2 *Preliminary, Primary, and Secondary Treatment*

Preliminary treatment includes screening to remove large objects, comminution (shredding) of large objects, and grit removal to remove abrasive materials. Primary treatment removes readily settleable solids and floating material, and commonly involves primary sedimentation (with or without chemical enhancement) and the use of fixed or rotary screens. Primary treatment typically removes approximately

25–50% of the incoming biochemical oxygen demand (BOD), 50–70% of the total suspended solids (TSS), and 65% of the oil and grease (Pescod 1992).

Secondary treatment involves the additional step of removal of residual soluble, colloidal, and suspended biodegradable solids by aerobic biological processes, and secondary sedimentation or clarification. Organic material biodegradation is typically performed using indigenous microorganisms in an aerobic environment. Secondary treatment processes vary primarily in how oxygen is supplied and the rate at which microorganisms metabolize the organic matter.

In the commonly used activated sludge process, organic matter biodegradation takes place in an aeration tank or basin containing a suspension of the wastewater and microorganisms. The contents of the aeration tank are mixed vigorously as oxygen is supplied using devices such as submerged diffusers that release compressed air and mechanical surface aerators that introduce air by agitating the liquid surface (Pescod 1992). Hydraulic retention time in the aeration tanks usually ranges from 3 to 8 h but can be higher with wastewaters with high BOD (Pescod 1992). In fixed-film reactors, the microorganism responsible for organic material biodegradation are attached to a stationary support media (trickling filters) or slowly rotating discs that are partially submerged in flowing wastewater (rotating biological contactors).

The microorganisms must be separated from the treated wastewater to produce clarified secondary effluent. Clarification is most commonly performed using sedimentation tanks, which are referred to a clarifiers or secondary clarifiers. High-rate biological treatment processes, in combination with primary sedimentation, typically remove 85% of the BOD and TSS originally present in the raw wastewater and some of the heavy metals and pathogens (Pescod 1992). Membrane bioreactors combine the conventional activated sludge process with a membrane system (instead of clarifiers) to remove suspended particles and pathogens from the treated effluent. Membrane bioreactor systems can produce much better quality water than conventional secondary treatment systems with reported BOD and TSS removals of greater than 98% (USEPA 2007).

22.3.3 Tertiary and Advanced Treatment

Tertiary treatment involves the addition of one or more treatment steps beyond secondary treatment to improve effluent quality. Tertiary and advanced treatments are typically applied to meet water quality requirements for wastewater reuse. Additional filtration is often performed to achieve a water quality that is suitable for reliable disinfection and to eliminate suspended solids on which bacteria and viruses can attach.

The term "advanced wastewater treatment" (AWT) is used when tertiary treatment includes nutrient (nitrogen and phosphorous) removal. The final treatment step is disinfection, which is performed to kill or deactivate any remaining pathogenic organisms present in the wastewater. Tertiary treatment and disinfection can include a wide variety of physical, biological, and chemical processes, and can result in the production of a very high-quality, safe water suitable for a wide variety of reuse applications. Advanced membrane treatment processes, which include microfiltration (MF), ultrafiltration (UF), nanofiltration (NF) and reverse osmosis (RO), are effective in reducing the concentrations of DOC, many (but not all) trace organic compounds, metals and salts, suspended solids, and pathogens.

Tertiary and advanced treatment processes include (National Research Council 1994; WEF and AWWA 1998; USEPA 2012):

- **Granular media (depth) filtration**: Sand, anthracite, garnet, and other media are used to remove suspended solids and reduce the turbidity of secondary effluent. Granular media filtration is important for effective disinfection.
- **Surface filtration**: Screens or fabrics (including disk filters) manufactured from nylon, polyester, acrylic, and stainless steel fibers are used to remove suspended solids.
- **Microfiltration and ultrafiltration**: MF uses membranes with pores between 0.10 and 10 μm and UF uses membranes with pores in the 0.001 (or 0.005) to 0.1 μm range. MF and UF are used to removal fine particles (including colloids), pathogens, and large molecular weight organic compounds.
- **Nanofiltration and reverse osmosis**: NF uses membranes with pores between 0.001 and 0.0001 μm and RO uses membranes with pores smaller than 0.0001 μm. NF and RO are used to remove organic chemicals and salts. Membrane treatment processes are less effective in the removal of low molecular weight ($\leq$500 Dalton) organic compounds.
- **Biofiltration (slow sand, rapid rate, and granular activated carbon [GAC] filtration)**: Granular media filters that are allowed to become biologically active for the purpose of removing biodegradable constituents.
- **Biological nitrification/denitrification**: Ammonia is oxidized to nitrate and subsequent denitrification converts the nitrate to atmospheric nitrogen. An anoxic heterotrophic bacterial process using the organic carbon of wastewater is employed for denitrification.
- **Coagulation, flocculation, and solid/liquid separation**: Coagulants and flocculants (e.g., alum [aluminum sulfate], ferric chloride, lime, polymers, prehydrolyzed aluminum or iron salts) are added to destabilize colloidal suspensions and clump the small, destabilized particles into larger aggregates that can be more easily separated by sedimentation or dissolved air flotation (DAF). Coagulation, flocculation, and solid/liquid separation remove suspended solids and reduce the concentrations of heavy metals, phosphates, organics, and microorganisms.
- **Ion exchange**: Softening process used to remove Ca and Mg.
- **Electrodialysis**: Membrane process used to remove ionized compounds (salts) and reduce the salinity of brackish waters with relatively low total dissolved solids (TDS) concentrations.
- **Chemical oxidation**: An oxidant (e.g., hydrogen peroxide [H_2O_2], ozone) is used to oxidize reduced inorganic compounds to less soluble oxidized forms and to oxidize organic molecules into compounds with a low toxicity and less objectionable characteristics.

- **Advanced oxidation processes (AOPs)**: Multiple oxidants, such as UV/H_2O_2, ozone/H_2O_2, and ozone/UV, are used to remove trace organic compounds that are not significantly removed during conventional water treatment processes. UV-based AOPs are frequently used to destroy nitrosamines (e.g., NDMA).
- **Adsorption with GAC**: Granular activated carbon (GAC) has an extremely high specific surface area that is available for removal of refractory and residual organic compounds by adsorption. GAC may also provide a medium for some fixed-film biological growth that may provide benefits for BOD and nitrate removal.
- **Air stripping**: Volatile organic compounds are removed by pumping water into the top of a tower packed with a media from which point it flows downward while air is blown upward.

22.3.4 Disinfection

Disinfection is the process of killing or inactivating pathogenic organisms. Water and wastewater systems typically use either chemicals or radiation (ultraviolet light) for disinfection. The target organisms are bacteria, viruses, protozoan parasites (e.g., *Giardia* and *Cryptosporidium*), and other parasites (e.g., nematodes such as *Ascaris*). Disinfection is generally taken to mean reducing the concentration of pathogenic microorganisms to a level such that they do not cause an unacceptable health risk. Disinfection does not imply sterilization, which is the destruction of all organisms. Pathogenic microorganism vary in their susceptibility to removal by different disinfectants. *Giardia* and *Cryptosporidium* oocysts, for example, tend to be resistant to chemical disinfectants. Disinfectants vary in their effectiveness, costs, safety, and formation of disinfectant byproducts (DBPs), some of which are known or suspected carcinogens. The World Health Organization (WHO 2006) has emphasized that the health risks from pathogens is much greater than those from disinfection byproducts and that concerns over DBPs should ***never*** preclude disinfection. Disinfection is addressed in virtually all water and wastewater treatment texts and numerous summary publications (e.g. National Drinking Water Clearinghouse 1996).

Chlorine is a very widely used for water disinfection because it is very effective at removing almost all pathogenic microorganisms. It can be supplied in the gaseous form, as a liquid (sodium hypochlorite), in a dry form (e.g., calcium hypochlorite), and generated on-site by passing an electrical current through a saline (sodium chloride) water solution (electrochlorine process). The effectiveness of chlorination is a function of dose and contact time. Chlorination offers relatively low protection against protozoa, has potential taste and odor objections (which can be overcome through accustomization), and produces disinfection byproducts (particularly trihalomethanes and haloacetic acids) that have demonstrated carcinogenic activity in laboratory animals.

The chloramination process is the addition of both chlorine and ammonia to water to form chloramines. Chloramines are weaker (slower acting) disinfectants than free chlorine but tend to persist longer. The major advantage of chloramination is that chloramines react with organic compounds less frequently than free chlorine and, as a result, produce fewer DBPs. Chloramination is used more frequently for secondary disinfection of drinking water to prevent biological growth in distribution systems.

Ozone (O_3) is a powerful oxidant that is generated on-site, produces no residuals, odor, or taste effects, and rapidly reverts to oxygen. In addition of inactivating pathogens, ozone is effective in removing organic compounds including trace organic compounds. Ozone rapidly decomposes in water and is therefore not suitable for secondary disinfection. Ozone may react with bromide present in water to form bromate, which is a suspected carcinogen in humans.

Ultraviolet (UV) disinfection damages the genetic material of organisms using UV radiation generated by a series of lamps. UV disinfection has the advantages of requiring no chemicals, producing no toxic residuals, and the equipment is safe and easy to operate. The effectiveness of UV disinfection is a function of residence time and the intensity of radiation absorbed by organisms. Intensity of absorbance is a function of distance, absorbance of the liquid, presence of particulates in the water (turbidity), and the transmittance of lamps and their enclosing quartz tube. Particles can shade target microorganisms from UV light. The main operational issues for the UV disinfection of wastewater are particle-associated microorganisms and the UV transmittance of wastewater. UV systems are poorly suitable for treating water with high suspended solids concentrations, color, and soluble organic matter. UV disinfection also does not provide a disinfectant residual and regrowth in the distribution system (or recharge well) is a concern.

22.3.5 Natural Wastewater Treatment Processes

Natural treatment processes include a variety of low-technology, low-cost wastewater treatment methods that have greatest implementation outside of developed countries and for small, decentralized facilities. Natural contaminant attenuation processes active in MAR systems can be effective in attenuating pathogens and chemical contaminants as part of multiple-barrier wastewater treatment systems (Chap. 7). Constructed wetlands (Sect. 12.7) are a proven water treatment technology that can also be used to directly recharge aquifers (leaky wetlands) or provide treatment for subsequent MAR. Constructed wetlands can reduce the concentrations of BOD, suspended solids, nitrogen, metals, trace organics, phosphorous and pathogens. Removal mechanisms include sedimentation, chemical precipitation, adsorption, microbial interactions, and uptake by vegetation. Soil-aquifer treatment systems (SAT; Chap. 19), rapid infiltration basins (Sect. 15.5), and slow-rate land application systems can treat water for subsequent uses and provide direct recharge.

Wastewater stabilization ponds (WSPs) utilize trains of ponds to treat wastewater in warm climates. WSP systems were reviewed by Ramadan and Ponce (n.d.). The basic WSP system design is a series of shallow holding basins or ponds that are used for secondary treatment of wastewater. The most common design consists of two parallel trains of ponds that may include, in order:

- anaerobic ponds,
- facultative ponds,
- maturation ponds (or constructed wetlands).

Anaerobic and facultative ponds are designed primarily for biochemical oxygen demand (BOD) removal and maturation ponds are designed for pathogen removal, although both processes occur in all three types of ponds. Anaerobic ponds are deep (commonly 3–5 m; 10–17 ft) to exclude dissolved oxygen (DO) and encourage the growth of anaerobic bacteria that breakdown organic matter in the effluent.

Facultative ponds are shallower ponds (1–2 m; 3–7 ft deep) that are designed for BOD removal through aerobic processes. Organic matter is metabolized by heterotrophic bacteria with DO provided by algae rather than by aeration equipment. The algae also remove nutrients. Maturation ponds are very shallow (<1 m deep; 3 ft deep) ponds that are designed to provide tertiary treatment. Constructed wetlands may also be used for maturation (i.e., polishing of the wastewater).

WSP systems should be constructed in areas where the surficial sediments have a low permeability (or the ponds should be lined), and designed and operated so that there is a steady flow of effluent, which has been shown to encourage the rapid and continuous growth of the bacteria involved in the biological breakdown of the effluent. The flow through the pond system should also be slow enough to provide a sufficient retention time for the biodegradation of organic matter and pathogen die-off. The systems should also not be overloaded with BOD.

WSPs can be a highly effective wastewater treatment method that can achieve >90% BOD removal, 70–90% nitrogen removal, 99.999% fecal coliform reduction, and 100% helminth removal (Ramadan and Ponce n.d.). They have the advantages of design, construction, and operational simplicity, relative low costs and skilled labor requirements, and a low energy requirement. The main disadvantages of WSPs are that they have large land requirements and need to be located and operated to minimize odor nuisance.

22.4 Wastewater Reuse Health Issues

The health hazards associated with biological and chemical contaminants in wastewater (and other waters) are a function of the concentration of contaminants in the water, the duration and level of exposure, and the dose-response function for each hazard. Risks associated with contaminant exposure depend upon both the duration and intensity of exposure with a fundamental distinction between acute and chronic exposure. An acute exposure is a one-time exposure to a substance or microorganism that is sufficient to cause a serious health impact. A one-time exposure to pathogenic microorganisms, such as consuming water that contains even small numbers of active *Cryptosporidium* or *Giardia* oocysts or enteric viruses, may be sufficient to cause serious illness.

Chronic exposure, on the contrary, involves repeated, continuous exposure to a hazardous substance over an extended period in which any single exposure is insufficient to cause serious harm. Most chemicals found in raw and treated wastewater would require long-term chronic exposure to induce illnesses. For example, some disinfection byproducts (e.g., trihalomethanes; THMs) have been linked to an increased risk of cancer, but the effect comes from long-term (multi-decadal) consumption of water containing THMs, rather than from a one-time exposure. USEPA lifetime health advisory (HA) levels are based on the exposure of a 70-kg adult consuming 2 L of water per day for life.

22.4.1 Pathogens

Microorganisms associated with waterborne disease are primarily enteric pathogens, have a fecal-oral or fecal-dermal route of infection, and survive in water (National Research Council 1998; Bos et al. 2010). A variety of protozoa, helminthes, trematodes, bacteria, and viruses have been identified as infectious agents in untreated municipal wastewater. However, in more than half of all reported outbreaks of waterborne disease, the disease-causing agent was never discovered, which suggests the presence of unrecognized pathogens (National Research Council 1998). The infectious agents present in wastewater depend upon the sources of the wastewater, the general health of the contributing population, the existence of "disease carriers" in the population, and the ability of the various infectious agents to survive in environments outside of their host (National Research Council 1994).

Risks from microbial contamination depend not only on the dose of microorganisms but also on the host's immune status. Limited data are available on the pathogen doses that are necessary to cause infection for most microorganisms and there is also a limited understanding of the relationship between infection and the various forms of illness (Macler and Merkle 2002). Sensitive populations, including children, the elderly, and people with compromised immune systems, stand a greater risk of severe outcomes (National Research Council 1998). Pathogens with a long persistence in the environment, low minimum infectious doses, that elicit little or no human immunity, and have long latency periods (e.g., helminthes) have a higher probability of causing infections than others (Bos et al. 2010).

The more commonly used treatment processes may not completely remove pathogens. It is not practical or affordable to test for all enteric viruses of concern on a routine basis, and such testing cannot be performed in real time (Asano and Cotruvo 2004). From a public health and process control perspective, the most critical group of pathogenic microorganisms is enteric viruses because of the potential for infection from low dose exposure and the lack of routine, cost-effective methods for their detection and quantification (Mujeriego and Asano 1999). It is not practicable or affordable to routinely test for all enteric viruses potentially present in an impaired water. Indicator organisms are, therefore, routinely used for pathogen monitoring. Indicator organisms provide some information on water quality, but the absence of indicator organisms does not guarantee that water is free of pathogens and is safe to ingest. The criteria for ideal indicators of drinking water pollution are that they (Standridge 2008; Asano et al. 2007):

- occur is proportion to the pollutant
- are never present in non-contaminated (safe) water
- are always present in contaminated water
- do not multiply in the environment (including water distribution system)
- die off in the environment more slowly than other pathogens
- can be easily detected in the laboratory
- are safe to work with
- have collection and analytical procedures that are not onerous and exceedingly expensive.

No ideal indicator organism has been identified (Asano et al. 2007). Fecal bacteria are commonly used as indicators of the presence of fecal contamination, particularly enteric pathogens. The presence or absence of enteric pathogens in water samples does not necessarily correlate with the presence of fecal contamination because only a small percentage of a population is infected at any given time and are excreting pathogens (Payment and Locas 2011). Fecal microbial indicators are not absolute indicators of the presence of pathogens but are rather an indicator of a probability of their co-occurrence (Payment and Locas 2011).

Total coliform bacteria have long been used as an indicator of fecal contamination because of the simplicity and low cost of their analysis. Total coliforms have been classically described as nonspore-forming, gram-negative rods capable of hydrolyzing lactose to acid and gas end products within 48 h at 35 °C (Standridge 2008). Total coliform bacteria do not necessarily have a fecal origin and there is strong evidence that total coliform and "fecal" coliform bacteria are poor indicators of the presence of fecal contamination (Standridge 2008). Total coliform bacteria are ubiquitous in natural waters and multiply in drinking water environments. The main argument for the common use of total and fecal coliform tests as fecal contamination indicators is that they are inexpensive and conservative, and are not "less protective of public health" (Standridge 2008). The conservatism of the tests comes at the expense of false positive detections of contamination.

Escherichia coli (*E. coli*) has been proposed as a better indicator of fecal contamination of drinking water (Edberg et al. 2000; Standridge 2008). Newly developed enzymatic methods for detection of *E. coli* are simple, rapid, specific, and sensitive (Standridge 2008). Nevertheless, *E. coli* testing has its limitations. Only certain variants of *E. coli* can cause serious illness, while many variants are harmless. *E. coli* can occur and multiply in environments without fecal contamination. *E. coli* as an indicator also has the limitation that it does not indicate the presence of non-fecal pathogens, such as *Legionella* and mycobacteria.

Gerba and Rose (2003) proposed that fecal coliform bacteria should be used as a treatment performance measurement and not as an indicator of virus or parasite performance and risk. International guidelines based solely on bacteria indicators and treatment requirements may not necessarily reflect the risks posed by the use of recycled waters for different purposes (Gerba and Rose 2003). Gerba and Rose (2003) recommended a multiple indicator approach including viruses. However, as previously noted, routine monitoring for all pathogens that could be present in wastewater is not feasible, would be highly expensive, and cannot be performed in real time (Asano and Cotruvo 2004).

The risks associated with wastewater reuse depend upon:

- the concentrations of pathogens in the recharged wastewater
- the probability of exposure to recharged wastewater (frequency, duration, and magnitude of exposure)
- dose-response relationships.

The health hazards associated with the reuse of wastewater can be evaluated through either epidemiological studies or risk assessments. Both epidemiological studies and risk assessments are complex (and thus time consuming and expensive) investigations that are typically not performed for individual wastewater reuse projects. Epidemiological studies and risk assessments require specialized professionals and are typically performed on a research or regulatory level to establish general water quality standards rather than to evaluate individual projects. Epidemiological studies compare the prevalence of a disease or infection in an exposed group compared to an unexposed control group. Epidemiological studies require a large amount of accurate data on the health history of the exposed and control groups to determine whether there is a statistically significant difference in their health. Such data may be very difficult to obtain in developing countries, particularly if the response is not severe, memorable, or normally recorded. For example, it is much more difficult to obtain data on the long-term incidence of a mild diarrheal disease from wastewater pathogens than for more severe diseases that result in hospitalization or death. The latter are more likely to be officially recorded.

Epidemiological studies are even more complicated when applied to chemical contaminants in which there is a long latency period. For example, chemically-induced cancers may not occur until decades after exposure. People exposed to carcinogenic chemicals may change jobs or move out of the area of exposure. It can be extraordinarily difficult, if not impossible, to locate workers who may have been exposed to a chemical at a farm or industrial facility decades earlier. Individuals may also be exposed to variety of carcinogenic chemicals over their lifetimes from different sources, adding considerable "noise" to the data.

Risk assessments can be defined broadly as the process of estimating the probability of occurrence of an event and the probable magnitude of adverse effects of the event on safely, health, and ecology over a specified time period (Asano et al. 2007). Risk assessment involves four main components:

(1) **Hazard identification**: recognition of microorganisms and chemicals that increase the incidence of a health condition.
(2) **Exposure assessment**: evaluation of exposure scenarios and the probability (and frequency) of exposure of an individual to a given chemical or pathogen dose over a specified time period.

(3) **Dose-response assessment**: quantification of the risk of disease or infection of an individual from a given chemical or pathogen dose.
(4) **Risk characterization**: combined exposure and dose-response assessments to estimate the incidence of a given adverse impact on a population.

The risk assessment process with respect to wastewater reuse can also include management tasks (NRMMC-EPHC-AHMC 2006) such as

- Identification of preventive measures to control identified hazards and to establish monitoring programs to ensure that preventive measures are operating efficiently.
- Verification that the management system consistently provides recycled water of a sufficient quality that is fit for its intended use.

As the National Research Council (2012) observed, "it is important to remember that risk is a function of hazard and exposure, and where there is no exposure, there is no risk". Exposure (dose) is a product of the concentration and amount of a medium to which a person is exposed. Routes of exposure to wastewater include direct ingestion, food, skin and eye contact, and inhalation. Exposure assessments should consider unintended and foreseeable plausible inappropriate uses of reclaimed water (National Research Council 2012).

Dose-response relationship is the most difficult component of risk assessments to accurately quantify. Dose-response relationships for some illnesses (e.g., cancer) may be linear in that any exposure may cause some assumed risk. Threshold (non-linear) dose-response relationships occur where multiple cells must be injured before an adverse effect is experienced and the injury must occur at a rate that exceeds the rate of repair (National Research Council 2012). On the wastewater reuse and MAR project level, dose-response relationships (where considered) are invariably taken from the literature rather than independently investigated.

The Australian risk assessments guidelines for MAR (NRMMC-EPHC–NHMRC 2009) include maximal and residual risk assessments. A maximal risk assessment identifies inherent risks in the absence of preventative measures. A residual risk assessment evaluates residual risks after consideration of potential preventative measures. For example, a maximal risk assessment of a reclaimed water ASR system would likely identify pathogens in the recovered water as a significant risk element. The residual risk assessment might consider risks remaining after natural attenuation during a planned storage period and post-treatment of the recovered water.

The USEPA (2012) made the distinction between high-income "resource-endowed" countries and low-income "developing" or "resource-constrained" countries. Resource-endowed countries have established human risk guidelines or standards that involve high-technology and high-cost approaches, whereas resource-constrained countries have considered adopting a more fit-for-purpose gradational process toward reducing health risks based on the WHO (2006) "Guidelines for the Safe Use of Wastewater, Excreta, and Greywater."

Numerical water quality standards for reuse, where promulgated, typically are based on the potential for and type of human exposure. In urban settings, the distinction is made between restricted and unrestricted irrigation, which differ based on whether exposure to reclaimed water is controlled or public exposure is likely. For agricultural irrigation, a distinction is made between the use of water on nonfood crops, crops eaten raw (i.e., not peeled, skinned, cooked or thermally processed), and crops not eaten raw. The USEPA (2012) recommended that for reclaimed water with no direct public or worker contact, fecal coliforms bacteria should not exceed 200/100 mL. For reclaimed water with direct or indirect public contact, no fecal coliforms should be detected per 100 mL. Where indirect potable reuse occurs, there should be no total coliform bacteria per 100 mL.

Wastewater (as opposed to reclaimed water) use is practiced in some countries, which is the "intentional or unintentional use of untreated, partially treated, or mixed wastewater that is not practiced under a regulatory framework or protocol designed to ensure the safety of the resulting water for the intended use" (USEPA 2012). The drivers for wastewater reuse is a combination of water scarcity and the dearth of effective collection and treatment systems. The latter reflects limited regulatory (including enforcement) and economic resources. WHO (2006) guidelines recommend that rather than relying on numerical water quality thresholds, multiple measures should be incorporated to reduce pathogen loads, such as produce disinfection, peeling, and cooking. The WHO (2006) approach involves:

(1) definition of a tolerable maximum burden of disease
(2) definition of a tolerable risk of disease and infection
(3) determination of the required pathogen reduction(s) needed to ensure that the tolerable disease and infection risks are not exceeded
(4) determination of how the required pathogen reductions can be achieved
(5) verification monitoring.

Quantitative Microbial Risk Assessment (QRA or QMRA) procedures were reviewed in detail by Haas et al. (1999) and their application to wastewater and storm water reuse is presented by Mara et al. (2007). QMRA, and similar risk assessment procedures used for chemical contaminants, are based on evaluations of the likelihood of exposure (frequency and extent) to a contaminant and the dose-response for the contaminant. The dose-response assessment may be either static or dynamic (Asano et al. 2007). Static assessments consider the dose-response function for a single exposure event. Dynamic assessments incorporate complexity associated with exposure, infection, and disease processes including indirect exposure (and multiple routes of exposure) and varying susceptibility to infection or illness. Dynamic modeling more closely resembles what occurs in nature but has much greater data requirements, which introduces additional uncertainty.

The most serious obstacle to the application of risk assessment to wastewater reuse is the very limited availability of dose-response data for most constituents of concern in wastewater (Asano et al. 2007). Both exposure scenarios and dose-response model equations include multiple parameters, the values of which are often poorly known. Consider, for example, the exposure of farm workers to reclaimed water used for irrigation. The exposure evaluation must consider the concentrations of pathogens in the reclaimed water, potential mechanisms of exposure (e.g., consumption of water and soil, inhalation of aerosols), and the likely frequency and magnitude of exposure.

Dose-response relationships may not be linear. For example, a threshold dose value may have to be exceeded for infection to occur. The response to a given pathogen dose will also vary between individuals. There will thus inherently be great uncertainty in the quantitative evaluation of the risks that farm workers face. Statistical techniques are available for dealing with the uncertainty in the values of the risk assessment parameters, such as stochastic methods (e.g., Monte Carlo simulations). A key point is that neither QMRA nor epidemiology is an exact science and any comparison of the two methods has to take this into account (Mara et al. 2007). In order for there to be a satisfactory agreement between the two methods, the values for the parameters used in QMRA and Monte Carlo simulations should be close to the actual values likely to occur in the field (Mara et al. 2007).

The results of epidemiological studies and risk assessments are a calculated risk of infection or disease for an individual exposed over a year, or other time period, which is evaluated relative to a "tolerable" risk. With respect to the use of wastewater for irrigation, the World Health Organization (WHO 2006) recommends that the tolerable impact from treated water consumption and water reuse in agriculture is $\leq 1 \times 10^{-6}$ disability-adjusted life year (DALY) per person per year. A DALY is equal to one year of healthy life lost and includes both time lost as the result of premature death and time spent disabled by disease. Health impacts are weighted in terms of severity within a range from 0 for no impacts to 1 for death (NRMMC-EPHC-AHMC 2006). DALYs are calculated as

$$\text{DALYs} = \text{YLL(years of life lost)} + \text{YLD(years lived with a disability or illness)} \tag{22.1}$$

For example, a mild diarrhea with a severity of 0.1 and duration of 7 days (7 days/365 days per year) results in a DALY of 0.002 per case (NRMMC-EPHC-AHMC 2006). DALYs per person per year are obtained by multiplying the frequency of disease (number of cases per year) by DALYs per case.

The threshold for what constitutes an acceptable or tolerable risk depends upon local circumstances and is not static. The acceptability of a given risk for a given activity depends upon (Asano et al. 2007):

- costs (human, social, and economic) associated with the risk
- costs of implementing measures to reduce the risk
- societal resources available that could be mobilized to reduce the risk
- other risks that could be reduced with available resources.

Public acceptance of risk in general is often based largely on emotion rather than objective analysis. Some members of the general public may strongly object to the notion of accepting any risks with respect to their water supply. Minute risks associated with drinking water quality may be considered much less acceptable than much greater risks routinely assumed, for example, by driving an automobile or smoking. The public tends to be more accepting of risks that they feel they understand and have some control over than risks that they do not fully understand and feel are being imposed upon them by others.

22.4.2 Chemical Contaminants

The National Research Council (1998) recognized three categories of chemical contaminants that are present in reclaimed water:

(1) inorganic chemicals and organic matter that are naturally present in the water supply
(2) chemicals created by industrial, commercial, and other human activities in the wastewater service area
(3) chemicals added or generated during water and wastewater treatment and distribution processes.

To this list can be added soluble microbial products (SMPs), which are organic compounds formed during the wastewater treatment process by the decomposition of organic matter (Drewes and Fox 1999).

Anthropogenic organic compounds, which includes trace organic compounds (TrOCs; also referred to as compounds of emerging concern), are of greatest concern because of the large number of compounds that might be present in wastewater, the inability to analyze for all of them, and the lack of toxicity information for most compounds. Virtually any chemical that is used in the service area of a wastewater treatment facility could potentially occur in the wastewater flow. Even if all organic chemicals in a wastewater water sample could be identified, there would be no basis for assigning risk to most of the identified compounds (National Research Council 1998). Most petroleum-based chemicals and solvents are at least partially removed in the sewage treatment process or are naturally attenuated in the groundwater environment.

The fate of TrOCs during MAR is discussed in Sect. 7.4. The National Research Council (2012) concluded that with respect to TrOCs in wastewater reuse that

> collectively while these findings are insufficient to ensure complete safety, toxicological and epidemiological studies provide supporting evidence that if there are any health risks associated with exposure to low levels of chemical substances in reclaimed water, they are likely to be small.

22.5 Wastewater MAR Issues

The use of treated wastewater and other impaired water in ASR and other MAR projects was reviewed by the National Research Council (1994, 1998, 2008) and Maliva and Missimer (2010, 2012). From a technical perspective, the design and operation of MAR systems using impaired water is not significantly different from that of potable-water MAR systems, other than that greater efforts may be required to manage clogging. Treated wastewater may have relatively high nutrient concentrations, which can result in a greater tendency for biological clogging.

Recharge of wastewater should not adversely impact potable water sources or otherwise create public health or environmental risks. In some countries, recharged water must meet promulgated numerical water quality standards. Wastewater MAR systems may contain unregulated chemicals (e.g., TrOCs) and microorganisms for which there are no promulgated standards, but yet might still impact public health and the environment. A fundamental distinction is whether an MAR project may involve indirect potable reuse. Environmental regulations and policies vary between countries and states on whether the water quality improvement benefits of MAR (i.e., natural aquifer treatment; Chap. 7) are recognized.

Depending upon the type of MAR system and local environmental regulations, wastewater may be required to received additional treatment (pretreatment) prior to recharge. Pretreatment requirements for MAR systems recharging treated wastewater are an additional construction and operational costs that has made some projects economically unviable. Reclaimed water often has a low economic value and, in some situations, there are costs associated with its disposal. The costs to treat wastewater to potable standards (where indirect potable reuse will not occur) and construct and operate an MAR system could substantially exceed the revenues generated from the sale of the reclaimed water. However, consideration should also be given to the economic value of freshwater conserved by the reuse of reclaimed water.

22.5.1 Pretreatment

MAR of wastewater differs from similar MAR systems using other types of water primarily by real and perceived health concerns and a greater susceptibility of wells and infiltration systems to clogging due to relatively high concentrations of nutrients. The concentrations of pathogens and chemical contaminants are significantly reduced in MAR systems by dilution and a variety of physical and biological contaminant attenuation processes. Hence, from a human and environmental health perspective, the pertinent issue is the concentrations in the water at the time of recovery (or discharge) rather than the concentration at the time of recharge. The contaminants present in recharged water and their natural attenuation in MAR systems are addressed in Chap. 7.

The pretreatment requirements for wastewater used for MAR depend upon (Asano and Cotruvo 2004):

- quality of the wastewater, particularly the presence and likely concentrations of pathogens, dissolved and suspended solids, heavy metals, and organic compounds of concern
- treatment processes that are technically available and economically feasible for removing contaminants of concern in the wastewater prior to recharge
- ambient groundwater quality, and present and potential future uses of the receiving aquifer
- natural contaminant attenuation process that occur during infiltration, percolation, and groundwater flow
- adverse fluid-rock interactions (e.g., arsenic leaching) that may occur after recharge

- spatial and temporal separation of the recharge location from potentially impacted sensitive receptors (i.e., maintaining sufficient travel times for natural attenuation processes to occur)
- amount of groundwater dilution and associated reductions in contaminant concentrations that will occur prior to recovery or the water reaching a sensitive receptor
- type, capability, and reliability of water treatment processes to remove contaminants that reach an extraction (recovery) point
- extent and type of human exposure to recharged water and known or potential health risks
- regulatory requirements
- public perception and cultural issues.

The last two factors may not be concordant with the actual health and environmental risks associated with MAR projects involving reclaimed water. For example, a very conservative (restrictive) approach is taken in the United States where all aquifers containing less than 10,000 mg/L of total dissolved solids (TDS) are considered to be underground sources of drinking water (USDWs), unless specifically exempted (which is uncommon and essentially not allowed in many states). The overriding requirement for injection well systems is that they do not endanger USDWs, which is defined as causing a violation of primary (heath-based) drinking water standards. Brackish USDW aquifers whose water cannot be directly consumed without advanced treatment (desalination) and/or will never be actually used for potable water supply are regulated in the same manner (i.e., give the same protections) as freshwater aquifers whose water is consumed directly with perhaps only disinfection. Over protection of brackish USDW aquifers has made some surface and reclaimed water ASR systems economically unviable.

In many countries, specific regulations for MAR in general, much less wastewater MAR, have not yet been promulgated and systems are regulated through existing environmental impact assessment procedures. The lack of specific criteria and guidelines governing aquifer recharge is currently hampering the implementation of large-scale groundwater recharge projects in some areas (Asano and Cotruvo 2004). Wastewater recharge may be just governed by the general mandate that projects should not cause significant environmental harm, which is open to individual interpretation, leading to higher costs, project delays, and project non-approval.

Public opposition may be aroused against wastewater MAR recharge projects under the perception that they will contaminate what are perceived to be pristine aquifers and the public water supply. The potential for public and political opposition is usually greatest for the first system(s) in an area. MAR using reclaimed water may require a substantial public outreach and education effort, especially for the first project in an area.

Pretreatment requirements typically have both regulatory and project technical elements. Clearly, recharged water must meet prevailing local regulatory requirements. However, it behooves professionals involved in MAR project to become involved, to the extent possible, in the rules and policy-making process.

As is the case for wastewater reuse in general, treatment standards for wastewater used in MAR necessarily must consider local socioeconomic conditions. Even if drinking water augmentation is not explicitly foreseen in an MAR project, drinking water quality standards are still commonly applied to the recharged water in many applications in developed countries (Wintgens et al. 2008). Arnold and Arnold (2009) observed that the developed world tends to overspend in the environmental area to achieve health benefits that are largely illusory and that in purchasing higher levels of water treatment or restricting the acceptable uses of reclaimed water, people may be paying for peace of mind as opposed to material health improvements.

From a purely technical perspective, health issues associated with the MAR of wastewater should consider natural attenuation processes (Chap. 7) along with the wastewater treatment prior to recharge and water treatment after recovery. MAR is thus one element (barrier) in a multiple-barrier approach to protecting public health and the environment.

22.5.2 Movement and Mixing of Recharged Treated Wastewater

The most basic question concerning MAR is where will the recharged water go. More precisely, what is the direction and rate of movement of recharged water and will its composition change along its flow path and by the time it reaches sensitive receptors (e.g., potable water supply wells). The concentrations of contaminants present in recharged will tend to decrease over time and aquifer flow due to dilution (mixing) and a variety of physical and biological attenuation processes addressed in Chap. 7. Pathogens are removed by physical processes (e.g., straining, filtration, and sorption) and inactivation. The latter depends upon the inactivation rate and subsurface residence time.

Accurate prediction of the movement and mixing of recharged wastewater requires numerical groundwater modeling based on a detailed and accurate aquifer characterization. A critical issue is aquifer heterogeneity. In highly heterogeneous aquifers in which flow is dominated by thin high-transmissivity strata, fractures, or karst conduits, water flow and solute transport can be very rapid and unpredictable. A key issue driving the extent of aquifer characterization and modeling is risks posed by a system. For example, if reclaimed water is being stored in an aquifer that is not used for potable supply and there is little potential for migration into an aquifer used for potable supply, then the risks posed by the system are low and the need to predict water movement is low. Conversely, recharge of reclaimed water into aquifers used for potable supply may require a detailed aquifer characterization and solute-transport modeling to demonstrate either that the water will not enter potable water supply wells, that there will be a sufficiently long residence time to allow for adequate pathogen attenuation, and/or that chemical contaminants will be sufficiently attenuated by dilution, sorption, and degradation processes.

The risks posed by MAR systems will also depend on the degree of treatment of the wastewater prior to recharge. Where wastewater is treated to essentially potable standards, then prediction of the movement and mixing of the recharged water is not a critical technical issue (i.e., health and safety concern) although it may still be a conservative regulatory concern. Predictive solute-transport modeling has considerable uncertainty in the absence of natural or anthropogenic tracer data for model calibration. The tracer data could be initial operational (or pilot system testing) data where recharged water has significant compositional differences with native groundwater.

22.5.3 *Monitoring*

It is easier to make regulations than to enforce them (WHO 2006). Establishing health-based targets can be a relatively simple matter, particularly if existing targets (e.g., World Health Organization or United States Environmental Protection Agency standards) are adopted. Within developed countries, existing governmental institutions typically exist that have the authority and resources to regulate large urban municipal wastewater treatment facilities and the use of reclaimed water, and the utilities themselves have technical and economic resources to perform monitoring. Monitoring serves three main purposes (WHO 2006):

(1) **Validation**: determination after construction that a system and its individual components are capable of meeting specific performance targets
(2) **Operational monitoring**: routine monitoring to determine if the treatment system is operating within design parameters and that health protection measures are working as designed
(3) **Verification**: determination if the end product meets treatment targets and ultimately health-based targets.

The parameters and sampling frequency for wastewater reuse systems are normally determined by the regulatory agency that has jurisdiction over the project and/or are formally established by rule. Parameters included in monitoring programs are usually those for which there is a promulgated water quality standard.

Monitoring can represent a significant fraction of the total system operational costs and can impact the economic feasibility of wastewater MAR projects. There is often a natural desire to have a very rigorous monitoring program to try to provide additional assurances that a reuse program does not pose an unacceptable health risk. Such monitoring may be necessary to gain support for projects. However, excessive monitoring results in additional costs and may provide no public or environmental health benefits. In general, more frequent analyses for pathogens are needed because a one-time exposure can cause health impacts. Chemical contaminants require less frequent analyses because their health risk are associated with long-term chronic exposure.

Too often a "one-size-fits-all" approach is taken, in which a standard monitoring program is required for every project irrespective of project-specific realities. Such an approach has the superficial benefits of apparent fairness and simplicity, but may result in considerable waste of effort and money. The challenge is for system owners and regulators to work together to develop "right sized" monitoring programs that provide necessary data on reclaimed water quality while not resulting in an excessive financial burden on system owners and operators.

Degree of regulatory compliance of wastewater reuse systems over long-periods of time is one of the current weak points in wastewater reuse in developing countries with a poorly developed regulatory system (Salgot et al. 2003). Reliability, with respect to wastewater treatment and reuse, is defined as the probability of successful performance of the unit and facility in satisfying specific operational conditions and producing effluent that meets minimum effluent water quality standards (Salgot et al. 2003). Reliability depends upon system design, operation, and maintenance, and requires frequent sampling to confirm compliance with minimum effluent quality standards. A key treatment facility operational issue is how plant upsets or malfunctions are handled to avoid discharges of inadequately treated effluent to the reclaimed water stream.

Reliability requirements also vary depending on the sensitivity of the intended use. Systems intended for direct or indirect potable reuse require higher degrees of reliability and more frequent sampling (continuous where possible) to ensure that the reclaimed water consistently meets water quality standards. Fail-safe procedures should be incorporated into the system design to ensure that water not meeting specifications cannot enter the water supply.

Additionally, controls are needed to prevent unintended and inappropriate uses of the reclaimed water after it leaves the wastewater treatment facility. Occasional monitoring is also recommended to confirm that unintended uses are not occurring. Different colored pipe are usually used to differentiate between reclaimed water and potable water lines. Reclaimed water pipe in the United States, Canada, Europe, Australia and much of the rest of the world is colored purple (lavender, lilac) to avoid cross-connections. However, despite such an obvious indicator, cross-connections of reclaimed water to residential potable water lines have still occurred in some rare instances (e.g., several homes in Cape Coral, Florida, in 2003), which highlights the need for vigilance. Warning signs are usually required where reclaimed water is used for public access area irrigation to discourage consumption.

22.6 Potable Reuse Basics, Health Issues, and Public Perceptions

22.6.1 Terminology

Indirect potable reuse was defined by the (United States) National Research Council (1998) as "the abstraction, treatment, and distribution of water for drinking from a natural source-water that is fed in part by the discharge of wastewater effluent." Indirect potable reuse includes storage of the treated wastewater in an environmental buffer, which results in some dilution and provides time for natural contaminant attenuation processes to occur, and for monitoring and storage to be initiated if an upset in the treatment system causes water to fail to meet applicable standards. The environmental buffer can be either a surface water body or an aquifer.

Direct potable reuse is defined by Asano et al. (2007) as the introduction of highly treated wastewater either directly into the potable water distribution system downstream of a water treatment plant or into the raw water supply immediately upstream of a water treatment plant. Direct potable reuse, which has been distastefully described by some project opponents as "toilet to tap," has historically been seldom seriously considered because of adverse public perception and opinion issues, but attitudes are changing in areas facing water scarcity.

The third type of potable reuse is "de facto" reuse, which is reuse that occurs but is not officially recognized as such. The term "de facto" reuse is recommended in preference to "unplanned," which presumes that water managers are unaware of the integrated nature of the nation's water supply (National Research Council 2012). A common type of de facto reuse occurs where wastewater is discharged to a river that is the water source for downstream communities. The water supply for downstream communities contains wastewater from all the communities located upstream of their intakes. The wastewater is diluted and naturally treated to vary degrees by the river flows and biological processes. De facto reuse also occurs by the recharge of aquifers used for potable supply by rapid infiltration basins and adsorption fields that are not specifically designated as indirect reuse systems. There is not a clear, widely accepted delineation between indirect and de facto reuse in terms of the fraction of wastewater in produced water or the travel distance and residence time. De facto reuse, since it widely occurs, is an important source of data on the potential health risks of potable reuse (National Research Council 2012).

The National Research Council (2012) rejected the use of "indirect" versus "direct" potable reuse because the terms are not related to produced water quality. However, this distinction is still widely used and is retained herein. A key conclusion of the National Research Council (2012) study is that potable reuse systems can be implemented with less risk than current water supplies that involve some de facto reuse. It was also concluded that there is no indication that health risks from using highly treated reclaimed water for potable purposes poses a greater risk than those using many existing water supplies.

The National Research Council (2012) noted that they were not aware of any situation in which the time delay provided a buffer has actually been used to respond to an unforeseen upset, and further concluded that "environmental buffers are not essential elements to achieve quality assurance in potable reuse projects." Nevertheless, environmental buffers have been, and appear to continue to be, crucial for public acceptance as they separate the water source from wastewater. An earlier National Research Council (1998) committee observed that the "loss of identity is an issue that seems more relevant to public relations that public health protection."

22.6.2 Potable Reuse Public Health Issues

Public health issues related to indirect potable reuse were reviewed in detail by the National Research Council (1998, 2012) and Rodriguez et al. (2009). The health risks from potable reuse lie in the potential for pathogens and chemical contaminants to enter the potable water supply. The National Research Council (2012) cautioned that the occurrence of a contaminant at a detectable level does not necessarily pose a significant risk and that although absolute safety is a laudable goal of society, some risk must be considered acceptable. Instead "it is appropriate to compare the risk of water produced by potable reuse projects with the risks associated with water supplies that are presently in use" (National Research Council 2012). With respect to pathogens, concentrations may be an order of magnitude less than that in some current drinking water supplies, particularly where existing systems involve some de facto reuse.

The National Research Council (1998) emphasized that drinking water standards cannot be relied upon as the sole standard for the safety of potable reuse of wastewater because the drinking water standards cover only a limited number of contaminants and they were intended for conventional, relatively uncontaminated sources of freshwater. Chemicals present in treated wastewater at low concentrations, such as TrOCs, may pose essentially no health risk for wastewater irrigation projects because human exposure (consumption) would be usually accidental, short-term, and likely of low volumes. Chemical contaminants present in water recovered from indirect potable reuse system are of greater concern because there is a greater likelihood for long-term consumption.

Rodriguez et al. (2009) reviewed the public health data from indirect potable reuse studies and concluded that the epidemiological and toxicological studies have not documented health risks. Epidemiological studies are constrained by diseases with long latency periods (cancer) being associated with many competitive risk factors and are complicated by limitations in the assessment of exposure. Epidemiological studies thus typically do not have the statistical power to detect the levels of risk considered significant from a population-based perspective (e.g., additional life-time cancer risk of 1:1,000,000; National Research Council 2012).

Toxicological studies of human health risks are generally based on extrapolation of toxicological analyses of animals, which are usually performed using concentrates of recycled water rather than individual compounds (Rodriguez et al. 2009). Standard toxicological studies measure the effects of exposures of a group of living organisms against a control group, which is subject to the same procedures but without exposure to the chemical or chemicals in question. Toxicological testing of individual constituents cannot predict health effects resulting from combinations of chemicals (synergistic effects). Two main types of toxicological tests are performed (WEF and AWWA 1998): in vitro testing of bacteria or mammalian tissue or cell cultures, and whole animal tests. Toxicological testing has the limitation that it may show evidence of a potential health risk but cannot be used to establish the absence of risk (National Research Council 2012).

Health criteria for long-term environmental exposure of the general population to TrOCs do not exist for these classes of compounds. The compounds are not regulated and developing toxicity criteria is time consuming and resource intensive (Snyder et al. 2010). The lack of drinking water guidelines for emerging contaminants (TrOCs) can lead to public and political rejection of otherwise sustainable reuse projects (Snyder et al. 2010). The general public may take greater comfort in that contaminant concentrations are below some numerical health-based drinking water standards (maximum contaminant levels), as opposed to concentrations reported to be very low and not believed to be harmful.

Snyder et al. (2010) developed a conservative screening level methodology that selects the lowest calculated level (most protective of public health) for several possible risk assessment schemes. For a pharmaceutical compound, the screening level is the lowest of:

- therapeutic dose divided by a default uncertainty factor (UF) of 3,000 and by another UF of 10 if the compound is a non-genotoxic carcinogen or endocrine disrupting compound (EDC).
- division of the literature-based no observed adverse effect level (NOAEL) by a default UF of 1,000 or the lowest observed adverse effect level (LOAEL) by a default UF of 3,000 and by another UF of 10 if the compound is a non-genotoxic carcinogen or endocrine disrupting compound (EDC).
- if the compound is a genotoxic carcinogen and tumor incidence data are available, develop a cancer slope factor and establish a comparison value assuming a de minimis cancer risk of 1 in 1,000,000.
- for genotoxic compounds with no tumor data available, the lower of the virtually safe dose (VSD) or threshold of toxicologic concern (TTC).

The actual threshold for adverse health impacts is expected to be much higher that the screening levels. However, it is highly unlikely that any laboratory test will ever fully establish the health risks in drinking water from any source (National Research Council 2012). Instead, the quality of reuse water should be compared to that of conventional drinking water supplies, which are assumed to be safe (National Research Council 2012).

Indirect potable reuse systems incorporate a multiple-barrier approach to protect public health, which may include some or all of the following elements (Asano et al. 2007; National Research Council 2012):

- source controls
- robust and redundant conventional wastewater treatment
- robust and redundant advanced wastewater treatment
- an environmental buffer that allows for additional natural treatment and dilution
- water treatment
- a robust monitoring program.

An important questioned raised by Asano et al. (2007) is whether to apply advanced treatment technologies (if necessary) to the wastewater or to the potable water treatment processes. For example, should reverse osmosis be applied to the secondary treated wastewater or to the water recovered for potable supply? Inasmuch as not all of the reclaimed water may enter the potable supply, applying advanced treatment to the water supply may be the more cost-effective option. Treating wastewater to potable standards can be a waste of resources if the aquifer water is of poor quality (non-potable) and recovered water will be subsequently treated to potable water standards at the water treatment plant (Hummer and Eden 2016).

The state-of-the-art advanced treatment (referred to as full advanced treatment or FAT), such as employed in the Groundwater Replenishment System in Orange County, California, includes microfiltration (MF) followed by reverse osmosis (RO) and an advanced oxidation process (AOP; ultraviolet light and hydrogen peroxide treatment), which produces a water of a quality far beyond that produced by many potable water systems (Markus 2009). AOP commonly includes a UV-based advanced oxidation process for removal of some organic compounds that passed through the RO membranes (e.g., NDMA) and inactivation of pathogens. Ozone treatment may be used either before the MF/UF units to prevent membrane fouling or downstream of the RO units for additional organic removal or degradation.

FAT pilot tested for an indirect potable reuse project in Clearwater, Florida, involved (Mercer et al. 2015):

- ultrafiltration as the primary process for pathogen removal
- reverse osmosis to reduce salinity and as a secondary process for pathogen removal
- advanced oxidation (UV and hydrogen peroxide) to reduce the concentrations of microconstituents that remain after RO
- remineralization (CO_2 and calcium carbonate)
- DO removal using membrane contactors and sodium bisulfide to prevent arsenic mobilization in the aquifer.

The pilot system removed all infectious pathogens and reduced the total organic carbon (TOC) concentration by 99% from about 10 mg/L to below the detection limit of 0.06 mg/L. The wastewater was tested for 62 microconstituents (TrOCs) of which 30 were detected. Only atenolol (a high blood pressure medication) was detected in the purified water, but with a 79% removal (decreased from 75 ng/L to 16 ng/L). The presence of atenolol may have been due to an underfeed of hydrogen peroxide on the day of sampling (Mercer et al. 2015). A reported operational issue is that residual hydrogen peroxide may use up sodium bisulfide added for DO removal. Some other projects in which FAT has been implemented, pilot tested, or is planned are

- San Diego Water Purification Demonstration Project (Steirer et al. 2012)
- Miami-Dade (Florida) South District Water Reclamation Plant; investigation of possible recharge of the Biscayne Aquifer at the Miami Zoo (Chalmers and Ferguson 2012)
- Singapore NEWater system (Seah and Woo 2012)
- Padre Dam Municipal Water District (San Diego County, California) Lau et al. (2016).

Toxicological testing performed on concentrates of the organic contaminants in treated wastewater are summarized below (WEF and AWWA 1998):

Whittier Narrows Groundwater Recharge Project: Stormwater and reclaimed water yielded the highest levels of mutagenicity with imported water the least. In more than half of the instances observed, chlorination lead to an apparent formation of mutagens.

Denver Potable Water Reuse Demonstration Project: RO and AOP treatment were employed. Except for nitrogen compounds, concentrations of constituents in the reclaimed water subject to high level treatment were equal to or less than the concentrations in Denver's existing supplies. No differences were observed in toxicity and carcinogenicity between the control and treated groups.

San Diego Total Resources Recovery Project: High-level treated wastewater and water from the City's raw water supply reservoir were tested. The results indicated that reclaimed water is unlikely to be more genotoxic or mutagenic than water from the current raw water supply source.

Tampa Water Resources Recovery Project: Pilot facility incorporated secondary treatment plus filtration and disinfection and advanced organics removal techniques (GAC, RO, UF). Treated water was compared to an existing water source (Hillsborough River). Results were reported to be uniformly negative for the product water from the AWT facility.

Gerrity et al. (2013) reviewed treatment trains used for potable reuse throughout the world. The widespread implementation of FAT is hindered by a number of sustainability issues including high capital and O&M costs, high energy consumption (and thus carbon footprint), practical limits on water recovery, and the need to discharge concentrated brine streams. FAT also has high technical skill requirements. As an alternative to FAT, a number of ozone-based alternatives are increasing in popularity throughout the world. These alternative systems involve some combination of membrane filtration, ozonation, and/or biological activated carbon (BAC). Primary limitations are the potential formation of bromate and NDMA during ozonation, inability to reduce TDS, and practical limits on TOC removal.

Stanford and Antolovich (2017) documented the pilot testing of a non-membrane advanced oxidation process at the Hollywood (Florida) Southern Regional Wastewater Treatment Plant (SRWWTP). The SRWWTP effluent has a greater than 1,000 mg/L chloride concentration. Its salinity is too high for either reuse or recharge of the shallow freshwater Biscayne Aquifer without costly RO desalination. The alternative option considered is to recharge the deeper confined Floridan Aquifer, which contains brackish water, by deep well injection. Pilot testing was performed to evaluate whether an alternative treatment system can meet Broward County requirements for the acceptable removal of emerging contaminants.

Secondary-treated effluent was further treated by deep-bed filtration, ion exchange to remove TOC and ammonia, biological activated carbon (BAC) and an advanced oxidation process with UV. Both membrane and non-membrane processes produced water that met Florida primary and secondary drinking water standards and achieved oxidation of emerging contaminants to target levels. The costs of the non-membrane treatment would be less than half the cost of traditional membrane-based treatment technology and result in substantially lower carbon emissions (Stanford and Antolovich 2017). Water recovered from the Floridan Aquifer typically requires RO treatment before potable reuse, so another independent barrier exists to protect public health.

22.6.3 Public Perception Issues

Numerous studies have been performed and papers published on the attitude of the public towards potable reuse and wastewater reuse in general. Public acceptance of reuse projects is vital for the future of wastewater reclamation, recycling, and reuse, as the end users must be willing to accept the water. Poor public perception (and associated loss of political support) has caused reuse projects to be abandoned. Highly treated wastewater may still be viewed as being just wastewater even though it may be of superior quality to currently used freshwater sources (Bixio et al. 2005). The "Law of Contagion" may apply in which once water has been in contact with contaminants it can be psychologically very difficult for people to accept that it has been purified (Khan and Gerrard 2006). Clearly, there is a need to educate the public that "Since all water is recycled in one way or another, the quality of the water at its point of use is much more important than its history" (Bouwer 2000).

The intentional addition of treated wastewater into the water supply may elicit stronger responses because it is viewed as a choice and is thus avoidable. The results of a survey in Australia revealed that the main perceived risk associated with potable reuse is to the safety of children and animals (Hurlimann 2007), which can elicit very strong reactions and opinions amongst the general public.

It has been observed by a number of workers that a key issue in the public perception of water reuse is the hydrological distance between the water's origin as waste and its use for personal activities, as well as the number of natural or artificial barriers along the way. Water reuse applications that consistently have had the least amount of community support have been those that involve recycling municipal wastewater into drinking water supplies, whereas projects that reuse treated wastewater for landscape and golf course irrigation usually face little public opposition. The passage of water through a natural environment, whether it be a river or aquifer, may reduce its taint as being wastewater and thus make reuse more acceptable. After recharge, wastewater becomes groundwater. Terminology can also impact public perception.

Describing water as "purified water" as opposed to "reclaimed water" or "treated wastewater" can impact public opinion.

Other factors that impact public perception of wastewater reuse include

(Bruvold 1985; Bahri and Brissaud 1996; Alhumoud et al. 2003; Gerba and Rose 2003; Daughton 2004; Marks 2004; Hartley 2006; Khan and Gerrard 2006; Hurlimann 2007; Ormerod and Scott 2013):

- whether the public believes that they will benefit from the project
- openness of the process; whether project development is being performed in the open and the public is having their concerns heard and addressed, as opposed to a project being secretive and perceived as being foisted upon them
- whether the public believes that water scarcity is real and there is an actual need for wastewater reuse
- confidence and trust in the organizations that will be implementing the reuse project and the regulatory agencies that will be overseeing the systems
- education; whether the public understands wastewater treatment and reuse processes and public health issues.

A WateReuse Foundation study by Haddad et al. (2009) noted than an implication of public perception issues is that unnecessary restrictions and conditions on approved projects can limit their beneficial service to the public. A region may place limits on a project's scope and/or require costly additional monitoring and reporting requirements that are not justifiable in terms of increased safety or achievement of other goals. Haddad et al. (2009) examined public perception of water reclamation and reuse from a judgment and decision making, and social psychology perspective. Several key observations are:

- people often have a gut feeling about particular issues and then construct reasons to support their feeling
- certain types of risks are exaggerated because of people's emotional reaction to them
- negative effects have a greater potency than equivalent positive events
- people tend to over-estimate near-term harm than long-term benefits; an example given is flu shots, which for some people the short-term discomfort has greater weight than the benefits of long-term immunity
- people have psychological traits that will affect they opinions on wastewater reuse.

People were found to vary in their "disgust sensitivity" (i.e., human emotions related to decay, foul odors, and body products), "contagion sensitivity" (i.e., belief that once something is in contact with a contagion it can never be made clean), and their degree of trust and cynicism. Haddad et al. (2009) observed that there was an increase in willingness to accept wastewater reuse with travel distance and time in the environment. An interesting result of their survey is that 35% of the respondents agreed that "if recycled water is part of my drinking water supply, as long as it is safe, I'd rather not know the details." Indeed, this "preference for ignorance" may explain the paucity of public concern expressed over widespread de facto indirect potable reuse.

Haddad et al. (2009) observed that opposition to wastewater reuse is not widespread and that a large majority of the population is either positively disposed or neutral to its practice. However, a small percentage (26%) agreed with the statement that "it is impossible for recycled water to be treated to a high enough quality that I would want to use it." A minority of the population will thus not be responsive to the most persuasive messages. A person most likely to reject "certified safe recycled water" is someone who is:

- disgust and contagion sensitive
- self-identified as not politically moderate
- less trusting in institutions and science
- less pro-technology
- more interested in knowing the history of the water he or she drinks
- less impressed by successful and more effective water treatment technologies.

Haddad et al. (2009) noted that their findings are inconsistent with the vehemence of some opposition to proposed wastewater reuse systems, which raises the question of whether the process of communication with program opponents and the public needs to be improved.

22.7 Wastewater MAR and Potable Reuse Experiences

Indirect potable reuse systems involving a groundwater buffer are a type of aquifer recharge and recovery (ARR) systems. The Hueco Bolson Recharge Project (Sect. 18.2.1) is a long operating system that recharges highly treated wastewater using wells and more recently infiltration basins, and recovery for potable water supply after additional conventional water treatment. The Belgium Dune Aquifer Recharge and Recovery (St-André System) recharges reclaimed water that is additionally treated by ultrafiltration pretreatment, reverse-osmosis, and ultraviolet disinfection into ponds and recovers the water using a battery of wells (Sect. 18.3.2). The recovered water is treated by aeration, slow sand filtration, chlorination, and UV treatment before being sent to the distribution system. Follows are summaries of some additional indirect potable reuse systems using MAR and non-potable MAR projects.

22.7.1 Montebello Forebay Groundwater Recharge Project

The Montebello Forebay Groundwater Recharge Project, located in southeastern Los Angeles County, California, is especially noteworthy because of its scale, long-operational history (it is the oldest indirect potable reuse project in California), and

Fig. 22.1 Aerial photograph of the Rio Hondo Spreading Grounds, Los Angeles County, California (December 2013, *Source* U.S. Geological Survey)

the considerable amount of study it has received. Recharge is performed by surface spreading at three facilities (Cook 2004):

- Rio Hondo Spreading Grounds (570 acres; 231 ha) (Fig. 22.1)
- San Gabriel Coastal Spreading Grounds (128 acres; 52 ha)
- Unlined San Gabriel River Channel (133 acres; 54 ha).

Recharge has been performed using storm water since 1938, imported surface water from Northern California and the Colorado River since 1953, and recycled water since 1962 (Cook 2004; Gasca et al. 2011; Johnson and Gagan 2011). The breakdown of water sources was reported to be (Johnson and Gagan 2011):

- rainfall and stormwater capture (~54,000 acre-feet/yr, AF/yr; 66.6 million m^3/yr; MCM/yr)
- Imported river water (~21,000 AF/yr; 25.9 MCM/yr)
- Tertiary treated recycled water (~50,000 AF/yr; 61.7 MCM/yr)

Johnson and Gagan (2011) report in 2011 that since 1938, a total of 7,335,063 AF (9,047 MCM) have been recharged at the Rio Hondo Spreading Grounds and San Gabriel Coastal Spreading Grounds, including 2,895,086 AF (3,571 MCM) of storm water (39%), 1,510,880 AF (1,864 MCM) of recycled water (21%) and 2,929,098 AF (3,613 MCM) of imported and other makeup water (40%). The use of imported water has been decreasing because of more limited supplies and continually rising costs. Additional stormwater recharge has been pursued to offset the reduction in imported water supply for recharge.

Use of recycled water began with the construction of the Whittier Narrows Water Reclamation Plant (WRP). The San Jose Creek and Pomona WRPs started to contribute water for recharge in the 1970s (Gasca and Hartling 2012). The WRPs were upgraded to tertiary treatment in the late 1970s, and the produced water meets federal and state drinking water standards for pesticides, major ions, heavy metals, trace organics, nitrogen, and radionuclides, and has extremely low levels of microorganisms (Gasca and Hartling 2012). The plants were further upgraded in the early 2000s to provide nitrification/denitrification. In the late 2000s sequential chlorination was added to minimize THMs and NDMA production and, in 2011, the Whittier Narrows WRP began using UV disinfection (Gasca and Hartling 2012). Since April 2009, the amount or recycled water that can be recharged is subject to a dilution-based limit of no more than 35% in any running five-year period (Gasca and Hartling 2012).

The National Research Council (1994) reported on toxicological and epidemiological studies performed on the system between 1969 and 1980. The epidemiological studies revealed no increased health impacts between households that received water containing the recharged reclaimed water and those that did not. A recognized limitation of the study was a short time period between first exposure to recovered water and study period relative to the long-latency periods (15 years of more) between first exposure and cancer diagnosis. The Ames test data for mutagenicity was determined not to be adequate to support a risk assessment of the source-water quality. The National Research Council (1994) concluded that the risks associated with the three sources of recharge water (imported water, stormwater, and reclaimed water) were not significantly different.

Epidemiological studies of the health impacts of the indirect potable reuse from the Montebello Forebay systems were performed by the Rand Corporation (Sloss et al. 1996, 1999). The study results indicated a higher rate of liver cancer in the population receiving the most reclaimed water, but this result was noted to be most likely due to factors other than reclaimed water or just chance. The study also did not find any evidence for a relationship between reclaimed water exposure and birth outcomes. If reclaimed water was impacting birth outcomes, then it was concluded that the effects were likely small.

The overall conclusion was that no evidence was found that populations consuming groundwater containing reclaimed water has higher risks of cancer, mortality, infectious diseases or adverse birth outcomes than those using water from other sources. There was no evidence that reclaimed water had an adverse impact on health. Sloss et al. (1996) cautioned that due to the limitations of the epidemiological methods, it is difficult to draw any definitive conclusions about the effect of reclaimed water on public health.

A basic constraint on the construction of new spreading facilities, such as the Montebello Forebay Groundwater Recharge Project, is the availability of potential sites in hydrogeologically favorable areas. Review of aerial photographs of Orange and Los Angeles County, California, reveals that new large-scale surface spreading facilities, such as the Rio Hondo Spreading Grounds, could not be constructed today in the urban area because of the very limited amounts of potentially available undeveloped land and its associated very high costs.

22.7.2 Town of Atlantis, South Africa

Most MAR systems are constructed within existing communities, which constrains design and operational options. The MAR system in the Town of Atlantis, located 50 km north of Cape Town, South Africa, is atypical in that it was implemented in a planned community (incepted in 1976). Water resources are very limited and groundwater resources are inadequate to meet the needs of the town. Aquifer recharge was recognized as being critical for a sustainable water supply. The history and operation of the Atlantis Water Resource Management Scheme (AWRMS) were discussed by Tredoux et al. (2002, 2012) and in considerable detail by the Department of Water Affairs (2010). The initial motivation of the scheme was to eliminate the need for a marine wastewater outfall, which had become cost prohibitive because of construction and monitoring requirements. Augmentation of the limited groundwater supplies subsequently became the primary objective of the scheme. About 30% of the town's water supplies are being augmented through artificial recharge. Importation of low-salinity water began in 1998, which decreases the groundwater and recycled water content of the final water supply (Department of Water Affairs 2010).

A diagram of the AWRMS layout is provided Fig. 22.2. Industrial and residential areas are separated, with potentially polluting activities located in an area of poor quality groundwater. Domestic and industrial wastewater are treated separately in twin wastewater treatment works. The domestic wastewater receives secondary treatment with nitrification-denitrification steps and polishing is a series of maturation ponds. The urban stormwater system was designed to control the flows of different salinities and to collect the best quality water for aquifer recharge. The low salinity stormwater and treated domestic wastewater from the maturation ponds are sent to two recharge basins (Basins #7 and Basin #12) located upgradient of the water supply production wells. Industrial treated wastewater and more saline stormwater

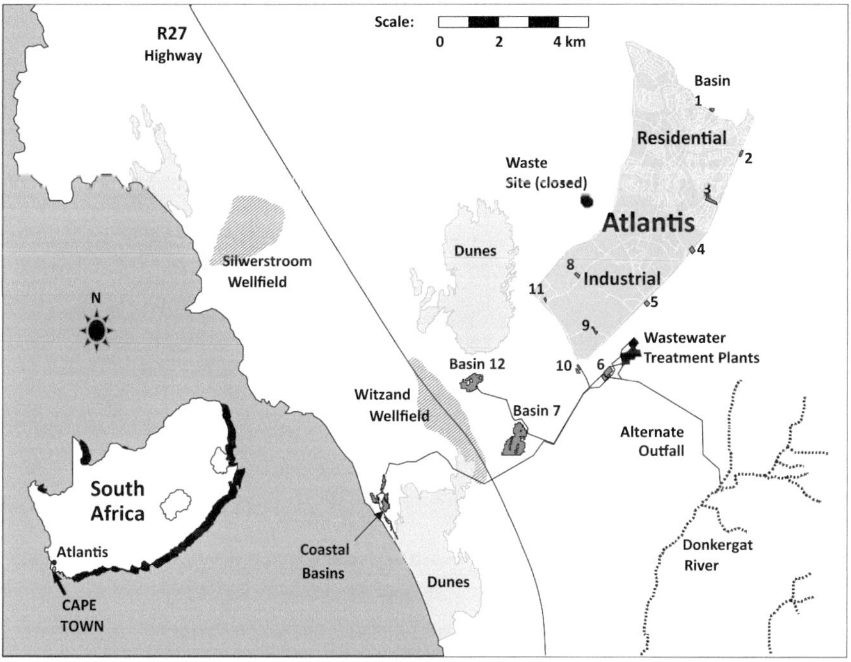

Fig. 22.2 Location and layout of the Atlantis, South Africa, water supply system (*Source* Tredoux et al. 2012, copyright IWA)

are discharged downgradient of the production wells. Infiltration rates in Basin #7 and Basin #12 were reported to be 0.01 m/d and 0.16 m/d, respectively.

Clogging of production wells was a serious operational challenge, which was caused by biological precipitation of iron deposits. Biological iron precipitation was believed to have been due to over pumping of the wells, which allowed oxygen to enter the aquifer. Introduced dissolved organic carbon may have also contributed to reducing conditions that allowed for the mobilization of iron (Tredoux et al. 2002, 2012).

Tredoux et al. (2012) summarized the performance of the AWRMS from a water quality perspective. The monitoring data for some wastewater-related parameters are summarized in Table 22.3. A total of 8 of about 40 tested micropollutants were detected in at least one production well. Traces of only two compounds were detected at ng/L levels in the final chlorinated drinking water: carbamazepine and its metabolite dihydrodihydroxycarbamazepine. The concentrations of the two detected micropollutants were well below internationally accepted norms (Tredoux et al. 2012).

The AWRMS system is noteworthy in that it pioneered the application of MAR as a water management tool for bulk water supply in southern Africa (Tredoux et al. 2012). It is also a successful example of indirect potable reuse not involving advanced water treatment of the wastewater. The Department of Water Affairs (2010) cautioned

Table 22.3 Summary of AWRMS water quality improvement

Parameter	Mean concentrations			
	Domestic secondary effluent	Stormwater	Abstracted groundwater	Final chlorinated drinking water
Dissolved organic carbon (mg/L)	8.8	8.8	4.2	3.5
Nitrate (mg/L)	11	2.5	0.6	0.6
Total Kjeldahl nitrogen (mg/l)	2.3	4.7	0.8	0.8
Total coliform bacteria (MPN/100 mL)	6.9×10^4	5.5×10^4	2	0
E. Coli (MPN/100 mL)	1.2×10^4	1.2×10^4	0	0
Enterococci (MPN/100 mL)	1,700	6.3×10^4	0	0
Clostridium spores (CFU/100 mL)	2.3×10^4	2.1×10^4	0	0
Bacteriophages (PFU/100 mL)	910	4.5×10^4	0	0

Source Tredoux et al. (2012), Table 8.4

that the longer-term sustainability of the AWRMS depends on proper maintenance of all the components for eliminating risks.

22.7.3 Bolivar, South Australia

The trial ASR system at the Bolivar Wastewater Treatment Plant site (WWTP), near the city of Adelaide, South Australia, is noteworthy as it is undoubtedly the most thoroughly investigated ASR system to date. The main objectives of the trial ASR project were to determine the technical feasibility, environmental sustainability, and economic viability of ASR using reclaimed wastewater from the Bolivar WWTP, and to confirm and demonstrate that any potential health risks associated with the practice can be controlled effectively within a strict quality regulation and monitoring regime (Martin et al. 2002). The primary goal of ASR in southern Australia is to store excess reclaimed and surface water (stormwater) during the winter for later non-potable uses in the dry summer period.

Commonwealth Scientific and Industrial Research Organisation (CSIRO) Land and Water took the technical lead in the scientific investigations for the Bolivar ASR project, which resulted in a large number of technical publications that have relevance for ASR using treated wastewater in general. Among the many technical issues investigated were:

- clogging of the ASR well and aquifer during injection (Pavelic et al. 2007a, 2007b)
- changes in the quality of injected water due to biogeochemical processes and fluid-rock interactions (Vanderzalm et al. 2002; Skjemstad et al. 2005; Greskowiak et al. 2005)
- presence and fate of pathogens (Barry et al. 2010)
- fate and transport of emerging contaminants (Ying et al. 2003, 2004)
- fate of disinfection byproducts (Nicholson et al. 2002; Pavelic et al. 2006a)
- hydraulic properties of the aquifer and their impact on system operation (recovery efficiency; Pavelic et al. 2006b).

The hydrogeology of the Bolivar ASR system site was discussed by Pavelic et al. (2006b). The storage zone is the "T2" aquifer of the Tertiary-aged Port Willunga Formation, which consists of fossiliferous and marly limestone to siliceous calcarenite. The T2 aquifer has a moderate transmissivity of approximately 180 m^2/d (1,940 ft^2/d) and the native groundwater salinity (TDS concentration) is approximately 2,100 mg/L at the project site. The Bolivar ASR system consists of a single ASR well and a network of 17 observation wells and piezometers. The ASR well is constructed with an 8-in. (20 cm) inner diameter fiberglass reinforced plastic casing set at 103 m (338 ft) below land surface (bls). The well is completed with a nominal 8-in. (20-cm) diameter open hole to the base of the T2 aquifer at 160 m (525 ft) below land surface (bls). Observation wells and piezometers include a transect of wells located at radial distances of 4 m (13 ft), 50 m (164 ft), 75 m (246 ft), 120 m (394 ft), 300 m (984 ft) and 600 m (1,968 ft) from the ASR well in the downgradient direction and a minor transect located perpendicular to the regional flow gradient.

Operational testing of the Bolivar ASR system began in October 1999. The injection rates ranged from 900 to 1,300 m^3/d (0.24–0.34 million gallons per day; MGD). Recovery was continued until the total dissolved solids reached the maximum permitted concentration of 1,500 mg/L for the first three operational cycles (Pavelic et al. 2006b; Barry et al. 2010). The recovery efficiency increased from 60 to 80% between the first and second cycles and was maintained at 80% in the third cycle. The recovery efficiency for the fourth cycle was 73% taking into account a reduced TDS threshold for recovery of approximately 1,300 mg/L (Barry et al. 2010). The very good performance of Bolivar ASR system can be attributed to a combination of:

- favorable aquifer hydrogeology, particularly a dominance of matrix flow
- modest native storage-zone salinity
- ability to recovery to a salinity close to that of the native storage-zone groundwater.

Management of well and aquifer clogging has been an important operational issue. Pavelic et al. (2007a), for example, evaluated the effectiveness of different redevelopment methods on the performance of the Bolivar ASR system and determined that daily surging (24 min) was the preferred redevelopment method. Short-term clogging due to particle retention had a low impact on long-term clogging rates as the organic particles were biodegraded. TOC concentration appeared to be the limiting factor for microbial growth and long-term biological clogging. A low pH allows for the unclogging of the aquifer by calcite dissolution (Pavelic et al. 2007a, b).

22.7.4 Reclaimed Water ASR in Florida (U.S.A.)

Southern and Central Florida experience a summer wet season, in which ample water is available, and a winter and spring dry season, which corresponds to the annual peak in population from tourism and seasonal residents. The state relies heavily on groundwater for both its potable and non-potable water supplies. Low dry season aquifer water levels are limiting further exploitation of groundwater resources and water shortages are being declared with increasing frequency. ASR is a highly logical water management solution in Florida. Excess water that would otherwise be lost to tide in the wet season could be stored underground for use in the dry season. Florida is a national leader in wastewater reuse, but further increases in reuse is hampered by inadequate dry season supplies. Potential reuse customers require a reliable year-round supply and tend to be unwilling to commit to a reuse system if they cannot be assured that they will receive water during dry periods when the water is most needed. Reclaimed water ASR also makes good sense in that excess reclaimed water that is available in the wet season (and can be a disposal challenge) could be stored to provide a more reliable supply in the dry season.

Implementation of ASR in Florida has been hampered by arsenic leaching into stored water (Sect. 6.5). The recharge of water with DO appears to result in the oxidative dissolution of arsenic-containing iron sulfide minerals, releasing the arsenic into stored water. Under federal and state Underground Injection Control (UIC) rules, injection that causes a violation of the arsenic drinking water standard (10 μg/L) in an aquifer considered a USDW (i.e., aquifers with TDS concentrations of less than 10,000 mg/L) is a regulatory violation. Injected reclaimed water must also not cause a violation of primary drinking water and groundwater standards for coliform bacteria and disinfection byproducts, particularly the total THMs standard of 80 μg/L.

Two strategies has been employed to address water quality issues applicable to reclaimed water ASR. Arsenic, coliform, THMs, and other water quality standards do not apply to non-USDW aquifers. Storage zones below the deepest USDW (i.e., aquifer zones with greater than 10,000 mg/l of TDS) have been utilized (e.g., Englewood South Regional WWTP and Collier County Reclaimed Water ASR systems) to avoid water quality restrictions associated with storing water in USDW aquifers. The disadvantages of this approach are greater well construction costs for the deeper wells and poorer recovery efficiencies due to the higher native groundwater salinity.

The Englewood South Regional WWTP Reclaimed Water ASR System was documented in a Monthly Operation Report submitted to the Florida Department of Environmental Protection to have had a 52.1% recover efficiency from 2001 through April 2016. This system is taken to be a successful example of a non-USDW ASR system. A recovery efficiency of greater than 50% is possible because recovery can continue up to 1,000 mg/L of chloride (compared to the 250 mg/L drinking water standard). Recovery efficiency is not critical because the stored water has a low value (i.e., the excess reclaimed water stored in the wet season would otherwise be disposed of).

Destin Water Users, Inc. (DWU), located on the western Florida panhandle, uses institutional controls to allow reclaimed water to be stored in a freshwater USDW aquifer (Maliva et al. 2013, 2018). The storage zone is lower part (main-producing zone) of the siliciclastic Sand-and-Gravel Aquifer, which is located between approximately 116 and 166 ft (35.7 and 50.6 m) below land surface (bls) at the DWU ASR system site. Under Florida reuse rules, ASR systems that utilize freshwater aquifers are essentially assumed to involve indirect potable reuse and are regulated as such. Meeting the required very onerous full treatment and disinfection requirements would have made the project economically not feasible. DWU was able to obtain a variance from the full treatment and disinfection requirements because a local ordinance restricts the use of the storage zone aquifer to irrigation. Institutional controls thus prevent potable use. Arsenic leaching has occurred in the system, but the concentrations have naturally declined over the operation of the system as the supply of leachable arsenic in the aquifer is being exhausted. The main operational challenge continues to be management of clogging of the screened ASR wells.

The important lesson of the DWU ASR systems is that the health risks associated with MAR of treated wastewater can be successfully managed by institutional and physical controls that prevent unintended and uncontrolled exposure to the water. It can be much less expensive to prevent unauthorized contact with recharged treated wastewater than treating the wastewater to such a high degree that there are no health risks in the often highly unlikely event that significant contact does occurs. The DWU is also an example of the aquifer zoning concept. In some areas (e.g., Destin barrier island), the optimal use of an aquifer may be for the storage of reclaimed water for non-potable uses rather than potable water supply. There is no realistic scenario in which the Sand-and-Gravel Aquifer in Destin would ever be used for municipal water supply because of its limited capacity and vulnerability to saline-water intrusion.

22.8 Direct Potable Reuse

Direct potable reuse is the introduction of highly treated wastewater either directly into the potable water distribution system downstream of a water treatment plant or into the raw water supply immediately upstream of a water treatment plant (Asano et al. 2007). Direct potable reuse does not incorporate an environmental buffer. Despite that advanced treated wastewater is cleaner than most bottled waters (Nagel 2015), many people still regard it as unclean and unwholesome through association with its source (Hummer and Eden 2016). An uniformed public may be the biggest obstacle to direct potable reuse becoming common practice (Hummer and Eden 2016). Public opinion can dramatically change through education. For example, potable reuse was opposed in San Diego, California, by a two to one majority in 2004, but a survey in 2014 had 79% support for potable reuse. Direct potable reuse practices and associated health risks were reviewed in detail by the Tchobanoglous et al. (2011), Natural Research Council (2012), USEPA (2012), and Lahnsteiner et al. (2018).

It is increasingly being observed that use of environmental buffers can be quite illogical from a technical perspective considering the high quality to which wastewater can be treated. The USEPA (2012) observed that:

> Research on the performance of several full-scale advanced water treatment operations indicates that some engineered systems can perform equally well or better than some existing environmental buffers in attenuating contaminants, and the proper use of indicators and surrogates in the design of reuse systems offers the potential to address many concerns regarding quality assurance.
>
> Implementation of technologies for increasingly higher levels of treatment for many of these IPR projects has led to questions about why reclaimed water would be treated to produce water with higher quality than drinking water standards, and then discharged to an aquifer or lake.

To date, there has been limited implementation of direct potable reuse. A very early temporary implementation of direct potable reuse occurred in Chanute, Kansas, as an emergency measure during an extreme drought in 1956–1957. The City's sole source of water, the Neosho River, ran dry. Secondary treated wastewater was stored in a pond creating by damming the river channel, retained for 17 days, and then sent to the water treatment plant (Hummer and Eden 2016). An epidemiological survey showed fewer cases of stomach and intestinal illnesses during the period of reclaimed use than in the following winter when Chanute resumed using river water, which received wastewater from upstream communities (Asano et al. 2007).

The City of Denver, Colorado, conducted the Direct Potable Water Reuse Demonstration Project from 1985 to 1992, which included whole animal testing of the highly treated water. Based on an assessment of the results of an extensive testing program, it was concluded that the reclaimed water met all health standards and was of equal and better quality than the city's drinking water. Nevertheless, direct potable reuse was not implemented in Denver.

The world's first permanent direct potable reuse system was commissioned in Windhoek, Namibia in 1968 (Haarhoff and Van der Merwe 1996; Law 2003; Du Pisani 2006; Asano et al. 2007). Direct potable reuse was implemented because of the lack of alternative economically viable water supply options. A cornerstone of the reclamation system is the separation of industrial wastes from domestic effluent, with the former being sent to a different wastewater treatment plant. The current facility, the New Goreangab Water Reclamation Plant, went into operation in August 2002 and includes the following series of processes: coagulation/flocculation, dissolved-air floatation, dual-media filtration, ozonation, granular activated carbon (GAC) filtration, membrane filtration (ultrafiltration), and chlorination (Law 2003; Du Pisani 2006).

Du Pisani (2006) emphasized that to obtain public confidence, quality monitoring and control are of the utmost importance, and that the most difficult challenge for emulating the Windhoek experience would be breaking down the fear-barrier to direct potable reuse. Du Pisani (2006) also observed that direct potable reuse may only be realistically considered in cases where no viable alternatives exist.

The Wichita Falls, Texas, temporary direct potable reuse project went online July 9, 2014 following extensive testing by the City of Wichita Falls and the Texas Commission on Environmental Quality (TCEQ). The system provided over 5 MGD (0.019 MCM/d) of water treated by (Wichita Falls Texas 2014; Water 360 n.d.):

(1) clarification with coagulant
(2) microfiltration
(3) reverse osmosis
(4) water release to a holding lagoon
(5) blending the reuse water with raw lake water from the City's water source lakes, Lake Arrowhead and Lake Kickapoo, on a 50–50 basis
(6) treating the blended water by conventional means

The blended water was treated through a conventional eight step process:

(1) treatment with chlorine dioxide
(2) pre-disinfection
(3) coagulation
(4) softening
(5) flocculation
(6) sedimentation
(7) re-stabilization
(8) fluoridation

The City of Wichita direct potable reuse system was implemented in response to a severe drought, which has since ended, and the city is now moving to convert operations into indirect potable reuse in which the treated wastewater will be discharged into Lake Arrowhead (Our Texas Water 2016).

Direct potable reuse at the Colorado River Municipal Water District Raw Water Production Facility (RWPF) at Big Springs, Texas, began in 2013 (Sloan 2012; Steinle-Darling et al. 2016). Secondary treated effluent receives full advanced treatment (FAT), consisting of MF and RO plus an advanced oxidation process (UV and H_2O_2). Less than 20% of the highly treated wastewater is blended with the existing surface water supply (Moss Creek Lake) and then sent to the water treatment plant. The FAT system achieves about a two orders of magnitude removal of TrOCs (pharmaceuticals and personal care products) by RO and slightly more by the advanced oxidation process (Steinle-Darling et al. 2016). TrOC concentrations are generally less than those in Moss Creek Lake Water. The treated water has some DBPs (THMs) but does not exceed regulatory standards. The water produced by the RWPF is nearly every way superior to the raw surface water with which it is being blended (Steinle-Darling et al. 2016).

Direct potable reuse is being investigated in a number of other communities facing water scarcity. The main issues facing implementation of potable reuse is gaining public support, treatment costs, and choosing between indirect and direct options.

References

Alhumoud, J. M., Behbehani, H. S., & Abdullah, T. H. (2003). Wastewater reuse practices in Kuwait. *The Environmentalist, 23,* 117–126.

Arnold, R. G., & Arnold, K. P. (2009). Integrated urban water management in the Tucson, Arizona metropolitan area. In L. M. Mays (Ed.), *Integrated urban water management: Arid and semi-arid regions* (pp. 113–138). Paris: UNESCO, and Leiden: Taylor & Francis.

Asano, T., & Cotruvo, J. A. (2004). Groundwater recharge with reclaimed wastewater; health and regulatory considerations. *Water Research, 38,* 1941–1951.

Asano, T., & Levine, A. D. (1996). Wastewater reclamation, recycling and reuse: Past, present, and future. *Water Science and Technology, 30*(11–11), 1–14.

Asano, T., Burton, F. L., Leverenz, H. L., Tsuchihashi, R., & Tchobanoglous, G. (2007). *Wastewater reuse. Issues, technologies and applications*. New York: McGraw-Hill.

Ayers, R. S., & Wescott, D. W. (1985). *Water quality for agriculture*. FAO Irrigation and Drainage Paper 29. Rome: Food and Agriculture Organization of the United Nations.

Bahri, A., & Brissaud, F. (1996). Wastewater reuse in Tunisia: Assessing a national policy. *Water Science and Technology, 33*(10–11), 87–94.

Barry, K., Vanderzalm. J., Pavelic, P., Regel, R., May, R., Dillon, P., Sidhu, J., & Levett, K. (2010). *Bolivar reclaimed water aquifer storage and recovery project: Assessment of the third and fourth ASR cycles*. Canberra: CSIRO: Water for a Healthy Country National Research Flagship.

Bixio, D., De heyder, B., Cikurel, H., Muston, M., Miska, V., Joksimovic, D., Schäfer, A.I., Ravazzini, A., Aharoni, A., Savic, D., & Thoeye, C. (2005). Municipal wastewater reclamation: Where do we stand? An overview of treatment technology and management practice. *Water Science & Technology: Water Supply, 5*(1), 77–85.

Bos, R., Carr, R., & Keraita, B. (2010). Assessing and mitigating wastewater-related health risks in low-income countries. In P. Dreschel, C. A. Scott, L. Raschid-Sally, M. Redwood, & A. Bahri (Eds.), *Wastewater irrigation and health. Assessing and mitigating risk in low income countries* (pp. 29–47). Ottawa: International Development Research Centre.

Bouwer, H. (1991). Groundwater recharge with sewage effluent. *Water Science and Technology, 23,* 2099–2108.

Bouwer, H. (2000). Integrated water management: Emerging issues and challenges. *Agricultural Water Management, 48,* 217–228.

Bruvold, W. H. (1985). Obtaining public support for reuse water. *Journal American Water Works Association, 77*(7), 72–77.

Chalmers, R. B., & Ferguson, J. (2012). South district water reclamation plant (SDWRP). In *Guidelines for water reuse,* EPA/600/R-12/618 (pp. D-80–D-82). Washington, DC: U.S. Environmental Protection Agency.

Cook, J. (2004). *Innovative applications in water reuse: Ten case studies*. Alexandra, VA: WateReuse Association.

Crites, R. W., Middlebrooks, E. J., Bastian, R. K., & Red, S. C. (2014). *Natural wastewater treatment systems* (2nd ed.). Boca Raton: CRC Press.

Daughton, C. G. (2004). Ground water recharge and chemical contaminants: Challenges in communicating the connections and collisions of two disparate worlds. *Ground Water Monitoring and Remediation, 24*(2), 127–138.

Department of Water Affairs. (2010). *The Atlantic water resources management scheme: 30 years of artificial groundwater recharge*. PRSA 000/00/11609/10-activity 17 (ASR.1). Pretoria: Department of Water Affairs.
Dillon, P. (2000). Water reuse in Australia: Current status, projections and research. In P. J. Dillon (Ed.), *AWA Water Recycling Forum, Proceedings of the First Symposium*, Adelaide, pp. 99–104.
Drewes, J. E., & Fox, P. (1999). Fate of natural organic matter (NOM) during groundwater recharge using reclaimed water. *Water Science and Technology, 40*(9), 241–248.
Drinan, J. E., & Spellman, F. R. (2013). *Water and wastewater treatment: A guide for the nonengineering professional* (2nd ed.). Boca Raton: CRC Press.
Du Pisani, P. L. (2006). Direct reclamation of potable water at Windhoek's Goreangab Reclamation Plant. *Desalination, 188,* 79–88.
Edberg, S. C. L., Rice, E. W., Karlin, R. J., & Allen, M. J. (2000). Escherichia coli: The best biological drinking water indicator for public health protection. *Journal of Applied Microbiology, 88*(S1), 106S–116S.
Friedler, E. (2001). Water reuse—An integral part of water resources management: Israel as a case study. *Water Policy, 3,* 29–39.
Gasca, M., Johnson, T., & Willardson, B. (2011). Keeping the water going: Challenges to recharge in the Montebello Forebay. In *Managed Aquifer Recharge Symposium*, January 25–26, 2011. Irvine, California.
Gasca, M., & Hartling, E. (2012). Montebello Forebay groundwater recharge project using recycled water, Los Angeles County, California. *Guidelines for water reuse* (pp. D42–D45). Washington, D.C.: USEPA.
Gerba, C. P., & Rose, J. B. (2003). International guidelines for water recycling: Microbiological considerations. *Water Science and Technology: Water Supply, 3*(4), 311–316.
Gerrity, D., Pecson, B., Trussell, R. S., & Trussell, R. R. (2013). Potable reuse treatment trains throughout the world. *Journal of Water Supply: Research and Technology-AQUA, 62*(6), 321–338.
Greskowiak, J., Prommer, H., Vanderzalm, J., Pavelic, P., & Dillon, P. (2005). Modeling of carbon cycles and biogeochemical changes during injection and recovery of reclaimed water at Bolivar. *South Australia. Water Resources Research, 41,* W10418.
Haarhoff, J., & Van der Merwe, B. (1996). Twenty-five years of wastewater reclamation in Windhoek, Namibia. *Water Science and Technology, 33,* 25–35.
Haas, C. N., Rose, J. B., & Gerba, C. P. (1999). *Quantitative microbial risk assessment*. New York: Wiley.
Haddad, B. M., Rozin, P., Nemeroff, C., & Slovic, P. (2009). *The psychology of water reclamation and reuse. Survey findings and research road map*. Alexandria, Virginia: WateReuse Foundation.
Hartley, T. W. (2006). Public perception and participation in water reuse. *Desalination, 187,* 115–126.
Hummer, N., & Eden, S. (2016). Potable reuse of water. *Arroyo*. University of Arizona Water Resources Research Center, Tucson, AZ. https://wrrc.arizona.edu/publications/arroyo-newsletter/arroyo-2016-potable-reuse-water.
Hurlimann, A. C. (2007). Is recycled water risky? An urban Australian community's perspective. *The Environmentalist, 27,* 83–94.
Jiménez, B., Mara, D., Carr, R., & Brissaud, F. (2010). Wastewater treatment for pathogen removal and nutrient conservation: Suitable systems for use in developing countries. In P. Dreschel, C. A. Scott, L. Raschid-Sally, M. Redwood, & A. Bahri (Eds.), *Wastewater irrigation and health. Assessing and mitigating risk in low income countries* (pp. 149–169). Ottawa: International Development Research Centre.
Johnson, T., & Gagan, M. (2011). The increasing role of storm water for groundwater recharge in the Central basin, Los Angeles Coastal Plain, southern California. In *Managed Aquifer Recharge Symposium*, January 25–26, 2011. Irvine, California.
Kariar, G. L., & Christian, R. A. (2013). *Wastewater treatment: Concepts and design approach* (2nd ed.). Delhi: PHI Learning Pvt. Ltd.

Khan, S. J., & Gerrard, L. E. (2006). Stakeholder communications for successful water reuse operations. *Desalination, 187,* 191–202.

Kivaisi, A. K. (2001). The potential for constructed wetlands for wastewater treatment and reuse in developing countries: A review. *Ecological Engineering, 16*(4), 545–560.

Lahnsteiner, J., van Rensburg, P., & Esterhuizen, J. (2018). Direct potable reuse—A feasible water management option. *Journal of Water Reuse and Desalination, 8*(1), 14–28.

Lau, A., Huston, P., & Pecson, B. (2016). Padre Dam's advanced water purification program: Building a better future inspired by creativity from the past. *Journal American Water Works Association, 108*(100), 68–76.

Law, I. B. (2003). Advanced reuse: From Windhoek to Singapore and beyond. *Water*, May, 44–50.

Lazarova, V., Cirelli, G., Jeffrey, P., Salgot, M., Icekson, N., & Brissaud, F. (2000). Enhancement of integrated water management and reuse in Europe and the Middle East. *Water Science and Technology, 42*(1–2), 193–202.

Levine, A. D., & Asano, T. (2004). Recovering sustainable water from wastewater. *Environmental Science and Technology, 38,* 201A–208A.

Macler, B., & Merkle, J. C. (2002). Current knowledge of groundwater microbial pathogens and their control. *Hydrogeology Journal, 8,* 29–40.

Maliva, R. G., Griswold, R. F., & Autrey, M. M. (2013). Prototype for a reclaimed water aquifer storage recovery system benefits and operational experiences. *Florida Water Resources Journal, 65*(3), 54–59.

Maliva, R. G., Autrey, M. M., Law, L., Manahan, E. S., & Missimer, T. M. (2018). Reclaimed water aquifer storage and recovery system: Update on a groundbreaking system in Florida. *Florida Water Resource Journal, 69*(2), 52–59.

Maliva, R. G., & Missimer, T. M. (2010). *Aquifer storage and recovery and managed aquifer recharge using wells: Planning, hydrogeology, design, and operation*. Houston: Schlumberger Corporation.

Maliva, R. G., & Missimer, T. M. (2012). *Arid lands water evaluation and management*. Berlin: Springer.

Mara, D. (2003). *Domestic wastewater treatment in developing countries*. Abingdon: Taylor & Francis.

Mara, D. D., Sleigh, P. A., Blumenthal, U. J., & Carr, R. M. (2007). Health risks in wastewater irrigation: Comparing estimates from quantitative microbial risk analyses and epidemiological studies. *Journal of Water and Health, 5,* 39–50.

Markus, M. R. (2009). Groundwater replenishment & water reuse: *The Water Report*, Issue 59 (January 15, 2009), 1–9.

Marks, J. S. (2004). Advancing community acceptance of reclaimed water. *Water* (August 2004), *31*(5), 46–51.

Martin, R., Clarke, D., Dennis, K., Graham, J., Dillon, P., Pavelic, P., & Barry, K. (2002). *Bolivar water re-use project.* Report DWLBC 2002/02. Australia: Department of Water, Land, and Biodiversity Conservation.

Mercer, T., Bennett, J., Fahey, P., Moore, E., MacNevin, D., & Kinslow, J. (2015). Groundwater replenishment performance and operations: Lessons learned during clearwater's one-year pilot. *Florida Water Resources Journal, 67*(2), 32–40.

Miller, G. W. (2006). Integrated concepts in water reuse: Managing global water needs. *Desalination, 187,* 66–75.

Mujeriego, R., & Asano, T. (1999). The role of advanced treatment in wastewater reclamation and reuse. *Water Science and Technology, 40*(4–5), 1–9.

Nagel, R. (2015). Making direct potable reuse a reality. *Journal American Water Works Association, 107*(7), 76–82.

National Drinking Water Clearinghouse. (1996). *Tech brief, disinfection*. Morgantown, WV: National Drinking Water Clearinghouse. http://www.nesc.wvu.edu/pdf/dw/publications/ontap/2009_tb/disinfection_DWFSOM50.pdf.

National Research Council. (1994). *Ground water recharge using waters of impaired quality*. Washington, DC: National Academy Press.
National Research Council. (1998). *Issues in potable reuse: The viability of augmenting drinking water supplies with reclaimed water*. Washington, DC: National Academy Press.
National Research Council. (2008). *Prospects for managed underground storage of recoverable water*. Washington, DC: National Academy Press.
National Research Council. (2012). *Water reuse: Potential for expanding the nation's water supply through the reuse of municipal wastewater*. Washington, DC: National Academy Press.
Nicholson, B. C., Dillion, P. J., & Pavelic, P. (2002). Fate of disinfection by-products during aquifer storage and recover. In P. J. Dillion (Ed.), *Management of aquifer recharge for sustainability* (pp. 155–160). Lisse: A.A. Balkema.
NRMMC-EPHC-AHMC. (2006). *National guideline for water recycling: Managing health and environmental risks (Phase 1)*. Natural Resource Management Ministerial Council, Environment Protection and Heritage Council, and Australian Health Ministers Conference.
NRMMC-EPHC-NHMRC. (2009). *Australian guidelines for water recycling: Managing health and environmental risks (Phase 2) managed aquifer recharge.* Natural Resource Management Ministerial Council, Environment Protection and Heritage Council, National Health and Medical Research Council.
Ormerod, K. J., & Scott, C. A. (2013). Drinking wastewater public trust in potable reuse. *Science, Technology and Human Values, 38*(3), 351–373.
Our Texas Water. (2016). *Wichita falls direct potable water reuse project*. Retrieved March 12, 2017 from https://ourtexaswater.org/projects/wichita-falls-direct-potable-water-reuse-project.
Pavelic, P., Dillon, P. J., & Nicholson, B. C. (2006a). Comparative evaluation of the fate of disinfection byproducts at eight aquifer storage and recovery sites. *Environmental Science and Technology, 40,* 501–508.
Pavelic, P., Dillon, P. J., & Simmons, C. T. (2006b). Multiscale characterization of a heterogeneous aquifer using an ASR operation. *Ground Water, 44,* 155–164.
Pavelic, P., Dillon, P. J., & Barry, K. E. (2007a). Management of clogging for reclaimed water in a carbonate aquifer. In P. Fox (Ed.), *Management of Aquifer Recharge for Sustainability: Proceedings of the 6th International Symposium on Managed Aquifer Recharge of Groundwater* (pp. 148–161). Phoenix: Acacia Publishing.
Pavelic, P., Dillon, P. J., Barry, K. E., Vanderzalm, J. L., Correl, R. L., & Rinck-Pfeiffer, S. M. (2007b). Water quality effects on clogging rates during reclaimed water ASR in a carbonate aquifer. *Journal of Hydrology, 334,* 1–16.
Payment, P., & Locas, A. (2011). Pathogens in water: Value and limitations of correlation with microbial indicators. *Ground Water, 49*(1), 4–11.
Pescod, M. E. (1992). *Wastewater treatment and use in agriculture*. FAO Irrigation and Drainage Paper 47. Rome: Food and Agriculture Organization of the United Nations.
Ramadan, H., & Ponce, V. M. (n.d.). *Design and performance of waste stabilization ponds, version 081218*. Retrieved May 31, 2018 from http://stabilizationponds.sdsu.edu/.
Riffat, R. (2013). *Fundamentals of wastewater treatment and engineering*. London: IWA Publishing and Boca Raton: CRC Press.
Rodriguez, C., Van Buynder, P., Lugg, R., Blair, P., Devine, B., Cook, A., et al. (2009). Indirect potable reuse: A sustainable water supply alternative. *International Journal of Environmental Research and Public Health, 6,* 1174–1209.
Salgot, M., Vergés, C., & Angelakis, A. N. (2003). Risk assessment in wastewater recycling and reuse. *Water Science and Technology: Water Supply, 3*(4), 301–309.
Seah, H., & Woo, C.-H. (2012). The multi-barrier safety approach for indirect potable reuse and direct nonpotable use of NEWWATER. In *Guidelines for water reuse,* EPA/600/R-12/618 (pp. E-99–E-101). Washington, DC: U.S. Environmental Protection Agency.
Sharma, S. K., Ernst, M., Hein, A., Jekel, M., Jefferson, B., & Amy, G. (2012). In C. Kazner, T. Wintgens, & P. Dillon P. (Eds.), *Water reclamation technologies for safe managed aquifer recharge* (pp. 239–258). London: IWA Publishing.

Skjemstad, J., Swift, R., & Hayes, M. (2005). Evaluation of changes in NOM during ASR at the Bolivar site. In P. Dillon & S. Toze (Eds.), *Water quality improvements during aquifer storage and recovery, volume 1: Water quality improvement processes,* Report 91056F (pp. 99-110). Denver: AWWA Research Foundation.

Sloan, D. W. (2012). Raw water production facility: Big Spring Plant. In *Guidelines for water reuse*, EPA/600/R-12/618 (pp. D-148–D-149). Washington, DC: U.S. Environmental Protection Agency.

Sloss, E. M., Geschwind, S. A., McCaffrey, D. F., & Ritz, B. R. (1996). *Groundwater recharge with recycled water: An epidemiologic assessment in Los Angeles, County, 1987–1991* (report prepared for the Water Replenishment District of Southern California). Santa Monica, CA: RAND.

Sloss, E. M., McCaffrey, D. F., Fricker, R. D., Geschwind, S. A., & Ritz, B. R. (1999). *Groundwater recharge with recycled water: Birth outcomes in Los Angeles County 1982–1993* (report prepared for the Water Replenishment District of Southern California). Santa Monica, CA: RAND.

Snyder, S. A., Stanford, B. D., Bruce, C. M., Pleus, R. C., & Drewes, J. E. (2010). *Identifying hormonally active compounds, pharmaceuticals, and personal care product ingredients of health concern for potential presence in water intended for indirect potable reuse*. Washington, DC: WateReuse Research Foundation.

Standridge, J. (2008). E. coli as public health indicator of drinking water quality. *Journal American Water Works Association, 100*(2), 65–75.

Stanford, B. D., & Antolovich, A. (2017). Advanced oxidation process, unique hydrogeology allow for innovative reuse strategy. *Journal American Water Works Association, 109*(2), 59–63.

Steinle-Darling, E., Sutherland, J., & Salveson, A. (2016). Sampled direct potable reuse water shows promising results. *Opflow, 42*(2), 20–22.

Steirer, M. A., Dorman, A., Van, A., & Quicho, J. (2012). City of San Diego—water purification demonstration project In *2010 guidelines for water reuse* (EPA/600/R-12/618) (pp. D-27–D-29). Washington, D.C.: U.S. Environmental Protection Agency.

Tchobanoglous, G., Leverenz, H., Nellor, M. H., & Crook, F. (2011). *Direct potable reuse: A path forward*. Alexandria, VA: WateReuse Foundation.

Tchobanoglous, G., Stensel, H. D., Tsuchihashi, R., & Burton, F. L. (2014). *Wastewater engineering: Treatment and resource recovery* (5th ed.). New York: McGraw-Hill.

Tredoux, G., Cavé, L. C., & Bishop, R. (2002). Long-term stormwater and wastewater infiltration in a sandy aquifer, South Africa. In P. J. Dillon (Ed.), *Management of aquifer recharge for sustainability* (pp. 41–48). Lisse: A.A. Balkema.

Tredoux, G., Genthe, B., Steyn, M., & Germanis, J. (2012) Managed aquifer recharge for potable reuse in Atlantis, South Africa. In C. Kazner, T. Wintgens & P. Dillon P. (Eds.), *Water reclamation technologies for safe managed aquifer recharge* (pp. 121–140). London: IWA Publishing.

USEPA. (2007). *Wastewater management fact sheet. Membrane bioreactors*. Washington, DC: U.S. Environmental Protection Agency.

USEPA. (2012). *Guidelines for water reuse* (EPA/600/R-12/618). Washington, DC: U.S. Environmental Protection Agency.

Vanderzalm, J. L., Le Gal La Salle, C., Hutson, J. L., & Dillon, P. J. (2002). Water quality changes during aquifer storage and recovery at Bolivar, South Australia. In P. J. Dillon (Ed.), *Management of aquifer recharge for sustainability* (pp. 82–88). Lisse: A.A. Balkema.

Vignesnaran, S., & Sundaravadivel, M. (2004). Recycle and reuse of domestic wastewater. In S. Vigneswaran (Ed.), *Wastewater recycle, reuse and reclamation, encyclopedia of life support systems (EOLSS)*. Oxford, UK: Eolss Publishers (http://www.eolss.net).

Water 360. (n.d.). *Recycled water for drinking: direct potable reuse a temporary solution for Wichita Falls, Texas*. Retrieved March 12, 2017 from http://legacywater360.server309.com/wp-content/uploads/2015/07/J003887-Wichita-falls-fact-sheet_2.pdf.

Water Environment Foundation. (2012). *Wastewater treatment plant design handbook*. Alexandria, VA: Water Environment Federation.

WEF and AWWA. (1998). *Using reclaimed water to augment potable water resources*. Alexandra, VA: Water Environment Foundation.

Wichita Falls Texas. (2014). *Direct potable reuse project*. Retrieved August 12, 2013 http://www.wichitafallstx.gov/index.aspx?nid=1595.

Wintgens, T., Salehi, F., Hochstrat, R., & Melin, T. (2008). Emerging contaminants and treatment options in water recycling for indirect potable reuse. *Water Science and Technology, 57*(1), 99–107.

WHO. (2006). *WHO guidelines for the safe use of wastewater, excreta, and greywater. Volume 1, Policy and regulatory aspects*. Geneva: World Health Organization.

Ying, G.-G., Kookana, R. S., & Dillon, P. (2003). Sorption and degradation of selected endocrine disrupting chemicals in aquifer material. *Water Research, 37*, 3785–3791.

Ying, G.-G., Kookana, R. S., & Dillon, P. (2004). Attenuation of two estrogen compounds in aquifer materials supplemented with sewage effluent. *Ground Water Monitoring and Remediation, 24*, 102–107.

Chapter 23
Low Impact Development and Rainwater Harvesting

23.1 Introduction and Definitions

Under predevelopment conditions, much of rainwater stayed in the area where it fell, either recharging the underlying shallow aquifer or slowly returning to the atmosphere via evapotranspiration. Land development activities increase the imperviousness of land surfaces, decreasing infiltration and increasing runoff. Increased imperviousness results in an increase in peak and total runoff and a decreased time to peak runoff and a shorter time of concentration. Traditional stormwater management seeks to collect and remove runoff from sites as quickly as possible with associated downstream impacts of water quality deterioration, erosion, and flooding. The alternative is to promote infiltration on adjacent pervious areas with the goals of decreasing runoff and increasing recharge (Holman-Dodds et al. 2003).

Stormwater management is a discipline unto itself. Many governmental jurisdictions (national, state, county, and city) have promulgated regulations for stormwater management that are applicable to new development activities. These regulations are in the form of rules, construction standards, and best management practices that are incorporated into a formal permitting process. Non-regulatory guidance documents are also widely available, now usually on-line.

Stormwater management practices serve multiple purposes. For example, the Florida Administrative Code (62-40.431(2,a)) states that

> The primary goals of the state's stormwater management program are to maintain, to the maximum extent practical, during and after construction and development, the pre-development stormwater characteristics of a site; to reduce stream channel erosion, pollution, siltation, sedimentation and flooding; to reduce stormwater pollutant loadings discharged to waters to preserve or restore designated uses; to reduce the loss of fresh water resources by encouraging the recycling of stormwater; to enhance ground water recharge by promoting infiltration of stormwater in areas with appropriate soils and geology; to maintain the appropriate salinity regimes in estuaries needed to support the natural flora and fauna; and to address stormwater management on a watershed basis to provide cost effective water quality and water quantity solutions to specific watershed problems.

R. G. Maliva, *Anthropogenic Aquifer Recharge*,
Springer Hydrogeology, https://doi.org/10.1007/978-3-030-11084-0_23

Stormwater management system design, in the very broad sense, commonly involves the following elements:

(1) delineation of drainage and subdrainage areas
(2) evaluation of the runoff generation potential of drainage and subdrainage areas using numerical and analytical techniques, such as the United States Soil Conservation Service curve number (SCS-CN) method (Natural Resources Conservation Service 2004)
(3) selection of design storm, rainfall event, or amount (e.g., first 1 in. of rainfall), which depends on stormwater management objectives
(4) determination of the required stormwater retention volume for the design storm
(5) design of a system to provide the target stormwater retention or detention.

Low Impact Development (LID) was pioneered in Prince George's County, Maryland, and is characterized as (PGCDER 1999):

> The low-impact development (LID) approach combines a hydrologically functional site design with pollution prevention measures to compensate for land development impacts on hydrology and water quality.
>
> The primary goal of low impact development methods is to mimic the predevelopment site hydrology by using site design techniques that store, infiltrate, evaporate, and detain runoff. Use of these technique helps to reduce off-site runoff and ensures adequate groundwater recharge.
>
> Stormwater is managed in small, cost-effective landscape features located on each lot rather than being conveyed and managed in large, costly pond facilities located at the bottom of the drainage area.

The U.S. Environmental Protection Agency (USEPA n.d.a) noted that the term "low impact development" refers to systems and practices that use or mimic natural processes that result in the infiltration, evapotranspiration or use of stormwater in order to protect water quality and associated aquatic habitats. Basic LID principles are:

- reduce impervious area
- disconnect impervious areas
- intercept stormwater before it comes in contact with impervious areas
- detain and infiltrate stormwater on site, as close as possible to the source.

The USEPA (n.d.b) defines the similar term "green infrastructure" (GI) as a

> cost-effective, resilient approach to managing wet weather impacts that provides many community benefits. While single-purpose gray stormwater infrastructure—conventional piped drainage and water treatment systems—is designed to move urban stormwater away from the built environment, green infrastructure reduces and treats stormwater at its source while delivering environmental, social, and economic benefits.

At both the site and regional scale, LID and GI practices are an approach to land development that works with nature to manage stormwater close to its source (USEPA n.d.a). LID is referred to as "sustainable drainage systems" (SuDS) in the United Kingdom and "water-sensitive urban design" (WSUD) in Australia. LID includes decentralized groundwater recharge systems, which manage stormwater on the scale of individual properties and developments (Stephens 2010; Stephens et al. 2012).

SuDS was described by the British Geological Survey (2017) as

> drainage solutions that provide an alternative to the direct channelling of surface water through networks of pipes and sewers to nearby watercourses. By mimicking natural drainage regimes, SuDS aim to reduce surface water flooding, improve water quality and enhance the amenity and biodiversity value of the environment. SuDS achieve this by lowering flow rates, increasing water storage capacity and reducing the transport of pollution to the water environment.

Water Sensitive Urban Design has been described as a philosophical approach to urban planning and design that aims to minimize the hydrological impacts of urban development on the surrounding environment (Lloyd et al. 2002). It is based on formulating plans that incorporate an integrated approach to the management of the urban water cycle (Wong 2006a). The guiding principles of Water Sensitive Urban Design (WSUD; Wong 2006a) are:

- reducing potable water demands through water efficient appliances and the use of alternative sources of water such as rainwater and treated wastewater
- minimization of wastewater generation and treatment of wastewater to standards suitable for reuse
- treating urban stormwater to meet quality objectives before discharge to surface waters or reuse
- using stormwater in urban settings to maximize visual and recreational amenity of developments.

Rainwater harvesting can be succinctly defined as the collection, storage, and reuse of rainwater. Rainwater harvesting is divided into passive and active methods. Passive rainwater harvesting consists of methods that slow the flow of water and allow it to infiltrate into the soil and underlying aquifers. Active methods involve the storage of rainwater commonly in tanks or cisterns. Rainwater harvesting is typically independently performed, small-scale activities, rather than elements of integrated water management systems.

The Ground Water Protection Council (2007) cautioned with respect to LID that "The real challenge will be to make these approaches standard practice at the low level and ensure that they are designed and maintained properly so that ground water is not degraded." They also cautioned that without considering groundwater, even LID technologies can allow polluted stormwater to impact groundwater, with the greatest risk occurring in locations with pollutant sources and sandy soils with a shallow water table. Natural filtering and sorption capacity of soils can be effective in removing some, but not all, contaminants. Factors affecting the potential for urban runoff to contaminate groundwater include:

- soil characteristics (sorption capacity, grain size, depth to the seasonal high water table)
- pollutant mobility

- pollutant abundance in urban runoff
- soluble fraction of pollutant
- pretreatment provided before infiltration.

From an anthropogenic aquifer recharge (AAR) perspective, LID and GI and rainwater harvesting include techniques that are intended to increase local groundwater recharge. Although the objectives of LID are laudable, it is important to recognize that some urban areas are experiencing rising shallow groundwater levels, with associated adverse impacts, and that increasing local recharge may exacerbate the problem (Sect. 24.2).

23.2 LID Approach

Preservation of the predevelopment hydrology of sites is the overall goal of LID (Dietz 2007). This is achieved by preserving as much of a site in an undisturbed condition as practical and where disturbance is necessary, reducing impacts to soils, vegetation, and aquatic systems. The USEPA (n.d.a) describes LID as employing

> principles such as preserving and recreating natural landscape features, minimizing effective imperviousness to create functional and appealing site drainage that treat stormwater as a resource rather than a waste product.

Stormwater management best management practices (BMPs) are defined by the USEPA as practices or combination of practices that are determined to be the most effective and practicable (including technological, economic, and institutional considerations) means of controlling point and nonpoint source pollutants at levels compatible with environmental quality goals. LID tools and techniques overlap with stormwater BMPs in that they serve both pollution treatment and water management objectives. The USEPA BMPs are defined based on pollution control, whereas LID practices also focus on infiltration and aquifer recharge.
The basic LID approach (PGCDER 1999) is an integrated process that includes the following elements:

- site planning to reduce, minimize, and disconnect the total impervious area at a site
- measures to minimize and mitigate the hydrological impacts of land use activities are implemented as close as possible to the source of runoff
- utilization of simple, non-structural methods
- creation of multifunctional landscape and infrastructure (e.g., bioretention cells can provide runoff control and treatment benefits plus ecological and aesthetic benefits)
- development is focused on areas that are less sensitive to disturbances and have lower value in terms of hydrological function (e.g., avoid developing areas with high infiltration rates or are environmentally sensitive, and locate impervious areas on less permeable soil types)

- increasing the time of concentration by lengthening flow paths, reducing lengths of runoff conveyance systems, increasing surface roughness (Manning's roughness coefficient), and diverting flow over pervious surfaces.

The State of Washington (WADOE 2013) listed key LID planning principles:

- preserving native vegetation
- protecting critical areas
- minimizing impervious surfaces
- minimizing grading and compaction of site soils
- preserving existing flow paths
- infiltrating stormwater runoff
- dispersing stormwater to vegetated facilities
- using naturalistic surface conveyance facilities
- utilizing small-scale, distributed LID BMPS.

The basic design of infiltration-based LID BMPs involves first evaluating site hydrogeology. Data are required on the depth to groundwater (including seasonal variations), the direction of groundwater flow and hydraulic gradients, and soil types present between land surface and the water table. Key soil considerations are the vertical hydraulic conductivity of surficial soils (infiltration rates) and the presence and depths of hydraulic restricting layers. Design considerations include (WADOE 2013):

- ability to meet site stormwater flow control and treatment requirements
- soils and subsurface hydrology
- site constraints (e.g., available space, existing facilities and vegetation)
- constructability
- ease of maintenance
- public acceptance

There is now a voluminous literature with respect to LID, green infrastructure, rainwater harvesting, and related subjects. In the United States, various state, and local agencies have developed LID manuals that both share common practices and take into consideration region-specific conditions including local climate, soil conditions, hydrogeology, and land development practices. The design of stormwater management systems must conform to local stormwater, roadway, utilities, and other engineering standards.

The key objective of LID is integrated water management systems. Considerable effort has been made toward development of modeling methods for LID. The goals of modeling are to provide insights on the performance of LID practices for achieving hydrological and water quality benefits and to serve as a guide for developing watershed planning and management strategies (Ahiablame et al. 2012). Models can also be used as design tools to evaluate various LID implementation strategies. Two main modeling approaches have been employed: process representation and practice representation (Ahiablame et al. 2012). Process representation modeling represents processes occurring in LIDs including infiltration, evapotranspiration, sedimentation, adsorption, and transformation of pollutants. A fundamental limitation of process representation (i.e., distributed parameter) modeling is that it is very data intensive and the required data are seldom available, especially for applied (as opposed to research) projects.

The practice representation approach uses an aggregation method to represent the LID practice as a whole (Ahiablame et al. 2012). The approach quantifies the effects of an LID practice by combining the complex processes that the practice can perform in one parameter. For example, the SCS-CN method can be used to represent the effects of porous pavement on infiltration and runoff volume by adjusting the value of the CN used for a site (or part of a site). Ahiablame et al. (2012) identified the following research opportunities to improve implementation of LID:

- characterization of runoff and water quality from different urban land uses
- need to continue data collection on the performance of LID systems over different spatial and temporal scales, and climate and environmental conditions
- enhancement of modeling techniques for evaluating the performance of LID practices
- scaling of the performance of LID practices from lot scales to watershed and regional scales
- development of easy-to-use decision support systems to facilitate incorporation of LID practices into new developments.

23.3 LID and Stormwater BMP Water Quality Improvements

Stormwater best management practices (BMPs), such as wet ponds, grass swales, infiltration basins, permeable pavements, bioretention systems, are widely implemented to manage the additional runoff generated by urbanization and to improve water quality. However, there is still uncertainty as to their effectives at achieving pollution reduction goals. Strecker et al. (2001) observed that "it is apparent that inconsistent study methods, lack of associated design information, and reporting protocols make wide-scale assessments difficult, if not impossible." The performance of an individual BMP system can be evaluated with some effort, but data have not been available on the general performance of systems.

A distinction is made between system "performance" and "effectiveness." The former refers to how well a BMP meets its intended water quality improvement goals for the water that actually enters the system, whereas the latter is a measure of how well the BMP meets its goal for all stormwater that reaches the BMP site, which includes bypass flow (Strecker et al. 2001). Performance and effectiveness can also be quantified in terms of how well the systems mitigate increased flows from urbanization (Strecker et al. 2001).

An additional parameter is efficiency, which is a measure of how well a BMP or BMP system removes pollutants (Strecker et al. 2001). Efficiency is often expressed as the percentage of a given pollutant that is removed by a BMP. The efficiency (percent removal) of a BMP with respect to a pollutant often depends on the concentration of the pollutant in the stormwater entering a BMP. Hence, removal percentages are not very useful for characterizing performance and effectiveness, unless looked at carefully (e.g., comparing data from only "dirty" sites; Strecker et al. 2001). Systems with relatively clean influent may have effluent with both low pollutant concentrations and low calculated percent removals.

As introduced by Strecker et al. (2001), the American Society of Civil Engineers (ASCE) and the U.S. Environmental Protection Agency (USEPA), under a cooperative agreement instituted in 1996, developed protocols for BMP monitoring and reporting, and an international BMP database. The International Stormwater Best Management Practices (BMP) Database project has a public-access website (www.bmpdatabase.org) that features a database of over 500 BMP studies, performance analysis results, tools for use in BMP performance studies, monitoring guidance, and other study-related publications. The overall purpose of the project is stated to provide scientifically sound information to improve the design, selection, and implementation of BMPs. In 2004, the project transitioned to a more broadly supported coalition of partners led by the Water Environment Research Foundation (WERF), including the Federal Highway Administration (FHWA), American Public Works Association (APWA), and the Environmental and Water Resources Institute (EWRI) of the ASCE (International Stormwater BMP Database n.d.).

A series of reports were published by the International Stormwater BMP Database that provide a statistical summary of the database. For example, a December 2014 report (International Stormwater BMP Database 2014) provides a statistical summary of influent and effluent solids, bacteria, nutrients, and metals data from eleven types of stormwater BMPs. The data have a great deal of scatter and are mixed as to whether statistically significant increases or decreases (or no change) occurred for given pollutants for given BMP categories. The overall data indicate that the BMPs tend to result in reductions in total suspended solids, bacteria, some metals (total cadmium, chromium, copper, iron, lead, nickel, and zinc) and total Kjeldahl nitrogen. Removals of total metals tends to be greater than dissolved metals reflecting the efficiency of the BMPs in removing particulate matter. Biofilter and bioretention systems had statistically significant increases in total phosphorous and orthophosphate. BMPS are not effective in removing salts.

Factors that affect the potential for a substance to cause groundwater contamination are (Clark et al. 2010):

- high mobility (low sorption potential in the vadose zone)
- high concentration in stormwater
- high solubility (as opposed to occurring in the particulate fraction).

Soil zone characteristics that impacts infiltration and pollutant transport through the vadose zone include (Clark et al. 2010):

- soil texture (porosity, pore size, permeability)
- soil organic content (higher contents favors sorption of pollutants)

- phosphorous content
- cation and anion exchange capacity
- depth of media and contact time
- pH
- ORP—anaerobic conditions can promote denitrification and the mobilization of metals
- pH-ORP—determines the speciation and thus mobility of metals
- sodium adsorption ratio of runoff (high SAR values promote clay dispersion and associated loss of permeability with montmorillonite, vermiculite, illite, and mica-derived clays being more sensitive to sodium than other clays).

Nieber et al. (2014) presented the results of a literature review and field investigation of potential impacts of stormwater infiltration at three infiltration sites in the Twin Cities Metropolitan Area, Minnesota. The sites consisted of an infiltration basin, a large rain garden, and an infiltration gallery in an industrial area. Contaminants of concern at the infiltration sites are chloride, nitrate, phosphorous, heavy metals, and petroleum hydrocarbons. Conclusions from the literature review include:

- surface-oriented practices have the capacity to capture some groundwater contaminants in the soil or infiltration media
- pathogens and suspended solids are generally filtered out by the soil and media through which they are infiltrated
- there is evidence of the biodegradation of petroleum hydrocarbons
- toxic metals will eventually have to be harvested, but the capacity of soils to retain toxic metals is substantial and hundreds of years may be required for saturation
- nitrate concentrations in stormwater are generally low and unlikely to cause high concentrations in drinking water
- subsurface infiltration systems have a greater potential for groundwater contamination with metals and petroleum hydrocarbons
- the potential for groundwater contamination with chloride exists in northern climates where salt is used for road deicing.

The field testing results include:

- 15% of surface water samples exceeded the USEPA chloride secondary drinking water maximum contaminant level (MCL) of 250 mg/L and no attenuation with depth occurred
- 4% of samples exceeded the USEPA nitrate primary drinking water MCL of 10 mg/L with no strong trend of attenuation with depth
- total phosphorous in the run off ranged between 5 and 1,400 μg/L
- lead exceeded the primary drinking water MCL (15 μg/L) in 8% of the samples
- copper, nickel, cadmium and zinc were detected but well below standards
- low metals and petroleum attenuation may occur in organic-poor soils and it was suggested that placing organic-rich soil around infiltration chambers may increase contaminant attenuation.

23.4 Low Impact Development Techniques Outline

LID is implemented on building and residential parcel, street, and neighborhood/development scales (Table 23.1). LID techniques associated with infiltration and thus aquifer recharge include (Hunt et al. 2010):

- infiltration basins (Chap. 15)
- vegetated swales
- infiltration wells and trenches (vadose zone recharge systems; Chap. 17)
- bioretention systems (including rain gardens and bioinfiltration systems)
- infiltration wetlands
- level spreaders and vegetated filter strips
- permeable pavements.

LID systems provide water quality improvement through filtration and biological treatment processes, including vegetation uptake and microbial processes (e.g., nitrification and denitrification). There is a large LID toolbox from which techniques can be selected as appropriate to meet site-specific objectives and physical and hydrogeological constraints.

Table 23.1 LID scales and techniques

Scale	LID techniques
Building	• Infiltration of roof stormwater runoff (above grade or below grade, e.g., dry wells)
Property/parcel	• Reduction of impervious area • Permeable pavements • Rain gardens/bioretention areas
Large commercial properties	• Infiltration areas • Pervious parking lots • Rain gardens/bioretention areas
Street level	• Cut curbs or elimination of curbs • Permeable pavement • Bioswales (linear bioretention facility) • Roadside infiltration trenches • Narrow streets
Development scale	• Clustered development and open space preservation • Treatment parks (constructed wetlands, infiltration ponds, rain gardens, filter strips) • Dry swales (grass swales) • Underground detention cells

Source UADOC (2010)

23.5 Infiltration Areas/Basins

Infiltration basins are a widely employed on-site stormwater recharge BMP on the development scale. The design of infiltration basins is addressed in Chap. 15. There has been relatively little investigation of the overall recharge benefits of infiltration basins systems. The retention and infiltration of water that would otherwise runoff would clearly be expected to enhance local groundwater recharge. However, there is little quantitative data on the effectiveness of infiltration basins towards maintaining or exceeding predevelopment recharge rates.

Miller (2006) documented an example of the water resources benefits of enhanced groundwater recharge from the capture of runoff. The study site is a 34 ha former mine site in New Mexico that was commercially developed. On-site runoff was routed to four earthen retention ponds. The site is located in a semiarid high-desert environment with an average annual precipitation of 330 mm/year and recharge rates estimated by previous investigators to be about only 1% of precipitation. The increased on-site recharge results in the developed of a perched groundwater mound. Current recharge was estimated using both a surface infiltration model (UNSAT-H), based on soil hydraulic properties and ET rates, and through calibration of a groundwater flow model using the MODFLOW-SURFACT code to be 40–60% of precipitation. This investigation demonstrates how the capture and retention of stormwater runoff from site development can substantially increase groundwater recharge rates.

23.6 Vegetated Swales

The USEPA (1999a) defined a vegetated swale as

> broad, shallow channel with a dense strand of vegetation covering the side slopes and bottom. Swales may be natural or manmade, and are designed to trap particulate pollutants (suspended solids and trace metals), promote infiltration, and reduce the flow velocity of storm water runoff.

Vegetated or grass swales (or channels) are shallow open channels that are used to treat and convey stormwater runoff. The base of the swale is located above the water table. A related feature is a wet swale, the base of which is at least seasonally below the water table. Swales, as linear features, are well suited to treat runoff from highways and roads, yards, small parking, areas and driveways.

Grass swales (Fig. 23.1a) are used primarily to reduce particulate pollutants by settling and filtration. Vegetative uptake and adsorption can also act to reduce dissolved pollutant concentrations. Particulate removal occurs when low flow velocities and shallow water depths allow for particle settling and the grass acts to filter runoff from the water-quality design storm. For larger storm events, the swale functions primarily as an alternative to traditional storm sewer systems. The runoff reduction performance of grass swales depends upon the permeability of the soil, which can be enhanced through the use of amendments (e.g., compost). As a conservative estimate,

Fig. 23.1 **a** Grassed dry swale, Sarasota, Florida. **b** Wet swale (treatment wetland), median of Daniels Parkway, Lee County, Florida

a properly designed vegetated swale may achieve a 25–50% reduction in particulate pollutants, including sediments and sediment-attached pollutants (USEPA 1999a). Swales are largely ineffective for removing dissolved pollutants.

Wet swales intersect the seasonal high water table (SHWT), which allows wetland vegetation to become established (Fig. 23.1b). Wetland swales can provide nutrient removal and resistance to flow.

Dry swales may completely infiltrate runoff from small storms. As storm intensity increases, a threshold is exceeded and attenuation of flow becomes greatly reduced. Swales must be sized to both convey the peak discharge of the design storm and have sufficient volume to retain and infiltrate flows from small storms. Check dams are commonly used to retain and slow the flow of stormwater (Fig. 23.2).

Basic vegetated swale design issues are (USEPA 1999a)

- A fine, close-growing water-resistant grass should be used because increasing the surface area of vegetation exposed to runoff increases its effectiveness.
- A trapezoidal cross-section with side slopes no steeper than 1:3 is recommended to maximize the wetted channel perimeter of the swale.

Fig. 23.2 Check dams in roadside dry swales. **a** Concrete check dams, U.S. 27 north of Avon Park, Florida. **b** Simple rock check dams, Gorges State Park, North Carolina

- Swales should be sized to treat the design flows and pass the peak hydraulic flows unless a bypass is provided.
- Check dams may be installed to promote additional infiltration, increase storage, and reduce flow velocities.

Where infiltration occurs, pollutant removal may occur by sorption reactions and biogeochemical processes. The water quality improvement function of vegetated swales is predicated upon their ability to maintain minimal flow velocities within the channel. Flow velocity within a channel is ultimately dictated by the channel cross-sectional geometry, roughness, longitudinal slope, and design discharge (VDOT 2013). Follows are some basic siting and design criteria for grass or vegetated swales (PGCDER 1999; Lowndes 2001; VDCR 2011; VDOT 2013):

- **Drainage area:** Vegetated swales have limited capacities constrained by target flow velocities. The suitable maximum drainage area is constrained by limits on the cross-sectional area of the channel. A maximum drainage area of 5 acres (2 ha) has been recommended (VDOT 2013).
- **Slope:** Grass swales function best when constructed with a slope as flat as practically possible (VDOT 2013). The longitudinal slope should not exceed 4%. Steeper slopes create runoff velocities that can cause erosion and do not have sufficient retention/contact time for pollutant removal. Where the topography is a relatively steep (greater than several degrees), the velocity of flow may be reduced through the construction of series of check dams across the swale.
- **Soils:** Soils should have sufficiently high infiltration rates so that swales completely drain within 72 h of a storm event to avoid undesirable marshy conditions and mosquito habitat (minimum 0.27 in./h, 0.69 cm/h; VDOT 2013). High infiltration rates (>5 in./h, 12/7 cm/h) provide little treatment capacity.
- **Depth to water table:** A minimum depth to the SHWT of 2 ft (0.6) below the proposed swale bottom was specified to allow for pollutant attenuation in the vadose zone (VDOT 2013).
- **Swale geometry:** It is essential to avoid concentration of flow within the channel. Trapezoidal cross-sections are preferred and parabolic and triangular geometries should be avoided.
- **Bottom width:** A range of 3–6 ft (0.9–1.8 m) was established to be acceptable (VDOT 2013). Other recommended ranges are 4–8 ft (1.2–2.4 m). Smaller widths are essentially non-constructible and greater widths result in channelization of flow. If the width is greater than recommended values, then check dams, level spreaders or other structures should be used to prevent braiding and erosion along the channel bottom (VDOT 2013).
- **Channel slide slopes:** Should be 3H: 1V or greater.
- **Channel depth:** The water quality (treatment) volume should flow at a height approximately equal to the grass height (usually 4 in., 10 cm). Overall depth should permit the conveyance of the design flood (e.g., ten-year flood event) while providing a minimum of six inches (12 cm) of freeboard.
- **Hydraulic capacity:** Should be sufficient to convey runoff from design storm events while not being erosive.
- **Flow velocity:** Should be as low as possible in order to achieve maximum pollutant removal. Swales should also be designed so that larger floods do not resuspend previously deposited sediments. For treatment purposes, the maximum flow velocity should not exceed 1 ft/s (0.3 m/s) for a 1-inch (2.5 cm) storm event. The VDOT (2013) recommended a maximum permissible velocity of 4 ft/s (1.2 m/s) for the two-year flow and 7 ft/s (2.1 m/s) for the ten-year flood.
- **Manning's "n" value:** Should be 0.2 for flow depths up to 4 in. (10 cm), decreasing to 0.03 at a depth of 12 in. (0.3 m; VDOT 2013).

- **Hydraulic residence time:** The time for water to flow the full length of the channel (or from every entry point) should be a minimum of 9 min.
- **Maximum drain time:** 24–72 h.
- **Grass:** Should be able to tolerate both wet and dry periods and have deep roots systems to resist scouring. Taller and denser grasses are preferred for particulate removal.

Various methods and software are available for the design of stormwater channels. Manning's equation is commonly used, which is an empirical equation in which the uniform flow rate in an open channel (Q) is a function of the channel flow (cross-sectional) area, hydraulic radius, slope, and roughness. Roughness is expressed using the empirical Manning's roughness coefficient (n). Manning's roughness coefficient can be estimated from the vegetation (grass) type, channel hydraulic radius (flow depth), and flow velocity using various published default value tables or iterative processes (VDOT 2013).

$$Q = \frac{1.00}{n}(AR^{2/3}S^{1/2}) \quad \text{(S.I units)} \tag{23.1}$$

$$Q = \frac{1.49}{n}(AR^{2/3}S^{1/2}) \quad \text{(U.S. Customary units)} \tag{23.2}$$

where

Q flow rate (m^3/s; ft^3/s)
n Manning's roughness coefficient, which depends upon grass height and thus moving program
A cross-sectional area of flow (m^2, ft^2)
R hydraulic radius (m, ft; cross-sectional area of flow/wetted perimeter)
S channel slope (m/m, ft/ft)

Limitations and potential negative impacts of swales include (USEPA 1999a):

- they are impractical in areas with very flat grades, steep slopes, or wet or poorly drained soils
- they are not effective and may even be eroded at high flow volumes and rates
- they can become drowning hazards, mosquito breeding areas, and emit odors
- suitable land may not be available
- infiltration may carry pollutants into the groundwater
- leaching from vegetation may increase the concentrations of nutrients and trace metals in the runoff.

Wet swales (linear treatment wetlands; Fig. 23.1b) are designed so that their bottoms are located below the water table, as least for part of the year, to facilitate the growth of wetland vegetation. Wet swales use retention time and vegetative growth to treat stormwater prior to discharge to a downstream surface water body. Wetlands are commonly discharge rather than recharge features. Infiltration wetlands are constructed in areas with permeable soils with high water tables. Infiltration (and thus recharge) occurs because water is ponded above the water table.

23.7 Infiltration Trenches and Wells

The USEPA defines a well as a bored, drilled, or driven shaft whose depth is greater than the largest surface dimension, or a dug hole whose depth is greater than the largest surface dimension. A trench is a linear excavation that is generally deeper than it is wide, and narrow compared with its length. Infiltration trenches and infiltration wells (also referred to as vadose or dry wells) discharge into the vadose zone. Infiltration wells and trenches, and other vadose zone recharge techniques, are discussed in Chap. 17. Infiltration trenches and wells may be constructed of a variety of materials and may be finished open or filled with gravel, aggregate, or other coarse material.

The major advantages of infiltration trenches and wells are that they have small surface footprints, low costs, and may, if site hydrogeological conditions are favorable, allow for recharge where surface-spreading infiltration methods are not feasible because surficial sediments have a low permeability.

Infiltration wells and trenches are used to capture and treat small amounts of runoff (first flush) but do not have capacity to control peak hydraulic flows (USEPA 1999b). Instead, infiltration trenches are used in conjunction with detention ponds to provide both water quality and peak flow control. The principle operational challenge impacting the operation of infiltration trenches is management of clogging and the potential for groundwater contamination. Runoff that contains high levels of sediments and hydrocarbons (oil and grease) that may clog a trench or well are often pretreated using, for example, grit chambers, sediment traps, swales, and vegetated filter strips. Trenches and wells also have the disadvantage that they cannot be rehabilitated by pumping as is commonly performed to rehabilitate phreatic wells (i.e., wells completed below the water table).

23.8 Bioretention Systems

23.8.1 Bioretention Basics: Definition, Benefits, and Limitations

Bioretention systems were developed in the early 1990s in Prince George's County in Maryland (PGCDER 1993). Bioretention was defined by the Prince George's County Department of Environmental Regulation (PGGDER 2007) as

> a terrestrial-based (upland as opposed to wetland), water quality and water quantity control practice using the chemical, biological, and physical properties of plants, microbes, and soils for removal of pollutants from stormwater runoff.

Designs and definitions of bioretention systems have evolved from the original Prince George's County application. Bioretention systems can be loosely defined as vegetated retention (infiltration) basins or may be more specifically defined as including

Fig. 23.3 Bioretention pond with irises in bloom, Fort Carson, Colorado. *Source* USEPA, https://www.epa.gov/region8/green-infrastructure

additional elements, particularly modifications of the soil and an underdrain system to improve contaminant removal and increase infiltration rates. The question arises as to the difference between a bioretention system and an infiltration basin that is covered with grass and other vegetation. A strict definition of bioretention systems would hold that they involve an engineered ecosystem designed for water quality improvement.

For example, the Atlanta Regional Commission (2001) defined a bioretention area as a "shallow stormwater basin or landscaped area that utilizes engineered soils and vegetation to capture and treat runoff". The State of Washington Department of Ecology (WADOE 2013) noted that the key features of bioretention facilities are that they:

- are engineered facilities
- are shallow landscaped depressions with a designed soil mix and plants adapted to the local climate and soil moisture conditions
- receive stormwater from small contributing areas
- have healthy soil structures and vegetation to improve the infiltration, storage, filtration, and slow release of stormwater flows
- may have underdrains.

Underdrain systems are used where soils have low infiltration rates and the water table is shallow. Where the soil drainage rate is high and the depth of groundwater is great, an underdrain system is not needed.

Bioretention systems utilize multiple processes, such as sedimentation, filtration, sorption, plant uptake, microbial activities, and volatilization, to improve the quality of infiltrated water. Bioretention facilities, as now commonly defined, are distinct from situations where stormwater is allowed to flow into natural low-lying vegetated areas. Bioretention basins are also referred to as "rain gardens" and "bioinfiltration systems," with the former term tending to be employed for smaller systems (e.g., systems constructed on residential lots). The term "rain garden" is alternatively used in some areas to describe a non-engineered landscape depression intended to capture stormwater from adjacent areas (WADOE 2013).

Bioretention systems are typically shallow depressions located in upland areas (as opposed to being located in wetlands) that are designed to retain runoff water in shallow (6–12-in.; 15–30-cm deep) ponds in which infiltration quickly occurs into the underlying prepared soil. Bioretention swales incorporate the same design features as bioretention basins or cells but are also designed to be part of a stormwater conveyance system when the maximum ponding depth is exceeded (WADOE 2013). Bioretention systems are designed to provide storm-water retention and pollutant removal, specifically treatment of the first-flush runoff from urban areas (Roy-Poirier et al. 2010).

Three basic types of bioretention systems vary depending on whether an underdrain system is employed (PGCDER 2007):

- **Infiltration/recharge facilities:** do not have an underdrain system and require that the underlying soils by sufficiently permeable (infiltration rates >0.5 in./h, 1.3 cm/h) to allow the system to quickly drain.
- **Filtration/partial recharge facilities:** have an underdrain system to increase the infiltration rate with collected water conveyed under gravity to other stormwater management elements.
- **Filtration only facilities:** Employ an underdrain and impermeable liner to reduce or eliminate the potential for groundwater contamination at "hot spots" (e.g., industrial sites).

Filtration/partial recharge systems can be designed and operated to provide pre-treatment for an MAR system. Bioretention systems are located to intercept runoff close to the source, thereby reducing the amount (and associated costs) of stormwater drainage infrastructure (PGCDER 2007). When properly located and designed, bioretention systems can mimic the preexisting hydrological conditions of sites by retaining and treating the additional runoff generated by increases in impervious area. As small-scale features, they have great flexibility for integration into the landscape and available space of sites.

Bioretention systems not only provide for water quantity and quality control, but add the many values of landscape diversity to developments, including provision of wildlife habitats and aesthetic benefits (Fig. 23.3; PGCDER 2007). Similarly, rain gardens constructed on residential lots can also be aesthetic features. Bioretention systems have the attraction that they can be integrated into urban neighborhoods and require little maintenance if properly designed and constructed (PGCDER 2007; Roy-Poirier et al. 2010). The original systems in Maryland were designed based on an upland terrestrial forest ecosystem, but the design concept has been expanded to include other design themes, such as a meadow and ornamental garden (PGCDER 2007). Other ecosystems are replicated in regions with different climatic conditions. Bioretention systems can also can serve to replace traditional high-water use landscaping with a no-water irrigation (Houdeshel et al. 2012).

Limitations and constraints of bioretention systems include (WADOE 2013):

- individual cells or basins should not be used to treat large drainage areas (i.e., it is a decentralized technology)
- systems could impact present and future site activities
- underdrain discharge may not be suitable for phosphorous-sensitive receiving water bodies
- systems may not be feasible where erosion, slope failure, or down-gradient flooding is a concern
- systems may not be feasible where rising groundwater levels are a concern (e.g., utilities and basement might be impacted)
- they may not be suitable in areas with known soil and groundwater contamination
- they may not be feasible in areas with thin or low permeability soils
- local jurisdictions may have mandatory or recommended setback requirements from buildings, drinking water wells, septic drainfields, and other facilities that could be impacted.

23.8.2 Bioretention System Design

The type bioretention system (Fig. 23.4) was developed in Prince George's County, Maryland (PGCDER 1993). The essential features are that water is conveyed as sheet flow to the treatment area, which consists of a grass buffer strip, sand bed, and ponding area (USEPA 1999c). The ponding area has an organic or mulch layer, planting soil, and plants. The area is graded with a depressed center and ponding depth of 15 cm (6 in.). Greater ponding depths can result in prolonged standing water and associated nuisances, such as mosquito breeding. Excess water is diverted away from the bioretention area. Design modifications include an underdrain system to enhance infiltration and to avoid prolonged periods of standing water.

Each component of a bioretention system serves a water treatment function (PGCDER 1993, 2007; USEPA 1999c):

- **Grass buffer:** reduces flow velocity and filters particulates
- **Sand bed:** reduces flow velocity, filters particulates, and spreads flow over the length of the bioretention system
- **Ponding area:** provides temporary storage
- **Organic or mulch layer:** filters particulates and provides an environment conducive for the growth of microorganisms and biodegradation of organic substances
- **Planting soil (filter medium):** provides storage, supports plant growth, and provides a suitable environment for the adsorption and biodegradation of contaminants.

The USEPA (1999c) recommended that the planting soil should be a sandy loam, loam sand, or loam with a clay content of 10–25%, an organic content of 1.5–3%, and an infiltration rate greater than 1.25 cm (0.5 in.) per hour. Too low of an infiltration

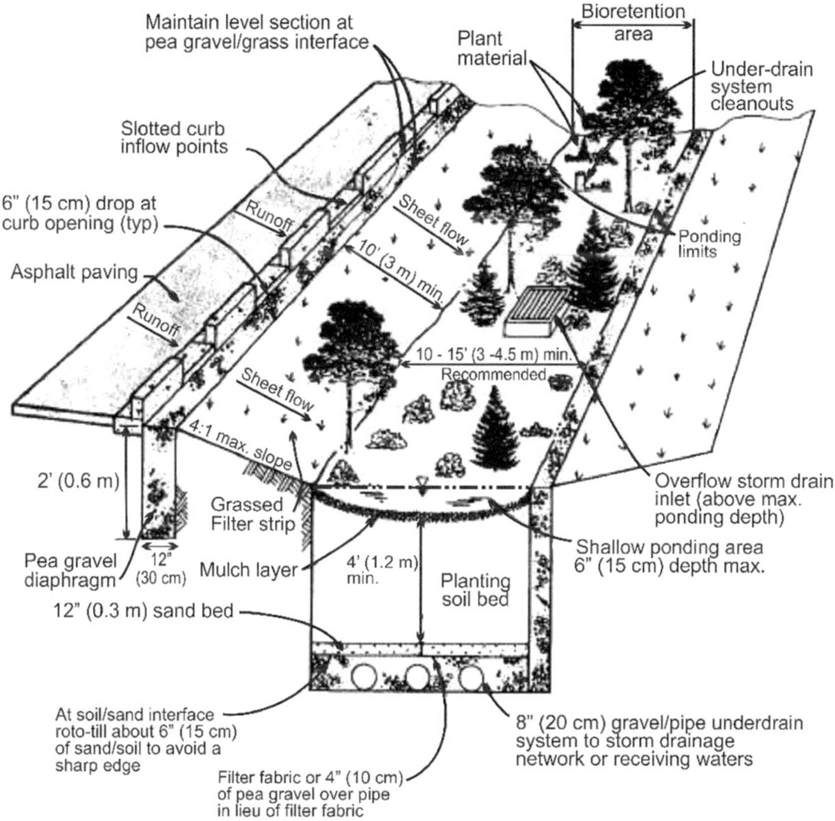

Fig. 23.4 Diagram of a bioretention system with an underdrain (Modified from PGCDER 1993)

rate may result in ponding. If the infiltration rate is too great, then water may flow through the system too quickly for adequate removal of contaminants.

The PGCDER (2007) recommends that the planting medium (bioretention soil) should be highly porous and permeable to allow for infiltration and provide for adsorption of organic nitrogen and phosphorous. The mix recommended by the PGCDER (2007) is 50–60% sand, 20–30% leaf compost, and 20–30% top soil. The bioretention soil media (BSM) should be selected or designed to (WADOE 2013):

- provide high enough infiltration rates to achieve desired surface water drawdown and system dewatering
- have infiltration rates that are not too high so as to optimize pollutant removal
- support long-term plant growth and soil health
- reduce or eliminate nutrient export during storm events
- optimize pollutant removal.

System design can enhance nutrient removal. Nitrification in the oxic soil environment can increase the concentration of nitrate. A solution for preventing excessive nitrate concentrations in systems with drains is to elevate the drain so that temporary anaerobic conditions develop below the drain for denitrification. Phosphorous removal can be promoted by increasing the Al, Fe, and Ca content of the soil and the thickness of the bioretention soil medium to 24–36 in. (0.6–0.9 m) to increase contact times with Al, Fe, and Ca components (WADOE 2013). Phosphorus removal can also be enhanced by maintaining healthy plant and soil microbial communities that are capable of rapid phosphorus uptake (WADOE 2013).

A key design parameter is the infiltration rate of the soils underlying the bioretention area. Some fines are needed for adsorption of contaminants, but too much fines can result in low infiltration rates. Measured infiltration rates may require a correction factor to account for site variability. It is recognized that professional judgement is required as to whether testing results are representative and, if not, the size of the correction factor that needs to be applied.

The mulch layer prevents erosion and protects the soil from excessive drying. PGCDER (2007) recommends 2–3 in. (5–8 cm) of fine shredded hardwood mulch or hardwood chips. A dense herbaceous ground cover (70–80% coverage) may be used instead of a mulch cover (PGCDER 2007). Vegetation should consist of native vegetation that resembles a multi-story terrestrial forest community ecosystem (USEPA 1999c). Plants should be selected that require minimum maintenance in the environment of the bioretention device. The vegetation should be able to tolerate a variable soil moisture regime and ponding fluctuations, and ecologically and aesthetically blend facilities into the environment (PGCDER 2007).

Bioretention systems are a recognized stormwater treatment technology and in the United States, regulatory requirements (construction standards) and guidance for their design and operation are included in state and local stormwater manuals. State and local construction standards and guidelines are commonly complete or partial adoptions of the USEPA guidelines or guidelines developed in other states (e.g., Prince George's County, Maryland). Design standards and guidance may reflect local climatic and hydrogeological conditions. While regulatory standards and guidelines for the most part are reasonable and serve their purpose of achieving more successful implementation of the technology, it must be stressed that there is often not a technical justification for why specific design criteria are considered optimal. Follows are some design recommendations or requirements for bioretention system from some American state stormwater manuals and technical publications, which were selected for their level detail and to illustrate some of the variation in approaches.

The Atlanta Regional Commission (2001) Georgia Stormwater Manual recommends that bioretention systems have

- approximately 5% of tributary impervious area
- no more than a 6% slope
- a minimum 2 ft (0.6 m) separation from the seasonal high water table.

Design parameters include:

- soil filter bed drain time of 48 h and a hydraulic conductivity of at least 9.5 ft/d (2.9 m/d)
- maximum ponding depth of 6 in. (0.15 m)
- planting soil depth of at least 4 ft (1.2 m) with 10–25% clay content, 1–3% organic matter, and <500 ppm solute salts)
- mulch layer thickness of 2–4 in. (5–10 cm)
- underlying sand layer approximately 12–18 in. (0.30–0.45 m) thick
- permeable filter fabric between the gravel layer (for underdrain) and planting soil
- outlet from underdrain
- emergency spillway (overflow) to safely convey flows that exceed the capacity of the facility
- vegetation should resemble a terrestrial forest ecosystem and preferably use native plants selected on the basis of climate and hydric tolerance
- pretreatment consisting of a grass filter strip below a level spreader.

The required area for bioretention facilities can be calculated using the equation (Atlanta Regional Commission 2001; PGCDER 2007):

$$A_f = \frac{V \cdot d_f}{K(h_f + d_f)t_f} \tag{23.3}$$

where

A_f area of ponding area (ft^2 or m^2)
V volume to be captured (ft^3 or m^3)
d_f filter bed depth (minimum 4 ft, 1.2 m)
K vertical hydraulic conductivity of filter bed (0.5 ft/d, 0.15 m/d for silt loam)
h_f average height of water above filter bed (ft; typically 0.25 ft inches or 7.6 cm; ½ of maximum ponding depth)
t_f filter bed drain time (days, 2 day max)

Wisconsin bioretention conservation practice standard (Wisconsin DNR 2014) includes the requirements:

- maximum ponding depth shall not exceed 12 in. (0.3 m)
- maximum drawdown time of ponded water shall not exceed 24 h
- a recommended soil mixture of 70–85% sand and 15–30% compost
- the engineered soil should be sufficient to support the rooting depth of the vegetation (minimum of 2 ft, 0.6 m)
- an underdrain with an underlying storage layer of sand and gravel is required unless the design infiltration rate of the native soil is determined to be at least 3.6 in./h (9.1 cm/h).
- the storage layer of sand and gravel shall be designed to achieve a total device drain time of 72 h
- an interface layer of 2–4 in. (5–10 cm) of sand is required above the native soil if the soil infiltration rates is less than 3.6 in./h (9.1 cm/h).

The State of Washington Department of Ecology (WADOE 2013) design recommendations address the minimum separation of systems from an underlying hydraulic restriction layer. The recommended minimum separation is 1 ft (0.3 m) where the contributing area is less than 5,000 ft^2 (465 m^2) of pollutant generating impervious surfaces or 10,000 ft^2 (929 m^3) of impervious surfaces. A minimum separation of 3 ft (0.9 m) is recommended when the pollutant-generating impervious surface is $\geq$5,000 ft^2 or total impervious surfaces are $\geq$10,000 ft^2.

Austin (2012) proposed the following design criteria for bioretention systems:

- The system area is typically 5–8% of catchment area, and varies depending on rainfall characteristics, catchment imperviousness, and stormwater management goals.
- Care should be taken to avoid compaction of basin floor. A maximum width of 25 ft (7.6 m) is suggested based on the ability to excavate a basin with heavy equipment located outside of the basin.
- The portion of a basin above land surface should be able to temporarily pond 70–75% of the design storm.
- The system surface should be covered with 3–4-inches (8–10 cm) of wood chips or mulch, which facilities microbial decomposition of some organic compounds (hydrocarbons).
- Basins should contain 24–48 in. (0.3–0.6 m) of sandy filter media (sandy loam). The filter media should consist of sand with variable amounts of silt, clay, and organic matter. High sand contents ($\geq$80%) may facilitate infiltration, but Austin (2012) observed that a sandy clay loam (54% sand, 26% silt, 20% clay) with 12.2% organic matter had better pollutant removal than a more sand-rich sandy loam soil.
- An underdrain layer of coarse gravel may improve infiltration.
- The subsoil (below filter media) should have an infiltration rate of 0.5 in./h (1.3 cm/h) or greater, otherwise an underdrain is used.
- An underdrain is used when the water table is seasonally within 3 ft (0.9 m) of the bioretention basin bottom.
- For nitrogen removal, saturated conditions should occur within system so that anaerobic conditions needed for denitrification can develop.
- A horizontal subsurface bed of soil carbon (wood chips) can be effective in increasing nitrate removal.
- Pretreatment is recommended using sedimentation basins, swales, or tanks.

23.8.3 Bioretention System Design in Arid Climates

States with xeric climates have the fastest population growths in the United States, inherently have limited water resources, and have native ecosystems that are less resilient to anthropogenic influences (Houdeshel et al. 2012). Bioretention system designs for wetter climates are therefore not appropriate for cold and warm desert regions (Houdeshel et al. 2012). As reviewed by Houdeshel et al. (2012), native plants in xeric climates are adapted to surviving in low-water environments. Plants employ two strategies for coping with water scarcity. Phreatophytes have large, deep (in some instances greater than 30 m) taproots to access groundwater at depth. Other plants utilize extensive wide, spreading shallow root networks to quickly capture and utilize small precipitation events. Houdeshel et al. (2012) recommended for bioretention systems in xeric climates, based on physiological traits and differences in water use, that a mixture of shallow-rooted cord grass and deeper-rooted shrubs will maximize the function and treatment benefits of the plant community. The plant species should be selected based on local and regional desert conditions.

Houdeshel et al. (2012) recommended an alternative bioretention system design in which the mulch layer is replaced by a 3–10 cm layer of decorative light-colored cobble or gravel. In arid (xeric) regions, mulch is rarely used as it becomes sun-faded, requires frequent maintenance and replacement, and dry conditions do not provide an environment that promotes decomposition (Houdeshel et al. 2012). The gravel layer is underlain by a 0.5 m-thick top soil layer and then a 0.6 m low-density media (gravel) layer, instead of a filtration layer. The gravel layer provides storage and allows for the slow infiltration of water into the underlying soil from which it can be accessed by selected plants that can root through the gravel layer. An additional modification is a gravel forebay at the entry point to facilitate downward flow into the gravel storage layer (Houdeshel and Pomeroy 2014).

Houdeshel and Pomeroy (2014) investigated at a test facility on the campus of the University of Utah, Salt Lake City, whether non-irrigated bioretention is feasible in cold arid and semiarid regions of the western United States (e.g., Salt Lake City, Utah; Boise, Idaho; Denver, Colorado). The concentration of run-off from impervious areas increases the total volume of water available to plants compared to natural environments, but the vegetation still experience prolonged periods of hot and dry conditions with no precipitation. The results of the investigation suggest that once the plants become established, native vegetation should be able to sustain itself under natural precipitation regimes without irrigation. An important implication is that correctly designed bioretention systems can be a zero-irrigation landscaping alternative with associated savings of scarce water.

23.8.4 System Construction

Brown and Hunt (2010) compared the infiltration rates of excavations for bioretention cells that were performed using the conventional "scoop" technique and the rake method. The former method smears the underlying soil surface, whereas the latter uses the teeth of the excavator's bucket to scarify the underlying soil surface. Infiltration rates were performed using the standard double-ring infiltrometer method before the excavation was backfilled with coarse aggregate. The different excavation techniques were applied to the bottom 30 cm (1 ft) of the excavation. The raking technique tended to yield more permeable and less compacted soil than the scoop method, but the difference was statistically significantly only for wet soil conditions.

Table 23.2 Summary of reported pollutant removal performances of bioretention systems

Parameter	% removal
Total suspended solids	97
Total phosphorous	35–65
Total nitrogen	33–66
Copper	36–93
Lead	24–99
Zinc	31–99
Oil and grease	99
Bacteria	70

Source PGCDER (2007)

The authors concluded that the use of the rake method under dry soil conditions is expected to increase long-term exfiltration rates from bioretention cells. For pure sand environments, with extremely high infiltration rates and hydraulic conductivities, excavation may be performed under either wet or dry conditions. However, for clay and loamy sand, excavation under dry soil conditions was recommended. The rake technique was recommended for other infiltration BMPS (e.g., trenches; Brown and Hunt 2010).

An important construction consideration for surface-spreading, bioretention, and other infiltration systems is avoiding compaction during construction, which can reduce porosity and permeability. Heavy equipment should not be driven on infiltration surfaces.

23.8.5 Bioretention System Performance

Although bioretention systems have relatively low maintenance requirements, successful performance depends on proper maintenance (PGCDER 2007). A main cause of failure is clogging and sedimentation. Vegetation also requires some maintenance including weeding and removal of exotic and invasive species. A number of papers have been published documenting laboratory and field studies of the performance of bioretention systems in terms of nutrient and pollutant removal, including several review papers. The then current state of knowledge as far as pollutant removal was summarized by the PGCDER (2007; Table 23.2).

Hunt and Lord (2006) examined the performance of bioretention systems at four locations in North Carolina. Total nitrogen removal ranged from 33 to 68%. Total phosphorous removal was related to the phosphorous content of the soil (P-index). The total phosphorus concentration in the effluent from a system with a high soil phosphorous concentration increased, whereas total phosphorous removals of up to 68% occurred in systems using soils with low phosphorous contents. The bioretention systems were found to be effective is removing particulate pollutants and metals.

Based on the experiences at the studied sites and studies by others, Hunt and Lord (2006) made the following recommendations:

- Phosphorous removal is enhanced with proper fill soil selection. Using low phosphorous (P-index) soils reduces phosphorus loads in the effluent.
- Nitrogen removal can be increased by retaining water in the bioretention cell. This can be achieved by an "interval water storage" (IWS) system in which the outlet pipe is higher than the drainage pipe and water has to flow "uphill". The top of the IWS system should be at least 18 in. (0.45 m) below the surface of the bioretention system.
- Recommended soil media recipe is 85–88% sand, 8–12% fines, 3–5% organics.
- Disturbed areas that are a source of sediments should be avoided.
- Pretreatment is recommended to avoid clogging (e.g., gravel verge, grass swale, forebay, small stilling basin).
- Maintenance required includes mowing and pruning of vegetation, and watering and spot fertilization during first year (and watering during droughts).
- Either a permeable fabric or transition layers may be used to prevent fines from the soil entering the underlying gravel.

Davis et al. (2009) and Roy-Poirier et al. (2010) reviewed the contaminant removal performance of bioretention systems. Bioretention systems tend to be efficient in the removal of oil and grease, most heavy metals, coliform bacteria and other pathogens. Removal of phosphate and nitrogen is more variable. Nitrogen removal depends on variations in redox (aerobic and anaerobic) conditions related to alternating wetting and drying. Aerobic conditions during dry periods can result in nitrification with associated produced nitrate accumulating in the soils and subject to release during subsequent storm events. Saturated conditions can result in anaerobic conditions conducive for denitrification and the removal of nitrate from infiltrated water. Phosphorus removal appears to be due mainly to sorption and is thus impacted by the phosphorous content of the soil. Soils with high phosphorus concentrations have low adsorptive capacities. Phosphorous in the soil may be mobilized and increase concentrations in the effluent.

Different states have varying design requirements or guidance for bioretention systems to achieve stormwater management and water quality improvement objectives. These objectives may be conflict. Higher infiltration rates are desirable for stormwater management but may be suboptimal for water quality improvements (Roy-Poirier et al. 2010).

23.9 Rain Gardens

Rain gardens are shallow planted depressions designed to capture and adsorb rainwater that runs off adjacent lawns and impervious areas (Fig. 23.5). Soil may be amended with sand and compost to enhance infiltration. Bioswales are swales planted with grasses or native vegetation that are designed infiltrate and filter water. Rain

Fig. 23.5 Rain garden constructed in a residential drainage swale, Lee County, Florida

gardens are intended to provide both stormwater treatment and landscaping. Some basic design principles for rain gardens are (Wisconsin DNR 2003; NRCS 2005):

- gardens should be located in low lying areas where stormwater will be intercepted
- soils should have adequate infiltration rates ($\geq$0.5 in./h; 1.3 cm/h)
- areas are typically 5–10% of the impervious area
- selected plants should be able to tolerate occasional standing water and the garden designed to be an attractive addition to a property and neighborhood
- impounded water should infiltrate within 12–24 h
- gardens should have sufficient volume to impound design rainfall events (e.g., 1 in; 2.5 cm)
- depths are typically 4–8 in. (10–20 cm)
- base should be level to avoid ponding of water on one side
- a downslope berm may be needed to retain stormwater
- gardens should be located away ($\geq$10 ft, 3 m) from buildings to avoid seepage into foundations
- gardens should not be constructed over septic systems.

23.10 Level Spreader and Vegetated Filter Strips

Level spreader and vegetated filter strip (LS-VFS) systems are constructed on sloped upland areas and are used to reduce stormwater flow, removal pollutants, and provide a diffuse (sheet) flow into riparian zones. Infiltration contributes to baseflow in nearby streams and rivers. LS-VFS systems have three main components (Hathaway and Hunt 2006; Winston and Hunt 2010; Knight et al. 2013; Fig. 23.6):

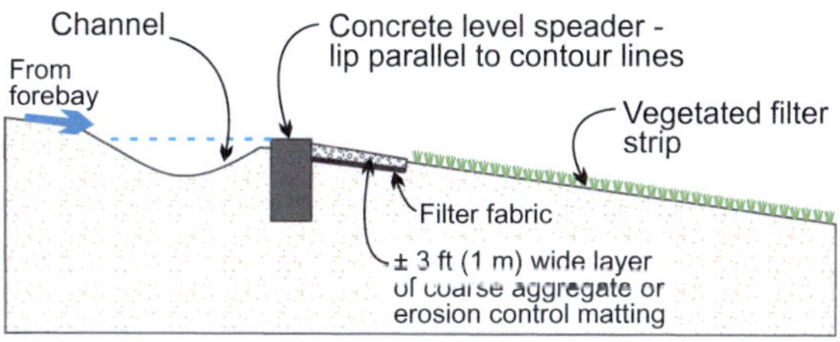

Fig. 23.6 Conceptual cross-section of level spreader and filter strip system

- **Forebay:** a depression, which may be lined with riprap, that acts to reduce runoff velocity and provide initial treatment of stormwater by allowing heavy sediment and debris to settle
- **Channel or swale:** a low dead-ended impoundment that fills to the level of the lower, downslope side on which a level spreader lip of uniform elevation is constructed of erosion resistant material
- **Vegetated filter strip (or riparian buffer):** a slope that is vegetated with trees, shrubs, or grasses.

The level spreader disperses stormwater along a topographic contour to reduce the potential for erosion and channeling of flow. The downslope vegetated filter strip provides a stable area for infiltration, sedimentation, and filtration to occur. Pollutants are also removed by biological and chemical processes within plants and soils. The main goal of filter strips is to allow water to remain longer in sheet flow than would occur in a naturally occurring landscape. Filter strips can be amended with compose or other materials to make the systems more permeable and enhance pollutant removal.

Design recommendations for LS-VFS systems include (Hathaway and Hunt 2006; Hunt et al. 2010; Winston et al. 2010; Winston and Hunt 2010):

- The level spreader lip should be constructed of concrete or metal and extend 3–6 in. above grade on the downslope side. Level spreaders with lips of earth or gravel (or both) eventually fail.
- Systems should be situated in areas away from natural swales, depressions, and mounds that could interfere with the diffuse (sheet) downslope flow of water.
- The best performing LS-VFSs systems have low slopes, dense vegetation, and small drainage areas.
- Erosion immediately downslope of the level spreader should be prevented by an erosion resistant zone, which could be constructed using erosion control matting or coarse aggregate placed atop a filter fabric.
- For a thick ground cover (e.g., grass), at least 10 ft (3 m) of level spreader should be provided for every cubic foot per second (cfs; 28.3 L/s) of flow. For a wooded (forested) filter zone, 50 ft (15 m) of level spreader per cfs should be provided.
- A flow-splitting device and bypass swale should be provided to handle runoff from rainfalls with intensities of greater than 1 in./h (2.5 cm/h).
- The maximum slope for the first 10 ft (3 m) of the VFSs should not exceed 4% and the overall slope for grass VFS should not exceed 8%.

- Periodic maintenance should be performed, which should include removal of debris and accumulated sediments in the forebay and spreader channel/swale, removal of trees and shrubs that grew on the level spreader lip or immediately down gradient, and mowing of the grass.
- The VFS soil may be amended with coarse-grained material (to increase infiltration rates) and nutrient (phosphate) adsorbing minerals.

The main advantages of LS-VFS systems are that they (Knight et al. 2013):

- are low cost to build and maintain
- are typically well received by landowners
- function in areas with a shallow water table.

Data from six LS-VFS systems at research sites in North Carolina indicated runoff reductions of 28–92% (Winston and Hunt 2010). Water quality data from LS-VFS research sites in North Carolina and Virginia indicate high reductions in total suspended solids concentrations (51–84%). Total nitrogen and phosphorus concentrations reductions were more variable with some systems showing increases in concentrations. However, total load reductions ranged from 49 to 62% for nitrogen and 32–48% for phosphorous.

Knight et al. (2013) evaluated the hydrological impacts and pollutant removal at four LS-VFSs and a grass swale at an experimental site in the Coastal Plain of North Carolina. Stormwater runoff filled a riprap lined forebay before being pumped off to the LS-VFSs and swale. The experimental system consisted for four pairs of VFSs, each with small (8 × 6 m) and large (20 × 6 m) VFS. One VSF of each size was amended with ViroPhos (70:30 sand:Virophos), a phosphorous sorptive aggregate manufactured by EnviRemed LLC. A down-gradient collection trough allowed for measurement of runoff volume from each VFS and sample collection.

The average reduction in runoff volume from the VFSs was reported to be 36–59%, compared to a reduction of 23% from the swale. There was no significant difference in runoff reduction between the amended and non-amended VFSs. Concentrations of total nitrogen (TN) were not significantly reduced when initial concentrations were less than 1 mg/L. All treatment systems marginally reduced the TN concentration of the discharge water, but the reductions were only statistically significant for the amended VFSs and swale. Total phosphorous concentration increased in the non-amended VFSs, small amended VFSs and the grass swale, with orthophosphate being the main contributor to the increase. The increase in total phosphorous concentration was attributed to solubilization of phosphorous in the soil or the decay of vegetation. A reduction in total phosphorous load (which considers the reduction in runoff volume) was achieved in all treatment systems except for the small non-amended VFS. The systems achieved average reductions in total nitrogen load of 38–69%.

23.11 Permeable Pavements

23.11.1 Introduction

Permeable pavements provide on-site water treatment and stormwater retention benefits. The pavements can be designed so that water is stored underground (harvested for reuse) or allowed to infiltrate. Storage occurs in a basal coarse aggregate layer. A great amount of literature on permeable pavements is available on line from government stormwater management and highway agencies, conference proceedings papers, and contractor literature. Ferguson (2005) identified the following permeable pavement types (alternative terms are provided):

- previous aggregate
- previous turf
- soft pervious surfacing (e.g., wood mulch and crushed shell)
- decks (elevated wooden structures)
- plastic geo-cells (plastic reinforced grid pavers) filled with sand/aggregate or turf
- open-jointed pavement blocks (permeable interlocking concrete pavements)
- open-celled paving grids (concrete grid pavers/pavements)
- pervious asphalt
- pervious concrete

A more general classification of permeable pavement types is:

- pervious asphalt
- pervious Portland cement concrete
- permeable interlocking concrete pavements (PICP)
- grid systems made of concrete or plastic with openings that are filled with either top soil and grass or permeable aggregate.

Pervious asphalt and concrete are similar in that fines are reduced or eliminated in the aggregate. The asphalt and concrete binds the aggregate at points of contact, retaining an open void network through which water can flow. Additional permeable pavement types are being developed. For example, KBI Flexi-Pave system (manufactured by K.B. Industries Inc.) uses rubber and stone granules and a proprietary epoxy binding agent (Fig. 23.7).

The main benefits of permeable pavement systems are that they can reduce runoff and increase local infiltration. Permeable pavements can thus reduce the need for stormwater management infrastructure. Developers could meet regulatory stormwater retention requirements while avoiding the land requirements (and associated costs) of infiltration/retention basins (Ferguson 2010). Permeable pavements also provide water treatment benefits. The main disadvantages of permeable pavements are (USEPA 1999d):

- many pavement engineers and contractor lack expertise with the technology
- pavements tend to become clogged if improperly installed or maintained

Fig. 23.7 Epoxy-bounded stone and rubber granules (KBI Flexi-Pave) used for a trail and parking lot, Wild Turkey Strand Preserve, Lee County, Florida

- reported high rates of failure
- risk of contamination of groundwater
- some building codes may not allow for their installation
- anaerobic conditions may develop in underlying soils, impeding microbiological decomposition of organic compounds, if the soils are unable to dry out between storms.

High failure rates reported by the USEPA (1999d) have been attributed to poor design, inadequate construction techniques, soils with very low permeability, heavy vehicular traffic, and resurfacing with nonporous pavement materials. Permeable pavements generally have low maintenance requirements, but some periodic maintenance may still be required to remediate clogging.

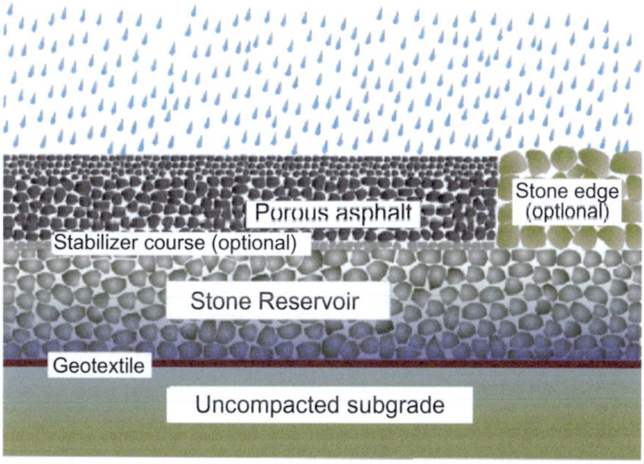

Fig. 23.8 Typical pervious asphalt pavement with stone reservoir. *Source* FHWA (2015a)

A common denominator for the installation of permeable pavement types, in general, is that their installation differs from that of typical pavements and it is important to involve a qualified professional for the selection of system type and design, and to contract installers with proper expertise and experience with the type of system to be installed. Installation should comply with applicable material specifications, national standards, and local regulatory requirements.

Permeable pavements experience losses of permeability from the entrainment of mineral and organic fines into the pores of pervious concrete and asphalt, and into aggregates used to fill joints and cells of PICP and grid pavement. Fines are derived from particulate emissions, traffic-caused abrasion, and organic substances and sediment from surrounding areas (Borgwardt 2006). Performance (infiltration rates) is highly affected by the age of the pavement.

23.11.2 Pervious (Porous) Asphalt

Pervious (porous) asphalt differs from conventional asphalt in that fines are reduced or eliminated in the aggregate. The asphalt binds the aggregate particles at their points of contact, retaining an open void (pore) network through which water can flow (FHWA 2015a). Previous asphalt pavements are typically built over an uncompacted subgrade to maximize infiltration through the soil (Fig. 23.8). The subgrade is overlain by a stone reservoir that serves both as a structural layer and for temporary storage of water. A geotextile fabric is placed over subgrade to prevent migration of fines into the stone reservoir.

Limitations of porous asphalt include (FHWA 2015a):

- higher construction costs, which may be offset by cost reductions for stormwater management infrastructure

- the vast majority of applications has been in areas with light automobile traffic and limited heavy loads; porous asphalt is typically recommended for parking areas and low-volume roadways
- the potential for clogging requires specialized maintenance, such as vacuuming and power washing
- it should not be used where there is a high risk of spills.

Site and design considerations include (FHWA 2015a):

- soil infiltration rates should be between 0.1 and 10 in./h (0.25–25 cm/h)
- minimum depth to bedrock or the seasonal high water table should be greater than 2 ft (0.6 m)
- the bottom of the infiltration bed (stone reservoir) should be flat
- for parking areas, slopes should be less than 5%
- overflows should be utilized where the pavement system is not designed to store and infiltrate the maximum anticipated precipitation at the site
- alternatives should be provided for stormwater to enter the stone reservoir (e.g., stone edge or drop inlet) in the case the surface becomes clogged
- clogging of the pavement surface and introduction of fine sediment into the stone reservoir should avoided during construction
- compaction of the subgrade should be avoided.

23.11.3 Pervious Concrete

Pervious (or porous) concrete contains little or no fine aggregates and the amount of water and cementitious material is carefully controlled (FHWA 2012). Pervious concrete is not a proprietary product, but is rather a recipe for concrete that can be made to order by any concrete batch plant. Pervious concrete has the same basic benefits of other permeable pavement types in that it can reduce runoff, recharge groundwater, eliminate or reduce the need for infiltration ponds and swales, capture the first flush of runoff (reducing contaminant concentration in runoff), and reduce the urban heat island effect (FHWA 2012). The main disadvantages pervious concrete are (FHWA 2012):

- limited use in high traffic areas (mostly applied to parking areas and low-volume roads)
- specialized construction procedures are required
- special attention and care is required for some soil types
- high construction costs

Tennis et al. (2004) provided an excellent introduction to pervious concrete pavements, which is summarized herein. The size distribution of the aggregate is kept narrow and the water content is limited to a narrow range to prevent the cement paste from closing the open structure. Typically, 15–25% voids (pores) are achieved in the hardened concrete and infiltration flow rates are typically in the range of

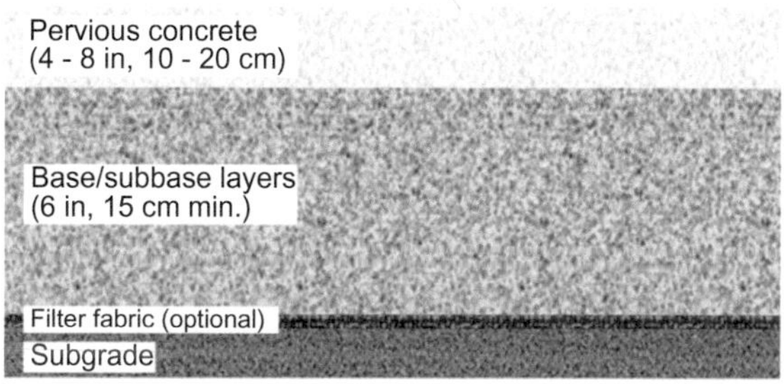

Fig. 23.9 Typical pervious concrete pavement cross section (Modified from FHWA 2012)

288–770 in./h (0.20–0.54 cm/s). Care must be taken during construction to avoid sealing the upper surface (e.g., smoothing with a trowel should not be performed). The mixture is typically thick when poured with an aggregate to cement ratio of 4.0–4.5 by mass.

The low mortar content and high porosity reduce the strength of pervious concrete, but it is still suitable for most low-volume pavement applications. It can achieve strengths in excess of 3,000 psi (20.5 MPa) and flexural strengths of more than 500 psi (3.5 MPa). The rougher surface texture makes the pavement more susceptible to abrasion and raveling (i.e., dislodgement of aggregate particles).

Pervious concrete pavements are installed on a crushed rock subbase (Fig. 23.9). Key design issues are the structural loads expected, the amount of rainfall expected and designed for, the required storage capacity of the aggregate and pavement, and the rate of infiltration into the underlying subsoil. The infiltration rate through the pavement may be much greater than that in the underlying soils. The high infiltration rates of the pavement result in the storage of water from rainfall events being the key design issue as water flows into the pavement and aggregate layer much more rapidly than it flows out. Tight, poorly drained soils (infiltration rates <0.5 in./h or 12 mm/h) and high water tables hinder the performance of pervious pavement systems. In such case, a drain pipe or channel may be installed in the aggregate layer.

Maintenance is performed by vacuuming or pressure washing. A key performance issue to minimizing the clogging of the porous structure. Hence it is important to minimize the washing of fine materials onto the pavements. A study of pervious concrete pavements in Florida indicated the pavements that were 10–15 years old were operating satisfactorily without significant clogging (Wanielista et al. 2007).

23.11.4 Permeable Interlocking Concrete Pavements (PICP)

Permeable interlocking concrete pavement (PICP), also referred to as concrete block permeable paving or pavement (CBPP), consists of solid concrete paving units with joints that create openings in the pavement surface when assembled in a pattern (FHWA 2015b). The joints are filled with permeable coarse sand or aggregate. The pavers are designed so that joints are wider than used in conventional pavers. PICPs

Fig. 23.10 Examples of permeable interlocking pavers

may be designed to include small apertures or have the traditional rectangular shape with slots or spacing lugs to keep pavers apart (Fig. 23.10).

Paving units are placed on a bedding layer of permeable aggregate (bedding course), which overlies a base and subbase reservoir (Fig. 23.11). The base and subbase reservoir store water and allow it to infiltrate into the underlying soil subgrade. Perforated underdrains are used in some pavements to remove water that does not infiltrate within a given design period. A geotextile layer may be placed between the subbase and subgrade to prevent fines from migrating into the subbase.

PICP share the general hydrological benefits of permeable pavements. PICP have the additional advantage of ease of maintenance. Paving units can be removed and

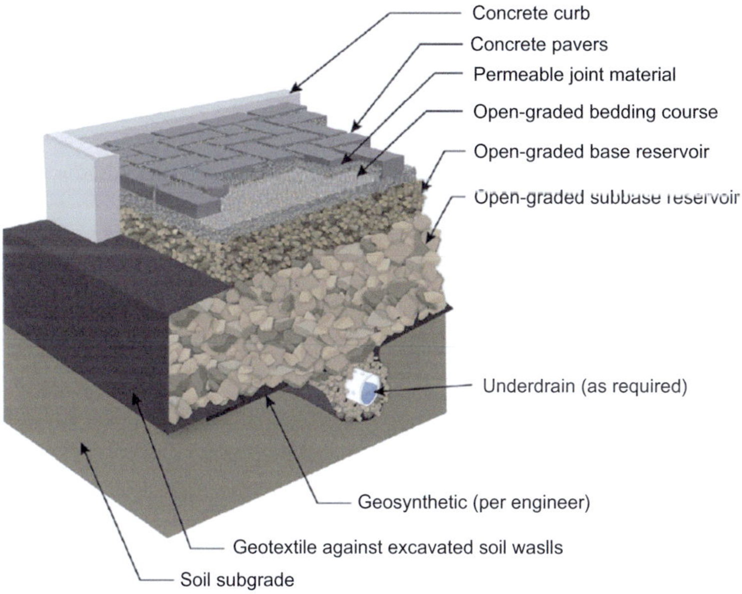

Fig. 23.11 Permeable interlocking concrete pavement design. *Source* FHWA (2015b)

reinstalled (if necessary). Additionally, PICP has aesthetic benefits in that they are considerably more attractive than conventional concrete and asphalt pavements. The limitations of PICPs include that they (FHWA 2015b):

- are intended for areas with less than 35 mph (50 kph) speed limits (e.g., walkways, driveways, parking lots, alleys, and low-speed roads)
- are generally not used in areas exposed to heavy loads
- should not be used in areas subject to the loading/unloading and storage of hazardous materials
- are generally not used where the seasonal high water table is less than 2 ft (0.6 m) below land surface
- should not be used at dirty sites where sediments can clog joints or apertures.

The Interlocking Concrete Pavement Institute (www.icpi.org) provides guidance documents, specifications, and training for PICP contractors. Most PICPs are machine installed and require specially trained contractors.

Surfaces should be tested for infiltration rate via ASTM (2013) Standard C1781, which is a single-ring (300-mm, 12.0-in diameter) test. The recommended minimum acceptable rate is 2,500 mm/h or 100 in./h (FHWA 2015b). PICP maintenance should include (FHWA 2015b):

- inspection 1–2 times per year
- infiltration tests via ASTM C-1781 if water ponds after rainstorms

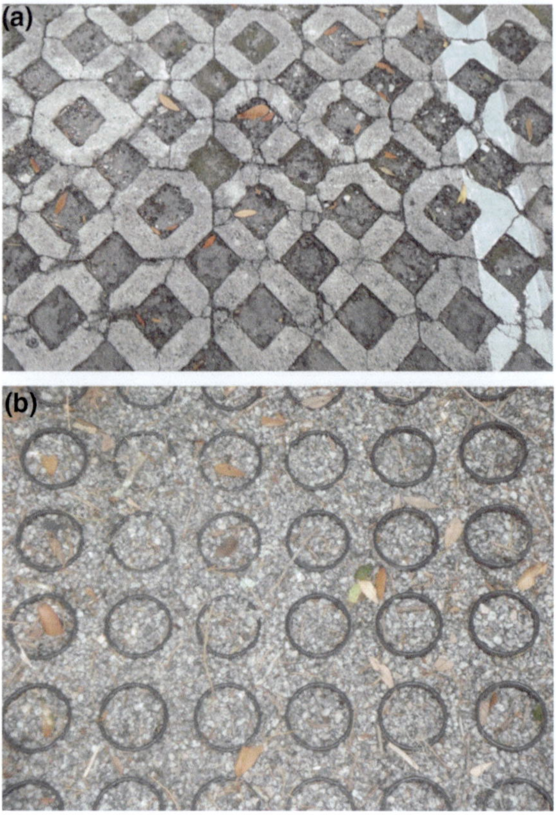

Fig. 23.12 **a** Concrete grid pavement used for parking spaces. **b** Plastic grid pavement used for a walking trail

- if the infiltration rate is less than 10 in./h (25 cm/h), then the sediment in the joints should be removed using a vacuum and be replaced with clean aggregate.

23.11.5 Plastic or Concrete Grid Systems

Grid systems consist of grids in which the cells are filled with gravel, sand, or soil, and may be either bare or planted with grass. The grid provides stability and structural support and act to keep the fill material in place (minimize erosion). Open-celled concrete grid pavers have large openings that are filled with gravel or grass (Fig. 23.12a) and are suitable only for areas with low speed limits (e.g., parking spaces and walkways). Plastic grid systems are also used primarily for driveways, parking areas, and foot paths (Fig. 23.12b).

23.11.6 Permeable Pavement Performance

Numerous studies have evaluated the performance of permeable pavements in terms of their initial infiltration rates, changes in infiltration rates over time (clogging), maintainability (ability to restore infiltration rates), and contaminant removal. Follows are summaries of some of the more pertinent studies.

Brattebo and Booth (2003) evaluated the long-term infiltration and water quality performance of permeable pavements as a stormwater management strategy. The field site, located in Renton, Washington state, consisted for four pairs of parking stalls, each constructed with a different permeable pavement system. The permeable pavement systems tested over the six-year study were:

- Grasspave2, a flexible plastic grid system filled with sand and planted with grass.
- Gravelpave2, a flexible plastic grid system filled with gravel.
- Turfstone, a concrete block lattice with about 60% impervious coverage.
- UNI eco-stone, small concrete blocks having about 90% impervious coverage with the inter-block spaces filled with gravel.

The permeable pavement systems showed varying, but generally only minor, signs of wear and tear after six years. Virtually all water infiltrated for every observed storm, with infrequent exceptions when cars covered the pavements and during the most prolonged periods of high-intensity rainfall. Surface runoff from the asphalt showed significantly higher concentrations of motor oil, copper, and zinc than in the infiltrated water. Brattebo and Booth (2003) noted that larger parking areas paved entirely with permeable pavement would almost certainly have sufficient uncovered areas to compensate for any local saturation that may occur around cars. A limitation of the study is that the Pacific Northwest of the United States generally has low rainfall intensities. Permeable pavements may have a lower effectiveness in areas underlain by less permeable soils (Brattebo and Booth 2003).

The USEPA, at the Edison Environmental Center parking lot (Edison, New Jersey), tested three permeable pavement types: interlocking concrete pavers, pervious concrete, and pervious asphalt. Borst et al. (2010) presented the results of the first six months of testing. Infiltration rates were measured using a 12-inch diameter cylinder in accordance with ASTM standard C1781 (ASTM 2013) using Neoprene sheeting for a bottom seal. All three pavement types had high infiltration rates that were greater than the reasonably expected rain events. The infiltration rates for the interlocking concrete pavers and pervious concrete (averaging between 776 and 2,219 cm/h) were an order of magnitude or more greater than the rates measured for the pervious asphalt (31–139 cm/h). The infiltration rates actually increased over time (from December 2009 to August 2010), which was likely due to a temperature effect (Borst et al. 2010).

Infiltration rates were measured at 40 permeable pavement sites in North Carolina, Maryland, Virginia, and Delaware (Bean et al. 2007). Pervious concrete, permeable interlocking concrete pavers, and concrete grid pavers were evaluated. Infiltration rates were measured using double-ring and single-ring infiltrometers in which the rings were sealed to the pavement using plumber's putty. Simulated maintenance tests were performed by replacing the void material to a depth of 13–19 mm. Key conclusions were:

- Maintenance is the key to sustaining high infiltration rates. Clogging can be controlled by regular maintenance, using a vacuum sweeper or pressure washer, or by replacing the upper layer of void-filling material.
- Permeable pavements should not be located near disturbed areas. Areas containing exposed and transportable fine soil particles should be avoided.
- Permeable pavers installed in sandy environments preserved high infiltration rates.

Collins et al. (2008) compared the hydrologic performance of standards asphalt and four permeable pavement types at an experimental parking lot in eastern North Carolina. The tested permeable pavements were:

- pervious concrete (PC),
- permeable interlocking concrete pavement (PICP) with small-sized aggregate in the joints and an open area of 12.9% (PICP1),
- PICP with an 8.5% open surface area (PICP2),
- concrete grid pavers (CGP) with a 28% open surface area filled with sand.

All tested permeable pavements provided statistically significantly better runoff reduction and peak flow mitigation than asphalt. The hydrological differences among the pavements were small in comparison to the overall improvements from asphalt (Collins et al. 2008). The PICP1 and CGP cells generated significantly lower total outflow volumes than the other tested pavements.

Yong et al. (2008) documented laboratory experiments in which three porous pavements were dosed with a semi-synthetic stormwater continuously over a period of 20 weeks, which simulated 20 years of real-life operation. Inflow rates were selected that are reflective of the temperate Melbourne and sub-tropical Brisbane, Australia climates. The porous pavement systems evaluated were:

- Monolithic pervious asphalt
- Hydrapave (Formpave), a modular concrete paver system
- Permapave, a modular polymer-fused aggregate product.

It was concluded that after 17 years of simulated continuous operation in Melbourne, Permapave is the only system that could cope with a 1-in-100 year event and that after the simulated 8.5 years in Brisbane, it could cope with a 1-in-5 year event. The tested pervious asphalt started to experience some clogging at the end of the simulation periods. Hydrapave was found to clog first and failed to cope with flooding conditions. All three of the tested systems showed a high rate of removal of suspended solids (close to 100%), but lesser removals of total phosphorous (~30%) and total nitrogen (~20%), with little difference between the systems.

Pavement effective life refers to both the number of years in service before hydraulic conductivity (infiltration) rates drop to unacceptable levels and/or pollutant (suspended solids) removal decreases to unacceptable levels (Pezzaniti et al.

2009). Pezzaniti et al. (2009) performed laboratory and field studies of the effective life of permeable paving systems. A laboratory rig was used to simulate 35 years of rainfall and associated suspended solids loading. Field testing was performed at four sites sides in Adelaide, South Australia. Three types of permeable pavement systems were tested:

PP1—Boral Formpave (PICP)
PP2—Rocla Ecoloc (PICP)
PP3—Grasspave (plastic molded grid system).

The Formpave system was tested with and without the equivalent of a yearly cleaning with a street-sweeping device. The laboratory results showed reductions in hydraulic conductivity of 59, 68, and 75%, respectively for PP1, PP2, and PP3. Average sediment retention for PP1, PP2, and PP3, were 94, 89, and 97%, respectively. At three of the sites, much greater reductions in infiltration rates due to clogging by sediment and organic matter were observed. It was concluded that prevention of this form of clogging is essential to maintaining the effective life of permeable pavements.

Jayasuriya and Kadurupokune (2010) compared the performance of three types of pavements, which were tested side by side in a car park in Melbourne, Australia:

- D&M Ecotrihex (PICP)
- Atlantis Turf Cell (reinforced grid)
- Conventional asphalt pavement.

The amount of runoff from each pavement was recorded. The average reduction in peak discharge (relative to conventional pavement) varied between 40 and 55% for the Ecotrihex pavement and between 45 and 60% for the Turf Cell pavement. Runoff volume decreased by 43–53% for the Ecotrihex pavement and 52–62% for the Turf Cell system. The permeable pavements reduced total suspended solids, total phosphorous, and total nitrogen loads by 70–100, 40–80, and 60–80%, respectively.

Lucke et al. (2013) compared the infiltration capacity of 18 permeable pavement (PICP) systems installed in the Netherlands and Australia with ages of 1–12 years. The pavements were evaluated relative to either the Australian performance criterion of being able to infiltrate a three month storm event or the European minimum infiltration rate of 92.2 mm/h. Infiltration rates are usually measured with single or double ring-type infiltrometers using a waterproof sealant at their base to prevent leakage. Larger square infiltrometers have the advantages of testing a larger area and many pavers. Lucke et al. (2013) used a square-shaped infiltrometer with an area of 1 m^2. Less than half of the sites were still performing satisfactorily with infiltration rates meeting or exceeding target rates. Infiltration capacity, in general, decreases with pavement age, probably due to clogging, but there is considerable variation in performance, which may be due to differences in operating conditions and maintenance.

Boogaard et al. (2014) evaluated the performance of eight permeable pavements in the Netherlands that had been in operation for seven years. The Netherlands requirements are a minimum initial infiltration capacity of 194 mm/h with maintenance recommended when rates fall below 0.50 m/d or 20.8 mm/h. The main types of permeable pavements used in Europe are:

- impermeable concrete pavers with wide joints or apertures
- porous concrete pavers with or without wide joints (permeable concrete interlocking pavers, PCIP)
- concrete or plastic grid pavers (CGP, PGP).

The limitation of small ring tests is that they evaluate only a small area the pavement, which may not be representative (Boogaard et al. 2014). Boogaard et al. (2014) proposed that a larger area ($\approx$50 m^2) test is more appropriate and presented a methodology in which small temporary dams are constructed on the pavement and a falling-head test performed. Soil or sand-filled plastic bags or cores were found to be the most effective dam materials and water levels were measured using pressure transducers and hand measurements. The infiltration rates in the eight pavements varied between 29 and 342 mm/h, with two higher than the 194 mm/h target value. In general, infiltration rates for permeable pavements start off very high with their performance deteriorating over time. Boogaard et al. (2014) recommended permeability testing after five years and maintenance after about 10 years of use.

The rainfall simulation infiltrometer (RSIT) was developed to better to replicate natural rainfall conditions. The system developed by Nichols et al. (2014) consists of a square steel frame with a series of PVC pipes on top. Rainfall was simulated by supplying water to the RSIT through a series holes (2.5 mm in diameter) drilled into the underside of the pipes. Water flowing from the holes in the pipe passes through two horizontal wire gauze sheets to break the flow into droplets. The maximum infiltration capacity of the pavement was calculated as the maximum flow rate before visibly ponding started to occur on the pavement surface outside of the framed area. A modified double-ring infiltration test was reported to produce surface infiltration results approximately 60% higher than the RSIT results (Nichols et al. 2014).

Wanielista et al. (2007) evaluated the performance of pervious concrete in eight parking lots in Florida, all of which had been operating for at least several years. It was recognized that standard surface infiltrometer tests are impacted by lateral flow within the concrete. In order to measure one-dimensional vertical flow, a new testing method was developed that involved using a 12-inch (30-cm) diameter coring bit to drill through the concrete. A steel tube is then inserted around the core and embedded into the underlying soil. Plumbers' putty was placed around the core to prevent side leakage. Specific-head tests (3-inches) were performed in which water was added at specified time intervals and the amount added recorded. The tests were stopped when the amount of water added each step stabilized. The cores were then recovered for laboratory analysis. The test results indicate the properly installed pervious concrete can continue to infiltrate even without routine maintenance. The four sites in Central Florida, in which no maintenance was performed, had average and median field infiltration rates of 9.89 in./h (25.1 cm/h) and 5.2 in./h (13.2 cm/h) after an average of 12.8 years of operation.

23.11.7 Maintenance

Hunt and Collins (2008) made the following observations concerning the clogging of permeable pavements:

- Permeable pavements clog, but clogging does not always mean sealing. Infiltration rates may be reduced but still be high enough to achieve stormwater infiltration design goals.
- Clogging increases with age and traffic use.
- Clogging impacts depend on the characteristics of nearby soils/sediment. The rate of surface infiltration tends to approach that of surroundings sediment. If pavement is located in a sandy area, then the clogging material will tend to be sand.
- Clogging cannot be avoided but can be managed through regular maintenance and avoiding areas with sources of clogging materials (disturbed areas). Hunt (2011) also observed that pavements located beneath tree canopies are more prone to clogging with vegetative debris and detritus.

Fines can fill gaps in PCIP systems creating a thin clogged "schmutzdecke" layer (Hunt 2011). Dietz (2007) recommended a combination of vacuuming, high-pressure washing, and replacement of interstitial gravel.

Hunt (2011) recommended as primary inspection and maintenance tasks:

- regular inspections; a simple qualitative test is to pour a bucket or large bottle of water on a pavement and examining how long it takes the water to infiltrate and the "water mark" that is left
- periodic preventative street sweeping (2–4 times per year)
- for PICP, replace the aggregate removed by sweeping and vacuuming.

The question remains as to whether periodic inspections and maintenance will be actually performed. Unless there is a strong regulatory driver, inspection and maintenance activities tend not to be performed unless there is an obvious problem.

23.12 Rainwater Harvesting and Water Harvesting

Rainwater harvesting is essentially the collection, storage, and use of rainwater. The closely related term "water harvesting" is commonly used to refer to the process of concentrating rainfall or runoff from a larger area for use in a smaller target area, typically for agricultural purposes (Oweis et al. 1999). Water harvesting was defined by the Food and Agricultural Organization of the United Nations (FAO) in its broadest sense as the "collection of runoff for its productive use" (Critchley and Siegert 1991). The term "rainwater harvesting" refers herein to small-scale activities that are performed to capture and use water that falls on individual parcels of land. Rainwater harvesting dovetails into stormwater management, which has a primary flood control objective and a greater (watershed) areal scope. Rainwater harvesting includes small-scale MAR techniques where the primary goal is local aquifer recharge.

Water harvesting has been used for millennia as either the primary or a supplemental source of water in arid and semiarid lands that do not have a reliable surface water source and local rainfall is inadequate to meet plant requirements. The limited rainfall often occurs in short-duration, high-intensity events. A solution to the water scarcity is to divide available land areas into catchment and smaller crop areas. Runoff collected from the catchment area combined with the rainfall that directly falls on the crop area becomes sufficient to meet crop water requirements (Stern 1979). The runoff from a catchment can be stored in a tank or pond, or be used to directly increase the soil moisture of a crop area. The basic principle is that in areas where precipitation for rainfed agriculture is inadequate, reasonable yields can still be achieved if the available rain is concentrated on a smaller area (Critchley et al. 1991).

Oweis et al. (1999) observed that

> The principle of agricultural rainwater harvesting is based on the concept of depriving part of the land of its share of precipitation, which is usually small and non-productive, and giving it to another part to increase the amount of water available to the latter part, which originally was not sufficient and to bring this amount closer to the crop water requirements so that an economical agricultural production can be achieved.

In sub-Saharan Africa, 95% of cultivated land is under rainfed agriculture and the rainfall-transpiration efficiency (percentage of precipitation used for transpiration) is low (<15%). Rainwater harvesting and management technologies have significant potential for improving rainfall-transpiration efficiency and sustaining (or increasing) rainfed agricultural production (Biazin et al. 2012).

Rainwater harvesting is part of LID and green infrastructure as it contributes to the goal of capturing precipitation before it leaves a property. Rainwater harvesting is often spontaneously implemented by property owners rather than being part of an integrated water management plan. Similarly, agricultural rainwater harvesting is also a decentralized water management strategy.

There is a rapidly growing popular and technical literature on rainwater harvesting. Rainwater harvesting methods have been reviewed by National Academy of Science (1974), Boers and Ben-Asher (1982), Bruins et al. (1986), Pacey and Cullis (1986), Waller (1989), Prinz and Singh (2000), Lancaster (2006, 2008), Waterfall (2006), Kinkade-Levario (2007), Downey (2010), Daily and Wilkins (2012), and Bickelmann (2013). Water harvesting for agriculture was reviewed by the UNEP (1984), Critchley and Siegert (1991), and Oweis et al. (1999, 2012). An internet search of "rainwater harvesting" gives a plethora of sites by government agencies, non-governmental organizations (NGOs), and a wide variety of contractors and commercial suppliers of rainwater collection supplies.

23.12.1 Rainwater Harvesting System Types

Rainwater harvesting systems are categorized as either earthwork (passive) or storage (active) systems. Earthwork systems are used to harvest and hold water in the soil or recharge the underlying aquifer (Lancaster 2006). Storage systems are used to harvest and store rainwater in tanks, cisterns, or reservoirs. Storage is critical for rainwater harvesting for year-round water supply, particularly in arid lands where rainfall is highly sporadic. There is great variation in the scale and technical sophistication of rainwater harvesting systems. Rainwater harvesting, for example, may range in scale from a simple barrel used to store roof runoff collected from gutters for irrigation use, or microcatchments constructed around a tree, to sophisticated systems involving treatment facilities, MAR, and distribution systems to serve multiple users. Rainwater harvesting is appropriate for developing countries as it has the advantages of being small-scale simple operations with high adaptability and low costs. Rainwater harvesting systems can also empower local communities to manage their own water resources.

In developed countries, rainwater harvesting is increasingly being implemented by individuals and organizations as part of a green ethic. Even in areas blessed with abundant rainfall, it is now common to see roof rain gutters connected to rain barrels for non-potable uses.

Earthworks systems are of concern herein because they can contribute to aquifer recharge. The objective of earthwork systems is to convert "convex" impervious landscapes where rainwater runs off into "concave" pervious landscapes that infiltrate water (Lancaster 2006). The goal is to slow the flow of water and spread harvested water over as much pervious area as possible to give the water maximum potential to infiltrate into the soil. Earthwork rainwater harvesting systems include (UNEP 1979; Lancaster 2006, 2008):

- **berms and basins:** shallow basins surrounded by a berm constructed perpendicular to the slope of the land
- **contour trenches**
- **terraces:** flat benches surrounded by a berms or low walls
- **French drains:** trenches or basins filled with porous material (gravel) with or without a perforated pipe
- **infiltration basins:** shallow excavated depressions with flat bottoms
- **permeable pavements**
- **modifications of land surface to reduce runoff:** e.g., contour farming and conservation tillage
- **diversion swales:** which intercept, infiltrate, and redirect the flow of water,
- **check dams:** low barriers placed perpendicular to the flow of water that slow and spread the flow or water
- **microcatchments**.

The simplest passive rainwater harvesting techniques involve the construction of structures or excavations that slow the flow of (or temporarily retain) runoff to

Fig. 23.13 Microcatchments capture and concentrate down-slope runoff

increase the amount of infiltration. Shallow depressions excavated on a slope or within or adjacent to an ephemeral stream channel can retain runoff and increase local infiltration. The local land surface should be contoured to direct water to the depression. Runoff may be diverted to collection areas using channels, swales, berms, and low walls. For example, a microcatchment is a contoured area with berms that is designed to capture and concentrate runoff into a small planting basin (Evenari et al. 1971; Fidelibus and Bainbridge 1994; Renner and Frasier 1995), which may contain a single tree (Fig. 23.13). Terraces constructed parallel to topographic contours (and thus perpendicular to the slope) have been used for millennia in some regions to capture runoff, control erosion, and create flat farm fields on sloped land (Fig. 23.14).

Ridges and furrows constructed parallel to topographic contours are effective in reducing runoff and increasing infiltration and soil moisture. Continuous contour trenches (CCT) and staggered contour trenches (SCT) are rectangular excavations constructed parallel to topographic contours to harvest runoff flowing down slope and recharge the shallow aquifer (Fig. 23.15). The trenches are kept empty to take full advantage of their storage capacity (Shinde et al. 2006). Sussman (2007) published a design manual for contour trenches. Contour trenches reduce flow velocity, promote infiltration, and prevent pollution from draining into water bodies. Soil excavated from the ditches is used to form berms on the downhill edge of the ditches. Berms are stabilized by planting with native vegetation or legumes. The distance between trenches is based on the slope of the field. The greater the slope, the lesser the distance. High slopes have greater flow velocities and thus flow has to be stopped more frequently. Sussman (2007) recommended a 10–12 m spacing for 0–4% slopes, an 8 m spacing for 4–8% slopes, and 6 m spacings for 8–15% slopes.

SCTs, which usually have lengths of 2–3 m, avoid the problem of lateral flow within poorly level continuous trenches. Downgradient trenches cover gaps in the line of upgradient trenches. The Gujarat State Watershed Management Agency (2011) Design Manual noted that

Fig. 23.14 Terraces. **a** Machu Picchu, Peru. **b** Jizan, Saudi Arabia (courtesy of Weixing Guo)

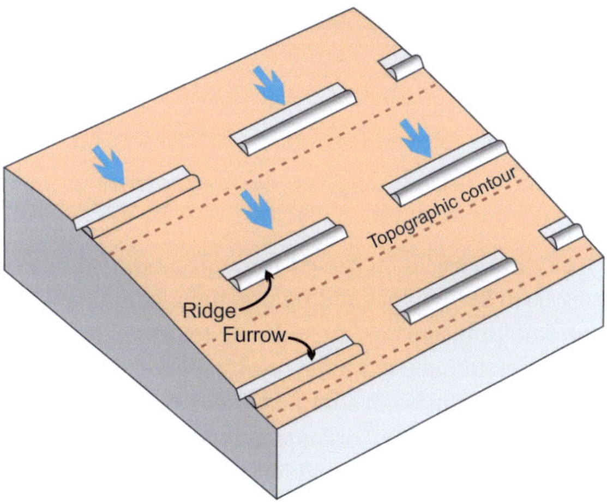

Fig. 23.15 Conceptual diagram of a staggered contour trench ridge-and-furrow system designed to capture and infiltration runoff

Fig. 23.16 Contour farming highlighted by corn stubble in snow, Waverly, Nebraska

it has been found that invariably errors have been made in contouring over long distances. If the contour trench is not level and by mistake sloped, then water starts to flow from the high point to the low point, cutting a path and increasing soil erosion. Therefore, instead of making trenches continuously, they should be made in a staggered, discontinuous manner.

Contour farming is defined as using ridges and furrows formed by tillage, planting, and other farming operations to change the direction of runoff from directly downslope to around the hillslope (USDA NRCS 2007; Fig. 23.16). This practice is applied to achieve one or more of the following objectives (USDA NRCS 2007):

- reduced sheet and rill erosion
- reduced transport of sediment, other solids, and the contaminants attached to them
- increased water infiltration.

23.12.2 Land Surface Modification

Infiltration can be increased through modifications of the land surface. It has very long been known that adding organic matter can improve the physical structure and water-holding capacity of soils. As reviewed by Martens and Frankenberger (1992), organic matter amendments result in increased aggregation of soil, although the specific mechanism is not completely understood. In an experimental study performed in Riverside, California, organic amendments (poultry manure, sewage sludge, barley straw, and alfalfa) were added to irrigated soil, which showed increased infiltration rates (18–25%), soil aggregate stability (22–59%), and soil moisture content (3–25%), and decreased bulk density (7–11%; Martens and Frankenberger 1992). The experimental data indicate that the increased infiltration rate was related to a stimulation of microbial activity that increased aggregate stability and decreased the bulk density of the soil in the tillage zone (Martens and Frankenberger 1992).

The USEPA supported a research project on the effects of urbanization and associated soil compaction on soil structure and infiltration rates, and the effectiveness of using compost as a soil amendment to increase rainwater infiltration and decrease runoff (Pitt et al. 1999). Approximately 150-double ring infiltrometer tests were performed in the Birmingham, Alabama area, which were divided into eight categories of soil conditions based on texture, moisture, and compaction. The field testing showed that compaction had the greatest effect on infiltration in sandy soils with non-compacted soils having an average infiltration rate of 441 mm/h (16.3 in./h) compared to an average rate of 64 mm/h (2.5 in./h) for compacted sandy soils. There was little detrimental effect associated with soil moisture in sandy soils. Compaction had about the same effect as moisture on clayey soils, with saturated compacted clayey soils having the lowest infiltration rates.

Field studies on the beneficial effects of using compost as a soil amendment to improve infiltration capacity and pollutant retention capacity were performed on glacial till soils in the Seattle, Washington, area (Pitt et al. 1999). Compost-amended soils were found to have significantly increased infiltration rates, but also increased concentrations of nutrients (phosphate, total phosphorus, ammonia-nitrogen, nitrate and total nitrogen) in surface and subsurface runoff from leaching of the compost. Surface runoff from the compost-amended soils had greater concentrations of almost all constituents compared to run off from soil-only test sites, with the exception of some cations (Al, Fe, Mn, Zn, Si) and toxicity. Due to decreased runoff volumes, the total surface water mass discharge of nutrient was reduced. Subsurface mass flow discharges of nutrients were expected to increase. Pitt et al. (1999) concluded that further research was needed to determine the optimum amount of compost amendment required to benefit urban soils without the associated leaching of nutrients.

Conservation tillage is any method of soil cultivation that leaves the previous year's crop residue (such as corn stalks or wheat stubble) on fields before and after planting the next crop to reduce soil erosion and runoff (Garg n.d.). Conservation tillage methods include no-till, strip-till, ridge-till, and mulch-till, with each requiring different types of specialized or modified equipment and adaptations in management (Garg n.d.). There is ample evidence in the tropics that conventional farming systems involving soil inversion using a plough and hoe contribute to soil erosion, desiccation, and accelerated oxidation of organic matter (Rockström et al. 2003). Field evidence demonstrates that conservation tillage can result in increased water productivity (Rockström et al. 2003).

The environmental benefits of conservation tillage include (Gard n.d.):

- reduction in soil erosion by as much as 60–90% depending on the conservation tillage method; pieces of crop residue shield soil particles from rain and wind until new plants produce a protective canopy over the soil
- improvement in soil structure and surface water quality by adding organic matter as the crop residue decomposes, which creates an open soil structure that increases infiltration and reduces runoff
- conservation of water by reducing evaporation at the soil surface
- conservation of energy due to fewer tractor trips across the field

- reduced potential air pollution from dust and diesel emissions
- crop residue provides food and cover for wildlife.

23.12.3 Downgradient Impacts and Legal Issues

Rainwater harvesting systems are intended to reduce runoff from properties. Depending upon the situation, reduction in runoff from upgradient properties can be beneficial or harmful to downgradient properties. Rainwater harvesting can be beneficial to downgradient users if it reduces flooding and increases groundwater levels. Adverse impacts may occur if captured runoff would otherwise become surface water that downgradient users rely upon. One person harvesting rainfall will have minimal impacts, but the impacts of widespread implementation might have significant impacts on downgradient properties and water users.

Rainwater harvesting is limited in several states in the western United States under the belief that it would impinge upon the rights of senior surface water rights holders. Depending upon where one lives, the rain that falls on one's roof and driveway may not belong to you. Surface water in the western United States is mainly governed under the prior appropriation doctrine in which senior water rights holders have priority over surface water flows. Captured runoff could result in decreased streamflows, adversely impacting holders of downstream surface water rights. However, Miller (2006) noted that much of the captured water will otherwise be lost to ET under natural conditions.

Under state of Colorado law, all precipitation is assumed to contribute to stream flow and, therefore, private properties (including rooftops) are considered to be tributaries to water bodies that have already been appropriated to senior water rights holders (Gaston 2010). A property owner that harvests rainwater is legally required to augment the streamflow for the lost water, which is an impractical requirement for household-scale systems (Gaston 2010). Colorado water law is thus a barrier to efficient water use as most (if not essentially all) of the water captured in a rooftop rainwater harvesting system would be put to a beneficial use, whereas only a very small fraction of the water that ran off from the same roof might ever reach a stream.

The state of Colorado somewhat relaxed the prohibition against rainwater harvesting with passage of bill HB 09-080 (effective July 1, 2009), which allows for limited rooftop rainwater harvesting where residents are legally entitled to a well and water is not available from a municipality or other water district. The water may only be collected from the roof of a building that is used primarily as a residence and may only be used in the same way as allowed in the well permit (Gaston 2010; Cummings 2012; Colorado Division of Water Resources 2015). If a well is permitted for household use, then the harvested rainwater may not be used for garden irrigation. Restrictions on rainwater/stormwater harvesting in Colorado are in conflict with Federal pollution control regulations (Cummings 2012). For example, stormwater retention/detention basins are elements of stormwater management and pollution control systems. Under current regulations, an individual can detain stormwater for later release, but may not use the water.

Colorado is losing its dubious distinction as the only state in the country to outlaw collecting rooftop runoff in rain barrels (Goodland 2016). In 2016, the Colorado Senate Agriculture, Natural Resources and Energy Committee voted to authorize rainwater collection. House Bill 16-1005 would allow Coloradans to use up to two 55-gallon (210 L) rain barrels to collect stormwater that runs off of roofs. Under the bill, harvested rainwater can only be used to water lawns or gardens (Goodland 2016).

Other states (e.g., Arizona and New Mexico) define appropriable water as water that flows in natural channels and does not include water tributary to the channels. Precipitation may therefore be harvested before it reaches a natural channel (Gaston 2010). Arizona actively encourages rainwater harvesting through tax incentives and ordinances have been passed that requires the practice. Most notable was the City of Tucson Ordinance 10597 adopted on October 14, 2008, which requires that 50% of the water used for landscaping on new commercial properties to come from harvested rainwater (Gaston 2010).

23.13 Soil Amendments for Improved Pollutant Removal

Infiltration basins and other stormwater management and LID systems are typically designed by just excavating the native soil and, depending upon the system, adding vegetation to the sides and bottom. There is increasing interest in engineering the systems to enhance infiltration rates and the removal of nutrients (nitrogen and phosphorus) and other contaminants. Alternative designs include replacing the soil below the basin with either new material or mixing the soil with various amendments. Chang et al. (2010b) summarized functionalized filter media for nutrient removal in stormwater retention basins. Media may be used in both natural and built environments to improve physicochemical and microbiological processes for nutrient removal. Such media can be made "green" by including recycled material, such a tire crumbs and saw dust, to increase treatment efficiency and effectiveness.

23.13.1 Organic Matter Amendments

Augmented heterotrophic denitrification systems use solid phase carbon sources (e.g., wood chips, saw dust, cardboard, paper, and various agricultural residues) and, depending on the system objectives, may require oxic-anoxic cycling (Chang et al. 2010c). Oxic conditions are required for nitrification and anoxic conditions are needed for denitrification. Kim et al. (2000) investigated a modification of the design of bioretention systems to improve nitrate removal by incorporating an underlying continuously submerged carbon layer that would promote denitrification. The design concept includes a drain to recover treated effluent for discharge to surface waters. Alternatively, the effluent could be used for aquifer recharge. Kim et al. (2000)

performed column studies of denitrification using two sets of organic substrates: alfalfa, newspaper, and leaf mulch compost (set no. 1) and sawdust, wood chips and wheat straw (set no. 2). Inorganic substrates evaluated were large sulfur particles, large sulfur particles with limestone, and small sulfur particles with limestone. The latter is referred to as sulfur-limestone autotrophic denitrification (SLAD; Kim et al. 2003). The columns were seeded with a supernatant of settled secondary effluent and fed anoxic synthetic stormwater. The goal of the investigation was to identify electron donor and carbon sources that promote significant denitrification and are stable for a long period of time in the subsurface.

Excellent (≥95%) nitrate removal occurred in the columns containing alfalfa, newspaper, wheat straw, wood chips, and sawdust. The greatest removal for an inorganic substrate occurred with the small (0.6–1.18 mm) sulfur and limestone particles. Wood chips and newspaper were identified as the best candidates in terms of both overall effluent quality and total nitrogen removal. The small sulfur particles and limestone had the best removal (~90%) presumably due to a higher surface area. The additional advantages of newspaper and wood chips are that they are a waste product (and thus inexpensive) and readily available.

Kim et al. (2003) reported on the results of subsequent phases of the investigation performed using the identified preferred substances (wood chips, newspaper and small sulfur and limestone particles):

- **Phase 2:** nitrate loading and flow rate optimization
- **Phase 3:** viability after long (30 and 84 days) dormant periods
- **Phase 4:** pilot-scale bioretention system study.

Nitrate and nitrite removals decreased approximately linearly as the nitrate loading increased. Newspaper was found to be the most promising electron donor for nitrate removal from stormwater runoff via denitrification (Kim et al. 2003). Dormant periods did have a short term (mostly <1 day) impact on nitrate removal. The column test studies showed a period of increasing nitrate concentrations (lesser removal) followed by rapidly decreasing concentrations to a greater than 90% removal. The results suggest that as the length of the dormant period increases, a greater the length of time is required for the system to recover. The pilot test studies confirmed the effectiveness of the proposed design for reengineered bioretention facilities, demonstrating mass nitrate removals of 70–80%.

23.13.2 Sorptive Media

Sorptive media are used to reduce contaminant concentrations in stormwater, wastewater and other impaired waters before discharge to surface waters or groundwater recharge. Although the media are often referred to as "sorptive," contaminant attenuation occurs as the result of a variety or processes including actual adsorption, cation exchange, precipitation, and microbially mediated reactions (Chang et al. 2010a). Chang et al. (2010a) provided a literature review of the various sorptive media that have been either experimentally tested and/or used at field sites. Media selection depends upon water treatment goals and multiple media may be mixed or used sequentially.

Phosphorous is the limiting nutrient for plant growth in many waters. Hence, phosphorous removal is an important objective for some stormwater management systems. Phosphorous can be removed in stormwater BMPs directly by plant uptake. Within soils, phosphorous removal occurs by precipitation with calcium, aluminum or iron, and adsorption onto iron and aluminum oxides and hydroxides. Erickson et al. (2007) documented batch and column studies of the effectiveness of a series of amendments for removing dissolved phosphorous from stormwater. The batch studies tested ASTM C33 sand, calcareous sand, limestone, aluminum oxide, steel wool, and three blast oxygen furnace (BOF) residues. The C33 sand was reported to removal 40% of the available phosphorous (0.485 mg PO_4^{-3}) after 10 h. Calcareous sand and washed calcareous sands removed phosphorous to below detected limits but increased the pH to between 9.8 and 10.1. BOF quickly removed phosphorus but increased pH to about 11. The addition of steel wool to the C33 sand increased phosphorus removal to 90% after 24 h.

Column studies were subsequently performed using C33 sand amended with calcareous sand, limestone, chopped granular steel wool, and steel wool fabric. C33 sand alone removed 2.1% of the dissolved phosphorous and its removal capacity was quickly exhausted. Steel-wool enhanced columns retained between 25 and 99% of the dissolved phosphorous without clogging the columns. The carbonate-enhanced columns also effectively removed phosphorous but the fine-grained precipitates clogged the filter fabric used and prevented the columns from draining. The results of these experiments demonstrated that addition of steel wool fabric to a sand filter can cost effectively increase phosphorous removal (Erickson et al. 2007).

Erickson et al. (2012) subsequently reported on column and field testing of filtration using sand amended with iron filings. As is the case for steel wool, oxidation (rusting) of the iron filings creates iron (oxy)hydroxides that bind phosphates by sorption. Iron filings are less expensive that steel wool and can be obtained with a similar size distribution as sand. Column tests using influent concentrations of approximately 0.313 mg PO_4^{-3}-P/L and iron filings concentrations (by weight) of 5, 2, 0.3 and 0% had median effluent concentrations of 0.036, 0.066, 0.271, and 0.328 mg PO_4^{-3}-P/L, respectively. After 100 m of treated depth, the phosphate removal capacity of the 0.3% iron-fillings sand was nearly exhausted.

Field testing was performed using two trenches installed in the City of Prior Lake, Minnesota. Sand mixtures with 7.2 and 10.7% by weight iron filings were tested. Phosphate removal efficiencies varied between 29 and 91%, and for most tests the removal was greater than 50%. Using a median reported nation-wide stormwater concentration of 0.12 mg PO_4^{-3}-P/L, the iron-enhanced sand filtration trenches would be expected to capture approximately 85–90% of the phosphate for most rainfall events (Erickson et al. 2012). Erickson et al. (2012) proposed other applications for their iron-enhanced sand filter (which they named the “Minnesota Filter”), such as a treatment layer below the planting media bed of bioretention systems.

O'Neill and Davis (2011a, b) documented batch and column testing results that demonstrated that an aluminum-based water treatment residual (WTR) amendment greatly improved the phosphorus removal of bioretention soil media (BSM).

Wanielista and Chang (2008) and Chang et al. (2010c) screened potential media based on their:

- relevance for nitrification and denitrification processes (or both) with documented literature effectiveness
- removal efficiency
- permeability
- cost
- availability in Florida
- additional environmental benefits.

Suitable sorption media identified were peat, sandy loam, sawdust, wood chips, tire crumbs, crushed limestone, and crushed oyster shell. Wanielista and Chang (2008) tested a functional sorption medium consisting of a mixture of sand (50%), limestone (20%), sawdust (15%) and tire crumb (15%). Removal efficiencies for a 5-h hydraulic retention time ranged from about 65–100% for ammonia, nitrate, orthophosphate, and total phosphorous. Based on isotherm analyses, the life expectancy of the sorption medium was estimated to be about 40 years for orthophosphate and much less for ammonia (0.25 year) and nitrate (2.11 years). However, the life expectancy analyses were based on sorption and did not consider the much greater microbiological removal of nitrogen compounds.

A key design issue is hydraulic retention time in the media. Wanielista and Chang (2008) recommended a minimum time on the order of 5 h or a design of 10 h to account for some preferential flow pathways in the media. An economic analysis indicates that the test functional sorption medium would be cost effective for treating stormwater because of the low cost of the medium (less than $150 per cubic yard; 2008 dollars) and little associated maintenance.

Chang et al. (2010c) documented column experiments using a medium consisting of 50% sand, 30% tire crumb, and 20% saw dust by weight. Nitrate removal ranged from 84.0 to 99.2% versus rates of 23–75.4% in control runs (sand without amendment). Phosphorous removal ranged from 78.8 to 93.9% versus 16.3 to 56.8% for control runs.

Field applications focusing on practical scale-uping of nutrient removal technologies and cost effectiveness assessments remain critical areas of concern (Chang et al. 2010a). Chang et al. (2010b) reviewed the design and application challenges of filter media for nutrient removal. The main design issues for sorptive media are:

- **performance:** efficiency of the processes (percent removal) and effluent water quality
- **capacity:** how much pollutant can be removed before the media needs to be replaced
- **side effects**.

Performance depends upon influent water quality (contaminant concentrations), media composition and properties (porosity, particle size, and permeability), media thickness, and hydraulic loading rate or retention time (HRT; Chang et al. 2010b). In addition, cost effectiveness is a major concern for any technology, which includes the initial cost to install the system and the costs to periodically replace the media (if necessary).

23.13.3 Biosorption Activated Media

Biosorption activated media (BAM) is a soil amendment technology designed to provide (Hood et al. 2013):

- inert filtration
- reactive filtration (adsorption and cation exchange)
- habitat for microorganisms to support biosorption and biological uptake.

Hood et al. (2013) performed field and laboratory experiments of BAM (trademarked Black and Gold product) to reduce total phosphorous and soluble reactive phosphorous (SRP) in stormwater. Black and Gold was reported to be composed of an uncompacted volume ratio of 75% expanded clay and 25% tire crumbs. The tire crumbs function like activated carbon and are effective in removing large organic molecules and non-polar compounds (Wanielista et al. 2014). A field-scale test bed was used that represented a highway and an adjacent grass-covered (Argentine bahia) swale (biofiltration system). Flow was spilt equally into two sides of the test bed, one containing 84 cm of Black and Gold and the other 84 cm of typical Florida sandy soil. Removal efficiencies were calculated from influent and effluent water samples. Column tests were also performed without the sod present to evaluate the effects of nutrient leaching from the soil. It was noted that the fresh Black and Gold and sandy soil used in the column test should have been minimally biologically reactive compared to the field test systems.

The results of the column tests indicate a total phosphorus removal of 60% for the Black and Gold versus 14% for the sandy soil. The total phosphorous removal in the field-scale tests was greater than 60% removal for Black and Gold (71% when leaching of phosphorous from the soil became negligible) versus no definite removal being evident in the sandy soil. The average soluble reactive phosphorous (SRP) removal by the Black and Gold was 95%. The results of the study demonstrated that BAM biofiltration systems are a feasible treatment method for removing phosphorous from highway runoff (Hood et al. 2013). BAM may also be used in engineered media filters to remove nutrients and other contaminants before discharge to surface water or groundwater recharge in MAR systems. For example, a layer of BAM was used for nutrient removal in an artificial wetland filter at Bok Tower Garden, Lake Wales, Florida.

Wanielista et al. (2014) presented column and field testing results on the effectiveness of BAM (Black and Gold media) for stormwater treatment in ultra-urban areas where more conventional treatment systems (e.g., bioretention) ponds are not feasible. Column testing was performed on 4-inch (10-cm) diameter, 24-inch (61-cm) high columns containing three mixtures that varied in their clay and tire crumb contents. Tests were performed of simulated 2-h and 24-h storm events. The Black and Gold medium had the highest average total nitrogen removals of 26 and 48%, respectively, for the 2-h and 24-h storm event tests. The greater contact time during the 24-h storm event promoted the anoxic conditions necessary for denitrification. Total phosphorous removal was also highest for the Black and Gold medium test, with efficiencies of 52 and 33%, respectively, for the 2-h and 24-h tests. SRP removal efficiencies were 57 and 66%, respectively for 2-h and 24-h tests.

BAM demonstration projects were performed using an off-line, up-flow filter in the City of Dunnelon, Florida, and an on-line, up-flow system in the City of Kissimmee, Florida. The off-line filter had average removal efficiencies of 60, 46 and 51%, respectively, for total nitrogen, total phosphorous, and total suspended solids. The on-line system had removal efficiencies of 45, 58, and 40%, respectively, for total nitrogen, total phosphorous, and total suspended solids. The water management districts in the state of Florida require a certification time of two years for the performance of treatment systems. Replacement times are a function of the amount of media used and influent water quality. Filtration and sedimentation of particulate matter before the sorption filter is necessary to extend the life of the filter media and to increase the performance of systems (Wanielista et al. 2014). Wanielista et al. (2014) provided a demonstration of the calculations for the media volumes needed to meet phosphorous goals.

O'Reilly et al. (2012) reported on the field testing of the use of a BAM in a full-scale infiltration basin located near Silver Springs and Ocala, north-central Florida. The test basin had a bottom area of 2,800 m^2, a total depth of 2.8 m, and a watershed of 22.7 ha. The basin was divided with a dike into two approximately equal area basins; a flood control basin and nutrient reduction basin. The surficial sediment of the latter was replaced, from the top down, with a 0.15 m layer of native top soil, a 0.30 M BAM layer, and a 0.10 M coarse sand filter layer. The BAM layer consisted of a 1.0:1.9:4.1 mixture (by volume) of tire crumb, fines (silt and clay) and sand. The top soil layer was intended to provide a source of organic carbon and the BAM layer provides increased sorption capacity and water retention.

Total dissolved phosphorous and orthophosphate concentrations were reduced in the nutrient reduction basin by greater than 70%. Minor nitrate removal was reported with the exception of one summer sample with a 45% reduction. O'Reilly et al. (2012) recommended a BAM recipe of 15% tire crumb, 25–50% fine-textured sediment (silt and clay), and the rest sand. The high moisture retention capacity of BAM and a greater frequency of stormwater storage contribute to conditions more favorable for the local formation of anoxic conditions in the BAM layer that enable the progression of biogeochemical processes toward denitrification.

23.14 Impediments to the Implementation of LID and Green Infrastructure

Goodwin et al. (2008) reported on the results of series of workshops on the barriers to the implementation of LID, which were conducted in three communities in the state of Oregon with different sizes, locations, and situations. Consistent themes emerged from the workshops:

- **Lack of basic understanding of planning and the impacts on growth:** There is a basic lack of understanding between land use and development decisions made today and future stormwater management and water quality consequences.
- **Need for active leadership:** It was recognized that there is a need for strong administrative support and direction to incorporate LID practices into codes and encourage developers to try LID projects.
- **Need for technical information and assistance:** A basic unfamiliarity with LID techniques and design was identified as an impediment to their implementation.
- **Funding, economics, and incentives:** Smaller jurisdictions often do not have the funding to develop, revise, and enforce new codes or regulations, or to educate builders and developers on LID techniques.

It was observed that existing codes often contain elements that may hinder the implementation of LID. For example, a requirement that streets have curbs and gutters may preclude the option of curbless streets draining to bioswales (Goodwin et al. 2008).

UNEP (2014) similarly observed that the impediments to the implementation of green infrastructure include:

- lack of awareness by decision makers
- regulatory or funding polices that stipulate traditional "gray" approaches
- lack of agreed upon methodologies for cost-benefit analyses that would enable a full comparison of gray and green infrastructure options
- perceived difficulties in quantifying the economic value of environmental benefits.

Wong (2006b) addressed some of the issues facing WSUD in Australia. From a technical perspective, construction and maintenance practices were identified as recurring impediments to the effective implementation of what are considered excellent conceptual designs. Socio-institutional dimensions of WSUD were identified as constraints in the implementation of WSUD. Considerable progress was reported in having WSUD philosophy, technology, and language adopted in industry standards and policies across all levels of government in Australia.

References

Ahiablame, L. M., Engle, B. A., & Chaubey, I. (2012). Effectiveness of low impact development practices: Literature review and suggestions for future research. *Water Air Soil Pollution, 223,* 4253–4273.

Atlanta Regional Commission. (2001). *Georgia stormwater management manual* (Vol. 2). Atlanta: Atlanta Regional Commission.

ASTM. (2013). *Standard test method for surface infiltration rate of permeable unit pavement systems (ASTM C1781/C1781M-13).* West Conshohocken, PA: ASTM International.

Austin, G. (2012). Design and performance of bioretention beds for removal of stormwater contaminants. *Journal of Green Building, 7*(1), 17–27.

Bean, E. Z., Hunt, W. F., & Bidelspach, D. A. (2007). Field survey of permeable surface infiltration rates. *Journal of Irrigation and Drainage, 133*(3), 249–255.

Biazin, B., Sterk, G., Temesgen, M., Abdulkedir, A., & Stroosnijder, L. (2012). Rainwater harvesting and management in rainfed agricultural systems in sub-Saharan Africa—A review. *Physics and Chemistry of the Earth, Parts A/B/C, 47,* 139–151.

Bickelmann, C. (2013). *Harvesting rainwater. Guide to water-efficient landscaping* (2nd ed.). Tucson: City of Tucson Water Department.

Boers, T. M., & Ben-Asher, J. (1982). A review of rainwater harvesting. *Agricultural Water Management, 5*(2), 145–158.

Boogaard, F., Lucke, T., van de Giesen, N., & van de Ven, F. (2014). Evaluating the infiltration performance of eight Dutch permeable pavements using a new full-scale infiltration testing method. *Water, 6*(7), 2070–2083.

Borgwardt, S. (2006). Long-term in-situ infiltration performance of permeable concrete block pavement. In: *8th International Conference on Concrete Block Paving*, November 6–8, 2006, San Francisco, CA.

Borst, M., Rowe, A. A., Stander, E. K., & O'Connor, T. P. (2010). *Surface infiltration rates of permeable surfaces: Six month update (November 2009 through April 2010).* Edison, NJ: U.S. Environmental Protection Agency, National Risk Management Research Laboratory, Water Supply and Water Resources Division.

Brattebo, B. D., & Booth, D. B. (2003). Long-term stormwater quantity and quality performance of permeable pavement systems. *Water Research, 37,* 4369–4376.

British Geological Survey. (2017). *Sustainable drainage systems (SuDS).* Retrieved June 2, 2018 from http://www.bgs.ac.uk/research/engineeringGeology/urbanGeoscience/SUDS/home.html.

Brown, R. A., & Hunt, W. F., III. (2010). Impacts of construction activity on bioretention performance. *Journal of Hydrologic Engineering, 15*(6), 386–394.

Bruins, H. J., Evenari, M., & Nessler, U. (1986). Rainwater-harvesting agriculture for food production in arid zones: The challenge of the African famine. *Applied Geography, 6*(1), 13–32.

Chang, N.-B., Hossain, F., & Wanielista, M. (2010a). Filter media for nutrient removal in natural systems and built environments: I—Previous trends and perspectives. *Environmental Engineering Science, 27*(9), 689–706.

Chang, N.-B., Wanielista, M., & Daranpob, A. (2010b). Filter media for nutrient removal in natural systems and built environments: II—Design and application challenges. *Environmental Engineering Science, 27*(9), 707–720.

Chang, N.-B., Wanielista, M., Moberg, M., Hossain, F., & Daranpob, A. (2010c). Use of functionalized filter medium for nutrient removal in stormwater ponds. In N.-B. Chang (Ed.), *Effects of urbanization on groundwater* (pp. 199–223). Reston, VA: ASCE.

Clark, S. E., Pitt, R., & Field, R. (2010). Groundwater contamination potential for infiltration of urban stormwater runoff. In N.-B. Chang (Ed.), *Effects of urbanization on groundwater* (pp. 119–164). Reston, VA: ASCE.

Collins, K. A., Hunt, W. F., & Hathaway, J. M. (2008). Hydrologic comparison of four types of permeable pavement and standard asphalt in eastern North Carolina. *Journal of Hydrologic Engineering, 13*(12), 1146–1157.

Colorado Division of Water Resources. (2015). *Rainwater collection in Colorado*. Denver: Colorado Division of Water Resources.

Critchley, W., & Siegert, K. (1991). *Manual for the design and construction of water harvesting schemes for plant production*. Rome: Food and Agriculture Organization of the United Nations.

Critchley, W., Siegert, K., & Chapman, C. (1991). *Water harvesting*. Rome: Food and Agriculture Organization of the United Nations.

Cummings, K. (2012). Adapting to water scarcity: A comparative analysis of water harvesting regulation in the four corner states. *Journal of Environmental Law and Litigation, 27,* 539–570.

Daily, C., & Wilkins, C. (2012). *Passive water harvesting*. Tucson: University of Arizona, College of Agriculture and Life Sciences Cooperative Extension.

Davis, A. P., Hunt, W. F., Traver, R. G., & Clar, M. (2009). Bioretention technology: Overview of current practice and future needs. *Journal of Environmental Engineering, 135*(3), 109–117.

Dietz, M. E. (2007). Low impact development practices: A review of current research and recommendations for future directions. *Water Soil Air Pollution, 186,* 351–363.

Downey, N. (2010). *Harvest the rain, how to enrich your life by seeing every storm as a resource*. Santa Fe, NM: Sunstone Press.

Erickson, A. J., Gulliver, J. S., & Weiss, P. T. (2007). Enhanced sand filtration for storm water phosphorus removal. *Journal of Environmental Engineering, 133*(5), 485–497.

Erickson, A. J., Gulliver, J. S., & Weiss, P. T. (2012). Capturing phosphates with iron enhanced sand filtration. *Water Research, 46*(9), 3032–3042.

Evenari, M., Shanan, L., & Tadmor, N. (1971). *The Negev. The challenge of a desert*. Cambridge, MA: Harvard University Press.

Ferguson, B. K. (2005). *Porous pavements*. Boca Raton: CRC Press.

Ferguson, B. K. (2010). Porous pavements in North America: Experience and importance. In *Proceedings NOVATECH 2010—7th International Conference on Sustainable Techniques and Strategies for Urban Water Management*, Lyon, France (28/06/2010–01/07/2010).

FHWA. (2012). *Pervious concrete (FHWA-HIF-13-009)*. Washington, DC: U.S. Department of Transportation Federal Highway Administration.

FHWA. (2015a). *Porous asphalt pavement with stone reservoirs (FHWA-HIF-15-009)*. Washington, DC: U.S. Department of Transportation Federal Highway Administration.

FHWA. (2015b). *Permeable interlocking concrete pavement (FHWA-HIF015-007)*. Washington, DC: U.S. Department of Transportation Federal Highway Administration.

Fidelibus, M. W., & Bainbridge, D. A. (1994). *Microcatchment water harvesting for desert revegetation*. Restoration Bulletin 5. San Diego: Soil Ecology and Restoration Group (SERG).

Garg, A., Kimball, B. A., Uprety, D. C., Hongmin, D., Upadhyay, J., & Dhar, S. (n.d.). *Conservation tillage*. Retrieved June 2, 2018 http://www.climatetechwiki.org/technology/conservation-tillage.

Gaston, T. L. (2010). Rainwater harvesting in the southwestern United States. A policy review of the Four Corners states. Unpublished research report.

Goodland, M (2016, March 30). Rainwater harvesting may no longer be outlawed in Colorado. *Colorado Independent*. http://www.coloradoindependent.com/158551/rainwater-harvesting-may-no-longer-be-outlawed-in-colorado.

Goodwin, D., Parry, B., Burris, F., Chan, S., & Punton, A. (2008). *Barriers and opportunities for low impact development: Case studies from three Oregon communities*. Corvallis: Oregon Sea Grant.

Ground Water Protection Council. (2007). *Ground water report to the nation: A call to action*. Oklahoma City: Ground Water Protection Council.

Gujarat State Watershed Management Agency. (2011). *Technical manual—Integrated water management program*. Government of Gujarat, India. Retrieved June 2, 2018 from http://gswma.gujarat.gov.in/pfile/Policies/Technical%20Manual%2018-06-2011.pdf.

Hathaway, J. M., & Hunt, W. F. (2006). *Level spreaders: Overview, design, and maintenance* (AGW-588-09), Urban Waterways Series. Raleigh, NC: North Carolina Cooperative Extension Services.

Holman-Dodds, J. K., Bradley, A. A., & Potter, K. W. (2003). Evaluation of hydrologic benefits of infiltration based urban stormwater management. *Journal American Water Resources Association, 39*(1), 205–215.

Hood, A., Chopra, M., & Wanielista, M. (2013). Assessment of biosorption activated media under roadside swales for the removal of phosphorus from stormwater. *Water, 5*(1), 53–66.

Houdeshel, C. D., Pomeroy, C., A., & Hultine, K. R. (2012). Bioretention design for xeric climates based on ecological principles. *Journal American Water Resources Association, 48*, 1178–1190.

Houdeshel, D., & Pomeroy, C. (2014). Storm-water bioinfiltration as no-irrigation landscaping alternative in semiarid climates. *Journal of Irrigation and Drainage Engineering, 140*(2), 06013004.

Hunt, W. F. (2011). *Maintaining permeable pavements* (AGW-588-23W). Urban Waterways Series. Raleigh, NC: North Carolina Cooperative Extension.

Hunt, W. F., & Collins, K. A. (2008). *Permeable pavement; Research update and design implications* (AGW-588-14). Urban Waterways series. Raleigh, NC: North Carolina Cooperative Extension.

Hunt, W., & Lord, W. (2006). *Bioretention performance, design, construction, and maintenance*. Urban Waterways. Raleigh, NC: NCSU A&T Cooperative Extension.

Hunt, W. F., Traver, R. G., Davis, A. P., Emerson, C. H., Collins, K. A., & Stagge, J. H. (2010). Low impact development practices: designing to infiltrate in urban environments. In N.-B. Chang (Ed.), *Effects of urbanization on groundwater* (pp. 308–343). Reston, VA: ASCE.

International Stormwater BMP Database. (n.d.). History. Retrieved June 2, 2018 from http://www.bmpdatabase.org/history.html.

International Stormwater BMP Database. (2014). International stormwater best management practices (BMP) database pollutant category statistical summary report. Solids, bacteria, nutrients, and metals, December 2014.

Jayasuriya, N., & Kadurupokune, N. (2010). Comparative performance of permeable and porous pavements. In *International Conference of Sustainable Built Environment*, December 13–14, 2010, Kandy, Sri Lanka.

Kim, H., Seagren, E. A., & Davis, A. P. (2000). Engineered Bioretention for removal of nitrate from stormwater runoff. In *Proceedings of the Water Environment Federation Technical Exhibition and Conference* (Vol. 2000, no. 13, pp. 623–632). Anaheim, California, October 14–18.

Kim, H., Seagren, E. A., & Davis, A. P. (2003). Engineered bioretention for removal of nitrate from stormwater runoff. *Water Environment Research, 75*(4), 355–367.

Kinkade-Levario, H. (2007). *Design for water: Rainwater harvesting, stormwater catchment, and alternate water reuse: Gabriola Island*. British Columbia: New Society Publishers.

Knight, E. M. P., Hunt, W. F., & Winston, R. J. (2013). Side-by-side evaluation of four level spreader–vegetated filter strips and a swale in eastern North Carolina. *Journal of Soil and Water Conservation, 68*(1), 60–72.

Lancaster, B. (2006). *Rainwater harvesting for drylands and beyond. Volume 1. Guiding principles to welcome rain into your life and landscape*. Tucson, AZ: Rainsource Press.

Lancaster, B. (2008). *Rainwater harvesting for drylands and beyond. Volume 2. Water-harvesting earthworks*. Tucson, AZ: Rainsource Press.

Lloyd, S. D., Wong, T. H., & Chesterfield, C. J. (2002). *Water sensitive urban design: A stormwater management perspective*. Victoria: Cooperative Research Centre for Catchment Hydrology.

Lowndes, M. (2001). *Grassed swales. The Wisconsin storm water manual* (G3691-7). Madison: University of Wisconsin Extension.

Lucke, T., Beecham, S., Boogaard, F., & Myers, B. (2013). Are infiltration capacities of clogged permeable pavements still acceptable? In *NovaTech 2013 Proceedings*.

Martens, D. A., & Frankenberger, W. T. (1992). Modification of infiltration rates in an organic-amended irrigated. *Agronomy Journal, 84*(4), 707–717.

Miller, M. (2006). Rainwater harvesting for enhanced groundwater recharge through capture of increased runoff from site development. In *Proceedings 2006 Universities Council on Water Resources (UCOWR) Conference*, July 18–20, 2006, Santa Fe, New Mexico.

National Academy of Sciences. (1974). *More water for arid lands. Promising technologies and research opportunities*. Washington, DC: National Academy of Sciences.

Natural Resources Conservation Service. (2004). Chapter 10. Estimation of direct runoff from storm rainfall, Part 630, hydrology, In *National engineering handbook*. Washington, DC: U.S. Department of Agriculture, Natural Resources Conservation.

Nieber, J. L., Arika, C. N., Lahti, L., Gulliver, J. S., & Weiss, P. T. (2014). *The impact of stormwater infiltration practices on groundwater quality*. Report prepared for the Metropolitan Council (Twin Cities, Minnesota). University of Minnesota Services.

Nichols, P. W. B., Lucke, Y., & Dierkes, C. (2014). Comparing two methods of determining infiltration rates of permeable interlocking concrete pavers. *Water, 6*, 2353–2366.

NRCS. (2005). *Rain gardens*. Washington, D.C.: U.S. Department of Agriculture Natural Resources Conservation Service. Retrieved June 2, 2018 from https://www.nrcs.usda.gov/Internet/FSE_DOCUMENTS/nrcs142p2_011366.pdf.

O'Neill, S. W., & Davis, A. P. (2011a). Water treatment residual as a bioretention amendment for phosphorus. I: Evaluation studies. *Journal of Environmental Engineering, 138*(3), 318–327.

O'Neill, S. W., & Davis, A. P. (2011b). Water treatment residual as a bioretention amendment for phosphorus. II: Long-term column studies. *Journal of Environmental Engineering, 138*(3), 328–336.

O'Reilly, A. M., Wanielista, M. P., Chang, N. B., Xuan, Z., & Harris, W. G. (2012). Nutrient removal using biosorption activated media: Preliminary biogeochemical assessment of an innovative stormwater infiltration basin. *Science of the Total Environment, 432*, 227–242.

Oweis, T., Hachum, A., & Kijne, J. (1999). *Water harvesting and supplementary irrigation for improved water use efficiency in dry areas*. Colombo, Sri Lanka: International Water Management Institute.

Oweis, T. Y., Prinz, D., & Hachum, A. Y. (2012). *Rainwater harvesting for agriculture in the dry areas*. Boca Raton: CRC Press.

Pacey, A., & Cullis, A. (1986). *Rainwater harvesting: The collection of rainfall and runoff in rural areas*. London: Intermediate Technology Publications.

Pezzaniti, D., Beecham, S. C., & Kandasamy, J. K. (2009). Influence of clogging on the effective life of permeable pavements. *Water Management, 162*(3), 211–220.

PGCDER. (1993). *Design manual for the use of bioretention in stormwater management*. Landover, MD: The Prince George's County Department of Environmental Resources, Environmental Services Division.

PGCDER. (2007). *Bioretention manual*. Landover, MD: The Prince George's County Department of Environmental Resources, Environmental Services Division.

PGCDER. (1999). *Low-impact development: An integrated design approach*. Landover, MD: Prince George's County, Maryland Department of Environmental Resources.

Pitt, R., Lantrip, J., Harrison, R., Henry, C. L., & Xue, D. (1999). *Infiltration through disturbed urban soils and compost-amended soil effects on runoff quality and quantity*, EPA 600/R-00/016. Cincinnati: U.S Environmental Protection Agency, Water Supply and Water Resources Division, National Risk Management Research Laboratory.

Prinz, D., & Singh, A. (2000). *Technological potential for improvements of water harvesting*. Contributing Paper for World Commission on Dams, Cape Town, Africa.

Renner, H. F., & Frasier, G. (1995). Microcatchment water harvesting for agricultural production: Part I: Physical and technical considerations. *Rangelands, 17*(3), 72–78.

Rockström, J., Barron, J., & Fox, P. (2003). Water productivity in rain-fed agriculture: Challenges and opportunities for smallholder farmers in drought-prone tropical agroecosystems. In J. W. Kijne, R. Barker, & D. Molden (Eds.), *Water productivity in agriculture: Limits and opportunities for improvement* (pp. 145–162). Wallingford: CAB International.

Roy-Poirier, A., Champagne, P., & Filion, Y. (2010). Review of bioretention system research and design: Past, present, and future. *Journal of Environmental Engineering, 136*(9), 878–889.

Shinde, M., Smot, I., & Gorantiwar, S. (2006). Assessment of water harvesting and groundwater recharge through continuous contour trenches. In *Recharge systems for protecting and enhancing groundwater resources*, Proceedings of the 5th International Symposium on Management of Aquifer Recharge, Berlin, Germany, June 11–16, 2005 (pp. 17–22). Paris: UNESCO.

Stephens, D. B. (2010). The role of decentralized artificial recharge systems in water resources management. In *55th Annual New Mexico Water Conference*, December 2–3, 2010. Corbett Center, New Mexico State University Las Cruces, New Mexico.

Stephens, D. B., Miller, M., Moore, S. J., Umstot, T., & Salvato, D. J. (2012). Decentralized groundwater recharge systems using roofwater and stormwater runoff. *Journal of the American Water Resources Association, 48*(1), 134–144.

Stern, P. H. (1979). *Small scale irrigation. A manual of lost cost water technology*. London: Intermediate Technology Publications.

Strecker, E. W., Quigley, M. M., Urbonas, B. R., Jones, J. E., & Clary, J. K. (2001). Determining urban storm water BMP effectiveness. *Journal of Water Resources Planning and Management, 127*(3), 144–149.

Sussman, D. (2007). *Design manual. Contour trenches*. Santa Barbara, CA: Bren School of Environmental Science and Management, University of California.

Tennis, P. D., Leming, M. L., & Akers, D. J. (2004). *Pervious concrete pavements*. Skokie, IL: Portland Cement Association and Silver Spring, MD: National Ready Mixed Concrete Association.

UADOC (University of Arkansas Community Design Center). (2010). *Low impact development. A design manual for urban development*. Fayetteville: University of Arkansas Press.

UNEP. (1979). *Rain and storm water harvesting for additional water supply in rural areas*. Nairobi: United Nations Environment Programme.

UNEP. (1984). *Rain and stormwater harvesting in rural areas: Dublin*. United Nations Environment Programme: Tycooly International Publishing.

UNEP. (2014). *Green infrastructure: Guide for water management: Ecosystem-based management approaches for water-related infrastructure projects*. United Nations Environment Programme (UNEP), UNEP-DHI Partnership—Centre on Water and Environment, International Union for Conservation of Nature (IUCN), The Nature Conservancy (TNC) and the World Resources Institute (WRI).

Usda, N. R. C. S. (2007). *Conservation practice standard, contour farming*. Washington, DC: U.S. Department of Agriculture, Natural Resources Conservation Service.

USEPA. (1999a). *Storm water technology fact sheet. Vegetated swales*. Washington, DC: U.S Environmental Protection Agency, Office of Water.

USEPA. (1999b). *Storm water technology fact sheet. Infiltration trench*. Washington DC: U.S. Environmental Protection Agency Office of Water (September 1999).

USEPA. (1999c). *Storm water technology fact sheet. Bioretention*, EPA 832-F-99-012. Washington, DC: United States Environmental Protection Agency Office of Water.

USEPA. (1999d). *Storm water technology fact sheet. Porous pavement*. Washington, DC: U.S Environmental Protection Agency, Office of Water.

USEPA. (n.d.a). *Urban runoff: Low impact development*. Retrieved June 2, 2018 from https://www.epa.gov/nps/urban-runoff-low-impact-development.

USEPA. (n.d.b). *What is green infrastructure*? Retrieved June 2, 2018 from https://www.epa.gov/green-infrastructure/what-green-infrastructure.

VDOT. (2013). Chapter 6—Vegetated water quality swale. In *Virginia DOT BMP manual of practice*. Richmond: Virginia Department of Transportation.

VDCR. (2011). Grass channels (Version 1.8). *Virginia DCR stormwater design specification no. 2*. Richmond: Virginia Department of Conservation and Recreation.

WADOE. (2013). *Eastern Washington low impact development guidance manual*. Lacey, WA: State of Washington Department of Ecology.

Waller, D. H. (1989). Rain water—An alternative source in developing and developed countries. *Water International, 14,* 27–36.

Wanielista, M., & Chang, N.-B. (2008). *Alternative stormwater sorption media for the control of nutrients*. Orlando: Stormwater Management Academy, University of Central Florida.

Wanielista, M., Chopra, M., Spence, J., & Ballock, C. (2007). *Hydraulic performance assessment of pervious concrete pavements for stormwater management credit*. Tallahassee: Florida Department of Transportation.

Wanielista, M., Chang, N.-B., Randall, A., Chopra, M., Hardin, M., Jones, J., et al. (2014). *Demonstration biomedia for ultra-urban stormwater treatment*. Orlando: Stormwater Management.

Waterfall, P. H. (2006). *Harvesting rainwater for landscape use*. Tucson: University of Arizona Cooperative Extension. https://extension.arizona.edu/sites/extension.arizona.edu/files/pubs/az1344.pdf.

Wisconsin DNR. (2014). *Bioretention for infiltration. Conservation practice standard*. Madison: Wisconsin Department of Natural Resources.

Winston, R. J., Hunt, W. F., Lord, W. G., & Lucas, A. C. (2010). *Level spreader update: Design, construction, and maintenance*, AGW-588-20W. Urban Waterways Series. Raleigh: North Carolina Cooperative Extension Services.

Winston, R. J., & Hunt, W.F. (2010). *Level spreader update: Performance and research* (AGW-588-21W). Urban Waterways Series. Raleigh: North Carolina Cooperative Extension Services.

Wisconsin DNR. (2003). *Rain gardens. A how-to manual for homeowners*. Madison: Wisconsin Department of Natural Resources and University of Wisconsin-Extension. Retrieved June 2, 2018 from http://dnr.wi.gov/topic/shorelandzoning/documents/rgmanual.pdf..

Wong, T. H. (2006a). An overview of water sensitive urban design practices in Australia. *Water Practice and Technology, 1*(1), wpt2006018.

Wong, T. H. (2006b). Water sensitive urban design—The journey thus far. *Australian Journal of Water Resources, 10*(3), 213–222.

Yong, C. F., Deletic, A., Fletcher, T. D., & Grace, M. R. (2008). The clogging behavior and treatment efficiency of a range of porous pavements. In *Proceedings 11th International Conference on Urban Drainage*, Edinburgh, Scotland.

Chapter 24
Unmanaged and Unintentional Recharge

24.1 Introduction

The bulk of this book is focused on managed aquifer recharge (MAR), which has been broadly defined as the "the purposeful recharge of water to aquifers for subsequent recovery or environmental benefits" (Dillon 2009). Unmanaged recharge (UMAR), which has also been referred to as "culturally modified recharge" (Stephens 1996), can be defined as recharge incidental to other human activities. The NRMMC, EPHC and NHMRC (2009) differentiated between unintentional recharge and unmanaged recharge. Unintentional recharge includes processes, such as leakage from water and wastewater mains, that are unplanned and often undesirable. Unmanaged recharge, as defined by the NRMMC, EPHC and NHMRC (2009), includes intentional activities and systems that have a primary disposal function, such as septic system leach fields. Aquifer recharge from irrigation return flows is also considered UMAR as it is the result of an intentional activity. Enhanced recharge as the result of changes in land use/land cover (LULC) also fall within the realm of anthropogenic aquifer recharge (AAR). Groundwater recharge can be increased either intentionally or unintentionally by changing LULC, particularly by decreasing evapotranspiration (ET) losses by removing or changing vegetation.

Unmanaged and unintentional recharge are important components of some local water budgets. Ironically, some cities in the Middle East, one of the driest areas of the world (e.g., Riyadh and Jeddah, Saudi Arabia; Kuwait City; Doha, Qatar), are experiencing rising shallow groundwater levels, which necessitates installation of drainage systems for the protection of building foundations, basements, and underground structures (Alhamid et al. 2007; Al-Sefry and Şen 2006). The distinction between unmanaged and unintentional recharge is not particularly relevant from an applied perspective, so long as the source of recharge is identified and quantified to some degree and water quality impacts understood and controlled.

R. G. Maliva, *Anthropogenic Aquifer Recharge*,
Springer Hydrogeology, https://doi.org/10.1007/978-3-030-11084-0_24

24.2 Urban Unmanaged Recharge

Sharp (2010) observed that "Although it is commonly stated, that groundwater recharge is reduced with urbanization because of the increase in impervious cover, the reverse is the more common condition—urbanization increases groundwater recharge." Factors that contribute to increased urban recharge and rising shallow groundwater levels include (Lerner 1986, 1990, 2002; Brassington and Rushton 1987; Foster 1990):

- leakage from utility mains and lines
- irrigation return flows from over irrigation of parks, gardens, and landscaped areas
- in situ sanitation in unsewered areas
- impervious covers that are more pervious than expected
- stormwater retention/detention/infiltration ponds
- discharges to losing (ephemeral) streams
- reductions in groundwater pumping after very long periods of pumping
- change from a vegetated land cover to an impervious cover with an associated reduction in evapotranspiration (ET).

The adverse impacts of rising groundwater levels in urban areas include (Al-Sefry and Şen 2006):

- flooding of house basements
- deterioration of roads and highways
- damage to building foundations
- contamination of soils
- public health impacts from ponded waters
- offensive odors
- breeding of mosquitos
- contamination of shallow aquifers that could be used as strategic storage reservoirs of artificially recharged freshwater for emergency situations.

Areas most likely to experience rising groundwater levels are low-lying areas with relatively low permeability aquifers, especially if the aquifers are not being exploited for groundwater supply (Al-Sefry and Şen 2006). Rising groundwater can cause flooding of cellars and basements, deep engineered structures (e.g., elevator shafts), and passenger transport (train) tunnels. Health hazards and nuisances may occur where the water table rises to land surface and results in standing water. In situ sanitation in unsewered areas increases groundwater recharge and is a source of pollution. In situ sewage disposal is a proven vector of pathogen transmission and increases in groundwater nitrate and dissolved organic carbon (including trace organic compounds) concentrations (Foster 1990).

Management of UMAR requires an identification of its sources and quantities, which usually involves an evaluation of utility operational data, (e.g., system water losses), water chemistry data, and groundwater modeling. Solutions to excessive UMAR in urban environments may include a program to reduce leakage from water and wastewater mains, and, in some instances, active groundwater pumping to lower water levels in shallow aquifers. Depending upon its quality, the recovered water may be suitable for some non-potable uses.

The complexity of urban environments means that identification of all of the very large number of point discharge locations is likely cost prohibitive. The objective is instead to determine whether sufficient individual sources (e.g., water mains leaks) exist to have an impact on overall urban recharge. Methods used to evaluate urban recharge include (Lerner 2002):

- piezometry—mapping of local rises in the water table
- tracers
  - inorganic tracers (major cations and anions) and trace elements
  - organic chemicals, including trace organic compounds related to specific sources (e.g., detergents may have domestic sewage sources)
  - microbiological parameters (e.g., fecal coliform bacteria)
  - stable isotopes (^{2}H, ^{15}N, ^{18}O, ^{35}S)
- water budget analysis
- unaccounted for water analysis
- minimum night flow analysis
- inverse groundwater modeling.

The cause(s) of groundwater rises may be inferred from the relationship between historic changes in groundwater levels, LULC changes, and the distribution of utility infrastructure. For example, rising groundwater that occurs preferentially in areas that are not yet connected to a sanitary sewer system could be related to on-site sewage disposal systems. Higher groundwater levels near parks and other irrigated areas might be due to excessive irrigation. The major limitation of the piezometry method is that it requires a time series of detailed groundwater level (i.e., potentiometric or piezometric surface) maps, which are often not available.

Tracers can be diagnostic of water sources. The ideal tracer (marker) species should be unique to a particular recharge source (irrespective of geographic location) and easily identifiable in a groundwater system (i.e., present at well above background concentrations; Barrett et al. 1997). Tracers indicative of sewage and other water sources were reviewed by Vengosh and Pankratov (1998), Barrett et al. (1997), and Gasser et al. (2010). Halides (Cl, Br, F) are effective and relatively inexpensive tracers, particularly where potable water supplies are fluoridated. Other, more sophistical tracers, such as nitrogen isotopes and refractory trace organic compounds (e.g., carbamazepine), may also provide useful information but are more expensive.

Tracer studies of urban recharge should start with a preliminary conceptual model that identifies potential sources of recharge, assesses potential groundwater geochemical conditions, and includes reasonable estimates concerning which tracers might display a broad range of concentrations that can be related to potential recharge sources. Mixing models are then used to determine the contributions of potential source waters to the groundwater present in studied wells. For each tracer, the average concentration in groundwater is a function of amount recharge from each source and the average concentration of each source water. Simple binary or multi-component mixing equations may be used:

$$C_i = \frac{1}{R}\sum R_j C_{ij} \tag{24.1}$$

where,

C_i concentration of a parameter in a groundwater sample
R_j amount of recharge from source "j"
C_{ij} concentration parameter "i" in water from source "j".

Mixing equation analysis become greatly complicated when aquifer water chemistry and recharge water compositions are spatially and temporally heterogeneous.

Unmanaged recharge amounts and locations may be evaluated through inverse-modeling, which is essentially the model calibration process in which model parameters (including recharge) are adjusted so that model outputs (simulated water levels) match field data. Inverse-modeling does not provide unique solutions and the accuracy of recharge estimates will depend upon the accuracy of other model parameters. If a model in poorly constrained (i.e., there is little data on aquifer hydraulic parameters and other elements of the water budget), then recharge values obtained through inverse modeling will have a low reliability. Nevertheless, high levels of urban recharge may be indicated if a model will not calibrate without additional local recharge added.

24.2.1 Potable Water Mains Leakage

Where potable water use is metered, losses from mains leakage are included in the unaccounted for water, which is the difference between produced water sent to the distribution system and metered water use. Unmetered water use includes both authorized uses (e.g., firefighting and unmetered customers) and unauthorized uses (e.g., illegal connections). The accuracy of estimates of leakage losses depends upon how well various unmetered water uses can be estimated and the accuracy of the metering.

Water losses are an economic burden to water utilities and various procedures have been developed to quantify leakage and locate leaks. Leakage losses can be roughly estimated from minimum night flow (MNF) analysis. MNF analysis is based on the concept that there is little actual water use very late at night (e.g., between 1:00 and 5:00 AM). Flows through the distribution network late at night is the sum of leakage and legitimate uses. Protocols have been developed for the correct performance and interpretation of MNF tests (e.g., Werner et al. 2011).

Lerner (1986) presented evidence of water main leakage being a significant source of groundwater recharge in Lima, Peru, and Hong Kong. Recharge rates from water main leaks were estimated to be about 50% of the produced water based on MNF results and evaluation of shallow aquifer water balances by groundwater model calibration. Lerner (1986) estimated that 40% of the average potable water supply for Lima at the time of the study was becoming recharge through leakage. The other 10% is on-premise leakage that enters the sewers. A groundwater model of the city would not calibrate unless the recharge from pipe leakage was included. Leakage was also determined to be the controlling factor for groundwater heads and slope stability in Hong Kong, but the actual recharge rate was not determined. Water main leakage in Hong Kong is evident by unnatural water levels and piezometric responses that can best be explained by leakage.

Leakage of water mains would be expected to contribute more water to recharge than from sewage mains because the former are pressurized. Sewers and storm drains are typically not pressurized (except for force mains), but usually receive less maintenance than water mains (Lerner 1986; Foster 1990).

24.2.2 Sewer Leaks—Exfiltration

Sewage mains are normally not metered and receive less monitoring and maintenance than potable water mains. Factors affecting the integrity of sewer pipes include (Ellis 2001):

- age; pipes in parts of some major European cities are over 100 years olds
- poor and/or outdated quality
- lack of sufficient maintenance
- inadequate funding relative to the high costs of replacement and rehabilitation
- geological conditions (tectonic movements and subsidence).

Current approaches used to identify and quantity exfiltration commonly involve water quality tracers, particularly parameters specific to sewage or present at distinctly difference concentrations in sewage than in groundwater including (Ellis 2001):

- standard ion chemistry
- fecal microorganisms (e.g., total and fecal coliforms, *E. coli*, fecal streptococcus, coliphage, fecal viruses)
- trace organics (pharmaceuticals, detergent constituents, industrial chemicals)
- stable isotopes (^{15}N).

A difference of opinion exists as to whether leakage (exfiltration) from sewer systems is an important source of groundwater recharge and contamination, or is unlikely to be a real problem because defects (leaks) in sewer systems are self-sealing due to the nature and constituents of sewage effluent (Blackwood et al. 2005). The self-sealing propensity is due to sediments and associated solids in sewage, and wall slimes and biofilm growth. Blackwood et al. (2005) provided experimental evidence that the gravel bed of sewer pipes contributes to more rapid and effective sealing of leaks. Testing results were also presented that suggest a rapid reduction in the concentration of fecal microorganisms in the soil beneath sewer pipes, which may be due to out competition by native organisms, soil type, and/or filtration (Blackwood et al. 2005). Ellis (2001) similarly concluded that the overall impact of sewer exfiltration on groundwater quality does not appears to be severe. However, it was noted that further work is needed to verify the nature and magnitude of long-term sewer exfiltration before it can be safely discounted as a potential diffuse source of groundwater pollution.

Closed-circuit television (CCTV) inspection of the Rastatt, Germany, sewer system revealed 31,006 defects within the 208 km sewer system (Wolf et al. 2006). The most common type of defects were damaged or improperly installed house connections (13,646) and joint displacements. The degree to which the defects were actually leaking could not be determined from the CCTV survey. The Rastatt sewer system would be considered a rather well-maintained system (Wolf et al. 2006), so greater defect rates may occur in other systems. The recharge rate to the urban aquifer from sewer leakage in the Rastatt sewer system was estimated to be 2.88–5.06 mm/year. Wolf et al. (2006) noted that development of a clogging layer may reduce exfiltration rates by sealing leaks. However, leakage rates may increase by an order of magnitude or more when the clogging layer is damaged by storms (Vollertsen and Hvitved-Jacobsen 2003; Wolf et al. 2006).

Wolf et al. (2006) also estimated leakage using tracers. Elevated boron concentrations were detected in the groundwater in the city center. Boron was widely used in detergents in the past. The iodated contrast media amidotrizoic acid and iothalamic acid were found to be present in both sewage and groundwater. Using the concentrations of the iodated contrast media as sewage tracers, it was estimated that there is 5–12% sewage in the urban aquifer of Rastatt.

Eiswirth and Hötzl (1999) investigated the impacts of sewer leaks on water quality at a constructed sewer test site in Rastatt. In Rastatt, 87% of the sewerage system is situated within the zone of fluctuation of the water table. Groundwater levels influence the exfiltration behavior of damaged sewers. Leakage (exfiltration) can occur when the sewers are located above the water table. Biodegradation and other attenuation processes within the unsaturated zone were found to reduce the concentrations of some contaminants (organic carbon and nitrate). The effects of sewer leaks were found to be strongly variable with impacts to water quality occurring only within a narrow zone next to the leaks.

Morris et al. (2006) investigated modern recharge of a Permo-Triassic sandstone aquifer (Sherwood Sandstone) in the Bessacarr-Cantley suburb of Doncaster, England, using environmental indicators. The tracers used included CFCs, SF_6, and microbiological fecal indicators (total and fecal coliforms, fecal streptococci, enteric viruses, and sulfite-reducing clostridia). CFC and SF_6 data indicate modern recharge has penetrated tens of meters below land surface, likely through fracture horizons. Fecal streptococci and sulfite-reducing clostridia spores were also detected at depth. The results indicate that urbanization of Bessacarr-Cantley over the previous 80 years had a slight impact on water quality, at least for the range of parameters examined in the study (Morris et al. 2006). The slight nature of the impacts was attributed in part to dilution from direct recharge through green space areas and local high storage in the poorly cemented shallow aquifer.

24.2.3 On-Site Septic Wastewater Treatment Systems

On-site septic wastewater-treatment systems (OWTS) can be a significant return flow to groundwater but are also a potential source of contamination from total dissolved solids (TDS), nitrates, pathogenic microorganisms, and a wide variety of organic chemicals. Recharge from OWTS can increase groundwater levels and stream baseflows. For example, a U.S. Geological Survey investigation in Gwinnett County, Georgia, during an extreme drought in October 2007 observed that the mean base-flow yield in high-density OWTS watersheds was 90% greater than that in a low-density OWTS watershed (Landers and Ankcorn 2008).

There is still very little quantitative data on the amount of actual recharge from individual systems or cumulatively from all the systems in a geographic area. McQuillan and Bassett (2009) reported that previous studies of the return flow from on-site septic systems in the City of Roswell, New Mexico, vicinity gave estimated return flows of 42.5 and 47% of the water delivered. The percentage of delivered water that recharges the water table aquifer via septic systems depends upon system design, local geology, and the fraction of household water use that does not enter the septic system (e.g., outdoors uses for landscape irrigation and car washing).

McQuillan and Bassett (2009) proposed that the recharge benefits of on-site wastewater systems could be increased by designing and locating systems to maximize recharge. Systems could be constructed deeper, below the zone of greatest ET, as opposed to the standard paradigm of constructing shallow systems to enhance natural treatment by soils. McQuillan and Bassett (2009) proposed that water quality issues associated with on-site dispersal systems may be of secondary concern to severe aquifer dewatering.

24.2.4 Increased Urban Imperviousness and Recharge

In urbanized areas, surficial soils become compacted by traffic and large portions of land surfaces are sealed by buildings and pavements, decreasing surface perviousness (Urbonas et al. 1992; Kennedy 2007; Kennedy et al. 2010; Stewart 2014) and generating more runoff than was generated prior to development. Herein, urban-enhanced infiltration is that volume of channel-bed focused infiltration that can be attributed to increased runoff from urban areas. Urban-enhanced groundwater recharge is the fraction of urban-enhanced infiltration that percolates to merge with the underlying saturated zone.

Recharge rates depend on the duration and areal extent of infiltration, and soil properties. In semiarid and arid lands where recharge occurs primarily (or only) where runoff is concentrated (i.e., recharge is indirect or focused), increased imperviousness can concentrate runoff and increase recharge. For example, recharging conditions in the Mojave Desert occur where urbanization has concentrated distributed runoff into a small number of fixed channels. Increased recharge is evident by the mobilization of chloride that had previously accumulated in the root zone (Izbicki et al. 2007).

An investigation of groundwater recharge in Coyote Wash, an 11,461 acre (4,638 ha) urban subwatershed of the Sierra Vista watershed, southeastern Arizona, showed that urban-enhanced channel recharge and stormwater basins may increase recharge rates by 200–300 acre feet/year (AF/year; 0.25–0.37 million cubic meters/year; MCM/year) above predevelopment levels (Milczarek e al. 2004). The increased recharge is attributed to an increase in impervious area and decreases in soil infiltration due to compaction and other impacts of site development. Many communities in Arizona are investigating, and even implementing, stormwater management approaches to take advantage of the increased runoff to increase groundwater recharge (Lohse et al. 2010). The potential tradeoff is that increased indirect recharge of stormwater may impact water quality. Modifications of receiving ephemeral channels, such as grass linings, may attenuate contaminants (Lohse et al. 2010).

Increases in impervious cover are thought to reduce local recharge, but "impervious" covers may have secondary porosity features that allow for infiltration. Wiles and Sharp (2008) examined the permeability of fractures and expansion joints in pavements and its role in urban recharge. Double-ring infiltrometer measurements were made in parking lots, roads, and concrete curb gutters in Austin, Texas. The equivalent hydraulic conductivity due to fractures and joints in the study area was calculated to be 5.9×10^{-5} cm/s $\pm$ 1.3×10^{-5} cm/s, which is equivalent to the values for fine-grained soils, sandstones, silts, and loams. Pavements may increase vertical infiltration and recharge if the ratio of the hydraulic conductivity of the urban surface to that of the dominant surface cover is much greater than 1, which appears to be the case in Austin, where the fine-grained alluvial soils have a reported average hydraulic conductivity of 1.07×10^{-5} cm/s (Wiles and Sharp 2008). Water that infiltrates through pavement may have a relatively high potential to become recharge due to the absence of evapotranspiration (Wiles and Sharp 2008). Wiles and Sharp (2008) estimated that 21% (170 mm) of the mean annual rainfall of (809 mm) is available as potential recharge. Further research is needed to determine the actual groundwater recharge.

24.2.5 Published Studies of Urban UMAR

UMAR was investigated in Barcelona, Spain, using Cl, SO_4^{-2}, ^{34}S, B, F, Br, EDTA, Zn, ^{18}O, D, total N, and residual alkalinity as tracers (Vázquez-Suñe et al. 2010). The basic challenges associated with using mixing models to ascertain the relative contribution of recharge sources are (Vázquez-Suñe et al. 2010):

- identification of conservative (non-reactive) tracer species that have markedly different concentrations in each recharge source
- identification of at least n − 1 tracer species, where n = number of potential recharge sources
- accurate determination of the concentrations of the tracer species in each source end-member, which may vary in both space and time.

The chemistry of water leaking from sewage lines may vary seasonally and between different locations in the collection network. A single set of values for the concentration of each tracer species may not adequately represent for example "sewage line leakage" recharge water. Mixing ratios are very sensitive to end-member concentrations.

Yang et al. (1999) examined urban recharge in the city of Nottingham (UK). Solute balances were used to supplement standard water balance and groundwater modeling studies. Groundwater modeling using the MODFLOW and MT3D96 codes was performed to estimate total recharge and solute data were intended to quantify the contributions of different sources of recharge. Three tracers were used (Cl, SO_4, and Total N) and three sources were considered (precipitation, leaking water mains, and sewage). A transient model was developed for a 145-year period (1850–1995), which was divided into 12 stress periods. Total recharge estimates were obtained for each of five years (1877, 1914, 1945, 1939, and 1965) in which land use was mapped and the expansion of the city is documented. The results of a sensitivity analysis indicate that recharge from water mains and sewage cannot be accurately measured due to the lack of good quality historical data and the long turnover time of the aquifer. Confidence intervals of ±40 and ±100% for mains and sewage recharge rates, respectively, were deemed to be appropriate. Total recharge was estimated to have decreased by about 8%, which is likely within the margin of error.

Kruse et al. (2013) demonstrated the use of model calibration to estimate urban recharge in the city of La Plata, Argentina. A groundwater model was developed using MODFLOW that was initially calibrated to 1940 condition. For year 2008, a simulation was first performed using an urban recharge rate of 0 m/d. Recharge was added until the modeled potentiometric surface matched the observed surface. Recharge in the urban area had to be increased by 60,000 m^3/d to adjust the simulated potentiometric surface to the observed surface. The simulated urban recharge is of a similar magnitude as the natural recharge by rainfall infiltration (Kruse et al. 2013).

24.3 Canal Seepage

Seepage from unlined conveyances can be an important source of unmanaged aquifer recharge. Canal recharge processes are analogous to the natural recharge resulting from transmission losses in streams. Natural groundwater recharge is favored along stream reaches because (Dagés et al. 2008)

- the amount of water available for recharge is large

- infiltration occurs under a positive pressure head
- prior water content of stream beds tend to be greater than nearby soils
- since the stream bed tends have a lower elevation, water has less distance to cover to reach the water table.

Recharge from transmission losses depend upon (Dagés et al. 2008)

- characteristics of the runoff events
- duration and area of inundation
- initial hydrologic conditions (e.g., soil moisture)
- sedimentation and erosion processes that modify the hydraulic properties of streambeds.

Dagés et al. (2008) performed an experimental study of seepage losses in a ditch during a typical Mediterranean runoff event. The experimental site was an isolated 10 m long segment of ditch in the Roujan catchment of the south of France. The depth to the water table during the test was reported to be 1.85 m (6.1 ft). Recharge processes were investigated using transects of piezometers and tracer (bromide) data. A total of 18 m^3 (2,113 gallons) was infiltrated. The greatest rise of the water table was 0.8 m (3.3 ft) and the hydraulic mound extended up to 16 m (52 ft) out from the ditch.

The piezometer response was rapid with an initial increase in head below the ditch detected 10 min after the start of recharge. After recharge stopped, the water table mounds dissipated almost as fast as they grew. The tracer migrated by piston-flow and the areal extent of the tracer plume was much less than the extent of the hydraulic mound. The fully saturated front reached the water table, which did not occur at some other study sites with a thick unsaturated zone and too dry conditions (Dagés et al. 2008; Izbicki et al. 2000).

Canal leakage can be intentionally taken advantage of as a source of recharge. Within the Alpujarra region of southern Spain, aquifer recharge is performed using unlined irrigation channels called "acequias," which were laid out in the 9th–15th centuries A.D. by the Moors, and possibly much earlier by the Romans (Pulido-Bosch and Sbih 1995). Dams divert snowmelt water in the spring to the channels, which gradually descend following topographic contours. The channels are 0.5–2.5 m (1.6–8.2 ft) wide and typically have lengths of up to 15 km (9.3 miles). The channels divert water to irrigated and recharge areas, and provide recharge through channel losses. The channel losses were reported to increase soil moisture, which supports downslope vegetation that is denser than would otherwise occur. Infiltration in recharge areas was also reported to support downgradient springs used for irrigation and domestic supplies. The local people of Alpujarra are conscious of the value of the recharge zones and acequias and make a special effort to preserve the recharge systems, which have been functioning for over a millennium (Pulido-Bosch and Sbih 1995).

UMAR can become part of the local hydrology and impact local water levels and ecosystems, particularly if it is locally a long-term phenomenon. The All-American Canal in the southern California is a good example of the transboundary issues that may arise when unmanaged seepage from an unlined canal becomes the local norm over time, even to the extent that local ecosystems have become dependent on it. The All-American Canal is an aqueduct that conveys Colorado River water to the Imperial Valley of Southern California. The name of the canal is derived from its being constructed entirely on the United States side of the United States-Mexico border. The canal, which was opened in 1942, was constructed unlined, and seepage recharge became an important component of the Mexicali-Imperial Valley aquifer water budget.

Users of the Mexicali-Imperial Valley aquifer had become accustomed to the seepage recharge. The reduction in recharge associated the canal lining would lower aquifer water levels, which would have detrimental impacts to farmers, cities (particularly city of Mexicali), and hydrologically connected wetlands. Of particular concern is the Andrade Mesa wetland area in Northern Mexico, which is an important feeding location for migratory birds.

However, the seepage losses from the All-American Canal are a loss of water to southern California water users. It was recognized that additional water would be available to southern California water users if the canal were lined to reduce the seepage losses. The San Diego County Water Authority agreed to pay part of the $285 million cost of a project to line the canal in exchange for the estimated 95 MCM (77,700 acre-feet) of water saved each year from the lining. Plans to line the canal developed into an international controversy that raised some basic issues concerning transnational groundwater management and law, which were discussed by Huber (2008) and Kibel (2008).

Mexican and American non-profit groups challenged the approval of the lining project in United States Federal District Court. The plaintiffs argued that although the canal lining would occur on United States soil, the environmental, social, and economic impacts to the Mexican side of the canal would revert back to the United States. For example, drying up of the Andrade Mesa wetlands would impact birds that spend part of their lives in the United States. The position of the Bureau of Reclamation was that the seepage water belonged to the United States as part of its Colorado River allocation under the 1944 Water Treaty and that its status remained unaffected by any conservation measures the United States should take. The litigation prompted the United States Congress to adopt legislation in December 2006 that exempted the project from compliance with United States environmental laws. The lining of 23 miles of canal was completed in 2009.

24.4 Irrigation Return Flows

24.4.1 Irrigation Basics

Irrigation can be defined as the provision of additional water to guarantee and increase crop production in areas where there is a water deficiency (Heathcote 1983). Irrigation can result in a tremendous increase in productivity versus rain-fed (dry land) agriculture and a greater variety of crops can be grown. Irrigation is essential for economically viable agriculture in arid and semiarid regions (Pescod 1992). There is little question that irrigation is vital for meeting current and future global food demands.

Irrigation has associated adverse impacts on water quantity and quality. Irrigation is by far the greatest human water use and has resulted in great stress on surface and groundwater resources in wide areas of the world, particularly in the arid and semiarid regions. Depending on local circumstances, irrigation can result in either a net input or output to local groundwater resources. In the case where an external water source is used, such as imported surface or reclaimed water, irrigation can result in a rise in the elevation of the water table. If a local shallow aquifer is used for water supply, irrigation can result in a lowering of aquifer water levels due to a net loss of water to evapotranspiration (ET).

Irrigation can also adversely impact soil and groundwater quality. Dissolved solids (e.g., salts, nutrients, and chemicals) in irrigation water will accumulate in the soil and shallow groundwater unless there is a corresponding outflow of the dissolved solids out of the groundwater basin. Salts present in imported surface water, reclaimed water, and groundwater from deeper confined aquifers can progressively build up within shallow groundwater unless measures are taken to manage the salt balance. Irrigation can also leach naturally existing salts and nutrients out of the soil. Improper irrigation and drainage practices can result in soil salinization, which has reached a critical stage in some irrigated areas, resulting in former arable land being taken out of production.

The following basic conditions should be met in order to make irrigated agriculture a success on a farm level (Pescod 1992):

- the required amount of water should be applied
- irrigation water should be of an acceptable quality
- water application should be properly scheduled
- appropriate irrigation methods should be used
- salt accumulation is the root zone as the result of plant ET should be prevented by means of leaching
- rises of the water table should be controlled by means of appropriate drainage
- plant nutrients should be managed in an optimal manner.

The amount of water required for irrigation depends on plant ET requirements, which must be adjusted for effective rainfall, leaching requirements, application losses (irrigation inefficiency) and other factors (Pescod 1992). Irrigation water applications should be matched to plant requirements. Water should be applied to crops before the soil moisture potential reaches a level for which the ET rate is likely to be reduced below its potential rate (Pescod 1992).

Irrigation applications in excess of plant requirements result in some deep percolation beyond plant root zones (i.e., return flows), which is an important local source of groundwater replenishment in irrigated arid regions (Bouwer 1978). Excessive irrigation and thus deep percolation is undesirable as it results in additional costs, wastes water, and can result in leaching of fertilizer and natural nutrients from soils. However, some deep percolation is necessary to prevent salt build-up in the plant root zone, which can adversely impact plant growth. Crop plants absorb essentially pure water leaving salts behind that accumulate in the root zone. The salt balance can be evaluation using the equation (Bouwer 1978):

$$C_i D_i = C_d D_d \tag{24.2}$$

where,

C_i TDS concentration of the irrigation water (mg/L)
D_i amount of irrigation water applied (mm)
C_d TDS concentration of deep-percolation water (mg/L)
D_d amount of deep percolation water (mm)

Leaching requirements depend upon the salt tolerance of the crops and the salinity of the irrigation water, with the amount of leaching necessary increasing with increasing plant sensitivity and irrigation water salinity.

Drainage is important to prevent salinization, which occurs when the water table rises close to land surface and the capillary rise of saline groundwater transports salts to the soil surface. As the water evaporates, salts are left behind at land surface. Drainage is performed to control the rise of the water table.

Irrigation methods can be divided into five main categories (Pescod 1992):

1. **flood irrigation**: water is applied over an entire field to infiltrate into the soil
2. **furrow irrigation**: water is applied to furrows between ridges upon which crops are grown
3. **sprinkler**: water is applied from above using sprinkler systems including solid-set, travelling, spray gun, and center-pivot
4. **sub-irrigation**: water is applied below the root zone in such a manner that it wets the root zone by capillary rise
5. **localized irrigation**: a variety of methods such as micro-irrigation, drip, bubblers, and microsprinklers are used to efficiently apply water to plants.

Irrigation methods vary in their "irrigation efficiency," which is defined as the ratio of the amount of water consumed by crops to the amount of water applied through irrigation. The applied water not used by crops is either lost to ET, recharges the underlying shallow aquifer, or is recovered by drainage systems. Furrow irrigation systems (Fig. 24.1a) have relatively low irrigation efficiencies, whereas sub-irrigation and localized recharge systems (e.g., Fig. 24.1b) have high efficiencies and thus tend to be used in water scare regions. However, some excess irrigation (irrigation inefficiency) is still needed to prevent salt build up in the root zone. High irrigation efficiencies and associated small amounts of deep percolation are preferred because (Bouwer 1987):

- less irrigation water is required
- less fertilizer is leached from the root zone

Fig. 24.1 (Top) Furrow irrigation near Phoenix, Arizona. (Bottom) Drip irrigation at a commercial aloe vera plantation in Curacao

- crop yields may be improved
- more salt is stored in the vadose zone as opposed to reaching the groundwater
- slower downward velocity of contaminants
- lesser potential for water logging of soils.

It has been observed that the behavioral response of irrigators to increased irrigation efficiency may result in increased actual water use through (Ward and Pulido-Velazquez 2008; Pfeiffer and Lin 2010), a shift to more profitable water-intensive crops, expansion of irrigated areas, greater irrigation rates, and more efficient plant use (i.e., greater evapotranspiration at the expense of return flows). For example, upon adoption of more efficient irrigation technologies, a farmer with a given supply of irrigation water would have the economic incentive to use the saved water to either irrigated a greater area or switch to a higher value, more water-intensive crop, as opposed to not using the water. The switch from "inefficient" furrow irrigation to drip irrigation will decrease water use, but also decreases groundwater recharge from return flows.

24.4.2 Remote Sensing Estimation of Irrigated Area and Water Use

Irrigation is by far the greatest consumptive use of water. Accurate estimation of irrigation and return flows is a technical challenge where water use is not accurately metered. In the absence of direct data on water use, irrigation water use can be estimated from irrigated area, crop water use requirements, and irrigation efficiencies. Remote sensing (RS) techniques have been demonstrated to be of value in quantifying irrigation water use, and there have been numerous published papers on the subject (e.g., Masoner et al. 2003; Ozdogan and Gutman 2008; Droogers et al. 2010; Romaguera et al. 2010). Summaries of several studies that illustrate some applications of RS to estimate irrigated area and water use are provided below.

Masoner et al. (2003) applied RS techniques to quantify irrigated areas and water use in the Lake Altus Drainage Basin, located in Oklahoma and Texas. Landsat Enhanced Thematic Mapper Plus (ETM+) imagery was used to map land use and irrigated cropland. Irrigated crop areas were determined using a ratio vegetation index consisting of a near infrared band (band 4) divided by a visible red band (band 3; Fig. 24.2). Identification of specific agricultural crops using satellite imagery requires knowledge of crop phenology (life cycles), climate for the particular growing season, and ground reference information on specific agricultural practices in the drainage basin (Masoner et al. 2003). Image date selection is vital for identification of many vegetation covers. Image dates should correspond to the time of peak greenness of crops (Masoner et al. 2003). For example, the best time to identify winter wheat crops in the Lake Altus Drainage Basin is in the spring when the crop is at its peak greenness (Masoner et al. 2003).

Irrigation water requirements were calculated using the following steps (Masoner et al. 2003):

- Reference ET rates were calculated from weather station data (method of Doorenbos and Pruitt was 1977 used).
- Crop ET rates were calculated as the product of the reference ET and crop coefficient.
- Effective precipitation (amount of precipitation available to meet the ET requirements of crops) was calculated from precipitation rates, crop ET rates, and a soil water storage factor.
- Irrigation requirements were calculated as difference between crop ET rates and effective precipitation.

Masoner et al. (2003) found that the irrigated areas (and thus irrigation water use) in the study area were significantly different from state reported irrigated areas.

A review by Velpuri et al. (2009) illustrates that single-date fine-resolution imagery acquired during critical growth stages is sufficient to identify irrigated areas, but multi-date time series are needed to distinguish between irrigated crop types and to derive irrigation intensity. To precisely identify irrigated areas, and also derive irrigation intensity and a cropping calendar, both fine-spatial and a time

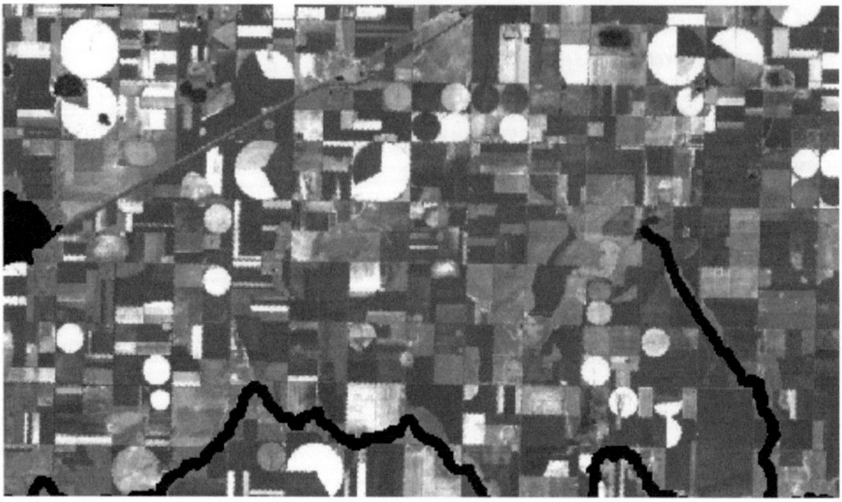

Fig. 24.2 Ratio classified image of part of Lake Altus Drainage Basin (Oklahoma and Texas). Brightness of pixels reflect ratio of Landsat ETM+ bands 3–4 with brighter colors indicating a higher ratio and healthier and greener vegetation (Masoner et al. 2003)

series of coarse-spatial resolution data sets are required (Thenkabail et al. 2005). Velpuri et al. (2009) evaluated the issue of resolution in mapping of irrigated areas in the Krishna River basin of India by comparing Advanced Very High Resolution Radiometer (AVHRR) 100,000 m, Moderate Resolution Imaging Spectroradiometer (MODIS) 500 m, MODIS 250 m, and LANDSAT ETM+ 30 m data. The results of the investigation demonstrated that the finer the resolution, the greater the irrigated area mapped. Fine-resolution techniques have a greater ability to detect fragmented irrigated areas.

Droogers (2002) reviewed past efforts at global irrigated area mapping. Droogers (2002) noted that there have been definition problems as to what constitutes an irrigated area. For example, there is large difference between supplemental irrigation when water is applied infrequently during irregular dry spells versus situations where there are annual applications throughout the growing season. Another obvious issue is that mapping irrigated areas in desert regions is relatively simple, as all green areas must be irrigated, but creating such a map for a relatively wet area is much more complicated (Droogers 2002).

A notable early RS global irrigated area mapping program is the USGS Global Land Cover Map. AVHRR data from April 1992 to March 1993 were used to calculate the normalized difference vegetation index (NDVI) using red and near infrared bands. Vegetation indices, such as NVDI, quantify the amount or condition of vegetation within a pixel from a green vegetation signal. Alternative vegetation indices are the Soil Adjusted Vegetation Index (SAVI) and Enhanced Vegetation Index (EVI).

Droogers (2002) proposed a new methodology to develop a Global Irrigated Area Map (GIAM). The methodology consists of the following steps:

- delineation of potentially irrigated areas based on climate data
- use of low-resolution monthly satellite images to determine vegetation indices (VIs)
- high-resolution satellite images are used to relate VIs to the vegetation cover (VC) of sample areas; high-resolution Advanced Spaceborne Thermal Emission and Reflection Radiometer (ASTER) images were used to classify irrigated areas down to the field level
- low-resolution VI images are converted to VC images using the VI-VC relationship (regression)
- irrigated areas are classified based on low-resolution VC images and potentially irrigated areas
- ground truthing is performed were high-resolution images are ambiguous.

Irrigation efficiency can be estimated from remote sensing-derived estimates of actual ET and data on water use. Wu et al. (2015) investigated the use of satellite ET data to evaluate irrigation water use efficiency in the middle reach of the Heihe River in northwestern China. Satellite (MODIS) based surface-energy balances may over estimate ET in arid and semiarid hydrological regimes where water availability limits ET (Seneviratne et al 2010; Wu et al. 2015 and references therein). Wu et al. (2015) incorporated soil moisture data derived from the Advanced Microwave Scanning Radiometer-Earth Observing System (AMSR-E) into the SEBS (Surface Energy Balance; Su 2002) model. This method remedied the shortcoming of overestimation of ET values in dry environments. Irrigation water volumes were calculated from gauging data at the entrance and exit of canals, subcanals, and field ditches. Average water consumption (ET) during the 2012 study period was 57% of the sum of total irrigation water and effective precipitation. The results showed that in some districts, over-irrigation exceeded the optimal amount (plant ET plus 150 mm/year for salt leaching) by more the 45% of the total ET demand.

24.4.3 Calculation of Return Flows

Methods of estimating irrigation return flows were review by Sammis et al. (1982) and Stephens et al. (2006). The basic methods and some key data requirements are

- soil-water balance residues, which require accurate site-specific data on evapotranspiration rates
- groundwater balance methods, which require accurate data on all elements of the water budget
- vadose zone and groundwater analyses of deep percolation, which requires data on unsaturated hydraulic conductivity, soil-water potential, and hydraulic gradient
- geochemical tracers (e.g., chloride, nitrate, tritium), which require accurate rainfall and soil concentration data
- soil temperature data analyses
- inverse modeling

Some of the methods are data intensive and, as a result, are not practical outside of a long-term research setting. The soil-water balance method is limited by the accuracy of actual ET rate measurements. Inverse modeling and geochemical methods, such as the chloride mass balance (CMB) method, are most commonly used to obtain long-term recharge rates.

Jiménez-Martinez et al. (2009) employed inverse-modeling of the root zone using the Hydrus-1D code to estimate irrigation return flows in the Campo de Cartagena area of southeastern Spain. ET rates were estimated from potential ET rates calculated using the Penman-Monteith method and crop coefficients, which vary with annual crop growth stage. The modeled return flows for the summer melon and fall lettuce crops were 22 and 65%, respectively. As is generally the case with inverse modelling, model results are limited by parametric uncertainty.

The groundwater balance method calculates recharge and return flows as the residual of the water budget. Change in storage may be estimated using the water table fluctuation method, provided that the specific yield of the surficial aquifer is known or can be accurately estimated. The data requirements can be reduced if some of the water balance elements are found to be negligible, for example, natural recharge during the dry season and evaporation from a deep water table (Maréchal et al. 2003). Ochoa et al. (2011, 2012) used water budget calculations and the water table fluctuation method to estimate return flows from canal seepage and irrigation in the Alcalde-Velarde Valley of New Mexico. The canal seepage rate was obtained from an inflow-outflow test. Deep percolation and canal seepage together accounted for 33.3% of total canal inflow.

Stovall and Rainwater (2002) presented the results of the calibration of a model (i.e., inverse modeling) of the Llano Estacado Region of the Southern High Plains of Texas. Recharge was estimated through the model calibration process to average 2.5 in./year (6.4 cm/year) with a range of <0.5–10.0 in./year (<1.3–25.4 cm/year). Recharge was substantially influenced by return flows and cultivation practices that limit runoff. Total recharge ranged from near zero in uncultivated areas to over 6 in./year (15 cm/year) in some irrigated areas.

Irrigation return flows were investigated at two agricultural (corn and sugar beets) fields irrigated using flood furrows and center-pivot sprinklers in Weld County, north-central Colorado, which is a semiarid region underlain by alluvial sediments (Arnold 2011). The depth to the water table was reported to be 10–30 ft (3–9 m) below land surface. Deep percolation was estimated using an unsaturated zone water-balance (UZMB) approach and the water table fluctuation method. The UZMB equates deep percolation to increases in soil-water storage below a plane in the unsaturated zone (zero-flux plane) that separates upward movement of soil water from ET and downward drainage of soil water toward the water table. The zero flux plane was assumed to be at the bottom of the root zone ($\approx$4 ft; 1.2 m bls).

The cumulative deep percolation beneath the flood-irrigated field was 7.7–8.8 in. (19.6–22.4 cm) during the monitoring period (June 13, 2008 to October 15, 2009), which corresponds to 40–52% of the irrigation water applied and 29–38% of irrigation water plus precipitation. The cumulative deep percolation beneath the sprinkler irrigated field was 1.2–3.2 in. (3.0–8.1 cm) during the monitoring period (June 13,

2008 to October 15, 2009), which corresponds to 5–14% of the irrigation water applied and 4–11% of irrigation water plus precipitation.

24.4.4 Return Flows from Wastewater Irrigation—Tula Valley, Mexico

The Tula Valley, Mexico, is perhaps the most studied example of groundwater recharge and unplanned reuse from wastewater irrigation. In Central Mexico, 60 m^3/s (2,119 ft^3/s) of untreated wastewater is transmitted through unlined canals from Mexico City 80 km (49.7 mi) north to the Tula Valley where it is used to irrigate 90,000 ha (347 mi^2) of farm land referred to as "El Mezquital." Wastewater has been sent to valley since 1789 and used for agricultural irrigation since 1896 (Jiménez 2010).

Water quality improves significantly during transit and the farmers benefit from its nutrient content. It is estimated that 25 m^3/s (883 ft^3/s) of the wastewater flow artificially recharges aquifers in the Tula Valley through a combination of canal seepage and irrigation return flows from very high irrigation rates (1.5–2.2 m/year; 4.9–7.2 ft/year; Jiménez and Chávez 2004). The consequences of the artificial recharge have been a rising water table (and associated flooding and salinization of some farm fields), increasing river flows, and the appearance of springs (Jiménez and Chávez 2004). Aquifer water levels rose 15–30 m from 1938 to 1990 and dozens of springs appeared. Tula River flow increased from 1.6 to 12.7 m^3/s between 1945 and 1995 (Jiménez 2010). The recharge created a local groundwater source for 500,000 people living in the Valley with only chlorination for treatment. The recovered water has been proven to be of acceptable quality due to natural contaminant attenuation processes (Jiménez and Chávez 2004; Jiménez-Cisneros 2012).

Jiménez (2008, 2009, 2010) reviewed groundwater quality data from the Tula Valley. Elevated concentrations of Cl, Fe, NO_3, SO_4 and fecal coliforms (which can be addressed through disinfection) were reported. Low levels of phenol chlorinated benzenes, and other identified and unidentified trace organic compounds were also detected but at much lower concentrations than in the wastewater. No urgent water quality issues were identified. The passage of the wastewater through the soil results in a considerable reduction in the concentrations of organic matter, metals, and nutrients, and a considerable increase in salts.

The Tula Valley aquifers were investigated as a potential drinking water source for Mexico City. Pumping of some of the artificially recharged water would allow for the recovery of some flooded and saline lands (Jiménez and Chávez 2004). Agricultural use and the canal seepage processes were found to be very effective in removing pollutants, but the recovered water may have elevated salinity and nutrients. Biological analyses of water from a newly formed spring indicated that the water was of very high quality. A major cost item would be constructing of a pipeline from the Tula Valley to Mexico City, and a water treatment system to treat the water to full potable standards.

24.5 Land Use/Land Cover Changes and Recharge

24.5.1 Introduction

Land use/land cover (LULC) changes can impact various components of water budgets including aquifer recharge rates. Evaluation of the hydrological impacts of land development on local water resources requires consideration of the direction and magnitude of all changes in water budget components. For example, it has been taken as a fact that land development and associated increased groundwater pumping in southwestern Florida was resulting in declining groundwater levels and adverse environmental impacts, such as reductions in the hydroperiods of wetlands. However, an evaluation of historical water level data revealed that in most areas, water levels in the shallow (water table) aquifer were either stable or have actually risen over time (Maliva and Hopfensperger 2007). Hydrological analyses that focused solely on groundwater pumping give a misleading picture of the condition of the shallow aquifer. Additional factors that impact local shallow groundwater levels include (Maliva and Hopfensperger 2007):

- reduction of ET resulting from the replacement of native vegetation with impervious surfaces (e.g., buildings, roads, driveways, and sidewalks)
- retrun flows from irrigation with externally derived water (e.g., desalinated water, reclaimed water, groundwater derived from distant wellfields)
- recharge from on-site sewage treatment and disposal systems (residential septic systems)
- stormwater management systems that retain and infiltrate water, and reduce runoff to tide.

Plant evapotranspiration is a usually the major outflow of the soil-water budget in vegetated areas and plays a controlling role in determining whether infiltrated water becomes recharge. It therefore stands to reason that changes in vegetation may significantly impact water budgets and recharge rates. Incidental increases in recharge rates associated with anthropogenic changes in LULC are considered a type of unmanaged and unplanned aquifer recharge. Where the changes in LULC are performed with the intended purpose of increasing recharge, the actions fall into the realm of MAR.

24.5.2 Vegetation Type and Groundwater Recharge

Petheram et al. (2002) evaluated the results of recharge studies across Australia with goal of developing simple empirical relationships for the impacts of land uses and covers on aquifer recharge. The primary controlling factors considered were land use, soil type, and climate. Land use was divided into three broad categories; annuals (shallow-rooted annual crops and pasture), perennials (perennial pastures and native herbaceous vegetation), and trees (very deep-rooted vegetation). Deep-rooted vegetation appears to result in dramatically lower recharge rates compared to both ground covered with shallow-rooted annual vegetation and bare ground (Gee et al. 1992; Petheram et al. 2002).

Kim and Jackson (2012) compiled and analyzed existing groundwater recharge data for different climates, soils, and vegetation types to determine the relationship between vegetation types and recharge. Vegetation was classified into five categories: cropland, grassland, woodland, scrubland, and non-vegetated. For given water inputs (precipitation plus irrigation) and potential ET (PET) rates, croplands have the highest recharge rates, followed by grasslands and then woodlands. It was concluded that agricultural conversion of grasslands and woodlands would probably bring about an increase in recharge. Woody plant invasion and afforestation of croplands and grasslands would probably reduce recharge. Increases in recharge associated with cultivation may pose a risk of salinization and degradation of groundwater quality. Kim and Jackson (2012) concluded that vegetation, and its interactions with other factors, has a strong effect on groundwater recharge, explaining about 24% of the global variation in recharge. The relative difference in recharge among vegetation types were greater in drier climates and clayey soils.

Vegetation adapted to water scare conditions (i.e., xerophytes) is highly adapted to extracting soil moisture (Stonestrom and Harrill 2007). Scanlon et al. (2003) investigated flow and transport in arid interdrainage (interfluve) areas in the southwestern United States using field measurements of matric potential, chloride concentration data, and modeling analyses. The studied areas were the High Plains site near Amarillo, Texas, the Eagle Flat and Hueco Bolson sites in the Chihuahuan Desert in west Texas, and the Amargosa Desert site in the Mojave Desert near Beatty, Nevada. Upward water potential gradients indicate that current water fluxes in the shallow subsurface at all four sites is upward. In general, there appears to be negligible current recharge in interdrainage desert regions. Lower chloride concentrations at depth indicate wetter conditions during the Pleistocene and earlier times then during the Holocene (Scanlon et al. 2003). Both field data and modeling results have demonstrated that a change in land cover from deep-rooted native xerophytic vegetation (trees and shrubs) to shallower-rooted agricultural crops and landscaping may increase groundwater recharge rates as the latter are less efficient at extracting soil moisture (Keese et al. 2005; Scanlon et al. 2005). Recharge rates have increased by one or two orders of magnitude in some areas with a natural or man-caused change in vegetation type (Scanlon et al. 2006).

Rangeland systems in the southwestern United States share similar characteristics as desert environments, particularly low matric potentials and upward potential gradients that are indicative of discharge rather than recharge conditions (Scanlon et al. 2005). Agricultural conversion affects key vegetation parameters, including fractional vegetation coverage, wilting point, and root depth (Scanlon et al. 2005, 2006). Wilting point is the minimum matric potential at which plants can take up water. The wilting point of native arid and semiarid region rangeland vegetation is typically much lower than that of typical agricultural crops (Scanlon et al. 2005). Native rangeland vegetation can draw much more water out of the soil than typical agricultural crops, which is an adaptation to water scarcity. The transition in LULC from rangeland vegetation to cultivated crops can result in an increase in groundwater recharge. The initiation of recharge is related in part to the addition of water through irrigation. However, the change from upward to downward water potential gradient can also occur as a result of the change in LULC to dryland agriculture. The recharge associated with dryland agriculture appears to be associated with reduced interception, reduced evapotranspiration, shallow rooting depths, fallow periods, and increased soil permeability caused by plowing (Scanlon et al. 2005, 2006).

Peck and Williamson (1987) documented how the clearing of native forest vegetation and its replacement with pasture or crops in Western Australia resulted in a substantial rise in the water table and the mobilization of salts in the soils. In areas that were cleared for agriculture, the potentiometric surface moved upward at more than 2.6 m/year (8.5 ft/year) as an average over several years. The potentiometric surface rise was equivalent to increased recharge estimated as 6–12% of rainfall, depending on the value of the specific yield used for the aquifer. Reforestation has been demonstrated to result in decreases in groundwater levels relative to areas still under pasture (Bell et al. 1990).

A number of other studies have similarly documented how removal or changes in vegetation resulted in increased recharge. Allison et al. (1990) reported that the clearing of native vegetation in a semi-arid region of southern Australia has led to increases in groundwater recharge of about two orders of magnitude. Nosetto et al. (2012) explored the hydrological impacts of replacing native grasslands and dry forests with herbaceous vegetation (crops) and eucalyptus plantations in the Entre Ríos province of Argentina. The study included satellite estimates of ET, soil-water modeling (using the Hydrus code), and soil sampling for moisture. Native dryland forest and eucalyptus plantations displayed similar annual average ET values that were about 50% greater than grassland and cropland plots. The decrease in ET associated a change to herbaceous covers was expected to increase groundwater recharge, resulting in shallower groundwater levels and eventually salt mobilization.

Leblanc et al. (2008) examined the hydrological impacts of land clearance in SW Niger, Sahal region of Africa. Most groundwater recharge in the region is indirect and occurs by deep infiltration in ponds and gullies. Land clearance, mostly performed to extend agricultural areas, was found to have resulted in a 2.5-fold increase in the drainage network, as manifested by an increase in the number and length of gullies. Groundwater level data display a widespread steady rise of the water table of approximately 4 m from early 1960 to 2005. The increased runoff to sites of indirect infiltration is believed to be responsible for the increased groundwater recharge.

Jobbagy and Jackson (2004) examined salinization and groundwater recharge in 20 paired grassland and adjacent afforested plots across ten sites in the Argentine Pampas. Two years of salinity and groundwater measurements at a 40 ha *Eucalyptus camaldulensis* plantation revealed that the plantation had reduced groundwater recharge and lowered the water table by 38 cm, on average, compared to the adjacent grassland.

Modern recharge in groundwater basins in the Trans-Pecos basin aquifer system of west Texas was investigated by Robertson and Sharp (2013, 2015). Trends of increasing nitrate concentration over time and the presence of CFCs in the groundwater are evidence for widespread modern recharge. The additional nitrate in the groundwater appears to be labile nitrate that was sequestered in the vadose zone beneath the root of native vegetation and subsequently leached due to increased recharge. Core data show partial to full flushing of naturally accumulated nitrate and chloride from the vadose zone beneath lands used for irrigated agriculture (currently and in the past), which is evidence that changes in LULC have impacted infiltration and recharge.

Increased recharge compared to pre-western settlement conditions was attributed to agricultural return flows and a change in land cover from thick, dense grasslands to sparser woody vegetation with shallow root depths and a greater bare ground area (Robertson and Sharp 2013, 2015). The change in vegetation was attributed to over-grazing, regional climate shift, and fire suppression. Non-irrigation related (widespread diffuse recharge) increases in infiltration is indicated by increased groundwater nitrate concentrations in non-irrigated areas. Net infiltration on the basin floors of the two modeled basins was estimated to contribute between 7 and 11.5% of the annual basin recharge (Robertson and Sharp 2015). Net infiltration below the root zone was modeled using the U.S Geological Survey INFIL 3.0.1 Code. The INFIL simulations estimated annual average net infiltration rates, not the absolute amount of water reaching the water table each year. The modeling results indicate that the change in vegetation regime increased net infiltration by as much as 48% from pre-western settlement vegetation scenarios.

The impacts of land use changes on water resources depend on numerous factors including the original vegetation to be replaced, the vegetation that is replacing it, whether the change is permanent or temporary, and associated land management practices involving alteration of drainage (Scanlon et al. 2007). Natural forests have greater ET rates than other types of vegetation, and the reduced ET in cultivated areas compared to converted forests provides more water for groundwater recharge and streamflow. The latter can result in increased stream baseflow. The conversion of native vegetation to rain-fed agriculture generally increases water quantity but decreases water quality (Scanlon et al. 2007). Water quality effects include the mobilization into the groundwater of natural salts and nutrients (e.g., nitrates) that have accumulated in the vadose zone, and soil salinization caused by the rising of the water table to close to land surface (Scanlon et al. 2007). Proposed or implemented forestation (reforestation) projects could reduce groundwater recharge, decrease runoff, and decrease stream sediment loads, the effects of which need to be considered in water resources management (Scanlon et al. 2007).

The change from discharge to recharge associated with the change in LULC from native rangeland to agriculture raises the possibility that removal of native vegetation can be used as a water-management tool (Scanlon et al. 2005, 2006). Potential water savings from vegetation removal and replacement arise from (Nagler et al. 2009):

- removal of vegetation and associated elimination of transpiration
- replacement of vegetation with high ET rates with plants with lower ET rates
- replacement of plant communities with vegetation that has an overall lower biomass and leaf area

- removal of deep-rooted vegetation where the water table is deep and their replacement with shallower rooted vegetation that cannot access the water table.

Removal of plants can result is lesser shading and, as a result, transpiration savings may be offset by increased direct evaporation from the ground surface. Large-scale water savings experiments in the western United States have not realized the expected increases in stream flow due to either (Nagler et al. 2009):

- absence of a significant difference in ET between native and non-native plants
- difficulty in accurately measuring small changes in stream flow
- complexities in the interaction of groundwater and surface water (which may not be well connected).

Nevertheless, disruption of native ecosystems has adverse ecological implications. Therefore, land cover and use conversions need to be evaluated within the context of overall land and water management, especially with consideration of water quality.

24.5.3 Phreatophyte Removal

As discussed by Meinzer (1923) and Robinson (1958), the flora in arid lands is divided mainly into two classes: xerophytes and phreatophytes. Xerophytes, which include cacti, depend on rains for their water supply and are thus adapted to small and irregular supplies of water. A phreatophyte was defined by Meinzer (1923) as "a plant that habitually obtains its water supply from the zone of saturation, either directly or through the capillary fringe." Evidence that phreatophytes utilize groundwater is provided by diurnal fluctuations of water levels in wells that tap the water table in areas of phreatophyte growth (Robinson 1958). Many phreatophytes are facultative in that they can also obtain water from soil moisture when available (e.g., after rains; Robinson 1958).

White (1932) proposed that phreatophyte ET rates can be quantified from diurnal fluctuations in the water table. The underlying assumption is that phreatophyte transpiration, and associated declines in the water table, only occurs or are significant during daylight hours. High-resolution hydrographs of wells completed (screened) across the water table record a phreatophyte transpiration signal and a background net inflow into or outflow from the water table aquifer (Fig. 24.3). The diurnal fluctuation method was reviewed by Loheide et al. (2005) who noted the following advantages:

- it provides daily estimates of ET
- it provides an integrated response to highly heterogeneous stresses
- it is a generic method that does not depend on a particular plant community
- low cost and simplicity.

The underlying equation is:

$$ET_{Phreatophyte} = S_y(\Delta s/t + R) \tag{24.3}$$

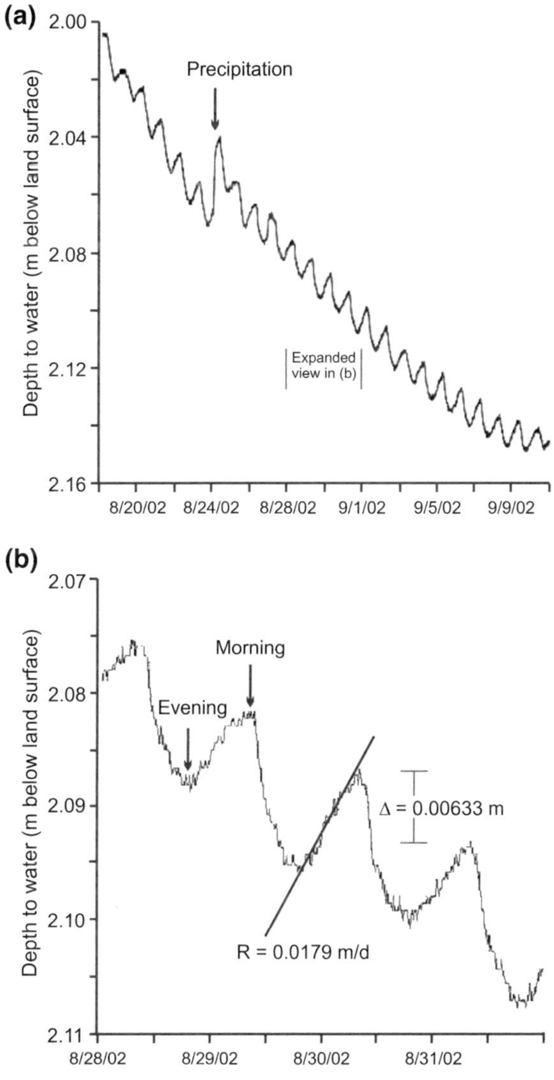

Fig. 24.3 Diurnal depth-to-water fluctuations recorded in a well in the riparian zone of the Arkansas River, near Larned, Kansas. Application of the White (1932) method to the data from August 30, 2002 gave a transpiration rate of 3.6 mm/d using a specific yield of 0.15 (from Loheide et al. 2005)

where

$ET_{Phreatophyte}$	daily average phreatophyte ET rate (mm/d)
S_y	specific yield (unitless)
Δs	daily change in storage (difference in morning daily maxima between successive days; mm)
t	time period of one day in rate time units (d)
R	net inflow (outflow) rate which is the slope of the night time (midnight to 4:00 AM) regression line (mm/d)

Loheide et al. (2005) suggested using an average R value for the day of interest and subsequent day. A modeling investigation by Loheide et al. (2005) confirmed the earlier observation that calculated ET rates are sensitive to the specific yield value used, which may be poorly known. ET rates will often be significantly overestimated in sediments with greater than 10% silts and clays because of the effects of drainage time and depth of the water table on specific yield. Measured specific yield is a property of the porous media, and also the depth to water, duration of drainage, and antecedent moisture conditions. Loheide et al. (2005) proposed that "readily available specific yield" (Meyboom 1967) be used in fine-grained sediments, which is defined as the amount of water released from the vadose zone during the time frame (<12 h) of the diurnal fluctuations. Loheide et al. (2005) provided trilinear diagrams and equations for estimating readily available specific yield values.

Butler et al. (2007) field tested the use of diurnal water-table fluctuations to quantify phreatophyte ET rates. The primary study site was the Larned Research Site (LRS), a highly instrumented site located adjacent to the Arkansas River, near Larned, Kansas. Three other auxiliary sites were also included in the investigation. Meteorological monitoring data confirmed that phreatophyte ET rates are controlled by global irradiance, vapor pressure deficit, wind speed, and air temperature. ET rates also vary with the depth of the water table relative to the root depth. If the water table is quickly lowered (e.g., by nearby groundwater pumping), then it may be out of reach, at least temporarily, of phreatophyte roots (Butler et al. 2007).

The presence of diurnal fluctuations in the water table should be considered an important indicator of groundwater consumption by phreatophytes (Butler et al. 2007). Diurnal fluctuations in the water table may not occur or be muted where plants obtain their water from the vadose zone. Butler et al. (2007) discussed some of the limitations and constraints of the use of diurnal water-table fluctuation data to estimate phreatophyte ET rates. An important consideration is that riparian vegetation can change substantially over a short distance and, therefore, data from a small number of wells may not be representative of phreatophyte water use throughout a given riparian zone.

Robinson (1952) noted that phreatophytes the 17 western states of the United States may "waste" as much as 20–25 million AF (24,700–30,800 MCM) of water into the atmosphere annually and that it may be possible to salvage a part of this water. It was estimated that in the state of Nevada, it may be practical to salvage about 25% of the water wasted annually, or about 400,000 AF (490 MCM).

Other workers reported that Robinson's (1952) numbers may be greatly exaggerated and served to create an alarmist interest in the "phreatophyte problem" (Van Hylckama 1980; Ritzi et al. 1985). From a decidedly anthropocentric perspective, Robinson (1958) noted that the transpirative draft of groundwater by phreatophytes is a "consumptive waste" as most phreatophytes have a low economic value and are heavy users of water. The groundwater used by most phreatophytes was noted to have a low beneficial use as far as man is concerned. The consumptively wasted water is available for "salvage" for beneficial consumptive use by man. Salvage may be achieved by rapidly lowering the water table beyond the root zone of phreatophytes or by the substitution of "nonbeneficial" phreatophytes with plants of high economic value (Robinson 1959).

Phreatophytes in the arid and semiarid areas of the American Southwest include native plants, such as mesquite (*Prosopis* spp.), willow (*Salix* spp.), and cottonwood (*Populus* spp.). Acacia trees are important phreatophytes in Africa, the Middle East, and South Asia. In the southwestern United States, native riparian-adapted species, including Fremont cottonwood (*Populus fremontii*), are being replaced by non-native species, such as saltcedar or tamarisk (*Tamarix* spp.) and Russian olive (*Elaeagnus augustofolia*). Saltcedar was introduced into the United States in the 1820s as an ornamental plant and was subsequently used to stabilize stream banks (Di Tomaso 1998). Similarly, introduced willow (*Salix* spp.) species have become widespread in the Murray-Darling Basin of Australia (Doody et al. 2011). Willow was reported to have been introduced into Australia in the 1800s by a European immigrant (Doody et al. 2011).

A provocative paper by Chew (2009) recounts how scientists recast tamarisks as "water-wasting foreign monsters." He specifically noted the

> By 1950 a permanent interagency bureaucracy had sprung up, focusing primarily on slaying the beast rather than demonstrating actual water salvage. It began to propagate the legend, as well as a few myths, that would keep its members busily cutting, bulldozing, spraying and reporting progress in terms of vegetation killed for another 20 years.

Saltcedar (Fig. 24.4) is able to outcompete native vegetation because of its ability to produce seeds continuously and tolerate extreme drought and flooding, salt tolerance, ability to recovery quickly after fire, and deep roots that gives it access to groundwater at depths of 10 m of more (Di Tomaso 1998; Hatler and Hart 2009). In addition to its use of water, saltcedar also increases flood risks by choking the normal overflow channels of streams (Robinson 1958). Saltcedar stands may have greater ET rates than native vegetation, which has been attributed to their substantially higher leaf area index (Nagler et al. 2003). Saltcedar removal is practiced for ecosystem restoration, fire control, and ET reduction. High ET loses from saltcedar (and other invasive phreatophyte species) is believed to lower groundwater levels and reduce stream baseflows.

Removal of phreatophytes has been demonstrated to result in the salvage of some water. For example, the U.S. Geological Survey compared ET before and after phreatophyte (mostly saltcedar and mesquite) removal along the Gila River flood plain of southeastern Arizona (Culler et al. 1982). ET was evaluated as the residual of a water budget equation, which was calculated for four contiguous reaches of the river. The removal of phreatophytes resulted in an average reduction in ET of 19 in. (480 mm) per year with a range of 14 in. (360 mm) to 26 in. (660 mm) per year. It was observed that a flood plain without phreatophytes is in an artificial condition and that the reduction in ET is temporary and would not apply after permanent replacement vegetation becomes established (Culler et al. 1982).

Welder (1988) evaluated the water salvage benefits of the removal of 19,000 acres (7,690 ha) of saltcedar trees along the Acme-Artesia reach of Pecos River of New Mexico. A reduction of transpiration was evident from the cessation of diurnal

Fig. 24.4 Dense strand of saltcedar, Havasu National Wildlife Refuge, Mohave County, Arizona. *Source* Garner and Truini (2011)

water-level fluctuations once the saltcedar was largely eradicated by root plowing. Groundwater level data suggested an increase in groundwater storage of about 6,000 AF (7.4 MCM), but the expected increase in stream baseflow was not realized. The effects of the saltcedar removal program on stream baseflow appears to have been masked by increased precipitation and decreased groundwater pumping.

Hart et al. (2005) evaluated the initial results of the "Pecos River Ecosystem Project" (Texas). Saltcedar was treated by helicopter applications of the herbicide Arsenal. Water losses from saltcedar strands was estimated by monitoring diurnal groundwater fluctuations in treated and untreated plots. Diurnal fluctuations were not evident at the investigated treated site during the growing season. The data shows a pronounced decrease in the total amount of annual ET loss and thus greater recharge after removal of the saltcedar. Hart et al. (2005) noted that native vegetation takes over some of the areas where saltcedar was treated, which could use some of the water salvaged by saltcedar control, and that there is a need for data on water use by native plants.

A subsequent paper on the Pecos River Ecosystem Project reported calculated water salvages of 0.13–0.68 m/year (Hatler and Hart 2009) based on natural revegetation with grasses, forbs, and saltcedar regrowth. The natural revegetation at the study site exhibited lower ET potential than the saltcedar that it replaced. Hatler and Hart (2009) emphasized that growth management is critical and long-term salt-cedar control is necessary to continue salvage benefits.

Moore and Owens (2012) evaluated the effects of saltcedar removal on ET losses in plots of mature cottonwood forest with a dense saltcedar understory. The study site is adjacent to the Middle Rio Grande River in central New Mexico. ET was quantified by sap flow measurements. The results of the investigation indicate that saltcedar and cottonwood (*Populus*) compete for water. Thinning of the saltcedar understory resulted in increased cottonwood transpiration and a minimal impact on the overall stand water balance. Moore and Owens (2012) questioned the hydrological value of water salvage programs involving the removal or thinning of exotic phreatophytic riparian vegetation. The key issue is the post-thinning response of native vegetation.

All other factors being equal, phreatophytes transpire less with increasing depth to groundwater (Ritzi et al. 1985). A water salvage alternative to eradication is to lower the water table by pumping along creeks (Ritzi et al. 1985). Instead of groundwater being lost to phreatophyte ET, it is put to a beneficial use. However, it was recognized that pumping rates will often be constrained by instream water requirements including surface water rights.

A review of saltcedar transpiration rates by Doody et al. (2011) from sap flow measurements gave a fairly wide range of 220–1,500 mm/year with a mean of 765 mm/year. Similar ET rates were obtained from soil moisture tower measurements. The ET rates of native riparian species were found to overlap the measured saltcedar range, with local values depending upon the depth to water, groundwater salinity, and climatic conditions. Doody et al. (2011) concluded that the water salvage potential for the removal of saltcedar in the western United States and willows in Australia is highly variable and site-specific. Permanent replacement of non-native species with native species adapted to similar ecological niches is unlikely to result in worthwhile water salvage (Doody et al. 2011). Doody et al. (2011) concluded that there was greater water salvage potential from willow removal in Australia because replacement by herbaceous plants could result in a substantial reduction in ET and willows removed from the permanently inundated zone of stream channels would not be replaced (i.e., willows occupy a vacant niche). A key conclusion of Doody et al. (2011) and other investigators is the need for accurate data on all elements of the water budget in order to accurately quantify water salvage.

References

Alhamid, A. A., Alfayzi, S. A., & Hamadto, M. A. (2007). A sustainable water resources management plan for Wadi Hanifa in Saudi Arabia. *Journal King Saud University Engineering Sciences, 19*(2), 209–222.

Allison, G. B., Cook, P. G., Barnett, S. R., Walker, G. R., Jolly, I. D., & Hughes, M. W. (1990). Land clearance and river salinisation in the western Murray Basin, Australia. *Journal of Hydrology, 119*(1–4), 1–20.

Al-Sefry, S. A., & Şen, Z. (2006). Groundwater rise problem and risk evaluation in major cities of arid lands—Jeddah case in Kingdom of Saudi Arabia. *Water Resources Management, 20,* 91–108.

Arnold, L. R. (2011). *Estimates of deep-percolation return flow beneath a flood—and a sprinkler-irrigated site in Weld County, Colorado, 2008–2009.* US. Geological Survey Scientific Investigations Report 2011–5001.

Barrett, M. H., Hiscock, K. M., Pedley, S., Lerner, D. N., Tellam, J. H., & French, M. J. (1997). Marker species for identifying urban groundwater recharge sources: A review and case study in Nottingham, UK. *Water Research, 33*(14), 3083–3097.

Bell, R. W., Schofield, N. J., Loh, I. C., & Bari, M. A. (1990). Groundwater response to reforestation in the Darling Range of Western Australia. *Journal of Hydrology, 115,* 297–317.

Blackwood, D. J., Gilmour, D. J., Ellis, J. B., Revitt, D. M., & Staines, A. (2005). Exfiltration from sewers—Is it a serious problem? In *10th International Conference on Urban Drainage, Copenhagen*, August 21–26, 2005 (pp. 1–8).

Bouwer, H. (1978). *Groundwater hydrology*. New York: McGraw-Hill.

Bouwer, H. (1987). Effect of irrigated agriculture on groundwater. *Journal of Irrigation and Drainage Engineering, 113*(1), 4–15.

Brassington, F. C., & Rushton, K. R. (1987). A rising water table in central London. *Quarterly Journal of Engineering Geology, London, 20,* 151–158.

Butler, J. J., Kluitenberg, G. J., Whittemore, D. O., Loheide, S. P., Jin, W., Billinger, M. A., & Zhan, X. (2007). A field investigation of phreatophyte-induced fluctuations in the water table. *Water Resources Research*, *43*(2). https://doi.org/10.1029/2005wr004627.

Chew, M. K. (2009). The monstering of tamarisk: How scientists made a plant into a problem. *Journal of the History of Biology, 42*(2), 231–266.

Culler, R. C., Hanson, R. L., Myrick, R. M., Turner, R. M., & Kipple, F. B. (1982). *Evapotranspiration before and after clearing phreatophytes, Gila River flood plain, Graham County, Arizona*. U.S. Geological Survey Professional Paper 655-P.

Dagès, C., Voltz, M., Lacas, J. G., Huttel, O., Negro, S., & Louchart, X. (2008). An experimental study of water table recharge by seepage losses from a ditch with intermittent flow. *Hydrological Processes, 22*(18), 3555–3563.

Dillon, P. (2009). Water recycling via managed aquifer recharge in Australia. *Boletín Geológico y Minero, 120*(2), 121–130.

Di Tomaso, J. M. (1998). Impact, biology, and ecology of saltcedar (*Tamarix* spp.) in the southwestern United States. *Weed Technology, 12*(2), 326–336.

Doody, T. M., Nagler, P. L., Glenn, E. P., Moore, G. W., Morino, K., Hultine, K. R., et al. (2011). Potential for water salvage by removal of non-native woody vegetation from dryland river systems. *Hydrological Processes, 25,* 4117–4131.

Doorenbos, J., & Pruitt, W. O. (1977). *Crop water requirements*. FAO Irrigation and Drainage Paper No. 24. Rome: Food and Agricultural Organization of the United Nations.

Droogers, P. (2002). *Global irrigated area mapping: Overview and recommendations (Paper 36)*. Colombo, Sri Lanka: International Water Management Institute.

Droogers, P., Immerzeel, W. W., & Lorite, I. J. (2010). Estimating actual irrigation application by remotely sensed evapotranspiration observations. *Agricultural Water Management, 97*(9), 1351–1359.

Eiswirth, M., & Hötzl, H. (1999). The impacts of leaking sewers on urban groundwater. In J. Chilton (Ed.), *Groundwater in the urban environment* (pp. 399–404). Rotterdam: A.A. Balkema.

Ellis, J. B. (2001). Sewer infiltration/exfiltration and interactions with sewer flows and groundwater quality. In *2nd International Conference Interactions Between Sewers, Treatment Plants and Receiving Waters in Urban Areas–Interurba II* (pp. 19–22).

Foster, S. S. D. (1990). Impacts of urbanization on groundwater. In *Hydrological processes and water management in urban areas* (Proceedings of the Duisberg Symposium, April 1988). IAHS Publication No. 198 (pp. 187–208).

Garner, B. D., & Truini, M. (2011). *Groundwater budgets for Detrital, Hualapai, and Sacramento Valleys, Mohave County, Arizona, 2007–08*. U.S. Geological Survey Scientific Investigations Report 2011–5159.

Gasser, G., Rona, M., Voloshenko, A., Shelkov, R., Tal, N., Pankratov, I., et al. (2010). Quantitative evaluation of tracers for quantification of wastewater contamination of potable water sources. *Environmental Science and Technology, 44*(10), 3919–3925.

Gee, G. W., Fayer, M. J., Rockhold, M. L., & Campbell, M. D. (1992). Variations in recharge at the Hanford Site. *Northwest Science, 66,* 237–250.

Hart, C. R., White, L. D., McDonald, A., & Sheng, Z. (2005). Saltcedar control and water salvage on the Pecos River, Texas, 1999–2003. *Journal of Environmental Management, 75*(4), 399–409.

Hatler, W. L., & Hart, C. R. (2009). Water loss and salvage in saltcedar (*Tamarix* spp.) stands on the Pecos River, Texas. *Invasive Plant Science and Management, 2*(4), 309–317.

Heathcote, R. L. (1983). *The arid lands: Their use and abuse*. London: Longman.

Huber, S. S. (2008). The legalities of lining the All-American Canal. In *CEMLP Annual Review*, The Centre for Energy, Petroleum, and Mineral Law and Policy, University of Dundee.

Izbicki, J. A., Johnson, R. U., Kulongoski, J., & Predmore, S. (2007). Ground-water recharge from small intermittent streams in the western Mojave Desert, California. In D. A. Stonestrom, J. Constantz, T. P. A. Ferre, & S. A. Leake (Eds.), *Ground-water recharge in the arid and semiarid southwestern United States* U.S. Geological Survey Professional Paper 1703G, (pp. 157–184).

Izbicki, J. A., Radyk, J., & Michel, R. L. (2000). Water movement through a thick unsaturated zone underlying an intermittent stream in the western Mojave Desert, southern California, USA. *Journal of Hydrology, 238*(3), 194–217.

Jiménez, B. (2008). Unplanned reuse of wastewater for human consumption: The Tula Valley of Mexico. In B. Jiménez, T. Asano, B. Ellis, L.-L. Bertrand-Krajewski, C. Binnie, & M. Kimber (Eds.), *Water reuse—An international survey: Contrasts, issues and needs around the world*. London: IWA Publishing.

Jiménez, B. (2009). Water and wastewater management in Mexico City. In L. W. Mays (Ed.), *Integrated urban water management: arid and semi-arid regions* (pp. 81–101). Paris: UNESCO, and Leiden: Taylor & Francis.

Jiménez, B. (2010). The unintentional and intentional recharge of aquifers in the Tula and the Mexico Valleys: The Megalopolis needs Mega solutions. In *Rosenberg Symposium*, Buenos Aires, Argentina.

Jiménez, B., & Chávez, A. (2004). Quality assessment of an aquifer recharged with wastewater for its potential use as drinking water source: "El Mezquital Valley" case. *Water Science and Technology, 50,* 269–276.

Jiménez-Cisneros, B. (2012). The planned and unplanned reuse of Mexico City's wastewater. In *Guidelines for water reuse, EPA/600/R-12/618* (pp. E-76–E-78). Washington, DC: U.S. Environmental Protection Agency.

Jiménez-Martínez, J., Skaggs, T. H., Van Genuchten, M. T., & Candela, L. (2009). A root zone modelling approach to estimating groundwater recharge from irrigated areas. *Journal of Hydrology, 367*(1), 138–149.

Jobbagy, E. G., & Jackson, R. B. (2004). Groundwater use and salinization with grassland afforestation. *Global Change Biology, 10*(8), 1299–1312.

Kennedy, J. R. (2007). *Changes in storm runoff with urbanization: The role of pervious areas in a semi-arid environment*. 2007. Master's thesis, Hydrology and Water Resources Dept. University of Arizona.

Kennedy, J. R., & Gungle, B. (2010). *Quantity and sources of base flow in the San Pedro River near Tombstone, Arizona*. U.S. Geological Survey Scientific Investigations Report 2010–5200.

Keese, K. E., Scanlon, B. R., & Reedy, R. C. (2005). Assessing controls on diffuse groundwater recharge using unsaturated flow modeling. *Water Resources Research, 41*, W06010. https://doi.org/10.1029/2004WR003841.

Kibel, P. S. (2008). A line drawn in water: Aquifers beneath the Mexico-United States border. *Water Law Review, 12,* 191–207.

Kim, J. H., & Jackson, R. B. (2012). A global analysis of groundwater recharge for vegetation, climate, and soils. *Vadose Zone Journal, 11*(1). https://doi.org/10.2136/vzj2011.0021ra.

Kruse, E., Carol, E., Mancuso, M., Laurencena, P., Deluchi, M., & Rojo, A. (2013). Recharge assessment in an urban area: A case study of La Plata, Argentina. *Hydrogeology Journal, 21*(5), 1091–1100.

Landers, M. N., & Ankcorn, P. D. (2008). *Methods to evaluate influence of onsite septic wastewater-treatment systems on base flow in selected watersheds in Gwinnett County, Georgia, October 2007.* U.S. Geological Survey Scientific Investigations Report 2008–5220.

Leblanc, M. J., Favreau, G., Massuel, S., Tweed, S. O., Loireau, M., & Cappelaere, B. (2008). Land clearance and hydrological change in the Sahel: SW Niger. *Global and Planetary Change, 61*(3), 135–150.

Lerner, D. N. (1986). Leaking pipes recharge ground water. *Ground Water, 24,* 654–662.

Lerner, D. N. (1990). Recharge due to urbanization. In D. N. Lerner, A. S. Issar, & I. Simmers (Eds.), *Groundwater recharge, a guide to understanding and estimating natural recharge, Contributions to Hydrogeology 8* (pp. 210–214). Kenilworth, UK: International Associations of Hydrogeologists.

Lerner, D. N. (2002). Identifying and quantifying urban recharge: A review. *Hydrogeology Journal, 10,* 143–152.

Loheide, S. P., Butler, J. J., & Gorelick, S. M. (2005). Estimation of groundwater consumption by phreatophytes using diurnal water table fluctuations: A saturated-unsaturated flow assessment. *Water Resources Research, 41*(7). https://doi.org/10.1029/2005wr003942.

Lohse, K. A., Gallo, E. L., & Kennedy, J. R. (2010). Possible tradeoffs from urbanization on groundwater recharge ad water quality. *Southwest Hydrology, 2010,* 18–32.

Maliva, R. G., & Hopfensperger, K. P. (2007). Impacts of residential development on humid subtropical freshwater resources: Southwest Florida Experience. *Journal American Water Resources Association, 43,* 540–1549.

Maréchal, J. C., Galeazzi, L., Dewandel, B., & Ahmed, S. (2003). Importance of irrigation return flow on the groundwater budget of a rural basin in India. In *Hydrology of the Mediterranean and Semiarid Regions*, Publication 278 (pp. 62–67). London: International Association of Hydrological Sciences.

Masoner, J. R., Mladinich, C. S., Konduris, A. M., & Smith, S. J. (2003). *Comparison of irrigation water use estimates calculated from remotely sensed irrigated acres and state reported irrigated acres in the Lake Altus drainage basin, Oklahoma and Texas, 2000 growing season.* U.S. Geological Survey Water-Resources Investigations Report 03-4155.

McQuillan, D., & Bassett, E. (2009). Return flow to ground water from onsite wastewater systems. In *18th Annual NOWRA Technical Conference and Expo*, April 6–9, 2009, Milwaukee, WI.

Meinzer, O. E. (1923). *Outline of ground-water hydrology, with definitions*. U.S. Geological Survey Water-Supply Paper 494.

Meyboom, P. (1967). Groundwater studies in the Assiniboine River drainage basin—Part II: Hydrologic characteristics of phreatophytic vegetation in south-central Saskatchewan. *Geological Survey Canada Bulletin* 139.

Milczarek, M., Yao, T.-M., Harding, J., Goodrich, D., & Levick, L. (2004). Predicting groundwater recharge rates in small urbanized watersheds. *Southwest Hydrology*, 6–7.

Moore, G. W., & Owens, M. K. (2012). Transpirational water loss in invaded and restored semiarid riparian forests. *Restoration Ecology, 20*(3), 346–351.

Morris, B. L., Darling, W. G., Cronin, A. A., Rueedi, J., Whitehead, E. J., & Gooddy, D. C. (2006). Assessing the impact of modern recharge on a sandstone aquifer beneath a suburb of Doncaster, UK. *Hydrogeology Journal, 14*(6), 979–997.

Nagler, P. L., Glenn, E. P., & Thompson, T. L. (2003). Comparison of transpiration rates among saltcedar, cottonwood and willow trees by sap flow and canopy temperature methods. *Agricultural and Forest Meteorology, 116*(1), 73–89.

Nagler, P. L., Shafroth, P. B., LaBaugh, J. W., Snyder, K. A., Scott, R. L., Merritt, D. M., & Osterberg, J. (2009). The potential water savings through the control of saltcedar and Russian olive. In P. B. Shafroth, C. A. Brown, & D. M. Merritt (Eds.), *Saltcedar and Russian olive control demonstration act science assessment* (pp. 35–47). U.S. Geological Survey Scientific Investigations Report 2009-5247.

Nosetto, M. D., Jobbágy, E. G., Brizuela, A. B., & Jackson, R. B. (2012). The hydrologic consequences of land cover change in central Argentina. *Agriculture, Ecosystems & Environment, 154,* 2–11.

NRMMC, EPHC, NHMRC. (2009). *Australian Guidelines for water recycling: managing health and environmental risks (Phase 2) managed aquifer recharge*. Australia: Natural Resource Management Ministerial Council, Environment Protection and Heritage Council, National Health and Medical Research Council.

Ochoa, C. G., Fernald, A. G., & Guldan, S. J. (2011). Caracterización del balance hídrico y la recarga por retorno de riego en un valle agrícola de una región semiárida de los Estados Unidos de América. *Estudios en la Zona no Saturada del Suelo, 10,* 337–340.

Ochoa, C. G., Fernald, A. G., Guldan, S. J., Tidwell, V. C., & Shukla, M. K. (2012). Shallow aquifer recharge from irrigation in a semiarid agricultural valley in New Mexico. *Journal of Hydrologic Engineering, 18*(10), 1219–1230.

Ozdogan, M., & Gutman, G. (2008). A new methodology to map irrigated areas using multi-temporal MODIS and ancillary data: An application example in the continental US. *Remote Sensing of Environment, 112*(9), 3520–3537.

Peck, A. J., & Williamson, D. R. (1987). Effects of forest clearing on groundwater. *Journal of Hydrology, 94,* 47–65.

Pescod, M. E. (1992). *Wastewater treatment and use in agriculture*, FAO Irrigation and Drainage Paper 47. Rome: Food and Agriculture Organization of the United Nations.

Petheram, C., Walker, G., Grayson, R., Thierfelder, T., & Zhang, L. (2002). Towards a framework for predicting impacts of land-use on recharge: 1. A review of recharge studies in Australia. *Australian Journal of Soil Research, 40*(3), 397–417.

Pfeiffer, L., & Lin, C. Y.-C. (2010). The effect of irrigation technology on groundwater use. *Choices, 25*(3).

Pulido-Bosch, A., & Sbih, Y. B. (1995). Centuries of artificial recharge on the southern edge of the Sierra Nevada (Granada, Spain). *Environmental Geology, 26*(1), 57–63.

Ritzi, R. W., Bouwer, H., & Sorooshian, S. (1985). Water resource conservation by reducing phreatophyte transpiration. In *First North American Riparian Conference, Riparian Ecosystems and Their Management: Reconciling Conflicting Uses*, Tucson, Arizona, April 16–18, 1985, Tucson, Arizona (pp. 191–196).

Robertson, W. M., & Sharp, J. M., Jr. (2013). Variability of groundwater nitrate concentrations over time in arid basin aquifers: Sources, mechanisms of transport, and implications for conceptual models. *Environmental Earth Sciences, 69*(7), 2415–2426.

Robertson, W. M., & Sharp, J. M. (2015). Estimates of net infiltration in arid basins and potential impacts on recharge and solute flux due to land use and vegetation change. *Journal of Hydrology, 522,* 211–227.

Robinson, T. W. (1952). Phreatophytes and their relation to water in Western United States. *Transactions, American Geophysical Union (EOS), 33,* 57–61.

Robinson, T. W. (1958). *Phreatophytes*. U.S. Geological Survey Water-Supply Paper 1423.

Robinson, T. W. (1959). *Phreatophyte research in Western United States, October 1958 to March 1959*. U.S. Geological Survey Circular 413.

Romaguera, M., Hoekstra, A. Y., Su, Z., Krol, M. S., & Salama, M. S. (2010). Potential of using remote sensing techniques for global assessment of water footprint of crops. *Remote Sensing, 2*(4), 1177–1196.

Sammis, T. W., Evans, D. D., & Warrick, A. W. (1982). Comparison of methods to estimate deep percolation rates. *Journal American Water Resources Association, 18*(3), 465–470.

Scanlon, B. R., Keese, K., Reedy, R. C., Simunek, J., & Andraski, B. J. (2003). Variations in flow and transport in thick desert vadose zones in response to paleoclimatic forcing (0–90 kyr): Field measurements, modeling, and uncertainties. *Water Resources Research, 39*(7), 1179.

Scanlon, B. R., Keese, K. E., Flint, A. L., Flint, L. E., Gaye, C. B., Edmunds, W. M., et al. (2006). Global synthesis of groundwater recharge in semiarid and arid regions. *Hydrological Processes, 20,* 3335–3379.

Scanlon, B. R., Reedy, R. C., Stonestrom, D. A., Prudic, D. E., & Dennehy, K. F. (2005). Impact of land use and land cover change on groundwater recharge and quality in the southwestern US. *Global Change Biology, 11,* 1577–1593.

Scanlon, B. R., Jolly, I., Sophocleous, M., & Zhang, L. (2007). Global impacts of conversions from natural to agricultural ecosystems on water resources: Quantity versus quality. *Water Resources Research, 43,* W03437. https://doi.org/10.1029/2006WR005486.

Seneviratne, S. I., Corti, T., Davin, E. L., Hirschi, M., Jaeger, E. B., Lehner, I., et al. (2010). Investigating soil moisture–climate interactions in a changing climate: A review. *Earth-Science Reviews, 99*(3), 125–161.

Sharp, J. M., Jr. (2010). The impacts of urbanization on groundwater systems and recharge. *Aqua Mundi, 1,* 51–56.

Stephens, D. B. (1996). *Vadose zone hydrology*. Boca Raton: CRCR Press.

Stephens, D. B., Blandford, T. N., Cattron, D., & Moore, S. (2006). Quantifying return flow to groundwater: What's in the tool box? In *Universities Council on Water Resources (UCOWR) Conference 'Increasing Freshwater Supplies'*, July 18–20, 2006, Santa Fe, New Mexico.

Stewart, A. M. (2014). Estimation of urban-enhanced infiltration and groundwater recharge, Sierra Vista subbasin, southeast Arizona USA. Ph.D. Dissertation. Tucson; University of Arizona.

Stonestrom, D. A., & Harrill, J. R. (2007). Ground-water recharge in the arid and semiarid southwestern United States—Climate and geologic framework. In D. A. Stonestrom, J. Constantz, T. P. A. Ferre, & S. A. Leake (Eds.), *Ground-water recharge in the arid and semiarid southwestern United States* U.S. Geological Survey Professional Paper 1703G, (pp. 1–27).

Stovall, J., & Rainwater, K. (2002). Evaluating recharge and irrigation return flows for the South Plains. In *Proceedings AWRA Summer Specialty Conference, Ground Water/Surface Water Interactions*, July 1–3, 2002, pp. 291–296.

Su, Z. (2002). The surface energy balance system (SEBS) for estimation of turbulent heat fluxes. *Hydrology and Earth System Sciences Discussions, 6*(1), 85–100.

Thenkabail, P. S., Schull, M., & Turral, H. (2005). Ganges and Indus river basin land use/land cover (LULC) and irrigated area mapping using continuous streams of MODIS data. *Remote Sensing of Environment, 95*(3), 317–341.

Urbonas, B. R., & Roesner L. A. (1992). Hydrologic design for urban drainage and flood control. In D. R. Maidment (Ed.), *Handbook of hydrology*, Chapter 28. New York: McGraw Hill.

Van Hylckama, T. E. (1980). *Weather and evapotranspiration studies in a saltcedar thicket, Arizona*. U.S. Geological Survey Professional Paper 491-F.

Vázquez-Suñe, E., Carrera, J., Tubau, I., X. Sánchez-Vila, X., & Soler, A. (2010). An approach to identify urban groundwater recharge. *Hydrology and Earth System Sciences, 14*, 2085–2097.

Velpuri, N. M., Thenkabail, P. S., Gumma, M. K., Biradar, C., Dheeravath, V., Noojipady, P., et al. (2009). Influence of resolution in irrigated area mapping and area estimation. *Photogrammetric Engineering & Remote Sensing, 75*(12), 1383–1395.

Vengosh, A., & Pankratov, I. (1998). Chloride/bromide and chloride/fluoride ratios of domestic sewage effluents and associated contaminated ground water. *Ground Water, 36*(5), 815–825.

Vollertsen, J., & Hvitved-Jacobsen, T. (2003). Exfiltration from gravity sewers: A pilot scale study. *Water Science and Technology, 47*(4), 69–76.

Ward, F. A., & Pulido-Velazquez, M. (2008). Water conservation in irrigation can increase water use. *Proceedings of the National Academy of Sciences, 105*(47), 18215–18220.

Welder, G. E. (1988). *Hydrologic effects of phreatophyte control, Acme-Artesia Reach of the Pecos River, New Mexico, 1967–1982*. U.S. Geological Survey Water-Resources Investigations Report 87-4148.

Werner, M., Maggs, I., & Petkovic, M. (2011). Accurate measurements of minimum night flows for water loss analysis. In *5th Annual WIOA NSW Water Industry Engineers & Operators Conference*. Water loss management program NSW, Exhileracing events centre, Newcastle (pp. 31–37).

White, W. N. (1932). *A method of estimating ground-water supplies based on discharge by plants and evaporation from soil: Results of investigations in Escalante Valley, Utah*. U.S. Geological Survey Water Supply Paper 659-A.

Wiles, T. J., & Sharp, J. M. (2008). The secondary permeability of impervious cover. *Environmental and Engineering Geoscience, 14*(4), 251–265.

Wolf, L. E. I. F., Eiswirth, M., & Hotzl, H. (2006). Assessing sewer-groundwater interaction at the city scale based on individual sewer defects and marker species distributions. *Environmental Geology, 49,* 849–857.

Wu, X., Zhou, J., Wang, H., Li, Y., & Zhong, B. (2015). Evaluation of irrigation water use efficiency using remote sensing in the middle reach of the Heihe river, in the semi-arid Northwestern China. *Hydrological Processes, 29*(9), 2243–2257.

Yang, Y., Lerner, D. N., Barrett, M. H., & Tellam, J. H. (1999). Quantification of groundwater recharge in the city of Nottingham, UK. *Environmental Geology, 38*(3), 183–198.

MIX
Papier aus verantwortungsvollen Quellen
Paper from responsible sources
FSC® C105338

If you have any concerns about our products,
you can contact us on
ProductSafety@springernature.com

In case Publisher is established outside the EU,
the EU authorized representative is:
Springer Nature Customer Service Center GmbH
Europaplatz 3, 69115 Heidelberg, Germany

Printed by Libri Plureos GmbH
in Hamburg, Germany